ANNUAL REVIEW OF
NEUROSCIENCE

EDITORIAL COMMITTEE (1994)

Responsible for the organization of Volume 17
(Editorial Committee, 1992)

ANNUAL REVIEW OF NEUROSCIENCE

VOLUME 17, 1994

W. MAXWELL COWAN, *Editor*

Howard Hughes Medical Institute

ERIC M. SHOOTER, *Associate Editor*

Stanford University School of Medicine

CHARLES F. STEVENS, *Associate Editor*

Salk Institute for Biological Studies

RICHARD F. THOMPSON, *Associate Editor*

University of Southern California

ANNUAL REVIEWS INC.   4139 EL CAMINO WAY   P.O. 10139   PALO ALTO, CALIFORNIA 94303-0897

ANNUAL REVIEWS INC.
Palo Alto, California, USA

International Standard Serial Number: 0147–006X
International Standard Book Number: 0–8243–2417-X

Annual Review and publication titles are registered trademarks of Annual Reviews Inc.

♾ The paper used in this publication meets the minimum requirements of American National Standard for Information Sciences—Permanence of Paper for Printed Library Materials, ANSI Z39.48-1984.

Typesetting by Kachina Typesetting Inc., Tempe, Arizona; John Olson, President; Jeannie Kaarle, Typesetting Coordinator; and by the Annual Reviews Inc. Editorial Staff

PRINTED AND BOUND IN THE UNITED STATES OF AMERICA

Ⱥ Annual Review of Neurosci.
Volume 17, 1994

CONTENTS

vi   Contents   *(Continued)*

SOME RELATED ARTICLES IN OTHER *ANNUAL REVIEWS*

From the *Annual Review of Cell Biology,* Volume 9 (1993):

The Role of GTP-Binding Proteins in Transport Along the Exocytic Pathway, Susan Ferro-Novick and Peter Novick

From the *Annual Review of Pharmacology and Toxicology,* Volume 34 (1994):

Permeation Properties of Neurotransmitter Transporters, Henry A. Lester, Sela Mager, Michael W. Quick, and Janis L. Corey

The Cholinesterases: From Genes to Proteins, Palmer Taylor and Zoran Radic

α_1-*Adregenic Receptor Subtypes,* Kenneth P. Minneman and Timothy A. Esbenshade

From the *Annual Review of Physiology,* Volume 56 (1994):

Modulation of Ion Channels by Protein Phosphorylation and Dephosphorylation, Irwin B. Levitan

Phosphoinositide Phospholipases and G Proteins in Hormone Action, J. H. Exton

Nucleus Tractus Solitarius—Gateway to Neural Circulatory Control, Michael C. Andresen and Diana L. Kunze

Synaptic Transmission in the Outer Retina, Samuel M. Wu

Ryanodine Receptor/Ca^{2+} Release Channels and Their Regulation by Endogenous Effectors, Gerhard Meissner

From the *Annual Review of Psychology,* Volume 45 (1994):

Images of the Mind: Studies of Verbal Response Selection, Marcus E. Raichle

Color Appearance: On Seeing Red—Or Yellow, Or Green, Or Blue, Israel Abramov and James Gordon

Chemical Senses, Linda M. Bartoshuk and Gary K. Beauchamp

Representations and Models in Psychology, P. Suppes, M. Pavel, and J.-Cl. Falmagne

Sexual Differentiation of the Human Nervous System, S. Marc Breedlove

Language Specificity and Elasticity: Brain and Clinical Syndrome Studies, Michael Maratsos and Laura Matheny

For the convenience of readers, a detachable order form/envelope is bound into the back of this volume.

Annu. Rev. Neurosci. 1994. 17:1–30

DEVELOPMENT OF MOTONEURONAL PHENOTYPE

Judith S. Eisen

Institute of Neuroscience, University of Oregon, Eugene, Oregon 97403

KEY WORDS: motoneuronal fate, pathfinding, induction, cell-cell interactions, zebrafish, chick, *Drosophila*

INTRODUCTION

The ability of animals to carry out their behavioral repertoires depends critically on the specificity of connections between motoneurons and the muscles they innervate. Learning how such specificity arises has been the goal of considerable research effort in vertebrates and invertebrates alike. The ultimate aim of this effort is to understand the molecular mechanisms by which motoneurons develop functionally appropriate synaptic connections, as well as other differentiated characteristics that define their phenotypes.

In adult organisms, neurons express characteristic features that allow us to recognize them. These features include location, shape, synaptic connectivity, and biochemical properties such as neurotransmitters, neurotransmitter receptors, and ion channels. A fundamental issue in neuronal development is understanding what factors influence the differentiation of a neuron so that it expresses its characteristic traits. At one extreme, all differentiated characteristics could be intrinsic properties with which a cell becomes endowed by its lineal antecedents. At the other extreme, each differentiated characteristic could be developed as the response to an extrinsic signal. In the first case, the entire developmental program of a neuron would be independent of its environment. In the second case, small changes in the environment could have profound effects, potentially rendering a neuron unrecognizable.

As will become clear in this review, motoneurons differ in their abilities to respond to environmental alterations, and these abilities may change over time. In some cases, individual characteristics may respond differently to environmental alterations, suggesting that they are independently regulated. For example, O'Brien et al (1990) found that when avian thoracic neural tube

1

was transplanted to the lumbar region, thoracic motoneurons developed both the soma organization and the peripheral synaptic connections appropriate for their new location; their central synaptic connections, however, remained appropriate for their site of origin. The basis of such differential regulation of individual characteristic is unknown. However, as we learn more about the molecular and genetic mechanisms that impinge on motoneuronal development, we should be able to build a picture of how the complex interplay between intrinsic cellular features and extrinsic environmental cues is manifest in the differentiated characteristics a motoneuron expresses.

This review focuses primarily on the development of three different groups of motoneurons: populations of motoneurons that innervate chick hindlimb, identified motoneurons that innervate zebrafish axial muscle, and identified motoneurons that innervate embryonic and larval *Drosophila* abdominal body wall muscles. Each of these systems has attributes that make it especially amenable to particular experimental approaches, and thus, each has provided a wealth of information on the mechanisms involved in motoneuronal development. To understand the mechanisms responsible for the development of motoneuronal phenotype, it is first necessary to define the criteria by which motoneurons are recognized. The first part of this review describes some of the characteristics by which motoneuronal phenotype can be identified; the second part focuses on the cellular, molecular, and genetic processes by which motoneurons develop their particular phenotypes. Over the past decade, numerous reviews have covered portions of this work, most of which concentrated on some aspect of development in only a single one of the experimental systems. Thus, rather than providing a comprehensive review, I will attempt to synthesize what is known about common features of development in these systems and to discuss what we can learn from apparent differences.

DEFINITION OF MOTONEURONAL PHENOTYPE

The major criterion used to identify motoneurons is the muscles they innervate. Learning the cellular processes by which motor growth cones find and recognize their appropriate muscle targets is a prerequisite for understanding the underlying molecular mechanisms. This portion of the review describes the organization of the neuromuscular systems under consideration and the cellular and molecular cues used during pathfinding and target recognition.

Neuromuscular System Organization

The skeletal muscles of the vertebrate trunk are derived from segmentally repeated blocks of mesoderm, called myotomes, that form along the rostrocaudal axis early in embryonic development. In some species, particularly anamniotes (see Fetcho 1986a) such as zebrafish, myotomes are retained

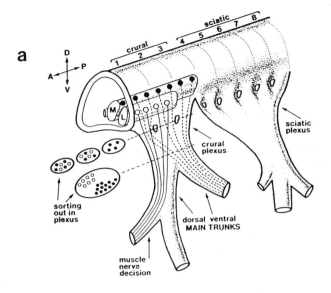

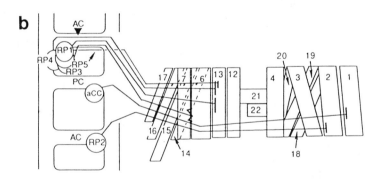

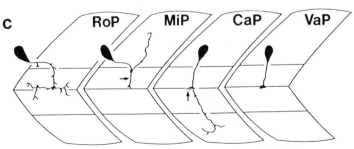

Figure 1 Neuromuscular system organization and identification of individual motoneurons. (*a*) Chick limb neuromuscular system. From Tang et al (1992) with permission of Cell Press. (*b*) *Drosophila* neuromuscular system showing identified motoneurons and muscle fibers. Adapted from Sink & Whitington (1991a) with permission of John Wiley & Sons, Inc. (*c*) Zebrafish identified motoneurons. Arrows denote branches that are later retracted.

in only slightly modified form in the adult as the major body musculature. In other species such as adult frogs, but particularly in amniotes (see Fetcho 1986b) such as chicks, myotomes undergo extensive rearrangements to contribute to paired limbs or to form complex axial musculature. Vertebrate trunk muscles are innervated by ipsilaterally located motoneurons in the ventral horn of the spinal cord; the regions of spinal cord containing motoneurons are bilaterally symmetric. Axial muscles are innervated by motoneurons in a medial motor column that extends the entire length of the spinal cord. In anamniotes, the medial motor column contains two distinct classes of motoneurons (Blight 1978, Myers 1985, Kimmel & Westerfield 1990): primary motoneurons, large cells that are few in number and develop early, and secondary motoneurons, smaller cells that are more numerous and develop later. Limb muscles are innervated by motoneurons in a lateral motor column, found only in the brachial and lumbosacral spinal cord regions. Within both medial and lateral motor columns, motoneurons innervating specific muscles are grouped into discrete pools at specific rostrocaudal locations. Thus, there is a direct matching between soma location within the spinal cord and innervation patterns in the periphery (Figure 1a).

The body wall muscles of abdominal segments 2–7 (A2–A7) of embryonic and larval *Drosophila* consist of a repeating pattern of about 30 identified muscle fibers on each side of each segment (Johansen et al 1989, Sink and Whitington 1991a; Figure 1b; one side of a segment is referred to as a hemisegment). A stereotyped, segmentally repeated set of about 34 motoneurons innervates the muscle fibers of each hemisegment. Some motoneurons are ipsilateral to the muscle fibers they innervate, whereas others are contralateral. Many *Drosophila* motoneurons have been individually identified by the muscle fibers they innervate and their soma positions (Figure 1b). Unlike vertebrates, *Drosophila* motoneuronal somata are not grouped together in a specific CNS region.

Motor Growth Cones Follow Stereotyped Pathways

Perhaps the most astonishing accomplishment of embryonic motoneurons is that their growth cones find and form synaptic connections with appropriate muscle targets, even though they are typically located at a distance of many cell diameters. This process, which will be described below for each system under consideration, involves at least two steps: growth cone extension to a region containing an appropriate target, often referred to as pathfinding, and formation of synapses with the correct type and number of target cells.

In chicks, each hindlimb muscle is innervated by hundreds of motoneurons whose somata are located in discrete motor pools in the lateral motor column (Landmesser 1992). Pathfinding by these cells appears to be precise from outset, with few obvious errors. Soon after leaving the CNS, motor growth cones collect into two plexuses, from which the major dorsal and ventral nerve trunks emerge (Lance-Jones & Landmesser 1981a). The growth cones of motoneurons destined to innervate different muscles intermingle extensively during axonal elongation. However, at the plexus region, growth cones extending toward specific muscles select divergent pathways that lead toward cell-specific targets (Figure 1a).

The segmentally iterated myotomes of embryonic zebrafish are innervated by a segmentally repeated set of three primary motoneurons, CaP, MiP, and RoP, each of which can be identified by its soma position in the spinal cord (Eisen et al 1986, Myers et al 1986; Figure 1c). Axonal outgrowth by these cells occurs in a stereotyped sequence in which CaP precedes MiP, which precedes RoP. The CaP growth cone extends ventrally from the spinal cord; when it reaches the region of the nascent horizontal septum, it pauses for about an hour, and then resumes ventral extension. The MiP and RoP growth cones initially extend caudally within the spinal cord. Where they encounter CaP, they reorient and follow the CaP axon to the nascent horizontal septum, where they also pause. When they resume extension, they select cell-specific divergent pathways that lead toward separate myotome regions. Zebrafish primary motoneurons appear to make few projection errors; however, both CaP (Eisen et al 1986, Liu & Westerfield 1990) and MiP (Eisen et al 1986) make a characteristic, ectopic axonal branch which is later retracted (Figure 1c). About half of the spinal hemisegments contain a fourth primary motoneuron, called VaP, which usually dies (Eisen et al 1990).

Studies in the grasshopper CNS (reviewed in Goodman et al 1984) and peripheral nervous system (PNS) (Keshishian 1980) were among the first to establish that pathfinding by identified neurons is highly precise from the outset. Homologues of many well-studied grasshopper neurons have been identified in *Drosophila* (Thomas et al 1984, Goodman et al 1984). Several of them are motoneurons that innervate abdominal body wall muscles, including aCC, U, and a group of neurons known as the RPs.

Pathfinding by individual members of the RP group has been extensively examined in *Drosophila* (Jacobs and Goodman 1989b, Halpern et al 1991, Sink & Whitington 1991b; Figure 2). The somata of RP1, RP3, RP4, and RP5 are contralateral to the identified muscle fibers they innervate, whereas the RP2 soma is ipsilateral. Initially, each contralateral motoneuron extends a growth cone that crosses the midline and fasciculates with the axon of its homologue. Most RP growth cones transiently contact the somata of their contralateral homologues or other RP motoneurons. They then turn, extend caudally along the contralateral longitudinal connective, and leave the CNS via the intersegmental nerve, previously pioneered by the growth cone of the aCC motoneuron. Where the segmental nerve joins the intersegmental nerve, a short distance outside the CNS, RP growth cones cross to the segmental

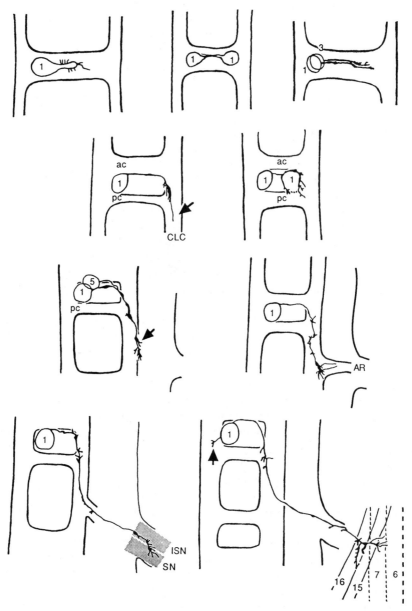

Figure 2 Pathfinding by identified RP motoneurons in *Drosophila*. Adapted from Sink &
Whitington (1991b) with permission of the Company of Biologists LTD. Abbreviations: ac,
anterior commissure; pc, posterior commissure; CLC, contralateral longitudinal connective; ISN,
intersegmental nerve; AR, anterior root of the ISN.

nerve. Although each RP motoneuron innervates a stereotyped set of muscle fibers, en route to those targets their growth cones make extensive contacts with inappropriate muscle fibers, which are later withdrawn. It is unclear whether these ectopic branches represent functional synaptic connections.

A striking feature of pathfinding in each of these systems is that after leaving the CNS, motor growth cones initially follow a common pathway from which they later diverge to select cell-type specific pathways toward their appropriate targets (see Figure 1). This pattern of pathfinding suggests the existence of both general guidance cues followed by all motor growth cones and guidance cues that may be specific for individual motor growth cones. That common pathways are delineated by general cues has been demonstrated in chick limbs, where surgically rerouted axons from inappropriate types of neurons follow the normal motor pathway (Lewis et al 1983), and in *Drosophila* wings, where growth cones from a variety of types of neurons transplanted to the wing follow a common sensory nerve pathway (Blair et al 1987). Experimental evidence for cell-specific pathways comes from studies in chick hindlimb in which motor growth cones correct their course within the plexus and project to appropriate muscles after the relationship between their somata and targets has been altered by spinal cord reversals (Lance-Jones & Landmesser 1980), limb bud reversals (Ferguson 1983, Stirling & Summerbell 1985), limb shifts (Lance-Jones & Landmesser 1981b), limb segment deletions (Whitelaw & Hollyday 1983), and limb segment duplications (Stirling & Summerbell 1988). Similarly, when single identified zebrafish primary motoneurons were transplanted to novel locations in the periphery, their growth cones selected appropriate cell-specific pathways toward their normal muscle targets (Gatchalian & Eisen 1992).

Environmental Cues Involved in Pathway Guidance and Target Recognition

What is the nature of the cues that guide motor growth cones to their appropriate targets, and where are they located? Interactions between navigating growth cones and their environments are likely to be mediated by appropriate temporal and spatial expression patterns of specific molecules, including diffusible factors, extracellular matrix (ECM) components, and cell-surface glycoproteins. Since the major criterion used to define motoneurons is the muscles they innervate, understanding the molecular basis by which motor growth cones find their targets is central to understanding the molecular basis of motoneuronal phenotype. Evidence for a role for each of the types of molecules mentioned above in guiding specific motor growth cones and establishing appropriate neuromuscular connections will be presented below. In addition to cues that attract growth cones or promote recognition between growth cones and their targets, some cues apparently act as barriers to growth cone advance. For example, chick motor growth cones seem actively to avoid some tissues, such as perinotochordal mesenchyme, pelvic girdle precursor (Tosney & Oakley 1990, Tosney 1992), and the posterior portion of somitic sclerotome (Keynes & Stern 1984). Such avoidance may result, at least in part, from growth cone collapse upon encountering particular cues (Kapfhammer & Raper 1987).

CUES THAT ACT AT A DISTANCE   Some growth cones are attracted to signals released by a target (Lumsden & Davies 1983, 1986; Heffner at al 1990) or by cells en route to a target (Dodd & Jessell 1988, Tessier-Lavigne et al 1988, Bovolenta & Dodd 1990). Is there evidence that such diffusible signals attract motor growth cones? McCaig (1986) found that when dissociated *Xenopus* neural tubes were cultured with somitic myoblasts or sheets of skin, presumptive motor neurites extended preferentially toward the myoblasts, suggesting that myoblasts released an attractive signal. This signal is likely to be diffusible, since both *Xenopus* (McCaig 1986) and chick (Henderson et al 1981, 1984) presumptive motor neurites were attracted by agar slabs conditioned by somitic myoblasts.

Muscle also appears to release a signal that attracts growth cones of axial motoneurons in vivo. Tosney (1987, 1988) found that removal of chick dermamyotome, the precursor of somitic muscles, affected formation of the motor nerve that normally innervated the missing axial muscle (Figure 3). When at least three segments of dermamyotome were deleted, the nerve that would normally innervate the muscle of the middle segment was absent (see

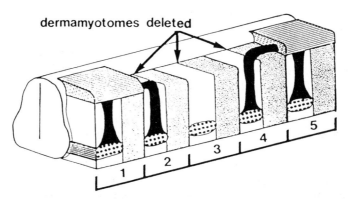

Figure 3   Effects of dermamyotome removals on axial nerve formation. Numbers denote somite levels; axial nerves are shown in black, dermamyotomes have horizontal lines, anterior sclerotome is white, and posterior sclerotome is stippled. From Tosney (1992) with permission of Raven Press.

also Phelan & Hollyday 1990). Nerves did form when muscle was present in the adjacent segments; however, they turned abnormally toward the nearest muscle, as though they were specifically attracted to it. Similarly, Eisen & Pike (1991) found that motor axons were stunted and developed abnormally in zebrafish embryos homozygous for the *spadetail* (*spt*) mutation which were deficient in trunk muscle (Kimmel et al 1989). Together, these studies suggest that muscle may normally provide a diffusible signal important for proper development of axial motoneurons.

In contrast, although extracts of chick limb muscle promote outgrowth by presumptive motoneurons in culture (Henderson et al 1981, 1982; Nurcombe & Bennett 1984), most evidence suggests that muscle is not required for proper pathfinding by chick limb motoneurons in vivo. Both major nerve trunks and muscle nerves develop following extirpation of somites to create muscleless forelimbs (Phelan & Hollyday 1990; but see Lewis et al 1981). Moreover, although chick limb motoneurons innervate muscles derived from the same segmental level as their somata (Keynes et al 1987, Lance-Jones 1988b), muscles are innervated normally following somite reversals which cause muscle to be derived from novel segments (Keynes et al 1987, Lance-Jones 1988a).

Despite these studies, it is clear that chick (Lance-Jones & Landmesser 1981b, Stirling & Summerbell 1988) and Japanese medaka fish (Okamoto & Kuwada 1991a,b) limb motoneurons can find their targets when they have been displaced. At least in chick, somatopleure-derived limb connective tissue appears to play an important role in this process. Lance-Jones & Dias (1991) found that when they transplanted chick somatopleure from the lumbosacral region to the thoracic region, the limb formed ectopically adjacent to the transplanted somatopleure. The somata of the motoneurons innervating the ectopic limb were located in their normal positions in the lumbosacral spinal cord, and their growth cones took altered routes to find the appropriate synaptic targets in their new locations. These results suggest that connective tissue provides cues involved in establishing proper connections between limb motor growth cones and their targets, although it is not clear whether it is the source of diffusible cues that attract motor growth cones.

CUES IN THE INTERSTITIAL SPACE   In chicks (Rogers et al 1986, Tosney et al 1986) and zebrafish (Westerfield 1987), ECM components, which are probably recognized by integrins and other receptors (reviewed in Reichardt and Tomaselli 1991), may provide both permissive and inhibitory cues for motor growth cone extension. The ECM component laminin is present on surfaces where motor growth cones extend, whereas another ECM component, fibronectin, is present in regions that motor axons appear to avoid. Moreover, in culture, zebrafish putative primary motoneurons extend neurites on laminin-coated dishes but not on dishes coated with fibronectin (Westerfield 1987). Although the temporal and spatial distributions of these ECM components are consistent with a role in delineating where motor growth cones can extend, they do not suggest a role in specific axonal guidance.

ECM is also present in developing insects, and laminin is expressed along the intersegmental and segmental nerves followed by motor growth cones (Montell & Goodman 1989). Enzymatic removal of the basal lamina in grasshopper limbs does not seem to affect pathway navigation by sensory growth cones (Condic & Bentley 1989), although whether ECM is necessary for motor growth cones has not been tested.

CUES ON THE CELL SURFACE   Studies on the development of identified neurons in grasshoppers (reviewed in Goodman et al 1984) led to the idea that their growth cones express selective affinities for specific axonal surfaces that are differentially labeled. There is now considerable evidence from both *Drosophila* and vertebrates that cues on the surfaces of axons and other cells play a major role in growth cone guidance. The existence of such cues was initially revealed by studies showing a guidance role for particular cells contacted by navigating growth cones. More recently, specific cell surface molecules involved in growth cone guidance have been isolated. In this section, I will first describe some of the cellular studies suggesting that cell surface cues guide motor growth cones, and then discuss some of the molecules that are thought to be involved.

Cellular evidence for surface cues   In both grasshoppers and *Drosophila,* identified glial cells are located in positions consistent with a role in guiding motor growth cones (Bastiani et al 1985, Bastiani & Goodman 1986, Jacobs & Goodman 1989a; Figure 4). Ablation of a specific grasshopper glial cell prevented formation of the intersegmental nerve (Bastiani & Goodman 1986), apparently by preventing the growth cones of the aCC and U motoneurons from leaving the CNS and extending into the periphery. Thus,

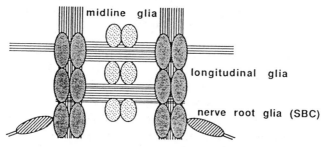

Figure 4   Location of glial cells in the *Drosophila* CNS and along the proximal nerve root. From Jacobs et al (1989) with permission of Cell Press.

contact with this glial cell appears to be required for proper motor growth cone guidance. Other glial cells are distributed along the peripheral nerve route (Klämbt et al 1991, Klämbt & Goodman 1991), but their capacity to function in guidance has not yet been tested.

The role of glial cells in guidance of vertebrate motor growth cones is somewhat less clear. In zebrafish embryos, neural crest cells that produce peripheral glia migrate along the same pathway as motor growth cones (Raible et al 1992); however, their migration along the relevant pathway region begins after motor growth cones have exited the CNS (S. H. Pike, E. Morin-Kensicki, D. W. Raible & J. S. Eisen, unpublished data), and they differentiate after motor axons have completed pathfinding (D. W. Raible & P. Henion, unpublished data). In contrast, putative Schwann cells have been observed ahead of the main limb and muscle nerves of the avian forelimb (Noakes & Bennett 1987). Removal of the neural crest has been reported to result in both formation of a normal pattern of motor nerves (Rickmann et al 1985) and failure of motor axons to enter the limb (Noakes et al 1988). In the mouse hindlimb, Schwann cells appear to be unnecessary for proper motoneuronal outgrowth, since hindlimb motor axons formed a normal pattern in embryos homozygous for the *Splotch* mutation (Grim et al 1992), in which lumbar Schwann cells are absent (Franz 1990).

Axonal surfaces provide important guidance cues for growth cones of *Drosophila* motoneurons in the CNS (see Goodman et al 1984). In all of the systems under consideration, at least some motor growth cones act as pioneers in establishing peripheral motor nerves, thus, they can clearly navigate in the absence of other axons. But what about later-developing motor growth cones? Sink & Whitington (1991b) found that although *Drosophila* RP motoneurons extended in the intersegmental nerve, they did not fasciculate with the pioneering aCC axon, suggesting that this axon was unnecessary for guidance. In embryonic zebrafish, ablation of any one of the primary motoneurons did not affect pathfinding by the others (Eisen et al 1989, Pike & Eisen 1990), indicating that their growth cones all have an independent ability to navigate in the periphery. However, given that ablation of primary motoneurons did affect pathfinding by later-developing secondary motoneurons (Pike et al 1992), primary motor axons are probably important in the guidance of secondary motor growth cones. Chick hindlimb motor axons do not have specific neighbor relationships during pathfinding (Tosney & Landmesser 1985), and they can find their appropriate muscles from aberrant locations (Lance-Jones & Landmesser 1981b), thus earlier axons may be unnecessary to guide later ones.

In insects, muscle cells evidently are important for the establishment of proper neuromuscular connections. Grasshoppers (Ho et al 1983, Ball et al 1985) and *Drosophila* (Leiss et al 1988, Johansen et al 1989, Bate 1990), possess a specific mesodermal cell, called a muscle pioneer, that forms a scaffold upon which each muscle is organized. Identified motor growth cones associate with particular muscle pioneers. Ball et al (1985) ablated a specific grasshopper limb muscle pioneer and found that growth cones that normally contacted it and specifically innervated the muscle it founded extended past the point where the muscle pioneer would normally be located, just what would be expected if contact with this cell were required for proper target recognition by motor growth cones. Similarly, Chiba et al (1993) found that *Drosophila* RP1 and RP3 motoneuron growth cones selected their appropriate muscle targets following deletion or duplication of identified muscle fibers, suggesting specific target recognition.

Molecular nature of cell surface cues   Two major types of cell surface molecules have been implicated in guiding growth cones of developing motoneurons in vertebrates: members of the immunoglobulin (I_g) superfamily of cell adhesion molecules, in particular the neural cell adhesion molecule NCAM (Edelman 1988), and calcium-dependent cadherins (Takeichi 1988). Molecules of both types have been shown to function in homophilic adhesion; however, this does not necessarily rule out participation in heterophilic interactions. Antibody blocking studies suggest that these molecules act synergistically during axonal extension on avian muscle cells in culture (Bixby et al 1987), and thus that multiple molecular cues may be involved in axonal guidance or target recognition. In vivo, the distribution of NCAM along chick hindlimb axons and the surfaces they contact is consistent with multiple roles for this molecule during axonal outgrowth and synapse formation (Tosney et al 1986), including mediating interactions both between motor growth cones and myotubes and among motor axons (Landmesser et al 1988). Another member of the I_g superfamily, L1 (Moos et al 1987), is probably involved in interactions among motor axons (Landmesser et al 1988). Blocking the function of either molecule with antibodies disrupts the pattern of axonal growth (Landmesser et al 1988).

More recently, Landmesser et al (1990) have implicated the level of polysialylation of NCAM in regulating motor nerve branch formation and thus the pattern of synaptic connections between chick hindlimb motoneurons and muscle. Removal of polysialic acid (PSA) caused changes in the nerve branching pattern in the fast region of the iliofibularis muscle, where axons typically branch without much fasciculation; in essence the branching pattern was altered to resemble that in the slow region, where axons typically show significant fasciculation. Since the innervation patterns of fast and slow regions are normally different (Dahm & Landmesser 1988), such changes in branching could lead to profound changes in muscle activity. Tang et al (1992) found differences in the distribution of PSA on motoneuronal somata and axons that projected to different muscle targets. Removal of PSA prior to axons sorting at the plexus resulted in some motoneurons projecting to incorrect muscles and all muscles receiving some incorrect projections. Thus, different levels of PSA could cause changes in the way distinct motoneurons recognize and/or respond to guidance cues in the plexus and limb. PSA has been found in *Drosophila* (Roth et al 1992), although whether it plays a similar role there is unknown.

Several glycoproteins thought to be important in pathfinding in the *Drosophila* CNS have been identified, some of which are members of the I_g superfamily (see Goodman et al 1992). One member of this group, fasciclin II (*fas II*), is expressed on the growth cone and axon of the aCC motoneuron within the intersegmental nerve (Grenningloh et al 1991). Another member of the I_g superfamily, *fas III* (Patel et al 1987), as well as two members of another group, the leucine rich repeat family (LRR) (Nose et al 1992), are also expressed on subsets of developing motoneurons and muscle cells, suggesting that they might be involved in motoneuronal target recognition. For example, somata and axons of RP1, RP3, and RP4 all express *fas III* (Halpern et al 1991). Identified muscle fibers 6 and 7, the targets of RP3, also express *fas III* transiently in the region where the neuromuscular synapses will form. However, identified muscle fibers 12 and 13, the targets of RP1 and RP4, do not express this molecule, raising the possibility that interactions between motor growth cones and their targets involve more than single homophilic adhesion molecules.

This idea is strengthened by the following experimental results. Synaptic sites on a subset of muscle fibers and the growth cones and axons of motoneurons that innervate them express *connectin,* a member of the LRR family (Nose et al 1992). One of the muscles expressing *connectin* is fiber 5, which is innervated by two motoneurons with ipsilateral somata (Cash et al 1992). Ablation of fiber 5 (Cash et al 1992) by laser-irradiation or mutations in the bithorax complex leads to ectopic neuromuscular endings on muscle fibers 12 or 4, which are probably displaced endings of the motoneurons that normally innervate fiber 5. Interestingly, neither fiber 12 nor 4 has been reported to express *connectin.*

Together, these studies in *Drosophila* suggest that matched expression of a single cell surface molecule on specific muscle fibers and motor growth cones is insufficient to explain the exquisitely precise pattern of neuromuscular specificity. Thus, it seems likely that proper formation of neuromuscular junctions requires the coordinate expression of multiple molecular cues, of which only a few have been identified. These cues are unlikely to be simply redundant, rather each one probably functions in some subtly different way (see Goodman et al 1992).

Is There Specific Recognition Between Motor Growth Cones and Muscles?

The studies described above suggest that individual *Drosophila* motor growth cones recognize specific muscle fibers. In contrast, there is little compelling evidence for specific recognition between motor growth cones and muscles in nonmammalian vertebrates. Motor growth cones show no preference for muscle of the same segmental level in vivo (Keynes et al 1987, Lance-Jones 1988a) or in vitro (McCaig 1986, Sanes & Poo 1989), nor do they show any preference in vitro between motoneurons and muscle cells of clonal origin (Sanes & Poo 1989; but see Moody & Jacobson 1983). Further, motoneurons can be forced to innervate inappropriate muscle targets stably in vivo (Landmesser & O'Donovan 1984; but see O'Brien et al 1990). Although perturbation experiments in chicks (Lance-Jones & Landmesser 1980, 1981b; Ferguson 1983) and zebrafish (Eisen et al 1989, Pike & Eisen 1990, Gatchalian & Eisen 1992) show that motor growth cones can find their appropriate targets under altered circumstances, most of these results can be attributed to finding appropriate pathways, rather than to specific target recognition. Despite the lack of evidence for target recognition in non-mammalian vertebrates, there is evidence from transgenic mice suggesting that mammalian muscle cells may acquire cell-autonomous differences based on their rostrocaudal positions (Donoghue et al 1992).

Role of Activity in Neuromuscular Specificity

The establishment and maintenance of appropriate synaptic connections often involve events that depend on neural activity. For example, proper connectivity in the vertebrate retina appears to require electrical activity (Shatz & Stryker 1988, Sretavan et al 1988; but see Stuermer et al 1990), and activity is important for normal synapse elimination at some neuromuscular junctions (reviewed in Jansen & Fladby 1990). In chick embryos, neuromuscular activity blockade produced striking differences in nerve branching patterns in fast and slow muscle regions (Dahm & Landmesser 1988) which were correlated with changes in the levels of PSA and were prevented by PSA removal (Landmesser et al 1990). These results suggest that activity may have profound effects on muscle innervation patterns. In contrast, in zebrafish, pathfinding, synaptogenesis, and the pattern of muscle innervation by primary motoneurons were normal following pharmacological blockade of sodium channels or acetylcholine receptors (Liu & Westerfield 1990) and in embryos homozygous for the *nic1* mutation, which lack functional acetylcholine receptors (Westerfield et al 1990). During initial pathfinding in *Drosophila*, many motoneurons make ectopic branches that are later retracted (Sink & Whitington 1991b). Mutations that change activity levels alter the density of motoneuronal arborizations, but their appropriateness appears to be unaffected (Budnick et al 1990, Zhong et al 1992). Thus, it is unclear whether activity plays a role in neuromuscular specificity in *Drosophila*.

Motor growth cones find and recognize their targets by interacting with specific environmental cues. Since pathfinding is precise from the outset, motoneurons must express at least some of the molecules that enable them to respond to pathfinding cues before encountering them; thus motoneurons appear to acquire cell-specific identities prior to axogenesis. The mechanisms involved in the acquisition of motoneuronal identity are considered below.

DETERMINATION OF MOTONEURONAL FATE

Cellular interactions appear to be important in determining cell fate in all multicellular animals (Stent 1985). The term *fate* is used here to refer to the differentiated cell type(s) normally formed by an embryonic cell or its progeny. To attain their fates, embryonic cells must respond to dynamically changing environmental signals that influence their development. This portion of the review describes some of the signalling processes thought to be involved in the generation of motoneurons and the specification of their identities.

Tissue Interactions Implicated in Induction of Vertebrate Motoneurons

Around the time of gastrulation, the dorsal region of vertebrate ectoderm receives an inducing signal from dorsal mesoderm and initiates neural development (Spemann 1938; reviewed in Gurdon 1987 and Papalopulu & Kintner 1992). Within the spinal cord, neurons develop a distinct dorsoventral pattern which has been proposed to depend on interactions with notochord, a derivative of dorsal mesoderm (Jessell & Dodd 1992). The first cells in the neural plate to show signs of overt differentiation are located at the ventral midline (Schoenwolf & Smith 1990), where they contact the notochord directly. These cells later become the spinal cord floor plate (Baker 1927, Kingsbury 1930), a specialized structure which may itself be involved in patterning the arrangement of other neural cell types (Jessell & Dodd 1992). Motoneurons are also located in the ventral spinal cord, near the floor plate but not in direct contact with it.

Jessell & Dodd (1992) and their associates (Jessell et al 1989; Yamada et al 1991; Placzek et al 1990, 1991) proposed that notochord induces floor plate and that both cell types may be involved in induction of motoneurons. Addition of a second notochord in chick embryos caused development of a second floor plate with motoneurons appropriately spaced around it; addition of only a floor plate also induced development of motoneurons. Surgical (Hirano et al 1991) or mutational (Bovolenta & Dodd 1991) ablation of notochord blocked formation of floor plate and motoneurons. Both medial and lateral motor column motoneurons appear to be affected in these experiments (J. Dodd & T. Jessell, personal communication). Although some experimental evidence suggests notochord and floor plate are able to induce motoneurons independently (Yamada et al 1991), they may normally work in concert.

In anamniotes such as *Xenopus* and zebrafish, the situation may be somewhat different. Notochordless *Xenopus* embryos produced by ultraviolet irradiation of fertilized eggs lack floor plates (Clarke et al 1991). Although the number of late-developing secondary motoneurons appeared to be considerably reduced in these embryos, the number of early-developing primary motoneurons was approximately normal, so that neither the notochord nor the floor plate was required for induction of this cell type in these experiments. This is consistent with studies showing that primary motoneurons have their birthdays during gastrulation (Lamborghini 1980), before overt appearance of the neural plate. Thus, these cells might be induced by earlier signals from other sources.

The zebrafish mutant *cyclops* (*cyc*) lacks a differentiated floor plate (Hatta et al 1991, Hatta 1992), yet both primary (Eisen 1991, Hatta 1992) and secondary (Hatta 1992) motoneurons are present; thus a differentiated floor plate is not required for motoneuronal induction in this species. In the *no tail* (*ntl*) mutant, which lacks a differentiated notochord, the floor plate is present and motoneurons appear essentially normal, so a differentiated notochord is also not required for induction of motoneurons in this species (Halpern et al 1993). However, cells of the notochord lineage are apparently present in *ntl* embryos and retain some of their early functions, such as floor plate induction.

These studies raise the issue of whether different types of vertebrate motoneurons (primary vs secondary, medial column vs lateral column, or amniote vs anamniote) are induced by different tissue interactions. As in *Xenopus*, zebrafish primary motoneurons have their birthdays during or soon after gastrulation; however, their final divisions occur near their final locations (C. B. Kimmel, R. M. Warga & D. A. Kane, personal communication), suggesting that motoneuronal precursors are in a position to receive signals from notochord and floor plate lineages, if they are present at the appropriate time. Learning whether motoneurons are present in zebrafish embryos in which both notochord and floor plate lineages are absent should help resolve this question.

In addition to the patterning role of notochord and floor plate in the organization of the spinal cord, work in avian embryos shows a patterning role for paraxial trunk mesoderm. This tissue clearly imposes a segmental arrangement on trunk motor nerves and sensory ganglia (Keynes & Stern 1984; Rickmann et al 1985; Tosney 1987, 1988) and affects the structure of the spinal cord by shaping segmental swellings (myelomeres) and by restricting anterior-posterior spread of clones derived from single neural tube cells (Lim et al 1991, Stern et al 1991). However, it is unclear whether segmented mesoderm influences specific neuronal cell types.

In zebrafish ventral spinal cord, the somata of primary motoneurons and at least one class of primary interneurons have an obvious segmental organization (Eisen & Pike 1991). Two kinds of evidence suggest that paraxial mesoderm may be involved in patterning the arrangement of these neurons. First, the segmental organization of paraxial mesoderm is altered in embryos homozygous for the recessive embryonic lethal mutation *spt* (Kimmel et al 1989, Eisen & Pike 1991). In these embryos, the arrangement of ventrally located, segmentally organized primary motoneurons and interneurons is altered in correspondence with the new pattern of segmented mesoderm, whereas the arrangement of dorsally located, nonsegmentally organized primary sensory neurons and interneurons remains unchanged (Eisen & Pike 1991). Because the *spt* mutation acts autonomously in paraxial mesoderm (Ho & Kane 1990) but not in motoneurons (Eisen & Pike 1991), it seems likely that the modified arrangement of motoneurons results from altered interactions with affected mesoderm. Second, heat shock alters the organization of segmented mesoderm and the arrangement of primary motoneurons in a corresponding fashion (Kimmel et al 1988). Whether this proposed patterning role of paraxial mesoderm on zebrafish primary motoneurons constitutes an inductive influence depends ultimately on timing. If paraxial mesoderm is appropriately positioned early enough, it might be involved in determining the fates of early-developing neurons in the ventral spinal cord. Alternatively, it may provide signals that allow ventral spinal neurons to complete differentiation along a developmental pathway already determined by interactions with signals from other sources.

Cellular and Genetic Interactions Involved in Formation of Drosophila Motoneurons

The neurectoderm of insect embryos consists of an apparently uniform sheet of cells, all of which apparently have the potential to develop as neuroblasts (Bate 1976, Doe & Goodman 1985a). Of the approximately 150 ectodermal cells in each hemisegment, only about 30 actually become neuroblasts, and only a specific subset of these produce motoneurons. Neuroblasts arise in several temporal waves (Campos-Ortega & Hartenstein 1985, Doe 1992a). Individual neuroblasts can be recognized by their positions, and each identified neuroblast produces a stereotyped set of progeny (Bate and Grunewald 1981, Goodman et al 1984, Doe & Goodman 1985a). Although each neuroblast forms at a specific location, it does not always arise from the same cell. Rather, interactions among a cluster of about four to six equivalent cells result in one of them becoming the neuroblast, while the others either differentiate into nonneuronal support cells or die (Doe & Goodman 1985b).

Genetic studies have identified two classes of *Drosophila* genes involved in neuroblast formation: the proneural genes (Ghysen & Dambly-Chaudiere 1989), which promote neurogenesis, and the neurogenic genes (Lehmann et al 1983), which suppress neurogenesis and facilitate epidermogenesis. The proneural genes include members of the *achaete-scute* complex (ASC) and *daughterless*, all of which encode proteins with a basic helix-loop-helix motif (reviewed in Jan & Jan 1990). Initially all cells of a cluster express the proneural genes; later, expression is retained only in the differentiating neuroblast. The ASC may be required for commitment of cells to a neuroblast fate, because loss of function mutations result in fewer neuroblasts (Jiménez & Campos-Ortega 1990). Additionally, individual proneural genes may be required only in specific subsets of neuroblasts, since when only one proneural gene is absent, only a few neuroblasts are missing (Martín-Bermudo et al 1991). The zygotic neurogenic genes include *Notch, Delta, big brain, neuralized, Enhancer of split,* and *mastermind. Notch* and *Delta* both encode membrane proteins with epidermal growth factor (EGF)-like repeats which may be involved in cell-cell interactions (see Greenspan 1992 and Greenwald & Rubin 1992). In embryos lacking any one of the first five of these genes, all cells in a cluster continue to express proneural genes and become neuroblasts (Skeath & Carroll 1992).

By the time neuroblasts begin to differentiate, the *Drosophila* blastoderm has already been divided into various domains by the action of pair rule and segment polarity genes (Nusslein-Volhard & Weischaus 1980). Two of these genes, *fushi tarazu* (*ftz*) and *even-skipped* (*eve*), may be involved in specifying several identified motoneurons (Doe et al 1988a,b); this role appears to be independent of their earlier functions. *ftz* is expressed in aCC, RP1, and RP2 motoneurons, as well as in the pCC interneuron. In the absence of neuronal *ftz* expression aCC and RP1 appear normal, but RP2 develops a morphology similar to RP1. *eve* is also expressed by aCC, pCC, and RP2, but not by RP1. In the absence of neuronal *eve* expression RP2 again develops a morphology similar to RP1, whereas aCC develops a morphology similar to pCC. These studies suggest that in the absence of expression of specific genes, the identities of particular neurons may be transformed.

Role of Positional Cues in Specifying Motoneuronal Identity

Three lines of evidence support the idea that neuroblast identity is positionally specified in insects. First, neuroblasts form a stereotyped array (Bate 1976, Hartenstein & Campos-Ortega 1984, Doe & Goodman 1985a, Doe 1992a) and each neuroblast produces a stereotyped set of progeny (Bate & Grunewald 1981, Goodman et al 1984, Doe & Goodman 1985a). Second, when a specific

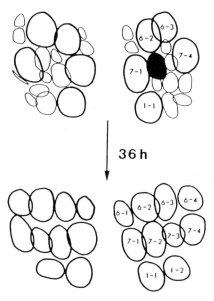

Figure 5 Early ablation of identified neuroblasts in the grasshopper embryo leads to their replacement by neighboring cells. The ablated neuroblast (*black*) in the right hand panel was replaced by one of the small cells in the left hand panel. From Doe & Goodman (1985b) with permission of Academic Press.

grasshopper neuroblast was ablated early in development, an adjacent ectodermal cell formed a new neuroblast which always developed the fate appropriate for its position (Doe & Goodman 1985b; Figure 5). Interestingly, if the neuroblast was replaced after it produced the first few cells of its lineage, the new neuroblast began the lineage again, producing duplicates of those cells (Doe & Goodman 1985a), suggesting that once a neuroblast is specified, it produces its cell-specific lineage invariantly. This is in contrast to studies that suggest that vertebrate neurons (reviewed in Sanes 1989), including motoneurons (Hartenstein 1989, Leber et al 1990, Eisen 1992a), do not arise by invariant lineages.

The third kind of evidence for positional specification of neuroblast fate comes from molecular genetic studies. Currently, the best candidates for providing positional cues along the anterior/posterior axis of the neurogenic region of the *Drosophila* blastoderm are the pair rule and segment polarity genes (Nusslein-Volhard & Weischaus 1980; for reviews of their role in positional specification of neuroblast fate see Doe 1992b and Goodman & Doe 1993). Pair rule gene products subdivide each segment into a series of molecularly distinct regions (Ingham et al 1988). Since each region may generate a single row of early-arising neuroblasts, these gene products may be involved in neuroblast specification (Goodman & Doe 1993); other gene products are likely to be involved in the specification of later-arising neuroblasts (Goodman & Doe 1993). Mutations in some but not all segment polarity genes also affect neuroblast specification (Patel et al 1989, Goodman & Doe 1993).

The molecular genetic studies in *Drosophila* are certainly consistent with positional specification of neuroblast fate, especially when considered with ablation studies from grasshoppers. However, it is clear from studies in zebrafish that cells that are specified, that is, developmentally distinct from their neighbors (Davidson 1990, Kimmel et al 1991), as assessed by expression of a cell-specific gene product, may be able to alter expression of that gene (Schulte-Merker et al 1992) and their fates (R. K. Ho, personal communication) when exposed to a new environment. Thus, cells may be specified without being irreversibly committed to develop their characteristic phenotypes (Kimmel et al 1991). Commitment can be tested when embryonic cells are challenged to differentiate in altered environments. Operationally, this requires that cells be cultured in vitro or transplanted to heterotopic locations in vivo.

Positional specification of chick hindlimb motoneuron precursors is suggested by studies in which three or four lumbosacral spinal cord segments were reversed about the rostrocaudal axis. Initially, it was found that motor axons followed aberrant pathways and projected to muscles appropriate for their original spinal cord locations (Lance-Jones & Landmesser 1980, 1981b). Because these transplants were done prior to motoneuronal birthdays, it appears that motoneuronal precursors were already committed to produce motoneurons that would project to specific muscles. When these experiments were repeated in younger embryos, the results were very different (Matise & Lance-Jones 1992). Motoneurons projected axons to muscles appropriate for their new spinal cord locations, suggesting that their precursors were uncommitted. These studies also suggest that positional cues are somehow involved in the commitment process, and that the time of exposure to these cues is critical. However, these population studies do not reveal whether individual motoneuronal precursors were committed, since interactions among cells within the transplant, so-called community effects (Gurdon 1988), may cause groups of transplanted cells to develop according to their site of origin, whereas individual cells of the same type will develop according to their new location, when they are transplanted to the same new environment (Ho 1992). Ultimately, cell commitment is best tested by culturing or transplanting single, identified cells.

Transplantation of individual zebrafish primary motoneurons to new spinal cord locations provides strong support for positional determination of

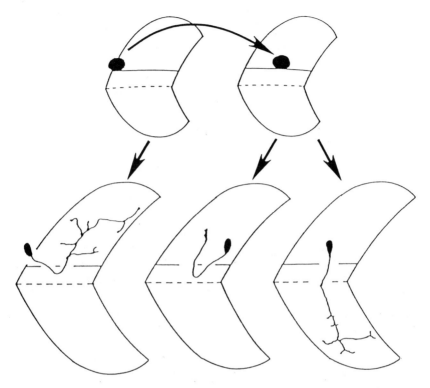

Figure 6   Zebrafish motoneurons develop position-specific axonal trajectories. Motoneurons transplanted from the MiP position of a donor embryo to the same position of a host develop a normal MiP morphology (*left*). Motoneurons develop a MiP axonal trajectory when transplanted from the MiP position to the CaP position about an hour before axogenesis (*middle*) and a CaP morphology when transplanted from the MiP position to the CaP position two to three hours before axogenesis (*right*).

motoneuronal identity (Eisen 1991; Figure 6). Replacement of a CaP with a MiP, or vice versa, about an hour before axogenesis resulted in the transplanted cell developing an axonal trajectory appropriate for its original position, suggesting that its axonal trajectory was already committed. Surprisingly, in some cases the somata of transplanted motoneurons actually moved from their new positions back to their original positions, suggesting that different spinal cord regions may have positional cues that motoneurons can recognize. When these experiments were repeated one to two hours earlier, as with the population studies of chick hindlimb motoneurons, a very different result was obtained. In this case, the transplanted motoneuron typically developed an axonal trajectory appropriate for its new position, suggesting that when it was moved, its axonal trajectory was uncommitted. Further, these results suggest that exposure to positional cues during a critical time period commits a cell to develop a specific axonal trajectory. A possible mechanism that could account for this result is that positional cues could induce a motoneuron in a specific location to express receptors that cause its growth cone to recognize only a particular pathway.

Role of Cell-Cell Interactions in the Expression of Motoneuronal Fate

Although positional cues are probably instrumental in specifying the identity of zebrafish primary motoneurons and *Drosophila* neuroblasts, interactions between specific neurons appear to determine how some cell fates are expressed. For example, initially the *Drosophila* aCC motoneuron and its sibling, the pCC interneuron, are indistinguishable; later, their growth cones make cell-specific pathway choices and they develop cell-specific fates (Kuwada & Goodman 1985). Studies have shown that these cells form an equivalence pair (see Kimble et al 1979 and Greenwald & Rubin 1992 for discussions of equivalence groups), since ablation of either cell within a few hours of becoming postmitotic resulted in the remaining cell developing into pCC. Thus, pCC is the primary fate and aCC is the secondary fate. Ablation at later stages, but still before axogenesis, resulted in the remaining cell developing into either aCC or pCC with equal probability, showing that these cells had become committed to different fates. Because the presence of both cells was required for one of them to adopt the secondary fate, interactions between them were likely involved in the process. Such interactions require a signal that tells the cells that the primary fate is assigned and thus allocates one of them to the secondary fate.

A similar situation occurs in the embryonic zebrafish, although this equivalence pair is formed by neurons that are not necessarily siblings (Eisen 1992b). As described earlier, about half of the zebrafish spinal hemisegments contain a VaP motoneuron, in addition to CaP, MiP, and RoP. CaP and VaP are initially indistinguishable; later, most VaPs die. Transplantation studies show that CaP and VaP are an equivalence pair in which CaP is the primary fate and VaP is the secondary fate. The fate choice of each cell appears to depend upon the presence and state of development of another CaP/VaP. Interestingly, ablation of CaP quite late allows VaP to develop into CaP, showing that these cells remain equivalent until nearly the time when VaP normally dies.

What is the basis of these interactions between individual neurons? The molecular genetic mechanisms underlying similar cell-cell interactions have been investigated in *Caenorhabditis elegans* and *Drosophila* (reviewed in Greenwald & Rubin 1992). Similar genes appear to be involved in both species: *lin-12* in *C. elegans* and *Notch* in *Drosophila*; both genes encode transmembrane proteins with EGF-like repeats that suggest a role in cell-cell signaling. Members of the *lin-12/Notch* family have also been found in vertebrates: *Xotch* in *Xenopus*, by Coffman et al (1990); *Notch1* and *Notch2* in rat, by Weinmaster et al (1991, 1992); and *TAN-1* in human, by Ellisen et al (1991). It will be interesting to learn whether similar gene products are involved in the interactions between CaP and VaP in the zebrafish.

Because so many motoneurons innervate each chick hindlimb muscle, it is not possible to examine whether there are equivalence pairs. However, interactions are clearly involved in determining the fates of these motoneurons. About half of the motoneurons initially generated eventually die. Like ones that survive, motoneurons that die extend growth cones and probably make contact with target muscles. It is thought that they die because they fail to receive neurotrophic support (Oppenheim et al 1988), and that this support may be based on neuromuscular activity (Oppenheim & Nuñez 1982, Pittman & Oppenheim 1979).

CONCLUSIONS

Both vertebrate and insect motoneurons appear to develop their specific phenotypes through a series of interactions with signals in the local environment. Motoneuronal precursors respond to these signals in ways that limit their ability to respond to other signals and thus restrict their potential to develop other fates. Transplantation studies in vertebrate embryos have been especially useful in elucidating cellular interactions involved in motoneuronal fate determination, and studies in *Drosophila* have begun to elucidate some of the underlying molecular and genetic mechanisms.

One of the intriguing discoveries of the past few years is that there are surprising similarities among some of the genes involved in regulating insect and vertebrate development. For example, homeobox-containing genes that control positional identities along the anterior/posterior body axis in *Drosophila* are conserved in vertebrates and appear to subserve a similar function (reviewed in McGinnis & Krumlauf 1992), mammalian homologues of *Drosophila* segment polarity genes function during nervous system development (Thomas & Capecchi 1990), and as in *Drosophila*, mammalian homologues of *achaete-scute* are expressed in subpopulations of neuronal precursors (Johnson et al 1990, Lo et al 1991). These similarities in gene expression patterns raise the exciting possibility that many of the molecular mechanisms underlying embryonic development in insects and vertebrates are also similar. Even if this is the case, however, considerably more work will be required to understand how patterns of gene expression regulate the phenotypes of individual motoneurons.

Although the motoneurons considered here show many similarities during development, they also reveal some intriguing differences. For example, distinct signals may be involved in specification of different types of vertebrate motoneurons, and even within a single species, different types of motoneurons may use distinct types of signals during pathfinding. These differences may reveal fundamental departures in underlying mechanisms, or they may simply represent subtle variations on a theme. Continued studies of motoneuronal development in a variety of species will be important to resolve this question.

ACKNOWLEDGMENTS

I thank Bruce Appel, Christine Beattie, Marnie Halpern, Robert Ho, David Raible, Monte Westerfield, and especially Charles Kimmel for comments on the manuscript; Jane Dodd, Chris Doe, Marnie Halpern, Paul Henion, Robert Ho, Tom Jessell, Don Kane, Charles Kimmel, Beth Morin-Kensicki, Sue Pike, David Raible, and Rachel Warga for sharing unpublished work; and Pat Edwards for typing. Original work from my laboratory was supported by NS23915, BNS8553146, NS01476, HD22486, a Searle Scholar Award, the Procter and Gamble Company, and the Murdock Charitable Trust.

Literature Cited

Baker RC. 1927. The early development of the ventral part of the neural plate of ambystoma. *J. Comp. Neurol.* 44:1–26

Ball EE, Ho RK, Goodman CS. 1985. Development of neuromuscular specificity in the grasshopper embryo: guidance of motoneuron growth cones by muscle pioneers. *J. Neurosci.* 5:1808–19

Bastiani MJ, Doe CQ, Helfand SL, Goodman CS. 1985. Neuronal specificity and growth cone guidance in grasshopper and *Drosophila* embryos. *Trends Neurosci.* 8:257–66

Bastiani MJ, Goodman CS. 1986. Guidance of neuronal growth cones in the grasshopper embryo. II. Recognition of specific glial pathways. *J. Neurosci.* 6:3542–51

Bate CM. 1976. Embryogenesis of an insect nervous system. I. A map of the thoracic and abdominal neuroblasts in *Locusta migratoria*. *J. Embryol. Exp. Morphol.* 35:107–23

Bate CM, Grunewald EB. 1981. Embryogenesis of an insect nervous system. II. A second class of neuron precursor cells and the origin of the intersegmental connectives. *J. Embryol. Exp. Morphol.* 61:317–30

Bate M. 1990. The embryonic development of larval muscles in *Drosophila*. *Development* 110:791–804

Bixby JL, Pratt RS, Lilien J, Reichardt LF. 1987. Neurite outgrowth on muscle cell surfaces involves extracellular matrix receptors as well as Ca^+-dependent and -independent cell adhesion molecules. *Proc. Natl. Acad. Sci. USA* 84:2555–59

Blair SS, Murray MA, Palka J. 1987. The guidance of axons from transplanted neurons through aneural *Drosophila* wings. *J. Neurosci.* 7:4165–75

Blight AR. 1978. Golgi-staining of "primary" and "secondary" motoneurons in the developing spinal cord of an amphibian. *J. Comp. Neurol.* 180:679–90

Bovolenta P, Dodd J. 1990. Guidance of commissural growth cones at the floor plate in embryonic rat spinal cord. *Development* 109:435–47

Bovolenta P, Dodd J. 1991. Perturbation of neuronal differentiation and axon guidance in the spinal cord of mouse embryos lacking a floor plate: analysis of Danforth's short-tail mutation. *Development* 113:625–39

Budnik V, Zhong Y, Wu C-F. 1990. Morphological plasticity of motor axons in *Drosophila* mutants with altered excitability. *J. Neurosci.* 10:3754–68

Campos-Ortega JA, Hartenstein V. 1985. *The Embryonic Development of Drosophila melanogaster*. New York/Berlin: Springer-Verlag

Cash S, Chiba A, Keshishian H. 1992. Alternate neuromuscular target selection following the loss of single muscle fibers in *Drosophila*. *J. Neurosci.* 12:2051–64

Chiba A, Hing H, Cash S, Keshishian H. 1993. Growth cone choices of *Drosophila* motoneurons in response to muscle fiber mismatch. *J. Neurosci.* 13:714–32

Clarke JDW, Holder N, Soffe SR, Storm-Mathisen J. 1991. Neuroanatomical and functional analysis of neural tube formation in notochordless *Xenopus* embryos: laterality of the ventral spinal cord is lost. *Development* 112:499–516

Coffman C, Harris W, Kintner C. 1990. *Xotch*, the *Xenopus* homologue of *Drosophila Notch*. *Science* 249:1438–41

Condic ML, Bentley D. 1989. Pioneer neuron pathfinding from normal and ectopic locations *in vivo* after removal of the basal lamina. *Neuron* 3:427–39

Dahm LM, Landmesser LT. 1988. The regulation of intramuscular nerve branching during normal development and following activity blockade. *Dev. Biol.* 130:621–44

Davidson EH. 1990. How embryos work: A comparative view of diverse modes of cell fate specification. *Development* 108:365–89

Dodd J, Jessell TM. 1988. Axon guidance and the patterning of neuronal projections in vertebrates. *Science* 242:692–99

Doe CQ. 1992a. Molecular markers for identified neuroblasts and ganglion mother cells in the *Drosophila* central nervous system. *Development* 116:855–63

Doe CQ. 1992b. The generation of neuronal diversity in the *Drosophila* embryonic central nervous system. In *Determinants of Neuronal Identity*, ed. M. Shankland, ER Macagno, pp. 119–54. San Diego: Academic Press. 528 pp.

Doe CQ, Goodman CS. 1985a. Early events in insect neurogenesis. I. Development and segmental differences in the pattern of neuronal precursor cells. *Dev. Biol.* 111:193–205

Doe CQ, Goodman CS. 1985b. Early events in insect neurogenesis. II. The role of cell interactions and cell lineage in the determination of neuronal precursor cells. *Dev. Biol.* 111:206–19

Doe CQ, Hiromi Y, Gehring WJ, Goodman CS. 1988a. Expression and function of the segmentation gene *fushi tarazu* during *Drosophila* neurogenesis. *Science* 239:170–75

Doe CQ, Smouse D, Goodman CS. 1988b. Control of neuronal fate by the *Drosophila* segmentation gene *even-skipped*. *Nature* 333:376–78

Donoghue MJ, Morris-Valero R, Johnson YR, et al. 1992. Mammalian muscle cells bear a cell-autonomous, heritable memory of their rostrocaudal position. *Cell* 69:67–77

Edelman GM. 1988. Morphoregulatory molecules. *Perspect. Biochem.* 27:3533–43

Eisen JS. 1991. Determination of primary motoneuron identity in developing zebrafish embryos. *Science* 252:569–72

Eisen JS. 1992a. Development of motoneuronal identity in the zebrafish. In *Determinants of Neuronal Identity*, ed. M Shankland, ER Macagno, pp. 469–96. San Diego: Academic Press. 528 pp.

Eisen JS. 1992b. The role of interactions in determining cell fate of two identified motoneurons in the embryonic zebrafish. *Neuron* 8:231–40

Eisen JS, Myers PZ, Westerfield M. 1986. Pathway selection by growth cones of identified motoneurons in live zebrafish embryos. *Nature* 320:269–71

Eisen JS, Pike SH. 1991. The *spt-1* mutation alters the segmental arrangement and axonal development of identified neurons in the spinal cord of the embryonic zebrafish. *Neuron* 6:767–76

Eisen JS, Pike SH, Debu B. 1989. The growth cones of identified motoneurons in embryonic zebrafish select appropriate pathways in the absence of specific cellular interactions. *Neuron* 2:1097–104

Eisen JS, Pike SH, Romancier B. 1990. An identified motoneuron with variable fates in embryonic zebrafish. *J. Neurosci.* 10:34–43

Ellisen LW, Bird J, West DC, et al. 1991. *TAN-1*, the human homolog of the *Drosophila Notch* gene, is broken by chromosomal translocations in T lymphoblastic neoplasms. *Cell* 66:649–61

Ferguson BA. 1983. Development of motor innervation of the chick following dorsoventral limb bud rotations. *J. Neurosci.* 3:1760–72

Fetcho JR. 1986a. The organization of the motoneurons innervating the axial musculature of vertebrates. I. Goldfish (*Carassius auratus*) and Mudpuppies (*Necturus maculosus*). *J. Comp. Neurol.* 249:521–50

Fetcho JR. 1986b. The organization of the motoneurons innervating the axial musculature of vertebrates. II. Florida water snakes (*Nerodia fasciata pictiventris*). *J. Comp. Neurol.* 249:551–63

Franz T. 1990. Defective ensheathment of motoric nerves in the splotch mutant mouse. *Acta Anat.* 138:246–53

Gatchalian CL, Eisen JS. 1992. Pathway selection by ectopic motoneurons in embryonic zebrafish. *Neuron* 9:105–12

Ghysen A, Dambly-Chaudiere C. 1989. Genesis of the *Drosophila* peripheral nervous system. *TIG* 5:251–55

Goodman CS, Bastiani MJ, Doe CQ, et al. 1984. Cell recognition during neuronal development. *Science* 225:1271–79

Goodman CS, Doe CQ. 1993. Embryonic

development of the *Drosophila* central nervous system. In *The Development of Drosophila*, ed. CM Bate, A Martinez-Arias. Cold Spring Harbor Press. In press

Goodman CS, Grenningloh G, Bieber AJ. 1992. Molecular genetics of neural cell adhesion molecules in *Drosophila*. In *The Nerve Growth Cone*, ed. PC Letourneau, SB Kater, ER Macagno, pp. 283–303. New York: Raven Press. 535 pp.

Greenspan RJ. 1992. Initial determination of the neurectoderm in *Drosophila*. In *Determinants of Neuronal Identity*, ed. M Shankland, ER Macagno, pp. 155–88. San Diego: Academic. 528 pp.

Greenwald I, Rubin GM. 1992. Making a difference: the role of cell-cell interactions in establishing separate identities for equivalent cells. *Cell* 68:271–81

Grenningloh G, Rehm EJ, Goodman CS. 1991. Genetic analysis of growth cone guidance in *Drosophila:* fasciclin II functions as a neuronal recognition molecule. *Cell* 67:45–57

Grim M, Halata Z, Franz T. 1992. Schwann cells are not required for guidance of motor nerves in the hindlimb in Splotch mutant mouse embryos. *Anat. Embryol.* 186:311–18

Gurdon JB. 1987. Embryonic induction—molecular prospects. *Development* 99:285–306

Gurdon JB. 1988. A community effect in animal development. *Nature* 336:772–74

Halpern ME, Chiba A, Johansen J, Keshishian H. 1991. Growth cone behavior underlying the development of stereotypic synaptic connections in *Drosophila* embryos. *J. Neurosci.* 11:3227–38

Halpern ME, Ho RK, Walker C, Kimmel CB. 1993. The zebrafish mutation *ntl* distinguishes between floor plate and muscle pioneer signalling properties of the notochord. *Cell* In Press

Hartenstein V. 1989. Early neurogenesis in *Xenopus:* the spatio-temporal pattern of proliferation and cell lineages in the embryonic spinal cord. *Neuron* 3:399–411

Hartenstein V, Campos-Ortega JA. 1984. Early neurogenesis in wild-type *Drosophila melanogaster*. *Roux's Arch. Dev. Biol.* 193: 308–25

Hatta K. 1992. Role of the floor plate in axonal patterning in the zebrafish CNS. *Neuron* 9:629–42

Hatta K, Kimmel CB, Ho RK, Walker C. 1991. The cyclops mutation blocks specification of the floor plate of the zebrafish central nervous system. *Nature* 350:339–41

Heffner CD, Lumsden GS, O'Leary DDM. 1990. Target control of collateral extension and directional axon growth in the mammalian brain. *Science* 247:217–20

Henderson CE, Huchet M, Changeux J-P. 1981. Neurite outgrowth from embryonic chicken spinal neurons promoted by media conditioned by muscle cells. *Proc. Natl. Acad. Sci. USA* 78:2625–29

Henderson CE, Huchet M, Changeux J-P. 1984. Neurite-promoting activities for embryonic spinal neurons and their developmental changes in the chick. *Dev. Biol.* 104:336–47

Hirano S, Fuse S, Sohal GS. 1991. The effect of the floor plate on pattern and polarity in the developing central nervous system. *Science* 251:310–13

Ho RK. 1992. Cell movements and cell fate during zebrafish gastrulation. *Development Suppl.* 65–73

Ho RK, Ball EE, Goodman CS. 1983. Muscle pioneers: large mesodermal cells that erect a scaffold for developing muscles and motoneurons in grasshopper embryos. *Nature* 301:66–69

Ho RK, Kane D. 1990. Cell-autonomous action of zebrafish *spt-1* mutation in specific mesodermal precursors. *Nature* 348:728–30

Ingham PW, Baker NE, Martinez-Arias A. 1988. Regulation of segment polarity genes in the *Drosophila* blastoderm by *fushi tarazu* and *even skipped*. *Nature* 331:73–75

Jacobs JR, Goodman CS. 1989a. Embryonic development of axon pathways in the *Drosophila* CNS. I. A glial scaffold appears before the first growth cones. *J. Neurosci.* 9:2402–11

Jacobs JR, Goodman CS. 1989b. Embryonic development of axon pathways in the *Drosophila* CNS. II. Behavior of pioneer growth cones. *J. Neurosci.* 9:2412–22

Jacobs JR, Hiromi Y, Patel NH, Goodman CS. 1989. Lineage, migration, and morphogenesis of longitudinal glia in the *Drosophila* CNS as revealed by a molecular lineage marker. *Neuron* 2:1625–31

Jan YN, Jan LY. 1990. Genes required for specifying cell fates in *Drosophila* embryonic sensory nervous system. *Trends Neurosci.* 13:493–98

Jansen JKS, Fladby T. 1990. The perinatal reorganization of the innervation of skeletal muscle in mammals. *Prog. Neurobiol.* 34: 39–90

Jessell TM, Bovolenta P, Placzek M, et al. 1989. Polarity and patterning in the neural tube: the origin and role of the floor plate. *Ciba Found. Symp.* 144:255–80

Jessell TM, Dodd J. 1992. Midline signals that control the dorso-ventral polarity of the neural tube. *Semin. Neurosci.* 4:317–25

Jiménez F, Campos-Ortega JA. 1990. Defective neuroblast commitment in mutants of the *achaete-scute* complex and adjacent genes of *D. melanogaster*. *Neuron* 5:81–89

Johansen J, Halpern ME, Keshishian H. 1989. Axonal guidance and the development of muscle fiber-specific innervation in *Drosophila* embryos. *J. Neurosci.* 9:4318–32

Johnson JE, Birren SJ, Anderson DJ. 1990. Two rat homologues of *Drosophila* achaete-scute specifically expressed in neuronal precursors. *Nature* 346:858–61

Kapfhammer JP, Raper JA. 1987. Collapse of growth cone structure on contact with specific neurites in culture. *J. Neurosci.* 7:201–12

Keshishian H. 1980. The origin and morphogenesis of pioneer neurons in the grasshopper metathoracic leg. *Dev. Biol.* 80:388–97

Keynes RJ, Stern CD. 1984. Segmentation in the vertebrate nervous system. *Nature* 310:786–89

Keynes RJ, Stirling RV, Stern CD, Summerbell D. 1987. The specificity of motor innervation of the chick wing does not depend upon the segmental origin of muscles. *Development* 99:565–75

Kimble J, Sulston J, White J. 1979. Regulative development in the post-embryonic lineages of *Caenorhabditis elegans*. In *Cell Lineage, Stem Cells and Cell Determination, INSERM Symp. No. 10*, ed. N Le Douarin, pp. 59–68. Amsterdam: Elsevier/North Holland Biomedical Press

Kimmel CB, Kane DA, Walker C, et al. 1989. A mutation that changes cell movement and cell fate in the zebrafish embryo. *Nature* 337:358–62

Kimmel CB, Schilling T, Hatta K. 1991. Patterning of body segments in the zebrafish embryo. *Curr. Top. Dev. Biol.* 25:77–110

Kimmel CB, Sepich DS, Trevarrow B. 1988. Development of segmentation in zebrafish. *Development Suppl.* 104:197–207

Kimmel CB, Westerfield M. 1990. Primary neurons of the zebrafish. In *Signals and Sense: Local and Global Order in Perceptual Maps*, ed. GM Edelman, WM Cowan, pp. 561–88. New York: Wiley Interscience

Kingsbury BF. 1930. The developmental significance of the floor-plate of the brain and spinal cord. *J. Comp. Neurol.* 50:177–206

Klämbt C, Goodman CS. 1991. The diversity and pattern of glia during axon pathway formation in the *Drosophila* embryo. *Glia* 4:205–13

Klämbt C, Jacobs JR, Goodman CS. 1991. The midline of the *Drosophila* central nervous system: a model for the genetic analysis of cell fate, cell migration, and growth cone guidance. *Cell* 64:801–15

Kuwada JY, Goodman CS. 1985. Neuronal determination during embryonic development of the grasshopper nervous system. *Dev. Biol.* 110:114–26

Lamborghini JE. 1980. Rohon-beard cells and other large neurons in *Xenopus* embryos originate during gastrulation. *J. Comp. Neurol.* 189:323–33

Lance-Jones C. 1988a. The effect of somite manipulation on the development of moto-neuron projection patterns in the embryonic chick hindlimb. *Dev. Biol.* 126:408–19

Lance-Jones C. 1988b. The somitic level of origin of embryonic chick hindlimb muscles. *Dev. Biol.* 126:394–407

Lance-Jones C, Dias M. 1991. The influence of presumptive limb connective tissue on motoneuron axon guidance. *Dev. Biol.* 143:93–110

Lance-Jones C, Landmesser L. 1980. Motoneuron projection patterns in the chick hind limb following early partial spinal cord reversals. *J. Physiol. (London)* 302:581–602

Lance-Jones C, Landmesser L. 1981a. Pathway selection by chick lumbosacral motoneurons during normal development. *Proc. R. Soc. London Ser. B* 214:1–18

Lance-Jones C, Landmesser L. 1981b. Pathway selection by embryonic chick motoneurons in an experimentally altered environment. *Proc. R. Soc. London Ser. B* 214:19–52

Landmesser LT. 1992. Growth cone guidance in the avian limb: a search for cellular and molecular mechanisms. In *The Nerve Growth Cone*, ed. PC Letourneau, SB Kater, ER Macagno, pp. 373–85. New York: Raven Press. 535 pp.

Landmesser LT, Dahm L, Schultz K, Rutishauser U. 1988. Distinct roles for adhesion molecules during innervation of embryonic chick muscle. *Dev. Biol.* 130:645–70

Landmesser LT, Dahm L, Tang J, Rutishauser U. 1990. Polysialic acid as a regulator of intramuscular nerve branching during embryonic development. *Neuron* 4:655–67

Landmesser LT, O'Donovan MJ. 1984. The activation patterns of embryonic chick motoneurones projecting to inappropriate muscles. *J. Physiol.* 347:205–24

Leber SM, Breedlove SM, Sanes JR. 1990. Lineage, arrangement, and death of clonally related motoneurons in chick spinal cord. *J. Neurosci.* 10:2451–62

Lehmann R, Jiménez F, Dietrich U, Campos-Ortega JA. 1983. On the phenotype and development of mutants of early neurogenesis in *Drosophila melanogaster*. *Roux's Arch. Dev. Biol.* 192:62–74

Leiss D, Hinz U, Gasch A, et al. 1988. β3 tubulin expression characterizes the differentiating mesodermal germ layer during *Drosophila* embryogenesis. *Development* 104:525–31

Lewis J, Al-Ghaith L, Swanson G, Khan A. 1983. The control of axon outgrowth in the developing chick wing. In *Limb Development and Regeneration, Part A*, ed. JF Fallon, AI Caplan, pp. 195–205. New York: Liss. 639 pp.

Lewis J, Chevallier A, Kieny M, Wolpert L. 1981. Muscle nerve branches do not develop in chick wings devoid of muscle. *J. Embryol. Exp. Morphol.* 64:211–32

Lim T-M, Jaques KF, Stern CD, Keynes RJ. 1991. An evaluation of myelomeres and segmentation of the chick embryo spinal cord. *Development* 113:227–38

Liu DWC, Westerfield M. 1990. The formation of terminal fields in the absence of competitive interactions among primary motoneurons in the zebrafish. *J. Neurosci.* 10:3947–59

Lo L-C, Johnson JE, Wuenschell CW, et al. 1991. Mammalian *achaete-scute* homolog 1 is transiently expressed by spatially restricted subsets of early neuroepithelial and neural crest cells. *Genes Dev.* 5:1524–37

Lumsden AGS, Davies AM. 1983. Earliest sensory nerve fibres are guided to peripheral targets by attractants other than nerve growth factor. *Nature* 306:786–88

Lumsden AGS, Davies AM. 1986. Chemotropic effect of specific target epithelium in the developing mammalian nervous system. *Nature* 323:538–39

Martín-Bermudo MD, Martínez C, Rodríguez A, Jiménez F. 1991. Distribution and function of the *lethal of scute* gene product during early neurogenesis in *Drosophila*. *Development* 113:445–54

Matise MP, Lance-Jones C. 1992. The timing of motoneuron commitment in the developing chick spinal cord. *Soc. Neurosci. Abstr.* 18:1111

McCaig CD. 1986. Myoblasts and myoblast-conditioned medium attract the earliest spinal neurites from frog embryos. *J. Physiol.* 375:39–54

McGinnis W, Krumlauf R. 1992. Homeobox genes and axial patterning. *Cell* 68:283–302

Montell DJ, Goodman CS. 1989. *Drosophila* laminin: sequence of B2 subunit and expression of all three subunits during embryogenesis. *J. Cell Biol.* 109:2441–53

Moody SA, Jacobson M. 1983. Compartmental relationships between anuran primary spinal motoneurons and somitic muscle fibers that they first innervate. *J. Neurosci.* 8:1670–82

Moos M, Tackle R, Scherer H, et al. 1987. Neural cell adhesion molecule L1 as a member of the immunoglobulin superfamily with binding domains similar to fibronectin. *Nature* 334:701–4

Myers PZ. 1985. Spinal motoneurons of the larval zebrafish. *J. Comp. Neurol.* 236:555–61

Myers PZ, Eisen JS, Westerfield M. 1986. Development and axonal outgrowth of identified motoneurons in the zebrafish. *J. Neurosci.* 6:2278–89

Noakes PG, Bennett MR. 1987. Growth of axons into developing muscles of the chick forelimb is preceded by cells that stain with

Schwann cell antibodies. *J. Comp. Neurol.* 259:330–47

Noakes PG, Bennett MR, Stratford J. 1988. Migration of Schwann cells and axons into developing chick forelimb muscles following removal of either the neural tube or the neural crest. *J. Comp. Neurol.* 277:214–33

Nose A, Mahajan VB, Goodman CS. 1992. Connectin: a homophilic cell adhesion molecule expressed on a subset of muscles and the motoneurons that innervate them in *Drosophila*. *Cell* 70:553–67

Nurcombe V, Bennett MR. 1982. Evidence for neuron-survival and neurite-promoting factors from skeletal muscle: their effects on embryonic spinal cord. *Neurosci. Lett.* 34:89–93

Nüsslein-Volhard C, Weischaus E. 1980. Mutations affecting segment number and polarity in *Drosophila*. *Nature* 287:795–801

O'Brien MK, Landmesser L, Oppenheim RW. 1990. Development and survival of thoracic motoneurons and hindlimb musculature following transplantation of the thoracic neural tube to the lumbar region in the chick embryo: functional aspects. *J. Neurobiol.* 21:341–55

Okamoto H, Kuwada JY. 1991a. Outgrowth by fin motor axons in wildtype and a finless mutant of the Japanese Medaka fish. *Dev. Biol.* 146:49–61

Okamoto H, Kuwada JY. 1991b. Alteration of pectoral fin nerves following ablation of fin buds and by ectopic fin buds in the Japanese Medaka fish. *Dev. Biol.* 146:62–71

Oppenheim RW, Haverkamp LJ, Prevette D, et al. 1988. Reduction of naturally occurring motoneuron death *in vivo* by a target-derived neurotrophic factor. *Science* 240:919–22

Oppenheim RW, Nuñez R. 1982. Electrical stimulation of hindlimb increases neuronal cell death in chick embryo. *Nature* 295:57–59

Papalopulu N, Kintner C. 1992. Induction and patterning of the neural plate. *Semin. Neurosci.* 4:295–306

Patel NH, Schafer B, Goodman CS, Holmgren R. 1989. The role of segment polarity genes during *Drosophila* neurogenesis. *Genes Dev.* 3:890–904

Patel NH, Snow PM, Goodman CS. 1987. Characterization and cloning of fasciclin III: a glycoprotein expressed on a subset of neurons and axon pathways in *Drosophila*. *Cell* 48:975–88

Phelan KA, Hollyday M. 1990. Axon guidance in muscleless chick wings: the role of muscle cells in motoneuronal pathway selection and muscle nerve formation. *J. Neurosci.* 10:2699–716

Pike SH, Eisen JS. 1990. Interactions between identified motoneurons in embryonic zebrafish are not required for normal motoneuron development. *J. Neurosci.* 10:44–49

Pike SH, Melancon EF, Eisen JS. 1992. Pathfinding by zebrafish motoneurons in the absence of normal pioneer axons. *Development* 114:825–31

Pittman R, Oppenheim RW. 1979. Cell death of motoneurons in the chick embryo spinal cord. V. Evidence that a functional neuromuscular interaction is involved in the regulation of naturally occurring cell death and the stabilization of synapses. *J. Comp. Neurol.* 187:425–46

Placzek M, Tessier-Lavigne M, Yamada T, et al. 1990. Mesodermal control of neural cell identity: floor plate induction by the notochord. *Science* 250:985–88

Placzek M, Yamada T, Tessier-Lavigne M, et al. 1991. Control of dorsoventral pattern in vertebrate neural development: induction and polarizing properties of the floor plate. *Development Suppl.* 2:105–22

Raible DW, Wood A, Hodsdon W, et al. 1992. Segregation and early dispersal of neural crest cells in the embryonic zebrafish. *Dev. Dyn.* 194:29–42

Reichardt LF, Tomaselli KJ. 1991. Extracellular matrix molecules and their receptors: functions in neural development. *Annu. Rev. Neurosci.* 14:531–70

Rickmann M, Fawcett JW, Keynes RJ. 1985. The migration of neural crest cells and the growth of motor axons through the rostral half of the chick somite. *J. Embryol. Exp. Morphol.* 90:437–55

Rogers SL, Edson KJ, Letourneau PC, McLoon SC. 1986. Distribution of laminin in the developing peripheral nervous system of the chick. *Dev. Biol.* 113:429–35

Roth J, Kempf A, Reuter G, et al. 1992. Occurrence of sialic acids in *Drosophila melanogaster*. *Science* 256:673–5

Sanes JR. 1989. Analysing cell lineage with a recombinant retrovirus. *Trends Neurosci.* 12:21–28

Sanes DH, Poo M-M. 1989. *In vitro* analysis of position- and lineage-dependent selectivity in the formation of neuromuscular synapses. *Neuron* 2:1237–44

Schoenwolf GC, Smith JL. 1990. Mechanisms of neurulation: traditional viewpoint and recent advances. *Development* 109:243–70

Schulte-Merker S, Ho RK, Herrmann BG, Nüsslein-Volhard 1992. The protein product of the zebrafish homologue of the mouse *T* gene is expressed in nuclei of the germ ring and the notochord of the early embryo. *Development* 116:1021–32

Shatz CJ, Stryker MP. 1988. Prenatal tetrodotoxin infusion blocks segregation of retinogeniculate afferents. *Science* 242:87–89

Sink H, Whitington PM. 1991a. Location and connectivity of abdominal motoneurons in the embryo and larva of *Drosophila melanogaster*. *J. Neurobiol.* 22:298–311

Sink H, Whitington PM. 1991b. Pathfinding in the central nervous system and periphery by identified embryonic *Drosophila* motor axons. *Development* 112:307–16

Skeath JB, Carroll SB. 1992. Regulation of proneural gene expression and cell fate during neuroblast segregation in the *Drosophila* embryo. *Development* 114:939–46

Spemann H. 1938. *Embryonic Development and Induction.* New York: Hafner. Reprinted in 1967

Sretavan DW, Shatz CJ, Stryker MP. 1988. Modification of retinal ganglion cell axon morphology by prenatal infusion of tetrodotoxin. *Nature* 336:468–71

Stent GS. 1985. The role of cell lineage in development. *Philos. Trans. R. Soc. London Ser. B* 312:3–19

Stern CD, Jaques KF, Lim T-M, et al. 1991. Segmental lineage restrictions in the chick embryo spinal cord depend on the adjacent somites. *Development* 113:239–44

Stirling RV, Summerbell D. 1985. The behavior of growing axons invading developing chick wing buds with dorsoventral or anteroposterior axis reversed. *J. Embryol. Exp. Morphol.* 85:251–69

Stirling RV, Summerbell D. 1988. Specific guidance of motor axons to duplicated muscles in the developing amniote limb. *Development* 103:97–110

Stuermer CAO, Rohrer B, Münz H. 1990. Development of the retinotectal projection in zebrafish embryos under TTX-induced neural-impulse blockade. *J. Neurosci.* 10:3615–26

Takeichi M. 1988. The cadherins: cell-cell adhesion molecules controlling animal morphogenesis. *Development* 102:639–55

Tang J, Landmesser L, Rutishauser U. 1992. Polysialic acid influences specific pathfinding by avian motoneurons. *Neuron* 8:1031–44

Tessier-Lavigne M, Placzek M, Lumsden AGS, et al. 1988. Chemotropic guidance of developing axons in the mammalian central nervous system. *Nature* 336:775–78

Thomas JB, Bastiani MJ, Bate M, Goodman CS. 1984. From grasshopper to *Drosophila*: a common plan for neuronal development. *Nature* 310:203–7

Thomas KR, Capecchi MR. 1990. Targeted disruption of the murine *int-1* protooncogene resulting in severe abnormalities in midbrain and cerebellar development. *Nature* 346:847–50

Tosney KW. 1987. Proximal tissues and patterned neurite outgrowth at the lumbosacral level of the chick embryo: deletion of the dermamyotome. *Dev. Biol.* 122:540–58

Tosney KW. 1988. Proximal tissues and patterned neurite outgrowth at the lumbosacral level of the chick embryo: partial and complete deletion of the somite. *Dev. Biol.* 127:266–86

Tosney KW. 1992. Growth cone navigation in the proximal environment of the chick embryo. In *The Nerve Growth Cone*, ed. PC Letourneau, SB Kater, ER Macagno, pp. 387–403. New York: Raven Press. 535 pp.

Tosney KW, Landmesser LT. 1985. Growth cone morphology and trajectory in the lumbosacral region of the chick embryo. *J. Neurosci.* 5:2345–58

Tosney KW, Oakley RA. 1990. The perinotochordal mesenchyme acts as a barrier to axon advance in the chick embryo: implications for a general mechanism of axonal guidance. *Exp. Neurol.* 109:75–89

Tosney KW, Watanabe M, Landmesser L, Rutishauser U. 1986. The distribution of NCAM in the chick hindlimb during axon outgrowth and synaptogenesis. *Dev. Biol.* 114:437–52

Weinmaster G, Roberts VJ, Lemke G. 1991. A homolog of *Drosophila Notch* expressed during mammalian development. *Development* 113:199–205

Weinmaster G, Roberts VJ, Lemke G. 1992. *Notch2*: a second mammalian *Notch* gene. *Development* 116:931–41

Westerfield M. 1987. Substrate interactions affecting motor growth cone guidance during development and regeneration. *J. Exp. Biol.* 132:161–75

Westerfield M, Liu DW, Kimmel CB, Walker C. 1990. Normal pathfinding by pioneer motor growth cones in mutant zebrafish lacking functional acetylcholine receptors. *Neuron* 4:867–74

Whitelaw V, Hollyday M. 1983. Thigh and calf discrimination in the motor innervation of the chick hindlimb following deletions of limb segments. *J. Neurosci.* 3:1199–215

Yamada T, Placzek M, Tanaka H, et al. 1991. Control of cell pattern in the developing nervous system: polarizing activity of the floor plate and notochord. *Cell* 64:635–47

Zhong Y, Budnik V, Wu C-F. 1992. Synaptic plasticity in *Drosophila* memory and hyperexcitable mutants: role of cAMP cascade. *J. Neurosci.* 12:644–51

Annu. Rev. Neurosci. 1994. 17:31–108

CLONED GLUTAMATE RECEPTORS

Michael Hollmann and Stephen Heinemann

Molecular Neurobiology Laboratory, The Salk Institute for Biological Studies, La Jolla, California 92037

KEY WORDS: NMDA receptor, AMPA receptor, kainate receptor, CPD receptor, metabotropic glutamate receptor, ionotropic glutamate receptor

INTRODUCTION

The application of molecular cloning technology to the study of the glutamate receptor system has led to an explosion of knowledge about the structure, expression, and function of this most important fast excitatory transmitter system in the mammalian brain. The first functional ionotropic glutamate receptor was cloned in 1989 (Hollmann et al 1989), and the results of this molecular-based approach over the past three years are the focus of this review. We discuss the implications of and the new questions raised by this work—which is probably only a glance at this fascinating and complex signaling system found in brains from the snails to man.

Glutamate receptors are found throughout the mammalian brain, where they constitute the major excitatory transmitter system. The longest-known and best-studied glutamate receptors are ligand-gated ion channels, also called ionotropic glutamate receptors, which are permeable to cations. They have traditionally been classified into three broad subtypes based upon pharmacological and electrophysiological data: α-amino-3-hydroxy-5-methyl-4-isoxazole propionate (AMPA) receptors, kainate (KA) receptors, and N-methyl-D-aspartate (NMDA) receptors. Recently, however, a family of G protein-coupled glutamate receptors, which are also called metabotropic glutamate or *trans*-1-aminocyclopentane-1,3-dicarboxylate (tACPD) receptors, was identified (Sugiyama et al 1987). (For reviews of the classification and the pharmacological and electrophysiological properties of glutamate receptors see Mayer & Westbrook 1987, Collingridge & Lester 1989, Honore 1989, Monaghan et al 1989, Wroblewski & Danysz 1989, Hansen &

31

Krogsgaard-Larsen 1990, Lodge & Johnson 1990, MacDonald & Nowak 1990, McDonald & Johnston 1990, Reynolds & Miller 1990, Sansom & Usherwood 1990, Schoepp et al 1990, Watkins et al 1990, Yoneda & Ogita 1991, and Zorumski & Thio 1992.)

Most plausible theories of learning, pattern recognition, and memory depend upon changes in the efficiency of chemical synapses (Cajal 1911, Hebb 1949, Hopfield 1982). The glutamate receptor has become the focus of much work because of several experimental findings that suggest a central role for glutamate receptors in learning and memory (for reviews see Collingridge & Bliss 1987, Morris 1989, Collingridge & Singer 1990, Etienne & Baudry 1990, Holmes & Levy 1990, Aroniadou & Teyler 1991, Collingridge et al 1991, Fields et al 1991, Izquierdo 1991, Racine et al 1991, Reid & Morris 1991, Advokat & Pellegrin 1992, Ben-Ari et al 1992, Bliss & Collingridge 1993). The NMDA receptor has attracted much attention because its properties make it an ideal candidate for a receptor involved directly in the learning process. The NMDA receptor is the only ligand-gated ion channel whose probability of opening depends strongly upon the voltage across the membrane, under physiological conditions. This property permits this receptor to operate as a "Hebbian" molecule (Hebb 1949). Additional evidence suggests that NMDA receptors play a role in the early stages of topographic map formation in the mammalian brain, where early experience during development is known to alter the maps (for review see Constantine-Paton et al 1990).

A large body of evidence indicates that glutamate receptors play a role in a number of brain diseases and the damage that occurs after head injuries. It also has been known for decades that glutamate is toxic to neurons in culture and in vivo, and many experiments implicate the glutamate receptor as a mediator of these toxic effects of glutamate. (For reviews see Meldrum 1985; Maragos et al 1987; Choi 1988; Olney 1989a,b; Klockgether & Turski 1989; Carlsson & Carlsson 1990; Choi & Rothman 1990; Cowburn et al 1990; Dingledine et al 1990a; Girault et al 1990; Meldrum & Garthwaite 1990; Penney et al 1990; Wachtel & Turski 1990; Advokat & Pellegrin 1992; Beal 1992; Storey et al 1992; and Appel 1993.) These observations have led investigators to suggest that many neurological accidents involving strokes in which there is a loss of oxygen and glucose or epileptic seizures result in brain damage because of over-stimulation by glutamate. It has also been proposed that degenerative diseases such as Alzheimer's disease, Huntington's disease, Parkinson's disease, and amyotrophic lateral sclerosis (ALS) may involve neuronal cell death caused by excessive activation of the glutamate receptor system. Thus, understanding the glutamate receptor systems is crucial for the understanding of basic brain functions such as learning and memory as well as for the rational treatment of disease.

Table 1 Structural and functional properties of cloned glutamate receptor genes

Gene	Synonyms	Species	Mature protein size (aa)	Molecular weight (Daltons)	Receptor type	Order of agonist potency at homomeric receptor[f]	N-glycosylation and phosphorylation sites[j]	References[g]	EMBL/ GenBank accession number
GluR1	GluR-K1	Rat	889	99,769	Ionotropic, AMPA receptor	QA>DOM~AMPA>GLU>KA	6/6/22/2/1	Hollmann et al 1989	X17184
	GluR-A	Rat	889					Keinänen et al 1990	M36418
	GluR-K1	Rat	889					Nakanishi et al 1990	M64752
	GluHI	Human	889					Puckett et al 1991	X58633
	KR4	Human	888					Potier et al 1992	X57497
	α1	Mouse	889					Sakimura et al 1990	M81886
	HBGR1	Human	888					Sun et al 1992	M85035
GluR2	GluR-B	Rat	862	96,400	Ionotropic, AMPA receptor	QA>AMPA>GLU>KA	5/6/22/1/1	Boulter et al 1990	M36419
	GluR-K2	Rat	862					Keinänen et al 1990	X54655
	α2	Rat	862					Nakanishi et al 1990	X57498
		Mouse	862					Sakimura et al 1990	M85036
GluR3	GluR-C	Rat	866	98,000	Ionotropic, AMPA receptor	QA>AMPA>GLU>KA	5/4/21/1/1	Boulter et al 1990	M36420
	GluR-K3	Rat	866					Keinänen et al 1990	X54656
		Rat	866					Nakanishi et al 1990	M85037
GluR4	GluR-D	Rat	881	101,034	Ionotropic, AMPA receptor	QA>AMPA>GLU>KA	6/5/27/1/1	Bettler et al 1990	M36421
		Rat	881					Keinänen et al 1990	
GluR5	GluR-5	Rat	875	102,785	Ionotropic, low affinity KA receptor	DOM>KA>QA~Glu>AMPA	8/3/19/3/0	Bettler et al 1990	M83561
	GluR-5	Rat	875				DOM>>QA~GLU>>AMPA[b]		Sommer et al 1992	Z11713
GluR6		Rat	877[h]	93,891	Ionotropic, low affinity KA receptor	DOM>KA>QA>GLU	8/4/21/3/0	Egebjerg et al 1991	Z11548
	β2	Mouse	(833)[a]			DOM>>QA>GLU>>AMPA[b]		Morita et al 1992	D10054
	GluR-6	Rat	877					Lomeli et al 1992	Z11715
GluR7	GluR-7	Rat	888	100,000	Ionotropic, low affinity KA receptor	DOM>KA>GLU>QA>AMPA[b]	9/3/19/1/0	Bettler et al 1992	M83552
	GluR-7	Rat	888					Lomeli et al 1992	Z11716

Table 1 (*Continued*)

Gene	Synonyms	Species	Mature protein size (aa)	Molecular weight (Daltons)	Receptor type	Order of agonist potency at homomeric receptor[f]	N-glycosylation and phosphorylation sites[j]	References[g]	EMBL/GenBank accession number
KA1	KA-1	Rat	936	105,000	Ionotropic, high affinity KA receptor	KA>QA>DOM>GLU>>AMPA[b]	9/2/10/1/0	Werner et al 1991	X59996
KA2	KA-2, γ, humEAA2	Rat, Mouse, Human	965, 965, 962	109,000, 107,176	Ionotropic, high affinity KA receptor	KA>QA>DOM>GLU>>AMPA[b]	11/2/11/1/0	Herb et al 1992, Sakimura et al 1992, Kamboj et al 1992	Z11581, D10011, S40369
KBP-chick		Chicken	464	51,800	KA binding protein	DOM~KA>>L-GLU>D-GLU >QA[b,c]	2/2/9/0/2	Gregor et al 1989	X17700
KBP-frog		Frog	470	52,500	KA binding protein	DOM>KA>QA~GLU[b]	2/2/7/1/1	Wada et al 1989	X17314
DGluR-I		*Drosophila*	964	108,482	Ionotropic, KA receptor	KA	11/2/12/0/0	Ultsch et al 1992	M97192
DGluR-II		*Drosophila*	983	101,790	Ionotropic, GLU receptor	GLU>QA,AMPA,KA	4/2/15/1/1	Schuster et al 1991	M73271
LymGluR		Pond snail	898	100,913	Orphan receptor (most probably an ionotropic GluR)	—	13/4/17/1/1	Hutton et al 1991	X60086
delta1	δ1	Mouse, Rat	994, 992	110,440	Orphan receptor (most probably an ionotropic GluR)	does not bind KA or AMPA	4/3/20/2/1[k]	Yamazaki et al 1992a, Lomeli et al 1993	D10171, Z17238
delta2		Rat	991	113,229	Orphan receptor (most probably an ionotropic GluR)	does not bind KA or AMPA	2/2/9/1/1	Lomeli et al 1993	Z17239

		Species	Residues	MW	Type	Pharmacology		Reference	Accession
NMDAR1		Rat	920	103,477	Ionotropic NMDA receptor	GLU>NMDA	10/4/18/2/2	Moriyoshi et al 1991	X63255
	ζ1	Mouse	920					Yamazaki et al 1992b	S88563
		Human	920					Karp et al 1993	D131515
NMDAR2A	NR2A	Rat	1445	163,267	Ionotropic NMDA receptor	—	18/1/12/1/0	Monyer et al 1992	M91561
	ε1	Mouse	1445					Meguro et al 1992	D10217
NMDAR2B	NR2B	Rat	1456	162,875	Ionotropic NMDA receptor	—	16/1/11/1/0	Monyer et al 1992	M91562
	ε2	Mouse	1456					Kutsuwada et al 1992	D10651
NMDAR2C	NR2C	Rat	1218[i]	133,513	Ionotropic NMDA receptor	—	5/1/17/1/0[k]	Monyer et al 1992	M91563
	ε3	Mouse	1220					Kutsuwada et al 1992	D10694
NMDAR2D	NR2D	Rat	1329[m]		Ionotropic NMDA receptor	—		Monyer & Seeburg 1992[l]	D13213[m]
	ε4	Mouse	1296	140,656			6/0/14/0/0[k]	Ikeda et al 1992	D12822
GBP		Rat	594	57,020	Insufficient data	Insufficient data[d]	0/0/1/0/0	Kumar et al 1991	S61973
mGluR1		Rat	1179	133,229	Metabotropic, ACPD receptor, linked to IP/Ca^{2+} signal transduction	QA>IBO~GLU>tACPD	8/5/37/2/0	Masu et al 1991	X57569
	Glu_GR	Rat	1179					Houamed et al 1991	M61099
mGluR2		Rat	854	95,770	Metabotropic, ACPD receptor, linked to inhibitory cAMP cascade	GLU~tACPD>IBO>QA	5/4/13/3/0	Tanabe et al 1992	M92075
mGluR3		Rat	857	98,960	Metabotropic, ACPD receptor, linked to inhibitory cAMP cascade	GLU~tACPD>IBO>QA	4/18/0/0	Tanabe et al 1992	M92076
mGluR4		Rat	880	101,810	Metabotropic, ACPD receptor, linked to inhibitory cAMP cascade	AP4>GLU>>tACPD QA is inactive	5/3/16/3/0	Tanabe et al 1992	M92077

Table 1 (*Continued*)

Gene	Synonyms	Species	Mature protein size (aa)	Molecular weight (Daltons)	Receptor type	Order of agonist potency at homomeric receptor[f]	N-glycosylation and phosphorylation sites[j]	References[g]	EMBL/GenBank accession number
mGluR5		Rat	1151	128,289	Metabotropic, ACPD receptor, linked to IP/Ca^{2+} signal transduction	QA>GLU~IBO>tACPD	8/12/38/0	Abe et al 1992	D10891
mGluR6		Rat	871[e]		Metabotropic, ACPD receptor linked to inhibitory cAMP cascade			Nakanishi 1992	

[a] The published sequence of β2 encodes a 833–amino acid (aa) protein; however, comparison of the cDNA sequence in the database to that of GluR6 suggests that the two proteins have the same size, and that a sequencing error (one missing base in a stretch of eight "A" residues) caused the truncated 833-aa protein.

[b] Order of potencies derived from inhibition of ligand binding experiments.

[c] Binding measured to partially purified protein fraction containing a prominent 49-kDa band plus additional proteins (Gregor et al 1988.

[d] GLU binding to a fusion protein partially purified from bacterial cultures by GLU affinity chromatography has been reported (Kumar et al 1991), but without controls for endogenous bacterial glutamate binding proteins. See p. 88 for discussion.

[e] Including signal peptide; sequence has not yet been published.

[f] See Table 4 for data and references for agonist potencies.

[g] All references are to the original cloning papers of the respective cDNA clones.

[h] In the original publication, the protein size was stated to be 853 aa, due to a sequencing error that was later corrected (Bettler et al 1992).

[i] In the original publication, the protein size was stated to be 943 aa, due to a 1-bp deletion in the sequenced clone, which introduced an early stop codon; the protein size was later corrected to 1218 aa (Burnashev et al 1992c).

[j] Number of consensus sites are for N-glycosylation (Marshall 1972), cAMP-dependent protein kinase A (Kennelly & Krebs 1991), protein kinase C (Kennelly & Krebs 1991), Ca^{2+}/calmodulin-dependent protein kinase II (Kennelly & Krebs 1991), and tyrosine kinase (Cooper et al 1984), respectively; only extracellular N-glycosylation sites and intracellular phosphorylation sites according to TMD topology model 2 (see Figure 4) are listed.

[k] Data from the mouse sequence.

[l] Sequence has not been published.

[m] Added in proof: sequence published by Ishii et al (1993).

Table 2 Amino acid sequence identities among glutamate receptor genes[a]

	AMPA receptors	KA receptors			Invertebrate receptors	GLU receptor	Orphan receptors	NMDA receptors		Status unknown	ACPD receptors
	GluR1–GluR4	Low affinity GluR5–GluR7	High affinity KA1, KA2	Binding proteins KBP-chick, KBP-frog	DGluR-I, LymGluR	Invertebrate receptor DGluR-II	delta1, delta2	NMDAR1	NMDAR2A–NMDAR2D	GBP	mGluR1–mGluR5
GluR1–GluR4	67.7–72.8	38.6–41.4	36.2–37.3	35.1–37.8	41.7–46.1	29.9–31.3	28.1–30.2	25.1–27.1	21.8–26.3	17.4–20.6	14.2–19.6
GluR5–GluR7		74.4–80.9	42.5–44.7	37.7–40.8	35.8–38.8	30.9–32.8	29.1–31.6	27.3–28.7	22.1–25.6	18.0–19.6	17.4–20.3
KA1, KA2			70.3	35.8–36.8	33.0–35.4	28.7–29.6	25.0–28.2	26.2–27.3	22.2–27.1	18.0–20.7	18.2–22.2
KBP-chick, KBP-frog				55.4	34.5–36.0	29.9–32.1	28.5–29.9	27.5–28.5	25.2–27.3	18.2–18.3	16.1–23.8
DGluR-I, LymGluR					40.2	29.1–29.6	26.0–29.1	26.2–28.0	21.7–25.0	15.5–18.1	15.3–20.2
DGluR-II						—	23.8–24.4	22.3	17.6–22.5	18.5	16.3–19.2
delta1, delta2							55.6	25.5–28.0	21.2–24.8	17.9–18.8	16.9–19.5
NMDAR1								—	25.6–27.5	18.0	18.2–20.2
NMDAR2A–NMDAR2D									42.2–55.9	16.7–19.2	15.0–21.5
GBP										—	16.2–19.5
mGluR1–mGluR5											41.8–68.0

[a] The program "gap" from the University of Wisconsin sequence analysis software (Devereux et al 1984) was used to calculate pairwise sequence identities. See Table 1 for references to sequences. Sequence identities were calculated as the percentage of the number of amino acid residues of the shorter one of the two clones compared in each case and are given as the range of identities found in the group of clones compared.

Table 3 Splice variants of glutamate receptor genes

Gene	Splice variants	Mature protein size (aa)	Location of variation	Type of splice variant	References	EMBL/GenBank accession number
GluR1	GluR1$_{flop}$ = GluR1	889	—	—	Sommer et al 1990	M36418
	GluR1$_{flip}$	889	In front of TMD IV	Alternate exon of same size	Sommer et al 1990	M38060
GluR2	GluR2$_{flop}$ = GluR2	862	—	—	Sommer et al 1990	M36419
	GluR2$_{flip}$	862	In front of TMD IV	Alternate exon of same size	Sommer et al 1990	M38061
GluR3	GluR3$_{flop}$ = GluR3	866	—	—	Sommer et al 1990	M36420
	GluR3$_{flip}$	866	In front of TMD IV	Alternate exon of same size	Sommer et al 1990	M38062
GluR4	GluR4$_{flop}$ = GluR4	881	—	—	Sommer et al 1990	M36421
	GluR4$_{flip}$	881	In front of TMD IV	Alternate exon of same size	Sommer et al 1990	M38063
	GluR4c	863	At C-terminus	Insertion of new exon, introduces new stop codon	Gallo et al 1992	S94371
GluR5	GluR5-1	890	At N-terminus	Insertion of new exon	Bettler et al 1990	M83560
	GluR5-2 = GluR5	875	—	—	Bettler et al 1990	M83561
	GluR-5-2a	826	At C-terminus	Insertion of new exon, introduces new stop codon	Sommer et al 1992	Z11712
	GluR-5-2b = GluR5-2 = GluR5	875	—	—	Sommer et al 1992	Z11713
	GluR-5-2c	904	At C-terminus	In-frame insertion of new exon	Sommer et al 1992	Z11714

	Name	No.	Location	Modification	Reference	Accession
mGluR1	mGluR1α = mGluR1	1179		—	Tanabe et al 1992	X57569
	mGluR1β	886	At C-terminus	Insertion of new exon, introduces new stop codon	Tanabe et al 1992	S78270[a]
	mGluR1a = mGluR1				Pin et al 1992	
	mGluR1b = mGluR1β	876	At C-terminus	Insertion of new exon, introduces new stop codon	Pin et al 1992	
	mGluR1c				Pin et al 1992	
NMDAR1	NMDAR1A = NMDAR1	920		—	Sugihara et al 1992	X63255
	NMDAR1B	941	At N-terminus	Insertion of new exon (5)	Sugihara et al 1992	S39217[a]
	NMDAR1C	883	At C-terminus	Deletion of exon 21	Sugihara et al 1992	S39218[a]
	NMDAR1D	904	At C-terminus	Partial deletion of exon 22	Sugihara et al 1992	S39219[a]
	NMDAR1E	867	At C-terminus	Deletion of exon 21, plus partial deletion of exon 22	Sugihara et al 1992	S39220[a]
	NMDAR1F	904	At N- and C-terminus	Insertion of new exon (5) plus deletion of 1 exon	Sugihara et al 1992	
	NMDAR1G	888	At N- and C-terminus	Insertion of new exon (5) plus deletion of exon 21 plus partial deletion of exon 22	Sugihara et al 1992	
	ζ1-2 = NMDAR1C	883	At C-terminus	Deletion of exon 21	Yamazaki et al 1992b	S37525
	NMDAR1-SL = NMDAR1	920			Anantharam et al 1992	
	NMDAR1-LL = NMDAR1B	941			Anantharam et al 1992	X65227
	NMDAR1-SS = NMDAR1C	883			Anantharam et al 1992	
	NMDAR1-LS = NMDAR1F	904			Anantharam et al 1992	

Table 3 (*Continued*)

Gene	Splice variants	Mature protein size (aa)	Location of variation	Type of splice variant	References	EMBL/GenBank accession number
	NMDA-R1a = NMDAR1	920			Nakanishi et al 1992	S44967[a]
	NMDA-R1b = NMDAR1B	940			Nakanishi et al 1992	
	NMDA-R1c = NMDAR1C	883			Nakanishi et al 1992	
	NMDAR1a = NMDAR1	920			Durand et al 1992	
	NMDAR1b = NMDAR1G	888			Durand et al 1992	
	NMDAR1c = NMDAR1E	867			Durand et al 1992	
	NMDAR1-1a = NMDAR1	920			Hollmann et al 1993	L08228[b]
	NMDAR1-1b = NMDAR1B	941			Hollmann et al 1993	L08228[b]
	NMDAR1-2a = NMDAR1C	883			Hollmann et al 1993	L08228[b]
	NMDAR1-2b = NMDAR1F	904			Hollmann et al 1993	L08228[b]
	NMDAR1-3a = NMDAR1D	904			Hollmann et al 1993	L08228[b]
	NMDAR1-3b	925	At N- and C-terminus	Insertion of new exon (5) plus partial deletion of exon 22	Hollmann et al 1993	L08228[b]
	NMDAR1-4a = NMDAR1E				Hollmann et al 1993	L08228[b]
	NMDAR1-4b = NMDAR1G				Hollmann et al 1993	L08228[b]

[a] Partial sequence.
[b] Genomic sequence.

Twenty-eight recombinant glutamate receptor (GluR) cDNAs plus a considerable number of splice variants thereof have been cloned since 1989, when the first functional glutamate receptor, GluR1, was isolated by expression cloning (Hollmann et al 1989). These 28 GluR genes include 22 members of the ionotropic subfamily as well as six metabotropic receptors (see Table 1), and there is little doubt that more genes remain to be discovered. In many cases, new GluR genes were identified independently and simultaneously in several laboratories and thus were given different names. Additionally, some authors suggested alternative names for previously published GluR genes, thus complicating the nomenclatural situation even further. For this review we use gene names consistent with those originally suggested for the first cloned representative of each subfamily, in order to avoid inconsistent names within a group of related genes.

Taking into account sequence homologies as well as pharmacological properties, the 28 GluR genes that so far have been characterized on a molecular basis can be grouped into 13 subfamilies, 10 for the ionotropic receptor class and 3 for the metabotropic receptor class (see Tables 2 and 7). A number of recent reviews (Barnard & Henley 1990, Betz 1990, Dingledine et al 1990b, Hollmann et al 1990, Gasic & Heinemann 1991, Barnes & Henley 1992, Darlison 1992, Gasic & Hollmann 1992, Henneberry 1992, Nakanishi 1992, Sommer & Seeburg 1992, Pin et al 1993) have discussed various subsets of these GluR genes. This review draws together and discusses current knowledge about the structural and functional properties of each of the currently known 28 receptor subunits as well as what is known about their distribution in the brain and their interactions with each other. More than 110 studies on recombinant glutamate receptors have been published since 1989. Data published prior to February 15, 1993, are included in this review as well as some in press data that have come to our attention.

IONOTROPIC RECEPTORS

AMPA Receptors: GluR1–GluR4

CLONING   In the traditional method for cloning receptors for neurotransmitters, researchers purify the receptor protein to obtain partial protein sequence information, which they then use to synthesize oligonucleotide probes for the screening of cDNA libraries. This approach has not worked for the cloning of functional glutamate receptors, mainly because high-affinity, high-specificity ligands suitable for affinity purification of glutamate receptor proteins were not available [with the exception of domoic acid (DOM), which was used to purify KA binding proteins; see below]. A new cloning technique, introduced by Masu et al in 1987, finally circumvented this problem. This technique exploits the capability of *Xenopus laevis* oocytes to functionally express neurotransmitter receptors from microinjected foreign mRNA. In vitro transcripts are made from pools of clones from cDNA libraries and injected into oocytes; then, utilizing a combination of molecular biological and electrophysiological techniques, researchers test the transcripts for the inherent electrophysiological responses of expressed ligand-gated ion channels or responses mediated by ion channels endogenous to the oocyte that couple to expressed foreign G protein-coupled receptors. We (Hollmann et al 1989) used this method, termed expression cloning, to search through a rat brain cDNA library and isolate the first recombinant glutamate receptor clone, GluR1. With the sequence information of this clone available, several groups (Boulter et al 1990, Keinänen et al 1990, Nakanishi et al 1990, Sakimura et al 1990) carried out standard homology screening and polymerase chain reaction (PCR)-mediated screening for related sequences and turned up the closely related receptor genes GluR2, GluR3, and GluR4 (see Table 1 for synonyms).

STRUCTURAL FEATURES   The four receptor subunits GluR1–GluR4 are of similar size (~900 amino acids), share 68–73% amino acid sequence identity, and show an overall structure that at first glance appears to be quite similar to that of other ligand-gated ion channels, such as the nicotinic acetylcholine receptor (nAChR), the γ-amino butyric acid receptor (GABAR), and the glycine receptor (GlyR). Like these receptors, GluR1–GluR4 contain a hydrophobic domain at the N-terminus that represents the signal peptide required for membrane insertion, plus four hydrophobic domains in the typical cluster-of-three-plus-one pattern (three hydrophobic regions near the middle of the protein and one in the C-terminal portion of the protein) (see Figures 1 and 2), which, in analogy to the nAChR, are interpreted as transmembrane domains (TMDs). Because the N-terminal receptor domain in front of the putative TMD I contains numerous consensus sites for N-glycosylation, it is believed to be extracellular. In GluR1, GluR3, and GluR4 all of the consensus sites for N-glycosylation present in these proteins are located in this N-terminal domain; because N-linked glycosylation has been demonstrated for these subunits (Rogers et al 1991, Hullebroeck & Hampson 1992), it can be concluded that the N-terminus in fact must be extracellular. This topology places the long segment between the putative TMD III and TMD IV on the cytoplasmic side, consistent with the presence of several consensus phosphorylation sites in this domain (see Table 1). The C-terminal domain downstream of TMD IV would be extracellular according to this model. In the absence of experimental evidence for the topology of the receptor subunits, hydrophobicity plots are generally used to assign TMDs. Although the very high hydrophobicity of these two regions apparently leaves little doubt about the

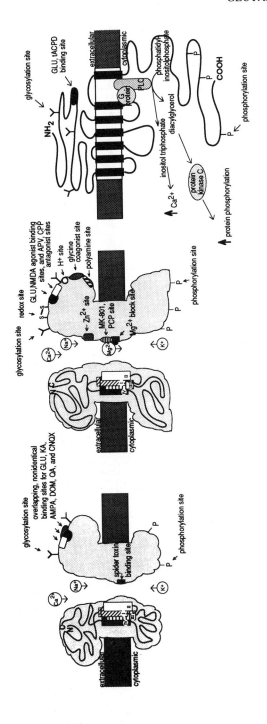

KA/AMPA receptor   **NMDA receptor**   **ACPD receptor**

Figure 1   Schematic representation of two ionotropic GLU receptor types (a KA/AMPA receptor and a NMDA receptor) and a metabotropic GLU receptor, indicating sites for various agonists, antagonists, and modulators, and for secondary modification. KA/AMPA and NMDA channels are shown with only two of the probably five subunits required for ion channel formation (see Figure 4). In the KA/AMPA and NMDA receptors, the subunit drawn on the left schematically indicates the coiling of the protein required to produce four TMDs as well as extracellular N- and C-termini according to model 2 (see Figure 4). PLC stands for phospholipase C.

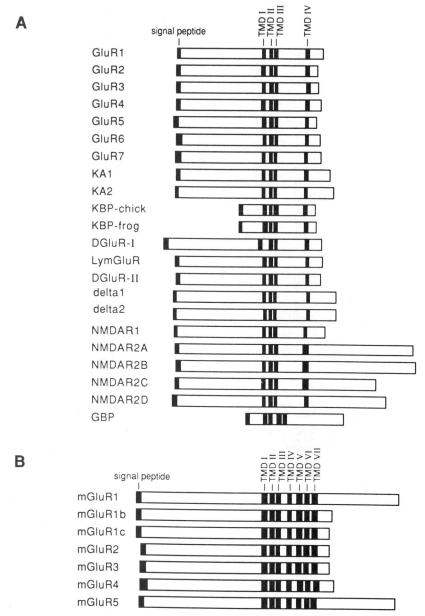

Figure 2 Bar graph representation of (*a*) the general structure of 21 ionotropic GluRs and the GBP, and (*b*) five metabotropic GluRs. For references to sequences, see Table 1. The signal peptide and the proposed TMDs of each subunit are indicated by black boxes at their respective positions in the sequence. Ionotropic GluRs have been lined up along their putative TMD II domains. Note the different distribution of putative TMDs in the GBP. For GluR5, the splice variant GluR5-2 is shown; NMDAR2D is represented by the mouse clone, ε4. Metabotropic GluRs have been lined up along their putative TMD I domains.

	TMD II		**TMD III**
GluR1	TTSDQSNE	FGIFNSLWFSLGAFM**Q**QGC	.DISPRSLSGRI
GluR2	S.SESTNE	FGIFNSLWFSLGAFM**R**QGC	.DISPRSLSGRI
GluR3	SPPDPPNE	FGIFNSLWFSLGAFM**Q**QGC	.DISPRSLSGRI
GluR4	S.DQPPNE	FGIFNSLWFSLGAFM**Q**QGC	.DISPRSLSGRI
GluR5[a,b]	S.DVVENN	FTLLNSFWFGVGALM**Q**QGS	.ELMPKALSTRI
GluR6[b]	S.DVVENN	FTLLNSFWFGVGALM**R**QGS	.ELMPKALSTRI
GluR7	S.EVVENN	FTLLNSFWFGMGSLM**Q**QGS	.ELMPKALSTRI
KA1	RCNLLVNQ	YSLGNSLWFPVGGFM**Q**QGS	.TIAPRALSTRC
KA2	RPHILENQ	YTLGNSLWFPVGGFM**Q**QGS	.EIMPRALSTRC
KBP-chick	EPKNEENH	FTFLNSLWFGAGALT**L**QGV	.TPRPKAFSVRV
KBP-frog	EPASEQNH	FTLLNSLWYGVGALT**L**QGA	.EPQPKALSARI
DGluR-I	VPPVPPNE	FTMLNSFWYSLAAFM**Q**QGC	.DITPPSIAGRI
LymGluR	HHSYIAND	FSISNSLWFSLGAFM**Q**QGC	.DISPRSMSGRI
DGluR-II	DPDELENI	WNVNNSTWLMVGSIM**Q**QGC	.DILPRGPHMRI
delta1	TQPRPSAS	ATLHSAIWIVYGAFV**Q**QGG	.ESSVNSVAMRI
delta2	S....MTS	TTLYNSMWFVYGSFV**Q**QGG	.EVPYTTLATRM
NMDAR1	SEEEEEDA	LTLSSAMWFSWGVLL**N**SGI	GEGAPRSFSARI
NMDAR2A	GKAPTGLL	FTIGKAIWLLWGLVF**N**NSV	PVQNPKGTTSKI
NMDAR2B	GREPGGPS	FTIGKAIWLLWGLVF**N**NSV	PVQNPKGTTSKI
NMDAR2C	GKKPGGPS	FTIGKSVWLLWALVF**N**NSV	PIENPRGTTSKI
NMDAR2D[c]	GKRPGGST	FTIGKSIWLLWALVF**N**NSV	PVENPRGTTSKI
GBP	FTFVGEVK	GFVRAQVWTYYVSYA**I**FFI	SLIVLSCCGDFR

Figure 3 Amino acid sequence comparison of the putative TMD II domains and adjacent regions of the 21 ionotropic glutamate receptor genes and the GBP. See Table 1 for references to sequences. Gaps (*periods*) were introduced to maximize homologies. The Q/R editing site of GluR1–GluR7 is shown in bold.

[a] Splice variant GluR5-2.

[b] Both edited (R) and unedited (Q) versions occur in native receptors.

[c] Sequence data from the mouse clone, ε4.

assignment of TMD III and TMD IV, some ambiguity remains in the domain preceding TMD III as to the assignment of TMD I and TMD II. The presence of three hydrophobic stretches in this region (residues 463–480, 521–539, and 596–614 of GluR1; all residues are numbered beginning with the first amino acid of the mature protein) has led to the proposal of two different models for the receptor topology (see Figure 4). Model 1 suggests the first two hydrophobic regions as TMD I and TMD II, respectively, disregarding the third hydrophobic region; model 2 assigns the second and third region as TMD I and TMD II, respectively, disregarding the first hydrophobic region (see Hollmann et al 1990 for a discussion). A third proposed model considers the first three plus the fifth hydrophobic regions to be true TMDs and places the fourth strongly hydrophobic domain intracellularly (Dingledine et al 1992). In this model (Figure 4), TMD II of model 2 becomes the third TMD and thus crosses the membrane from extracellular to intracellular, rather than from intracellular to extracellular. This change was suggested in order to bring a site critical for channel properties (the Q/R site; see below) close to the intracellular mouth of the channel, where the selectivity filter in nAChRs has been located (see Dingledine et al 1992). A definite clarification of the topology will probably not be possible until crystal structure analysis of the receptor protein is available; however, sequence information from additional homologous genes and data from mutagenesis studies on the putative TMD II (see below) have accumulated evidence in favor of model 2, which will be used throughout this review. However, one should remember that both models are based on the assumption that GluRs have the same basic structure as nAChRs, which may turn out to be untrue, especially given the lack of sequence homology between GluRs and other ligand-gated ion channels (see next paragraph) and some very recent immunological evidence that the C-terminus of GluR1 may be intracellular (Molnar et al 1993). However, prediction of receptor topology based on immunological evidence has been proven problematic in the case of the nAChR (Lindstrom et al 1984, Ratnam & Lindstrom 1984, Young et al 1985), and thus should be viewed with caution.

SEQUENCE HOMOLOGIES WITH OTHER GENES   In spite of the apparent structural homology with other ligand-gated ion channels, the overall amino acid sequence identity of GluR1–GluR4 with nAChRs, GABARS, and GlyRs is only ~20% and thus not significantly higher than for two random sequences. Furthermore, a region in the N-terminal domain that was regarded as a ligand-gated ion channel signature structure because of its conservation among all nAChR, GABAR, and GlyR subunits (Barnard et al 1987) is conspicuously absent in GluRs. Also, although in GluRs the intracellular loop between TMD III and TMD IV is highly conserved between different subunits, this is the most variable domain in other ligand-gated ion channels (Boulter et al 1990). The reverse is true, on the other hand, for the N-terminal domain. Thus, the superfamily hypothesis, which suggests that all ligand-gated ion channels are evolutionarily related to a common ancestor gene, does not appear to be valid for the glutamate receptor family. GluR1–GluR4, however, share significant localized amino acid sequence identity with the *Escherichia coli* periplasmic glutamine-binding protein in the region immediately preceding TMD I (which may be involved in ligand binding; see below) and in the intracellular domain in front of TMD IV (Nakanishi et al 1990). This may indicate that the ligand binding site of GluRs was acquired through exon shuffling from a prokaryotic gene (Nakanishi et al 1990). The glutamate receptors are highly conserved between mammals; the rat, mouse, and human GluR1 are 96–97% identical at the amino acid level.

MECHANISMS CREATING RECEPTOR DIVERSITY   Each of the GluR1–GluR4 subunits exists in two different forms created by alternative splicing of a 115–base pair (bp) region immediately preceding TMD IV (Sommer et al 1990; Table 3). The two alternate exons have been termed "flop" and "flip" (although flip is not the reverse of flop, as the name unfortunately suggests), and their signature sequences, which are invariant in GluR1–GluR4, are N-$(X)_{21}$-GGGD and S-$(X)_{21}$-KDSG for flop and flip, respectively. The forms are equally abundant but show different regional distributions in the brain, particularly in CA3 pyramidal cells (only flip) and dentate gyrus granule cells (more flop than flip) of the hippocampus (Sommer et al 1990). Their expression in cerebellar Purkinje cells is differentially regulated (Lambolez et al 1992), as is their expression during development (Monyer et al 1991). The two splice variants do not confer different pharmacological properties to the receptors, but they do differ notably in the efficacy of glutamate (GLU) in activating the receptor; flip channels are more efficient than flop channels. Sommer et al (1990) have speculated that an adaptive switch from flop to flip receptors might play a role in the maintenance of long-term potentiation (LTP). The observed developmental switch in the rat from predominantly flip variants before birth to flip plus flop variants after birth might reflect a need for the more efficient flip receptors during synaptogenesis, which later in develop-ment are tuned down by the addition of flop variants, similar to the switch in nAChRs from fetal to adult forms (Monyer et al 1991).

Another means of creating receptor diversity is differential splicing at the C-terminus, which has been reported for GluR4. In a cDNA clone named GluR4c, a 113-bp alternate exon inserted downstream of TMD IV introduces an alternate stop codon and creates a shorter version of GluR4 (Gallo et al 1992; Table 3). This splice variant has a higher amino acid sequence identity (72%) with GluR1–GluR3 at the C-terminus than with the original GluR4, suggesting the possibility that C-terminal splice variants for GluR1–GluR3 may exist with C-termini matching the original GluR4. The variant C-terminus of GluR4c, which can occur with either the flop or the flip version of the receptor, does not appear to change the functional properties of the receptor. However, the splice variant shows a distribution in the brain that overlaps with but is different from GluR4 and is developmentally regulated in cerebellar granule cells, where it appears to parallel the maturation of these cells (Gallo et al 1992). Gallo et al (1992) have speculated that the variant C-terminus may serve to stabilize the receptor or localize it to a specific region within the cell. Alternatively, the C-terminus could determine which subunits go together to form heteromeric receptors.

An additional, quite amazing mechanism of creating receptor diversity was discovered in Peter Seeburg's laboratory (Sommer et al 1991). A glutamine (Q) residue in the putative TMD II is encoded in the genes for GluR1–GluR4, but nevertheless all GluR2 cDNA clones from adult animals ($>$ 1000 were analyzed) actually contain an arginine (R) instead at this position (residue 586 in GluR2). Because only one gene exists for each of the four receptors and no alternate exons are present for TMD II, this observation is best explained by RNA editing, a mechanism that can specifically alter a single codon in a messenger RNA (Sommer et al 1991). This editing mechanism appears to be developmentally regulated; low levels of unedited GluR2 RNA are present in fetal but not in adult brain (Burnashev et al 1992b). This RNA editing of a single codon is especially noteworthy because it changes an amino acid that crucially affects ion channel function by determining the rectification properties of the channel and by controlling the divalent cation permeability of GluR1–GluR4 (Hume et al 1991, Verdoorn et al 1991; see below). Editing at this position has not been observed in GluR1, GluR3, or GluR4, and it remains to be explained how the amazing specificity for GluR2 is generated, given the almost identical nucleotide sequences in GluR1–GluR4 in the vicinity of the editing site. Sommer et al (1991) have speculated that some feature of the secondary structure of the RNA may confer specificity of editing.

FUNCTIONAL PROPERTIES   The GluR1–GluR4 subunits each form homomeric ion channels when expressed in oocytes or in transfected cells. Agonist potencies follow the order quisqualate (QA) $>$ DOM ~AMPA $>$ GLU $>$ KA. AMPA, GLU, and QA act as partial agonists, and KA and DOM appear to be full agonists eliciting maximal responses (Hollmann et al 1989, Nakanishi et al 1990, Sakimura et al 1990). The same order of agonist potency for GluR1–GluR4 (Table 4) was found in [^{3}H]AMPA ligand binding experiments (Keinänen et al 1990, Kawamoto et al 1991) and led to the designation of these receptors as AMPA-preferring receptors (also called AMPA receptors, AMPA/KA receptors, or KA/AMPA receptors). Although AMPA is the most potent specific agonist, KA does elicit the largest responses, and consequently most expression studies use KA as the standard agonist for GluR1–GluR4. Both agonists act competitively at the same site (Lambolez et al 1991). KA (and DOM) steady-state currents are larger because these agonists do not desensitize the receptors. GLU, QA, and AMPA currents, on the other hand, are smaller because of fast desensitization. The desensitization time constants for GluR2 and GluR4 are ~36 ms and ~8 ms, respectively (Burnashev et al 1992a). After an initial large peak current, desensitization rapidly reduces the current 2.5- to 8-fold to a steady-state level (Verdoorn et al 1991). The finding that both AMPA and KA can activate the same receptor was unexpected, given the differential distribution of AMPA and KA binding sites in the brain (Monaghan et al 1989), which had led to the conclusion that AMPA and KA receptors were separate entities. However, unitary KA/AMPA receptors had been postulated previously from electrophysiological data of cultured neurons that showed cross-desensitization between the two agonists (Kiskin et al 1990). The apparent discrepancy between the pharmacological properties of GluR1–GluR4 currents and the divergent regional distribution patterns for [^{3}H]AMPA and [^{3}H]KA binding is partly explained by the intrinsic technical impossibility of measuring tritiated ligand binding at the very low affinity levels that are still sufficient to produce electrophysiological responses. That apparent discrepancy furthermore suggests the existence of another class of KA receptors insensitive to or with only very low affinity for AMPA, which would account for the [^{3}H]KA binding sites found in the brain (see section on GluR5–GluR7 and KA1–KA2, below).

Among GluR1–GluR4, GluR2 stands out by producing only very small currents as a homomeric channel. Also, GluR2 shows a linear (or slightly outwardly rectifying) current/voltage (I/V) relation, whereas the other receptors display strong inward rectification (Boulter et al 1990, Nakanishi et al 1990, Verdoorn et al 1991). These I/V relations are probably the result of the conductance properties of the channel rather than a gating phenomenon (Verdoorn et al 1991). Although no single-channel studies of recombinant AMPA channels have been published, a conductivity of < 1 pS has been claimed for homomeric (Keinänen et al 1990) as well as heteromeric (Sommer et al 1990) channels, and different conductance states for AMPA vs KA-activated channels have been noted (Stein et al 1992).

Coexpression of two or more subunits does not alter the order of agonist potency, although potencies are generally lower in heteromeric receptors (Boulter et al 1990, Nakanishi et al 1990), approaching potencies seen when total rat brain RNA is expressed in oocytes (Dawson et al 1990). Also, current amplitudes are much larger for most subunit combinations than for the respective homomeric receptors, indicating subunit interaction in a functional heteromeric receptor complex (Boulter et al 1990, Nakanishi et al 1990, Sakimura et al 1990). Responses are fast; decay time constants are ~16 ms for GluR1/GluR2 and ~6 ms for a homomeric GluR1 channel (Verdoorn et al 1991). Heteromeric complexes have inwardly rectifying I/V relations unless GluR2 is present. If GluR2 is present, the I/V relation becomes linear or outwardly rectifying, demonstrating that the channel properties of the GluR2 subunit dominate those of the entire heteromeric complex (Boulter et al 1990, Nakanishi et al 1990, Sakimura et al 1990). Both homomeric and heteromeric channels show cooperativity, which suggests a multimeric structure of the native channel (Nakanishi et al 1990).

Table 4 EC$_{50}$S of various glutamate receptor agonists, and K$_D$S and K$_I$S of agonists in competitive ligand binding assays with membranes from transfected cell lines

GluR subunit, or combination of subunits	EC$_{50}$ values (μM) for receptor activation by various agonists									K$_D$ of ligand binding		K$_I$ values (nM) for inhibition of [³H]KA binding					
	GLU	GLYC	QA	AMPA	KA	DOM	NMDA	IBO	tACPD	[³H]KA (nM)	[³H]AMPA (nM)	GLU	QA	AMPA	KA	DOM	CNQX
GluR1	9.2[b] 3.4[d]		0.15[b] 0.1[e]	1.3[c]	39[a] 36[b] 32[c] 34[d] 50[e] 45[f]	1.8[a] 1.3[d] 1.9[f]				++[a,f]	++[ac,af]						
GluR1/GluR2					110[h]												
GluR1/GluR2$_{flip}$				2.2[c]	110[c]												
GluR1$_{flip}$/GluR2	5.7[b]		0.12[b]		91[b]												
GluR1$_{flip}$/GluR2$_{flip}$	6.2[g]			3.3[g]	57.5[g]												
GluR1/GluR3$_{flip}$				24[c]	55[c]												
GluR1/GluR2$_{flip}$/ GluR3$_{flip}$				13[c]	100[c]												
GluR2										12[ac]		490[ac]	9[ac]		9000[ac]		320[ac]
GluR2$_{flip}$/GluR3$_{flip}$				19[c]	110[c]												
GluR2$_{flip}$/GluR4$_{flip}$	32.3[g]			5.0[g]	64.6[g]												
GluR3																	
GluR3$_{flip}$				36[c]	130[c]						++[ac]						
GluR4																	
GluR4c	1.8[i]			44.3[i]							++[ac]						

GluR5-2a	631[i]		3000[i]	33.6[j]	1.2[j]	73.3[j]	290[j]	280[j]	~3000[j]	2.1[j]	2000[j]
GluR6	31[k] 273[ak]	11[k]		1.0[k] 1.4[am]	0.14[j]	95[l] 36[ad]	3100[l] 1080[j]	1100[l] 403[j]	>10000j	59[j] 8.6[j]	3150[j]
β2	50[m]	34[m]		1.6[m]							
β2/KA2	14[m]	14[m]		1.5[m]							
GluR7						77[l] 62.5[j]	1100[l] 2194[j]	6900[l] 7022[j]	>10000[j] 63[l]	12[l] 36.6[j]	340[l] 820[j]
KA1						4.7[ac] —[ac]	200[ae]	18[ae]	>5000[ae]	40[ae]	
KA2						15[ag]	480[ag]	58[ag]	>10000[ag]	275[ag]	
KA2 (humEAA2)						2.6[ah]	51.3[ah]	10.1[ah]		12.9[ah]	3010[ah]
KBP-chick						560[ai]	20000[ai]	316000[ai]	280[ai]	250[ai]	
KBP-frog						5.5[aj]	800[aj]	590[aj]	13.2[aj]	2.7[aj]	830[aj]
DGluR-I	35000[o]										
DGluR-II				75[n]							
NMDAR1	0.72[p] 1.6[q] 0.64[r] 1.7[s]	0.077[f]				27[p] 12[r] 13.2[s]					
NMDAR1B	1.1[r]	0.120[f]				26[r] 30[t]					
NMDAR1G	5.6[s]					67[s]					
NMDAR1/ NMDAR2A	1.7[v]	2.1[v] GLYC alone: 2.0[u]									

Table 4 (*Continued*)

GluR subunit, or combination of subunits	EC$_{50}$ values (μM) for receptor activation by various agonists									K$_D$ of ligand binding		K$_I$ values (nM) for inhibition of [^{3}H]KA binding					
	GLU	GLYC	QA	AMPA	KA	DOM	NMDA	IBO	tACPD	[^{3}H]KA (nM)	[^{3}H]AMPA (nM)	GLU	QA	AMPA	KA	DOM	CNQX
NMDAR1/ NMDAR2B	0.8[v]	0.3[v]															
NMDAR1/ NMDAR2C	0.7[v]	0.2[v]															
NMDAR1/ NMDAR2D	0.4[w]	0.09[w]															
mGluR1	9[x]		0.2[x]					6[x]	50[x]								
	12[y]		0.7[y]					32[y]	380[y]								
mGluR1c	13[z]		0.75[z]					60[z]	130[z]								
mGluR2	4[aa]							35[aa]	5[aa]								
mGluR5	10[ab]		0.3[ab]					10[ab]	50[ab]								

[a] Hollmann et al 1989; [b] Sakimura et al 1990; [c] Nakanishi et al 1990; [d] Dawson et al 1990; [e] Lambolez et al 1991; [f] Sun et al 1992; [g] Stein et al 1992; [h] Keller et al 1993; [i] Gallo et al 1992; [j] Sommer et al 1992; [k] Egebjerg et al 1991; [l] Bettler et al 1992; the published EC$_{50}$ for DOM (14 μM (J. Egebjerg, pers. commun.); [m] Sakimura et al 1992; [n] Ultsch et al 1992; [o] Schuster et al 1991; [p] Moriyoshi et al 1991; [q] Sekiguchi et al 1992; [r] Nakanishi et al 1992; [s] Durand et al 1992; [t] Sugihara et al 1992; [u] Meguro et al 1992; [v] Kutsuwada et al 1992; [w] Ikeda et al 1992; [x] Masu et al 1991; [y] Houamed et al 1991; [z] Pin et al 1992; [aa] Tanabe et al 1992; [ab] Abe et al 1992; [ac] Keinänen et al 1990, IC$_{50}$ values except for CNQX; [ad] Lomeli et al 1992; [ae] Werner et al 1991; [af] Kawamoto et al 1991; [ag] Herb et al 1992; [ah] Kamboj et al 1992; [ai] Gregor et al 1988, IC$_{50}$ values measured for partially purified protein fraction; [aj] Wada et al 1989; [ak] Raymond et al 1993; [al] Gregor et al 1992; [am] Wang et al 1993.

Heteromeric receptors containing either only flop or only flip splice variants are indistinguishable in pharmacological profile, desensitization properties, I/V relation, and Ca^{2+} permeability (Sommer et al 1990, Hollmann et al 1991). However, current amplitudes of both peak and steady-state currents are four- to fivefold smaller for GLU (but not KA) at flop receptors compared with flip receptors. This could enable neurons to switch from "low gain" to "high gain" glutamate receptors simply by alternative splicing of transcripts that are being produced anyway (Sommer et al 1990).

Selective inhibitors for the various GluR subunits would be valuable tools. However, 6-cyano-7-nitroquinoxaline-2,3-dione (CNQX) and 2,3-dihydroxy-6-nitro-7-sulfamoyl-benzo(f)quinoxaline (NBQX), the potent antagonists of native non-NMDA receptors that are also potent inhibitors of GluR1–GluR4, show no (CNQX) or only low (NBQX) selectivity for different subunits (Nakanishi et al 1990, Lambolez et al 1991, Stein et al 1992). The dye Evans Blue blocks GluR1–GluR4 and all possible heteromeric combinations of these

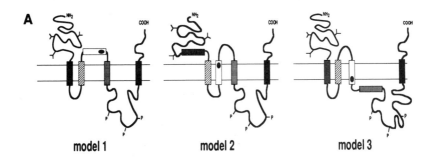

Figure 4 (*a*) Schematic representation of the three different models proposed to date for the topology of ionotropic GluRs. The five different hydrophobic domains that have been considered candidates for TMD regions in one or more of the three models are shown. It has to be kept in mind, however, that other hydrophobic regions exist in each of the receptor subunits and potentially could form additional TMDs that would suggest a different topology. The black dot in the third hydrophobic domain represents the Q/R editing site (see text). Model 2 is currently favored by most workers. (*b*) Schematic model of the proposed pentameric structure of glutamate receptor channels. Note that TMD II in each subunit is believed to face the ion pore.

receptors, with the exception of GluR3 homomeric receptors (Keller et al 1993; B Keller, personal communication). The dye also does not block GluR6, and may provide a starting point in the search for subunit-specific inhibitors. A toxin from the spider *Nephila clavata,* joro spider toxin (JSTX), has an even more striking subunit-specificity. It acts as a use-dependent channel blocker for those GluRs with rectifying I/V curves but not those with linear I/V relations (Blaschke et al 1993). Changing the rectification properties of the channel through mutagenesis at the Q/R site of TMD II (see below, and Figures 3 and 4) also changes the channel block by JSTX, suggesting that JSTX binds at or close to the site that determines the rectification properties.

DIVALENT CATION PERMEABILITY   Non-NMDA receptors in most neurons have traditionally been regarded as rather impermeable to divalent cations, in particular to Ca^{2+} (Mayer & Westbrook 1987), although in certain neurons Ca^{2+}-permeable KA receptors were observed (Murphy et al 1987, Iino et al 1990, Gilbertson et al 1991). It was therefore quite surprising to find that all recombinant AMPA-preferring receptor subunits and their heteromeric combinations, except for GluR2 and combinations containing GluR2, were quite permeable to various divalent cations, including Ca^{2+} and Mg^{2+} (Hollmann et al 1991), as demonstrated by I/V curve shifts to positive potentials in high divalent/low monovalent cation solutions. Ca^{2+} permeability of AMPA-preferring receptors was later also demonstrated directly with fluorescence-monitoring of Ca^{2+} influx into GluR3-expressing oocytes upon channel activation in physiological solutions (Keller et al 1992).

Just as GluR2 dominates the rectification properties of heteromeric channels, it also determines their Ca^{2+} permeability, which is almost abolished in the presence of GluR2. These observations suggested that GluR2 is different from GluR1, GluR3, and GluR4 at some key position, most likely in the channel domain. Because the most obvious site difference between GluR2 and the other subunits is the Q/R editing site in TMD II (Figure 3), this site was investigated through mutagenesis. It was found to be the critical determinant not only of the rectification properties of the channels (Hume et al 1991, Verdoorn et al 1991, Curutchet et al 1992) but also of their divalent cation permeability (Hume et al 1991, Burnashev et al 1992b). A mutant GluR2 containing Q instead of R in TMD II is made Ca^{2+}-permeable with a rectifying I/V curve, whereas a mutant GluR3 (or GluR1 or GluR4) containing R instead of Q has extremely low Ca^{2+}-permeability and exhibits a linear I/V relation. Although these data seemed to suggest that rectification properties and divalent cation permeability were coupled, further mutagenesis studies showed that the two are not invariably linked. If a histidine (H) or an asparagine (N) is introduced at the Q/R site of the inwardly rectifying,

Ca^{2+}-permeable GluR1, GluR3, or GluR4, Ca^{2+}-permeability is maintained but the subunit now exhibits a linear I/V relation (Burnashev et al 1992a, Curutchet et al 1992, Dingledine et al 1992). A similar effect occurs when an aspartic acid residue that is four amino acids C-terminal of the Q/R site is replaced with N (Dingledine et al 1992). Moreover, mutant receptors containing N at the Q/R site show a differential effect of this mutation on Ca^{2+}- and Mg^{2+}-permeability: these receptors prefer Ca^{2+} over Mg^{2+} (Burnashev et al 1992a). Thus, these mutations cause functional properties similar to those seen at the NMDA receptor channel in the absence of magnesium, which is interesting because NMDA receptors naturally contain an asparagine residue at a position equivalent to the Q/R site, and they also prefer Ca^{2+} over Mg^{2+} (see below).

Additional mutagenesis experiments showed that the smaller positively charged lysine at the Q/R site gives a channel with low divalent permeability, whereas a large tryptophan residue, or negatively charged residues, block the channel for all ions (Dingledine et al 1992). The serine residues in TMD II of GluR1–GluR4 appear not to be involved in channel function (Dingledine et al 1992), unlike their role in nAChRs. This finding argues against a strict structural homology of the channel-lining domains in nAChRs and GluRs. Taken together, the available mutagenesis data argue strongly that the Q/R site is close to the ion path of the channel, which makes it likely that TMD II is the channel-lining domain of GluRs.

The relative permeability ratios of GluR1–GluR4 channels for divalent cations versus monovalent cations (P_{div}/P_{monov}) was estimated to be 0.01–0.05 for Ca^{2+}-impermeable receptors and 0.8–3.0 for Ca^{2+}-permeable receptors (Hume et al 1991, Burnashev et al 1992a). This compares well with reported ratios of <0.2 and 5 for Ca^{2+}-impermeable Type I KA receptors and Ca^{2+}-permeable Type II KA receptors, respectively, in cultured hippocampal neurons (Iino et al 1990). It also demonstrates that Ca^{2+}-permeability of GluR1, GluR3, and GluR4 receptors and their heteromeric combinations is substantial, in the same range as that of nAChRs ($P_{div}/P_{monov} = 1$–2), and not very much smaller than the Ca^{2+}-permeability ($P_{div}/P_{monov} = 8$–10) reported for the NMDA receptor (Dingledine et al 1992).

GluR2 may serve an important function in regulating the Ca^{2+}-permeability of GluR1–GluR4 receptors in vivo, during development as well as during certain abnormal conditions such as might occur in neurodegenerative disorders. GluR2 in neocortex, striatum, and cerebellum has a developmental pattern different from that of GluR1 and GluR3. GluR2 levels increase monotonically relative to those of GluR1 plus GluR3, presumably leading to the formation of progressively fewer Ca^{2+}-permeable receptors (Pellegrini-Giampietro et al 1992).

EXPRESSION OF RECEPTOR RNA   In situ hybridization studies have revealed widespread but differential distribution of the GluR1–GluR4 RNAs (Table 5) as well as their independent regulation during development. The near-ubiquity of GluR2 RNA matches well with electrophysiological data indicating that most native KA/AMPA-receptors in the brain that have been studied electrophysiologically have a linear I/V relation. Notable differences in subunit expression exist in cerebellar granule cells, which contain GluR2 and GluR4 RNA but not GluR3 and little if any GluR1 RNA, cerebellar Bergmann glia, which contain GluR1 and GluR4 but not GluR2 and GluR3 RNA, and brainstem, which apparently contains solely GluR4 RNA. In the hippocampus, GluR4 RNA, in contrast to GluR1–GluR3 RNA, is less abundant in CA3 and dentate gyrus than in CA1 pyramidal cells (Bettler et al 1990, Keinänen et al 1990). In the retina, GluR1 and GluR2 RNA is distributed throughout the inner nuclear layer (INL) and the ganglion cell layer (GCL), whereas GluR3 and GluR4 RNA is found only in the inner third of the INL and in the GCL (Hughes et al 1992, Müller et al 1992). The distribution of RNA in the brain largely matches the distribution of [^{3}H]AMPA binding sites, but some discrepencies are evident, especially in the cerebellum, the reticular thalamic nucleus, and the hippocampal CA3 region, where [^{3}H]AMPA binding is low but GluR1–GluR4 RNA is abundant (Keinänen et al 1990). Interestingly, immunological detection of GluR1 protein revealed lower expression in CA3 than in CA1–CA2 (Blackstone et al 1992b), which correlates better with [^{3}H]AMPA binding data than with GluR1 RNA expression. For differential expression of flop and flip splice variants see the discussion above.

Expression of the GluR1–GluR4 genes during development is seen as early as embryonal day 10 (E10) in both the central nervous system (CNS) and the peripheral nervous system (PNS) and is most pronounced in areas where neurons differentiate and assemble into nuclei (Bettler et al 1990). Postnatally, expression of the GluR1–GluR4 RNA in general increases gradually. A transient elevation relative to the adult level has been reported for GluR1 and GluR3 RNAs between postnatal days 14 (P14) and P21 (Pellegrini-Giampietro et al 1992).

PROTEIN-BIOCHEMICAL CHARACTERIZATION OF RECOMBINANT RECEPTORS   Researchers have used antibodies generated against either bacterial fusion proteins of GluR1–GluR4 or synthetic C-terminal peptides to label 105- to 108-kDa proteins on Western blots of rat and human synaptic membranes and membranes from GluR1-transfected cell lines (Wenthold et al 1990b; Rogers et al 1991; Blackstone et al 1992a,b; Hullebroeck & Hampson 1992). Also, GluR1 is enriched in synaptic membrane preparations, particularly in postsynaptic density preparations, and the receptor protein is glycosylated, as expected given the presence of five (GluR2, GluR3) and six (GluR1, GluR4)

Table 5 Regional distribution[a] of glutamate receptor RNAs as revealed by in-situ hybridization histochemistry

GluR gene	1	2	3	4	5	6	7	8	9	10	11	12	13	14	15	16	17	18	19	20	21	22	23	24	25	26	27	28	29	References
GluR1	l	h	l	h	h	h	h	h	h	h	l	h	l/a	l	h		l	h	h/a	a				h	h	a/l	1	1	h	a,b,c,d,e,f,g,h,i,ae,af
GluR2	l	h	l	h	h	h	h	h	h	h	l	h	a	l	h		l	h	h/a	a				h	a	h	a	h	h	a,b,c,d,e,f,g,h,i,ae
GluR3	l	h	h	h	h	h	h	h	h	h	l	h	a	l	h		l	h	l/a	a				h	a	h	a	h	h	a,b,c,d,e,g,h,i,ae
GluR4	h	h	h	h	l	l	l	l	l	h	a	h	a	h	h	h	h	h	l	h				h/a	h	h/a	a	h/1	1	b,c,d,e,j,k,ae
GluR5	h	h	l	l	l	l	l	l	l	h	a	a	a	h	h	a	l	h	h	h				h	h	h/a	1	h	h	c,d,j
GluR6	h	l	h	h	h	h	h	l	h		l	a		l			l	l	l	1				h	h	h	1	h		l
GluR7	h	h	h	l	l	a	a	a	l			h	1				h	1/h	h	h				h	a	a			l	m,n
KA1	l	l	a	l	l	a	a	h	1		a		l	a			a	a	a					1	l		1/h			o,p
KA2	h	h	h	h	h	a	a	h	h	h		h	a			h	l	1	a	h				h	a	1				p
KBP-chick										h						h								h	h	1				q
delta1				1	1	1	1	1	1									1						h	h	a				ad
delta2				l	a	a	a	l	a	a								a	a					h	a	h	a			ad
NMDAR1	h	h	h	h	h	h	h	h	a		h		a	h	h		1	h	h	h				h	h	h		h		r,s,t,u
NMDAR2A	h	h	a	h	h	h	h	h				h			h		a/1	a	a	a				a	h	h				s,v
NMDAR2B	h	h	a	h	h	h	h	h	a		a	a/1	a/1	a	a/1		1	a/1	a	a				h						v,w
NMDAR2C	l	a	a	a	a	a	a	a	a	a	a	a	a	a	a		l	a	a	a	a			h						v,w
NMDAR2D	h																		h					h						ac
mGluR1	h	h	h	h	l	h	h	h	h	h	l	h	h	h	l	a	h	h	l	h				h	h	h	1	1	h	x,y,z
mGluR2	h	h		h		a	h	h	h			h						1		h				a			h			aa
mGluR3		h						h																1						aa
mGluR4				h																1			1	a		a				aa
mGluR5	h	h		h	h	h	h	h	h		h			h			h			1				1	a				l	ab

[a] Listed are 29 brain regions mentioned explicitly in the literature cited below as being labeled by probes for glutamate receptor subunits. Labeling intensity is marked as either high ("h"), which includes strong and medium labeling, low ("l"), or absent ("a"). No entry in the table means brain region has not been mentioned explicitly in the literature. The 29 regions are 1) olfactory bulb, 2) neocortex, 3) entorhinal cortex, 4) hippocampus, 5) CA1 pyramidal cell layer, 6) CA2 pyramidal cell layer, 7) CA3 pyramidal cell layer, 8) dentate gyrus, 9) subiculum, 10) piriform cortex, 11) striatum, 12) caudate-putamen, 13) globus pallidus, 14) septum, 15) amygdala, 16) medial habenula, 17) thalamus, 18) reticular thalamic nucleus, 19) hypothalamus, 20) brainstem, 21) pons, 22) substantia nigra, 23) cerebellum, 24) Bergmann glia, 25) Purkinje cell layer, 26) cerebellar granule cell layer, 27) cerebellar molecular layer, 28) spinal cord, 29) retina. References included in this table are, [a] Boulter et al 1990; [b] Kreinen et al 1990; [c] Hughes et al 1992; [d] Müller et al 1992; [e] Lambolez et al 1991; [f] Monyer et al 1991; [g] Pellegrini-Giampietro et al 1992; [h] Pellegrini-Giampietro et al 1993; [i] Sommer et al 1990; [j] Bettler et al 1990; [k] Gallo et al 1992; [l] Egebjerg et al 1991; [m] Bettler et al 1992; [n] Lomeli et al 1992; [o] Werner et al 1991; [p] Herb et al 1992; [q] Eshhar et al 1992; [r] Moriyoshi et al 1991; [s] Meguro et al 1992; [t] Shigemoto et al 1992a; [u] Nakanishi 1992a; [v] Monyer et al 1992; [w] Kutsuwada et al 1992; [x] Masu et al 1991; [y] Pin et al 1992; [z] Shigemoto et al 1992b; [aa] Tanabe et al 1992; [ab] Abe et al 1992; [ac] Nakanishi 1992; [ad] Lomeli et al 1993; [ae] Henley et al 1993; [af] Gall et al 1990.

potential N-glycosylation sites (see Table 1) in the extracellular domains (Rogers et al 1991, Blackstone et al 1992b, Hullebroeck & Hampson 1992). GluR1–GluR3 (GluR4 was not investigated) have been shown to carry a variety of carbohydrate side chains [an estimated 4.8% (5.1 kDa) of the molecular mass consists of carbohydrates] and are also O-glycosylated, but do not contain any sialic acid residues. The receptors bind lectins directly, as has been shown on Western blots of antisera-purified proteins (Hullebroeck & Hampson 1992). Lectins decrease or prevent native AMPA receptor desensitization (Mayer & Vyklicky 1989, Huettner 1990), and direct interaction of lectins with the receptor subunit probably causes this desensitization, possibly by preventing an agonist-induced conformational change that desensitizes the receptor, or by preventing subunit dissociation of the multimeric receptor complex (Hullebroeck & Hampson 1992).

Stimulation of KA-gated currents by a cAMP has been observed for GluR1 and GluR3 coexpressed in oocytes (Keller et al 1992), suggesting modulation by a cAMP-dependent protein kinase (PKA). This finding is in agreement with observations in retinal horizontal cells, where KA currents are enhanced by PKA (Liman et al 1989). In vitro phosphorylation studies of GluR1 expressed in transfected cell lines demonstrated basal phosphorylation of the receptor by protein serine/threonine kinases. Phosphorylation above the basal level could not be achieved with either PKA or protein kinase C (PKC) when the kinases were cotransfected into the cells (Moss et al 1993), possibly because of already saturating levels of basal phosphorylation in these cells. However, cotransfection with v-src tyrosine kinase produced phosphorylation of GluR1 at a tyrosine (Y) residue (Moss et al 1993). Because the only consensus site for tyrosine phosphorylation in GluR1 is located at residue Y655 in the domain between TMD III and TMD IV, these data support the topological assignment of this region as an intracellular domain.

Receptor complexes were solubilized in nondenatured form from rat brain (Wenthold et al 1990b, 1992; Blackstone et al 1992b) and GluR-transfected cell line membranes (Blackstone et al 1992b, Wenthold et al 1992), and identified with subunit-specific anti-GluR antibodies. Complexes from rat brain were analyzed for ligand binding and found to contain two different [³H]AMPA binding sites, with affinities of 4.6 nM and 323 nM, but no [³H]KA binding sites (Wenthold et al 1990b). Antibodies specific for each of the subunits GluR1–GluR4 coprecipitated all four subunits from solubilized, nondenatured membranes, suggesting that each of the subunits GluR1–GluR4 can form heteromeric complexes with all other subunits of this subfamily (Wenthold et al 1992). No additional, accessory proteins were detected in these receptor complexes (Wenthold et al 1992). Analysis of solubilized receptor complexes from rat brain and transfected cell lines with gel-exclusion chromatography and sucrose gradient density centrifugation showed that a single fraction contained [^{3}H]AMPA binding activity as well as GluR1 immunoreactivity. The complex was identified as an aggregate of 8.2-nm Stoke's radius with a sedimentation coefficient of 18.0 S, leading to a molecular weight estimate of 610 kDa (Blackstone et al 1992b). In another study, the largest nondenatured receptor complexes immunopurified from solubilized rat brain membranes with antibodies specific for GluR1, GluR2, GluR3, and GluR4 were estimated at 590 kDa in each case (Wenthold et al 1992). Membranes chemically cross-linked prior to solubilization yielded the same complexes. These molecular weights are best fit by assuming a pentameric receptor stoichiometry, the same stoichiometry that has been reported for other ligand-gated ion channels, such as nAChRs (Cooper et al 1991) and GlyRs (Langosch et al 1988).

In a study of the CG4 glial precursor cell line, which expresses GluR2, GluR3, and GluR4 subunits (and additionally, GluR6, GluR7, KA1, and KA2), a monoclonal, subunit-specific anti-GluR2 antibody directed against the N-terminal domain of GluR2 coprecipitated AMPA receptor subunits, but not the low-affinity KA receptor subunits GluR6 and GluR7, nor the high-affinity KR receptor subunit KA2, from solubilized, nondenatured cell membranes. This observation indicates that GluR2 forms native heteromeric complexes with AMPA receptor subunits, but not with GluR6, GluR7, or KA2 (RB Puchalski, J-C Louis, N Brose, et al, submitted). In the same cell line, complexes of GluR6 and/or GluR7 and the high-affinity KA receptor subunit KA2 were detected (see section on GluR5–GluR7), demonstrating that receptor complexes of different subunit composition can coexist in the same cell (RB Puchalski, J-C Louis, N Brose, et al, submitted).

Researchers have used immunostaining with GluR1-specific antisera to show that the protein distribution in the brain (with the exception of a few areas such as cerebellar Purkinje cells, hippocampal CA3 pyramidal cells, and the medial habenula) matches well with the RNA distribution observed with in situ hybridization (Rogers et al 1991, Blackstone et al 1992b, Petralia & Wenthold 1992). Staining was especially intense in several regions of the limbic system, including the hippocampus, lateral septum, central nucleus of the amygdala (CNA), and lateral bed nucleus of the stria terminalis (BNST). Because these regions are preferentially affected by KA neurotoxicity and are also reciprocally interconnected by excitatory projections (e.g. CNA and BNST), glutamatergic feedback loops may exist that could explain focal seizure activity induced by amygdala stimulation (kindling) or by direct KA injection into the amygdala, a method that serves as a model for temporal lobe epilepsy in humans (for a discussion, see Rogers et al 1991). Immunostaining did not provide convincing evidence for presynaptic localization of GluR1–GluR4 (Hampson et al 1992, Petralia & Wenthold 1992, Martin et al 1993). Most of the immunostaining seen at the electronmicroscopic level was clearly postsynaptic, at postsynaptic densities or in the dendritoplasm, with some label in the cytoplasm (Hampson et al 1992, Petralia & Wenthold 1992). Excitotoxic lesion of the amygdala by DOM injection, which is known to cause cell death of hippocampal CA3 neurons while sparing the mossy fibers that provide the presynaptic input from granule cells, caused a reduction in the protein levels of GluR1, GluR2, and GluR3 in the CA3 area of the hippocampus 30 days after the lesion, suggesting postsynaptic localization of the receptors (Hampson et al 1992). Localization of GluR1–GluR4 subunits in glial cells was apparently restricted to the Bergmann glia in the cerebellum (Hampson et al 1992, Petralia & Wenthold 1992).

Molnar et al (1993) reported in an electronmicroscopic study that antibodies to a C-terminal peptide of GluR1 labeled the intracellular face of the postsynaptic membrane, whereas antibodies to an N-terminal peptide labeled the extracellular face. This suggests that, contrary to all the topology models suggested (see Figure 4), the C-terminus may be intracellular. More direct evidence will be necessary, however, to definitely determine the topology of GluRs.

Vanderklish et al (1992) have investigated the possible involvement of GluR1 subunits in LTP induction in cultured hippocampal slices. This group found that repeated transfection of the cultured slice for several days with an antisense oligonucleotide to GluR1 (but not with a sense oligonucleotide) reduced the level of GluR1 protein expression as seen on Western blots, without affecting levels of GluR2 or GluR3 protein expression. Induction of what was believed to be the equivalent of the LTP phenomenon seen in standard hippocampal slices was significantly reduced in the antisense oligonucleotide-treated cultured slices. (LTP was seen in only 29% of the pathways vs 70% of the sense oligonucleotide control.) This result was interpreted as an indication that AMPA-preferring receptors lacking the GluR1 subunit may not be competent to express LTP (Vanderklish et al 1992).

GENE STRUCTURE AND CHROMOSOMAL LOCALIZATION   The complete gene structure of GluR1–GluR4 is unknown. It has been shown, however, that the gene portion encoding the receptors' TMD I through TMD III region consists of two exons separated by a large intron; the first exon encodes both TMD I and TMD II, and the second exon encodes TMD III (Sommer et al 1991). The exons for flop and flip variants are adjacent, separated by an 810-bp intron, and flanked by introns of 810 bp and 1020 bp (Sommer et al 1990). Data from the cloning of a human GluR1 gene indicate that this gene (which is 97% identical with the rat GluR1) is at least 50 kilo-base pairs (kbp) and up to 250 kbp in size (Puckett et al 1991).

The genes for GluR1–GluR4 are located on four different chromosomes (Table 6), unlike nAChR and GABAR subunit genes, some of which occur in gene clusters (McNamara et al 1992). GluR1 is localized on chromosome 5q33. This region is syntenic with portions of mouse chromosomes 11 and 5 to which a number of genes for neurological disorders have been mapped, such as shaker-2 (*sh-2*), spasmodic (*spd*), vibrator (*vb*), tipsy (*ti*), trembler (*Tr*), ataxia (*ax*), and bouncy (*bc*) (Puckett et al 1991, McNamara et al 1992). GluR2 is located on chromosome 4q32–33, in a region to which no known neurological disorders have been mapped. The gene for Huntington's disease, a long-time candidate as a glutamate receptor-linked disorder, is localized on chromosome 4, but positioned far away at p16.3. GluR3 is localized on chromosome Xq25–26, where genes for the oculocerebralrenal disorder syndrome of Lowe, Turner syndrome, and a form of congenital nerve deafness and albinism have been mapped (McNamara et al 1992). GluR4 is located on chromosome 11q22–23, in an area to which ataxia telangiectasia, a type of schizophrenia, and tuberous sclerosis have been mapped; this position is syntenic with a region on mouse chromosome 9 that carries the gene for an epileptic mouse mutant, *El-1* (McNamara et al 1992).

Thus, none of the known major neurological disorders that have been mapped in humans appear to be linked with GluR1–GluR4, although several minor disorders are potential candidates. However, to prove the involvement of GluR1–GluR4 in any of these diseases, linkage studies are not sufficient.

Table 6 Chromosomal localization of human glutamate receptor genes

Gene	Chromosome	References
GluR1	5q33	Puckett et al 1991
	5	Potier et al 1992
	5q32–33	McNamara et al 1992
	5q31.3–33.3	Sun et al 1992
GluR2	4q32–33	McNamara et al 1992
	4q25–34.3	Sun et al 1992
GluR3	Xq25–26	McNamara et al 1992
GluR4	11q22–23	McNamara et al 1992
GluR5	21q21.1–22.1	Eubanks et al 1993
	21q22	Potier et al 1993
GluR7	1	JO McNamara, pers. comm.
NMDAR1	9q34.3	Karp et al 1993
	9q35	J Eubanks, pers. comm.

For a firm link of receptor malfunction with any of these diseases, analysis of GluR1–GluR4 cDNAs or genes isolated from patient tissue will be necessary.

Low-Affinity KA Receptors: GluR5–GluR7

CLONING AND STRUCTURAL FEATURES   Using low-stringency hybridization screening with GluR1–GluR4 probes and PCR-mediated DNA amplification with degenerate primers made to regions of high homology between GluR1–GluR4, researchers have cloned a group of three receptors, GluR5, GluR6, and GluR7, that share 75–80% amino acid sequence identity with each other but only ~40% with GluR1–GluR4 (Table 2; Figure 5), setting them apart as a different subfamily of receptors (Bettler et al 1990, Egebjerg et al 1991, Bettler et al 1992, Morita et al 1992, Sommer et al 1992). GluR5–GluR7 also apparently share the typical overall structure of ligand-gated ion channels (see Figure 2). Each has a large N-terminal, presumably extracellular domain containing eight N-glycosylation consensus sites, which are conserved among the three receptors (Table 1); four putative TMDs arranged in the cluster-of-three-plus-one pattern with a large, putatively intracellular domain between TMD III and TMD IV; and a presumably extracellular C-terminal domain. The hydropathy pattern used to derive the TMD topology is even less clear for GluR5–GluR7 than for GluR1–GluR4, and several algorithms predict up to seven potential TMDs. However, by homology with other glutamate receptor subunits, four TMDs appear to be the most likely configuration

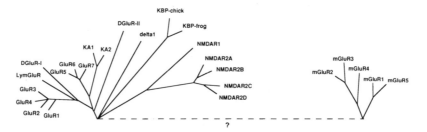

Figure 5 Phylogenetic tree of ionotropic and metabotropic glutamate receptors, constructed from pairwise sequence comparisons by using the algorithm of Feng & Doolittle (1987). Common branching points reflect shared ancestor sequences. The total branch length between any two genes indicates the relative evolutionary distance between them. The angles in the tree are arbitrary and have no meaning. Although some low-level sequence homology between ionotropic and metabotropic glutamate receptors has been discussed (Masu et al 1991), it appears unlikely that the two families of receptors share a common ancestor gene. Any sequence similarity may reflect evolutionary convergence rather than evolutionary relatedness.

(Bettler et al 1990). Like GluR1–GluR4, but in striking contrast to nAChR, GABAR, and GlyR subunits, GluR5–GluR7 share higher amino acid sequence identity C-terminal of their TMD I ($\sim$82%) than N-terminal of it ($\sim$74%).

MECHANISMS CREATING RECEPTOR DIVERSITY   The same editing mechanism acting on the codon for the Q/R site in TMD II of GluR2 as discussed above has also been found to operate at the homologous site in GluR5 and GluR6, but not in GluR7 (Sommer et al 1991, 1992; Bettler et al 1992; Lomeli et al 1992). In contrast to the editing of GluR2, however, that of GluR5 and GluR6 is incomplete, so that both edited (R) and unedited (Q) forms coexist. Researchers have found 39% of GluR5 cDNAs and 75% of GluR6 cDNA clones to be in the edited form (Sommer et al 1991). As is the case for GluR2, editing of GluR5 or GluR6 changes the amplitude of the response and the rectification properties of the ion channel (Sommer et al 1992, see below). Köhler et al (1993) recently reported two additional editing sites for the TMD I of GluR6. In this case, an isoleucine (I) and a tyrosine (Y) that are encoded in the gene are edited to become valine (V) and cysteine (C), respectively. The edited receptor differs in its ion permeability properties, suggesting that, unexpectedly, TMD I is involved in determining the properties of the ion channel (Köhler et al 1993).

Several splice variants have been found for GluR5 (Table 3). They differ in the presence or absence of an exon encoding a 15-amino acid insert in the N-terminal domain (GluR5-1 and GluR5-2, respectively) and in the presence (GluR5-2a, GluR5-2c) or absence (GluR5-2b) of additional exons in the C-terminal domain. Of the two C-terminal alternate exons, one introduces a stop codon and creates a truncated but functional receptor (GluR5-2a), whereas the other exon represents an in-frame insertion that creates a functional receptor with a larger C-terminal domain (GluR5-2c). N- and C-terminal splice variants can occur together and can be either edited or unedited (Sommer et al 1992).

FUNCTIONAL PROPERTIES   GluR5(Q), which is the unedited form of GluR5, when expressed in oocytes gives only tiny responses to GLU (Bettler et al 1990). In transfected cell lines, GluR5(Q) produces larger responses and can be activated by DOM, KA, GLU, and AMPA (Sommer et al 1992); the order of potency is DOM > KA > GLU > AMPA (Table 4). The difference in response amplitudes between oocytes and cell lines may be explained by a rapid desensitization of the response (see below) that cannot be resolved in oocytes, and by the use of different splice variants in the two studies: whereas unedited GluR5 (=GluR5-2b) was tested in oocytes, unedited GluR5-2a was used for the cell-line study. In [^{3}H]KA ligand binding experiments, GluR5-2a showed a five- to tenfold larger b_{max} than GluR5, indicating more efficient expression of the shorter splice variant GluR5-2a (Sommer et al 1992). The edited form of GluR5-2a produces no responses by itself, but upon coexpression with the unedited form changes the I/V relationship from inwardly rectifying to linear, thereby demonstrating that the edited form is a functional receptor (Sommer et al 1992).

GluR6(R), the edited form of GluR6, responds to KA, QA, DOM, and GLU but, quite surprisingly, not to AMPA, and is blocked by CNQX (Egebjerg et al 1991). The order of agonist potency at GluR6 is DOM > KA > QA > GLU (Table 4). The absence of responses to AMPA demonstrates that AMPA is an agonist at only a subset of ionotropic non-NMDA receptor subunits. Upon pretreatment of GluR6-expressing oocytes with the lectin concanavalin A (ConA), which is known to prevent receptor desensitization (Mayer & Vyklicky 1989, Huettner 1990), amplitudes are increased by 75- to 150-fold. The I/V curves of GluR6(R) channels are linear, as expected given the presence of R at the Q/R site, whereas the GluR6(Q) form exhibits an inwardly rectifying I/V curve (Egebjerg et al 1991, Dingledine et al 1992, Morita et al 1992, Egebjerg & Heinemann 1993). GluR6(R) and GluR6(Q) are both permeable to Ca^{2+}, with P_{div}/P_{monov} = 0.47 and 1.2, respectively, when measured in oocytes (Egebjerg & Heinemann 1993). Although GluR6(R) is approximately three times less permeable to divalent cations than GluR6(Q), its properties are quite different from GluR1–GluR4, where R at the Q/R site efficiently prevents divalent cation permeation (see above). GluR6(R) thus provides another example where a linear I/V relationship of the channel is not coupled to divalent impermeability, and also indicates that the Q/R site is probably not the only determinant of Ca^{2+}-permeability of the ion channel (Egebjerg & Heinemann 1993).

GluR7 does not produce any detectable responses in oocytes or transfected cell lines, not even after ConA pretreatment (Bettler et al 1992, Lomeli et al 1992). This could be the result of either extremely rapid desensitization or the absence of an essential additional subunit. [3H]KA ligand binding experiments with membranes of transfected cell lines demonstrated that KA binds to GluR5, GluR6, and GluR7 with similar affinity (K_Ds are ~70 nM; see Table 4). Competition experiments with various agonists revealed an order of affinity (Table 1) that was slightly different for GluR5, GluR6, and GluR7, with respect to the relative affinities of GLU and QA. All three receptors, however, bind DOM with the highest affinity of all agonists tested, and they show no significant AMPA binding (Bettler et al 1992, Lomeli et al 1992, Sommer et al 1992). Both electrophysiological and ligand binding data indicate that GluR5, GluR6, and GluR7 represent a glutamate receptor subfamily pharmacologically distinct from the AMPA-preferring receptors GluR1–GluR4. Because their [3H]KA binding affinities match very well with the lower-affinity class of the two [3H]KA binding affinity sites found in brain membranes as well as in brain receptor autoradiography (K_D ~50 nM, as opposed to ~5 nM for the high-affinity site), and because AMPA is not an agonist (or, for GluR5, only an extremely weak agonist), GluR5–GluR7 appear to represent the classic low-affinity KA receptor (Bettler et al 1992).

Contrary to GluR1–GluR4, both GLU and KA (and also DOM) desensitize GluR5 and GluR6 receptors. The desensitization time course for GluR5 shows biphasic kinetics (except for DOM), with desensitization constants (τ) of $\tau1$ = 8.9 ms and $\tau2$ = 68.6 ms for 1 mM GLU, $\tau1$ = 15.3 ms and $\tau2$ = 281 ms for 300 mM KA, and τ = 2.2 s for 50 mM DOM. GluR5 and GluR6 have no detectable steady-state currents, but desensitize completely (Sommer et al 1992), quite different from GluR1–GluR4, which show a substantial current even after prolonged agonist application. The desensitization properties of GluR5 and the order of agonist potency (notably, the much higher potency of DOM over KA) are reminiscent of KA responses from C-fibers of dorsal root ganglion cells (Huettner 1990). Because GluR5 is expressed in the spinal cord (see below), it may constitute the C-fiber glutamate receptor, or at least be part of a heteromeric receptor complex in the C-fiber.

GluR6 responses can be modulated (~50% increase in current amplitude) with phosphorylation by PKA, which phosphorylates GluR6 at serines 666 and 684. This modulatory effect of PKA does not alter the affinity for GLU, the I/V relation, or the desensitization properties (Raymond et al 1993, Wang et al 1993), and thus probably results from a change in single-channel conductance, open frequency, mean open time, or number of active receptors. Initial data suggest an increase in channel open frequency (Wang et al 1993). Mutagenesis experiments demonstrated that phosphorylation of Ser-684 alone (which is located at a very strong PKA consensus site) is sufficient to bring about the maximal stimulatory effect of PKA, whereas phosphorylation at Ser-666 (located at a very weak PKA consensus site) by itself produces only 50% of that effect (Wang et al 1993), suggesting that Ser-684 may be the preferred phosphorylation site in vivo.

Coexpression of GluR5, GluR6, or GluR7 with GluR1–GluR4 does not lead to any detectable differences in the responses compared with the respective single subunits, supporting the hypothesis that GluR5–GluR7 do not form heteromeric complexes with GluR1–GluR4 (Bettler et al 1990, 1992; Sommer et al 1992). GluR5 and GluR6 subunits do, however, coassemble into heteromeric complexes with KA2, a member of another glutamate receptor subfamily (see below). Unexpectedly, no interaction of GluR7 with either KA2 or KA1 could be demonstrated (Lomeli et al 1992).

EXPRESSION OF RECEPTOR RNA In situ hybridization studies reveal a pronounced differential distribution of GluR5, GluR6, and GluR7 RNAs in many regions of the brain (Table 5). In addition, the distribution pattern is quite different from that of GluR1–GluR4 RNA, being generally more restricted, and appears to be developmentally regulated. Levels of expression are generally lower than those of GluR1–GluR4, particularly for GluR5 RNA (Bettler et al 1990).

The most prominent differences in the expression of GluR5, GluR6, and GluR7 RNAs are in the hippocampus, where GluR7 is not detected (except for low levels in the dentate gyrus), GluR5 shows only low levels, and GluR6 is highly expressed (Bettler et al 1990, Egebjerg et al 1991, Lomeli et al 1992). Expression of GluR6 RNA is more pronounced in CA3 pyramidal cells than in CA1 cells, which fits well with electrophysiological data that indicates a KA receptor with nanomolar affinities on CA3 but not CA1 neurons (Robinson & Deadwyler 1981, Ben-Ari & Gho 1988). In cerebellar granule cells, GluR6 RNA expression is high, GluR5 is low, and GluR7 is absent, whereas in the caudate-putamen GluR7 RNA is high, GluR6 is low, and GluR5 is absent. In the retina GluR6 RNA is expressed in the ganglion cell layer, and GluR5 is mainly seen in the inner nuclear layer (Egebjerg et al 1991, Hughes et al 1992, Müller et al 1992).

The expression patterns of GluR6 and GluR7 transcripts are clearly different from the distribution pattern of $[^3H]$AMPA binding sites (Cotman et al 1987) in the brain, but taken together they overlap very well with most of the $[^3H]$KA binding sites. The combined expression patterns of GluR6 and GluR7 RNAs also correlate very well with areas known to be particularly vulnerable to KA- and DOM-induced neurotoxicity, such as the reticular thalamic nucleus, striatum, hippocampus, cerebral cortex, piriform cortex, and cerebellum (Bettler et al 1992). Thus, receptors containing GluR6 and/or GluR7 could potentially be responsible for KA- and DOM-induced neuronal degeneration in those brain regions.

Expression of GluR5 RNA during development starts at E10 in postmitotic neurons. It is most pronounced where neurons differentiate, and later, during postnatal development, becomes spatially more restricted, with generally lower levels of expression. The most pronounced change occurs in cerebellar granule cells, which express GluR5 RNA highly until P12. Then, the expression in the granule cells decreases during development compared with Purkinje cells, which keep their high levels of GluR5 RNA expression. In the PNS, expression of the GluR5 gene starts at E11, steadily increases until E16, and then stabilizes, showing a distribution pattern that is very different (except in dorsal root ganglia) from that of GluR1–GluR4 RNA (Bettler et al 1990). GluR7 RNA expression starts at E15 and reaches the adult level and pattern by the time of birth (Lomeli et al 1992).

PROTEIN-BIOCHEMICAL CHARACTERIZATION OF RECOMBINANT RECEPTORS
Researchers have generated a polyclonal antiserum directed against a C-ter-

minal synthetic peptide of GluR6 and have shown it to immunoprecipitate a 112-kDa protein from solubilized cell membranes of GluR6-transfected cells. A 99-kDa protein was also observed and demonstrated to represent a nonglycosylated precursor (Raymond et al 1993). The anti-GluR6 antibody cross-reacted with GluR7 but not GluR1–GluR5 or KA1–KA2. The affinity-purified antiserum coprecipitated GluR6 and/or GluR7 with KA2 from solubilized, nondenatured membranes of the CG4 glial precursor cell line, but it did not precipitate GluR2–GluR4 subunits that are also present in these cells (RB Puchalski, J-C Louis, N Brose, et al, submitted). This observation shows that GluR6 and/or GluR7 form native heteromeric receptor complexes with the high-affinity KA receptor subunit KA2 but not with the AMPA-pre-ferring subunits GluR2–GluR4. This observation is consistent with electro-physiological data that suggest heteromeric receptor formation between GluR6 and KA2 (see below).

Co-transfection of cells with GluR6 and the catalytical subunit of PKA induced strong phosphorylation at two PKA consensus sites (residues Ser-666 and Ser-684) in the putatively intracellular loop between TMD III and TMD IV, demonstrating directly that this domain in fact is located intracellularly (Raymond et al 1993, Wang et al 1993).

GENE STRUCTURE AND CHROMOSOMAL LOCALIZATION   The gene structure of GluR5–GluR7 is mostly unknown, except for the part that encodes the region of TMD I through TMD III of GluR5 and GluR6. In both genes the three TMDs are encoded by three different exons, unlike in the GluR1–GluR4 genes, where TMDs I and II are encoded in the same exon. Also, the splice site of the intron that separates TMD II and III in GluR5–GluR7 is different from the respective site in GluR1–GluR4 (Sommer et al 1991).

GluR5 has been mapped to chromosome 21q22 (Table 6), close to the locus for familial ALS, a disorder that has long been thought to be linked to a malfunction of glutamatergic neurotransmission. Interestingly, GluR5 RNA is localized in the ventral horn of the spinal cord, where the ALS-susceptible neurons reside; these neurons have also been shown in explant culture to be selectively vulnerable to non-NMDA receptor-mediated GLU toxicity (Eubanks et al 1993). Researchers are currently testing the possible link between the GluR5 subunit and ALS by analyzing the GluR5 gene of ALS patients.

GluR7 is located on chromosome 1 (Table 6); its exact position is currently being mapped (JO McNamara, personal communication).

High-Affinity KA Receptors: KA1 and KA2

CLONING AND STRUCTURAL FEATURES   In several laboratories, researchers have cloned two receptors, KA1 and KA2, by using PCR-mediated DNA amplification with degenerate primers made to regions of high homology in the GluR1–GluR4 sequences and low-stringency hybridization screening with probes derived from GluR1–GluR4 cDNAs. The receptors share 70% amino acid sequence identity with each other but only ~37% with GluR1–GluR4 and ~43% with GluR5–GluR7 (Table 2; Figure 5), suggesting that they form a separate subfamily of glutamate receptors (Werner et al 1991, Herb et al 1992, Kamboj et al 1992, Sakimura et al 1992).

KA1 and KA2 share the same overall structure as GluR1–GluR7 (see Figure 2). Amino acid sequence identity with other GluRs is highest in the region ~130 residues upstream of TMD I through TMD IV, whereas both N- and C-terminal domains show only low homology. There are numerous N-glycosylation consensus sites in the N-terminal domains of both KA1 and KA2, and both receptors have a variety of protein kinase consensus sites in their putative intracellular domains between TMD III and TMD IV (Table 1). The region preceding TMD I (~130 amino acids) is highly homologous to GluR1–GluR7, and it probably contains the ligand binding site. The mouse, rat, and human KA2 subunits exhibit 98–99% amino acid sequence identity.

MECHANISMS CREATING RECEPTOR DIVERSITY   No splice variants have been reported for KA1 and KA2. Also, there are no indications of editing at the Q/R site in TMD II, which contains a Q in both receptors (see Figure 3).

FUNCTIONAL PROPERTIES   No electrophysiological responses have been detected when either KA1 or KA2 was expressed as a homomeric protein in oocytes or in transfected cell lines (Werner et al 1991, Herb et al 1992, Kamboj et al 1992, Sakimura et al 1992). Failure to detect responses may result from very rapid desensitization, a very small unitary conductance, or a low open probability of the homomeric channels. However, [^{3}H]KA ligand binding studies with membranes of transfected cell lines demonstrated the presence of high-affinity binding sites for KA (K_{DS} were 2–15 nM; see Table 4), whereas [^{3}H]AMPA binding was undetectable (Werner et al 1991, Herb et al 1992, Kamboj et al 1992). The affinity of KA1 and KA2 for KA (binding of which is inhibited by CNQX) is significantly higher than that of GluR5–GluR7, suggesting that KA1 and KA2 may represent the high-affinity [^{3}H]KA receptor seen in ligand binding studies of synaptic membranes and in autoradiographic studies of brain sections. In these studies, [^{3}H]KA binding sites with K_{DS} of ~5 nM and ~50 nM have been detected, and while GluR5–GluR7 match the low-affinity 50-nM sites (see above), KA1 and KA2 may represent the high-affinity 5-nM sites. The sequence of agonist potencies, KA > QA > DOM > GLU >> AMPA (Table 4), also demonstrates that KA1 and KA2 are pharmacologically distinct from GluR5–GluR7 and are high-affinity KA receptors. It has not been possible to demonstrate that KA1 and KA2 can form functional homomeric receptors, so it may be that the high-affinity [³H]KA binding sites detected by receptor autoradiography do not represent functional KA receptors linked to ion channels, but rather KA1 or KA2 subunits waiting for coassembly with other subunits in order to form functional heteromeric channels, which may then have lower KA binding affinity.

When the KA1 subunit was coexpressed with GluR1–GluR4, no changes in functional properties were observed, indicating that interaction between these receptor subunits is unlikely (Werner et al 1991). Coexpression of KA1 with GluR5–GluR7 has not been reported. For KA2, however, coexpression with GluR5 and GluR6 (experiments with GluR1–GluR4 and GluR7 have not been reported) produced receptors with functional properties not present in the respective homomeric receptors (Herb et al 1992, Sakimura et al 1992). Both edited (R) and unedited (Q) versions of GluR5 and GluR6 produced functional receptors upon coexpression with KA2—even the combination GluR5(R)/KA2, which is remarkable because neither of these two subunits alone expresses functional channels. The desensitization of the GluR5/KA2 channels was biphasic and fairly rapid, with time constants of $\tau 1 = 15$ ms and $\tau 2 = 690$ ms (Herb et al 1992). The most impressive functional difference seen upon coexpression of GluR6 with KA2 was the responsiveness of the heteromeric receptor to AMPA, a property not seen with GluR6 alone. The AMPA response of GluR6/KA2 was nondesensitizing, which is quite different from AMPA currents of GluR1–GluR4 (Herb et al 1992, Sakimura et al 1992). The rectification properties of the heteromeric GluR5/KA2 and GluR6/KA2 channels in each case were determined by the presence (linear I/V curve) or absence (inwardly rectifying I/V curve) of an edited subunit containing R at the Q/R site in TMD II (Herb et al 1992).

If heteromeric assembly of receptors in vivo is the rule, it is possible that pharmacologically "pure" KA receptors (nonresponsive to AMPA) do not exist in the brain at all, although it cannot be ruled out that in some specific subpopulation of cells GluR6 receptors are expressed as homomeric complexes. Likewise, no data support the existence of pure AMPA receptors that cannot be activated by KA. These findings obviously call into question the prevailing nomenclature, which seeks to distinguish KA receptors from AMPA receptors, and point out the necessity to develop a more rational nomenclature.

In addition to electrophysiological differences, heteromeric receptors also display altered [³H]KA ligand binding properties. For example, GluR5(R)/KA2 receptors bind KA with a K_D of 90 nM and are competitively inhibited by DOM with a K_I of 665 nM. GluR5(R) and KA2 subunits alone have K_Ds for KA of 73 nM and 15 nM, respectively, and DOM inhibits binding with K_Is of 2.1 nM and 275 nM, respectively (Herb et al 1992). Thus, the formation of heteromeric receptors alters not only channel properties, but also the ligand binding site. We therefore can speculate that the ligand binding site is formed at the interface of two or more subunits, as has been suggested for nAChRs. Should this turn out to be the case, the possibility exists that many different ligand binding sites may be formed through different combinations of subunits. Coassembly of KA2, GluR6, and/or GluR7 into native heteromeric receptor complexes has been demonstrated in immunoprecipitation studies (RB Puchalski, J-C Louis, N Brose, et al, submitted), as discussed above in the section on GluR5–GluR7.

EXPRESSION OF RECEPTOR RNA   KA1 and KA2 RNAs are distributed quite differently in the brain. Whereas KA1 RNA is generally expressed at low levels and is abundant in only two cell types, CA3 pyramidal cells and dentate granule cells of the hippocampus, KA2 RNA is widely expressed in most parts of the brain, reminiscent of GluR2 (Herb et al 1992). Notably, KA1 RNA is absent from, and KA2 RNA is only weakly expressed in the reticular thalamic nucleus, an area of high levels of $[^3H]KA$ binding in autoradiographic studies, and one of the target areas for KA-induced neurotoxicity. Therefore, other glutamate receptor subunits, possibly GluR5–GluR7, are likely to mediate excitotoxicity in this particular brain area. The high expression of KA1 RNA in hippocampal granule cells has been interpreted as evidence for the possible presynaptic localization of KA1 (Werner et al 1991), because when the mossy fibers, the projections of granule cells to the CA3 region, are interrupted, CA3 cells lose their high-affinity $[^3H]KA$ binding sites (Represa et al 1987).

During development, KA1 and KA2 RNAs can be detected as early as E14. KA1 RNA has a very restricted expression pattern and shows different developmental regulation in different regions. In hippocampus, expression develops at E19 and reaches adult levels at P1, whereas in subiculum, striatum, and cortex expression in the adult is reduced (or even absent, as in the striatum) when compared with the early postnatal days. KA2 RNA is highly and nearly ubiquitously expressed in embryonic CNS and some regions of the PNS, such as dorsal root ganglia (Werner et al 1991, Herb et al 1992).

As discussed above, KA1 and KA2 represent a kainate receptor subfamily functionally distinct from GluR5–GluR7. However, receptors with properties different from either class can be formed upon coexpression of members of these two distinct subfamilies in in vitro systems. The same may be true in vivo, and in fact RNA distribution patterns suggest the possibility of coexpression of the two classes in the same neuron, particularly in cerebellar granule cells, hippocampal pyramidal and dentate granule cells, and the caudate-putamen (GluR6/KA2); the medial habenula, amygdala, and E17 dorsal root ganglia (GluR5/KA2); and hippocampal pyramidal and dentate granule cells (GluR6/KA1) (Herb et al 1992).

KA Binding Proteins: KBP-chick and KBP-frog

CLONING   The isolation of cDNAs of two KA binding proteins (KBPs) was reported simultaneously with the publication of the cloning of the GluR1 subunit. These genes were isolated by conventional cloning procedures, which included purification of the protein, sequencing of proteolytic or chemical cleavage fragments, and screening of cDNA libraries with oligonucleotide probes synthesized according to the partial sequences obtained from the purified peptides. The two KBPs isolated appear to represent the homologous genes of chick (Gregor et al 1989) and frog (Wada et al 1989).

STRUCTURAL FEATURES   The two KBPs, herein named KBP-chick and KBP-frog, are surprisingly small proteins (464 and 470 amino acid residues, respectively) when compared with other GluRs (Table 1; Figure 2). Their general structure, nevertheless, appears to be similar to that of other GluRs, with a N-terminal domain (which contains several N- and O-glycosylation sites) followed by four putative TMDs in the cluster-of-three-plus-one pattern typical of ligand-gated ion channels, and a presumably extracellular C-terminal domain. Sequence comparison reveals considerable overall homology (~30% amino acid sequence identity) with other GluR subfamilies and 55% identity between the two proteins (Table 2). The major difference between other GluRs and KBP-chick and KBP-frog is the KBPs' short N-terminal domains, which extend for only 147–148 amino acid residues upstream of the putative TMD I; in other GluRs this domain is generally 530–550 amino acid residues in size. Interestingly, however, this short N-terminal domain is just long enough to include the entire region of ~130 amino acids upstream of TMD I, which is homologous to the *E. coli* periplasmic glutamine-binding protein that has been suggested to contain the ligand binding site of GluRs (see discussion of GluR1–GluR4).

No splice variants have been discovered for either of the two genes, and it has been shown that the gene of KBP-chick does not contain flop/flip exons (Eshhar et al 1992, Gregor et al 1992). No mammalian homologues have been reported.

FUNCTIONAL PROPERTIES   KBP-chick and KBP-frog when expressed in oocytes or transfected cells did not produce detectable ion channel activity (Wada et al 1989, Gregor et al 1992). Additionally, antisense oligonucleotides specific for KBP-frog when coinjected with total frog brain RNA did not inhibit KA responses (Wada et al 1989). Thus, despite their structural features that are similar to other ligand-gated ion channels, the function of these two proteins remains a mystery.

[3H]KA ligand binding studies with membranes from cell lines transfected with KBP-frog revealed the same high affinity (K_D = 5.5 nM) as was found for the purified protein and for frog brain membranes. As in frog brain membranes, DOM had an even higher affinity than KA, and QA, GLU, and CNQX were 200- to 300-fold less potent in displacing KA binding than was DOM (Table 4). However, native frog brain membranes display a second [^{3}H]KA binding site with lower affinity (K_D = 35 nM), which is not seen in the recombinant KBP-frog (Wada et al 1989). This could indicate that a second subunit may be required in order to assemble a properly functional ion channel. KPB-chick when expressed in transfected cells binds [^{3}H]KA with a K_D of 560 nM (Gregor et al 1992), a much lower affinity than seen with the KBP-frog, but consistent with the affinity of KA for the KBP from chick cerebellum, which is much lower than that of the frog KBP.

EXPRESSION OF RECEPTOR RNA   The expression of KBP-chick RNA in the cerebellum is localized exclusively and abundantly in the Bergmann glia cells, which tightly surround the Purkinje cells (Eshhar et al 1992, Gregor et al 1992). There is also some expression in cerebral hemispheres. This distribution was confirmed by immunostaining with a monoclonal antibody against the isolated 49-kDa protein that was used to obtain partial sequence information for the cloning. The same distribution was also seen with [^{3}H]KA receptor autoradiography, which identified a single binding site of very high density, ~100-fold higher than usually found for neurotransmitter receptor proteins (Ortega et al 1991a). Cultured Bergmann glia are stained by the monoclonal anti-KBP antibody, confirming the in situ hybridization data (Gregor et al 1989, Ortega et al 1991a). The restricted distribution of KBP-chick RNA mainly on Bergmann glia cells, together with the close association of Bergmann glia and Purkinje cells, was interpreted by some workers as indicating a possible role for KBP-chick subunits in the modulation of the efficacy of Purkinje cell synapses, possibly by some yet to be demonstrated ion channel activity of KBP-chick (Gregor et al 1988, Ortega et al 1991a,b). During development, KBP-chick RNA is expressed several days after Bergmann glia have developed, when granule cells start to migrate. This may indicate a role for the KBP-chick subunit on Bergmann glia cells in coordinating or regulating granule cell migration (Ortega et al 1991a).

KBP-frog RNA is found abundantly in neurons in the optic tectum, telencephalon, and cells lining the lateral ventricles. This pattern is identical with that found with immunostaining and [^{3}H]KA receptor autoradiography (Wada et al 1989). [^{3}H]KA receptor autoradiography was performed with 8-nM and 80-nM [^{3}H]KA in order to differentiate between the high- and low-affinity binding sites. Both were found to share exactly the same distribution pattern, rendering it likely that they are coexpressed in the same cells and may actually be part of the same receptor complex (Wenthold et al

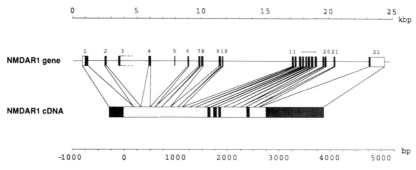

Figure 6 Exon/intron structure of the NMDAR1 gene (modified from Hollmann et al 1993) and position of the 22 exons in the cDNA clone. Exons shown in black indicate coding region, and exons shown in white encode untranslated portions of the cDNA (stippled ends of the cDNA graph). The four putative TMDs are indicated by black boxes on the cDNA. Exon 3 was found only in a truncated version of NMDAR1, and exon 5 is present only in a subset of splice variants of NMDAR1 (see Table 3 for details).

1990a). At the electronmicroscopic level, the native KBP subunit of the frog was found to be localized mainly extrasynaptically, although some labeling was seen in postsynaptic densities (Wenthold et al 1990a). The extrasynaptic localization, lack of function, and high density of expression do not support a typical neurotransmitter receptor role for the frog KBP (Wenthold et al 1990a). Its function remains to be elucidated and might be very different from that of other GluRs. Although the frog KBP appears to be evolutionarily related to other GluRs, it probably diverged from them early in phylogeny (Figure 5).

PROTEIN-BIOCHEMICAL CHARACTERIZATION   The purified KBP from chick migrates as a 49- to 50-kDa protein on denaturing protein gels, and as a 350-kDa complex on nondenaturing gels. This was interpreted as suggesting a heptameric subunit composition if the receptor is assembled exclusively from KBP-chick subunits (Ortega et al 1991b). However, detergent micelles contributing to the size of the receptor complex probably distorted this estimate of the actual complex size. Bergmann glia membranes show a cross-reacting protein band at ~93 kDa, which may represent a related but independent subunit, possibly one of the other cloned GluRs that are known to be present in Bergmann glia cells, for example GluR1 and GluR4.

Affinity-purified KBP from chick can be phosphorylated in vitro with the catalytical subunit of PKA, in accord with two PKA consensus sites found in the proposed intracellular domain between TMD III and TMD IV. Receptor phosphorylation has been suggested to abolish desensitization of KA-induced Na^+ influx through the ion channel of the chick KBP in cultured Bergmann glia cells (Ortega et al 1991a,b); however, the well-known presence of other

GluR subunits in these cells, at least in the rat, prevents a simple interpretation of these experiments. Because the agonist KA inhibits in vitro phosphorylation of KBP-chick, it has been speculated that agonist binding may lead to a conformational change that buries the phosphorylation site (Ortega & Teichberg 1990).

Monoclonal antibodies made against DOM affinity-purified KBP from frog recognize an ~48-kDa protein on Western blots of frog CNS membranes, plus a minor ~99-kDa band, and the antibodies immunoprecipitate both high- and low-affinity [³H]KA binding sites. These antibodies also recognize GluR1, GluR2, and GluR3 on Western blots, but not the KBP subunit from chick. This cross-reactivity may explain the strong labeling of an ~99-kDa protein band (but no 49-kDa band) seen on Western blots of rat brain membranes (Wenthold et al 1990a).

GENE STRUCTURE   The gene that encodes the KBP-chick subunit is 11.2 kbp in size, with 11 exons (32–248 bp in length) interrupted by 10 introns (0.2–2.4 kbp in length). All introns have perfectly conserved GT/AG splice sites, and the promoter region apparently contains a typical RNA polymerase II transcription initiation element. There are no differences in the gene and cDNA sequences, except for three nucleotide exchanges that did not change the amino acid sequence (Eshhar et al 1992, Gregor et al 1992). The four putative TMD are encoded on four different exons. The splice sites in front of TMD I, after TMD I, and after TMD II are precisely conserved among KBP-chick, GluR5, and GluR6, whereas only the site in front of TMD I is conserved between GluR1–GluR4 and KBP-chick (Gregor et al 1992). This indicates that KBP-chick is more closely related to the low-affinity KA receptors than to the AMPA-preferring receptors (see also Figure 5).

Invertebrate Glutamate Receptors: DGluR-I, DGluR-II, and LymGluR

CLONING AND STRUCTURAL FEATURES   Three invertebrate GluRs have been cloned to date, two from *Drosophila melanogaster,* named DGluR-I (Ultsch et al 1992) and DGluR-II (Schuster et al 1991), and one from the pond snail *Lymnaea stagnalis,* which was called "Lym" in the original publication (Hutton et al 1991), a name we have emended to LymGluR in this review. All three genes were initially identified through PCR-mediated DNA amplification from genomic libraries with degenerate primers made to conserved sequences in GluR1 and the KBPs. Partial genomic clones were then used to screen cDNA libraries, and several overlapping partial clones were spliced together to obtain full-length clones.

All three invertebrate genes code for proteins with structures very similar to GluR1–GluR7, KA1, and KA2 (see Figure 2). DGluR-I, DGluR-II, and LymGluR have large, presumably extracellular N-terminal domains of 585, 533, and 539 amino acids, respectively, which contain numerous N-glycosylation consensus sites (Table 1). The N-terminal domain is followed by a core region of 304, 280, and 281 amino acids, respectively, each with four hydrophobic regions in the cluster-of-three-plus-one pattern likely to represent TMDs I through IV. The C-terminal domains, which are presumably extracellular, consist of 75, 170, and 78 amino acids, respectively, and contain potential N-glycosylation sites in DGluR-I and LymGluR. The putative TMD II domains contain a Q at the Q/R site in all three gene products (Figure 3), and the presumably intracellular loops between TMD III and TMD IV contain several consensus sites for protein kinases (Table 1). The ~130–amino acid region preceding TMD I, which shows significant sequence homology with *E. coli* periplasmic glutamine-binding protein and has been implicated in ligand binding in other GluR genes, is conserved in all three invertebrate genes as well.

There is a high amino acid sequence homology between DGluR-I and LymGluR (~40% overall identity), whereas DGluR-I and DGluR-II, and DGluR-II and LymGluR share only ~29% identity. Thus, it appears that LymGluR may be the *L. stagnalis* homologue of the *D. melanogaster* DGluR-I gene, whereas DGluR-II is only a distantly related gene (Figure 5). This notion is supported by sequence comparisons with other GluR genes (Table 2) that show that both DGluR-I and LymGluR share ~33–46% amino acid sequence identity with AMPA and KA receptor genes, whereas DGluR-II is only 29–31% identical with these genes. Similarly, the homology of DGluR-I and LymGluR with NMDA receptors is higher than that of DGluR-II (Table 2).

FUNCTIONAL PROPERTIES   DGluR-I and DGluR-II (Schuster et al 1991, Ultsch et al 1992) but not LymGluR (Hutton et al 1991) form functional ion channels when expressed in oocytes.

DGluR-I is activated by KA with an EC_{50} of ~75 μM. CNQX reversibly blocks the response, and philanthotoxin partially blocks it. The ion channel shows a marked inward rectification similar to that seen in GluR1, GluR3, and GluR4, which is in accordance with the presence of Q at the Q/R site of TMD II. DGluR-I cannot be activated by other glutamatergic agonists, such as AMPA, QA, NMDA, ibotenate (IBO), L-homocysteic acid (L-HCA), or L-aspartate (L-ASP), and, quite surprisingly, DGluR-I is not activated by GLU (Ultsch et al 1992). This raises the question of what the endogenous ligand of this receptor might be, given that the sole identified agonist, KA, naturally occurs only in some algae and not in animals. It is possible, however, that GLU desensitizes the receptor extremely rapidly, or that it has a much lower affinity for this invertebrate receptor, which could be an essential feature owing to the high concentration of glutamate constantly present in *D. melanogaster* hemolymph. Alternatively, another receptor subunit may be required in order to assemble a fully functional receptor with the correct pharmacology.

DGluR-II has been reported to be activated by GLU and L-ASP, and with significantly smaller currents ($<$ 10 nA in oocytes) by QA, AMPA, and KA. However, unphysiologically high agonist concentrations were required (the EC_{50} for GLU was 35 mM, for L-ASP 50 mM), which leaves some doubts about the specificity of this receptor activation. Also, the response to GLU is insensitive to CNQX, DNQX, and argiotoxin, all of which inhibit non-NMDA responses of vertebrate GluRs (Schuster et al 1991). Should this pharmacological profile be confirmed, DGluR-II would appear to represent a receptor subfamily with unique properties, for which no mammalian homologues have been found so far. The reversal potential of the DGluR-II response is -10 mV; no data are available on divalent cation permeability or the rectification properties, although the presence of Q at the Q/R site in TMD II would predict an inwardly rectifying I/V curve. DGluR-II is expressed in somatic musculature (see below), which has been shown to express two types of glutamate receptors, an excitatory, cation-conducting receptor (D-type) and an inhibitory, anion-conducting receptor (H-type). DGluR-II appears to represent a D-type receptor, although its agonist potencies are much lower than those described for other invertebrate (crustacean) glutamate receptors. These low agonist potencies, however, may result from the absence of an additional subunit present in vivo or from rapid desensitization in the oocyte expression system (Schuster et al 1991).

Although several different LymGluR clones have been tested, no functional ion channels could be detected (Hutton et al 1991). It is possible that the LymGluR subunit needs to coassemble with a second subunit for function. It has also been speculated that the oocyte expression system may be incapable of expressing *L. stagnalis* glutamate receptor, but this seems unlikely given the fact that *D. melanogaster* receptors are readily expressed.

Ligand binding experiments have not been reported for any of the three proteins.

EXPRESSION OF RECEPTOR RNA   DGluR-I RNA is expressed in esophageal ganglia and the ventral cord of *D. melanogaster,* starting from late embryonal stages. The message then decreases during larval development (as revealed by Northern blot analysis) to become undetectable in late larvae. Gene expression is resumed in early pupal stages and gradually increases until the adult stage is reached (Ultsch et al 1992).

DGluR-II RNA shows an entirely different pattern of expression. Unlike DGluR-I RNA, which is expressed in the *D. melanogaster* nervous system,

DGluR-II RNA is expressed in somatic musculature with a striking, typically segmental organization of the expression pattern. Both the expression pattern and the level of expression are developmentally regulated. Expression is first seen in embryonal stages, then decreases in first larval stages, but resumes during larval growth. In pupae and adults, level of expression is strongly decreased (Schuster et al 1991). For LymGluR, no RNA could be detected with a genomic probe on Northern blots of RNA from *L. stagnalis* egg mass, adult nervous system tissue, or adult muscle tissue, indicating a very low abundance of this receptor (Hutton et al 1991). The expression pattern of LymGluR RNA thus remains to be determined.

CHROMOSOMAL LOCALIZATION   DGluR-I resides on *D. melanogaster* chromosome 3L, position 65C (Ultsch et al 1992), whereas DGluR-II maps to chromosome 2L, position 25F, a region that is characterized by several lethal mutations (Schuster et al 1991). The chromosomal localization of LymGluR is unknown.

A single DGluR-I gene was found in *D. melanogaster* and a single exon was found to encode the region encompassing TMD I, II, and III, including some sequence downstream of TMD III. This exon is separated by an intron from the exon that encodes the rest of the putative cytoplasmic loop between TMD III and TMD IV, which, after another intron, is followed by the exon that encodes TMD IV. No alternate exons were found for the region homologous to the flop/flip domain in GluR1–GluR4, although the intron/exon border is conserved in this region. The gene structures of D-GluR-II and LymGluR are unknown.

Orphan Receptor Subunits: delta1 and delta2

CLONING AND STRUCTURAL FEATURES   Screening of a mouse forebrain library by low-stringency hybridization with GluR1–GluR4 probes led to the isolation of a cDNA with low but significant homology (~26–30% amino acid sequence identity; see Table 2) to other GluR subfamilies (Table 2; Figure 5). The gene was named δ1 (herein spelled delta1) to indicate that it represents a different subfamily of GluRs (Yamazaki et al 1992a). Later, another receptor subunit 56% identical in sequence with delta1 was isolated and named delta2 (Lomeli et al 1993). Despite the low sequence homology with other GluRs, delta1 and delta2 have all the major structural features of GluR1–GluR7, KA1, and KA2. Their overall structure shows large (~550 amino acids), presumably extracellular N-terminal domains with two to four N-glycosylation consensus sites, an ~290–amino acid core region with four hydrophobic stretches in the cluster-of-three-plus-one pattern likely to represent TMD I through TMD IV, and a C-terminal domain of ~160 amino acids (Figure 2). The putative TMD

II domains contain Q residues at the Q/R site (Figure 3), and the presumably intracellular loop between TMD III and TMD IV has several consensus sites for protein kinases (Table 2).

FUNCTIONAL PROPERTIES   Neither delta1 nor delta2 when transfected into cell lines expressed functional ion channels in response to GLU, KA, AMPA, or the neurotransmitter candidates L-HCA and L-cysteinesulfinic acid (L-CSA) (Lomeli et al 1993). Moreover, in contrast to other subunits that lack homomeric ion channel function, such as KA1, KA2, GluR7, or the KBPs, no specific [^{3}H]KA or [^{3}H]AMPA binding could be detected in cell membranes, which suggests that the delta subunits either belong to a class of receptors pharmacologically unrelated to other GluRs, represent modulatory or structural subunits lacking ligand binding sites, or are not expressed in the cells tested. When delta2 was cotransfected into cells with GluR2, GluR5, KA1, NMDAR1, NMDAR2A, or NMDAR2C, no change in the functional properties of the respective homomeric receptor was evident. When delta1 was coexpressed with GluR7, no agonist-induced responses were detectable. Thus, the electrophysiological and pharmacological properties of these receptors, which have all the hallmarks of a glutamate-gated ion channel, remain a mystery.

EXPRESSION OF RECEPTOR RNA   The expression of delta1 RNA in adult rat brain is low and diffuse, mainly in hippocampal pyramidal and dentate granule cell layers. This gives delta1 the distinction of being the mammalian GluR gene with the lowest level of RNA expression (Lomeli et al 1993). In contrast, delta2 RNA is highly and specifically expressed in cerebellar Purkinje cells, and is expressed at low levels in the cingulate cortex and hippocampal dentate granule cells.

The delta1 gene is much more prominently expressed in young animals than in adults, with higher levels in caudate-putamen at P0 and in parts of the thalamus at P12, which then decline to adult levels (Lomeli et al 1993). This may indicate a role for delta1 in development. The delta2 gene does not appear to be developmentally regulated (Lomeli et al 1993).

NMDA Receptors Forming Functional Homomeric Channels: NMDAR1

CLONING   As had been the case for the first non-NMDA receptor, GluR1, it was the powerful method of expression cloning that led to the isolation of the first cDNA clone for an NMDA receptor subunit, NMDAR1 (Moriyoshi et al 1991). By screening a rat forebrain library made from size-selected RNA of 3–5 kbp, and by testing initial pools of ~3000 cDNA clones at a time, in an admirable effort Nakanishi's group isolated a cDNA that showed low but significant homology (22–29% amino acid sequence identity) with other GluR genes (Table 2; Figure 5). When expressed by itself in homomeric form in oocytes, this cDNA induced responses that had all the hallmark features of NMDA receptors (see below). Isolation of a mouse NMDAR1 clone was later reported by another group, who successfully exploited the low sequence homology of NMDAR1 with other GluRs by using low-stringency hybridization screening with GluR1, GluR2, GluR6, and KA2 probes (Yamazaki et al 1992b). A third group later reported the cloning of NMDAR1 through a combination of a hybrid depletion assay with expression cloning and low-stringency hybridization screening (Nakanishi et al 1992). PCR-generated probes homologous to conserved GluR sequences in the TMD III region were tested for depletion of NMDA receptor responses expressed in oocytes from total RNA; a probe identified with this technique was then used to isolate a full-length NMDAR1 cDNA clone.

STRUCTURAL FEATURES   Despite the low overall homology with non-NMDA glutamate receptors (25–29% amino acid sequence identity), NMDAR1 shares the typical structure of ligand-gated ion channels (see Figure 2). NMDAR1 has a large N-terminal, presumably extracellular domain containing 10 N-glycosylation sites: 4 putative TMDs arranged in the cluster-of-three-plus-one pattern with a large, putatively intracellular domain between TMD III and TMD IV, and a presumably extracellular C-terminal domain. The ~160-residue domain N-terminal of TMD I in NMDAR1 is highly homologous with the other GluRs and *E. coli* periplasmic glutamine-binding protein, except that this region is larger (it is only ~130 residues in other GluRs), owing to extra sequence in NMDAR1 that is absent from the other GluRs. This region may be involved in ligand binding to the NMDAR1 receptor. In addition, as in other GluRs, a region N-terminal of TMD IV in the putative intracellular loop shows significant homology with the *E. coli* periplasmic glutamine-binding protein, but the function of this region is unknown (Moriyoshi et al 1991). The amino acid sequence of the human NMDAR1 is 99% identical with the rat NMDAR1 (Karp et al 1993).

MECHANISMS CREATING RECEPTOR DIVERSITY   A total of eight splice variants have been reported for NMDAR1 (Table 3). They are created by all possible combinations of three different, independently occurring NMDAR1 splicing events: the insertion of one extra exon (exon 5 in Figure 6) of 63 bp in the N-terminal domain, the deletion of exon 21 (111 bp) in the C-terminal domain, and the use of an alternate splice acceptor site in the C-terminal exon 22 resulting in the deletion of 356 bp. Use of the alternate splice site in exon 22 removes the original stop codon and converts a stretch of 66 bp of untranslated sequence of NMDAR1 into coding sequence, with a new stop codon (Anantharam et al 1992, Durand et al 1992, Nakanishi et al 1992, Sugihara et al 1992, Yamazaki et al 1992b, Hollmann et al 1993). Estimates by Sugihara et al (1992) indicate that the relative abundance of the various splice variant RNAs is such that NMDAR1 is prevalent with ~67%, splice variants with the N-terminal insertion amount to ~15%, and those with a C-terminal deletion (but no N-terminal insertion) make up ~18% of all cDNA clones. A PCR study of relative abundance, which, however, did not include splice variants containing the 356-bp deletion, found a slightly different distribution: 48%, 20%, and 32%, respectively (Anantharam et al 1992).

Splice variants that contain the highly charged N-terminal insertion of 21 residues (which carries six positive and three negative charges) display different agonist and antagonist potencies and produce larger currents than do receptors lacking the insertion (Durand et al 1992, Nakanishi et al 1992, Hollmann et al 1993). NMDAR1 receptors lacking the N-terminal insertion are more sensitive to proton inhibition (S Traynelis, personal communication) and are selectively potentiated (by up to 175%) by micromolar concentrations of Zn^{2+}, an action of Zn^{2+} quite different from the inhibition seen with higher Zn^{2+} concentrations at all NMDA receptors studied so far (Hollmann et al 1993). Splice variants with C-terminal deletions show no differences in their basic responses, but exhibit a higher susceptibility to modulation by PKC (Durand et al 1992). This observation raises the possibility that, in contrast to the prevailing model for receptor topology, at least part of the C-terminus may be intracellular.

In addition to the eight functional splice variants, an apparently nonfunctional receptor truncated in the N-terminal domain after only 181 amino acids has been reported, but its significance remains to be elucidated (Sugihara et al 1992). It has been suggested that the truncated receptor may serve some regulatory role in NMDAR1 expression. All variants were shown to represent true splice variants of the NMDAR1 gene, which consists of 22 exons and 21 introns (see Figure 6) (Hollmann et al 1993).

FUNCTIONAL PROPERTIES   When expressed in oocytes, NMDAR1 responds in the presence of GLY to GLU, NMDA, QA, IBO, L-ASP, and L-HCA, but it does not respond to KA, AMPA, or tACPD, or to GLY by itself. Omission of GLY abolishes responses to NMDA completely (Nakanishi et al 1992) or reduces them greatly (Moriyoshi et al 1991, Yamazaki et al 1992b). A dipeptide endogenously present in the brain, N-acetylaspartylglutamate (NAAG), also activates the receptor in a GLY-dependent manner, with an EC_{50} of 185 μM (Sekiguchi et al 1992). This indicates that NAAG is ~100-fold less potent than GLU ($EC_{50} = 0.6–1.7$ μM) and ~10-fold less potent than NMDA ($EC_{50} = 12–27$ μM; see Table 4), but adds another candidate to the list of endogenous ligands for NMDA receptors that includes L-GLU, L-ASP, L-HCA, and L-CSA.

The I/V curve of NMDAR1 in the absence of Mg^{2+} is linear with a reversal potential of 0 mV, whereas in the presence of Mg^{2+} responses at potentials more negative than -20 mV are blocked (Moriyoshi et al 1991, Nakanishi et al 1992). The ion channel is highly permeable to Ca^{2+} (Yamazaki et al 1992b). Interestingly, an asparagine is located in the putative TMD II at the site equivalent to the Q/R site in other GluRs (see Figure 3), where it has been shown to influence divalent cation permeability and rectification properties. An asparagine introduced into this position in GluR3 or GluR4 by mutagenesis kept these receptors permeable to Ca^{2+}, made them less permeable to Mg^{2+}, and led to linear rectification properties, thus producing NMDA receptor-like properties. It was therefore not too surprising to find an asparagine residue at this position in an authentic NMDA receptor. Several mutagenesis studies carried out at this site of NMDAR1 confirm its involvement in the divalent cation permeability properties of the channel (Mori et al 1992, Burnashev et al 1992c, Sakurada et al 1993). Because these mutants have generally been analyzed in heteromeric receptors containing various NMDAR2 subunits, we discuss them in the section on NMDAR2 subunits (see below).

The competitive NMDA receptor antagonists D-2-amino-5-phosphono-valerate (D-APV), 3-(2-carboxypiperazine-4-yl)propylphosphonate (CPP), and cis-4-(phosphonomethyl)piperidine-2-carboxylate (CGS 19755) all inhibit NMDAR1 responses, as does 7-chloro-kynurenate (7-Cl-KYNA), a competitive antagonist at the glycine site. The channel blockers (+)5-methyl-10,11-dihydro-5H-dibenzo[a,d]cyclohept-5,10-imine maleate [(+)MK-801], Zn^{2+}, and phencyclidine also inhibit the receptor, and Mg^{2+} shows the voltage-dependent channel block typical for NMDA receptors (Moriyoshi et al 1991, Durand et al 1992). NMDAR1 is sensitive to protons, which inhibit the receptor with an IC_{50} of $\sim$50 nM, i.e. pH 7.3 (S Traynelis, personal communication). In the only ligand binding study of NMDAR1-transfected cells reported to date, [^{3}H]MK-801 was found to bind to cell membranes with a K_D of 8.5 nM, similar to the K_D of 5.0 nM found in rat brain membranes (Chazot et al 1992). The receptor density in transfected cell line membranes, however, was $\sim$20-fold less (113 fmol/mg protein) than what is usually found in brain membranes (1930 fmol/mg), which may indicate that coexpression of additional subunits (possibly NMDAR2) increases the efficacy of receptor expression. This may also explain the considerable increase in current seen when NMDAR1 is coexpressed with NMDAR2 subunits (see below).

Spermine has been reported to slightly ($\sim$1.4-fold) increase the responses of NMDAR1, whereas PKC activators such as phorbol 12-myristate 13-acetate (PMA) produce a 3- to 20-fold stimulation of the receptor response, depending on the splice variant analyzed, an effect inhibited by staurosporine, a specific blocker of PKC. PKA activators such as 8-bromo-cAMP or forskolin, on the other hand, did not alter the receptor response to agonist (Durand et al 1992, Yamazaki et al 1992b).

The data show that the functional properties reported for NMDA receptors from studies in other systems, such as tissue slices or cell culture, including the voltage-dependent block by Mg^{2+} which is the distinguishing feature of NMDA receptors, can all be found in NMDAR1. Thus, all the different sites responsible for interaction with the various agonists, antagonists, or modulators reside on the same protein. Heteromeric subunit assembly is obviously not required in order to obtain a fully functional receptor; however, current amplitudes at homomeric NMDAR1 receptors are fairly small, smaller than responses from total brain RNA expressed in oocytes. This indicates that other subunits are probably necessary to give the receptor its full in vivo efficacy.

EXPRESSION OF RECEPTOR RNA   NMDAR1 RNA is expressed in almost all neuronal cells throughout the CNS as well as the PNS (Table 5), with particularly high levels in the cerebellum, hippocampus, cerebral cortex, and olfactory bulb (Moriyoshi et al 1991, Shigemoto et al 1992b). RNAse protection assays have shown that the expression of NMDAR1 splice variants is differentially regulated. In the cerebellum, variants containing the N-terminal insertion are five times more abundant than those without the insertion, whereas in other brain areas, such as the cerebral cortex, hippocampus, pons/medulla, hypothalamus, striatum, midbrain, and olfactory bulb, the ratio is reversed (Nakanishi et al 1992).

PROTEIN-BIOCHEMICAL CHARACTERIZATION OF RECOMBINANT RECEPTORS   Rat brain membranes as well as membranes from NMDAR1-transfected cell lines showed a 117-kDa band on Western blots with a polyclonal antiserum made to a synthetic C-terminal peptide of NMDAR1 (Chazot et al 1992). In membranes from transfected cells, an additional 97-kDa band was observed which was demonstrated to represent a nonglycosylated receptor by the ability of N-glycanase to convert the 117-kDa protein into the 97-kDa protein (Chazot et al 1992). Thus, the in vivo receptor appears to be quite substantially glycosylated, with 17% of the total molecular mass representing carbohydrate side chains.

GENE STRUCTURE AND CHROMOSOMAL LOCALIZATION   The gene for NMDAR1 spans at least 25 kbp and consists of 22 exons and 21 introns. Most exons are ~100 bp in size (the smallest is 63 bp), the only exceptions being exon 1 (473 bp), exon 22 (1248 bp), and exon 3 (> 2700 bp), which is only found in the truncated splice variant. Intron sizes vary widely, between

100 bp and almost 5500 bp. All introns feature the typical GT/AG splice site consensus sequence (see Figure 6, p. 73; Hollmann et al 1993). The gene is located on chromosome 9q34.3 (Table 6) and is close to the locus for tuberous sclerosis and idiopathic torsion dystonia (Karp et al 1993; J Eubanks, personal communication). A region on mouse chromosome 2 that is syntenic to the human locus 9q34.3 encodes two neurological mutants, *lethargic* (*lh*) and *wasted* (*wst*) (Karp et al 1993).

NMDA Receptors Functional Only as Heteromeric Channels: NMDAR2A–2D

CLONING   With the sequence of NMDAR1 at hand, primers were designed to regions conserved between NMDAR1 and other GluRs. A region 60 bp upstream of TMD I and another region at the end of TMD III were used for PCR-mediated DNA amplification. This approach, combined with low-stringency hybridization screening, led to the isolation of four more NMDAR cDNAs with low homology (21–27% amino acid sequence identity) to other GluRs, including only 26–27% with NMDAR1 (Ikeda et al 1992, Kutsuwada et al 1992, Meguro et al 1992, Monyer et al 1992). The four NMDAR2 subunits, which were named NMDAR2A, NMDAR2B, NMDAR2C, and NMDAR2D (see Table 1 for synonyms), share considerable homology with each other (42–56% amino acid sequence identity), thereby establishing a new subfamily of GluR genes (Figure 5).

STRUCTURAL FEATURES   Although the four receptor subunits NMDAR2A, NMDAR2B, NMDAR2C, and NMDAR2D have the same basic structure as NMDAR1 (see above) and other GLU-gated ion channels, they differ in possessing strikingly large C-terminal domains of 627, 644, 404, and 461 amino acids, respectively. Because the same TMD topology is assumed for NMDAR2 subunits as for all other GluRs, these huge C-terminal domains are thought to be extracellular (Monyer et al 1992). However, this topology is an assumption and is not based on any data; it remains to be shown whether the C-termini are in fact extracellular. Recent immunological evidence suggesting the C-terminus of GluR1 may be intracellular (Molnar et al 1993; but see caution mentioned in section on GluR1) and the finding that C-terminal splice variants of NMDAR1 vary in their modulation by PKC (Durand et al 1992) may indicate that a reevaluation of GluR receptor topology could be necessary and that at least part of the domain C-terminal of TMD IV could be intracellular. In that case, the C-termini could well play some intracellular role as the target domains of modulatory or accessory proteins, in promoting receptor assembly, or in sorting or targeting NMDA receptor channels. Alternatively, different C-termini may cause different conformations of the intracellular domain between TMD III and TMD IV and thereby modulate receptor function. The large size of the NMDAR2 subunits, which are 133–163 kDa prior to any secondary modification, compares surprisingly well with data from radiation inactivation experiments that implicated a 209-kDa protein in [3H]CPP binding as well as a 121-kDa protein responsible for [3H]GLU binding (Honore et al 1989).

The Q/R site in the putative TMD II is occupied by asparagine (N), just as in the NMDAR1 subunit, making this a feature common to and exclusively found in all subunits of the NMDA receptor subfamily (see Figure 3). The region N-terminal of TMD I that in all ionotropic GluRs shows homology with the *E. coli* periplasmic glutamine binding protein is also conserved in the four NMDAR2 subunits and is ~160 amino acids in size, just as in NMDAR1 (Meguro et al 1992). A sequence distantly related to a zinc finger has been noted in the large C-terminal domains of both NMDAR2A and NMDAR2B (Monyer et al 1992), but the significance of this observation remains to be shown.

FUNCTIONAL PROPERTIES   None of the four NMDAR2 subunits assembles into functional ion channels when expressed as homomeric receptors in oocytes or transfected cell lines (Ikeda et al 1992, Kutsuwada et al 1992, Meguro et al 1992, Monyer et al 1992). Coexpression of NMDAR2A with either NMDAR2B or NMDAR2C, and NMDAR2B coexpressed with NMDAR2C also did not produce functional channels (Monyer et al 1992). However, when coexpressed with NMDAR1 each of the four NMDAR2 subunits coassemble with NMDAR1 into functional heteromeric receptors that have properties different from those of the homomeric NMDAR1 receptor. Most notably, current amplitudes for NMDAR1/NMDAR2D, NMDAR1/ NMDAR2C, NMDAR1/NMDAR2A, and NMDAR1/NMDAR2B are ~5-, ~20-, ~40-, and ~60-fold larger, respectively, than NMDAR1 homomeric currents (Ikeda et al 1992, Kutsuwada et al 1992, Meguro et al 1992). In

Table 7 Amino acid sequence identity comparison of metabotropic receptor subfamilies[a]

MGluR genes	mGluR1, mGluR5	mGluR2 mGluR3	mGluR4
mGluR1, mGluR5	63.0	43.9–45.6	41.8–42.3
mGluR2, mGluR3		68.0	45.4–46.5
mGluR4			100

[a] The program "gap" from the University of Wisconsin sequence analysis software (Devereux et al 1984) was used to calculate sequence identities. See Table 1 for references to sequences.

addition, agonist potencies are slightly higher for heteromeric receptors (see Table 7).

A completely unexpected finding was that the coagonist GLY alone is able to activate NMDAR1/NMDAR2A, NMDAR1/NMDAR2B, and NMDAR1/NMDAR2C receptors, in the absence of GLU or NMDA (Kutsuwada et al 1992, Meguro et al 1992). The action of GLY as an agonist in its own right appears to be specific, because it can be inhibited by APV and occurs with an EC_{50} of 2 μM, which is roughly the same affinity that GLY possesses as a coagonist with GLU or NMDA (Table 4). This property of GLY, which has never been observed for NMDA receptors in vivo, raises some doubts as to whether heteromeric NMDAR1/NMDAR2 receptors represent the actual in vivo subunit combination, despite the large responses these subunit combinations can produce in expression systems. Thus, additional subunits may be required to make up the native NMDA receptor complex.

Just as is the case for the homomeric NMDAR1 receptor, heteromeric combinations of NMDAR1 and NMDAR2 subunits are highly permeable to Ca^{2+} and are blocked by Zn^{2+} and (+)MK-801, and by Mg^{2+} in a voltage-dependent manner (Ikeda et al 1992, Kutsuwada et al 1992, Monyer et al 1992). The blocking action of these inhibitors is stronger for some combinations (NMDAR1/NMDAR2A, NMDAR1/NMDAR2B) than for others (NMDAR1/NMDAR2C). In particular, Mg^{2+} is only a very weak blocker for NMDAR1/NMDAR2C channels, which remain open even at 1 mM Mg^{2+} at − 70 mV. Because NMDAR2C RNA is almost exclusively expressed in the cerebellar granule cell layer, it has been suggested that NMDAR1/NMDAR2C may constitute the native NMDA receptor in mossy fiber/granule cell synapses in the cerebellum (Kutsuwada et al 1992); however, there is no electrophysiological evidence for a NMDA receptor that is not blocked by Mg^{2+} in granule cells.

The rise time of currents produced by NMDAR1/NMDAR2A receptors as well as NMDAR1/NMDAR2C receptors is fairly slow (12–13 ms), in accordance with the slow rise time of NMDA currents in vivo. The decay time of the currents is very different between NMDAR1/NMDAR2A (119 ms) and NMDAR1/NMDAR2C (382 ms) receptors, showing that the properties of the different heteromeric channels are not equivalent (Monyer et al 1992).

Competitive inhibition by APV at the GLU binding site and by 7-Cl-KYNA at the GLY binding site, although seen with all heteromeric combinations, differs quantitatively among them. The rank order of susceptibility to inhibition by APV is NMDAR1/NMDAR2A > NMDAR1/NMDAR2B > NMDAR1/NMDAR2C > NMDAR1/NMDAR2D, whereas that for 7-Cl-KYNA is NMDAR1/NMDAR2C > NMDAR1/NMDAR2B > NMDAR1/NMDAR2A ~NMDAR1/NMDAR2D (Ikeda et al 1992, Kutsuwada et al

1992). PKC activators potentiate responses of NMDAR1/NMDAR2A and NMDAR1/NMDAR2B, but not NMDAR1/NMDAR2C receptors, suggesting selective regulation by phosphorylation of certain heteromeric NMDA receptors. As expected, PKC potentiation is sensitive to the PKC inhibitor staurosporine; it lasts for ~20 and ~60 min for NMDAR1/NMDAR2A and NMDAR1/NMDAR2B, respectively (Kutsuwada et al 1992).

Single-channel properties of heteromeric NMDAR1/NMDAR2 have recently been reported from expression studies in oocytes. NMDAR1/NMDAR2A and NMDAR1/NMDAR2B are very similar in that each has two conductance levels, a main level of 50 pS with a mean open time of ~2.6 ms and a sublevel of 38 pS with a mean open time of ~0.6 ms (Stern et al 1992). These properties are similar to native receptors found in hippocampal CA1, dentate gyrus, and cerebellar granule cells. NMDAR1/NMDAR2C receptors are quite different, having a main conductance level of 36 pS and a sublevel of 19 pS, both with mean open times of 0.61 ms, similar to channels found in certain cultured cerebellar neurons (Stern et al 1992).

MUTAGENESIS STUDIES   Mutagenesis studies have been undertaken to elucidate the role of the putative channel-lining domain TMD II and its immediate surroundings (Figure 3). In the case of the NMDAR1 subunit, a peculiar stretch of six negatively charged amino acid residues (EEEEED) is present N-terminal of TMD II, at the putative intracellular mouth of the channel. No other GluR subunits share this sequence, and it has been suggested that this unique region might be responsible for the voltage-dependent Mg^{2+} block. However, when the six residues were replaced with TSDQSN, which is the homologous sequence in GluR1, coexpression of this mutant with NMDAR2B did not produce any differences in block by Mg^{2+}, Zn^{2+}, or MK-801, indicating that this site is not likely involved in channel block (Mori et al 1992).

When the asparagine at the Q/R site of TMD II of NMDAR1 is mutated to a glutamine (N $\rightarrow$ Q mutation), heteromeric channels formed with either NMDAR2A, NMDAR2B, or NMDAR2C show lower permeability for Ca^{2+} than the wildtype (Burnashev et al 1992c, Mori et al 1992, Sakurada et al 1993). The permeability to Mg^{2+} remains low and the mutant channel is still blocked by Mg^{2+}, although to a lesser extent compared with the wildtype. For NMDAR1/NMDAR2A, the IC_{50} for Mg^{2+} is 20.6 μM at the mutant channel and 3.6 μM at the wildtype channel (Burnashev et al 1992c). When the same N $\rightarrow$ Q mutation is introduced into one of the NMDAR2 subunits, coexpression with wildtype NMDAR1 produces channels that show the same high permeability for Ca^{2+} as the wildtype, whereas their permeability to Mg^{2+} is significantly increased while the blocking potency of Mg^{2+} is considerably decreased; for NMDAR1/NMDAR2A(N595Q), the IC_{50} for

Mg^{2+} is 78.5 μM. When all the NMDAR1 and the NMDAR2 subunits of a heteromeric receptor carry the N $\rightarrow$ Q mutation, the resulting channels are permeable to both Ca^{2+} and Mg^{2+} and are no longer blocked by Mg^{2+} [IC_{50} for Mg^{2+} is $>$ 100 μM for NMDAR1(N598Q)/NMDAR2A(N595Q)], similar to the Q versions of GluR1–GluR4 channels (Burnashev et al 1992c, Mori et al 1992). These data show that the asparagine residues in the TMD II domains of NMDAR1 and NMDAR2 subunits are not equivalent for channel function in heteromeric receptor complexes. The asparagine in NMDAR1 strongly affects the Ca^{2+} permeability of heteromeric receptors, but the Mg^{2+} permeability is only moderately affected; the same residue in NMDAR2 subunits strongly affects the Mg^{2+} permeability but not the Ca^{2+} permeability.

Native NMDA receptors are only slightly blocked by Ca^{2+}; the same is true of recombinant receptors, e.g. NMDAR1/NMDAR2A, which in 1.8 mM Ca^{2+} solution carries 75% of the current carried in Ca^{2+}-free solution. When the N $\rightarrow$ Q mutation is introduced into either NMDAR1 or NMDAR2A, the channel block by Ca^{2+} increases, leaving only ~45% of the control current observed in Ca^{2+}-free solution (Burnashev et al 1992c). A simple interpretation is that the asparagine residues in both the NMDAR1 and the NMDAR2 subunits are in the path of ion permeation and can interact with Ca^{2+}, in addition to Mg^{2+}. Thus, it is likely that the asparagine residue is part of two different sites that interact with divalent cations in the channel, one site that affects divalent permeability and another site that independently produces channel block. This conclusion is supported by data obtained from a mutant of NMDAR1 in which the asparagine was replaced by arginine. This mutant, coexpressed with NMDAR2A, forms a receptor with no divalent cation permeability that is not blocked by Mg^{2+} at all, possibly because divalent ions cannot enter this channel (Burnashev et al 1992c). Channel block by Zn^{2+} and MK-801 is differentially affected by the N $\rightarrow$ Q mutation (Sakurada et al 1993). This mutation does not significantly affect the Zn^{2+} block, regardless of whether it is introduced into NMDAR1, NMDAR2B, or both subunits. However, it abolishes the block by MK-801 when NMDAR2B or both subunits are mutated, but not when NMDAR1 alone carries the mutation (Mori et al 1992). This suggests that the Zn^{2+} site and the MK-801 site are separate entities and do not overlap. By the same argument, the mutagenesis data suggest that the MK-801 site and the Mg^{2+} site do overlap.

Taken together the data suggest that the asparagine residue is located in the ion permeation path and is part of the selectivity filter of the channel that includes the site for the Mg^{2+} and MK-801 block (Burnashev et al 1992c, Mori et al 1992, Sakurada et al 1993). However, because NMDAR1/NMDAR2C channels are less sensitive to Mg^{2+} block than are NMDAR1/NMDAR2A channels, residues other than the asparagine must participate in forming this site.

EXPRESSION OF RECEPTOR RNA   The expression patterns of NMDAR2A–NMDAR2D RNAs are quite different (Table 5). The regional distribution of NMDAR2A RNA is widespread and largely overlaps that of NMDAR1, whereas NMDAR2B RNA shows a more restricted pattern (e.g. it is not expressed in thalamus, amygdala, hypothalamus, brainstem, or cerebellum). NMDAR2C RNA, on the other hand, is almost exclusively expressed in the granule cell layer of the cerebellum, with some low-level expression in olfactory bulb and thalamic nuclei (Kutsuwada et al 1992, Meguro et al 1992, Monyer et al 1992). NMDAR2D RNA appears to be most prominently expressed in brainstem, cerebellum, and olfactory bulb (Nakanishi 1992).

The Glutamate Binding Protein: GBP

CLONING AND STRUCTURAL FEATURES   When the report of the expression cloning of NMDAR1 was published, it coincided with the report of the cloning of a cDNA for a protein suggested to represent the glutamate binding subunit of the NMDA receptor (Kumar et al 1991). Kumar et al (1991) isolated this cDNA clone, named GBP for glutamate binding protein, by screening a rat hippocampal expression library with a monoclonal antibody and polyclonal antisera made against a 71-kDa glutamate binding protein that had been isolated and purified by protein-biochemical methods (Chen et al 1988, Ly & Michaelis 1991). The isolated protein had been reported to be one component of a partially purified set of four to six proteins (31, 36, 41, 43, 62, and 70 kDa) that could be eluted with NMDA from a GLU affinity column. The whole set of proteins (but not the 70-kDa protein alone) was reported to show NMDA-displacable [^{3}H]GLU binding, MK-801–displaceable [^{3}H]N-(1-[thienyl]cyclohexyl) piperidine ([^{3}H]TCP) binding, and strychnine-insensitive [^{3}H]GLY binding. The set of proteins when reconstituted into liposomal membranes was also reported to produce ion channels with some properties of NMDA receptors, such as block by APV of NMDA- or GLU-evoked cationic currents (Ly & Michaelis 1991). The GBP antibody was reported to block GLU activation of ion flux in liposomes and to inhibit NMDA neurotoxicity in primary hippocampal cultures.

The cDNA clone that was isolated from a bacterial colony immunoreactive to the GBP antibody encodes a protein of 594 amino acids with a putative signal peptide of 22 amino acids. The GBP has no significant sequence homology with any of the GluR genes (amino acid sequence identity is 15–20%) and has no homology with any other ligand-gated ion channel or with the metabotropic gluatamate receptors. The domain N-terminal of TMD I, conserved between all other GluRs and the E. coli periplasmic glutamine binding protein and suspected to contain the ligand binding domain, is absent from GBP. Of the 23 amino acids that are absolutely conserved in all 21 ionotropic GluR genes, only four are conserved in the GBP. Also, the N-terminal, putatively extracellular domain, in contrast to all other GluRs, does not contain any consensus N-glycosylation sites. There are some rather unique sequence homologies with other genes that have not been detected in ionotropic or metabotropic GluR genes. At the N-terminus, high amino acid sequence identity is found at residues 60–95 with the N-terminal segment of pro-α-(I) collagen, and two short regions (one of them located within the proposed TMD II domain) show up to 60% amino acid sequence identity with γ-glutamyl transpeptidase.

A four-TMD topology was proposed for the GBP based on the hydrophobicity plot of the sequence, which contains at least six domains that qualify equally well as potential TMDs (Kumar et al 1991). The four proposed TMDs of the GBP differ from the putative TMDs of all other GluRs in not being arranged in the cluster-of-three-plus-one pattern that appears to be a hallmark of all GluRs, and in fact of all ligand-gated ion channels. Instead, the four TMDs in the GBP are closely clustered in the N-terminal half of the cDNA clone, and there is no large intracellular loop between TMD III and TMD IV (see Figure 2). Additionally, all four TMDs in the GBP contain one charged amino acid (TMD I even has two); in all other ionotropic GluRs the only charged residue is the arginine at the Q/R sites of GluR2, GluR5, and GluR6, which is introduced into the proteins through an RNA editing mechanism. Moreover, there is virtually no sequence homology in the putative channel-lining domain TMD II between GBP and the other GluRs (Figure 3). Of 19 residues in this domain, 13 are unique in the GBP when compared with all 21 ionotropic GluRs, four residues are shared by a single receptor, one residue is shared by two other receptors, and only one residue is conserved among all the receptors. Given that none of the structural features of the GBP resemble those of any other GluR, the GBP represents either a unique new family of glutamate receptors or an unrelated protein that was picked up by cross-reactivity of the anti-GBP antibody used in the isolation procedure. Direct functional evidence will be required before the recombinant GBP can be safely regarded as a glutamate receptor subunit (see below).

FUNCTIONAL PROPERTIES   No data have been published on the expression of GBP in oocytes or transfected cell lines. However, preliminary evidence suggests that the GBP does not assemble into functional ion channels when expressed as a homomeric protein in oocytes (Michaelis et al 1992). GBP also does not cause any detectable differences in the properties of GluR1–GluR6 upon coexpression with these subunits in oocytes (M Hollmann, unpublished data; R Dingledine, personal communication). Furthermore, when antisense GBP transcripts are coinjected with total rat brain RNA into oocytes, responses to NMDA remain unchanged as compared with controls injected only with total rat brain RNA (M Hollmann, unpublished data).

It has been reported that a fusion protein of the cloned GBP and β-galactosidase could be partially purified from GBP-expressing bacterial cultures on a GLU affinity column. This partially purified protein fraction was shown to bind GLU with a K_D of 263 nM, similar to the affinity of the GBP purified through protein-biochemical procedures from rat brain membranes (Chen et al 1988). However, because bacteria contain many glutamate binding proteins that could potentially copurify with the GBP fusion protein on an affinity column (in fact, an additional band not reactive with the GBP antibody was seen in the affinity column eluate; see Kumar et al 1991), more direct evidence will be necessary to establish that the cloned GBP gene is identical with the GBP isolated by biochemical means. Once this has been shown, it must still be rigorously demonstrated that the GBP is a subunit of the NMDA receptor complex that was reconstituted in liposomes.

EXPRESSION OF RECEPTOR RNA   Northern blot analysis showed expression of the GBP gene in hippocampus, cerebellum, and cortex, with much lower expression in brainstem. No expression was found in liver, lung, or muscle tissue (Kumar et al 1991).

Metabotropic Receptors

CLONING   The cloning of the first metabotropic glutamate receptor, mGluR1, marked another triumph of the expression cloning technique (Houamed et al 1991, Masu et al 1991). There is virtually no sequence homology between mGluR1 and other G protein-coupled receptors, which explains why metabotropic glutamate receptors have not been found among the many previously cloned seven-TMD receptors and underscores the importance of the expression cloning approach in finding new genes. The lack of sequence homology with any other receptor places mGluR1 in a new class of G protein-coupled receptors. With the sequence information of the first mGluR in hand, researchers who used low stringency hybridization screening with mGluR1 probes soon discovered a whole family of additional related genes, which have been consecutively named mGluR2 through mGluR6 (Abe et al 1992, Nakanishi 1992, Tanabe et al 1992).

STRUCTURAL FEATURES   All six mGluR cDNAs are large proteins (854–1179 amino acids; see Table 1) whose structures consist of a large hydrophilic N-terminal domain of ~550 residues; a core region of ~250 amino acids containing seven hydrophobic stretches thought to represent the seven-TMD domain characteristic of G protein-coupled receptors; and a large, hydrophilic C-terminal domain of varying length in the different receptor subunits. The N-terminal domain is believed to be extracellular and contains numerous consensus sites for N-glycosylation (see Table 1), whereas the C-terminal domain, in which several consensus sites for phosphorylation are present, is intracellular, if the assumed seven-TMD topology is correct. In mGluR1–mGluR5, 21 cysteine residues are absolutely conserved, 17 of which are located in the N-terminal domain and one each in the first and second extracellular loop and in TMDs V and VI, respectively (Tanabe et al 1992). This feature, which is not found in the ionotropic GluRs, may indicate that a rigidly controled tertiary structure is important for mGluRs.

SEQUENCE HOMOLOGIES WITH OTHER GENES   mGluR1 and mGluR5 are more closely related to each other than to other mGluRs, and the same is true for mGluR2 and mGluR3 (Table 7; Figure 5). The largest differences among mGluRs are in the C-terminal domains. mGluR4 is equally distantly related to all other mGluRs, whereas for mGluR6 no sequence data is available, yet. The different sequence homology levels within the mGluR family are reflected in differences in functional properties among these proteins (see below). Although there is no overall sequence homology between mGluRs and ionotropic GluRs, two stretches in the N-terminal domain (residues 215–352 and 453–597 of mGluR1) have been reported to show some low-level homology with GluR1–GluR4 (Masu et al 1991). However, when randomly shuffled permutations and the original sequences of these two sequence stretches are compared for their homologies with GluR1–GluR4, only the first region (residues 215–352) shows homology (with GluR1, GluR2, and GluR4) slightly higher than random. The significance of this extremely low homology has not been investigated, but may reflect conserved sequences within the ligand binding domains of these receptors.

The mGluRs are considerably larger than any other G protein-coupled receptor, particularly those for small ligands, to which they have no overall sequence homology. Additionally, almost none of the conserved amino acids of G protein-coupled receptors are conserved in mGluRs (Masu et al 1991). Several stretches of proline/glutamine and glutamate/aspartate in the C-terminal domain can also be found in adrenergic and muscarinic receptors; however, these sequences are normally present in the third intracellular loop of those receptors (Houamed et al 1991). The third intracellular loop between TMD V and TMD VI, which is putatively cytoplasmatic and is thought to represent the G protein-coupling domain in other receptors, is unusually short in mGluRs and highly conserved among all subunits, irrespective of their signal trans-duction pathways (Houamed et al 1991, Tanabe et al 1992). This suggests that G protein coupling in mGluRs involves a different or an additional domain. Thus, just as the ionotropic glutamate receptors were found not to fit into the superfamily of previously analyzed ligand-gated ion channels, metabotropic glutamate receptors do not fit into the superfamily of seven-TMD receptors but appear to define a new class.

Houamed et al (1991) have noted that a stretch of residues (amino acids 133–237) in the N-terminal domain of mGluR1 shows 29% amino acid sequence identity with sea urchin guanylate cyclase, a peptide receptor with a single TMD and an intracellular guanylate cyclase domain, but the significance of this observation is unclear.

MECHANISMS CREATING RECEPTOR DIVERSITY   Two splice variants have been reported for mGluR1, named mGluR1b and mGluR1c (see Table 3 for synonyms), which arise from the insertion of an additional exon and the use of an alternate exon at the C-terminal domain, respectively (Tanabe et al 1992, Pin et al 1992). Both variations in splicing create shortened versions of mGluR1 that are functional but show differences in their response kinetics (see below).

Receptors Linked to IP/Ca^{2+} Signal Transduction: mGluR1 and mGluR5

FUNCTIONAL PROPERTIES   When expressed in oocytes, both mGluR1 and mGluR5 respond to L-GLU, QA, IBO, and tACPD, producing large, long-lasting oscillating currents (Houamed et al 1991, Masu et al 1991, Pin et al 1992). These currents are typical for receptors coupled to phospholipase C, which releases Ca^{2+} from internal stores via inositol phosphates (IP); the released Ca^{2+} then activates endogenous Ca^{2+}-activated Cl^--channels. Consistent with this interpretation, current responses reversed at the Nernst potential for Cl^- currents (-22 mV in oocytes) and did not depend on extracellular Ca^{2+}, but were suppressed by the injection of EGTA. EGTA injection selectively inhibited mGluR1 responses (and not coexpressed Ca^{2+}-independent K-channels, which responded to membrane depolarization), thus demonstrating that mere membrane depolarization is not sufficient to trigger mGluR signal transduction pathways (Masu et al 1991, Pin et al 1992). L-ASP, KA, AMPA, and 2-amino-4-phosphonobutyrate (AP4) all elicit responses that are 100-fold smaller than those induced by GLU, whereas NMDA is entirely inactive. The rank order of potency for both mGluR1 and mGluR5 is QA > IBO ~GLU > tACPD (Table 1). Pertussis toxin (PTX), which inhibits a subpopulation of phospholipase C–activating G proteins, partially inhibits agonist-stimulated IP formation (Houamed et al 1991, Pin et al 1992). D,L-AP3 (2-amino-3-phosphonopropionate) and L-AP4 do not inhibit either mGluR1 or mGluR5, indicating that at least one additional mGluR subunit must exist that is coupled to IP/Ca^{2+} signal transduction. This elusive metabotropic glutamate receptor has been demonstrated to exist in the hippocampus, is blocked by both D,L-AP3 and L-AP4, and, unlike mGluR1 and mGluR5, prefers IBO over QA (Abe et al 1992).

The short splice variants of mGluR1 have the same pharmacology as mGluR1a, but produce longer lasting, delayed oscillations that peak at 14.2 s after agonist application (mGluR1c) as opposed to 5.8 s for mGluR1a (Pin et al 1992). It has been speculated that the differently sized C-termini could be responsible for different intracellular targeting of the splice variants, thus generating different Ca^{2+} responses in different parts of the cell (Pin et al 1992). A mutant mGluR1a receptor truncated at the C-terminus gives responses with characteristics similar to those of the C-terminal splice variants. This suggests that the C-terminal domain may be involved in coupling to the signal transduction pathway and is not likely to participate in agonist binding (Pin et al 1992).

Successful mGluR expression in transfected cell lines was found to require constant enzymatical removal of glutamate from the medium, to avoid excitotoxic effects that lead to selective pressure against mGluR-expressing cells. In stable cell lines obtained through this method, mGluR1 and mGluR5 were shown to stimulate inositol phosphate (IP) formation, with different time courses for inositol monophosphate (IP1), inositol biphosphate (IP2), and inositol triphosphate (IP3) (Abe et al 1992, Aramori & Nakanishi 1992). Also, intracellular free Ca^{2+} increased in these cells upon receptor stimulation in both Ca^{2+}-containing and Ca^{2+}-free medium. In addition, agonists stimulated cAMP formation and arachidonic acid release in cells transfected with mGluR1, with the same order of agonist potency found for stimulation of IP formation. Thus, mGluR1 couples to at least three different signal transduction pathways (Aramori & Nakanishi 1992). There is, however, no indiscriminate interaction with all G proteins, because activation of mGluR1 (or mGluR5) does not inhibit forskolin-induced cAMP formation and thus is not coupled to the inhibitory cAMP cascade (Aramori & Nakanishi 1992). PTX differentially affects IP formation (partial block), cAMP formation (stimulation), and arachidonic acid release (dose-dependent block), demonstrating that different G proteins are likely to be involved in the activation of the three different signal transduction pathways to which mGluR1 can couple, at least in transfected cell lines (Aramori & Nakanishi 1992). In support of this interpretation, the PKC activator PMA also displayed a differential effect on IP formation (strong block), cAMP formation (slight block), and arachidonic acid release (stimulation).

Although mGluR1 and mGluR5 are very similar in their functional properties, one difference has been noted: the blocking action of PTX (Abe et al 1992) is very weak for mGluR5 (~15% blocked) compared with mGluR1

($\sim$40% blocked). This is in agreement with observations that the PTX-sensitivity of metabotropic glutamate responses differs in different brain regions. Another functional difference is that mGluR5 does not couple to the stimulatory cAMP cascade (Abe et al 1992). To summarize these data, mGluR1 and mGluR5 are both linked to IP/Ca^{2+} signal transduction via a PTX-sensitive G protein. In addition, mGluR1 is coupled to cAMP signal transduction and arachidonic acid release, via separate G proteins, whereas mGluR5 is not linked to cAMP signal transduction.

EXPRESSION OF RECEPTOR RNA   Both mGluR1 and mGluR5 RNAs are widely expressed in the brain (see Table 5), overlapping in most regions but also showing some clear differences, e.g. in the substantia nigra, mitral and tufted cells of the olfactory bulb, and the superior colliculus, where mainly mGluR1 RNA is expressed, and in cerebral cortex, subiculum, CA1, nucleus accumbens, inferior colliculus, anterior olfactory nucleus, and internal granule cells of the main olfactory bulb, where mainly mGluR5 RNA is expressed. In the cerebellum, mGluR1 RNA is expressed in Purkinje cells and granule cells whereas mGluR5 RNA is absent from both cell types except during development. Notably, mGluR1 (but not mGluR5) in CA1 shows low RNA expression, whereas CA2/CA3 pyramidal cell layers of the hippocampus have high GluR1 and GluR5 expression (Masu et al 1991). mGluR1 RNA expression develops gradually starting at E18 and reaches adult levels at P11, paralleling maturation of neuronal elements (Shigemoto et al 1992a). The high level of expression of mGluR1 RNA in CA3 pyramidal neurons of the hippocampus and cerebellar Purkinje cells suggests that mGluR1 may be involved in LTP at the mossy fiber-CA3 synapse and in long-term depression (LTD) at parallel fiber-Purkinje cells synapses, both of which are PTX-sensitive phenomena (Masu et al 1991). Studies with cultured cerebellar granule cells showed that one of the mechanisms regulating the expression of mGluR1 RNA is an agonist-induced down-regulation, a mechanism that has also been observed for other G protein-coupled receptors (Bessho et al 1993). Most of the mGluR1 RNA expression in the brain is in glutamatergic target areas, in accordance with a postsynaptic localization of the receptor, although mitral and tufted cells of the olfactory bulb and Schaffer collateral terminals on CA3 neurons may represent sites of presynaptic localization. No expression was seen in glial cells, although metabotropic glutamate receptors coupled to the IP/Ca^{2+} signal transduction pathway have been shown to be present in astrocytes (Shigemoto et al 1992a). Thus, the existence of another mGluR subunit is likely, a conclusion that has also been reached from the pharmacological evidence discussed above.

The expression pattern of mGluR1 RNA, which is very similar to the immunostaining pattern obtained with a subunit-specific anti-mGluR1 antiserum (Martin et al 1992), overlaps in many areas of the brain with that of the ionotropic GluRs, suggesting the possibility of functional cooperation. Such cooperation has been proposed for arachidonic acid release from cultured striatal neurons and for LTD induction in cultured Purkinje cells (Shigemoto et al 1992a). Immunostaining with antibodies specific for mGluR1, GluR1, GluR2/GluR3, and GluR4 has shown colocalization of mGluR1 with GluR2 and/or GluR3, but not GluR1 or GluR4, in Purkinje cell bodies and dendritic shafts (Martin et al 1992). This suggests a possible cooperation of mGluR1 and GluR2/GluR3 during LTD, given that climbing fiber input to Purkinje cells is thought to be crucial for LTD generation, and climbing fiber terminals synapse on dendritic shafts of Purkinje cells (Martin et al 1992). However, it was later noted that the anti-GluR2/GluR3 antibody could possibly cross-react with GluR4c, a short splice variant of GluR4 (Martin et al 1993); thus, the association of mGluR1 and GluR2/GluR3 on Purkinje cell dendritic shafts needs to be reexamined with more specific antibodies.

Immunostaining also indicated that mGluR1 is localized predominantly postsynaptically, in accordance with the data obtained from in situ hydridization studies (Martin et al 1992).

PROTEIN-BIOCHEMICAL CHARACTERIZATION OF RECOMBINANT RECEPTORS   A polyclonal antibody made against a synthetic peptide from the C-terminal domain of mGluR1 detects a 142-kDa band on Western blots of membranes from human and rat brain tissue (Blackstone et al 1992a, Martin et al 1992). Another antiserum made against a bacterial fusion protein of mGluR1 has been used to detect mGluR1 in membranes from mGluR1-transfected cell lines (Aramori & Nakanishi 1992).

GENE STRUCTURE   The complete gene structure of mGluR1 has not been reported. However, Houamed et al (1991) have shown that the seven-TMD region is encoded by a single exon and flanked by two introns of 50 bp each. The presence of introns in the mGluR1 gene is in contrast to many other G protein-coupled receptors that generally do not have introns. This could be taken as an indication that mGluR1 may have evolved through exon shuffling of the seven-TMD domain, which transferred this domain from some other, probably unrelated protein into the context of a glutamate recognition site.

Receptors Linked to cAMP Inhibition: mGluR2–mGluR4 and mGluR6

FUNCTIONAL PROPERTIES   None of the four receptors mGluR2, mGluR3, mGluR4, and mGluR6 produces detectable responses when expressed in oocytes (Nakanishi 1992, Tanabe et al 1992). Upon expression in transfected cell lines, mGluR2 was shown to only slightly affect IP formation, but to markedly inhibit forskolin-induced cAMP formation, an effect that is inhibited by PTX in a dose-dependent manner. Thus, mGluR2 appears to be coupled to the inhibitory cAMP cascade via a PTX-sensitive G protein (Tanabe et al 1992). The rank order of agonist potency is GLU ~tACPD > IBO > QA (Nakanishi 1992). mGluR3, mGluR4, and mGluR6 are similarly coupled to the inhibitory cAMP cascade. mGluR3 and mGluR2 share the same rank order of agonist potency, whereas the rank order for mGluR4 is quite different, AP4 > GLU ~IBO > tACPD; the rank order for mGluR6 has not been reported (Nakanishi 1992). Also, a recently developed mGluR agonist, 2-(carboxycyclopropyl)glycine (CCG), can be used to distinguish between responses mediated by mGluR2 and mGluR4. One of eight stereoisomers, L-CCG-I, activates mGluR2 with a 10-fold higher potency compared with GLU, whereas at mGluR4, GLU and L-CCG-I are equipotent (Hayashi et al 1992). Thus, the different pharmacological properties are correlated with the level of amino acid sequence homology between the mGluR subunits, which is high (68% identity) between mGluR2 and mGluR3 and only ~46% between mGluR4 and both mGluR2 and mGluR3 (Table 7). The high potency of AP4 at mGluR4 may mean that mGluR4 represents or is part of the elusive, long sought-after AP4 receptor that has been found in the retina and in the hippocampus (Koerner & Johnson 1992). The finding that mGluRs can be coupled to the inhibitory cAMP cascade is especially interesting given the long prevailing view that all metabotropic glutamate receptors are coupled to IP/Ca^{2+} signal transduction (Tanabe et al 1992). Because receptors coupled to the inhibitory cAMP cascade are involved in inhibitory neurotransmission, this finding may suggest a new and expanded role for glutamate receptors; however, it is possible that these receptors are located in inhibitory neurons and thus ultimately are excitatory.

EXPRESSION OF RECEPTOR RNA   All mGluR subunit RNAs are differentially but widely distributed in the brain. mGluR2 RNA is prominently expressed in Golgi cells of the cerebellum, granule cells of the dentate gyrus, most neuronal cells of the cerebral cortex, and intrinsic neurons of the olfactory bulb. mGluR3 RNA is highly expressed in dentate gyrus and cerebral cortex neurons, and also in glial cells throughout the brain. Although mGluR3 RNA was found in glial cells, it is unlikely to represent the glial metabotropic glutamate receptor previously studied (as discussed above), given that this receptor is coupled to the IP/Ca^{2+} signal transduction pathway. mGluR4 RNA is prominent in granule cells of the cerebellum (Tanabe et al 1992).

THE LIGAND BINDING SITE OF GluRs

Sequence comparison of all ionotropic GluRs listed in Table 1 (excluding the GBP gene, which has not yet been shown convincingly to belong to the glutamate receptor family) reveals a total of 23 amino acids that are absolutely conserved among all of the 21 genes (only four of these 23 residues can be lined up with the GBP). The distribution of the 23 residues across the receptor subunit is such that 11 occur in the N-terminal domain, one in TMD I, one in TMD II, three in TMD III, five in the intracellular loop in front of TMD IV, and two at TMD IV. Of the 11 residues absolutely conserved in the N-terminal domain, 10 are located in a region of ~130 (in non-NMDA receptors) or ~160 (in NMDA receptors) residues immediately preceding the putative TMD I. This region and the domain in front of TMD IV that contains five absolutely conserved residues are identical with the two regions that are homologous with the *E. coli* periplasmic glutamine binding protein (Nakanishi et al 1990). This observation suggests that a good candidate for the agonist binding site is the conserved region N-terminal of TMD I. The equivalent domain in nAChR has been shown to contain a ligand binding domain in front of TMD I in that receptor. Assuming that the topology of model 2 is correct, the second consensual region is intracellular, and thus is unlikely to harbor the ligand binding site.

In one mutagenesis study of GluR1 a number of charged residues in this putative ligand binding domain were mutated and analyzed for their effect on channel activity and agonist potencies (Uchino et al 1992). It was concluded that the glutamate at position 398, the asparagine at 443, and the lysine at 445 are involved in ligand binding, because changing them to oppositely charged residues lowered the potency of agonists without changing the maximal current amplitude. Mutations at positions 398 and 443 also produced a differential effect on the potencies of different agonists (Uchino et al 1992).

Thus, it appears that the 130- to 160–amino acid region homologous to *E. coli* permease located N-terminal of TMD I of ionotropic glutamate receptors is at least part of the ligand binding site. It is necessary, however, to use more direct methods to characterize the glutamate binding site. For the metabotropic glutamate receptors, no experimental data on the localization of the ligand binding site are available.

RECOMBINANT GLUTAMATE RECEPTORS IN DISEASE STATES

Several studies have investigated possible links of various disease states with glutamate receptor expression levels. Most of these studies were very limited in scope and investigated only one or very few of the cloned receptor subunits, and contradictory results were frequently reported. However, it should be kept in mind that any correlation of a complex biological phenomenon such as a neurological disorder with the expression level of a particular gene should be interpreted very cautiously, because in most cases it is difficult to tell which is the cause and which is the effect. We nevertheless want to mention the data that have been discussed as evidence for the possible involvement of particular GluR receptor subunits in certain diseases.

It has been reported that limbic seizures induced by electrolytic lesions in the hilus of the hippocampal dentate gyrus caused a delayed, pronounced reduction in GluR1 RNA expression in the granule cells of the contralateral dentate gyrus, which returned to control levels one month after the lesion. The treatment also reduced expression in CA1 and CA3 pyramidal layers as well as in superficial layers of the neocortex and piriform cortex. These results were interpreted as an indication that epileptogenic patterns of physiological activity may downregulate the expression of GluR1 (Gall et al 1990). Another study reported, however, that hippocampal kindling specifically increased GluR1$_{flip}$ and GluR2$_{flip}$ expression in hippocampal dentate granule cells, without altering the expression of flop subunits. Expression levels returned to control values one month after kindling. The researchers speculated that the switch to flip splice variants, which carry larger currents, might underlly neuronal injury (Kamphuis et al 1992).

One study of GluR1 expression in the hippocampus of Alzheimer's patients found increased levels of GluR1 RNA, which the researchers attributed to upregulation of the receptor in response to glutamatergic deafferentiation (Harrison et al 1990). However, another study showed no difference in hippocampal GluR1 RNA levels of Alzheimer's disease patients and normal controls (Potier et al 1992).

In an animal model for ischemia, GluR2 RNA levels were found to be reduced 24 hrs post-ischemia relative to GluR1 and GluR3 RNA levels (Pellegrini-Giampietro et al 1993), paralleled by an increase in the Ca^{2+} content of the ischemic tissue during the same period. This was interpreted as evidence that reduced expression of the GluR2 subunit, which when present in heteromeric receptors conveys low Ca^{2+} permeability, may lead to a pathological increase in Ca^{2+} permeability resulting in excitotoxic cell death (Pellegrini-Giampietro et al 1993).

In a study of GluR1 RNA expression in hippocampal subfields of schizophrenia patients, a reduction of GluR1 RNA (but not total RNA or RNA for a GTP-binding protein used as a control) was seen in all subfields of hippocampi from patients compared with normal individuals. The decrease in expression was particularly pronounced in the CA3 area and the dentate gyrus, but only the reduction of GluR1 RNA in the CA3 area was statistically significant (Harrison et al 1991). A decrease in GluR1 protein expression in telencephalic but not cerebellar or brainstem tissue of aged mice (25 months old) vs young mice (3 months old) has been reported (Bahr et al 1992), and was found to parallel a reduction in [^{3}H]AMPA binding sites. Because the levels of dopamine, GABA, and serotonin receptor proteins did not change, it was suggested that glutamatergic neurotransmission mediated by AMPA-preferring GluRs may be selectively impaired during aging.

NOMENCLATURAL CONSIDERATIONS

A unified, consistent, and comprehensive nomenclature is desirable, but it is hard to conceive while the total complexity of GluR genes and their functional relationships with each other remain largely unknown. The pitfalls of trying to create a consistent nomenclature are well exemplified by the use of Greek letters for sequence-related groups of GluR genes (Meguro et al 1992, Sakimura et al 1992). The Greek letter system, as originally conceived for the nAChR receptor and extended to GABAR and GlyR, carries the implicit assumption that the different receptor classes represent subunits that can form heteromeric receptors with each other. Although this is true for other ligand-gated ion channels, it certainly is not true for the glutamate receptor subunit family, where at present there is no evidence, for instance, for the formation of heteromeric receptors containing both non-NMDA and NMDA subunits. Another point of concern is that the system fails to address the question of how to place those nonmammalian genes such as the KA binding proteins from chick and frog, and invertebrate GluRs such as those from *D. melanogaster* and *L. stagnalis*. For the time being it therefore seems to be prudent to abstain from inventing a new nomenclature for GluR genes. In due time, when more information becomes available on additional genes and particularly on the interaction of the various receptor subunits in vivo, a comprehensive and truly meaningful nomenclatural system will become possible.

OUTLOOK

What have we learned from the flurry of molecular cloning activity directed towards glutamate receptors over the past four years, and what remains to be done? It has become clear that glutamate receptors, like most of the other brain receptors, are encoded in the mammalian genome by a large family of genes coding for proteins with significant sequence homology presumably produced by gene duplication during evolution. Surprisingly, the ionotropic glutamate receptors have little sequence homology with the other ligand-gated ion channels, such as the nAChR, GABAR, GlyR, and serotonin receptors. Despite this, the hydrophobicity profile of the ionotropic glutamate receptors suggests that they are structurally related to the other ionotropic receptors. Similarly, the metabotropic glutamate receptors, which show little sequence homology with other members of the large superfamily of metabotropic receptors, nevertheless appear to be seven-TMD receptors that couple to specific G proteins.

A real mystery is why both the ionotropic glutamate receptors and the metabotropic glutamate receptors, which are structurally unrelated, seem to have extraordinarily large extracellular domains to bind the small ligand glutamate. This finding suggests that the large extracellular domains of both the ionotropic and metabotropic families of glutamate receptors serve some additional function. Perhaps they bind other modulatory ligands that are yet to be discovered. The extracellular domain must extend some significant distance into the synaptic cleft and contribute to the extracellular environment through which the transmitter and retrograde signals must pass. It is even possible that this large extracellular domain is itself a ligand and sends a signal back to the presynaptic neuron during synaptic signaling.

The discovery that in addition to NMDA receptors many of the cloned ionotropic non-NMDA glutamate receptors are quite permeable to calcium ions raises the question of their role in regulating the level of calcium during plastic phenomena such as LTP and LTD. The role of glutamate receptors in neurological disorders and degenerative brain diseases, which may involve calcium-induced cell death, must also be reexamined for each of the receptor subunits.

The unexpected finding that RNA editing determines the structure of some of the non-NMDA receptors raises many questions that remain to be answered. What is the mechanism of RNA editing, and why is this mechanism used for glutamate receptor biosynthesis? Are there diseases that result from a defect in RNA editing?

A pressing issue is the three-dimensional structure of the receptor proteins. Without reliable structural data, for instance, it is very hard to interpret mutagenesis experiments directed at such important unsolved questions as where the ligand binding site is, which domain(s) make up the ion channel, and which domains are intracellular targets for phosphorylation or interaction with potential modulatory proteins. Another important problem to be solved is the stoichiometry of the receptor complex, for which everything from tetramers to heptamers has been proposed. Also, it will be necessary to determine which subunits are expressed in which cell types, and which subunits actually form heteromeric receptors in vivo. Related to this is the basic question of what determines whether a given subunit can coassemble with another one, and how the coassembly process is regulated. The mystery of delta subunit function and the role of the KBPs remains to be solved, and without a doubt more genes are waiting to be discovered in the glutamate receptor family.

The final frontier then will be to understand how the interplay of the various glutamate receptors contributes to synaptic plasticity, memory storage, and cognition in the brain.

ACKNOWLEDGMENTS

We thank Anne-Marie Quinn for help with the construction of the phylogenetic tree; Ray Dingledine, Jim Eubanks, James McNamara, Ralph Puchalski, and Stephen Traynelis for permission to include some of their data prior to publication; and Jan Egebjerg and Stephen Traynelis for their comments on the manuscript.

Literature Cited

Abe T, Sugihara H, Nawa H, et al. 1992. Molecular characterization of a novel metabotropic glutamate receptor mGluR5 coupled to inositol phosphate/Ca^{2+} signal transduction. *J. Biol. Chem.* 267:13361–68

Advokat C, Pellegrin AI. 1992. Excitatory amino acids and memory—evidence from research on Alzheimer's disease and behavioral pharmacology. *Neurosci. Biobehav. Rev.* 16:13–24

Anantharam V, Panchal RG, Wilson A, et al. 1992. Combinatorial RNA splicing alters the surface charge on the NMDA receptor. *FEBS Lett.* 305:27–30

Appel SH. 1993. Excitotoxic neuronal cell death in amyotrophic lateral sclerosis. *Trends Neurosci.* 16:3–5

Aramori I, Nakanishi S. 1992. Signal transduction and pharmacological characteristics of a metabotropic glutamate receptor, mGluR1, in transfected CHO cells. *Neuron* 8:757–65

Aroniadou VA, Teyler TJ. 1991. The role of NMDA receptors in long-term potentiation (LTP) and depression (LTD) in rat visual cortex. *Brain Res.* 562:136–43

Bahr BA, Godshall AC, Hall RA, Lynch G. 1992. Mouse telencephalon exhibits an age-related decrease in glutamate (AMPA) receptors but no change in nerve terminal markers. *Brain Res.* 589:320–26

Barnard EA, Darlison MG, Seeburg P. 1987. Molecular biology of the GABA$_A$ receptor: The receptor/channel superfamily. *Trends Neurosci.* 10:502–8

Barnard EA, Henley JM. 1990. The non-NMDA receptors—types, protein structure and molecular biology. *Trends Pharmacol. Sci.* 11:500–7

Barnes JM, Henley JM. 1992. Molecular characteristics of excitatory amino acid receptors. *Prog. Neurobiol.* 39:113–33

Beal MF. 1992. Mechanisms of excitotoxicity in neurologic diseases. *FASEB J.* 6:3338–44

Ben-Ari Y, Aniksztejn L, Bregestovski P. 1992. Protein kinase-C modulation of NMDA currents—an important link for LTP induction. *Trends Neurosci.* 15:333–39

Ben-Ari Y, Gho M. 1988. Long-lasting modification of the synaptic properties of rat CA3 hippocampal neurones induced by kainic acid. *J. Physiol.* 404:365–84

Bessho Y, Nawa H, Nakanishi S. 1993. Glutamate and quisqualate regulate expression of metabotropic glutamate receptor messenger RNA in cultured cerebellar granule cells. *J. Neurochem.* 60:253–59

Bettler B, Boulter J, Hermans-Borgmeyer I, et al. 1990. Cloning of a novel glutamate receptor subunit, GluR5—expression in the nervous system during development. *Neuron* 5:583–95

Bettler B, Egebjerg J, Sharma G, et al. 1992. Cloning of a putative glutamate receptor—a low affinity kainate-binding subunit. *Neuron* 8:257–65

Betz H. 1990. Ligand gated channels in the brain: the amino acid receptor superfamily. *Neuron* 5:383–92

Blackstone CD, Levey AI, Martin LJ, et al. 1992a. Immunological detection of glutamate receptor subtypes in human central nervous system. *Ann. Neurol.* 31:680–83

Blackstone CD, Moss SJ, Martin LJ, et al. 1992b. Biochemical characterization and localization of a non-N-methyl-D-aspartate glutamate receptor in rat brain. *J. Neurochem.* 58:1118–26

Blaschke M, Keller BU, Rivosecchi R, et al. 1993. A single amino-acid determines the subunit-specific spider toxin block of AMPA/kainate receptor channels. *Proc. Natl. Acad. Sci. USA* 90:6528–32

Bliss TVP, Collingridge GL. 1993. A synaptic model of memory—long-term potentiation in the hippocampus. *Nature* 361:31–39

Boulter J, Hollmann M, O'Shea-Greenfield A, et al. 1990. Molecular cloning and functional expression of glutamate receptor subunit genes. *Science* 249:1033–37

Burnashev N, Khodorova A, Jonas P, et al. 1992a. Calcium-permeable AMPA-kainate receptors in fusiform cerebellar glial cells. *Science* 256:1566–70

Burnashev N, Monyer H, Seeburg PH, Sakmann B. 1992b. Divalent ion permeability of AMPA receptor channels is dominated by the edited form of a single subunit. *Neuron* 8:189–98

Burnashev N, Schoepfer R, Monyer H, et al. 1992c. Control by asparagine residues of calcium permeability and magnesium blockade in the NMDA receptor. *Science* 257: 1415–19

Cajal SR. 1911. *Histologie du systeme nerveux de l'homme et des vertebres.* Paris: Moloine

Carlsson M, Carlsson A. 1990. Interactions between glutamatergic and monoaminergic systems within the basal ganglia—implications for schizophrenia and Parkinsons disease. *Trends Neurosci.* 13:272–76

Chazot PL, Cik M, Stephenson FA. 1992. Immunological detection of the NMDAR1 glutamate receptor subunit expressed in human embryonic kidney-293 cells and in rat brain. *J. Neurochem.* 59:1176–78

Chen JW, Cunningham MD, Galton N, Michaelis EK. 1988. Immune labeling and purification of a 71-kDa glutamate-binding protein from brain synaptic membranes. *J. Biol. Chem.* 263:417–26

Choi DW. 1988. Glutamate neurotoxicity and diseases of the nervous system. *Neuron* 1: 623–34

Choi DW, Rothman SM. 1990. The role of glutamate neurotoxicity in hypoxicischemic neuronal death. *Annu. Rev. Neurol.* 13:171–82

Collingridge GL, Blake JF, Brown MW, et al. 1991. Involvement of excitatory amino acid receptors in long-term potentiation in the Schaffer collateral commissural pathway of rat hippocampal slices. *Can. J. Physiol. Pharmacol.* 69:1084–90

Collingridge GL, Bliss TVP. 1987. NMDA receptors—their role in long-term potentiation. *TINS* 10:288–93

Collingridge GL, Lester RAJ. 1989. Excitatory amino acid receptors in the vertebrate central nervous system. *Pharmacol. Rev.* 41:143–210

Collingridge GL, Singer W. 1990. Excitatory amino acid receptors and synaptic plasticity. *Trends Pharmacol. Sci.* 11:290–96

Constantine-Paton M, Cline HT, Debski E. 1990. Patterned activity, synaptic convergence, and the NMDA receptor in developing visual pathways. *Annu. Rev. Neurosci.* 13:129–54

Cooper E, Couturier S, Ballivet M. 1991. Pentameric structure and subunit stoichiometry of a neuronal nicotinic acetylcholine receptor. *Nature* 350:235–38

Cooper JA, Esch FS, Taylor SS, Hunter T. 1984. Phosphorylation sites in enolase and lactate dehydrogenase utilized by tyrosine kinases in vivo and in vitro. *J. Biol. Chem.* 259:7835–41

Cotman CW, Monaghan DT, Ottersen OP, Storm-Mathisen J. 1987. Anatomical organization of excitatory amino acid receptors and their pathways. *TINS* 10:273–79

Cowburn RF, Hardy JA, Roberts PJ. 1990. Glutamatergic neurotransmission in Alzheimers disease. *Biochem. Soc. Trans.* 18: 390–92

Curutchet P, Bochet P, Decarvalho LP, et al. 1992. In the GluR1 glutamate receptor subunit a glutamine to histidine point mutation suppresses inward rectification but not calcium permeability. *Biochem. Biophys. Res. Commun.* 182:1089–93

Darlison MG. 1992. Invertebrate GABA and glutamate receptors—molecular biology reveals predictable structures but some unusual pharmacologies. *Trends Neurosci.* 15:469–74

Dawson TL, Nicholas RA, Dingledine R. 1990. Homomeric GluR1 excitatory amino acid receptors expressed in Xenopus oocytes. *Mol. Pharmacol.* 38:779–84

Devereux J, Haeberli P, Smithies O. 1984. A comprehensive set of sequence analysis programs for the VAX. *Nucleic Acids Res.* 12:387–95

Dingledine R, Hume RI, Heinemann SF. 1992. Structural determinants of barium permeation and rectification in non-NMDA glutamate receptor channels. *J. Neurosci.* 12: 4080–87

Dingledine R, McBain CJ, McNamara JO. 1990a. Excitatory amino acid receptors in epilepsy. *Trends Pharmacol. Sci.* 11:334–38

Dingledine R, Myers SJ, Nicholas RA. 1990b. Molecular biology of mammalian amino acid receptors. *FASEB J.* 4:2636–45

Durand GM, Gregor P, Zheng X, et al. 1992. Cloning of an apparent splice variant of the rat N-methyl-D-aspartate receptor NMDAR1 with altered sensitivity to polyamines and activators of protein kinase C. *Proc. Natl. Acad. Sci. USA* 89:9359–63

Egebjerg J, Bettler B, Hermans-Borgmeyer I, Heinemann S. 1991. Cloning of a cDNA for a glutamate receptor subunit activated by kainate but not AMPA. *Nature* 351:745–48

Egebjerg J, Heinemann SF. 1993. Ca^{2+} permeability of unedited and edited versions of the kainate selective glutamate receptor GluR6. *Proc. Natl. Acad. Sci. USA* 90:755–59

Eshhar N, Hunter C, Wenthold RJ, Wada K. 1992. Structural characterization and expression of a brain specific gene encoding chick kainate binding protein. *FEBS Lett.* 297: 257–62

Etienne P, Baudry M. 1990. Role of excitatory amino acid neurotransmission in synaptic plasticity and pathology—an integrative hypothesis concerning the pathogenesis and evolutionary advantages of schizophrenia-related genes. *J. Neural Transm.* 39–48

Eubanks JH, Puranam RS, Kleckner NW, et

al. 1993. The gene encoding the glutamate receptor subunit GluR5 is located on human chromosome 2lq2l.1–22.1 in the vicinity of the gene for familial amyotrophic lateral sclerosis. *Proc. Natl. Acad. Sci. USA* 90: 178–82

Feng A-F, Doolittle RF. 1987. Progressive sequence alignment as a prerequisite to correct phylogenetic trees. *J. Mol. Evol.* 25: 351–60

Fields RD, Yu C, Nelson PG. 1991. Calcium, network activity, and the role of NMDA channels in synaptic plasticity in vitro. *J. Neurosci.* 11:134–46

Gall C, Sumikawa K, Lynch G. 1990. Levels of mRNA for a putative kainate receptor are affected by seizures. *Proc. Natl. Acad. Sci. USA* 87:7643–47

Gallo V, Upson LM, Hayes WP, et al. 1992. Molecular cloning and developmental analysis of a new glutamate receptor subunit isoform in cerebellum. *J. Neurosci.* 12: 1010–23

Gasic GP, Heinemann S. 1991. Receptors coupled to ionic channels: the glutamate receptor family. *Curr. Opin. Neurobiol.* 1: 20–26

Gasic GP, Hollmann M. 1992. Molecular neurobiology of glutamate receptors. *Annu. Rev. Physiol.* 54:507–36

Gilbertson TA, Scobey R, Wilson M. 1991. Permeation of calcium ions through non-NMDA glutamate channels in retinal bipolar cells. *Science* 251:1613–15

Girault JA, Halpain S, Greengard P. 1990. Excitatory amino acid antagonists and Parkinsons disease. *Trends Neurosci.* 13:325–26

Gregor P, Eshhar N, Ortega A, Teichberg VI. 1988. Isolation, immunochemical characterization and localization of the kainate subclass of glutamate receptor from chick cerebellum. *EMBO J.* 7:2673–79

Gregor P, Mano I, Maoz I, et al. 1989. Molecular structure of the chick cerebellar kainate-binding subunit of a putative glutamate receptor. *Nature* 342:689

Gregor P, Yang XD, Mano I, et al. 1992. Organization and expression of the gene encoding chick kainate binding protein, a member of the glutamate receptor family. *Mol. Brain Res.* 16:179–86

Hampson DR, Huang XP, Oberdorfer MD, et al. 1992. Localization of AMPA receptors in the hippocampus and cerebellum of the rat using an anti-receptor monoclonal antibody. *Neuroscience* 50:11–22

Hansen JJ, Krogsgaard-Larsen P. 1990. Structural, conformational, and stereochemical requirements of central excitatory amino acid receptors. *Med. Res. Rev.* 10:55–94

Harrison PJ, Barton AJL, Najlerahim A, Pearson RCA. 1990. Distribution of a kainate/AMPA receptor mRNA in normal and Alzheimer brain. *NeuroReport* 1:149–52

Harrison PJ, McLaughlin D, Kerwin RW. 1991. Decreased hippocampal expression of a glutamate receptor gene in schizophrenia. *Lancet* 337:450–52

Hayashi Y, Tanabe Y, Aramori I, et al. 1992. Agonist analysis of 2-(carboxycyclopropyl) glycine isomers for cloned metabotropic glutamate receptor subtypes expressed in chinese hamster ovary cells. *Br. J. Pharmacol.* 107:539–43

Hebb DO. 1949. *The organization of behavior.* New York: Wiley

Henley JM, Jenkins R, Hunt SP. 1993. Localisation of glutamate receptor binding sites and messenger RNAs to the dorsal horn of the rat spinal cord. *Neuropharmacology* 32:37–41

Henneberry RC. 1992. Cloning of the genes for excitatory amino acid receptors. *Bioessays* 14:465–71

Herb A, Burnashev N, Werner P, et al. 1992. The KA-2 subunit of excitatory amino acid receptors shows widespread expression in brain and forms ion channels with distantly related subunits. *Neuron* 8:775–85

Hollmann M, Boulter J, Maron C, et al. 1993. Zinc potentiates agonist-induced currents at certain splice variants of the NMDA receptor. *Neuron* 10:943–54

Hollmann M, Hartley M, Heinemann S. 1991. Ca^{2+} permeability of KA-AMPA gated glutamate receptor channels depends on subunit composition. *Science* 252:851–53

Hollmann M, O'Shea-Greenfield A, Rogers SW, Heinemann S. 1989. Cloning by functional expression of a member of the glutamate receptor family. *Nature* 342:643–48

Hollmann M, Rogers SW, O'Shea-Greenfield A, et al. 1990. The glutamate receptor GluR-K1—structure, function, and expression in the brain. *Cold Spring Harbor Symp. Quant. Biol.* 55:41–55

Holmes WR, Levy WB. 1990. Insights into associative long-term potentiation from computational models of NMDA receptor-mediated calcium influx and intracellular calcium concentration changes. *J. Neurophysiol.* 63: 1148–68

Honore T. 1989. Excitatory amino acid receptor subtypes and specific antagonists. *Med. Res. Rev.* 9:1–23

Honore T, Drejer J, Nielsen EO, et al. 1989. Molecular target size analyses of the NMDA-receptor complex in rat cortex. *Eur. J. Pharmacol.* 172:239–47

Hopfield JJ. 1982. Neural networks and physical systems with emergent collective computational abilities. *Proc. Natl. Acad. Sci. USA* 79:2554–58

Houamed KM, Kuijper JL, Gilbert TL, et al. 1991. Cloning, expression, and gene structure of a G-protein-coupled glutamate receptor from rat brain. *Science* 252:1318–21

Huettner JE. 1990. Glutamate receptor channels in rat DRG neurons—activation by kainate and quisqualate and blockade of desensitization by Con-A. *Neuron* 5:255–66

Hughes TE, Hermans-Borgmeyer I, Heinemann S. 1992. Differential expression of glutamate receptor genes (GluR1–5) in the rat retina. *Visual Neurosci.* 8:49–55

Hullebroeck MF, Hampson DR. 1992. Characterization of the oligosaccharide side chains on kainate binding proteins and AMPA receptors. *Brain Res.* 590:187–92

Hume RI, Dingledine R, Heinemann SF. 1991. Identification of a site in glutamate receptor subunits that controls calcium permeability. *Science* 253:1028–31

Hutton ML, Harvey RJ, Barnard EA, Darlison MG. 1991. Cloning of a cDNA that encodes an invertebrate glutamate receptor subunit. *FEBS Lett.* 292:111–14

Iino M, Ozawa S, Tsuzuki K. 1990. Permeation of calcium through excitatory amino acid receptor channels in cultured rat hippocampal neurones. *J. Physiol. (London)* 424: 151–65

Ikeda K, Nagasawa M, Mori H, et al. 1992. Cloning and expression of the epsilon4 subunit of the NMDA receptor channel. *FEBS Lett.* 313:34–38

Ishii T, Moriyoshi K, Sugihara H, et al. 1993. Molecular characterization of the family of the N-methyl-D-aspartate receptor subunit. *J. Biol. Chem.* 268:2836–43

Izquierdo I. 1991. Role of NMDA receptors in memory. *Trends Pharmacol. Sci.* 12:128–29

Kamboj RK, Schoepp DD, Nutt S, et al. 1992. Molecular structure and pharmacological characterization of humEAA2, a novel human kainate receptor subunit. *Mol. Pharmacol.* 42:10–15

Kamphuis W, Monyer H, Derijk TC, Dasilva FHL. 1992. Hippocampal kindling increases the expression of glutamate receptor-A flip and receptor-B flip messenger RNA in dentate granule cells. *Neurosci. Lett.* 148:51–54

Karp SJ, Masu M, Eki T, et al. 1993. Molecular cloning and chromosomal localization of the key subunit of the human N-methyl-D-aspartate receptor. *J. Biol. Chem.* 268: 3728–33

Kawamoto S, Onishi H, Hattori S, et al. 1991. Functional expression of the alpha1 subunit of the AMPA-selective glutamate receptor channel, using a baculovirus system. *Biochem. Biophys. Res. Commun.* 181:756–63

Keinänen K, Wisden W, Sommer B, et al. 1990. A family of AMPA-selective glutamate receptors. *Science* 249:556–60

Keller BU, Blaschke M, Rivosecchi R, et al. 1993. Identification of a subunit-specific antagonist of α-amino-3-hydroxy-5-methyl-4-isoazolepropionate/kainate receptor channels. *Proc. Natl. Acad. Sci. USA* 90:605–9

Keller BU, Hollmann M, Heinemann S, Konnerth A. 1992. Calcium influx through subunits GluR1/GluR3 of kainate/AMPA receptor channels is regulated by cAMP dependent protein kinase. *EMBO J.* 11:891–96

Kennelly PJ, Krebs EG. 1991. Consensus sequences as substrate specificity determinants for protein kinases and protein phosphatases. *J. Biol. Chem.* 266:15555–58

Kiskin NI, Krishtal OA, Tsyndrenko AY. 1990. Cross-desensitization reveals pharmacological specificity of excitatory amino acid receptors in isolated hippocampal neurons. *Eur. J. Neurosci.* 2:461

Klockgether T, Turski L. 1989. Excitatory amino acids and the basal ganglia—implications for the therapy of Parkinsons disease. *Trends Neurol.* 12:285–86

Koerner JF, Johnson RL. 1992. Excitatory amino acid receptors: design of agonists and antagonists. In *L-AP4 Receptor Ligands,* ed. P. Krogsgaard-Larsen JJ Hansen, pp. 308–330. West Sussex, UK: Ellis Horwood

Köhler M, Burnashev N, Sakmann B, Seeburg PH. 1993. Determinants of Ca^{2+} permeability in both TM1 and TM2 of high-affinity kainate receptor channels: Diversity by RNA editing. *Neuron* 10:495–500

Kumar KN, Tilakaratne N, Johnson PS, et al. 1991. Cloning of cDNA for the glutamate-binding subunit of an NMDA receptor complex. *Nature* 354:70–73

Kutsuwada T, Kashiwabuchi N, Mori H, et al. 1992. Molecular diversity of the NMDA receptor channel. *Nature* 358:36–41

Lambolez B, Audinat E, Bochet P, et al. 1992. AMPA receptor subunits expressed by single Purkinje cells. *Neuron* 9:247–58

Lambolez B, Curutchet P, Stinnakre J, et al. 1991. Electrophysiological and pharmacological properties of GluR1, a subunit of a glutamate receptor-channel expressed in Xenopus oocytes. *Neurosci. Lett.* 123:69–72

Langosch D, Thomas L, Betz H. 1988. Conserved quaternary structure of ligand gated ion channels: the postsynaptic glycine receptor is a pentamer. *Proc. Natl. Acad. Sci. USA* 85:7394–98

Liman ER, Knapp AG, Dowling JE. 1989. Enhancement of kainate-gated currents in retinal horizontal cells by cyclic AMP-dependent protein kinase. *Brain Res.* 481:399–402

Lindstrom J, Criado M, Hochschwender S, et al. 1984. Immunochemical tests of acetylcholine receptor subunit models. *Nature* 311:573–75

Lodge D, Johnson KM. 1990. Noncompetitive excitatory amino acid receptor antagonists. *Trends Pharmacol.* 11:81–86

Lomeli H, Sprengel R, Laurie DJ, et al. 1993. The rat delta-1 and delta-2 subunits extend the excitatory amino acid receptor family. *FEBS Lett.* 315:318–22

Lomeli H, Wisden W, Köhler M, et al. 1992. High-affinity kainate and domoate receptors in rat brain. *FEBS Lett.* 307:139–43

Ly AM, Michaelis EK. 1991. Solubilization, partial purification, and reconstitution of glutamate-activated and N-methyl-D-aspartate-activated cation channels from brain synaptic membranes. *Biochemistry* 30:4307–16

MacDonald JF, Nowak LM. 1990. Mechanisms of blockade of excitatory amino acid receptor channels. *Trends Pharmacol.* 11: 167–72

Maragos WF, Greenamyre JT, Penney JB, Young AB. 1987. Glutamate dysfunction in Alzheimer's disease: an hypothesis. *TINS* 10:65–68

Marshall RD. 1972. Glycoproteins. *Annu. Rev. Biochem.* 41:673–702

Martin LJ, Blackstone CD, Huganir RL, Price DL. 1992. Cellular localization of a metabotropic glutamate receptor in rat brain. *Neuron* 9:259–70

Martin LJ, Blackstone CD, Huganir RL, Price DL. 1993. The striatal mosaic in primates: striosomes and matrix are differentially enriched in ionotropic glutamate receptor subunits. *J. Neurosci.* 13:782–92

Masu M, Tanabe Y, Tsuchida K, et al. 1991. Sequence and expression of a metabotropic glutamate receptor. *Nature* 349:760–65

Masu Y, Kazuhisa N, Tamaki H, et al. 1987. cDNA cloning of bovine substance-K receptor through oocyte expression system. *Nature* 329:836–38

Mayer ML, Vyklicky L Jr. 1989. Concanavalin A selectively reduces desensitization of mammalian neuronal quisqualate receptors. *Proc. Natl. Acad. Sci. USA* 86:1411–15

Mayer ML, Westbrook GL. 1987. The physiology of excitatory amino acids in the vertebrate central nervous system. *Prog. Neurobiol.* 28:197–276

McDonald JW, Johnston MV. 1990. Physiological and pathophysiological roles of excitatory amino acids during central nervous system development. *Brain Res. Rev.* 15:41–70

McNamara JO, Eubanks JH, McPherson JD, et al. 1992. Chromosomal localization of human glutamate receptor genes. *J. Neurosci.* 12:2555–62

Meguro H, Mori H, Araki K, et al. 1992. Functional characterization of a heteromeric NMDA receptor channel expressed from cloned cDNAs. *Nature* 357:70–74

Meldrum B. 1985. Possible therapeutic applications of antagonists of excitatory amino acid neurotransmitters. *Clin. Sci.* 68:113–22

Meldrum B, Garthwaite J. 1990. Excitatory amino acid neurotoxicity and neurodegenerative disease. *Trends Pharmacol. Sci.* 11:379–87

Michaelis EK, Michaelis ML, Kumar KN, et al. 1992. Purification, reconstitution, and cloning of an NMDA receptor-ion channel complex from rat brain synaptic membranes—implications for neurobiological changes in alcoholism. In *Neurobiology of Drug and Alcohol Addiction*, ed. PW Kalivas, HH Samson, pp. 7–18. New York: N. Y. Acad. Sciences

Molnar E, Baude A, Richmond SA, et al. 1993. Biochemical and immunocytochemical characterisation of antipeptide antibodies to a cloned GluR1 glutamate receptor subunit—cellular and subcellular distribution in the rat forebrain. *Neuroscience* 53:307–26

Monaghan DT, Bridges RJ, Cotman CW. 1989. The excitatory amino acid receptors: their classes, pharmacology, and distinct properties in the function of the central nervous system. *Annu. Rev. Pharmacol. Toxicol.* 29:365

Monyer H, Seeburg P. 1992. Developmental expression of NMDA receptor subtypes. *Soc. Neurosci. Abstr.* 18:#172.1

Monyer H, Seeburg PH, Wisden W. 1991. Glutamate-operated channels—developmentally early and mature forms arise by alternative splicing. *Neuron* 6:799–810

Monyer H, Sprengel R, Schoepfer R, et al. 1992. Heteromeric NMDA receptors—molecular and functional distinction of subtypes. *Science* 256:1217–21

Mori H, Masaki H, Yamakura T, Mishina M. 1992. Identification by mutagenesis of a Mg^{2+}-block site of the NMDA receptor channel. *Nature* 358:673–75

Morita T, Sakimura K, Kushiya E, et al. 1992. Cloning and functional expression of a cDNA encoding the mouse beta2 subunit of the kainate-selective glutamate receptor channel. *Mol. Brain Res.* 14:143–46

Moriyoshi K, Masu M, Ishii T, et al. 1991. Molecular cloning and characterization of the rat NMDA receptor. *Nature* 354:31–37

Morris RGM. 1989. Synaptic plasticity and learning: selective impairment of learning in rats and blockade of long-term potentiation in vivo by the N-methyl-D-aspartate receptor antagonist AP5. *J. Neurosci.* 9:3040–57

Moss SJ, Blackstone CD, Huganir RL. 1993. Phosphorylation of recombinant non-NMDA glutamate receptors on serine and tyrosine residues. *Neurochem. Res.* 18:105–10

Müller F, Greferath U, Wässle H, et al. 1992. Glutamate receptor expression in the rat retina. *Neurosci. Lett.* 138:179–82

Murphy SN, Thayer SA, Miller RJ. 1987. The effects of excitatory amino acids on intracellular calcium in single mouse striatal neurons. *J. Neurosci.* 7:4145–58

Nakanishi N, Axel R, Shneider NA. 1992. Alternative splicing generates functionally distinct N-methyl-D-asparte receptors. *Proc. Natl. Acad. Sci. USA* 89:8552–56

Nakanishi N, Shneider NA, Axel R. 1990. A family of glutamate receptor genes—evidence for the formation of heteromultimeric receptors with distinct channel properties. *Neuron* 5:569–81

Nakanishi S. 1992. Molecular diversity of glutamate receptors and implications for brain function. *Science* 258:597–603

Olney JS. 1989a. Excitotoxicity and N-methyl-D-aspartate receptors. *Drug Dev. Res.* 17: 299

Olney JW. 1989b. Excitatory amino acids and neuropsychiatric disorders. *Biol. Psychiatry* 26:505–25

Ortega A, Eshhar N, Teichberg VI. 1991a. Properties of kainate receptor/channels on cultured Bergmann glia. *Neuroscience* 41: 335–49

Ortega A, Lamed Y, Gregor P, Teichberg VI. 1991b. The chick cerebellar kainate binding protein—structure and function. In *Excitatory Amino Acids*, ed. BS Meldrum, F Moroni, RP Simon, JH Woods, 5:97–103. New York: Raven. 780 pp.

Ortega A, Teichberg VI. 1990. Phosphorylation of the 49-kDa putative subunit of the chick cerebellar kainate receptor and its regulation by kainatergic ligands. *J. Biol. Chem.* 265:21404–6

Pellegrini-Giampietro DE, Bennett MVL, Zukin RS. 1992. Are Ca^{2+}-permeable kainate/AMPA receptors more abundant in immature brain? *Neurosci. Lett.* 144:65–69

Pellegrini-Giampietro DE, Zukin RS, Bennett MVL, et al. 1993. Switch in glutamate receptor subunit gene expression in CA1 subfield of hippocampus following global ischemia in rats. *Proc. Natl. Acad. Sci. USA* 89:10499–503

Penney JB, Maragos WF, Greenamyre JT, et al. 1990. Excitatory amino acid binding sites in the hippocampal region of Alzheimer's disease and other dementias. *J. Neurol. Neurosurg. Psychiatry* 53:314–20

Petralia RS, Wenthold RJ. 1992. Light and electron immunocytochemical localization of AMPA-selective glutamate receptors in the rat brain. *J. Comp. Neurol.* 318:329–54

Pin J-P, Fagni L, Bockaert J. 1993. Metabotropic glutamate receptors: targets for new neuropharmacologically active drugs. *Curr. Drugs* In press

Pin JP, Waeber C, Prezeau L, et al. 1992. Alternative splicing generates metabotropic glutamate receptors inducing different patterns of calcium release in xenopus oocytes. *Proc. Natl. Acad. Sci. USA* 89:10331–35

Potier M-C, Dutriaux A, Lambolez B, et al. 1993. Assignment of the human glutamate receptor gene GluR5 to 21q22 by screening a chromosome 21 YAC library. *Genomics* 15:696–97

Potier M-C, Spillantini MG, Carter NP. 1992. The human glutamate receptor cDNA GluR1: cloning, sequencing, expression and localization to chromosome 5. *DNA Seq.* 2:211–18

Puckett C, Gomez CM, Korenberg JR, et al. 1991. Molecular cloning and chromosomal localization of one of the human glutamate receptor genes. *Proc. Natl. Acad. Sci. USA* 88:7557–61

Racine RJ, Moore KA, Wicks S. 1991. Activation of the NMDA receptor—a correlate in the dentate gyrus field potential and its relationship to long-term potentiation and kindling. *Brain Res.* 556:226–39

Ratnam M, Lindstrom J. 1984. Structural features of the nicotinic acetylcholine receptor revealed by antibodies to synthetic peptides. *Biochem. Biophys. Res. Commun.* 122: 1225–33

Raymond LA, Blackstone CD, Huganir RL. 1993. Phosphorylation and modulation of recombinant GluR6 glutamate receptors by cAMP-dependent protein kinase. *Nature* 361:637–41

Reid IC, Morris RGM. 1991. N-methyl-D-aspartate receptors and learning—a framework for classifying some recent studies. In *Excitatory Amino Acids*, ed. B. S. Meldrum F Moroni RP Simon JH Woods, 5:521–32. New York: Raven. 780 pp.

Represa A, Tremblay E, Ben-Ari Y. 1987. Kainate binding sites in the hippocampal mossy fibres: localization and plasticity. *Neuroscience* 20:739–48

Reynolds IJ, Miller RJ. 1990. Allosteric modulation of N-methyl-D-aspartate receptors. *Adv. Pharmacol.* 21:101–26

Robinson JH, Deadwyler SA. 1981. Kainic acid produces depolarization of CA3 pyramidal cells in the in vitro hippocampal slice. *Brain Res.* 221:117–27

Rogers SW, Hughes TE, Hollmann M, et al. 1991. The characterization and localization of the glutamate receptor subunit GluR1 in the rat brain. *J. Neurosci.* 11:2713–24

Sakimura K, Bujo H, Kushiya E, et al. 1990. Functional expression from cloned cDNAs of glutamate receptor species responsive to kainate and quisqualate. *FEBS Lett.* 272:73–80

Sakimura K, Morita T, Kushiya E, Mishina M. 1992. Primary structure and expression of the gamma2 subunit of the glutamate receptor channel selective for kainate. *Neuron* 8:267–74

Sakurada K, Masu M, Nakanishi S. 1993. Alteration of Ca^{2+} permeability and sensitivity to Mg^{2+} and channel blockers by a single amino acid substitution in the N-methyl-D-aspartate receptor. *J. Biol. Chem.* 268:410–15

Sansom MSP, Usherwood PNR. 1990. Single channel studies of glutamate receptors. *Int. Rev. Neurobiol.* 32:51–106

Schoepp D, Bockaert J, Sladeczek F. 1990. Pharmacological and functional characteristics of metabotropic excitatory amino acid receptors. *Trends Pharmacol. Sci.* 11:508–15

Schuster CM, Ultsch A, Schloss P, et al. 1991. Molecular cloning of an invertebrate glutamate receptor subunit expressed in drosophila muscle. *Science* 254:112–14

Sekiguchi M, Wada K, Wenthold RJ. 1992. N-acetylaspartylglutamate acts as an agonist upon homomeric NMDA receptor (NMDAR1) expressed in Xenopus oocytes. *FEBS Lett.* 311:285–89

Shigemoto R, Nakanishi S, Mizuno N. 1992a. Distribution of the messenger RNA for a metabotropic glutamate receptor (mGluR1) in the central nervous system—an in situ hybridization study in adult and developing rat. *J. Comp. Neurol.* 322:121–35

Shigemoto R, Ohishi H, Nakanishi S, Mizuno N. 1992b. Expression of the messenger RNA for the rat NMDA receptor (NMDAR1) in the sensory and autonomic ganglion neurons. *Neurosci. Lett.* 144:229–32

Sommer B, Burnashev N, Verdoorn TA, et al. 1992. A glutamate receptor channel with high affinity for domoate and kainate. *EMBO J.* 11:1651–56

Sommer B, Keinänen K, Verdoorn TA, et al. 1990. Flip and flop: a cell-specific functional switch in glutamate-operated channels of the CNS. *Science* 249:1580–85

Sommer B, Kohler M, Sprengel R, Seeburg PH. 1991. RNA editing in brain controls a determinant of ion flow in glutamate-gated channels. *Cell* 67:11–19

Sommer B, Seeburg PH. 1992. Glutamate receptor channels—novel properties and new clones. *Trends Pharmacol. Sci.* 13:291–96

Stein E, Cox JA, Seeburg PH, Verdoorn TA. 1992. Complex pharmacological properties of recombinant alpha-amino-3-hydroxy-5-methyl-4-isoxazole propionate receptor subtypes. *Mol. Pharmacol.* 42:864–71

Stern P, Behe P, Schoepfer R, Colquhoun D. 1992. Single-channel conductances of NMDA receptors expressed from cloned cDNAs—comparison with native receptors. *Proc. R. Soc. London Ser. B* 250:271–77

Storey E, Kowall NW, Finn SF, et al. 1992. The cortical lesion of Huntington's disease—further neurochemical characterization, and reproduction of some of the histological and neurochemical features by N-methyl-D-as-

partate lesions of rat cortex. *Ann. Neurol.* 32:526–34

Sugihara H, Moriyoshi K, Ishii T, et al. 1992. Structures and properties of 7 isoforms of the NMDA receptor generated by alternative splicing. *Biochem. Biophys. Res. Commun.* 185:826–32

Sugiyama H, Ito I, Hirono C. 1987. A new type of glutamate receptor linked to inositol phospholipid metabolism. *Nature* 325:531–33

Sun W, Ferrer-Montiel AV, Schinder AF, et al. 1992. Molecular cloning, chromosomal mapping, and functional expression of human brain glutamate receptors. *Proc. Natl. Acad. Sci. USA* 89:1443–47

Tanabe Y, Masu M, Ishii T, et al. 1992. A family of metabotropic glutamate receptors. *Neuron* 8:169–79

Uchino S, Sakimura K, Nagahari K, Mishina M. 1992. Mutations in a putative agonist binding region of the AMPA-selective glutamate receptor channel. *FEBS Lett.* 308:253–57

Ultsch A, Schuster CM, Laube B, et al. 1992. Glutamate receptors of Drosophila melanogaster: cloning of a kainate-selective subunit expressed in the central nervous system. *Proc. Natl. Acad. Sci. USA* 89:10484–88

Vanderklish P, Neve R, Bahr BA, et al. 1992. Translational suppression of a glutamate receptor subunit impairs long-term potentiation. *Synapse* 12:333–37

Verdoorn TA, Burnashev N, Monyer H, et al. 1991. Structural determinants of ion flow through recombinant glutamate receptor channels. *Science* 252:1715–18

Wachtel H, Turski L. 1990. Glutamate—a new target in schizophrenia. *Trends Pharmacol.* 11:219–20

Wada K, Dechesne CJ, Shimasaki S, et al. 1989. Sequence and expression of a frog brain complementary DNA encoding a kainate-binding protein. *Nature* 342:684

Wang L-Y, Taverna FA, Huang X-P, et al. 1993. Phosphorylation and modulation of a kainate receptor (GluR6) by cAMP-dependent protein kinase. *Science* 259:1173–75

Watkins JC, Krogsgaard-Larsen P, Honore T. 1990. Structure activity relationships in the development of excitatory amino acid receptor agonists and competitive antagonists. *Trends Pharmacol.* 11:25–33

Wenthold RJ, Hampson DR, Wada K, et al. 1990a. Isolation, localization, and cloning of a kainic acid binding protein from frog brain. *J. Histochem. Cytochem.* 38:1717–23

Wenthold RJ, Hunter C, Wada K, Dechesne CJ. 1990b. Antibodies to a C-terminal peptide of the rat brain glutamate receptor subunit, GluR-A, recognize a subpopulation of

AMPA binding sites but not kainate sites. *FEBS Lett.* 276:147–50

Wenthold RJ, Yokotani N, Doi K, Wada K. 1992. Immunochemical characterization of the non-NMDA glutamate receptor using subunit-specific antibodies—evidence for a hetero-oligomeric structure in rat brain. *J. Biol. Chem.* 267:501–7

Werner P, Voigt M, Keinänen K, et al. 1991. Cloning of a putative high-affinity kainate receptor expressed predominantly in hippocampal CA3 cells. *Nature* 351:742–44

Wroblewski JT, Danysz W. 1989. Modulation of glutamate receptors: molecular mechanisms and functional implications. *Annu. Rev. Pharmacol. Toxicol.* 29:441–74

Yamazaki M, Araki K, Shibata A, Mishina M. 1992a. Molecular cloning of a cDNA encoding a novel member of the mouse glutamate receptor channel family. *Biochem. Biophys. Res. Commun.* 183:886–92

Yamazaki M, Mori H, Araki K, et al. 1992b. Cloning, expression and modulation of a mouse NMDA receptor subunit. *FEBS Lett.* 300:39–45

Yoneda Y, Ogita K. 1991. Neurochemical aspects of the N-methyl-D-aspartate receptor complex. *Neurosci. Res.* 10:1–33

Young EF, Ralston E, Blake J, et al. 1985. Topological mapping of acetylcholine receptor: evidence for a model with five transmembrane segments and a cytoplasmic COOH-terminal peptide. *Proc. Natl. Acad. Sci. USA* 82:626–30

Zorumski CF, Thio LL. 1992. Properties of vertebrate glutamate receptors—calcium mobilization and desensitization. *Prog. Neurobiol.* 39:295–336

Annu. Rev. Neurosci. 1994. 17:109–132

HOX GENES AND REGIONALIZATION OF THE NERVOUS SYSTEM

Roger Keynes

Department of Anatomy, University of Cambridge, Downing Street, Cambridge CB2 3DY, United Kingdom

Robb Krumlauf

Laboratory of Developmental Neurobiology, National Institute for Medical Research (NIMR), The Ridgeway, Mill Hill, London NW7 1AA, United Kingdom

KEY WORDS: *Hox* code, segmentation, hindbrain, rhombomere, homeobox genes

INTRODUCTION

The *Hox* genes form a network of transcription factors implicated in the regulation of axial patterning in vertebrate development. Their ordered patterns of expression in different tissues suggest that they may be used independently in a variety of embryonic sites. A major site of their expression is the developing nervous system. The combination of detailed studies at both cellular and molecular levels provides an opportunity to correlate the expression and function of *Hox* genes with the generation of anatomical diversity. The main purpose of this review is to explore the relationship of the *Hox* genes to regionalization and segmental pattern formation in the CNS, with particular emphasis on the hindbrain.

HOX GENES: EVOLUTION AND AXIAL PATTERNING

The homeodomain defines a 60 amino acid motif containing sequence-specific DNA binding activity and comprising part of a larger protein that regulates gene transcription (reviewed by Gehring 1987, 1992, Hoey & Levine 1988,

Scott et al 1989). Structural studies using X-ray crystallography and nuclear magnetic resonance spectroscopy have shown that the homeodomain configuration is related to the helix-turn-helix motif of several yeast and prokaryotic DNA-binding proteins (Otting et al 1988, 1990, Qian et al 1989, Kissinger et al 1990, Gehring 1992).

Homeodomain proteins and their genes have been further divided on the basis of sequence homology into various classes (Scott et al 1989). *Antp* class homeodomains share approximately 60% homology with the homeodomain encoded by the *Drosophila* homeotic gene *Antennapedia*, including a conserved sequence motif encoded in the exon immediately 5′ of the homeodomain exon, termed the *YPWM box* or *hexapeptide* (Mavilio et al 1986, Krumlauf et al 1987). In addition this class shares an identical core sequence of 12 amino acids within α-helix III shown by structural studies to encode the part of the protein that contacts DNA (Otting et al 1988, 1990, Qian et al 1989). The latter region is suggested to play an important role in determining the DNA sequence specificity of the homeodomain, and its presence raises the possibility of cross-regulatory interactions between *Antp* class proteins and their targets. All the *Antp* class genes of the fly map to the *Antennapedia* and *Bithorax* complexes on chromosome 3, and these are referred to collectively as the *HOM-C* (Homeotic) complex.

The discovery of the homeobox led quickly to the recognition that *Antp* class homeobox genes, designated *Hox* genes, are also present in the vertebrate genome, where their organization and deployment is remarkably similar to the *HOM-C* of the fly (Akam 1989, Duboule & Dollé 1989, Graham et al 1989, Boncinelli et al 1991, Scott 1992). Unlike *HOM-C,* the mouse *Hox* genes are grouped in four complexes, presumably derived by duplication of an ancestral cluster (Kappen et al 1989, Schugart et al 1989, Krumlauf 1992, McGinnis & Krumlauf 1992). Sequence analysis demonstrated that specific genes in each complex are structurally homologous to specific members of *HOM-C,* based principally upon identity in the homeodomain and its flanking regions (Figure 1). Furthermore the order of the homologous *HOM-C/Hox* genes along the chromosome is in an identical register (Duboule & Dollé 1989, Graham et al 1989, Boncinelli et al 1991). These structural relationships between *HOM-C* and *Hox* argue that there was an ancestral cluster in the evolutionary progenitor of the arthropods and vertebrates that contained at least five *Antp* class homeobox genes (Krumlauf 1992, McGinnis & Krumlauf 1992).

In his pioneering analysis of homeotic mutations in the Bithorax complex in *Drosophila*, Lewis (1978) proposed a colinearity between the spatial activity of genes along the anteroposterior (A-P) axis of the embryo and their position within the complex. He postulated that such organization could form a combinatorial code at the genetic level that would facilitate the specification

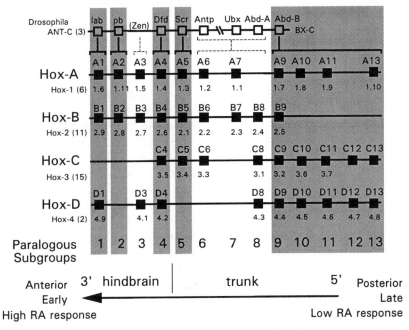

Figure 1   Organization and evolutionary relationships between the *Hox* and *HOM-C* homeotic complexes. The *Drosophila* complex is at the top and the four vertebrate complexes below. The new *Hox* nomenclature (Scott 1992) is above the line and the old below. Vertical rows imply an evolutionary homology or identity. The grey shaded boxes delineate the five different members of *HOM-C* that have clearly defined vertebrate homologues, for example, *Dfd* and *Hox-a4, b4, c4* and *d4*. The number of each paralogous group is noted below each vertical row. The direction of transcription of all genes is 5' to the right and 3' to the left. The colinear properties of the complexes with respect to A-P level, time, and retinoic acid sensitivity are indicated at the bottom of the diagram. [Modified from McGinnis & Krumlauf (1992) and Scott (1992).]

of axial diversity and segment identity. Colinear and ordered domains of expression have also been observed for the *Hox* genes in a variety of developing tissues, in particular the central and peripheral nervous system (Duboule & Dollé 1989, Giampaolo et al 1989, Graham et al 1989, Wilkinson et al 1989b, Hunt et al 1991a,c), neural crest and branchial arches (Hunt et al 1991a,c), mesoderm (Gaunt et al 1988, Dressler & Gruss 1989, Duboule & Dollé 1989, Kessel & Gruss 1991), and limb buds and gonads (Dollé et al 1989 1991, Izpisua-Belmonte et al 1991a,b). This conservation of colinear patterns in vertebrates suggests the operation of some evolutionary selective pressure related to an important functional property of the complexes. By analogy to the combinatorial code proposed by Lewis (1978), it is believed that the vertebrate *Hox* genes provide a molecular system for the specification of regional diversity along the body axis (Hunt et al 1991a, Kessel & Gruss 1991, Duboule 1992).

SEGMENTATION IN THE DEVELOPING NERVOUS SYSTEM

Regional subdivision by morphological repeat units, or segments, is a common feature of animal development in many phyla and is widely held to provide a convenient basis for the generation of regional diversity during development and evolution. In the nervous system of higher vertebrates, macroscopic bulges of the early neural tube, neuromeres, have been recognized for many years (Orr 1887, Vaage 1969) but were of uncertain significance until recently. Detailed cellular studies in the chick embryo have shown that the neuromeres of the developing hindbrain, rhombomeres, are arranged as morphological repeat-units, each pair innervating one adjacent branchial arch (Lumsden & Keynes 1989). The trigeminal (fifth cranial) branchiomotor nerve, for example, originates from rhombomeres 2 and 3 (r2,3) and innervates all muscles derived from the 1st branchial arch; the facial (7th cranial) branchiomotor nerve arises from r4,5 and innervates 2nd arch musculature and the glossopharyngeal nerve arises from r6,7 and innervates 3rd arch musculature (see Figure 2). Such striking registration between the rhombomeres and innervation patterns of the branchial arches argues that rhombomeres are indeed true segmental units of construction.

A lineage study in the chick embryo hindbrain, which used a fluorescent marker to follow clones of cells derived from individual labeled parent cells, has shown that rhombomeres are also units of lineage restriction (Fraser et al 1990). When parent cells are labeled before the appearance of boundaries between rhombomeres, their descendent clones straddle the boundaries. However, when cells are labeled at or after the appearance of boundaries the clones never cross a rhombomere boundary. Each rhombomere can therefore be regarded as a cell grouping or compartment whose constituent cells are unable to mix with the cells of neighboring rhombomeres. This would allow the opportunity for early establishment and maintenance of cell phenotype based on local information within the rhombomeric compartment.

The neuroepithelial cells at rhombomere boundaries are distinguished by a relatively low rate of proliferation and by the expression of a distinct set of cell surface molecules (Lumsden 1990, Guthrie et al 1991). They also form a site of preferential axon growth in a transverse plane (Lumsden & Keynes, 1989). Further evidence that there is an opportunity for cell-cell communication within a rhombomere, but not between rhombomeres, comes from a study on gap-junctional communication in the chick hindbrain (Martinez et al 1992). Iontophoretic injection of biocytin or Lucifer Yellow into single cells within

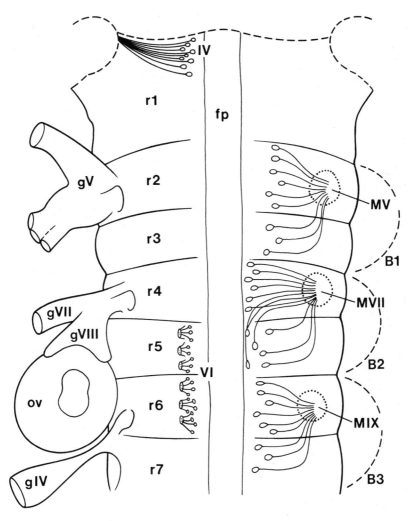

Figure 2   Schematic diagram of the developing hindbrain of a 3-day chick embryo. r1-r7, rhombomeres; gV-gIX, cranial sensory ganglia; B1-B3, branchial arches; MV-MIX, branchiomotor
nerves; ov, otic vesicle; fp, floor plate. [Modified from Figure 4, Lumsden & Keynes (1989).]

a rhombomere is followed by transjunctional diffusion to neighboring cells
only within that rhombomere.

It is believed that, as for insect compartments, the segregation of the
hindbrain epithelium into segmental units arises as adjacent groups of cells
acquire different cell surface properties. These properties are likely to be

expressed in an alternating or two-segment repeat pattern along the A-P axis of the hindbrain. Support for this comes from donor-to-host grafting experiments in the chick embryo. Normal boundaries arise when rhombomeres from adjacent positions, or positions three rhombomeres distant from one another, are juxtaposed. They do not appear, however, when the juxtaposed rhombomeres have identical or alternate positions of origin (Guthrie & Lumsden 1991). Thus cells from even numbered rhombomeres will mix with those from other even numbered rhombomeres, and odd with odd, but cells with even and odd origins will not mix.

Hindbrain segmentation also influences the development of pattern in the adjacent peripheral nervous system and branchial arch structures. The neural crest cells migrate from the dorsal neural epithelium and invade the neighboring branchial arches (Le Douarin 1983). Grafting experiments in chick embryos, which place 1st arch crest in the position of the 2nd or 3rd arch crest, have shown that the pattern of crest-derived skeletal and connective tissues is specified according to the A-P level of origin of the crest cells from the neural epithelium (Noden 1983, 1988). This has recently been extended to the neural derivatives of the cranial crest in the hindbrain (Kuratani & Eichele 1993). These experiments imply that the cranial neural crest is prepatterned before it migrates into the branchial arches, and it is tempting to speculate that the neural crest receives this information early during the process of hindbrain segmentation. Moreover, recent experiments in the chick embryo (Lumsden et al 1991, Sechrist et al 1993), using the fluorescent, lipid-soluble dye DiI to follow the pathways of crest migration from individual rhombomeres, show that there is segmental variation in the patterns of crest generation and/or migration. The 1st, 2nd, and 3rd arches are filled by crest from r2, 4, and 6, respectively. Rhombomeres 3 and 5, however, appear to release fewer crest cells and are associated with increased levels of cell death in the dorsal midline (Lumsden et al 1991, Jeffs et al 1992).

PATTERNS OF *HOX* EXPRESSION IN THE DEVELOPING NERVOUS SYSTEM

Analysis of expression patterns in mouse embryos has shown that the *Hox* genes are expressed in overlapping domains along the A-P axis of the developing hindbrain and spinal cord (Gaunt 1988, Gaunt et al 1988, Holland & Hogan 1988, Duboule & Dollé 1989, Giampaolo et al 1989, Graham et al 1989, Wilkinson et al 1989b, Hunt et al 1991a). As mentioned earlier, there is a colinear relationship between the register of genes in the *Hox* complexes and their distinct A-P boundaries of expression in the CNS. In vertebrates all the genes in a *Hox* complex are oriented in the same 5' to 3' direction with respect to transcription, such that the *abd-B* related members are the most 5′ in a complex (see Figure 1). Like the *HOM-C* genes in *Drosophila*, the genes most closely related to *abd-B* correlate with the most posterior regions of the embryo. The anterior limits of *Hox* expression for genes located sequentially 3′ in the cluster are positioned successively in an anterior direction along the A-P axis. Members related to the *labial* gene (paralogous group 1) are the only exception to this colinearity. One member, *Hox-d1*, is the only gene not expressed in the CNS, and the remaining two members, *Hox-a1* and *b1*, have A-P limits caudal to paralogous group 2.

Within the hindbrain the expression boundaries of paralogous groups 1–4 coincide with the segment (rhombomere) boundaries of the neural epithelium. For example, the anterior boundaries for members of groups 2, 3, and 4 show a two-segment repeat pattern, lying at rhombomere boundaries r2/3, r4/5 and

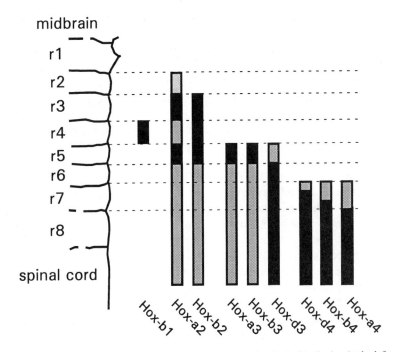

Figure 3   Segment boundaries and variations in *Hox* expression in the hindbrain. At the left of the drawing is a diagram of the mouse hindbrain at 9.5 dpc with the rhombomere (r) boundaries demarcated. At the right are bars marking the anterior rhombomere boundaries of expression (each *Hox* paralogue is identified at the bottom of each bar). The solid regions denote highest levels of expression and the grey shaded regions lower levels of expression. Note that while the A-P boundaries of paralogues can be similar, segment-specific variations occur between different members. [Based on Figure 4 in Hunt et al (1991b).]

r6/7, respectively (Figure 3). While the expression patterns for the *labial* paralogous group follow a different trend, they still reflect hindbrain segmentation (Murphy et al 1989, Wilkinson et al 1989b, Frohman et al 1990, Hunt et al 1991a,c, Lufkin et al 1991, Murphy & Hill 1991). Both *Hox-A1* and *b-1* have anterior expression limits in early embryos (8.0 dpc) that map to the future r3/4 boundary. In later stages (9.5 dpc) *Hox-b1* expression is up-regulated in r4, and down-regulated in more posterior regions, while *Hox-a1* is only down-regulated. This is again in contrast to other *Hox* genes that remain active in the nervous system through to adulthood.

The patterns of *Hox* expression in the hindbrain are not static throughout development, and another feature of their segmental expression is a relative up-regulation of activity in specific segments or anterior regions of their expression domains (Wilkinson et al 1989b, Hunt et al 1991a). In this second phase, the anterior boundaries do not change, but selected rhombomeres have distinctly elevated levels of expression which can differ between paralogous members; this pattern is illustrated in Figure 3. For example, in addition to the up-regulation of *Hox-b1* in r4, noted above, levels of *Hox-b2* increase in r3, r4 and r5, and *Hox-b3* is expressed at relatively high levels in r5. This may suggest that the final domains of activity of these genes, where critical threshold dosage levels for conferring regional identity are attained, coincide with specific segments.

Offsets in the two-segment periodicity between clusters could provide a system for generating a great degree of regional diversity and specification at the level of individual rhombomeres. However, one of the surprising findings to come from comparisons of segmental gene expression between *Hox* genes is that paralogous members have similar anterior boundaries (Figure 3; Hunt et al 1991a). The early patterns therefore delineate pairs of rhombomeres, and another system must be involved in helping to distinguish between the two members of a pair. This could happen very early, at the level of specification of differences between odd and even numbered rhombomeres. The odd and even numbered members of a pair might then have a differential ability to interpret the same *Hox* code. Alternatively it may be that the later variable levels of *Hox* expression in specific segments are more important for determining individual rhombomeric identities. Figure 3 shows that in the later phase each rhombomere does have a different combination of *Hox* genes that are active at a high level.

The overlap in paralogous genes raises the possibility that there might be some functional redundancy or compensation between members. However, despite the similarity in A-P boundaries of expression there are other temporal changes in the patterns of *Hox* expression in the nervous system which show that there is not a complete concordance between the paralogous genes. When

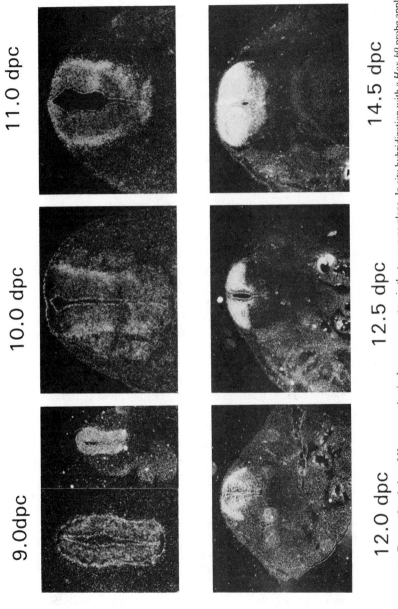

9.0dpc   **10.0 dpc**   **11.0 dpc**

12.0 dpc   **12.5 dpc**   **14.5 dpc**

Figure 4   Temporal variation of *Hox* expression in the nervous system in the transverse plane. In situ hybridization with a *Hox-b9* probe applied to transverse sections of mouse embryos from six different stages. Dorsal is to the top in all sections. This pattern is duplicated by all members of the *Hox-B* complex and is different from members of other complexes. Expression is initially uniform at 9.0 dpc and begins to increase in lateral regions and decrease in ventral regions between 10.0 and 11.0 dpc. There is a progressive dorsal restriction as indicated by the sharp boundary of expression at 12.5 dpc. Expression changes again by 14.5 dpc where ventral regions again have increased levels. These patterns mirror the birth of major classes of neurons in the spinal cord. [The figure is based on Figure 4 in Krumlauf et al (1991) and Graham et al (1991).]

viewed in the transverse plane *Hox* genes display dynamic patterns characteristic for each complex (Gaunt 1991, Graham et al 1991). For example all of the members of the *Hox-B* complex initially have a uniform distribution, which decreases ventrally as the motor neurons form and increases laterally as the commissural neurons differentiate (Figure 4). Expression then becomes restricted to dorsal sensory regions for a period before reappearing in ventral regions. These changes mirror the birth of specific classes of neurons in the CNS and suggest that the *Hox* genes could provide positional addresses for differentiating neurons (Graham et al 1991). Members of the other *Hox* complexes have distinct dorsoventral (D-V) distributions (Gaunt et al 1990, Gaunt 1991), suggesting that they could be patterning different neuronal classes. Therefore, genes from the four complexes could work independently to pattern subclasses of cells at the same A-P level, and they could combine to provide the complete positional information required in the CNS in later stages of neurogenesis. Other genes, such as *Pax* genes and those encoding growth factor homologues, are implicated in phenotype specification in the D-V axis (Gruss & Walther 1992, Jessell & Melton 1992), and the *Hox* genes presumably interact with these to define A-P identities.

How early is the identity of a rhombomere established, and are *Hox* genes linked with the process? Grafting experiments in the chick embryo suggest that the cell autonomous identity of a rhombomere is indeed established very early (Guthrie et al 1992, Kuratani & Eichele 1993). At the six somite stage, before rhombomeres become cellular compartments and their boundaries form, prospective r4 regions have been grafted to ectopic locations in the developing hindbrain. Rhombomeres do form in the ectopic locations and properly express the *Hox-b1* gene in an r4-like pattern typical of the site of origin and not the ectopic location (Guthrie et al 1992, Krumlauf et al 1993). Furthermore, peripheral nerves derived from the ectopic rhombomere also have an r4 associated identity (Krumlauf et al 1993). These experiments demonstrate that well before compartment boundaries form, there has been a regionalization of the developing hindbrain sufficient to allow a programme of cell autonomous gene expression and morphogenesis characteristic of a specific rhombomere. The tight coupling of gene expression to segmental pattern formation, together with early autonomy of expression, suggests that the *Hox* genes, like their *HOM-C* counterparts, may play a role in the specification of segment phenotype.

A further observation consistent with the coupling of hindbrain *Hox* gene expression to segmentation comes from a detailed study in the chick embryo using an antibody raised against the protein coded by the upstream, non-homeobox exon of the gene equivalent to mouse *Hox-b1* (Sundin & Eichele 1990). Shortly after the formation of r4, expression is excluded here from the ventral floor plate cells, corresponding to the finding that this region of the developing hindbrain contains no barriers to clonal expansion (Fraser et al 1990).

Hox Expression Patterns in the Branchial Arches Reflect Hindbrain Segmentation

Rhombomeres and neural crest contain sufficient prepatterning information to generate their characteristic properties in ectopic sites (Noden 1983, 1988, Kuratani & Eichele 1993). Since the neural crest migrates from the dorsal hindbrain epithelium, could its positional specification be linked with that of the hindbrain through the *Hox* genes? Because the chick grafting experiments (Guthrie et al 1992, Kuratani & Eichele 1993) have shown that cell autonomous properties of the future rhombomeres are already established before the onset of neural crest migration, this is certainly possible. There is support for the idea in that *Hox* expression in the neural crest also displays ordered and colinear domains (Hunt et al 1991a,c). For example members of paralogous group 2 (*Hox-a2 and b2*) are expressed in the surface ectoderm, mesenchymal derivatives, and sensory nerves in all branchial arches below the first arch. Similarly, group 3 genes are expressed in the same cell types in arches below the second arch, and group 4 genes in tissues below the third arch (Hunt et al 1991a,c, Hunt & Krumlauf 1992). Initially during early emigration from the rhombomeres, all neural crest cells from a specific rhombomere appear to express the same paralogous genes. However, at later stages within an arch, different paralogues have different regional expression patterns (Hunt et al 1991a). This suggests that individual genes are patterning different subsets of cells in a branchial arch, and has important implications for the interpretation of phenotypes in *Hox* mutants (see below).

It is interesting that this pattern in the branchial arches is offset in a posterior direction by one rhombomere from that in the hindbrain itself. We believe that this is due to the different properties of neural crest emigrating from odd and even numbered rhombomeres. No expression of *Hox* genes is observed adjacent to r3 and r5 (Hunt et al 1991a,c, Sham et al 1993), corresponding to the reduced crest emigration and increased levels of cell death noted in these same areas (Lumsden et al 1991, Jeffs et al 1992). As with the crest it is important to note that the A-P boundaries of *Hox* expression in other tissues, such as paraxial and lateral plate mesoderm, also fail to coincide with those of the neural tube. In all cases the boundary in the CNS lies most anterior. It is not yet clear whether these offsets result simply from morphogenetic movements following an early axial registration of *Hox* expression (in response, for example, to the same signals upstream of *Hox* expression), or whether instead they reflect independent pathways of development that are set up during the period of gastrulation.

EXPERIMENTALLY INDUCED *HOX* MUTATIONS

In *Drosophila*, loss-of-function mutations in *HOM-C* result in anteriorization of posterior segments, and gain-of-function mutations generate posterior transformations. This is precisely what is seen in the mouse following mutations in *Hox* genes created by transgenesis or targeted disruption in embryonic stem (ES) cells. In the axial skeleton, loss-of-function alleles have resulted in anterior homeotic transformations of vertebrae. In *Hox-c8⁻* mice the first lumbar vertebra is transformed into a thoracic vertebra resulting in an extra pair of ribs (Le Mouellic et al 1992), and in *Hox-b4⁻* mutants the second cervical vertebra is partially converted into an additional atlas or first cervical vertebra (Ramirez-Solis et al 1993). Furthermore, ectopic expression of the *Hox-d4* gene in transgenic mice results in the posterior transformation of the occipital bones in the skull (Lufkin et al 1992), and alterations to *Hox-c6* and *Hox-c8* expression produce extra ribs by transforming the first lumbar vertebra (Jegalian & De Robertis 1992, Pollock et al 1992). Together, these findings argue strongly that a *Hox* code analogous to that in *Drosophila* is operating during the development of the vertebrate axial skeleton to regulate the identity of specific vertebrae.

In the nervous system the evidence that a *Hox* code may function is less clear. Mutations in *Hox-a1* and *Hox-a2* have dramatic effects in the branchial region of the head (Chisaka & Capecchi 1991, Lufkin et al 1991, Chisaka et al 1992). Some but not all neurogenic and mesenchymal components derived from the neural crest are altered in the arches, as well as endodermally and mesodermally derived cells. The phenotypes correlate with the patterns of expression observed in the branchial arches and confirm that the genes are important in craniofacial morphogenesis. However, no obvious transformations in branchial tissues or the CNS have been observed in these mutants. The primary phenotypes described represent abnormal patterning and loss of multiple structures or cell populations. This suggests that the *Hox* genes are involved in patterning multiple tissues which ultimately need to interact to form the proper head structures, and one might not expect simple transformations if such an interdependent patterning cascade were perturbed.

We have argued that segmentation in the hindbrain might be important outside the CNS for imparting positional information to the branchial arches by way of the neural crest. The neural crest defects in the mouse mutations seem to support this idea, but it was somewhat surprising that no alterations in the CNS or hindbrain rhombomeres were observed in such mutants. This raised the possibility that *Hox* expression and function in neural crest could be mediated independent of hindbrain segmentation. Recently, however, a more detailed analysis of *Hox-a1⁻* mice has revealed that there are early defects in the generation of specific rhombomeres (Dollé et al 1993). In situ hybridization studies in homozygous mutant embryos, with *Krox-20* as a marker for r3 and r5 and *Hox-b1* as a marker for r4, have shown that r5 is almost completely absent and r4 is drastically reduced in size. While r4 and r5 are abnormal, the other surrounding rhombomeres appear unaffected and express the normal distributions of *Hox* genes. This does not reflect a transformation in the CNS, as it appears that r4 and r5 attempt to form but the lack of functional *Hox-a1* protein prevents their proper growth and elaboration. The alterations to expression suggest that *Hox-a1* may be required directly or indirectly to maintain the expression of *Hox-b1* and *Krox-20* by cross-regulatory interactions. Regardless, these data confirm that *Hox* genes are important in generating segmental patterns in the CNS.

HOX GENES AND RETINOIC ACID

Another source of support for *Hox* genes specifying rhombomeric identities in the CNS comes from experiments with retinoic acid (RA). RA is a potent teratogen that induces craniofacial abnormalities (Lammer et al 1985) and defects in the central nervous system (reviewed in Maden & Holder 1992, Morriss-Kay 1993; see also review by Jessell in this volume). There are marked effects on the CNS in many vertebrate embryos which suggest that RA may be both directly and indirectly causing truncations and transformations in axial structures (Durston et al 1989, Sive et al 1990, Holder & Hill 1991, Papalopulu et al 1991, Ruiz Altaba & Jessell 1991a,b, Sive & Cheng 1991). In cell culture RA is known to activate *Hox* genes in a dose dependent colinear manner (Simeone et al 1990, 1991, Papalopulu et al 1991), and this also occurs in *Xenopus* embryos (Papalopulu et al 1991, Dekker et al 1993). In addition RA treatment of vertebrate embryos leads to changes in the spatial distribution of *Hox* and *Krox-20* expression in the hindbrain (Morriss-Kay et al 1991, Papalopulu et al 1991, Sive & Cheng 1991, Conlon & Rossant 1992, Marshall et al 1992, Sundin & Eichele 1992). The close parallels in patterns of *Hox* expression in tissues altered by RA, the ability of RA to alter *Hox* expression domains, and the detection of retinoids in the embryo (Hensen's node, floor plate) (Hornbruch & Wolpert 1986, Wagner et al 1990, Hogan et al 1992) have led to the speculation that RA is an in vivo morphogen that plays a role in generating the proper boundaries of *Hox* expression. It is not clear if such activation would be directly mediated by the RA nuclear receptors, although binding sites for an RAR-β have been identified in the *Hox-a1* gene (Langston & Gudas 1992).

Recently, detailed analysis on retinoic acid–induced changes to the patterns of *Hox* expression and hindbrain neuroanatomy have revealed homeotic transformations (Marshall et al 1992, Kessel, 1993). Exposure of mouse embryos at 7.5 dpc (pre-headfold stage) results in a rapid expansion of r4

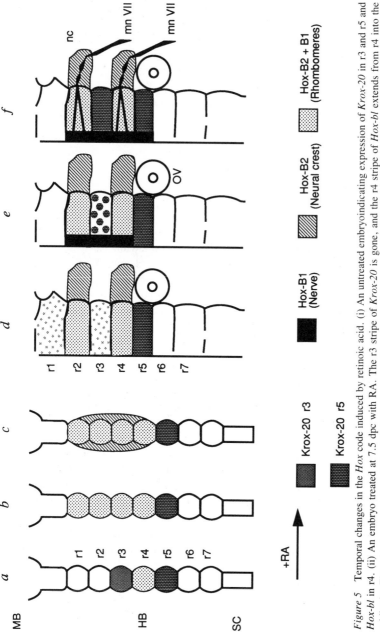

Figure 5 Temporal changes in the *Hox* code induced by retinoic acid. (i) An untreated embryoindicating expression of *Krox-20* in r3 and r5 and *Hox-b1* in r4. (ii) An embryo treated at 7.5 dpc with RA. The r3 stripe of *Krox-20* is gone, and the r4 stripe of *Hox-b1* extends from r4 into the midbrain. There is also expression of *Hox-b2* which is expanded into the midbrain. (iii) Neural crest expressing the *Hox-b2* gene begins to emerge from the expanded domain expressing *hox-b1*. (iv) Expression of *Hox-b1* begins to decline in r1 and r3, (marked by the asterisks)leaving two strong stripes in r2 and r4. (v) Expression of *Hox-b1* is completely absent in r1 and r3, and *Krox-20* is reactivated in r3 (small round dots). Midline expression is observed in r2/3/4, corresponding to the cell bodies of the moter nerves, which are just beginning to form. (vi) Final pattern at 10.5 dpc where there is a duplication in r2/3 of the markers normally expressed in r4/5, representing a transformation of r2/3 into the r4/5 identity. [This figure is based on Figure 4 in Langston & Gudas (1992). MB, midbrain; HB, hindbrain; SC, spinal cord; OV, otic vesicle; nc, neural crest; mn VII, facial branchiomotor nerve.]

markers such as *Hox-b1* into anterior regions which are subsequently modified (Marshall et al 1992). The end result of these modifications to *Hox* expression is that genes normally expressed in r4 and r5 are now also expressed in r2 and r3. Associated with these changes in gene expression is a transformation of the trigeminal nerve into a facial nerve identity, based on the locations of axon projections and cell bodies (Marshall et al 1992, Kessel 1993).

Figure 5 summarizes the temporal changes in segmental gene expression observed in the mouse hindbrain upon exposure of embryos to RA. One striking conclusion from these studies is that elevating the expression of *Hox-b1* into anterior regions does not permanently transform these regions into an r4 identity. The gene is down-regulated in alternate rhombomeres and remains active only in r2 and r4. This suggests that even in the early pre-headfold stage of mouse embryos (which is the time of RA treatment) there must be alternating properties within the hindbrain that can account for the later differential response of *Hox* genes to RA. This could be at the level of specifying odd versus even numbered rhombomeres, with the implication that segmental regionalization within the CNS is initiated much earlier than anticipated from the grafting and cell lineage studies. However, these experiments do demonstrate that *Hox* genes are linked with transformations in the CNS and support the idea that a *Hox* code for segment identity could be operating in the hindbrain.

OTHER REGIONS OF THE CNS

We have argued that segmentation processes in the hindbrain involving *Hox* genes play an influential role in regional patterning. Whether the same holds true for other axial regions is less clear. In the spinal cord of higher vertebrates, neuromeres (myelomeres) are observed, but with the possible exception of the projection patterns of the preganglionic sympathetic axons (Ezerman et al 1990), there is no clear indication of segmented patterns of neuronal development (Lim et al 1991). On the other hand, lineage studies of the spinal cord neural epithelium, using a fluorescent marker dye, have shown the presence of lineage restriction boundaries aligned with the boundaries that separate the adjacent somites into anterior and posterior halves (Stern et al 1991). The maintenance of these boundaries depends on some influence, mechanical or chemical, of the adjacent somites, for lineage restriction is disrupted when the somites are excised surgically (Stern et al 1991). There is also good evidence for segmentation of the neural pattern in the spinal cord of lower vertebrates. In the case of the zebrafish, mutation-induced disruptions of the somite pattern cause parallel disruptions in the pattern of primary motor neurons (Eisen & Pike 1991). Together with the influence of the somite mesoderm on the chick spinal cord lineage restrictions, this would suggest that spinal cord segmentation is not intrinsic, but arises instead from influences from neighboring paraxial mesoderm.

The boundaries of *Hox* expression that map to the spinal cord are sharp. In view of the correlation between rhombomere boundaries and *Hox* expression in the hindbrain it will be interesting to examine whether boundaries in the spinal cord map to somite-imposed lineage restrictions. If so, it is possible that segmentation and *Hox* gene expression were co-opted in early vertebrates in the development of both spinal cord and hindbrain.

In contrast to the spinal cord, a system of segmentation directly analogous to that of the hindbrain appears to be operating in the developing forebrain. The four diencephalic neuromeres have also been shown to be units of cell lineage restriction in the chick embryo (Figdor & Stern 1993). The fact that none of the *Hox* genes are expressed in the developing forebrain implicates other, distinct sets of genes in the positional specification and patterning of this brain region and its subdivisions, the diencephalon and telencephalon. Members of the *Dlx*, *Nkx* and *Wnt* multigene families, which have homologues in *Drosophila*, are expressed in the developing diencephalon in a spatial pattern whose boundaries align with those of the diencephalic neuromeres (Price et al 1991, 1992, Robinson et al 1991, Roelink & Nusse 1991).

The most anterior head structures in *Drosophila* are patterned by several homeobox genes which are not part of the *HOM-C* complex. It is interesting that vertebrate homologues of these genes, *Emx-1 & 2* and *Otx-1 & 2*, have been identified, and they are expressed in nested domains during the early development of the mouse telencephalon, diencephalon and mesencephalon (Simeone et al 1992a,b, 1993). The domains appear to correspond to single or multiple neuromeric regions, again suggesting a match between segmental morphogenesis and homeobox gene expression in the CNS. Further candidates for roles in forebrain regionalization include another distal-less homologue, *Tes-1* (Porteus et al 1991), and the homeodomain proteins of the POU subfamily, which are widely expressed in the developing forebrain and midbrain (Treacy & Rosenfeld 1992).

REGIONALIZATION CASCADE

The large body of evidence we have presented argues strongly that *Hox* genes have a pivotal role in specification of positional values, analogous to the homeotic genes in *Drosophila*. They clearly represent one of the components involved in regionalizing the nervous system, and the task ahead is to determine how they fit into a larger-scale cascade of interactions leading to the final neural pattern.

Upstream of the Hox Genes

We have seen above that regional autonomy of *Hox* expression is already established in late neural plate stages, and it is an open question as to how early this pattern is specified. Classical experiments on amphibian embryos have shown that axial mesoderm induces the overlying ectoderm to adopt a neural fate, and that axial mesoderm from different A-P levels induces corresponding A-P characteristics in the CNS (Spemann 1938, Hamburger 1988, Kintner 1992). In the trunk the somites appear to be have a major influence on segmental patterning in the adjacent spinal cord (Lim et al 1991, Stern et al 1991), so one might predict that regionalization of the hindbrain at the level of rhombomeres could also arise from segmental patterning in the underlying axial mesoderm, rather than originating intrinsically within the neural ectoderm.

Apart from the fact that there is no evidence for segmentation in axial mesoderm in the head, there are experimental data which argue strongly that interactions within the ectoderm itself can generate regional patterning (Doniach 1992, Doniach et al 1992, Ruiz i Altalba 1992). In *Xenopus*-derived exogastrulae (Kintner 1992) and Keller sandwiches (Keller & Danilchik 1988), in which there is no involution of organizer (mesoderm) tissue beneath the ectoderm, not only does neural tissue form but regional markers such as *Krox-20* and *Engrailed-2* are expressed in appropriate restricted domains. The implication of these experiments is that planar signals within the ectoderm are capable of establishing fine-grained regionalization (Doniach 1992, Doniach et al 1992, Ruiz i Altalba 1992). It is still clear, nevertheless, that vertical signals from the mesoderm are required to generate the complete pattern. It seems likely that the genes which control the early segment-restricted patterns of *Hox* expression in the hindbrain are being activated during this process of induction and segmentation, whose molecular basis remains unknown.

Segmentation

Could *Hox* genes be segmentation genes themselves? Although we have argued by analogy with *Drosophila* that they are likely to define segment identity, there are early phases of expression in the mouse which could suggest such a role. *Hox-a1*, *b1*, and *B2* are expressed with sharp anterior boundaries in early headfold mouse embryos, at a time when the segments should be forming. Furthermore, in null mutants of *Hox-a1* (Lufkin et al 1991, Chisaka et al 1992) rhombomeres r4 and r5 fail to develop properly and are almost completely absent based on the lack of r5 stripe of *Krox-20*, and r4 stripe of *Hox-B1* (Dollé et al 1993). This suggests that *Hox-a1* is required to generate or elaborate these particular segments, but it could be equally necessary for their maintenance rather than their initial construction. The question must therefore remain open, pending further proof.

Perhaps a more promising candidate as a segmentation gene is *Krox-20* (Chavrier et al 1988). This zinc-finger containing gene is restricted to r3 and r5 before the up-regulation of *Hox* genes in specific rhombomeres (Wilkinson et al 1989a,b, Sham et al 1993). Recently it has been shown that *Krox-20* is a positive regulator of *Hox-b2* during hindbrain segmentation (Sham et al 1993). *Krox-20* activates the expression of *Hox-b2* in r3 and r5, which would be consistent with a putative role as a segmentation gene. Furthermore the *Hox-b2* paralogue, *Hox-a2*, is expressed in high levels in r3 and r5, suggesting another target for regulation by *Krox-20*. *Krox-20* could therefore be functioning to regulate multiple *Hox* genes positively or negatively. To date no other zinc finger genes related to *Krox-20* have been found to have rhombomere-restricted expression domains, so it is uncertain if *Krox-20* is a member of a family of segmentation genes with similar functions in other rhombomeres. The only other gene found to display segment-restricted patterns of expression in the early developing hindbrain is *sek*, encoding a novel receptor tyrosine kinase (Gilardi-Hebenstreit et al 1992, Nieto et al 1992). This gene is also expressed in the future r3 and r5 domains, and detailed temporal studies are needed to determine if it is expressed earlier or later than the *Krox-20* and *Hox* genes.

Based on the current evidence the most convincing role for the *Hox* genes is in determining regional identities. Perhaps the best evidence comes from phenotype transformations that arise in the axial skeleton when *Hox* expression has been altered by gain or loss of function (Balling et al 1989, Kessel et al 1990, Jegalian & De Robertis 1992, Le Mouellic et al 1992, Lufkin et al 1992, Pollock et al 1992, Ramirez-Solis et al 1993). While there is less evidence for the CNS, retinoic acid does induce changes in *Hox* expression that result in transformation of r2/3 to an r4/5 phenotype (Marshall et al 1992). Targeted disruptions in mice have clearly demonstrated that the *Hox* genes have a normal role in patterning the neural crest, but transformation of crest-derived structures has not been observed (Chisaka & Capecchi 1991, Lufkin et al 1991, Chisaka et al 1992). Rather, the phenotypes have been described in terms of an absence or disorganization of crest-derived tissues.

Although we have emphasized the similarities between *HOM-C* and *Hox* it should be stressed that a *Hox* code in vertebrates may operate in a different manner to that proposed for *HOM-C*. In *Drosophila* a particular combination of homeotic genes confers an identity. In vertebrates, however, the multiple paralogues which arose due to duplication and divergence add an extra layer of complexity. Is the identity determined when all paralogues are expressed in the same cell, or is it a function of differential expression of paralogues in subsets of cells at the same segmental level, as seen in the branchial arches?

We favor the latter hypothesis, because it explains the phenotypes in the mouse *Hox* mutants and in an evolutionary sense provides a mechanism for greater phenotypic diversity.

Downstream of the Hox Genes

Other than the *HOM-C/Hox* genes themselves, very little is known about the target genes they activate to initiate the cascade of interactions that modulate regional diversity. Genetic screens in *Drosophila* for genes involved in embryonic pattern formation have failed to identify *HOM-C* targets, presumably because mutations in such genes do not yield distinct pattern formation phenotypes. This implies that the downstream genes are required for a variety of purposes in morphogenesis and differentiation, and they would be most likely to yield lethal phenotypes. Logical candidates would include cell and substrate adhesion molecules, growth factors and their receptors, cytoskeletal proteins and cell signalling molecules. In this regard it is interesting that one immediate downstream target of *Ubx* in *Drosophila* is *connectin*, which encodes a cell adhesion molecule (Gould & White 1992, Nose et al 1992, White et al 1992). In vertebrates *Hox* genes activate and repress the promoter activity of another cell adhesion molecule (N-CAM) in cell culture (Jones et al 1992). Even though the target genes have not been identified, gene swap experiments suggest that human and *Drosophila* genes can activate similar downstream genes (Malicki et al 1990, McGinnis et al 1990). Ectopic expression experiments in *Drosophila*, using heat shock promoters and the vertebrate *Hox-d4* and *Hox-b6*, give similar phenotypes to their *Drosophila* homologues *Dfd* and *Antp*. Defining the nature and function of these downstream targets will continue to be one of the most active areas of *Hox* research in the coming years.

CONCLUDING REMARKS

Following the discovery of the homeobox, progress in identifying vertebrate genes involved in regionalization of the nervous system has been remarkably rapid. It is now clear that the *Hox* genes are fundamental components of the molecular mechanisms that create and specify diversity in tissues along the A-P axis, including the nervous system. The striking structural and functional parallels between the arthropod and vertebrate in *HOM/Hox* complexes has led to the speculation that this conservation reflects selection based on a common role in segmentation. However, we feel it is more likely that these genes were part of an axial system, perhaps with a primary role in A-P patterning in the nervous system, that was later co-opted for segmentation. The intensive search for the most primitive organism with a *Hox* complex should shed some light on the original function of such an A-P system. Many other gene families and their products have also been implicated in regional patterning of the nervous system (*Pax, En, POU, Wnt, Otx, Emx, Evx, BMP, β-TGF,* etc), and understanding how they interact to produce the extreme complexity of the nervous system presents a major challenge for future research.

Literature Cited

Akam M. 1989. Hox and HOM: homologous gene clusters in insects and vertebrates. *Cell* 57:347–49

Balling R, Mutter G, Gruss P, Kessel M. 1989. Craniofacial abnormalities induced by ectopic expression of the homeobox gene Hox-1.1 in transgenic mice. *Cell* 58:337–47

Boncinelli E, Simeone A, Acampora D, Mavilio F. 1991. HOX gene activation by retinoic acid. *Trends Genet.* 7:329–34

Chavrier P, Zerial M, Lemaire P, et al. 1988. A gene encoding a protein with zinc fingers is activated during G0/G1 transition in cultured cells. *EMBO J.* 7:29–35

Chisaka O, Capecchi M. 1991. Regionally restricted developmental defects resulting from targetted disruption of the mouse homeobox gene *hox*1.5. *Nature* 350:473–79

Chisaka O, Musci T, Capecchi M. 1992. Developmental defects of the ear, cranial nerves and hindbrain resulting from targeted disruption of the mouse homeobox gene *Hox*-1.6. *Nature* 355:516–20

Conlon R, Rossant J. 1992. Exogenous retinoic acid rapidly induces anterior ectopic expression of murine *Hox*-2 genes *in vivo. Development* 116:357–68

Dekker E-J, Pannese M, Houtzager E, et al. 1993. Colinearity in the *Xenopus laevis Hox*-2 complex. *Mech. Dev.* 40:3–12

Dollé P, Izpisua-Belmonte J-C, Brown JM, et al. 1991. *Hox*-4 genes and the morphogenesis of mammalian genitalia. *Genes Dev.* 5:1767–76

Dollé P, Izpisua-Belmonte JC, Falkenstein H, et al. 1989. Coordinate expression of the murine *Hox*-5 complex homoeobox-containing genes during limb pattern formation. *Nature* 342:767–72

Dollé P, Lufkin T, Krumlauf R, et al. 1993. Local alterations of *Krox*-20 and *Hox* gene expression in the hindbrain of *Hox-a1(Hox-1.6)* homozygote null mutant embryos. *Development.* In press

Doniach T. 1992. Induction of anteroposterior neural pattern in *Xenopus* by planar signals. *Development* (Suppl.):183–93

Doniach T, Phillips CR, Gerhart JC. 1992. Planar induction of anteroposterior pattern in the developing central nervous system of *Xenopus laevis. Science* 257:542–45

Dressler GR, Gruss P. 1989. Anterior boundaries of Hox gene expression in mesoderm-derived structures correlate with the linear gene order along the chromosome. *Differentiation* 41:193–201

Duboule D. 1992. The vertebrate limb:A model system to study the *Hox/HOM* gene network during development and evolution. *BioEssays* 14:375–84

Duboule D, Dollé P. 1989. The structural and functional organization of the murine HOX gene family resembles that of Drosophila homeotic genes. *EMBO J.* 8:1497–505

Durston A, Timmermans J, Hage W, et al. 1989. Retinoic acid causes an anteroposterior transformation in the developing central nervous system. *Nature* 340:140–44

Eisen J, Pike S. 1991. The *spt*-1 mutation alters segmental arrangement and axonal development of identified neurons in the spinal cord of the embryonic zebrafish. *Neuron* 6:767–76

Ezerman EB, Glover JC, Forehand CJ. 1990. Segmental organisation of thoracic preganglionic neurons in chick and rat embryos. *Abstr. Soc. Neurosci.* 16:145.10.

Figdor M, Stern CD. 1993. Segmental organisation of embryonic diencephalon. *Nature.* 363:630–34

Fraser S, Keynes R, Lumsden A. 1990. Segmentation in the chick embryo hindbrain is defined by cell lineage restrictions. *Nature* 344:431–35

Frohman MA, Boyle M, Martin GR. 1990. Isolation of the mouse *Hox*-2.9 gene; analysis of embryonic expression suggests that positional information along the anterior-posterior axis is specified by mesoderm. *Development* 110:589–607

Gaunt SJ. 1988. Mouse homeobox gene transcripts occupy different but overlapping domains in embryonic germ layers and organs: a comparison of *Hox*-3.1 and *Hox*-1.5. *Development* 103:135–44

Gaunt SJ. 1991. Expression patterns of mouse *Hox* genes: Clues to an understanding of developmental and evolutionary strategies. *BioEssays* 13:505–13

Gaunt SJ, Coletta PL, Pravtcheva D, Sharpe PT. 1990. Mouse Hox-3.4: homeobox sequence and embryonic expression patterns compared with other members of the *Hox* gene network. *Development* 109:329–39

Gaunt SJ, Sharpe PT, Duboule D. 1988. Spatially restricted domains of homeogene transcripts in mouse embryos: relation to a segmented body plan. *Development* 104 (Suppl.):169–79

Gehring WJ. 1987. Homeoboxes in the study of development. *Science* 236:1245–52

Gehring WJ. 1992. The homeobox in perspective. *Trends Biochem. Sci.* 17:277–80

Giampaolo A, Acampora D, Zappavigna V, et al. 1989. Differential expression of human HOX-2 genes along the anterior-posterior axis in embryonic central nervous system. *Differentiation* 40:191–97

Gilardi-Hebenstreit P, Nieto A, Frain M, et al. 1992. An EPH-related receptor protein-tyrosine kinase gene segmentally- expressed in the developing mouse hindbrain. *Oncogene* 7:2499–506

Gould AP, White RAH. 1992. *Connectin,* a target of homeotic gene control in Drosophila. *Development* 116:1163–74

Graham A, Maden M, Krumlauf R. 1991. The murine Hox-2 genes display dynamic dorsoventral patterns of expression during central nervous system development. *Development* 112:255–64

Graham A, Papalopulu N, Krumlauf R. 1989. The murine and *Drosophila* homeobox clusters have common features of organisation and expression. *Cell* 57:367–78

Gruss P, Walther C. 1992. *Pax* in development. *Cell* 69:719–22

Guthrie S, Lumsden A. 1991. Formation and regeneration of rhombomere boundaries in the developing chick hindbrain. *Development* 112:221–29

Guthrie S, Muchamore I, Kuroiwa, A, et al. 1992. Neuroectodermal autonomy of *Hox-2.9* expression revealed by rhombomere transpositions. *Nature* 356:157–59

Guthrie SC, Butcher M, Lumsden A. 1991. Patterns of cell division and interkinetic nuclear migration in the chick embryo hindbrain. *J. Neurobiol.* 22:742–54

Hamburger V. 1988. *The Heritage of Experimental Embryology: Hans Spemann and the Organizer.* Oxford: Oxford Univ. Press

Hoey T, Levine M. 1988. Divergent homeobox proteins recognize similar DNA sequences in Drosophila. *Nature* 332:858–61

Hogan BLM, Thaller C, Eichle G. 1992. Evidence that Hensen's node is a site of retinoic acid synthesis. *Nature* 359:237–41

Holder N, Hill J. 1991. Effects of retinoic acid on zebrafish embryos: modulation of *engrailed* protein expression at the midbrain-hindbrain border and development of cranial ganglia. *Development* 113:1159–70

Holland P, Hogan B. 1988. Expression of homeo box genes during mouse development: a review. *Genes Dev.* 2:773–82

Hornbruch A, Wolpert L. 1986. Positional signalling by Hensen's node when grafted to the chick limb bud. *J. Embryol Exp. Morphol.* 94:257–65

Hunt P, Gulisano M, Cook M, et al. 1991a. A distinct *Hox* code for the branchial region of the head. *Nature* 353:861–64

Hunt P, Krumlauf R. 1992. *Hox* codes and positional specification in vertebrate embryonic axes. *Annu. Rev. Cell Biol.* 8:227–56

Hunt P, Whiting J, Nonchev S, et al. 1991b. The branchial *Hox* code and its implications for gene regulation, patterning of the nervous system and head evolution. *Development* 113(Suppl. 2):63–77

Hunt P, Wilkinson D, Krumlauf R. 1991c. Patterning the vertebrate head: murine Hox 2 genes mark distinct subpopulations of premigratory and migrating neural crest. *Development* 112:43–51

Izpisua-Belmonte J-C, Falkenstein H, Dollé P, et al. 1991. Murine genes related to the *Drosophila AbdB* homeotic gene are sequentially expressed during development of the posterior part of the body. *EMBO J.* 10: 2279–89

Izpisua-Belmonte J-C, Tickle C, Dollé P, et al. 1991. Expression of homeobox *Hox-4* genes and the specification of position in chick wing development. *Nature* 350:585–89

Jeffs P, Jaques K, Osmond M. 1992. Cell death in cranial neural crest development. *Anat. Embryol.* 185:583–88

Jegalian BG, De Robertis EM. 1992. Homeotic transformation in the mouse induced by overexpression of a human *Hox-3.3* transgene. *Cell* 71:901–10

Jessell T, Melton DA. 1992. Diffusible factors in vertebrate embryonic induction. *Cell* 68: 257–70

Jones FS, Prediger EA, Bittner DA, et al. 1992. Cell adhesion molecules as targets for *Hox* genes: Neural cell adhesion molecule promoter activity is modulated by co-transfection with *Hox-2.5* and *-2.4. Proc. Natl. Acad. Sci. USA* 89:2086–90

Kappen C, Schugart K, Ruddle F. 1989. Two steps in the evolution of antennapedia-class vertebrate homeobox Genes. *Proc. Natl. Acad. Sci. USA* 86:5459–63

Keller R, Danilchik M. 1988. Regional expression, pattern and timing of convergence and extension during gastrulation of *Xenopus laevis*. *Development* 103:193–209

Kessel M. 1993. Reversal of axonal pathways from rhombomere 3 correlates with extra *Hox* expression domains. *Neuron* 10:379–93

Kessel M, Balling R, Gruss P. 1990. Variations

of cervical vertebrae after expression of a *Hox 1.1* transgene in mice. *Cell* 61:301–8

Kessel M, Gruss P. 1991. Homeotic transformations of murine prevertebrae and concomitant alteration of Hox codes induced by retinoic acid. *Cell* 67:89–104

Kintner C. 1992. Molecular bases of early neural development *Xenopus* embryos. *Annu. Rev. Neurosci.* 15:251–84

Kissinger C, Liu B, Martin-Blanco E, et al. 1990. Crystal structure of an engrailed homeodomain-DNA complex at 2.8Å resolution: A framework for understanding homeodomain-DNA interactions. *Cell* 63: 579–90

Krumlauf R. 1992. Evolution of the vertebrate *Hox* homeobox genes. *BioEssays* 14:245–52

Krumlauf R, Holland P, McVey J, Hogan B. 1987. Developmental and spatial patterns of expression of the mouse homeobox gene, *Hox 2.1*. *Development* 99:603–17

Krumlauf R, Hunt P, Graham A, Wilkinson D. 1991. Patterning regional identity: spatially-restricted and dynamic expression patterns of *Krox-20* and *Hox* genes in the developing nervous system. *Semin. Dev. Biol.* 2:375–84

Krumlauf R, Hunt P, Sham, M-H, et al. 1993. *Hox* genes: A molecular code for patterning regional diversity in the nervous system and branchial structures. *Restorative Neurol. Neurosci.* 5:10–12

Kuratani SC, Eichele G. 1993. Rhombomere transposition repatterns the segmental organization of cranial nerves and reveals cell-autonomous expression of a homeodomain protein. *Development* 117:105–17

Lammer E, Chen D, Hoar R, et al. 1985. Retinoic acid embryopathy. *New. Engl. J. Med.* 313:837–41

Langston AW, Gudas LJ. 1992. Identification of a retinoic acid responsive enhancer 3' of the murine homeobox gene *Hox-1.6*. *Mech. Dev.* 38:217–28

Le Douarin N. 1983. *The Neural Crest*. Cambridge: Cambridge Univ. Press

Le Mouellic H, Lallemand Y, Brulet P. 1992. Homeosis in the mouse induced by a null mutation in the homeo-gene *Hox-3.1*. *Cell* 69:251–64

Lewis E. 1978. A gene complex controlling segmentation in *Drosophila*. *Nature* 276: 565–70

Lim TM, Jaques KF, Stern CD, Keynes RJ. 1991. An evaluation of myelomeres and segmentation of the chick embryo spinal cord. *Development* 113:227–38

Lufkin T, Dierich A, LeMeur M, et al. 1991. Disruption of the *Hox-1.6* homeobox gene results in defects in a region corresponding to its rostral domain of expression. *Cell* 66:1105–19

Lufkin T, Mark M, Hart C, et al. 1992. Homeotic transformation of the occipital

bones of the skull by ectopic expression of a homeobox gene in transgenic mice. *Nature* 359:835–41

Lumsden A. 1990. The cellular basis of segmentation in the developing hindbrain. *Trends Neurosci.* 13:329–35

Lumsden A, Keynes R. 1989. Segmental patterns of neuronal development in the chick hindbrain. *Nature* 337:424–28

Lumsden A, Sprawson N, Graham A. 1991. Segmental origin and migration of neural crest cells in the hindbrain region of the chick embryo. *Development* 113:1281–91

Maden M, Holder M. 1992. Retinoic acid and development of the central nervous system. *BioEssays* 14:431–38

Malicki J, Schughart K, McGinnis W. 1990. Mouse Hox-2.2 specifies thoracic segmental identity in Drosophila embryos and larvae. *Cell* 63:961–67

Marshall H, Nonchev S, Sham, M-H, et al. 1992. Retinoic acid alters the hindbrain *Hox* code and induces the transformation of rhombomeres 2/3 into a rhombomere 4/5 identity. *Nature* 360:737–41

Martinez S, Geijo E, Sanchez-Vives MV, et al. 1992. Reduced junctional permeability at interrhombomeric boundaries. *Development* 116:1069–76

Mavilio F, Simeone A, Giampaolo A, et al. 1986. Differential and stage-related expression in embryonic tissues of a new human homeo box gene. *Nature* 324:664–67

McGinnis W, Krumlauf R. 1992. Homeobox genes and axial patterning. *Cell* 68:283–302

McGinnis N, Kuziora M, McGinnis W. 1990. Human Hox-4.2 and Drosophila Deformed encode similar regulatory specificities in Drosophila embryos and larvae. *Cell* 63: 969–76

Morriss-Kay G. 1993. Retinoic acid and craniofacial development: molecules and morphogenesis. *BioEssays* 15:9–15

Morriss-Kay G, Murphy P, Hill R, Davidson D. 1991. Effects of retinoic acid on expression of Hox 2.9 and Krox 20 and on morphological segmentation in the hindbrain of mouse embryos. *EMBO J.* 10:2985–96

Murphy P, Hill RE. 1991. Expression of mouse *labial*-like homeobox-containing genes, *Hox 2.9* and *Hox 1.6*, during segmentation of the hindbrain. *Development* 111:61–74

Murphy P, Davidson DR, Hill RE. 1989. Segment-specific expression of a homeobox-containing gene in the mouse hindbrain. *Nature* 341:156–59

Nieto A, Gilardi-Hebenstreit P, Charnay P, Wilkinson D. 1992. A receptor protein tyrosine kinase implicated in the segmental patterning of the hindbrain and mesoderm. *Development* 116:1137–50

Noden DM. 1983. The role of the neural crest in patterning of avian cranial skeletal, connective, and muscle tissues. *Dev. Biol.* 96: 144–65

Noden DM. 1988. Interactions and fates of avian craniofacial mesenchyme. *Development* 103(Suppl.):121–40

Nose A, Mahajan VB, Goodman CS. 1992. Connectin: A homophilic cell adhesion molecule expressed on a subset of muscles and the motoneurons that innervate them in *Drosophila. Cell* 70:553–67

Orr H. 1887. Contribution to the embryology of the lizard. *J. Morphol.* 1:311–72

Otting G, Qian YQ, Billeter M, et al. 1990. Protein—DNA contacts in the structure of a homeodomain—DNA complex determined by nuclear magnetic resonance spectroscopy in solution. *EMBO J.* 9:3085–92

Otting G, Qian YQ, Muller M, et al. 1988. Secondary structure determination for the Antennapedia homeodomain by nuclear magnetic resonance and evidence for a helix-turn-helix motif. *EMBO. J.* 7:4305–9

Papalopulu N, Lovell-Badge R, Krumlauf R. 1991. The expression of murine *Hox-2* genes is dependent on the differentiation pathway and displays collinear sensitivity to retinoic acid in F9 cells and *Xenopus* embryos. *Nucleic Acids Res.* 19:5497–506

Pollock RA, Jay G, Bieberich CJ. 1992. Altering the boundaries of *Hox-3.1* expression: evidence for antipodal gene regulation. *Cell* 71:911–24

Porteus MH, Bulfone A, Ciaranello RD, Rubenstein JLR. 1991. Isolation and characterisation of a novel cDNA clone encoding a homeodomain that is developmentally regulated in the ventral forebrain. *Neuron* 7: 221–29

Price M, Lazzaro D, Phol T, et al. 1992. Regional expression of the homeobox gene *Nkx-2.2* in the developing mammalian forebrain. *Neuron* 8:241–55

Price M, Lemaistre M, Pischetola M, et al. 1991. A mouse gene related to *Distal-less* shows a restricted expression in the developing forebrain. *Nature* 351:748–51

Qian YQ, Billeter M, Otting G, et al. 1989. The structure of the Antennapedia homeodomain determined by NMR spectroscopy in solution: comparison with prokaryotic repressors. *Cell* 59:573–80

Ramirez-Solis R, Zheng H, Whiting J. 1993. *Hox-B4* (*Hox- 2.6*) mutant mice show homeotic transformation of a cervical vertebra and defective closure of the sternal rudiments. *Cell* 73:279–94

Robinson G, Wray S, Mahon K. 1991. Spatially restricted expression of a member of a new family of murine *Distal-less* homeobox genes in the developing forebrain. *New Biol.* 3:1183–94

Roelink H, Nusse R. 1991. Expression of two members of the *Wnt* family during mouse

development - restricted temporal and spatial patterns in the developing neural tube. *Genes Dev.* 5:381–88

Ruiz Altaba A, Jessell T. 1991a. Retinoic acid modifies mesodermal patterning in early *Xenopus* embryos. *Genes Dev.* 5:175–87

Ruiz Altaba A, Jessell T. 1991b. Retinoic acid modifies the pattern of cell differentiation in the central nervous system of neurula stage *Xenopus* embryos. *Development* 112:945–58

Ruiz i Altalba A. 1992. Planar and vertical signals in the induction and patterning of the *Xenopus* nervous system. *Development* 116: 67–80

Schughart K, Kappen C, Ruddle F. 1989. Duplication of large genomic regions during the evolution of vertebrate homeobox genes. *Proc. Natl. Acad. Sci. USA* 86:7067–71

Scott MP. 1992. Vertebrate homeobox gene nomenclature. *Cell* 71:551–53

Scott MP, Tamkun JW, Hartzell GW, 1989. The structure and function of the homeodomain. *Biochim. Biophys. Acta* 989:25–48

Sechrist J, Serbedzija GN, Scherson T, et al. 1993. Segmental migration of the hindbrain neural crest does not arise from its segmental generation. *Development* 118:691–703

Sham MH, Vesque C, Nonchev S, et al. 1993. The zinc finger gene *Krox-20* regulates *Hox-B2* during hindbrain segmentation. *Cell* 72: 183–96

Simeone A, Acampora D, Arcioni L, et al. 1990. Sequential activation of HOX2 homeobox genes by retinoic acid in human embryonal carcinoma cells. *Nature* 346: 763–66

Simeone A, Acampora D, Gulisano M, et al. 1992a. Nested expression domains of four homeobox genes in developing rostral brain. *Nature* 358:687–90

Simeone A, Acampora D, Mallamaci A, et al. 1993. A vertebrate gene related to orthodenticle contains a homeodomain of the bicoid class and demarcates anterior neuroectoderm in the gastrulating mouse embryo. *EMBO J.* In press

Simeone A, Acampora D, Nigro V, et al. 1991. Differential regulation by retinoic acid of the homeobox genes of the four HOX loci in human embryonal carcinoma cells. *Mech. Dev.* 33:215–27

Simeone A, Gulisano M, Acampora D, et al. 1992b. Two vertebrate homeobox genes realted to the *Drosophila empty spiracles* gene are expressed in the embryonic cerebral cortex. *EMBO J.* 11:2541–50

Sive H, Cheng P. 1991. Retinoic acid perturbs the expression of *Xhox.lab* genes and alters mesodermal determination in *Xenopus laevis. Genes Dev.* 5:1321–32

Sive H, Draper B, Harland R, Weintraub H. 1990. Identification of a retinoic acid-sensitive period during primary axis formation in *Xenopus laevis*. *Genes Dev*. 4:932–42

Spemann H. 1938. *Embryonic Development and Induction*. New York: Hafner

Stern C, Jaques K, Lim T, et al. 1991. Segmental lineage restrictions in the chick embryo spinal cord depend on the adjacent somites. *Development* 113:239–44

Sundin O, Eichele G. 1990. A homeo domain protein reveals the metameric nature of the developing chick hindbrain. *Genes Dev*. 4: 1267–76

Sundin O, Eichele G. 1992. An early marker of axial pattern in the chick embryo and its respecification by retinoic acid. *Development* 114:841–52

Treacy MN, Rosenfeld MG. 1992. Expression of a family of POU-domain protein regulatory genes during development of the central nervous system. *Annu. Rev. Neurosci*. 15: 139–65

Vaage S. 1969. The segmentation of the primitive neural tube in chick embryos (*Gallus domesticus*). *Adv. Anat. Embryol. Cell Biol*. 41:1–88

Wagner M, Thaller C, Jessell T, Eichele G. 1990. Polarising activity and retinoid synthesis in the floor plate of the neural tube. *Nature* 345:819–22

White RAH, Brookman JJ, Gould AP, et al. 1992. Targets of homeotic gene regulation in *Drosophila*. *J. Cell Sci. Suppl*. 16:53–60

Wilkinson DG, Bhatt S, Chavrier P, et al. 1989a. Segment-specific expression of a zinc finger gene in the developing nervous system of the mouse. *Nature* 337:461–64

Wilkinson DG, Bhatt S, Cook M, et al. 1989b. Segmental expression of hox 2 homeobox-containing genes in the developing mouse hindbrain. *Nature* 341:405–9

Annu. Rev. Neurosci. 1994. 17:133–51

THE EARLY REACTIONS OF NON-NEURONAL CELLS TO BRAIN INJURY

Dennis M. D. Landis

Departments of Neurology and Neurosciences, School of Medicine, Case Western Reserve University, Cleveland, Ohio 44106

KEY WORDS: astrocyte, macrophage, endothelial, potassium, cell swelling, brain homeostasis, microglia

INTRODUCTION

One of the great goals in neuroscience research has been to improve the recovery of the mammalian central nervous system after damage. Considerable effort has gone into learning about changes that occur in the brain after injury, and the extent to which these changes are protective or injurious in themselves. This review deals with the responses of non-neuronal cells in the central nervous system that occur from the first seconds to hours after brain injury. This is a very large topic, and it has been necessary to select only some aspects for consideration. Emphasis is placed on three cell populations: astrocytes, brain macrophages, and endothelial cells. For each cell type, we can distinguish responses related to properties existing before the lesion, and responses reflecting the expression of new properties. Where possible, interactions between these cell types are discussed, but the complex and important interactions with the immune system are mentioned only briefly. Different classes of injury are associated with different arrays of cellular responses; the injury of ischemic stroke is discussed in sufficient detail to emphasize the complexity and interactions of cellular responses. Only some of the cellular responses to injury have obvious benefits, and others may fail to support or may inhibit recovery.

133

ASTROCYTES

Astrocytes normally have important roles in maintaining the composition of the extracellular fluid of the brain, and the initial astrocytic responses to injury are closely related to these normal roles. The unusual demands of injury may overwhelm the capacity of astrocytes to restore homeostasis and may trigger astrocytic changes that exacerbate the original injury. It is convenient in description to distinguish *immediate responses* (those utilizing preexisting cellular properties), *early responses* (those involving modification of preexisting properties and acquisition of new ones), and *late responses* (such as change in cell shape, migration, or proliferation).

Astrocytes Are Not All the Same

Astrocytic processes partially or entirely invest all neuronal structures, and entirely invest all vascular structures, but the precise distribution of astrocytic processes and the nature of the functions they subserve probably vary from brain region to brain region. For example, the shape of astrocytes in white matter is quite different from that of astrocytes in the vicinity of neurons. Heterogeneity in shape and structure of astrocytes in adult brain probably reflects the fact that several cell lineages contribute to the cell population. At least two lineages have been recognized in the optic nerve, and several more are likely to exist in spinal cord (Miller et al 1989, Miller & Szigeti 1991). Despite the range in morphology, several homeostatic functions are presumed to be performed by all astrocytes, especially those related to the normal regulation of the composition of the brain's extracellular fluid.

In both gray and white matter, astrocyte membrane composition is not uniform over the cell surface. Freeze fracture techniques have shown that astrocytic processes investing vascular structures and forming the surface of the brain have high concentrations of a peculiar specialization of intramembrane structures, termed *assemblies* (Landis & Reese 1974, Landis & Reese 1981b, Landis & Reese 1982). Fewer assemblies are present on astrocytic processes investing cell bodies and synapses, and virtually none are present on the cell body. Astrocytes are thus polarized, with a membrane composition that varies with the nature of the adjacent cellular process. It is important to note that astrocytic processes investing vascular structures do not constitute the blood-brain barrier; that barrier is imposed by tight junctions linking endothelial cells. Astrocytes are extensively interconnected by gap junctions, and so the myriad astrocytic processes in the brain approximate a syncytium through which ions and small molecules may move freely.

Astrocyte Functions

Three normal homeostatic functions carried out by astrocytes have special importance because of their relation to the initial responses of astrocytes in the context of brain injury: K^+ homeostasis, CO_2 metabolism, and neurotransmitter metabolism. In each instance, loss or distortion of the function in damaged brain may actually extend the initial damage.

POTASSIUM HOMEOSTASIS   *Normal potassium homeostasis*   In normal brain, astrocytes take up K^+ from regions of extracellular space where the concentration is relatively high; they then redistribute the K^+ through their cytoplasmic volume, including the syncytium formed by gap junctions. K^+ leaves the astrocytic compartment to rejoin the extracellular space in regions where extracellular K^+ concentration is relatively low (Gardner-Medwin 1981, Gardner-Medwin 1983). This "spatial buffering" of K^+ concentration in the brain's extracellular space occurs in addition to K^+ diffusion in the extracellular space.

If neuronal activity is diffusely increased, K^+ concentration may be increased over a relatively large volume of extracellular space, and redistribution by spatial buffering may be insufficient to restore normal K^+ concentrations. In such a situation, another astrocytic mechanism may emerge: the movement of K^+ from the extracellular space into the vascular compartment (Paulson & Newman 1987). In this instance, K^+ exits from the astrocytic compartment via perivascular astrocytic processes. It is taken up by endothelial cells, moved across their cytoplasm, and dumped into the vascular space (Goldstein & Betz 1983). Presumably, K^+ lost from the brain via this mechanism of "potassium siphoning" is replaced in cerebrospinal fluid secreted by the choroid plexus.

Another aspect of "potassium siphoning" is pertinent to normal metabolism and to the initial response after injury. If the concentration of extracellular K^+ increases in the vicinity of brain arteriolar smooth muscle, then the smooth muscle relaxes and blood flow is increased (Paulson & Newman 1987). Neuronal activity is thus transduced into a K^+ signal, and the vasculature responds by increasing flow in tissue where neurons are active.

Disordered potassium homeostasis   In pathological circumstances, the K^+ load from increased extracellular concentrations may be more than spatial buffering and siphoning can manage. Experimentally, this might be induced by applying concentrated KCl to a cortical surface. In human pathology, one aspect of an intracerebral hemorrhage is the release of large amounts of K^+ from ruptured erythrocytes. In either setting, astrocytes swell. This is presumably due to passive flux of K^+ through astrocytic channels, as in normal buffering or siphoning, plus K^+ and Cl^- entry via separate voltage-gated channels (Giulian & Lachman 1985), KCl co-transport, and possibly Na^+, K^+, Cl^- co-transport as well. Astrocytic swelling per se may alter astrocyte function—for example, by causing the release of intracellular glutamate (Kimelberg et al 1990).

Potassium metabolism also might be disrupted by anything which will uncouple gap junctions. Both spatial buffering and siphoning operate through the syncytium of astrocytic processes. If gap junction communication is disrupted, the isolated astrocytes will swell if they have insufficient apposition to vascular elements to dump K^+. It is not clear, though, that an isolated loss of gap junction communication actually occurs. A more realistic and worse alteration of K^+ homeostasis would be a combination of uncoupling of gap junctions and cessation of local blood flow. In this instance, cell-to-cell syncitial communication would be lost, and the isolated cells would be unable to dump their K^+ into the vascular sink.

CO_2 METABOLISM   *Normal CO_2 metabolism*   Neuronal metabolism is normally aerobic. The CO_2 formed by normal neuronal metabolism diffuses into astrocytic processes and is converted to bicarbonate (and a proton) by carbonic anhydrase (Kimelberg & Bourke 1982). The bicarbonate diffuses to the vicinity of blood vessels, and there it is exchanged for extracellular Cl^-. The proton similarly moves and is exchanged for extracellular Na^+. Outside the cell, CO_2 is regenerated and diffuses into endothelial cells and then into the vascular space. The system of bicarbonate transport may have evolved in part to defend extracellular pH. If CO_2 produced by neuronal metabolism simply diffused to the vascular space, there would be a gradient from origin to vessel, and an accompanying pH gradient. The role of astrocytes in bicarbonate metabolism serves to maintain a relatively uniform extracellular pH.

Pathological changes   In the bicarbonate metabolism described above, the proton and bicarbonate ions are exchanged for a Na^+ and a Cl^- ion. The Na^+ is extruded by the Na^+, K^+-ATPase. If, however, energy supply becomes limited and ATP declines, the activity of the Na^+, K^+-ATPase may decrease to the point that intracellular Na^+ accumulates. This osmotically active ion is accompanied by water, so there is a commensurate increase in cell volume. Note that the carbonic anhydrase is not ATP-dependent and so can continue to add protons and bicarbonate, while the Na^+, K^+-ATPase is unable to compensate. Again, astrocytic swelling may be damaging itself. Extensive swelling, for example, may lead to local increase in tissue pressure, and a local decrease in vascular perfusion.

NEUROTRANSMITTER METABOLISM   *Normal uptake and responses*   Astrocytes take up several neurotransmitters, including glutamate and GABA. Astrocytic uptake is probably especially important in glutamatergic neurotransmission, in which it serves to remove glutamate from the vicinity of the synapse after desensitization has ended the burst of ligand-gated channel openings.

Astrocytes do more than simply mop up neurotransmitters in the extracellular space, however. Increasing evidence indicates that astrocytes respond to transmitters. Glutamate analogs, for example, appear to depolarize astrocytes via ligand-gated receptors (Bowman & Kimelberg 1984, Kettenmann et al 1984, Tang & Orkand 1986). Glutamate receptors on astrocytes are not activated by NMDA (Giulian et al 1993). There are $GABA_A$-activated chloride channels in cortical astrocytes (Giulian & Baker 1985, Kettenmann et al 1984). Catecholamines evoke changes in second messenger systems, such as cAMP (Hirata et al 1983). These astrocytic responses to neurotransmitter are probably not involved in information processing but may be integral to the coordination of astrocytic and neuronal metabolism.

Pathology resulting from altered uptake   Uptake of glutamate by astrocytes is a Na^+-dependent process, in which Na^+ and glutamate enter and K^+ is extruded (Brew & Attwell 1987). If extracellular K^+ is increased, uptake of glutamate is inhibited, because the gradient favoring K^+ extrusion is decreased (Barbour et al 1988). The glutamate transporters utilized by astrocytes appear to be members of a new gene family with Na^+ and K^+ ionic dependence (Pines et al 1992, Storck et al 1992). Recently, physiological studies of glutamate transport in salamander retinal glial cells have found that the uptake of a single molecule of glutamate is accompanied by influx of two Na^+ ions, efflux of a single K^+ ion, and efflux of a OH^- ion (Bouvier et al 1992). This stoichiometry has special significance in the setting of anoxia, in which the combination of high extracellular K^+, decreased Na^+, and altered pH in the extracellular space may inhibit astrocytic uptake of glutamate to the extent that the extracellular concentration may rise to as much as $370\mu M$ (from a normal of $0.6\mu M$) and precipitate the well-described excitotoxic cascade (Choi 1988). Neurons with NMDA-type glutamate receptors suffer a prolonged and abnormal influx of ionized calcium, which leads to cell death if sufficiently severe.

Persistent increases in the concentration of extracellular glutamate may directly injure the astrocyte. Glutamate applied to cultured astrocytes causes transient increases in intracellular ionized calcium, recruited from intracellular stores and from the extracellular space (Cornell-Bell et al 1990). Whether this may be toxic in the setting of injury is uncertain, but it is likely that the astrocytic vulnerability to extracellular glutamate is substantially less than that of neurons.

Initial Responses

The initial astrocytic responses to injury are simply normal astrocytic properties operating in an abruptly altered environment. If the injury is accompanied by increase in extracellular K^+, astrocytic spatial buffering and siphoning mechanisms will attempt to restore normal concentrations—if this fails, local blood flow regulation will also be lost, and there will be astrocytic swelling. If the local oxygen supply declines, the continued operation of bicarbonate formation in astrocytes will eventually lead to astrocytic swelling. The nature of glutamate transport is such that when extracellular K^+ is increased, and neurons therefore depolarized, astrocytes are least able to take up glutamate and so may fail to prevent excitotoxicity.

ASTROCYTIC SWELLING   Several processes result in swelling of astrocytes in the first seconds to minutes of injury:

1. Potassium uptake
   passive uptake
   via voltage-gated channels
   KCl co-transport
   Na^+, K^+, Cl^- co-transport
   active uptake (Na^+, K^+-ATPase)
2. Sodium uptake
   Na^+/H^+ antiport
3. Glutamate uptake
4. Lactate uptake

These should all be regarded as normal processes, occurring to an abnormal extent in the injured tissue. All are osmotically active, and so the swelling reflects water entering with the solute.

Freeze fracture electron microscopic methods have revealed that distortion of an intramembrane particle (assemblies) may be correlated with lack of oxidative phosphorylation, and astrocytic swelling (Landis & Reese 1981a). This change in membrane structure might represent the formation of a nonselective cation channel; the effect would be additive to those listed above.

Astrocytic swelling has at least two significant sequelae. First, astrocytic swelling in a region of the brain can increase the volume of the tissue, which may distort it grossly and contribute to mechanical injury. If the increase in volume is great enough, intracerebral pressure will increase, and vascular perfusion may decrease. The second class of effect, though, may be on the astrocyte itself. Swelling may cause the release of glutamate and lactate, exacerbating the metabolic disarray in the injured tissue (Kimelberg et al 1990), and may also cause the influx of ionized calcium, leading in the worst instance to astrocytic destruction.

ENERGY METABOLISM   Glycogen is present in astrocytic cytoplasm. Astrocytes probably break down glycogen and provide glucose for neuronal and astrocytic metabolism when blood supply of energy substrates is limited. When oxygen supply declines, astrocytes can switch to glycolytic metabolism. This creates lactate in astrocytic cytoplasm, and it is likely that the intracellular pool equilibrates with lactate in the extracellular space via a membrane transporter.

The switch to anaerobic glycolysis has significant effects on cell pH (Kraig & Chesler 1990) and on cell volume (Staub et al 1990). Two lactate molecules are generated per glucose consumed, resulting in an increase in cell osmolality. This, in turn, contributes to astrocytic swelling. Some of the protons generated by glycolytic metabolism render the cytoplasm acidotic, while others are transported via a Na^+/H^+ antiport exchanger to the extracellular space. The Na^+ accumulation contributes to cell swelling, and the proton extrusion contributes to acidosis in the extracellular fluid.

Early Changes

Early changes refer to astrocytic responses that involve the modification of pre-existing cell properties or the acquisition of new properties. These emerge within hours of injury.

EXPRESSION OF GFAP   The most extensively studied aspect of astrocytic early changes is an increase in GFAP (glial fibrillary acidic protein) immunoreactivity, especially as detected with immunocytochemical methods (Lindsay 1986). Despite a large number of investigations, the regulation of this response and its functional significance are incompletely understood. Mechanical injury to adult brain almost invariably results in increased GFAP immunoreactivity locally, within one to three days of the injury. The astrocytic change in that setting may reflect in part responses to blood-borne factors or factors released by macrophages. However, an increase also occurs in GFAP immunoreactivity in astrocytic processes in the dorsal horn of the spinal cord after sciatic nerve lesions. This change in GFAP expression in the absence of local axonal degeneration suggests that changes in axonal function may be sufficient to elicit the astrocytic response (Murray et al 1990).

Even with focal injury, the volume of brain containing astrocytes with increased GFAP immunoreactivity is large. This may be taken as evidence of communication among astrocytes, perhaps mediated in part by the gap junction syncytium. A focal cortical injury may evoke altered GFAP expression among cortical astrocytes at some distance, but less change in the astrocytes in the underlying white matter. Thus the increased expression of GFAP is not simply a response to a diffusable signal in the extracellular space, because such a diffusable signal would be expected to reach deep into the white matter. Alternatively, GFAP regulation, or more generally the response to injury, may be different in white matter astrocytes.

Altered GFAP expression is easily detected, but its functional significance is still uncertain. It is often assumed that this increase in the expression of intermediate filaments is related to preserving or enhancing structural integrity of brain tissue. Increased GFAP immunoreactivity can occur in the absence of change in astrocyte cell number or perikaryonal size (Wells et al 1992).

EXPRESSION OF MHC PROTEINS   Astrocytes in normal brain do not express the cell surface proteins associated with the major histocompatibility genes. Type I and type II antigens may appear following injury, especially that associated with viral infection (Olsson et al 1987, Suzumura et al 1986). The type II antigens (Ia in rodents) can function in the presentation of processed antigen to competent lymphocytes in vitro (Fontana et al 1984), and they may have a similar role in the setting of injury.

The precise stimuli that lead to the expression of type I and type II MHC antigens following injury in vivo have yet to be defined, but there is evidence that a variety of cytokines (including interferon-γ, interleukins, tumor necrosis factors α and β, and transforming growth factor β-1) can modulate MHC expression in cultured cells (Hertz et al 1990, Johns et al 1992).

Late Changes

Late astrocytic responses emerge hours to days after the injury. While this review has been largely restricted to earlier responses, the late changes deserve mention because they have a special clinical importance. It may be hard to find tools that modify the immediate and early astrocytic changes in the initial seconds to hours after injury, but late changes may be sufficiently delayed in their onset that one can hope to influence them.

HYPERTROPHY   Astrocytes surviving the initial injury may increase in size, including perikaryon and processes. Usually such increases in cell size are detected with methods such as GFAP immunoreactivity, and the cell size is presumed to be related to a role in maintaining brain structure. In some instances, though, increased cell size may be related to a requirement for increased astrocytic metabolic activity, such as the hypertrophy seen in association with hepatic encephalopathy.

PROLIFERATION   After an injury that destroys neuronal tissue, an increase occurs in the proportion of the residual brain occupied by astrocytes. This reflects the loss of the neuronal volume and may reflect both hypertrophy of preexisting astrocytes and the proliferation of astrocytes. Double-labeling methods have been used to show that cells with astrocytic characteristics (e.g. GFAP immunoreactivity) undergo mitosis after brain injury (Latov et al 1979, Miyake et al 1988, 1992).

The signal for astrocytic division after injury is still uncertain, and the nature of the dividing cell is not yet known. Astrocytes may arise in injured brain from the division of previously differentiated astrocytes, from the division of a precursor population, or from both. Astrocytes (or astrocytic precursors) probably proliferate at low rates in normal adult brain; this is likely to be accompanied by a balancing rate of astrocyte cell death (Korr 1986). There is virtually no information about the cessation of division after injury, except that astrocytes in injured brain do not continue to divide to the point that their volume exceeds that of the original local brain volume. Similarly, the rate of astrocyte cell death after injury (not attributable to the injury itself) has yet to be defined.

MIGRATION   Cultured astrocytes can be injected into the brains of normal recipients and will migrate throughout the brain (Goldberg & Bernstein 1988a, Goldberg & Bernstein 1988b, Jacque et al 1986). They seem to move preferentially in parallel with the course of axons in white matter tracts and along the brain or spinal cord surface. Whether the same phenomenon occurs in injured brain is still unknown. In organotypic cultures, astrocytes are generated in the vicinity of the ventricular zone and migrate throughout the slice to form an abnormal glial limitans (Del Rio et al 1991). This migration probably does not follow the pathways that support normal development in the hippocampus, but the migration provides indirect evidence that astrocytes in an abnormal environment are able to migrate.

Roles in Regeneration

Astrocytes have important roles in guiding neuronal migration and growing axons in the normal development of the mammalian central nervous system. This capacity, though, seems to disappear when cellular movements and initial connectivity are established. For example, axons normally grow into the spinal cord from neurons in dorsal root ganglia. If, however, axon ingrowth is delayed experimentally, one can define an age after which the axons of dorsal ganglia neurons cannot penetrate into the spinal cord. The growing tips of the axons appear to stop in the vicinity of their initial contact with astrocytic processes. This may indicate that normal astrocytes in mature tissue do not support axon ingrowth, or that they contribute to actively inhibit that ingrowth.

Similarly, studies of injured spinal cord in adult animals consistently show a virtually complete failure of axons to regrow past the astrocytes persisting at the site of the original injury. This has contributed to the idea that reactive astrocytes are inimical to brain recovery (Liuzzi & Lasek 1987). The idea has never been fully tested, though, because techniques for selective removal of astrocytes reacting to injury have not been available.

On the other hand, astrocytes in injured brain have the capacity to make several of the NGF family of neurotrophic factors. These glia-derived factors may be very important in limiting neuronal cell death in the context of certain classes of injury, especially axon transection.

MACROPHAGES

Cells with the capacity for phagocytosis of cellular debris are present in normal nervous system, and this capacity is promptly expressed in the setting of injury, which results in cell death or the degeneration of cell processes. In addition, widespread damage can attract the blood-derived macrophages. Brain macrophages responding to injury acquire type I and II MHC markers and so are probably competent to interact with cells of the immune system. The influx of blood-derived macrophages and the activation of resident macrophages in the brain may have the undesirable consequence of nonselective killing of "bystander" cells.

Microglia and Perivascular Macrophages

The term "ramified microglia" has been applied to macrophages with a characteristic shape in brain parenchyma. Studies of developing animals suggest that these microglial cells are initially derived from a wave of invading blood-derived macrophages (Ling & Wong 1993). Microglia, though, may be only one component of the brain macrophage population. There is a population of perivascular macrophages, separated from the brain parenchyma by a basal lamina. Immunocytochemical probes indicate that perivascular macrophages have an immunophenotype distinct from that of microglia (Graeber et al 1990, Graeber et al 1989, Streit et al 1989). These perivascular macrophages are derived from blood-borne monocytes which enter and exit the brain throughout adult life (Hickey & Kimura 1988), but which do not appear to become ramified microglia. Perivascular macrophages are competent to serve as antigen-presenting cells in experimental autoallergic encephalomyelitis (Hickey & Kimura 1988, Lassmann et al 1993), and so may be a population specialized for interaction with the immune system.

Following focal injury such as injection of the excitotoxin kainic acid, there is a prompt and widespread change in the ramified microglia. The reactive microglia stain more intensely with the *Griffonia simplicifolia* isolectin B4, and the cells change shape to become globular or to develop pseudopodia; these changes occur before blood-borne monocytes invade the lesion and its surround (Marty et al 1991). The activated microglia may have the capacity for mitosis (Streit & Kreutzberg 1988), though this is difficult to prove in the absence of markers that clearly distinguish activated microglia and blood-derived monocytes.

Injury and Blood-Derived Macrophages

After injury that creates cellular debris, blood-derived macrophages enter the brain and presumably contribute to the phagocytosis and removal of debris. The brain macrophages thus derive from three distinct populations: ramified microglia present in brain parenchyma since the early postnatal period, perivascular macrophages that had taken up position prior to injury, and monocytes that enter the brain from the blood after injury. It becomes difficult to distinguish the populations in the context of injury, because microglia may change shape and may migrate to the site of extracellular debris in a fashion indistinguishable from that of the blood-derived populations.

Activated microglia may contribute to the signaling that attracts blood-derived macrophages. Activated microglia can release cytokines such as interleukin 1, which in turn may act on endothelial cells to increase their adhesion to circulating macrophages and leukocytes (Pober & Cotran 1990).

Brain macrophages certainly have the capacity to take up cellular debris but may also have the capacity to injure normal cells or to kill them (Giulian et al 1993). The capacity for damaging apparently normal cells is evident in macrophages that strip axonal boutons from the surface of chromatolytic neurons—in this instance, the postsynaptic cell had been altered by axotomy, but it is the presynaptic axons that are destroyed by the macrophages (Blinzinger & Kreutzberg 1968). Brain macrophages can lyse oligodendrocytes in vitro, apparently by the generation of free radicals and induction of lipid peroxidation in the oligodendrocytes (Merrill & Zimmerman 1991).

Signaling by Brain Macrophages

The expression of type I and type II MHC antigens by reactive microglia and blood-derived macrophages in the brain is probably one aspect of the interactions of these cells with cells of the immune system. Brain macrophages may also release substances that influence astrocytic and/or endothelial cell properties involved in their responses to brain injury (Giulian & Baker 1985, Giulian et al 1986, Giulian & Lachman 1985, Giulian et al 1993, Giulian & Young 1986).

ENDOTHELIAL CELLS

The role of vascular endothelial cells in the blood-brain barrier is well established: The tight junctions between endothelial cells form the barrier to movement of ions and solutes in the extracellular space from blood to brain, and from brain to blood (Brightman & Reese 1969). Further, carrier-mediated transport systems in endothelial cells form the basis of selective movement of sugars and amino acids from blood to brain, down their concentration gradients.

Brain endothelial cells may also take part in the formation of brain extracellular fluid, perhaps by transport of Na^+ (Betz & Goldstein 1986) (this might also account for the large mitochondrial capacity). Surgical extirpation of the choroid plexus reduces but does not eliminate cerebrospinal fluid production (Milhorat et al 1971). There appears to be fluid formation at the level of capillaries and, even in normal situations, a bulk flow from extracellular space in brain parenchyma to the cerebrospinal fluid in the ventricular system (Cserr 1988, Cserr et al 1977).

Endothelial cells in the brain may have the capacity to alter vascular permeability: Permeability is increased following perfusion with fluid of high osmotic activity and subsequently is restored to normal.

Expression of cell surface proteins on the lumenal membrane of endothelial cells may have a critical role in attracting the attention of circulating immune cells and supporting their entry into brain parenchyma.

Ion Homeostasis and Permeability Changes

As discussed above, brain endothelial cells have a high concentration of Na^+, K^+-ATPase on the ablumenal membrane (Betz 1986), and so they are able to move extracellular K^+ from the brain extracellular space into the endothelial cytoplasm, and then into the vascular lumen. Mitochondrial volume in cerebral endothelial cells has been estimated to be about four times that of the mitochondrial volume in non-neural endothelial cells (Oldendorf et al 1977). The comparatively large volume of mitochondria presumably reflects high energy requirements, perhaps including the Na^+, K^+- ATPase. In the setting of injury involving loss of oxygen, endothelial energy metabolism may be insufficient to support the continued uptake of K^+. This would provide yet another block in the removal of K^+ from the brain extracellular space and might contribute to continued neuronal depolarization. It is a peculiar feature of endothelial cells in adult brain that they lack gap junctions. Potassium entering one cell has to be handled by that cell and is not redistributed to adjacent endothelial cells.

Mechanical or hypoxic injury sufficient to kill endothelial cells will obviously result in opening of the blood-brain barrier. It has been difficult to determine whether reversible endothelial cell injury can be associated with loss and then recovery of normal permeability function. In normal brain, transient perfusion of the vascular system with a solution containing high osmotic activity will result in a loss of the blood-brain barrier to extracellular solutes. This change in permeability is short-lived. It has been viewed as an "opening" of the tight junction system and then "resealing," but the actual mechanism of the permeability change may be a bypass of the tight junction system by transcellular channels. These experiments reveal the capacity of endothelial cells to recover and reestablish in minutes to hours normal vascular permeability.

The permeability characteristics of endothelial cells in brain vasculature may be influenced by an interaction with astrocytes. For example, the endothelial cells in the vasculature of metastatic tumors do not form a continuous belt of tight junctions, and the vessels are therefore leaky to extracellular solutes. The endothelial cells in these tumors are presumably derived from brain endothelial cells, but they are surrounded by tumor cells, not astrocytes. Similarly, endothelial cells in poorly differentiated glial tumors may also fail to establish normal brain-like permeability characteristics in the absence of nearby normal astrocytes. At least one set of experiments suggests that rodent astrocytes can induce brain-like permeability in surrounding vessels when transplanted to chick allantoic membrane (Janzer & Raff 1987). Transplantation of embryonic brain tissue to the abdominal cavity (of chicks) results in the invasion of the neural tissue by mesoderm-derived endothelial cells, but these cells acquire all the permeability characteristics of the blood-brain barrier (Stewart & Wiley 1981). Moreover, astrocyte conditioned medium in the presence of an inducer of cAMP will support the formation of tight junctions by cultured brain-derived endothelial cells. One aspect of the astrocytic response to injury in adult brain may be to support the recovery of the blood-brain barrier permeability characteristics of endothelial cells.

Astrocyte interactions with endothelial cells may involve specific factors (Goldstein 1988). Endothelial cells make platelet-derived growth factor, which appears to act on astrocytes but not on endothelial cells (Heldin et al 1981). Astrocytes may produce endothelial cell growth factor, which may act on endothelial cells to influence proliferation, and also on astrocytes, to influence differentiation (Morrison et al 1985). The mitogenic effect on endothelial cells of endothelial cell growth factor can be antagonized by transforming growth factor-β (Schroder et al 1986, Sporn et al 1986).

Cerebral endothelial cells are characterized by a high content of gamma-glutamyl transpeptidase activity. Co-culture of astrocytes and brain-derived endothelial cells will support the development of similar high activity in vitro (cell contact seems to be required, since conditioned medium is ineffective) (DeBault 1981).

Signaling by Endothelial Cells

In peripheral tissues, endothelial cells are dynamic components of the inflammatory response. For example, exposure to tachykinins will cause peripheral endothelial cells to increase local permeability to large blood-derived proteins. A similar process in vasculature of the dura may be associated with human migraine. However, there is no evidence that brain endothelial cells respond to circulating peptides by changing permeability characteristics.

Endothelial cells in brain can generate signals. The expression of a specific membrane-associated protein, for example, seems to be involved in the penetration into the brain of immune system cells and macrophages in the course of immune demyelination (Yednock et al 1992). Brain endothelial cells in the context of injury may similarly alter their surface proteins to promote local entry of circulating macrophages and immune cells. To some extent, this change in endothelial cell adhesion may be induced by cytokines released by activated microglia (Pober & Cotran 1990).

Modification of the Extracellular Matrix

After recovery from injury, the basal lamina around blood vessels may contain components not evident in normal brain. For example, laminin immuno-reactivity may be markedly increased in blood vessels in the vicinity of injury. The functional significance of this is still uncertain, as is the nature of the cell forming the laminin. The composition of the perivascular basal lamina and brain extracellular matrix components may have a role in the failure of neural regeneration after injury.

AN EXAMPLE OF BRAIN INJURY: ISCHEMIC STROKE

The nature and extent of the cellular responses to injury reflect the nature of the alterations in the cellular environment. Ischemic stroke is a complex and common injury, and it provides a useful example of injury that evokes many of the cellular responses detailed above.

A blood clot lodging at the bifurcation of a blood vessel will cause an abrupt cessation of blood flow in the vessels beyond that point. Arterial blood vessels in mammalian brain have anastomoses, so some of the tissue distal to the site of the embolic occlusion will be supported by anastomotic supply, while some of the tissue has little or no anastomotic supply. The damage caused by the clot is therefore not uniform—the ischemic insult will be worst in the center of the lesion, and less in a surround that receives some blood supply.

The most complex and fascinating cellular responses occur in the zone between normal tissue and the tissue simply destroyed by the complete cessation of blood supply.

In the first minute after block of blood supply, there is an abrupt increase in extracellular K^+ and a dramatic drop in cellular ATP. At least some of this early rise in K^+ comes from neuronal activity. Briefly, neurons continue to fire action potentials, and K^+ released during the repolarization phase accumulates in the extracellular space. The normal regulation of extracellular K^+ fails. Spatial buffering and siphoning are ineffective when there is a widespread, synchronous release of K^+ from the hypoxic neurons. Flux of K^+ through endothelial cells into the vascular lumen fails as the energy supply of endothelial cells becomes inadequate to support continued function of the Na^+,K^+-ATPase.

Until the supply of oxygen is exhausted, cells continue to generate CO_2, which enters the astrocytic compartment and is converted to bicarbonate. Removal of bicarbonate, however, becomes ineffective. There is no sink for CO_2 to diffuse into, because the blood flow becomes stagnant. Moreover, when lack of oxygen limits aerobic ATP production, astrocytes may fail to support sufficient Na^+, K^+-ATPase activity to remove the Na^+ and Cl^- which enters during normal CO_2 metabolism. The cell accumulates Na^+ and Cl^- and swells.

Ineffective clearance of K^+ from the extracellular space continues to depolarize neurons; synapses that utilize glutamate as a neurotransmitter will release it into the extracellular space. Unfortunately, the astrocytes are now inefficient at taking up the glutamate, and the transmitter accumulates. The combination of neuronal depolarization and persistent increase in extracellular glutamate concentration is sufficient to open the NMDA type of ligand-gated neuronal glutamate receptors, and it causes an influx of ionized calcium into the neuronal spines and dendrites. This, in turn, triggers the excitotoxic cascade of neuronal injury.

Normal brain tissue is supported by aerobic metabolism. When oxygen abruptly declines, there is a switch to glycolytic metabolism and a consequent rise in the production of lactate by astrocytes. Glycolytic production of lactate is accompanied by the generation of protons and a significant decline in intracellular pH. During ischemia a decline also occurs in extracellular pH, and movement of protons from the astrocytic compartment to the extracellular space via a Na^+/H^+ antiport exchanger may contribute to the extracellular acidosis.

The first minutes of injury thus result in anaerobic metabolism, decreased extracellular pH, increased extracellular K^+, and neuronal excitotoxicity. Astrocytes swell with accumulating K^+ and with accumulating Na^+ related to bicarbonate metabolism and to glycolytic production of lactate.

After the first minutes, depletion of oxygen and the metabolic disarray are sufficiently severe to kill cells in a central region of tissue. Surrounding this, though, are cells damaged, but not killed, by limited energy metabolism and abrupt metabolic loads. Here, the normal vascular response to extracellular K^+ is advantageous—smooth muscle relaxes, and resistance arterioles dilate to deliver as much blood supply as persisting vasculature can support. An equilibrium is established between continuing lactate production and the capacity of astrocytes to clear extracellular lactate. While gap junctions are certainly uncoupled in the center of the lesion, in the periphery they function to redistribute K^+ and deliver it to the vicinity of perfused vasculature, ultimately restoring normal extracellular K^+.

Early changes emerge in cells on the periphery of the lesion and extend outward from it. GFAP immunoreactivity is increased in astrocytes. Microglial macrophages respond to the presence of extracellular debris with shape change, migration, and phagocytosis. Endothelial cells may begin to express the cell surface signals that attract blood-derived macrophages. It is possible, but not yet clearly demonstrated, that endothelial cells damaged by the initial metabolic chaos and which consequently lost their barrier function are now able to recover and restore barrier function.

In the hours to days after the embolic occlusion of the blood vessel, the clot usually breaks up, and the block to blood flow is cleared. This, however, creates a surprising set of new problems. In areas of brain where the vasculature itself was damaged by the ischemic insult, restoration of blood flow and arterial blood pressure may result in rupture of damaged blood vessels, and extravasation of blood into the brain parenchyma. At the cellular level, however, restitution of blood flow improves K^+ clearance and all the energy-dependent astrocytic uptake mechanisms. Moreover, some of the stimulus to anaerobic metabolism is elimina d, and the extracellular lactate load decreases. Most of the macrophages in \ e dead tissue have been derived from circulating cells and return to the circulation after ingesting debris.

Forty-eight hours after the initial circulatory block, the damage is largely done. Astrocytes have begun to enlarge and proliferate. Endothelial cells, too, begin to proliferate, to restore the population diminished by the ischemic insult. Newly formed vessels, especially capillaries, acquire a basal lamina composition different from that of normal brain.

Why Is Recovery After Injury Incomplete?

The immediate, early, and later cellular responses discussed above have been presented from the perspective of restoration of normal brain homeostasis. After the injury, though, neuronal recovery is incomplete. In mature brain, there is no generation of new neurons by mitosis, even after injury. Recovery is therefore limited to restoration of connectivity where it has been interrupted, and resumption of normal function by transiently injured cells. Preexisting connectivity, though, is never fully restored. The problem is most obvious in the long axons of the spinal cord. Though the cells of origin and the target neurons may have been spared direct injury, transected axons fail to regrow through damaged spinal cord.

The failure of transected axons to regrow is not simply an inability to elongate. If suitable paths and substrates are provided, transected mammalian axons can grow comparatively long distances. From such observations, it would seem that one aspect of the failure of regrowth after damage is the absence of a cellular or extracellular environment that supports axon elongation. It has been suggested repeatedly that astrocytes reacting to injury are somehow involved in this failure to provide a supportive environment. If this is true, then removal or modification of reactive astrocytes holds enormous promise for restoration of function after injury.

The failure of axon regrowth is not due simply to some characteristics of reactive astrocytes. Proteins normally present on oligodendrocytes appear to decrease the extent of axon elongation through an injury (Schnell & Schwab 1990), an effect which would be additive to any property of reactive astrocytes. It is possible that positional and directional cues exist transiently during development and serve then to guide growing axons. Those guides are not available after injury, and axon regrowth may fail in their absence. Finally, it may be that the normal adult mammalian brain, including adult astrocytes, suppresses axon regrowth after the initial connections have been established during development. Substantial evidence suggests the capacity of neuronal cell bodies and dendrites to remodel in adult tissue, but none for the formation of new axonal connections. The absence of axon regrowth after injury may thus indicate that the normal brain environment has been efficiently restored.

ACKNOWLEDGMENT

Our work has been supported by NIH NS 22614 and a grant from the American Paralysis Association.

Literature Cited

Barbour B, Brew H, Attwell D. 1988. Electrogenic glutamate uptake in glial cells is activated by intracellular potassium. *Nature* 335: 433–35

Betz AL. 1986. Transport of ions across the blood-brain barrier. *Fed. Proc.* 45: 2050–54

Betz AL, Goldstein GW. 1986. Specialized properties and solute transport in brain capillaries. *Annu. Rev. Physiol.* 48:241–50

Blinzinger K, Kreutzberg G. 1968. Displacement of synaptic terminals from regenerating motoneurons by microglial cells. *Z. Zellforsch. Mikrosk. Anat.* 85:145–57

Bouvier M, Szatkowski M, Amato A, Attwell D. 1992. The glial cell glutamate uptake carrier countertransports pH-changing ions. *Nature* 360:471–74

Bowman CL, Kimelberg HK. 1984. Excitatory amino acids directly depolarize rat brain astrocytes in primary culture. *Nature* 311: 656–59

Brew H, Attwell D. 1987. Electrogenic gluta-

mate uptake is a major current carrier in the membrane of axolotl retinal glial cells. *Nature* 327:707–9

Brightman MW, Reese TS. 1969. Junctions between intimately apposed cell membranes in the vertebrate brain. *J. Cell Biol.* 40:648–77

Choi D. 1988. Glutamate neurotoxicity and diseases of the nervous system. *Neuron* 1: 623–34

Cornell-Bell AH, Finkbeiner SM, Cooper MS, Smith SJ. 1990. Glutamate induces calcium waves in cultured astrocytes: long-range glial signaling. *Science* 247:470–73

Cserr HF. 1988. Role of secretion and bulk flow of brain interstitial fluid in brain volume regulation. *Ann. NY Acad. Sci.* 529:9–20

Cserr HF, Cooper DN, Milhorat TH. 1977. Flow of cerebral interstitial fluid as indicated by the removal of extracellular markers from rat caudate nucleus. *Exp. Eye Res.* 25 (Suppl.):461–73

DeBault LE. 1981. Gamma-glutamyltranspeptidase induction mediated by glial foot process to endothelium contact in co-culture. *Brain Res.* 220:432–35

Del Rio JA, Heimrich B, Soriano E, et al. 1991. Proliferation and differentiation of glial fibrillary acidic protein-immunoreactive glial cells in organotypic slice cultures of rat hippocampus. *Neuroscience* 43:335–47

Fedoroff S, Vernadakis A, eds. 1986. *Astrocytes. Cell Biology and Pathology of Astrocytes,* Vol. 3. Orlando, Fla: Academic

Fontana A, Fierz W, Wekerle H. 1984. Astrocytes present myelin basic protein to encephalitogenic T-cell lines. *Nature* 307:273–76

Gardner-Medwin AR. 1981. Possible roles of vertebrate neuroglia in potassium dynamics, spreading depression and migraine. *J. Exp. Biol.* 95:111–27

Gardner-Medwin AR. 1983. Analysis of potassium dynamics in mammalian brain tissue. *J. Physiol.* 335:393–426

Giulian D, Baker TJ. 1985. Peptides released by ameboid microglia regulate astroglial proliferation. *J. Cell Biol.* 101:2411–15

Giulian D, Baker TJ, Shih LC, Lachman LB. 1986. Interleukin 1 of the central nervous system is produced by ameboid microglia. *J. Exp. Med.* 164:594–604

Giulian D, Lachman L. 1985. Interleukin 1 stimulation of astroglial proliferation after brain injury. *Science* 228:497–99

Giulian D, Vaca K, Corpuz M. 1993. Brain glia release factors with opposing actions upon neuronal survival. *J. Neurosci.* 13:29–37

Giulian D, Young DG. 1986. Brain peptides and glial growth. II. Identification of cells that secrete glia-promoting factors. *J. Cell Biol.* 102:812–20

Goldberg WJ, Bernstein JJ. 1988a. Fetal cortical astrocytes migrate from cortical homografts throughout the host brain and over the glial limitans. *J. Neurosci. Res.* 20:38–45

Goldberg WJ, Bernstein JJ. 1988b. Migration of cultured fetal spinal cord astrocytes into adult host cervical cord and medulla following transplantation into thoracic spinal cord. *J. Neurosci. Res.* 19:34–42

Goldstein GW. 1988. Endothelial cell-astrocyte interactions. A cellular model of the blood-brain barrier. *Ann. NY Acad. Sci.* 529:31–39

Goldstein GW, Betz AL. 1983. Recent advances in understanding brain capillary function. *Ann. Neurol.* 14:389–95

Graeber MB, Streit WJ, Kiefer R, et al. 1990. New expression of myelomonocytic antigens by microglia and perivascular cells following lethal motor neuron injury. *J. Neuroimmunol.* 27:121–32

Graeber MB, Streit WJ, Kreutzberg GW. 1989. Identity of ED2-positive perivascular cells in rat brain. *J. Neurosci. Res.* 22:103–6

Heldin CH, Westermark B, Wateson A. 1981. Special receptors for PDGF on cells from connective tissue and glia. *Proc. Natl. Acad. Sci. USA* 78:3664–68

Hertz L, McFarlin D, Waksman B. 1990. Astrocytes: auxiliary cells for immune responses in the central nervous system. *Immunol. Today* 11:256–68

Hickey WF, Kimura H. 1988. Perivascular microglial cells of the CNS are bone marrow-derived and present antigen in vivo. *Science* 239:290–92

Hirata H, Slater NT, Kimelberg HK. 1983. α-Adrenergic receptor-mediated depolarization of rat neocortical astrocytes in primary culture. *Brain Res.* 270:358–62

Jacque CM, Suard IM, Collins VP, Raoul MM. 1986. Interspecies identification of astrocytes after intracerebral transplantation. *Dev. Neurosci.* 8:142–49

Janzer RC, Raff MC. 1987. Astrocytes induce blood-brain barrier properties in endothelial cells. *Nature* 325:253–57

Johns LD, Babcock G, Green D, Freedman M, Sriram S, Ransohoff R. 1992. Transforming growth factor-B1 differentially regulates proliferation and MHC class-II antigen expression in forebrain and brainstem astrocyte primary cultures. *Brain Res.* 585:229–36

Kettenmann H, Backus KH, Schachner M. 1984. Aspartate, glutamate, and gamma-aminobutyric acid depolarize cultured astrocytes. *Neurosci. Lett.* 52:25–29

Kimelberg HK, Bourke RS. 1982. Anion transport in the nervous system. In *Handbook of Neurochemistry,* ed. A Lajtha, 1:31–67. New York: Plenum

Kimelberg HK, Goderie SK, Higman S, et al. 1990. Swelling-induced release of glutamate, aspartate, and taurine from astrocyte cultures. *J. Neurosci.* 10:1583–91

Korr H. 1986. Proliferation and cell cycle parameters of astrocytes. See Fedoroff & Vernadakis 1986, pp. 77–127

Kraig RP, Chesler M. 1990. Astrocytic acidosis in hyperglycemic and complete ischemia. *J. Cereb. Blood Flow Metab.* 10:104–14

Landis DMD, Reese TS. 1974. Arrays of particles in freeze-fractured astrocytic membranes. *J. Cell Biol.* 60:316–20

Landis DMD, Reese TS. 1981a. Astrocyte membrane structure: Changes after circulatory arrest. *J. Cell Biol.* 88:660–63

Landis DMD, Reese TS. 1981b. Membrane structure in mammalian astrocytes: A review of freeze-fracture studies in adult, developing, reactive and cultured astrocytes. *J. Exp. Biol.* 95:35–48

Landis DMD, Reese TS. 1982. Regional organization of astrocytic membranes in cerebellar cortex. *Neuroscience* 7:937–50

Lassmann H, Schmied M, Vass K, Hickey WF. 1993. Bone marrow derived elements and resident microglia in brain inflammation. *Glia* 7:19–24

Latov N, Nilaver G, Zimmerman EA, et al. 1979. Fibrillary astrocytes proliferate in response to brain injury. A study combing immunoperoxidase technique for glial fibrillary acid protein and radioautography of tritiated thymidine. *Dev. Biol.* 72:381–84

Lindsay RM. 1986. Reactive gliosis. See Fedoroff & Vernadakis 1986, pp. 231–62

Ling E-A, Wong W-C. 1993. The origin and nature of ramified and amoeboid microglia: A historical review and current concepts. *Glia* 7:9–18

Liuzzi FJ, Lasek RJ. 1987. Astrocytes block axonal regeneration in mammals by activating the physiological stop pathway. *Science* 237:642–45

Marty S, Dusart I, Pechanski M. 1991. Glial changes following an excitotoxic lesion in the CNS—I. Microglia/macrophages. *Neuroscience* 45:529–39

Merrill JE, Zimmerman RP. 1991. Natural and induced cytotoxicity of oligodendrocytes by microglia is inhibitable by TGFβ. *Glia* 4:327–31

Milhorat TH, Hammock MD, Fenstermacher JD, et al. 1971. Cerebrospinal fluid production by the choroid plexus and brain. *Science* 173:330–32

Miller RH, ffrench-Constant C, Raff MC. 1989. The macroglial cells of the rat optic nerve. *Annu. Rev. Neurosci.* 12:517–34

Miller RH, Szigeti V. 1991. Clonal analysis of astrocyte diversity in neonatal rat spinal cord cultures. *Development* 113:353–62

Miyake T, Hattori T, Fukuda M, et al. 1988. Quantitative studies on proliferative changes of reactive astrocytes in mouse cerebral cortex. *Brain Res.* 451:133–38

Miyake T, Okada M, Kitamura T. 1992. Reactive proliferation of astrocytes studied by immunohistochemistry for proliferating cell nuclear antigen. *Brain Res.* 590:300–2

Morrison RS, de Vellis J, Lee YL, et al. 1985. Hormones and growth factors induce the synthesis of glial fibrillary acidic protein in rat brain astrocytes. *J. Neurosci. Res.* 14:167–76

Murray M, Wang SD, Goldberger ME, Levitt P. 1990. Modification of astrocytes in the spinal cord following dorsal root or peripheral nerve lesions. *Exp. Neurol.* 110:248–57

Oldendorf WH, Cornford ME, Brown WJ. 1977. The large apparent work capacity of the blood-brain barrier: a study of the mitochondrial content of capillary endothelial cells in brain and other tissues of the rat. *Ann. Neurol.* 1:409–17

Olsson T, Maehlen J, Löve A, et al. 1987. Induction of class I and class II transplantation antigens in rat brain during fatal and non-fatal measles virus infection. *J. Neuroimmunol.* 12:265–77

Paulson OB, Newman EA. 1987. Does the release of potassium from astrocyte endfeet regulate cerebral blood flow? *Science* 237:896–98

Pines G, Danbolt NC, Bjøras M, et al. 1992. Cloning and expression of a rat brain L-glutamate transporter. *Nature* 360:464–67

Pober JS, Cotran RS. 1990. Cytokines and endothelial cell biology. *Physiol. Rev.* 70:427–51

Schnell L, Schwab ME. 1990. Axonal regeneration in the rat spinal cord produced by an antibody against myelin-associated neurite growth inhibitors. *Nature* 343:269–72

Schroder M, Muller G, Birchmeier W, Bolen P. 1986. Transforming growth factor-beta inhibited endothelial cell proliferation. *Biochem. Biophys. Res. Commun.* 37:295–302

Sporn MB, Roberts AB, Wakefield LM, Assoian RK. 1986. Transforming growth factor-β: biological function and chemical structure. *Science* 233:532–34

Staub F, Baethmann A, Peters J, et al. 1990. Effects of lactic acidosis on glial cell volume and viability. *J. Cereb. Blood Flow Metab.* 10:866–76

Stewart PA, Wiley MJ. 1981. Developing nervous tissue induces formation of blood-brain barrier characteristics in invading endothelial cells: A study using quail-chick transplantation chimeras. *Dev. Biol.* 84:184–92

Storck T, Schulte S, Hofmann K, Stoffel W. 1992. Structure, expression, and functional analysis of a Na+-dependent glutamate/aspartate transporter from rat brain. *Proc. Natl. Acad. Sci. USA* 89:10955–59

Streit WJ, Graeber MB, Kreutzberg GW. 1989. Expression of Ia antigen on perivascular and microglial cells after sublethal and lethal motor neuron injury. *Exp. Neurol.* 105:115–26

Streit WJ, Kreutzberg GW. 1988. Response of endogenous glial cells to motor neuron degeneration induced by toxic ricin. *J. Comp. Neurol.* 268:248–63

Suzumura A, Lavi E, Weiss S, Silberberg D. 1986. Coronavirus infection induces H-2 antigen expression on oligodendrocytes and astrocytes. *Science* 232:991–93

Tang C-M, Orkand RK. 1986. Glutamate depolarization of glial cells in Necturus optic nerve. *Neurosci. Lett.* 63:300–4

Wells J, Vietje BP, Wells DG, Paradee J. 1992. Isomorphic activation of astrocytes in the somatosensory thalamus. *Glia* 5:154–60

Yednock TA, Cannon C, Fritz LC, et al. 1992. Prevention of experimental autoimmune encephalomyelitis by antibodies against alpha4β1 integrin. *Nature* 356:63–66

Annu. Rev. Neurosci. 1994. 17:153–83

NITRIC OXIDE AND SYNAPTIC FUNCTION

Erin M. Schuman* and Daniel V. Madison

Dept. of Molecular and Cellular Physiology, Stanford University Medical Center, Stanford, California 94303; *present address: Division of Biology, California Institute of Technology, Pasadena, California 91125

KEY WORDS: plasticity, long-term potentiation, long-term depression, NMDA receptor, hippocampus

INTRODUCTION

The free radical gas nitric oxide (NO) is a recently identified neuronal messenger that carries out diverse signaling tasks in both the central and peripheral nervous systems. Whereas most neurotransmitters are packaged in synaptic vesicles and secreted in a Ca^{2+}-dependent manner from specialized nerve endings, NO is an unconventional transmitter which is not packaged in vesicles, but rather diffuses from its site of production in the absence of any specialized release machinery. The lack of a requirement for release apparatus raises the possibility that NO can be released from both pre- and postsynaptic neuronal elements. In addition, because NO is gaseous and extremely membrane permeant, it can bypass normal signal transduction routes involving interactions with synaptic membrane receptors. Although the targets of NO have not yet been completely described, it is known that NO can bind to the iron contained in heme groups, leading to conformational changes in associated proteins, such as guanylyl cyclase.

NO as an Intercellular Signaling Molecule

The idea that NO may participate in modulating neuronal function originally arose from the discovery that it is an important intercellular signal that maintains vascular tone and resistance. It had long been known that acetylcholine (Ach), as well as many other neurotransmitters and neuromodulators, when applied to arteries or veins was capable of producing relaxations of the smooth muscle. In 1980, Furchgott & Zawadski reported that the ACh-induced

153

relaxation of rabbit aorta required the presence of endothelial cells. In an elegant series of bioassays, endothelial cells were removed from the intimal strip of rabbit aorta, and no relaxation could be elicited. The relaxation of the muscle was then restored by the addition of exogenous endothelial cells. A diffusible factor produced in endothelial cells, endothelial-derived relaxing factor (EDRF), was proposed to account for the observed smooth muscle relaxation. Further studies showed that the relaxation produced by ACh and other agents was Ca^{2+}-dependent (Griffith et al 1986). In addition, the relaxation was thought to be mediated by rises in cGMP that were shown to occur in the muscle but not in the endothelial cells (Rapoport et al 1983). It was also known that several nitrovasodilators (agents that generate NO, e.g. glyceryl trinitrate and sodium nitroprusside) did not require the presence of endothelial cells to elicit relaxation. Thus, it was proposed that EDRF is NO, based on the following observed similarities of NO and EDRF: both agents are extremely labile (half-life = 4–6 s), the relaxations induced by both substances are blocked by hemoglobin (which binds NO) or by generators of O_2^-, and the effects of both NO and EDRF are enhanced by superoxide dismutase, which scavenges superoxide ions. In 1987, two groups (Ignarro et al 1987, Palmer et al 1987) directly demonstrated that the vascular endothelium actually releases NO in quantities sufficient to account for the biological activity of EDRF.

Since this initial discovery, NO has been implicated in several other systems, including macrophage cytotoxicity (Marletta 1989), nonadrenergic noncholinergic intestinal relaxation (Desai et al 1991), penile erection (Rajfer et al 1992), neurotoxicity (Dawson et al 1991b), and plasticity in the hippocampus (Bohme et al 1991, O'Dell et al 1991, Schuman & Madison 1991, Haley et al 1992) and cerebellum (Crepel & Jaillard 1990, Shibuki & Okada 1990). The first demonstration of NO acting as a neuronal messenger came from studies in cerebellar granule cells by Garthwaite and colleagues (1988). These investigators demonstrated that the application of NMDA to granule cells resulted in rises in cGMP levels that were blocked by both NO synthase (NOS) inhibitors and hemoglobin, suggesting that NO was functioning as an intercellular messenger. These studies drew a significant amount of attention to the signal transduction pathway involving NMDA receptors, NOS, and guanylyl cyclase, and no doubt served as an impetus to many future inquiries, in particular those regarding the role of NO in synaptic plasticity. Although NO has recently been shown to function in a wide variety of central and peripheral processes, this paper is limited to a brief review of NO and NO synthase as well as a discussion of NO's role in the modulation of synaptic function in the following areas: NMDA receptor currents, neurotoxicity, secretion, long-term depression and potentiation, and animal learning.

NITRIC OXIDE AND NITRIC OXIDE SYNTHASES

NOS Isoforms

Nitric oxide is produced by an NO synthase (NOS). To date, several different nitric oxide synthases have been identified: one or more inducible NOSs present in macrophages, neutrophils, hepatocytes, and possibly glial cells, and at least two different constitutive forms present in endothelial cells and neurons. Four distinct isoforms of NOS have been cloned thus far: a brain NOS (Bredt et al 1991c), an endothelial NOS (Lamas et al 1992, Marsden et al 1992, Sessa et al 1992), a macrophage NOS (Lowenstein et al 1992, Xie et al 1992), and a hepatocyte NOS (Geller et al 1993). The different classes of cloned enzymes share about 50% identity in their amino acid sequences. All forms of NOS characterized thus far require several electron donors [flavin adeninedinucleotide (FAD), flavin mononucleotide (FMN), nicotinamide adenine dinucleotide phosphate (NADPH), and tetrahydrobiopterin] and produce NO by oxidizing one of the terminal guanidino nitrogens of L-arginine, resulting in the stoichiometric production of L-citrulline (Figure 1).

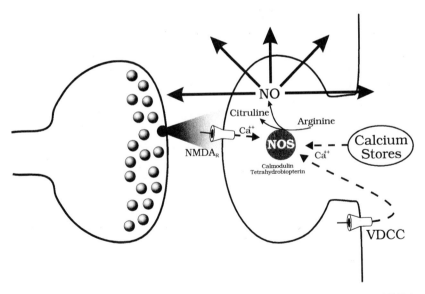

Figure 1   Diagram of NO production at synapses. The calcium signal derived from the NMDA receptor-channel, voltage-dependent Ca^{2+} channels (VDCC), or intracellular stores binds calmodulin and activates the nitric oxide synthase (NOS). Activated NOS produces NO from L-arginine. NO can then diffuse from its site of production to influence all nearby synapses.

The amino acid sequence of a brain NOS, originally purified from rat cerebellum (Bredt & Snyder 1990), encodes a protein of 160 kD that contains several recognition sites for required cofactors, including a basic amphipathic α helix calmodulin-binding consensus site, a cAMP-dependent protein kinase phosphorylation consensus sequence, a NADPH-binding domain, and potential binding sites for FMN and FAD (Bredt et al 1991a,c). The C-terminal half of NOS shows substantial homology to rat cytochrome P-450 reductase and sulphite reductase (Bredt et al 1991c), which also contain binding sites for NADPH, FMN, and FAD. The full-length cDNA was inserted into an expression vector and transfected into human kidney 293 cells. The expressed NOS protein exhibited catalytic activity with properties corresponding to those observed with the native NOS in cerebellum, namely a dependence on Ca^{2+} and NADPH and inhibition by calmodulin antagonists.

A family of endothelial NOSs has also been cloned, including a bovine endothelial NOS (Lamas et al 1992, Sessa et al 1992) and a human NOS (Janssens et al 1992, Marsden et al 1992). The deduced amino acid sequences of both the bovine and human endothelial NOSs encode a protein of approximate molecular mass 133 kDa, consistent with observed molecular weight (135,000) of the purified protein (Pollock et al 1991). The sequences of the two isoforms are highly homologous with one another (90%) and exhibit 50 and 60% homology with the cloned macrophage and brain NOS, respectively. Like the brain NOS, the endothelial NOS also contains binding regions for calmodulin, NADPH, FMN, and FAD, as well as a consensus sequence for phosphorylation by cAMP-dependent protein kinase. Interestingly, the endothelial NOS sequence also contains a consensus sequence for myristylation at the amino terminus; as discussed below, this may account for particulate localization of the endothelial NOS. The cloned cDNAs have also been inserted into expression vectors and transiently expressed in COS cells (Lamas et al 1992, Sessa et al 1992) or NIH3T3 cells (Janssens et al 1992). The transfected cells exhibited Ca^{2+}-dependent conversion of L-arginine to NO and citrulline that was sensitive to the NOS inhibitor L-NARG or L-NAME.

Two apparently distinct isoforms of inducible NOS have been cloned: one from macrophages (Lowenstein et al 1992, Lyons et al 1992, Xie et al 1992) and one from human hepatocytes (Geller et al 1993). These two inducible NOSs possess 80% amino acid sequence homology. The sequences of both forms encode smaller proteins than the brain NOS; the approximate molecular mass is 130 kD. As in other NOS isoforms, recognitions sites for FMN, FAD, and NADPH are present. Although the activity of the macrophage NOS has been observed to be largely Ca^{2+}- and calmodulin-independent, the enzyme contains a recognition site for calmodulin binding. In human 293 kidney cells transfected with the macrophage cDNA, the expression and activity of macrophage NOS was markedly enhanced by treatment with lipopolysaccharide but was not affected by Ca^{2+} chelators (Lowenstein et al 1992). The hepatocyte NOS also contains a calmodulin-binding domain. In contrast to the macrophage NOS, however, the cloned hepatocyte NOS expressed in human 293 kidney cells displayed an activity that was significantly attenuated by both Ca^{2+} chelation and calmodulin antagonists (Geller et al 1993). Further studies are obviously needed to explore the potential Ca^{2+}- and calmodulin-dependence of inducible NOS activity and to determine whether these two cofactors might also play a role in modulating the expression of inducible NOS.

Enzyme Activation

The expression of the inducible NOS requires protein synthesis and is initiated by various cytokines and microbial products (Hibbs et al 1987, Stuehr & Marletta 1987). Following induction, NO is produced in large quantities (nanomoles) for several hours. In contrast, the constitutive NOS found in blood vessels and in brain remains active for relatively short periods of time and produces smaller quantities of NO (picomoles). This NOS is activated by Ca^{2+} that is bound to calmodulin (Knowles et al 1989, Bredt & Snyder 1990). In the periphery, the primary source of Ca^{2+} may be agonist-induced phosphoinositide (PI) hydrolysis resulting in inositol-triphosphate-mediated release of Ca^{2+} from intracellular stores. In the brain, the flux of Ca^{2+} through the NMDA receptor-channel has been implicated as the source of Ca^{2+} in many systems, although it seems possible that influx of Ca^{2+} via voltage-sensitive Ca^{2+} channels or release of Ca^{2+} from intracellular stores via neurotransmitter-induced PI hydrolysis (e.g. the metabotropic glutamate receptor) may also activate the brain NOS (Figure 1).

Localization

The purification and cloning of the various NOS isozymes has prompted the development of antibodies and antisense oligonucleotides that in turn have permitted immunohistochemical mapping of NOS and localization of the NOS mRNA by in situ hybridization. In addition, the histochemical marker nitrotetrazolium blue (NTB) reacts with NOS-containing neurons (Dawson et al 1991a, Hope et al 1991). This reaction is accounted for by the redox activity of NOS, which reduces nitrotetrazolium blue to NADPH diaphorase. In situ hybridization (Bredt et al 1991b) for the brain NOS (originally purified from cerebellum) reveals a high density of silver grains in the cerebellum, olfactory bulb, and the pedunculopontine tegmental nucleus. In the pedunculopontine tegmental area NOS-positive cells also contain choline acetyltransferase (Dawson et al 1991a). Strong hybridization is also apparent in the hippocampus (dentate gyrus), supraoptic nucleus, and superior and inferior colliculus.

Isolated NOS-containing neurons have been observed in the cerebral cortex and the corpus striatum (Bredt et al 1991, Dawson et al 1991). These cells also stain positive for somatostatin and neuropeptide Y (Vincent et al 1983, Dawson et al 1991).

A matter of considerable debate concerns the localization of NOS in the hippocampus and the cerebellum, two areas where NO has been implicated in synaptic plasticity. Several studies (Bredt et al 1990, 1991b; Valtschanoff et al 1993) have noted a lack of NOS immunoreactivity in rat CA1 pyramidal neurons, the site where NO has been proposed to be produced during the induction of long-term potentiation (O'Dell et al 1991, Schuman & Madison 1991). However, recent studies have reported that CA1 pyramidal neurons of the hippocampus stain for NADPH diaphorase (Wallace & Fredens 1992) or an NOS antibody (Schweizer et al 1993). Likewise, staining for NOS and NADPH diaphorase has been observed in cerebellar granule cells and basket cells, but has not been detected in Purkinje cells, where modification of postsynaptic glutamate receptors has been proposed to underlie long-term depression (LTD). This has led to the suggestion that during LTD, NO is generated in other NOS-containing neurons in the cerebellar circuit. Of course, the possibility remains that Purkinje cells express lower levels of NOS, which have not been detected; alternatively, it is also possible that a different isoform of brain NOS may be present in Purkinje neurons and CA1 pyramidal cells.

Regulation

The NOS isoforms may be regulated by several posttranslational forms of modification, including phosphorylation and myristylation. Purified brain NOS can be phosphorylated by cAMP-dependent protein kinase (Brune & Lapetina 1991, Bredt et al 1992), protein kinase C (Nakane et al 1991, Bredt et al 1992), and Ca^{2+}/calmodulin-dependent protein kinase II (Nakane et al 1991, Bredt et al 1992, Schmidt et al 1992). The phosphorylation by all three kinases occurs primarily on serine residues; each kinase predominantly phosphorylates a distinct residue (Bredt et al 1992). The effects of phosphorylation on NOS activity appear to be controversial: PKC has been reported to both increase (Nakane et al 1991) and decrease (Bredt et al 1992) NOS activity. cAMP-dependent protein kinase has been reported to have no effect on NOS activity (Brune & Lapetina 1991, Bredt et al 1992). Phosphorylation by Ca^{2+}/calmodulin-dependent protein kinase has been reported to decrease NOS activity (Nakane et al 1991, Schmidt et al 1992) or to have no effect (Bredt et al 1992). The apparent discrepancies between these findings may result from differing basal conditions used in the assays, including the presence or absence of Ca^{2+} and/or calmodulin in the reaction mixtures. More studies are needed to clarify the effects of phosphorylation on physiological NOS activity and to explore the possibility that NO may in turn modulate the activity of protein kinases.

Evidence indicates that the endothelial isoform of NOS can be myristylated (Pollock et al 1992). Co- or posttranslational modification of proteins by myristylation is thought to confer membrane association. Whereas the macrophage and brain forms of NOS appear to be located primarily in soluble fractions, the endothelial form of NOS is located predominantly in particulate fractions (Forstermann et al 1991). None of the cloned NOS isoforms appear to have hydrophobic signal sequences that would correspond to membrane-associated regions. However, at the amino terminal portion, the endothelial NOS contains a consensus sequence for N-myristyl transferase, an enzyme that catalyzes myristylation (Kaplan et al 1988). The brain and macrophage forms of NOS lack this consensus sequence (Bredt et al 1991c, Xie et al 1992). Incubation of bovine endothelial cells with [^{3}H] myristate results in the incorporation of myristate into the endothelial NOS (Pollock et al 1992). Thus, the fatty acid acylation of the endothelial NOS may serve as a membrane anchor. Site-directed mutagenesis of the N-myristyl transferase consensus sequence is needed to determine whether this type of modification is required for the localization of endothelial NOS to particulate fractions. Since there are a few reports (Forstermann et al 1992, Hiki et al 1992) of insoluble forms of both the macrophage and brain NOS enzymes, it will be interesting to see whether myristylation may also target other forms of NOS to the membrane.

NO Donors, Scavengers and Inhibitors

A variety of pharmacological tools have been used to elucidate the functions of NO. Many of the classic nitrovasodilators exert their actions by releasing NO. These types of compounds, including sodium nitroprusside, hydroxyl-amine, isosorbide dinitrate, 3-morpholino-sydnonimine (SIN-1), and S-ni-troso-N-penicillamine (SNAP) can been used to assess the sufficiency of NO as a signaling molecule in various systems. These agents release NO by different mechanisms: some compounds that are presumably membrane impermeant, such as SIN-1 and SNAP, release NO in the extrasynaptic space; others, such as hydroxylamine and isosorbide dinitrate, are thought to release NO from intracellular locations, since they likely require cellular enzymes such as catalases and cytochromes to release NO. It is important to note that higher concentrations of NO donors may be needed when working with intact tissue (e.g. brain slices), since it has been shown that the concentrations of donors required to elevate cGMP in slices are orders of magnitude higher than those required to stimulate guanylyl cyclase in broken cell preparations (Southam & Garthwaite 1991). This observation indicates that intact tissue may possess mechanisms for rapidly inactivating NO.

In addition, several competitive inhibitors of the NOS are available, including L-arginine derivatives such as NG-monomethyl-L-arginine (L-NMMA, L-Me-Arg), NG-nitro-L-arginine (NARG), and L-nitro arginine methyl ester (L-NAME). Many of these compounds also have D-isomeric forms that do not inhibit NOS and thus serve as useful controls. Another compound that has proven to be particularly useful in blocking NO's action is hemoglobin. NO and other putative messengers, such as carbon monoxide (CO), bind avidly to the iron in the heme group of hemoglobin. Because hemoglobin is a large protein, it is unlikely to cross cellular membranes. Thus, when applied extracellularly hemoglobin may provide information regarding NO's function as an intercellular, rather than intracellular, messenger.

NO Effectors

The major effector of NO identified in many tissues is a soluble guanylyl cyclase (Arnold et al 1977, Miki et al 1977, Murad et al 1978). The soluble guanylyl cyclase is a heterodimer that contains a heme, the region responsible for NO activation of the cyclase. When NO binds to Fe^{2+} in the porphyrin ring of heme, this interaction pulls the Fe^{2+} out of the plane of the porphyrin ring, resulting in a conformational change and activation of the guanylyl cyclase (Wolin et al 1982). The resulting rises in cGMP levels can then affect ion channel or phosphodiesterase activity, or activate a cGMP-dependent protein kinase. In smooth muscle cells, the NO-induced rises in cGMP may activate a cGMP-dependent protein kinase that is ultimately responsible for muscle relaxation (Rapoport et al 1983). Alternatively, cGMP has been observed to decrease intracellular Ca^{2+} levels, which may also contribute to relaxation (Rashatwar et al 1987). Carbon monoxide, also recently identified as a potential messenger molecule (Verma et al 1993), also activates guanylyl cyclase (Brune & Ullrich 1987), although much less potently than NO (Furchgott & Jothianandan 1991).

NO can also combine with superoxide anions to form peroxynitrite. Peroxynitrite ultimately decomposes to hydroxide and NO_2 free radicals, which are believed to be the bactericidal and tumoricidal effectors of activated macrophages and neutrophils (Beckman et al 1990). NO may also exert its cytotoxic effects by binding to the iron-sulphur centers of enzymes involved in mitochondrial transport electron chain (Granger et al 1980), the citric acid cycle (Drapier & Hibbs 1986), and DNA synthesis (Nakaki et al 1990). If the inducible NOS exists in noncultured astrocytes, then the NO generated by the inducible NOS may also contribute to the neuronal damage associated with cerebral ischemia (Nowicki et al 1991).

NO may also produce its effects by stimulating the ADP-ribosylation of proteins (Brune & Lapetina 1989). ADP-ribosylation involves the covalent attachment of ADP-ribose to substrate proteins; this reaction is usually catalyzed by cellular ADP-ribosyltransferases. Brune & Lapetina (1989) demonstrated that sodium nitroprusside induced the ADP-ribosylation of a 39-kD protein in platelets. Later studies have identified this 39-kD protein as glyceraldehyde 3' phosphate dehydrogenase (GAPDH) (Dimmeler & Brune 1992, Kots et al 1992, Zhang & Snyder 1992) and have indicated that NO promotes the auto-ADP-ribosylation of GAPDH, rather than activating a distinct ADP-ribosyltransferase. NO first stimulates the S-nitrosylation of a cysteine residue adjacent to the NAD-binding site in the catalytic region of GAPDH (Molina y Vedia et al 1992). The subsequent auto-ADP-ribosylation of GAPDH is thought to occur on this S-nitrosylated cysteine residue (Dimmeler & Brune 1992, Zhang & Snyder 1992). The ADP-ribosylation of GAPDH results in a reduction of the normal dehydrogenase activity of GAPDH (Dimmeler et al 1992, Zhang & Snyder 1992). However, in addition to stimulating auto-ADP-ribosylation of GAPDH, NO apparently may also modulate the activity of endogenous cellular ADP-ribosyltransferases. Several groups have also described NO-stimulated ADP-ribosylation of distinct neuronal proteins (Williams et al 1992) that lack NAD^+-binding domains, including transducin (Ehret-Hilberer et al 1992) and other putative GTP-binding proteins (Duman et al 1991). The ADP-ribosylation of these proteins has been proposed to be mediated by a distinct ADP-ribosyltransferase. Thus, it appears that NO may stimulate both auto- and ADP-ribosyltransferase–mediated covalent modifications. Future studies should be aimed at identifying additional substrates for NO-stimulated ADP-ribosylation as well as delineating the functional consequences of this form of covalent modification in neurons.

NO AND SYNAPTIC FUNCTION

N-methyl-D-aspartate Receptor-Channel

One role of NO in the brain may be as a neuromodulatory substance, analogous to some neurotransmitters. One example of NO's modulatory function may be it's reported ability to influence ion currents through the N-methyl-D-aspartate (NMDA) receptor channels. The NMDA receptor-channels are a rather unique class of glutamate receptor channels that usually require depolarization (Mayer & Westbrook 1987, Nowak et al 1984) to flux Ca^{2+} (MacDermott et al 1986, Jahr & Stevens 1987, Ascher & Nowak 1988). Thus, by modulating current flow through this particular channel, NO could potentially influence many Ca^{2+}-regulated neuronal processes that utilize this receptor, such as synaptic transmission, plasticity, neurotoxicity, and some aspects of development.

Several different NO-donating compounds (sodium nitroprusside, nitroglycerin, S-nitrosocysteine, and SIN-1) have been shown to reduce NMDA currents (Lei et al 1992, Manzoni et al 1992a). The use of several different NO donors is important, since it has been shown that sodium nitroprusside can exert effects on NMDA currents that can be reversed by hemoglobin, but are apparently unrelated to NO since these effects are not shared by other NO-donating compounds (East et al 1991, Manzoni et al 1992b). Manzoni et al (1992a) demonstrated that the SIN-1-induced reduction in the NMDA current was accompanied by an attenuation of NMDA-mediated rises in intracellular Ca^{2+}, as revealed by measurements of Fura-2 fluorescence (see also Hoyt et al 1992). The effects of NO could be blunted by simultaneous application of hemoglobin, and were absent when NO-depleted SIN-1 was applied. On the basis of these results, it was suggested that NO may play a role as a feedback modulator. According to this scheme, when the NMDA receptor is activated the resulting entry of Ca^{2+} into the cell activates the NO synthase, leading to the reduction of subsequent NMDA currents.

Much has been learned about the possible mechanisms of the observed NO modulation of NMDA currents. Early studies had shown that NMDA currents could be influenced by the redox state of a site on the receptor channel complex (Aizenman et al 1989, 1990; Lazarewicz et al 1989). A pair of closely-spaced cysteine residues thought to reside on the extracellular side of the channel may form a disulfide bond that constitutes the redox site of the NMDA complex. Reducing this site with agents such as dithiothreitol (DTT) increases the current flow through the channel, whereas oxidizing the redox site, by using 5,5-dithio-bis-2-nitrobenzoic acid (DNTB), decreases the current flow (Lei et al 1992). DNTB can reverse the DTT-induced potentiation of current flow, but this reversal can be prevented by treatment with the irreversible sulfhydryl alkylating agent N-ethylmalemide (NEM). Such treatments similarly affect the synaptic currents through the NMDA channels (Tauck 1992).

The effect of NO on NMDA currents appears to be mediated through this redox site. It has been proposed that the free sulfhydryl groups on the NMDA channel complex are oxidized in the presence of NO to form S-nitrosothiols (Lei et al 1992). In support of this idea, after treatment with NO-donating compounds the oxidant DNTB had no further significant effect (Lei et al 1992). The reducing agent DTT could substantially reverse the effects of NO. In addition, the action of nitrogylcerine was blocked by treatment with NEM. Cyclic GMP apparently does not mediate the effect of NO at the redox site, as application of cGMP did not have any effect on the current (East et al 1991, Kiedrowski et al 1992). Indeed, the SIN-1-induced decrease in the NMDA current can still be recorded in isolated outside-out patches of membrane, suggesting that NO's effect is not mediated by a soluble messenger.

Reduction of NMDA currents by NO may have significance for many cellular processes involving the NMDA receptor. However, at present this inhibition appears to be essentially a feedback mechanism, since the NMDA receptor must first be activated (to flux Ca^{2+} that activates the NO synthase) before the currents can be reduced. In light of this consideration, some processes—those that depend on very brief activation of the NMDA receptor—may not be affected by the subsequent inhibition of the NMDA current. An example of such a process may be long-term potentiation (see below), in which NMDA receptors are transiently activated during the induction of LTP, but are not required for the maintenance or expression of LTP. Thus, this feedback inhibitory mechanism may be of more relevance to processes that involve prolonged activation of NMDA receptors, such as neurotoxicities.

Neurotoxicity

Considerable controversy surrounds the role of NO in various forms of neurotoxicity that possess different etiologies. One possible consequence of the NO-mediated reduction in NMDA currents discussed above could be attenuation of NMDA-mediated neurotoxicity. In addition, it has been known for some time that neurons that stain positive for NADPH diaphorase (and thus presumably contain NOS) are particularly resistant to neurological insults (Ferrante et al 1985, Beal et al 1986, Koh et al 1986, Hyman et al 1992). Conversely, the documented participation of NO in macrophage-mediated cell killing (Hibbs et al 1988) suggests that NO might be involved in promoting glutamate-mediated cell death. The data examining NO's role in glutamate-induced toxicity often lead to different conclusions; some results are consistent with NO acting as a neuroprotective agent while others suggest that NO is neurotoxic.

Dawson et al (1991b) have shown that inhibition of NO production can have profound effects in preventing glutamate and NMDA-mediated death of cultured cortical neurons. Using a trypan blue exclusion assay, they demonstrated that the application of NOS inhibitors attenuated the neurotoxic effects of glutamate. This inhibition of toxicity could largely be reversed by the application of excess L-arginine. In agreement, Izumi et al (1992) have shown that in hippocampal slices NOS inhibitors can prevent glutamate and NMDA-mediated cell death. As in the culture system, this inhibition is reversed by the addition of excess L-arginine. These results suggest that NO promotes and is necessary for glutamate-mediated cell death.

In contrast to the above results, other data show that NO can prevent NMDA-mediated cell death. Lei et al (1992) have shown that application of sodium nitroprusside or nitroglycerine prevents the toxicity of NMDA in cultured cortical neurons. This study showed a parallel reduction in the NMDA current, as well as diminished intracellular Ca^{2+} levels with NO. These changes were proposed to underlie the observed protective effect of NO against the NMDA-mediated toxicity.

Many studies examining the relationship between NO and glutamate/ NMDA-mediated toxicity have failed to support any role for NO, either protective or toxic. Kiedrowski et al (1991) showed that sodium nitroprusside could prevent toxicity, as in Lei et al (1992). However, it was concluded that the protective effect of sodium nitroprusside was not mediated by NO, since it was not reproduced by another NO donor, SNAP, but could be reproduced by ferricyanide (a ferricyanide group is present in sodium nitroprusside). It should be noted, however, that the results of Lei et al (1992) did not rely entirely upon the use of sodium nitroprusside, but were also obtained with another NO donor, nitroglycerine. In addition, in rats chronically treated with NOS inhibitors, no decrease in NMDA toxicity could be detected (Lerner-Natoli et al 1992). In other studies NOS inhibitors did not decrease the toxicity caused by glutamate, NMDA, or other agonists in cultures of cerebellar granule cells (Puttfarcken et al 1992), in neurons cultured from whole rat brain (Demerle-Pallardy et al 1991), or in glial free neuronal cultures (Pauwels & Leysen 1992).

Thus, experimental results support a role for NO in neurotoxicity ranging from protective to no role to toxic. The discrepancies between these studies could potentially arise from several different sources. First, they may simply represent the outcomes of nonstandardization of technique and preparation. Different cell populations may be differentially sensitive to NO. In addition, differing ratios of cell types (neuronal vs glial) in different preparations may contribute to the discrepancies. Also, the methods and durations of applications of NO donors and NOS inhibitors may also be important. Second, there may be multiple pathways that mediate glutamate-induced toxicity. NO may participate in neurotoxicity, but neurodegeneration could still proceed in the absence of NO via a parallel redundant mechanism. Third, NO itself may exert multiple, perhaps opposing actions, depending on the timing or concentration of its application. A more precise definition of the role of NO in cytotoxicity mediated by glutamate or other factors awaits further experiments that consider the above issues. One result that does emerge clearly at this time is that cGMP production stimulated by NO does not appear to play any role in producing toxicity, since direct application of cGMP is not toxic to cells (see Lustig et al 1992).

Secretion

In several different brain regions NO has been shown to modulate synaptic function by altering the release of neurotransmitter from presynaptic nerve endings. Using a push-pull cannula, Prast and Phillipu (1992) examined the ability of the NO donor SIN-1 to modulate the basal release of acetylcholine in the basal forebrain, an area where NOS-containing neurons are colocalized with choline acetyltransferase (Dawson et al 1991a). Introduction of the NO donor SIN-1 into the superfusate induced a near doubling of the basal release of acetylcholine. Superfusion of the tissue with the NOS inhibitor NARG reduced the basal release of acetylcholine by roughly 40%, suggesting that there is continuous NO production that regulates secretion in this system. In hippocampal slices, the NO donor hydroxylamine stimulated the efflux of [^{3}H] norepinephrine and [^{14}C] acetylcholine (Lonart et al 1992). The hydroxylamine-stimulated release was attenuated by hemoglobin, suggesting that the effect was mediated by NO. Extracellularly applied EGTA also abolished the NO-stimulated release, suggesting that NO exerts its effect by modulating Ca^{2+}-dependent exocytosis.

Two groups have examined the effects of NO on both the basal and evoked release of dopamine from rat striatal slices. Zhu & Luo (1992) observed that basal dopamine release increased up to 330% of baseline following the addition of sodium nitroprusside. L-arginine also produced a large enhancement of basal release when added to the bathing medium; this potentiation of release could be blocked by the coadministration of the NOS inhibitor L-Me-Arg. Hanbauer et al (1992) have reported an NMDA-evoked release of [^{3}H] dopamine that is sensitive to NOS inhibitors and hemoglobin. Exogenous application of NO also elicited increases in the basal release of [^{3}H] dopamine from primary cultures of ventral mesencephalic neurons.

In addition to several examples of NO-induced stimulation of release, in at least one system NO has also been shown to have an inhibitory influence on secretion. The magnocellular secretory neurons of the paraventricular and supraoptic nuclei in the hypothalamus stain intensely with antibodies to NOS (Bredt et al 1990). The paraventricular nucleus is the major source of the hypophysiotropic factor corticotropin-releasing hormone (CRH) (Kawano et al 1988). L-arginine, NO donors, or NOS inhibitors had no effect on the basal secretion of CRH in hypothalamic explants, as measured by radioimmunoassay (Costa et al 1993). However, the release of CRH induced by depolarization (40 mM K^+) or the cytokine interleukin 1B was potently reduced by either L-arginine or an NO donor. The inhibitory effect of L-arginine was blocked when the NOS inhibitor L-Me-Arg or hemoglobin was coincidently applied. These results suggest that NOS activation upon depolarization may function as inhibitory feedback contributing to hypotension by reducing CRH release.

Thus there are several examples of NO modulating the release of a variety of secretory substances. It appears that in different systems NO is capable of either increasing or decreasing neurotransmitter release. In theory, this bipotential control could be accomplished by NO's acting on different downstream enzymes or secretory targets, or by a common target molecule or enzyme whose modulation state determines whether release will be augmented or depressed. As such, it will be interesting to examine the molecular mechanisms by which NO modulates neurotransmitter release.

Because previous studies have shown that NO can modulate Ca^{2+} influx (Lei et al 1992), it is possible that NO alters presynaptic Ca^{2+} influx or homeostasis. Alternatively, NO may alter the function of various synaptic vesicle proteins implicated in secretion.

Long-Term Potentiation

Long-term potentiation (LTP), which has been observed in many brain areas, has proven to be a powerful system for the study of the molecular mechanisms that underlie activity-dependent enhancement of synaptic strength. At the CA1-Schaffer collateral synapses of the hippocampus, LTP occurs when the excitatory synapses are stimulated such that the depolarization of postsynaptic CA1 neurons is coincident with the release of neurotransmitter from the presynaptic CA3 nerve terminals. This is usually accomplished through the delivery of high-frequency stimulation to presynaptic axons (100 Hz; tetanus) or through the pairing of postsynaptic depolarization produced by current injection with low-frequency stimulation of presynaptic axons (pairing). Most of our understanding of the molecular processes responsible for LTP has to do with those events that underlie the initiation, or induction, of LTP. Studies from several laboratories have highlighted a cascade of postsynaptic events that initiate LTP, including postsynaptic depolarization (Malinow & Miller 1986), glutamate binding to the NMDA receptor-channel (Collingridge et al 1983), and Ca^{2+} influx (Lynch et al 1983, Malenka et al 1988). The rise in Ca^{2+} has been proposed to activate any one or combination of postsynaptically located Ca^{2+}-dependent enzymes, including protein kinase C (Lovinger et al 1987; Malinow et al 1988, 1989), Ca^{2+}/calmodulin-dependent protein kinase II (Malenka et al 1989, Malinow et al 1989, Silva et al 1992), calpain (Lynch & Baudry 1984), phospholipase A_2 (Williams et al 1989), and NOS (Bohme et al 1991, O'Dell et al 1991, Schuman & Madison 1991a, Haley et al 1992), all of which have been implicated in LTP to some extent.

In contrast to the possible exclusive role of the postsynaptic neuron in the induction of LTP, several lines of evidence suggest that the presynaptic neuron may also participate in the longer lasting aspects of LTP, known as maintenance and expression. Quantal analyses of synaptic transmission before and after LTP have often concluded that at least part of the increase in synaptic strength observed following LTP induction results from an increase in the release of neurotransmitter (Bekkers & Stevens 1990, Malinow 1991, Malinow & Tsien 1990, Kullman & Nicoll 1992, Malgaroli & Tsien 1992; but see Foster & McNaughton 1991, Manabe et al 1992). Thus, LTP is induced postsynaptically but may be expressed, at least in part, presynaptically. This shift of locus requires that the presynaptic cell receive a signal from the postsynaptic cell that indicates that LTP induction is occurring. This postsynaptically generated retrograde signal would then be responsible for bringing about increases in neurotransmitter release. An early candidate for this signal was arachidonic acid, generated by a Ca^{2+}-sensitive phospholipase A_2. Extracellularly applied inhibitors of PLA_2 have been shown to block LTP (Williams et al 1989), and extracellular application of arachidonic acid coupled with weak tetanic stimulation can enhance synaptic transmission (Williams et al 1989), although the onset of this enhancement is much slower than the onset of LTP.

More recently, several labs have queried the possibility that NO may function as retrograde signal in LTP. Given the Ca^{2+}- and calmodulin-dependence of the brain NOS and the established role of both of these molecules in LTP induction (Madison et al 1991), NO appears at the outset to be particularly well suited to perform the functions of a retrograde signal. In initial experiments we (Schuman & Madison 1991a) and others (Bohme et al 1991, O'Dell et al 1991, Haley et al 1992) showed that extracellular application of NOS inhibitors prevents tetanus-induced LTP. The inhibition of LTP produced by competitive NOS inhibitors can be reversed by the addition of L-arginine, as would be expected if the actions of the inhibitors are on the NOS. It appears that NO production is necessary only during LTP induction, since NOS inhibitors applied 20–30 min after high-frequency stimulation do not reverse established LTP (O'Dell et al 1991, Haley et al 1992).

The above experiments utilized extracellular bath application of NOS inhibitors to demonstrate NO's importance in the production of LTP. However, these experiments do not indicate the synaptic site of NO generation. By definition, the retrograde signal must be produced in the postsynaptic neuron. Indeed, this appears to be the case for NO, since NOS inhibitors injected into the postsynaptic neuron will block LTP induced by pairing postsynaptic depolarization with low frequency stimulation of afferents (O'Dell et al 1991, Schuman & Madison 1991). Injection of the D-isomers of NOS inhibitors has no effect on LTP production. Hemoglobin applied extracellularly also attenuates LTP (Bohme et al 1991, O'Dell et al 1991, Schuman & Madison 1991, Haley et al 1992), whereas methemoglobin, which has a much lower affinity for NO, has no effect. The reduction of LTP by hemoglobin is consistent with the idea that NO functions as an intercellular signal, traveling from the post- to the presynaptic neuron.

If postsynaptically released NO interacts with the presynaptic terminal to bring about LTP, then the exogenous application of NO coupled with presynaptic activity should be sufficient to induce synaptic potentiation. The ability of NO or NO donors to increase synaptic strength in hippocampal slices has not been readily observed, possibly because of the extreme lability of NO and the difficulty of reaching appropriate concentrations in the depth of the tissue (Southam & Garthwaite 1991). However, a few groups have had some success. Bohme and colleagues (Bohme et al 1991, Bon et al 1992) have shown that extracellular application of two different NO donors, hydroxylamine and sodium nitroprusside, can potentiate synaptic transmission in a manner that occludes normal synaptically-induced LTP. Direct application of NO has also been demonstrated to augment synaptic transmission: in cultured hippocampal neurons NO increases the frequency of spontaneous miniature synaptic events (O'Dell et al 1991). In hippocampal slices bathed in the NMDA receptor antagonist AP5, NO induces potentiation when paired with a weak tetanus (50 Hz), but not when it is applied in the absence of presynaptic activity (Zhou et al 1993). One caveat regarding this study concerns the relatively high frequency of presynaptic stimulation required to elicit the NO-induced potentiation; in theory, frequencies as low as 1 Hz should be sufficient to produce potentiation, given that this frequency of presynaptic stimulation can induce LTP when paired with postsynaptic depolarization. Nonetheless, the observed activity-dependence of the NO-induced potentiation is noteworthy, since it may provide an explanation for how NO can mediate the input-specific nature of LTP: only those synapses that are active during LTP induction become potentiated (Barrionuevo & Brown 1983).

The existence of a diffusible retrograde signal in LTP raises interesting possibilities regarding the specific synapses that will be influenced by its generation. In the absence of a precise targeting mechanism or an extremely efficient breakdown pathway, it is possible that a diffusible signal, such as NO, will interact with nearby synapses that have not participated directly in its production, resulting in a non-Hebbian heterosynaptic potentiation. Indeed, Bonhoeffer and colleagues have observed this type of potentiation in cultured hippocampal slices (Bonhoeffer et al 1989) and visual cortex slices (Kossel et al 1990). In these studies, pairing postsynaptic depolarization of an individual neuron with low-frequency stimulation resulted in a decrease in spike latency in the paired cell as well as a nearby cell. We (Schuman & Madison 1991b, 1993) have also observed that in acute hippocampal slices, LTP induced by pairing in one CA1 pyramidal can spread to nearby ($\sim$ 100 μm), but not spatially remote ($>$ 500 μm) synapses. These results are consistent with the postsynaptic generation of a diffusible factor, such as NO, that spreads to influence nearby synapses.

How does NO bring about the increase in synaptic strength that underlies LTP? An early study suggested that NO may activate a guanylyl cyclase, since membrane permeant analogues of cGMP (coapplied with an NOS inhibitor) partially reversed the inhibition of LTP normally observed with NOS inhibitors (Haley et al 1992). Sweatt and colleagues have also observed that tetanic stimulation results in large rises in cGMP that are blocked by NOS inhibitors (Chetkovich et al 1993). However, if guanylyl cyclase is the target of NO, then application of membrane permeant analogues of cGMP in conjunction with low-frequency presynaptic stimulation should be sufficient to potentiate synaptic transmission. This is not what has been experimentally observed (Schuman et al 1992). Under most stimulation parameters, extracellular application of cGMP analogues has no effect on baseline levels of synaptic transmission (Haley et al 1992). High-frequency stimulation delivered in the presence of cGMP analogues (and an NMDA receptor antagonist) usually results in a transient depression (Schuman et al 1992). Additionally, cGMP depresses a Ca^{2+} current in hippocampal neurons (Doerner & Alger 1988); at first glance, this result is the opposite of what might be expected if cGMP were involved in increasing neurotransmitter release.

Another potentially interesting NO target in LTP may be a cytosolic ADP-ribosyltransferase (ADPRT) (Brune & Lapetina 1989; Figure 2). An earlier study (Goh & Pennefather 1989) showed that slices from rats pretreated with pertussis toxin, a bacterial ADP-ribosyltransferase, failed to exhibit LTP. Recent preliminary results suggest that LTP can be prevented by extracellular application of ADP-ribosyltransferase inhibitors (Schuman et al 1992). Postsynaptic injections of an ADPRT inhibitor did not prevent LTP, consistent with a presynaptic requirement for ADPRT activity. However, the link between NO production and ADP-ribosyltransferase activity in LTP is still untested. It will be interesting to see if NO-induced increases in synaptic strength are mediated by ADP-ribosyltransferase activity. Also, it remains to be determined whether LTP-inducing high-frequency stimulation results in the NO-dependent ADP-ribosylation of specific proteins.

In sum, evidence from several laboratories supports a role for NO as a signaling molecule in LTP. However, the involvement of NO in LTP is not without its mysteries or disputes. As mentioned previously, one caveat is the failure of most histological studies to detect NOS in the CA1 pyramidal cell region of the hippocampus. Recent studies, however, have identified NOS-immunopositive (Schweizer et al 1993) and NADPH diaphorase-staining (Wallace & Fredens 1992) CA1 pyramidal neurons. Also, since only one brain NOS isoform has been identified thus far (Bredt et al 1991c), it is possible that additional, as yet unidentified isoforms, not recognized by the available antibodies or oligonucleotides, may also be present in these neurons.

Another area of controversy concerns reports of NOS inhibitor-insensitive forms of LTP that occur under certain stimulation parameters (Gribkoff et al 1992, Chetkovich et al 1993) or at higher temperatures (Li et al 1992, Chetkovich et al 1993). A recent study (Haley et al 1993) has examined both of these issues. These investigators found that at temperatures above 30°C, NOS inhibitors block LTP induced by short (2 × 100 Hz for 0.25 s), but not long (2 × 100 Hz for 0.5 s) duration tetanic stimulation. These results suggest that at physiological temperatures stronger stimulation parameters may activate alternative biochemical pathways (see also Chetkovich et al 1993).

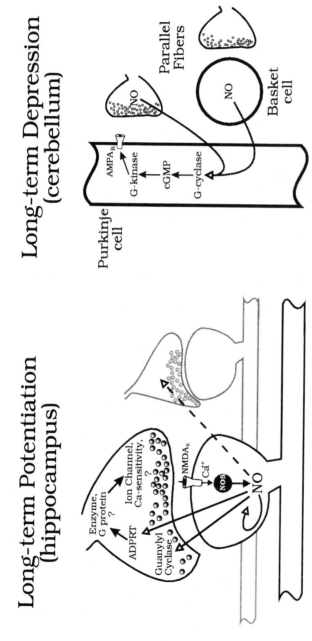

Figure 2 Diagram of possible schemes for NO action in long-term potentiation and depression. In long-term potentiation (*left*) Ca²⁺ influx through the NMDA channel may activate the NOS, resulting in NO production. NO would then diffuse to the presynaptic terminal of the same synapse to interact with a guanylyl cyclase and/or an ADPRT. An NO-stimulated ADPRT may then ADP-ribosylate a GTP-binding protein or other enzyme, which would in turn alter ion channel activity or change the Ca²⁺ sensitivity of the neurotransmitter release process such that more neurotransmitter is released during LTP. In long-term depression (*right*) NO may be produced by Ca²⁺ signals in the parallel fibers or basket cells. NO may then diffuse to Purkinje cells to activate guanylyl cyclase and elevate cGMP levels. The cGMP could then activate a cGMP-dependent protein kinase that phosphorylates α-amino-3-hydroxy-5-methyl-4-isoxazole proprionic acid (AMPA) receptors, or associated molecules, resulting in decreased postsynaptic responsiveness.

NOS inhibitors have also been reported to block NMDA-mediated inhibition of LTP (Izumi et al 1992). Taken together, these findings suggest previously unappreciated complexities in the interactions between NO and the experimental paradigms used in LTP studies. Thus, in future endeavors, much more attention should be paid to the various LTP induction procedures and experimental conditions that are used in different labs. The breadth of biochemical pathways implicated in LTP is ever expanding. The task of future experiments will be to understand how various putative signaling pathways (e.g. arachidonic acid, carbon monoxide, and nitric oxide) may interact to ultimately orchestrate the observed increases in synaptic strength. In addition, the possibility that given biochemical pathways may be selectively invoked based on the stimulation parameters used to induce LTP (Gribkoff et al 1992, Haley et al 1993) or the preexisting history of the synapse (Izumi et al 1992, Larkman et al 1992) needs to be explored further.

Long-Term Depression

Certain patterns of neuronal activity can also lead to persistent decreases in synaptic strength, or long-term depression (LTD). Like LTP, LTD has been documented in many brain areas, including visual cortex, the hippocampus, and the cerebellum (Ito 1989). In the cerebellar cortex, the Purkinje cells possess two separate sets of excitatory synapses, one from the parallel fibers (granule cells axons) and one from the climbing fibers (inferior olive axons). Both the parallel fiber and the climbing fiber synapses use glutamate as a neurotransmitter. Conjunctive stimulation (1–4 Hz, for 25 s to 10 min) of the parallel fibers and the climbing fibers produces a long-lasting depression of synaptic transmission at the synapses between the parallel fibers and Purkinje cells. The observed decrease in synaptic strength results from a reduction in the sensitivity of postsynaptic AMPA receptors (Ito et al 1982, Crepel & Krupa 1988, Hirano 1991, Linden et al 1991).

Studies aimed at elucidating the molecular mechanisms that underlie LTD have outlined a series of events that includes glutamate binding to postsynaptic receptors, rises in intracellular Ca^{2+}, and activation of protein kinase(s). Glutamate released from climbing fibers is thought to activate primarily the AMPA class of receptor. Because each climbing fiber possesses multiple synapses with each Purkinje cell, climbing fiber activity can potently depolarize Purkinje neurons, resulting in the activation of voltage-dependent Ca^{2+} channels and the influx of Ca^{2+}. The glutamate released from parallel fibers has been proposed to activate both AMPA and metabotropic classes of glutamate receptors. Activation of both of these receptors is a necessary step in the induction of LTD, because antagonists to either class of receptor will block LTD (Linden et al 1991). Activation of the metabotropic receptor activates phospholipase C, resulting in generation of diacylglycerol and an

IP$_3$-mediated rise in intracellular Ca^{2+}. The Ca^{2+} signal (also derived from climbing fiber–mediated activation of voltage-sensitive Ca^{2+} channels), as well as the generation of diacylglycerol, serves to activate protein kinase C. PKC activity is required for LTD (Linden & Connor 1991), and LTD can be mimicked by the application of phorbol esters (Crepel & Krupa 1988). In addition to PKC, activation of cGMP-dependent protein kinase has also been implicated in LTD, as discussed in further detail below. The activity of these protein kinases has then been proposed to bring about alterations in the sensitivity of AMPA-type glutamate receptors by directly phosphorylating channel subunits or associated molecules.

Where might NO fit into the induction cascade outlined above? The brain NO synthase that was originally purified from cerebellum (Bredt & Snyder 1990) is found in the granule cells as well as the inhibitory basket cells, but has not been detected in the Purkinje cells (Bredt et al 1991b). Thus, the Ca^{2+} signal in Purkinje cells that is required for LTD most likely does not function as an activator of NOS (unless an as yet unidentified isoform of NOS exists in Purkinje cells). However, NOS may be activated by an influx of Ca^{2+} in the granule cells or by granule cell-induced excitation of the basket cells (Figure 2). If NO plays a role in LTD, then guanylyl cyclase is a likely target, since cGMP concentrations are much higher in the cerebellum than in other brain areas. Immunohistochemical studies indicate that guanylyl cyclase (Nakane et al 1983) and cGMP-dependent protein kinase (Lohmann et al 1981) are present in Purkinje cell bodies, dendrites, and axons. In contrast, cGMP is found primarily in Bergmann fibers and cell bodies, and in astroglial cells in the granular layer and white matter, but appears to be absent from Purkinje cells (de Vente et al 1989). In addition, several different agonists, including glutamate and kainate, as well as the NO donor sodium nitroprusside, fail to elevate cGMP in Purkinje neurons (Garthwaite & Garthwaite 1987). These findings may indicate that the guanylyl cyclase present in Purkinje cells is stimulated by an as yet unidentified signal transduction cascade.

Is there any direct evidence that NO participates in LTD? Crepel & Jaillard (1990, Daniel et al 1993) demonstrated that extracellular application of the NOS inhibitor L-NMMA blocks LTD produced by pairing parallel fiber-mediated EPSPs with postsynaptic Purkinje cell Ca^{2+} spikes in cerebellar slices. However, LTD is not blocked when the NOS inhibitor is included in the whole-cell recording pipette, suggesting that NO production is not required in Purkinje cells (Daniel et al 1993). In contrast, inclusion of an NO donor in the Purkinje cell recording pipette resulted in a progressive decline in the amplitude of parallel fiber-mediated EPSPs (Daniel et al 1993). This NO-mediated decrease in the EPSP prevented the subsequent induction of LTD by pairing parallel fiber stimulation with Ca^{2+} spikes. LTD can be prevented by the extracellular application of methylene blue, which has been reported to inhibit guanylyl cyclase (Crepel & Jaillard 1990). In addition, bath or intracellular application of 8-bromo-cGMP also depressed the Purkinje cell EPSP. These results are consistent with a role for NO in LTD in which NO is produced by parallel fiber stimulation and then diffuses into the Purkinje cells to activate guanylyl cyclase and depress the EPSP.

A role for NO has been explored in another LTD paradigm. Ito & Karachot (1990) have documented a quisqualate (QA)-induced desensitization of Purkinje cell glutamate receptors in grease gap recordings from Purkinje cell axons in cerebellar slices (Ito & Karachot 1989). QA is believed to induce the observed desensitization by acting upon two classes of glutamate receptors, both the ionotropic AMPA type and the metabotropic type. Application of AMPA alone does not induce the desensitization (presumably because it does not act at the metabotropic receptor), but the coapplication of AMPA and the NO donor sodium nitroprusside or a membrane permeant cGMP analogue will produce desensitization (Ito & Karachot 1990). Also, prior incubation with either the NOS inhibitor L-NMMA or hemoglobin blocked the QA-induced desensitization of responses. These results can be contrasted to the findings of Linden & Connor (1992), who showed that in cultured Purkinje neurons, NO is not important for the depression glutamate currents produced by conjoint depolarization and glutamate iontophoresis. The differences in these two findings may be accounted for by the different preparations (cerebellar slices vs cultured Purkinje neurons) or the different induction procedures (quisqualate applications vs glutamate iontophoresis coupled with depolarization) used.

Another study has shown that a correlate of LTD, the alteration of extracellular K^+ concentration ($[K^+]_o$), is also influenced in a manner consistent with a role for NO. When the parallel fibers are stimulated in the molecular layer of a cerebellar slice, an increase in ($[K^+]_o$) can be recorded with an ion-sensitive electrode (Shibuki & Okada 1990). LTD, produced by conjunctive stimulation, is accompanied by a depression of the parallel fiber–elicited K^+ response (Shibuki & Okada 1990). The conjunctive stimulation–induced decrease in ($[K^+]_o$) is blocked when cerebellar slices are bathed in the NOS inhibitor L-NMMA or hemoglobin (Shibuki & Okada 1991). In addition, sodium nitroprusside or a cGMP analogue paired with parallel fiber stimulation significantly depressed the K^+ response. This study also showed that an NO-sensitive probe inserted in the molecular layer was able to detect increases in NO concentration following conjunctive stimulation.

Thus, evidence from several studies suggests that NO functions as a important signal in the cellular events that underlie LTD. How might NO be incorporated into the anatomical and cellular circuitry important for LTD? The most parsimonious transduction scheme would most likely begin with NO generation in the basket or granule cells, induced by parallel fiber stimulation (Figure 2). NO would then diffuse to Purkinje cells to activate guanylyl cyclase, increase cGMP levels, and potentially activate a cGMP-dependent protein kinase. A cGMP-dependent protein kinase is one kinase that has been proposed to mediate the decreased postsynaptic responsiveness by phosphorylating a postsynaptic AMPA receptor or associated molecule (Ito 1989). A caveat to the above sequence of events is that researchers have been unable to observe NO stimulation of Purkinje cell guanylyl cyclase (Garthwaite & Garthwaite 1987). In addition, the molecular underpinnings of the proposed down-regulation of the AMPA channels need to be further explored. It will be interesting to see whether the particular AMPA receptor subunits expressed in Purkinje cells possess consensus sites for phosphorylation by cGMP protein kinase or PKC, and whether NO can stimulate the phosphorylation of the receptor subunits by either of these kinases.

Animal Learning

The involvement of NO in LTP and LTD has prompted several investigators to explore the role of NO in the acquisition and retention of learned behavioral tasks, such as the Morris Water Maze, a radial arm maze, classical conditioning of the eyeblink response, and passive avoidance learning. The Morris Water Maze (Morris 1984) is a spatial learning task that requires an animal to find a platform submerged in a pool of opaque water based on spatial cues provided by the surrounding environment. During training trials, animals are placed at random positions in the pool and the amount of time it takes them to find the submerged platform is measured (escape latency). Previous work has shown that the hippocampus (Morris et al 1990) and NMDA receptor activity (Morris et al 1986) are required for animals to learn this task. A recent study suggests that one target of the NMDA-mediated Ca^{2+} influx that is necessary for learning this spatial task may be a NOS. Rats that received systemic injections of an NOS inhibitor (L-NAME; 75 mg/kg) prior to training had significantly longer escape latencies than control animals (Chapman et al 1992). The effect of the NOS inhibitor was abolished when L-arginine was coadministered. When NOS inhibitors were injected after animals learned the task, the animals retained their ability to navigate to the platform, implying that NOS activity is required during the acquisition but not the retention of the memory.

Bohme and colleagues (1993) have also implicated NO in another test of spatial learning, performance in a radial arm maze. In this task, rats were required to make one visit to each arm of an eight-arm radial maze, in order to obtain a food reward. An error was recorded when a rat entered a previously visited arm within a given training period. Vehicle-injected controls can successfully navigate ($<$ 2 errors/session) in the maze by the third day of training. Rats that received injections of the NOS inhibitor L-NARG (100 mg/kg, i.p.) for four days preceding the initiation of the training mastered this task much more slowly than other animals. This study also showed that the same injection of L-NARG prevented LTP in hippocampal slices prepared ex vivo. Lower doses (25 mg/kg) that were ineffective in blocking LTP from ex vivo slices also did not impair maze learning.

Classical conditioning of eyeblink responses involves pairing a tone (conditioned stimulus) with an air puff to the eye (unconditioned stimulus), which normally elicits an eyeblink. After days of training, the tone alone will elicit an eyeblink (conditioned response). Lesion studies in rabbits indicate that the acquisition of this learned behavior requires the cerebellum (McCormick & Thompson 1984). Given the marked presence of the NOS in cerebellar granule cells, NO seems at the outset to be a good candidate for a mediator of this type of learning. Indeed, the acquisition of this classically conditioned response was blocked in rabbits that received daily injections (10 mg/kg) of L-NAME prior to training (Chapman et al 1992). However, on subsequent days when the injections were switched to D-NAME, the animals showed normal acquisition of the conditioned response. Interestingly, animals that had received D-NAME injections learned successfully, and their retention of the conditioned response could not be attenuated by subsequent injections of L-NAME. One caveat to the above study concerns the inverse dose-response relationship observed: lower doses (10 mg/kg) of L-NAME were effective in preventing the conditioned response, whereas higher doses (75 mg/kg) were ineffective.

Two studies have examined the requirement for NOS activity in different passive avoidance learning tasks. In a chick one-trial passive avoidance paradigm, chicks that initially peck at a bead coated with a bitter substance subsequently avoid dry, uncoated beads. A previous study suggested that this type of learning has been shown to rely on activation of NMDA receptors (Burchuladze & Rose 1990). Holscher & Rose (1992) found that chicks that received i.p. injections of NARG prior to training exhibited an initial disgust avoidance of the bitter bead, but did not avoid the bead during the test phase. The initial display of avoidance towards the bitter bead suggests that the NARG injections do not block learning by simply altering taste perception. These investigators also noted a failure of NOS inhibitors to alter established memory. In contrast, in a one-trial shock avoidance learning task, NOS inhibitors appeared not to interfere with learning (Bohme et al 1993). In this study, rats that received NOS inhibitor injections (100 mg/kg) learned as rapidly as control animals to avoid a dark chamber where they had previously experienced an electric shock. Thus, a requirement for NO in passive avoidance learning appears to depend on the species or the particulars of the experimental protocol, which may include the sensory modalities utilized during the tasks. It may be the case that NO is important for learning tasks that involve olfactory systems, since NOS inhibitors have also been shown to be important for another form of olfactory memory (Bohme et al 1993).

Thus several studies suggest that NO may participate in the acquistion of learned behaviors. However, a general caveat that must be applied to these studies involves potential systemic effects of blocking NO production. Alterations in blood pressure could alter an animal's ability to learn for a variety of reasons including alterations in motivation or activation of compensatory physiological systems that oppose learning. In addition, with the modes of inhibitor administration used in these studies (systemic or intaperitoneal injections) it is impossible to ascertain the site, neural or peripheral, where the inhibitor is action. A more informative approach might involve direct injections of NOS inhibitors into brain structures previously implicated in the behavioral changes. Nonetheless, keeping the above considerations in mind, these studies make a promising start toward the elucidation of how NO may modulate complex behavioral phenomena like learning and memory.

PERSPECTIVES

We have summarized data that suggests that NO is an important signaling molecule in a variety of physiological and pathophysiological processes. The observation that neuronal NOS requires both Ca^{2+} and calmodulin for its activity raises the possibility that NO may function in many other systems where rises in intracellular Ca^{2+}, particularly those contributed by NMDA receptors, are known to act as a triggering step. Thus, the examples of NO-induced modulation discussed in this review are, no doubt, just a beginning.

In the various behavioral and cellular models of plasticity where the role of NO has been explored, it appears that NO functions as an early signal, responsible for the acquisition of information rather than its maintenance or long-term storage. This idea is suggested by the demonstrations that NOS inhibitors are without effect when injected after animals have learned either the Morris Water Maze, classical conditioning of eyeblink responses, or a passive avoidance task. These observations nicely parallel the finding that NOS inhibitors applied after the induction of LTP do not affect the enhanced synaptic transmission. Thus, continuous production of NO does not appear to underlie the long-lasting phases of synaptic or behavioral plasticity. Indeed, short-lived production of NO is what might be expected, given what is known about the activity of constitutive NOSs. However, it remains to be determined whether the activity of constitutive NOSs can be modified to produce NO for longer durations. In addition, it will be interesting to see if inducible NOSs, which produce NO for prolonged periods of time, may also be present in the central nervous system (see Galea et al 1992).

Although it has not been explored in much detail, a role for NO appears promising (Gally et al 1990, Montague et al 1991) in the development and the stabilization of synaptic connections. Both neuronal activity and bidirectional synaptic signaling have been proposed to underlie the remodeling and refinement of many developing synapses (Kandel & O'Dell 1992, Goodman & Shatz 1993). NO may be well-suited to mediate some of these functions, since it is diffusible and optimally positioned to detect neural activity by virtue of the Ca^{2+}-dependence of the NOS. In addition, NOS has also been localized to neurons known to play an important role in development. For example, in the cortex, NOS is localized to a small population of interneurons that are dispersed through layers II–VI as well as in the subcortical white matter (Mizukawa et al 1988). These interneurons in the subcortical white matter are derived from the population of subplate neurons (Chun & Shatz 1989), which are known to pioneer the development of cortical connections (Ghosh et al 1990).

The notion of diffusible gaseous messengers raises the problem of how signaling specificity can be achieved. As mentioned above, one way specificity can be accomplished is to require the messenger production to coincide with synaptic activity. Indeed, this is what has been observed experimentally in the case of NO-induced increases in synaptic strength in the hippocampus (Zhou et al 1993). This concept is also highlighted by a recent report demonstrating NO-induced enhancement of immediate early gene expression (Peunova & Enikolopov 1993). These investigators observed that NO can substantially augment the Ca^{2+}-induced increases in c-fos expression, although NO alone was without effect. The facilitatory effect of NO required strict temporal contiguity of the Ca^{2+} and NO signals, indicating that coincidence detection can occur at the level of transcriptional regulation. Similar mechanisms to confer specificity may be employed in other systems that utilize diffusible messengers. Achieving a molecular understanding of NO's interaction with synaptic activity is an important area of future investigation.

Finally, although this review has dealt exclusively with potential functions of NO, it appears that there may be other forms of small diffusible signaling molecules, including CO (Verma et al 1993) and OH (Zoccarato et al 1989). The diffusibility of these messengers allows for coordinated molecular communication between ensembles of neurons, a feature not provided by conventional neurotransmitters. In addition, the extent and duration of each messenger's influence can be controlled by the different half-lives of the molecules as well as different diffusion constants. In future studies it will be interesting to see how the enzymatic pathways that make these new messenger molecules can be regulated and how these signals may ultimately interact to modulate the activity of synapses.

Literature Cited

Aizenman E, Lipton SA, Loring RH. 1989. Selective modulation of NMDA responses by reduction and oxidation. *Neuron* 2:1257–63

Aizenman E, Hartnett KA, Reynolds IJ. 1990. Oxygen free radicals regulate NMDA receptor function via a redox modulatory site. *Neuron* 5:841–46

Arnold WP, Mittal CK, Katsuki S, Murad F. 1977. Nitric oxide activates guanylate cyclase and increases guanosine 3′:5′-cyclic monophosphate levels in various tissue preparations. *Proc. Natl. Acad. Sci. USA* 74: 3203–7

Ascher P, Nowak L. 1988. The role of divalent cations in the N-methyl-D-aspartate responses of mouse central neurones in culture. *J. Physiol. (London)* 399:247–66

Barrionuevo G, Brown TH. 1983. Associative long-term potentiation in hippocampal slices. *Proc. Natl. Acad. Sci. USA* 80:7347–51

Beal MF, Kowall NW, Ellison DW, et al. 1986. Replication of the neurochemical characteristics of Huntington's disease by quinolinic acid. *Nature* 321:168–71

Beckman JS, Beckman TW, Chen J, et al. 1990. Apparent hydroxyl radical production by peroxynitrite: implications for endothelial injury from nitric oxide and superoxide. *Proc. Natl. Acad. Sci. USA* 87:1620–24

Bekkers JM, Stevens CF. 1990. Presynaptic mechanism for long-term potentiation in the hippocampus. *Nature* 346:724–29

Bohme GA, Bon C, Lemaire M, et al. 1993. Altered synaptic plasticity and memory formation in nitric oxide inhibitor treated rats. *Proc. Natl. Acad. Sci. USA* In press

Bohme GA, Bon C, Stutzmann JM, et al. 1991. Possible involvement of nitric oxide in long-term potentiation. *Eur. J. Pharmacol.* 199: 379–81

Bon C, Bohme GA, Doble A, et al. 1992. A role for nitric oxide in long-term potentiation. *Eur. J. Neurosci.* 4:420–24

Bonhoeffer T, Staiger V, Aertsen A. 1989. Synaptic plasticity in the rat hippocampal slice cultures: local "Hebbian" conjunction or pre- and postsynaptic stimulation leads to distributed synaptic enhancement. *Proc. Natl. Acad. Sci. USA* 86:8113–17

Bredt DS, Ferris CD, Snyder SH. 1992. Nitric oxide synthase regulatory sites: phosphorylation by cyclic AMP dependent protein kinase, protein kinase C, calcium/calmodulin protein kinase; identification of flavin and calmodulin binding sites. *J. Biol. Chem.* 267:10976–81

Bredt DS, Glatt CE, Hwang PM, et al. 1991b. Nitric oxide synthase protein and mRNA are discretely localized in neuronal populations of mammalian central nervous system together with NADPH diaphorase. *Neuron* 7:615–24

Bredt DS, Hwang PH, Glatt C, et al. 1991c. Cloned and expressed nitric oxide synthase structurally resembles cytochrome P-450 reductase. *Nature* 351:714–18

Bredt DS, Hwang PH, Snyder SH. 1990. Localization of nitric oxide synthase indicating a neural role for nitric oxide. *Nature* 347:768–70

Bredt DS, Snyder SH. 1990. Isolation of nitric oxide synthetase, a calmodulin-requiring enzyme. *Proc. Natl. Acad. Sci. USA* 87:682–85

Brune B, Lapetina EG. 1989. Activation of a cytosolic ADP-ribosyltransferase by nitric oxide generating agents. *J. Biol. Chem.* 264:8455–58

Brune B, Lapetina EG. 1990. Properties of a novel nitric oxide-stimulated ADP-ribosyltransferase. *Arch. Biochem. Biophys.* 279: 286–90

Brune B, Lapetina EG. 1991. Phosphorylation of nitric oxide synthase by protein kinase A. *Biochem. Biophys. Res. Commun.* 181:921–26

Brune B, Ullrich V. 1987. Inhibition of platelet aggregation by carbon monoxide is mediated by activation of guanylate cyclase. *Mol. Pharmacol.* 32:497–504

Burchuladze R, Rose SPR. 1990. Memory formation in the chick depends on membrane bound protein kinase C. *Brain Res.* 535:131–38

Chapman PF, Atkins CM, Allen MT, et al. 1992. Inhibition of nitric oxide synthesis impairs two different forms of learning. *NeuroReport* 3:567–70

Chetkovich DM, Klann E, Sweatt JD. 1993. Nitric oxide synthase-independent long-term potentiation in area CA1 of hippocampus. *NeuroReport* 4:919–22

Chun JJM, Shatz CJ. 1989. The earliest generated neurons of the cat cerebral cortex: characterization by MAP2 and neurotransmitter immunohistochemistry during fetal life. *J. Comp. Neurol.* 282:555–69

Collingridge GL, Kehl SJ, McLennan H. 1983. Excitatory amino acids in synaptic transmission in the Schaffer collateral-commisural pathway of the rat hippocampus. *J. Physiol.* 334:33–46

Costa A, Trainer P, Besser M, Grossman A. 1993. Nitric oxide modulates the release of corticotropin-releasing hormone from rat hypothalamus in vitro. *Brain Res.* 605:187–92

Crepel F, Jaillard D. 1990. Protein kinases, nitric oxide, and long-term depression of synapses in the cerebellum. *NeuroReport.* 1:133–36

Crepel F, Krupa M. 1988. Activation of protein kinase C induces a long-term depression of glutamate sensitivity of cerebellar Purkinje cells. An in vitro study. *Brain Res.* 458:397–401

Daniel H, Hemart N, Jaillard D, Crepel F. 1993. Long-term depression requires nitric oxide and guanosine 3′:5′ cyclic monophosphate production in cerebellar purkinje cells. *Eur. J. Neurosci.* 5:1079–82

Dawson TM, Bredt DS, Fotuhi M, et al. 1991a. Nitric oxide synthase and neuronal NADPH diaphorase are identical in brain and peripheral tissues. *Proc. Natl. Acad. Sci. USA* 88:7797–801

Dawson VL, Dawson TM, London ED, et al. 1991b. Nitric oxide mediates glutamate neurotoxicity in primary cortical culture. *Proc. Natl. Acad. Sci. USA* 88:6368–71

Demerle-Pallardy C, Lonchampt M-O, Chabrier P-E, Braquet P. 1991. Absence of implication of L-arginie/Nitric oxide pathway on neuronal cell injury induced by l-glutamate or hypoxia. *Biochem. Biophys. Res. Commun.* 181:456–64

Desai KM, Sessa WC, Vane JR. 1991. Involvement of nitric oxide in the reflex relaxation of the stomach to accomodate food or fluid. *Nature* 351:477–79

de Vente J, Bol JGJM, Steinbusch HWM. 1989. Localization of cGMP in the cerebellum of the adult rat: an immunohistochemical study. *Brain Res.* 504:332–37

Dimmeler S, Brune B. 1992. Characterization of a nitric oxide-catalyzed ADP-ribosylation of glyceraldehyde-3-phosphate dehydrogenase. *Eur. J. Biochem.* 210:305–10

Dimmeler S, Lottspeich F, Brune B. 1992. Nitric oxide causes ADP-ribosylation and inhibition of glyceraldehyde-3-phosphate dehydrogenase. *J. Biol. Chem.* 267:16771–74

Doerner D, Alger BE. 1988. Cyclic GMP depresses hippocampal Ca^{2+} current through a mechanism independent of cGMP-dependent protein kinase. *Neuron* 1:693–99

Drapier JC, Hibbs JB. 1986. Murine cytotoxic activated macrophages inhibit aconitase in tumor cells. Inhibition involves the iron-sulfer prosthetic group and is reversible. *J. Clin. Invest.* 78:790–97

Duman RS, Terwilliger RZ, Nestler EJ. 1991. Endogenous ADP-ribosylation in brain: initial characterization of substrate proteins. *J. Neurochem.* 57:2124–32

East SJ, Batchelor AM, Garthwaite J. 1991. Selective blockade of N-methyl-D-asparate receptor function by the nitric oxide donor, nitroprusside. *Eur. J. Pharmacol.* 209:119–21

Ehret-Hilberer S, Nullans G, Aunis D, Virmaux N. 1992. Mono ADP-ribosylation of transducin catalyzed by rod outer segment extract. *FEBS Lett.* 309:394–98

Ferrante RJ, Kowall NW, Richardson EP, et al. 1985. Selective sparing of a class of striatal neurons in Huntington's disease. *Science* 230:561–63

Forstermann U, Schmidt HHHW, Kohlhass KL, Murad F. 1992. Induced RAW 267.4 macrophages express soluble and particulate nitric oxide synthase: inhibition by transferring growth factor-beta. *Eur. J. Pharmacol.* 225:161–65

Forstermann U, Pollock JS, Schmidt HHHW, et al. 1991. Calmodulin-dependent endothelium-derived relaxing factor/nitric oxide synthase activity is present in the particulate and cytosolic fractions of bovine aortic endothelial cells. *Proc. Natl. Acad. Sci. USA* 88:1788–92

Foster TC, McNaughton BL. 1991. Long-term enhancement of CA1 synaptic transmission is due to increased quantal size, not quantal content. *Hippocampus* 1:79–91

Furchgott RF, Jothianandan D. 1991. Endothelium-dependent and independent vasodilation involving cGMP: relaxation induced by nitric oxide, carbon monoxide, and light. *Blood Vessels* 28:52–61

Furchgott RF, Zawadski JV. 1980. The obligatory role of endothelial cells in the relaxation of arterial smooth muscle by acetylcholine. *Nature* 288:373–76

Galea E, Feinstein DL, Reis DJ. 1992. Induction of calcium-independent nitric oxide synthase in primary rat glial cultures. *Proc. Natl. Acad. Sci. USA* 89:10945–49

Gally JA, Montague PR, Reeke GN, Edelman GM. 1990. The NO hypothesis: possible effects of a short-lived, rapidly diffusible signal in the development and function of the nervous system. *Proc. Natl. Acad. Sci. USA* 87:3547–51

Garthwaite J, Charles SL, Chess-Williams R. 1988. Endothelium-derived relaxing factor release on activation of NMDA receptors suggests role as intracellular messenger in the brain. *Nature* 336:385–88

Garthwaite J, Garthwaite G. 1987. Cellular origins of cGMP responses to excitatory amino acid receptor agonists in rat cerebellum in vitro. *J. Neurochem.* 48:29–39

Geller DA, Lowenstein CJ, Shapiro RA, et al. 1993. Molecular cloning and expression of inducible nitric oxide synthase from human hepatocytes. *Proc. Natl. Acad. Sci. USA* 90:3491–95

Ghosh A, Antonini A, McConnell SK, Shatz CJ. 1990. Requirement for subplate neurons in the formation of thalamocortical connections. *Nature* 347:179–81

Goh JW, Pennefather PS. 1989. A pertussis toxin-sensitive G protein in hippocampal long-term potentiation. *Science* 244:980–83

Goodman CS, Shatz CJ. 1993. Developmental mechanisms that generate precise patterns of neuronal connectivity. *Neuron* 10:77–98

Granger DL, Taintor RR, Cook JL, Hibbs JB. 1980. Injury of neoplastic cells by murine macrophages leads to inhibition of mitochondrial respiration. *J. Clin. Invest.* 65:357–60

Gribkoff VK, Lum-Ragan JT. 1992. Evidence for nitric oxide synthase inhibitor-sensitive and insensitive hippocampal synaptic potentiation *J. Neurophysiol.* 68:639–42

Griffith TM, Edwards DH, Newby AC, et al. 1986. Production of endothelium-derived relaxant factor is dependent on oxidative phosphorylation and extracellular calcium. *Cardiovasc. Res.* 20:7–12

Haley JE, Wilcox GL, Chapman PF. 1992. The role of nitric oxide in long-term potentiation. *Neuron* 8:211–16

Haley JE, Malen PL, Chapman PF. 1993. Nitric oxide synthase inhibitors block LTP induced by weak, but not strong, tetanic stimulation at physiological brain temperatures in rat hippocampal slices. *Neurosci. Lett.* In press

Hanbauer I, Wink D, Osawa Y, et al. 1992. Role of nitric oxide in NMDA-evoked release from [^{3}H]-dopamine striatal slices. *NeuroReport* 3:409–12

Hibbs JB, Vavrin Z, Taintor RR. 1987. L-arginine is required for expression of the activated macrophage effector mechanism causing selective metabolic inhibition in target cells. *J. Immunol.* 138:550–56

Hibbs JB, Taintor RR, Vavrin Z, Rachlin EM. 1988. Nitric oxide: a cytotoxic activated macrophage effector molecule. *Biochem. Biophys. Res. Commun.* 157:87–94

Hiki K, Hattori R, Kawai C, Yui Y. 1992. Purification of insoluble nitric oxide synthase from rat cerebellum. *J. Biochem.* 111:556–58

Hirano T. 1991. Differential pre- and postsynaptic mechanisms for synaptic potentiation and depression between a granule cell and a Purkinje cell in rat cerebellar culture. *Synapse* 7:321–23

Holscher C, Rose SPR. 1992. An inhibitor of nitric oxide synthesis prevents memory formation in the chick. *Neurosci. Lett.* 145:165–67

Hope BT, Michael GJ, Knigge KM, Vincent SR. 1991. Neuronal NADPH diaphorase is a nitric oxide synthase. *Proc. Natl. Acad. Sci. USA* 88:2811–14

Hoyt KR, Tang L-H, Aizenman E, Reynolds IJ. 1992. Nitric oxide modulates NMDA-induced increases in intracellular Ca^{2+} in cultured rat forebrain neurons. *Brain Res.* 592:310–16

Hyman BT, Marzloff K, Wenniger JJ, et al. 1992. Relative sparing of nitric oxide syn-thase-containing neurons in the hippocampal formation in Alzheimer's disease. *Ann. Neurol.* 32:818–20

Ignarro LJ, Buga GM, Wood KS, et al. 1987. Endothelium-derived relaxing factor produced and released from artery and vein is nitric oxide. *Proc. Natl. Acad. Sci. USA* 84:9265–69

Ito M. 1989. Long-term depression. *Annu. Rev. Neurosci.* 12:85–102

Ito M, Karachot L. 1989. Long-term desensitization of quisqualate-specific glutamate receptors in Purkinje cells investigated with wedge recordings from rat cerebellar slices. *Neurosci. Res.* 458:397–401

Ito M, Karachot L. 1990. Messengers mediating long-term desensitization in cerebellar Purkinje cells. *NeuroReport* 1:129–32

Ito M, Sakurai M, Tongroach P. 1982. Climbing fiber induced depression of both mossy fibre responsiveness and glutamate sensitivity of cerebellar Purkinje cells. *J. Physiol.* 324:113–24

Izumi Y, Clifford DB, Zorumski CF. 1992. Inhibition of long-term potentiation by NMDA-mediated nitric oxide release. *Science* 257:1273–76

Jahr CE, Stevens CF. 1987. Glutamate activates multiple single channel conductances in hippocampal neurones. *Nature* 325:522–25

Janssens SP, Shimouchi A, Quertermous T, et al. 1992. Cloning and expression of a cDNA encoding human endothelium-derived relaxing factor/nitric oxide synthase. *J. Biol. Chem.* 267:14519–22

Kandel ER, O'Dell TJ. 1992. Are adult learning mechanisms also used for development? *Science* 258:243–45

Kaplan JM, Mardon G, Bishop JM, Varmus HE. 1988. The first seven amino acids encoded by the v-src oncogene act as a myristylation signal: lysine 7 is a critical determinant. *Mol. Cell. Biol.* 8:2435–41

Kawano H, Daikoku S, Shibasaki T. 1988. CRF-containing neuron systems in the rat hypothalamus: retrograde tracing and immunohistochemical studies. *J. Comp. Neurol.* 272:260–68

Kiedrowski L, Costa E, Wroblewski JT. 1992. Sodium nitroprusside inhibits N-methyl-D-aspartate-evoked calcium influx via a nitric oxide- and cGMP independent mechanism. *Mol. Pharmacol.* 41:779–84

Kiedrowski L, Manev H, Costa E, Wroblewski JT. 1991. Inhibition of glutamate-induced cell death by sodium nitroprusside is not mediated by nitric oxide. *Neuropharmacology* 30:1241–43

Knowles RG, Palacios M, Palmer RMG, Moncada S. 1989. Formation of nitric oxide from L-arginine in the central nervous system: a transduction mechanism for stimulation of soluble guanylate cyclase. *Proc. Natl. Acad. Sci. USA* 86:5159–62

Koh J-Y, Peters S, Choi DW. 1986. Neurons containing NADPH-diaphorase are selectively resistant to quinolate toxicity. *Science* 234:73–76

Kossel A, Bonhoeffer T, Bolz J. 1990. Non-Hebbian synapses in rat visual cortex. *NeuroReport* 1:115–18

Kots AY, Skurat AV, Sergienko EA, et al. 1992. Nitroprusside stimulates the cysteine-specific mono(ADP-ribosylation) of glyceraldehyde-3-phosphate dehydrogenase from human erythrocytes. *FEBS Lett.* 300:9–12

Kullmann DM, Nicoll RA. 1992. Long-term potentiation is associated with increases in quantal content and quantal amplitude. *Nature* 357:240–44

Lamas S, Marsden PA, Li GK, et al. 1992. Endothelial nitric oxide synthase: molecular cloning and characterization of a distinct constitutive enzyme isoform. *Proc. Natl. Acad. Sci. USA* 89:6348–52

Larkman A, Hannay T, Stratford K, Jack J. 1992. Presynaptic release probability influences the locus of long-term potentiation. *Nature* 360:70–73

Lazarewicz JW, Wroblewski JT, Palmer ME, Costa E. 1989. Reduction of disulfide bonds activates NMDA-sensitive glutamate receptors in primary cultures of cerebellar granule cells. *Neurosci. Res Commun.* 4:91–97

Lei SZ, Pan Z-H, Aggarwal SK, et al. 1992. Effect of nitric oxide production on the redox modulatory site of the NMDA receptor-channel complex. *Neuron* 8:1087–99

Lerner-Natoli M, Rondouin G, deBock F, Bockaert J. 1992. Chronic NO synthase inhibition fails to protect hippocampal neurones against NMDA toxicity. *NeuroReport* 3:1109–12

Li YG, Errington ML, Williams JH, Bliss TVP. 1992. Temperature-dependent block of LTP by the NO synthase inhibitor L-NARG. *Soc. Neurosci. Abstr.* 18:342

Linden DJ, Connor JA. 1991. Participation of postsynaptic PKC in cerebellar long-term depression in culture. *Science* 254:1656–59

Linden DJ, Connor JA. 1992. Long-term depression of glutamate currents in cultured cerebellar purkinje neurons does not require nitric oxide signalling. *Eur. J. Neurosci.* 4:10–15

Linden DJ, Dickinson MH, Smeyne M, Connor JA. 1991. A long-term depression of AMPA currents in cultured cerebellar Purkinje neurons. *Neuron* 7:81–89

Lohmann SM, Walter U, Miller PE, et al. 1981. Immunohistochemical localization of cyclic GMP-dependent protein kinase in mammaliam brain. *Proc. Natl. Acad. Sci. USA* 78:653–57

Lonart G, Wang J, Johnson KM. 1992. Nitric oxide induces neurotransmitter release from hippocampal slices. *Eur. J. Pharmacol.* 220: 271–72

Lovinger DM, Wong KL, Murakami K, Routtenberg A. 1987. Protein kinase C inhibitors eliminate long-term potentiation. *Brain Res.* 436:177–83

Lowenstein CJ, Glatt CS, Bredt DS, Snyder SH. 1992. Cloned and expressed macrophage nitric oxide synthase contrasts with the brain enzyme. *Proc. Natl. Acad. Sci. USA* 89:6711–15

Lustig HS, von Brauchitsch KL, Chan J, Greenberg DA. 1992. cGMP modulators and excitotoxic injury in cerebral cortical cultures. *Brain Res.* 577:343–46

Lynch G, Baudry M. 1984. The biochemistry of memory: a new and specific hypothesis. *Science* 224:1057–63

Lynch G, Larson J, Kelso S, et al. 1983. Intracellular injections of EGTA block induction of hippocampal long-term potentiation. *Nature* 305:719–21

Lyons CR, Orloff GJ, Cunningham JM. 1992. Molecular cloning and functional expression of an inducible nitric oxide synthase from a murine macrophage cell line. *J. Biol. Chem.* 267:6370–74

MacDermott AB, Mayer ML, Westbrook GL, et al. 1986. NMDA-receptor activation increases cytoplasmic calcium concentration in cultured spinal cord neurones. *Nature* 321:519–22

Madison DV, Malenka RC, Nicoll RA. 1991. Mechanisms underlying long-term potentiation of synaptic transmission. *Annu. Rev. Neurosci.* 14:379–97

Malenka RC, Kauer JA, Perkel DJ, et al. 1989. An essential role for postsynaptic calmodulin and protein kinase activity in long-term potentiation. *Nature* 340:554–57

Malenka RC, Kauer JA, Zuker RJ, Nicoll RA. 1988. Postsynaptic calcium is sufficient for potentiation of hippocampal synaptic transmission. *Science* 242:81–84

Malgaroli A, Tsien RW. 1992. Glutamate-induced long-term potentiation of the frequency of miniature synaptic currents in cultured hippocampal neurones. *Nature* 357: 134–39

Malinow R. 1991. Transmission between pairs of hippocampal slice neurons: quantal levels, oscillations and LTP. *Science* 252:722–24

Malinow R, Madison DV, Tsien RW. 1988. Persistent protein kinase activity underlies long-term potentiation. *Nature* 335:820–24

Malinow R, Miller JP. 1986. Postsynaptic hyperpolarization during conditioning reversibly blocks induction of long-term potentiation. *Nature* 320:529–30

Malinow R, Schulman R, Tsien RW. 1989. Inhibition of postsynaptic PKC or CAMKII blocks induction but not expression of LTP. *Science* 245:862–66

Malinow R, Tsien RW. 1990. Presynaptic enhancement shown by whole-cell recordings of long-term potentiation in hippocampal slices. *Nature* 346:177–80

Manabe T, Renner P, Nicoll RA. 1992. Postsynaptic contribution to long-term potentiation revealed by the analysis of miniature synaptic currents. *Nature* 355:50–55

Manzoni O, Prezeau L, Deshager S, et al. 1992a. Sodium nitroprusside blocks NMDA receptors via formation of ferrocyanide ions. *NeuroReport* 3:77–80

Manzoni O, Prezeau L, Marin P, et al. 1992b. Nitric oxide-induced blockade of NMDA receptors. *Neuron* 8:653–62

Marletta MA. 1989. Nitric oxide: biosynthesis and biological significance. *Trends Biol. Sci.* 14:488–92

Marsden PA, Schappert KT, Chen HS, et al. 1992. Molecular cloning and characterization of human endothelial nitric oxide synthase. *FEBS Lett.* 307:287–93

Mayer ML, Westbrook GL. 1987. Permeation and block of N-methyl-D-aspartic acid receptor channels by divalent cations in mouse central neurones. *J. Physiol.* 394:501–28

McCormick DA, Thompson RF. 1984. Cerebellum: essential involvement in the classically conditioned eyelid response. *Science* 223:296–99

Miki N, Kawabe Y, Kuriyama K. 1977. Activation of cerebral guanylate cyclase by nitric oxide. *Biochem. Biophys. Res. Commun.* 75:851–56

Molina y Vedia L, McDonald B, Reep B, et al. 1992. Nitric oxide-induced S-nitrosylation of glyceraldehyde-3-phosphate dehydrogenase inhibits enzyme activity and increase endogenous ADP-ribosylation. *J. Biol. Chem.* 267:24929–32

Mizukawa K, Vincent SR, McGeer PL, McGeer EG. 1988. Ultrastructure of reduced nicotinamide dinucleotide phosphate (NADPH) diaphorase- positive neurons in the cat cerebral cortex, amygdala and caudate putamen. *Brain Res.* 452:286–92

Montague PR, Gally JA, Edelman GM. 1991. Spatial signaling in the development and function of neural connections. *Cereb. Cortex* 1:199–220

Morris RGM. 1984. Development of a water maze procedure for studying spatial learning in the rat. *J. Neurosci. Meth.* 11:47–60

Morris RGM, Anderson E, Lynch GS, Baudry M. 1986. Selective impairment of learning and blockade of long-term potentiation by an N-methyl-D-aspartate receptor antagonist AP5. *Nature* 319:774–76

Morris RGM, Schenk F, Tweedie F, Jarrard L. 1990. Ibotenate lesions of the hippocampus and/or subiculum: dissociating components of allocentric spatial learning. *Eur. J. Neurosci.* 2:1016–28

Murad F, Mittal CK, Arnold WP, et al. 1978. Guanylate cyclase: activation by azide, nitro compounds, nitric oxide, and hydroxyl radical and inhibition by hemoglobin and myoglobin. *Adv. Cyclic Nucl. Res.* 9:145–58

Nakaki T, Nakayama M, Kato R. 1990. Inhibition by nitric oxide and nitric oxide-producing vasodilators of DNA synthesis in vascular smooth muscle cells. *Eur. J. Pharmacol.* 189:347–53

Nakane M, Ichikawa M, Deguchi T. 1983. Light and electron microscopic demonstration of guanylate cyclase in rat brain. *Brain Res.* 273:9–15

Nakane M, Mitchell J, Forstermann U, Murad F. 1991. Phosphorylation by calcium-calmodulin-dependent protein kinase II and protein kinase C modulates the activity of nitric oxide synthase. *Biochem. Biophys. Res. Commun.* 180:1396–402

Nowak L, Bregestovski P, Ascher P, et al. 1984. Magnesium gates glutamate-activated channels in mouse central neurones. *Nature* 307:462–65

Nowicki JP, Duval D, Poignet H, Scatton B. 1991. Nitric oxide mediates neuronal death after focal cerebral ischemia in the mouse. *Eur. J. Pharmacol.* 204:339–40

O'Dell TJ, Hawkins RD, Kandel ER, Arancio O. 1991. Tests on the roles of two diffusible substances in LTP: evidence for nitric oxide as a possible early retrograde messenger. *Proc. Natl. Acad. Sci. USA* 88:11285–89

Palmer RMJ, Ferrige AG, Moncada S. 1987. Nitric oxide accounts for the biological activity of endothelium-derived relaxing factor. *Nature* 327:524–26

Pauwels P, Leysen JE. 1992. Blockade of nitric oxide formation does not prevent glutamate-induced neurotoxicity in neuronal cultures from rat hippocampus. *Neurosci. Lett.* 143:27–30

Peunova N, Enikolopov G. 1993. Amplification of calcium-induced gene transcription by nitric oxide in neuronal cells. *Nature.* 364:450–53

Pollock JS, Förstermann U, Mitchell JA, Warner TA, Schmidt HHHW, et al. 1991. Purification and characterization of particulate endothelium-derived relaxing factor synthase from cultured and native bovine aortic endothelial cells. *Proc. Natl. Acad. Sci. USA* 88:10480–84

Pollock JS, Klinghofer V, Forstermann U, Murad F. 1992. Endothelial nitric oxide synthase is myristylated. *FEBS Lett.* 309:402–4

Prast H, Phillipu A. 1992. Nitric oxide releases acetylcholine in the basal forebrain. *Eur. J. Pharmacol.* 216:139–40

Puttfarcken PS, Lyons WE, Coyle JT. 1992.

Dissociation of nitric oxide generation and lainiate-mediated neuronal degeneration in primary cultures of rat cerebellar granule cells. *Neuropharmacology* 31:565–75

Rajfer J, Aronson WJ, Bush PA, et al. 1992. Nitric oxide as a mediator of the corpus cavernosum in response to nonadrenergic non cholinergic transmission. *N. Engl. J. Med.* 326:90–94

Rapoport RM, Draznin MB, Murad F. 1983. Endothelium-dependent relaxation in rat aorta may be mediated through cGMP-dependent phosphorylation. *Nature* 306:174–76

Rashatwar SS, Cornwell TL, Lincoln TM. 1987. Effects of 8-bromo-cGMP on Ca^{2+} levels in vascular smooth muscle cells: possible regulation of Ca^{2+}-ATPase by cGMP-dependent protein kinase. *Proc. Natl. Acad. Sci. USA* 84:5685–89

Schmidt HHHW, Pollock JS, Nakane M, et al. 1992. Ca^{2+}/calmodulin-regulated nitric oxide synthases. *Cell Calcium* 13:427–34

Schuman EM, Madison DV. 1991a. A requirement for the intercellular messenger nitric oxide in long-term potentiation. *Science* 254:1503–6

Schuman EM, Madison DV. 1991b. An inhibitor of nitric oxide synthase prevents long-term potentiation (LTP). *Soc. Neurosci. Abstr.* 17:2

Schuman EM, Madison DV. 1993. Long-term potentiation induced in a single CA1 pyramidal neuron can enhance nearby, but not distant, synapses. *Soc. Neurosci. Abstr.* In press

Schuman EM, Meffert MK, Schulman H, Madison DV. 1992. A potential role for an ADP-ribosyltransferase in hippocampal long-term potentiation. *Soc. Neurosci. Abstr.* 18:761

Schweizer FE, Wendland B, Ryan TA, et al. 1993. Evidence for the presence of nitric oxide synthase in rat hippocampal pyramidal cells. *Soc. Neurosci. Abstr.* In press

Sessa WC, Harrison JK, Barber CM, et al. 1992. Molecular cloning and expression of a cDNA encoding endothelial cell nitric oxide synthase. *J. Biol. Chem.* 267:15274–76

Shibuki K, Okada D. 1990. Long-term synaptic changes in rat cerebellar slices reflected in extracellular K+ activity. *Neurosci. Lett.* 113:34–39

Shibuki K, Okada D. 1991. Endogenous nitric oxide release required for long-term synaptic depression in the cerebellum. *Nature* 349:326–28

Silva AJ, Stevens CF, Tonegawa S, Wang Y. 1992. Deficient hippocampal long-term potentiation in α-calcium-calmodulin kinase II mutant mice. *Science* 257:201–6

Southam E, Garthwaite J. 1991. Comparative effects of some nitric oxide donors on cyclic GMP levels in rat cerebellar slices. *Neurosci. Lett.* 130:107–11

Stuehr DJ, Marletta MA. 1987. Induction of nitrite/nitrate synthesis in murine macrophages by BCG infection, lymphokines, or interferon γ. *J. Immunol.* 139:518–25

Tauck DL. 1992. Redox modulation of NMDA receptor-mediated synaptic activity in the hippocampus. *NeuroReport* 3:781–84

Ueda K, Hayaishi O. 1985. ADP-ribosylation. *Annu. Rev. Biochem.* 54:73–100

Valtschanoff JG, Weinberg RJ, Kharazia VN, et al. 1993. Neurons in the rat hippocampus that synthesize nitric oxide. *J. Comp. Neurol.* 330:1–11

Verma A, Hirsch DJ, Glatt CE, et al. 1993. Carbon monoxide: a putative neural messenger. *Science* 259:381–84

Vincent SR, Johansson O, Hokfelt T, et al. 1983. NADPH-diaphorase: a selective histochemical marker for striatal neurons containing both somatostatin- and avian pancreatic polypeptide (APP)-like immunoreactivities. *J. Comp. Neurol.* 217:252–63

Wallace MN, Fredens K. 1992. Activated astrocytes of the mouse hippocampus contain high levels of NADPH-diaphorase. *NeuroReport* 3:953–56

Williams JH, Errington ML, Bliss TVP. 1989. Arachidonic acid induces a long-term activity-dependent enhancement of synaptic transmission in the hippocampus. *Nature* 341:739–42

Williams MB, Li X, Gu X, Jope RS. 1992. Modulation of endogenous ADP-ribosylation in rat brain. *Brain Res.* 592:49–56

Wolin MS, Wood KS, Ignarro LJ. 1982. Guanylate cyclase from bovine lung. A kinetic analysis of the regulation of unpurified soluble enzyme by protoporphyrin IX, heme, and nitrosyl-heme. *J. Biol. Chem.* 257:11312–20

Xie Q, Cho HJ, Calayeay J, et al. 1992. Cloning and characterization of inducible nitric oxide synthase from mouse macrophages. *Science* 256:225–28

Zhang J, Snyder SH. 1992. Nitric oxide stimulates auto-ADP-ribosylation of glyceraldehyde-3-phosphate dehydrogenase. *Proc. Natl. Acad. Sci. USA* 89:9382–85

Zhou M, Small SA, Kandel ER, Hawkins RD. 1993. Nitric oxide and carbon monoxide produce long-term enhancement of synaptic transmission in the hippocampus by an activity-dependent mechanism. *Science* 260:1946–49

Zhu XZ, Luo LG. 1992. Effect of nitroprusside (nitric oxide) on endogenous dopamine release from rat striatal slices. *J. Neurochem.* 59:932–35

Zoccarato F, Deana R, Cavallini L, Alexandre A. 1989. Generation of hydrogen peroxide by cerebral-cortex synaptosomes. *Eur. J. Biochem.* 180:473–78

Annu. Rev. Neurosci. 1994. 17:185–218

THE SUBPLATE, A TRANSIENT NEOCORTICAL STRUCTURE: Its Role in the Development of Connections between Thalamus and Cortex

Karen L. Allendoerfer

Neurosciences Program, Stanford University, Stanford, California 94305; present address: Division of Biology, California Institute of Technology, Pasadena, California 91125

Carla J. Shatz

Division of Neurobiology, Department of Molecular and Cell Biology, University of California, Berkeley, California 94720

KEY WORDS: neurogenesis, pioneer neurons, waiting period, cell death, ocular dominance columns

INTRODUCTION

The functioning of the mammalian brain depends upon the precision and accuracy of its neural connections, and nowhere is this requirement more evident than in the neocortex of the cerebral hemispheres. The neocortex is a structure that is divided both radially, from the pial surface to the white matter into six cell layers, and tangentially into more than 40 different cytoarchitectural areas (Brodmann 1909). For instance, within the cerebral hemispheres, sets of tangential axonal connections link neurons within a given cortical layer to each other and also link neurons of different cortical areas; sets of radial connections link neurons of different layers together. In addition, the major input to the neocortex arises from neurons in the thalamus, which in turn receive a reciprocal set of connections from the cortex. These connections are highly restricted: In the radial domain, thalamic axons make their major projection to the neurons of cortical layer 4, and the neurons of cortical layer 6 project back to the thalamus. Connections are also restricted tangentially, in that neurons located in specific subdivisions of the thalamus send their axons to specific cortical areas. For instance, neurons in the lateral geniculate nucleus (LGN) of the thalamus connect with primary visual cortex, whereas those situated in the ventrobasal complex connect with somatosensory cortex. There are also local patterns of connections within a given cortical area, for example, the ocular dominance columns in primary visual cortex of higher mammals, or the barrels in rodent somatosensory cortex (Woolsey & van der Loos 1970). The ocular dominance columns are based on the fact that the inputs of LGN axons representing the two eyes are segregated from each other in layer 4 and their terminal arbors are clustered together in patches (LeVay et al 1980).

A primary question is how these sets of connections form during development. The purpose of this review is to consider this question as it pertains specifically to the formation of connections between thalamus and cortex [for a more general review of the formation of connectivity, see Goodman & Shatz (1993)]. Several major steps are involved in this developmental process. First, the constituent neurons of the thalamus and cortex must be generated. Next, axons must grow along the appropriate pathways and select the appropriate targets. In the visual system, this means that LGN axons must grow up through the internal capsule, bypass many other inappropriate cortical areas, and then select visual cortex. Finally, the axons must enter the cortical plate, recognize and terminate within layer 4, and segregate to form ocular dominance columns. Thus, in addition to the general problems of pathfinding and target selection faced by all developing neurons, thalamic neurons are faced with a series of tangential and radial decisions as they form the final pattern of connections within neocortex: they must choose the correct cortical area and the correct layer, and must restrict the extent of their terminal arbors. In addition, similar problems must be solved by the neurons of cortical layer 6 as they grow towards and invade their thalamic targets.

A growing body of evidence suggests that the formation of connections between thalamus and cortex requires the presence of a specific and transient cell type, subplate neurons. These neurons are present early in development, but by adulthood the majority have disappeared. Here we consider their life history and review the evidence for their role in the patterning of connections.

DEVELOPMENT AND MATURATION OF THE SUBPLATE NEURONS

The Preplate is Comprised of Subplate and Marginal Zone Cells

In the development of the cerebral cortex, an early germinal zone, the ventricular zone (VZ), gives rise, through successive rounds of cell division and migration, to the postmitotic neurons that comprise the adult cortical layers. As Figure 1 shows, during development the histology of the cerebral wall is very different from that of the adult (Boulder Committee 1970). Early in development, in addition to the VZ there is a cellular zone located immediately below the pial surface, which has been termed the preplate (Rickmann et al 1977, Stewart & Pearlman 1987) [or primordial plexiform layer (Marin-Padilla 1971)] (Figure 1b). The Golgi studies of Marin-Padilla (1971) showed that the preplate is filled with loosely packed, polymorphic cells with a neuronal morphology. As the cortex matures, a zone of densely-packed pyramidal cells appears in the middle of the preplate, and this was termed the cortical plate (Figure 1c). These observations led Marin-Padilla to propose that the primordial plexiform layer is split apart by later-forming neurons of the cortical plate. In particular, he suggested that in the adult, the outer neurons form layer 1, the inner neurons form layer 7, and the number of neurons in these two layers remains unchanged during the subsequent maturation and growth of the cortex (reviewed in Marin-Padilla 1988).

In more recent years, ^{3}H-thymidine labeling experiments have confirmed and extended this idea, and demonstrated that some revision is necessary. For example, in the development of the cat cerebral cortex, such experiments have demonstrated directly that neurons belonging to the preplate are the earliest-generated neurons of the cerebral cortex. By using ^{3}H-thymidine birthdating and autoradiography, Luskin & Shatz (1985b) determined that genesis of cells in the occipital pole begins after E21 in the cat. When fetuses are ^{3}H-thymidine–labeled at E24 (but not earlier, at E21) and analyzed at E31, labeled cells, the first postmitotic cells of the visual cortex, can be observed in the preplate. However, if this autoradiographic analysis is performed later, at E40, the early-generated population (labeled at E24) can be seen to have split into two zones, as shown in Figure 1c (Luskin & Shatz 1985b). The deeper of the two zones is called the subplate (SP), a zone first defined by Kostovic & Molliver (1974) that is situated below the cortical plate. The other zone, termed the marginal zone (MZ) (Boulder Committee 1970), is located immediately below the pial surface. In a parallel study focusing on the cortical plate, Luskin & Shatz (1985a) showed that the neurons that eventually comprise the adult cortical layers 2–6 are generated next. Those neurons constituting the deepest cortical layer (layer 6) are generated first, and those occupying the most superficial layers (layers 2 and 3) are generated last (Luskin & Shatz 1985a), as had previously been observed in rodents by Angevine & Sidman (1961).

Taken together, these observations demonstrated that in the cat the early-generated preplate is split in two by the incoming migrating cortical plate neurons, thereby creating a cellular framework consisting of the MZ and SP. That the early-generated population of cells consists of neurons has been confimed by means of ^{3}H-thymidine birthdating combined with immuno-histochemistry for neuronal markers (Chun et al 1987; Chun & Shatz 1988a,

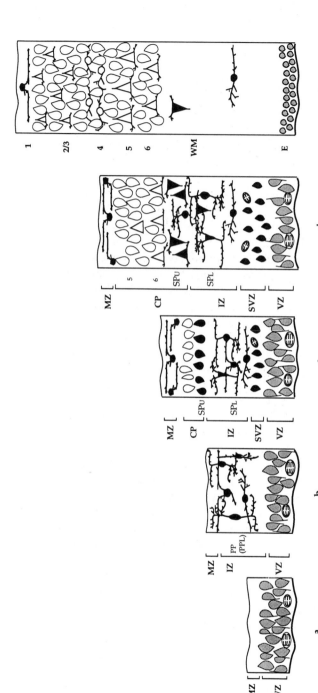

Figure 1 Schematic diagram of the histological changes in the cerebral wall during neocortical development, based on ³H-thymidine labeling studies combined with morphological studies including Golgi impregnations, DiI labeling, and immunohistochemistry. Diagram updated and modified from Boulder Committee (1970). (*a*) Initially, the cerebral wall is comprised of a germinal zone, the ventricular zone (VZ), and a marginal zone (MZ). (*b*) As the first postmitotic cells migrate from the VZ, they settle below the MZ to form the preplate (PP), or primordial plexiform layer (PPL). This is a zone of loosely-packed cells (*black*), many of which have neuronal morphologies, that is situated within the intermediate zone (IZ) just below the MZ. (*c*) With ensuing neurogenesis and migration, a cell-dense zone, the cortical plate (CP), forms. PP neurons are split into two populations that comprise the MZ and the SP. In higher mammals, ³H-thymidine labeling studies show that the SP consists of two subdivisions, upper (SP$_U$) and lower (SP$_L$). Cells residing in the SP$_U$ form the base of the cortical plate as histologically defined, but the fact that the majority disappear by adulthood, coupled with their early birthdates, indicates that the SP$_U$ cells are part of the subplate neuron population. Transient cells of the SP$_U$ are shown in black, and permanent neurons that comprise the base of layer 6 are shown in white. (*d*) At even later times in development, the cortical plate thickens as layers 5 and 6 form, and the MZ and SP neurons achieve maturity. (*e*) By adulthood, the majority of neurons in the MZ and in both the SP$_U$ and SP$_L$ have disappeared, leaving scattered interstitial neurons in the white matter (WM) and Cajal-Retzius neurons in layer 1; the VZ and subventricular zone (SVZ) have also disappeared, leaving an ependymal layer (E).

1989a); such experiments reveal striking similarities between immunostained early-generated cells and the Golgi drawings of Marin-Padilla (1971, 1988).

In addition, the ^{3}H-thymidine labeling studies suggest that at least in the cat some of the subplate neurons condense into a layer of cells in the middle of the preplate, as shown in Figure 1c. The appearance of this clear histological layer has traditionally been interpreted as signaling the formation of the cortical plate. Thymidine birthdating, however, reveals that this first accumulation of neurons into a "plate" does not go on to form the base of layer 6 in the adult. Rather, like the neurons that initially reside in the preplate, very few survive (Figure 1e); for example, neurons generated at E24 condense to form the cortical plate by E30, but few if any can be found in adult layer 6 (Luskin & Shatz 1985b) (see Figure 1 and section titled "Death of Subplate Neurons," below). Instead, adult layer 6 is formed by neurons generated later (after E30) that take up positions in the cortical plate directly above the upper subplate (Luskin & Shatz 1985b). Thus, these experiments suggest that subplate neurons reside in two zones, a loosely packed zone below the cortical plate called the lower subplate (Luskin & Shatz 1985b) and a more densely packed zone at the base of the cortical plate called the upper subplate (Luskin & Shatz 1985b) (see Figure 1c, d). A similar subdivision of the subplate into upper and lower portions has also been noted in primates (Kostovic & Rakic 1990) through histological criteria.

Thymidine-labeling studies in rodents also indicate the presence of an early-generated population of neurons that forms a preplate. However, the exact relationship between these early-generated neurons and the permanent neurons of the cortical plate is harder to determine. In higher mammals such as cats, monkeys, and ferrets, where the period of neurogenesis is prolonged, a single injection of ^{3}H-thymidine can label a subset of neurons within a single cortical layer, whereas in rodents, because of the rapid pace of cortical neurogenesis (see Table 1), a similar injection typically labels cells in multiple layers. Therefore, the ^{3}H-thymidine labeling technique can be used reliably in higher mammals to identify subplate neurons based on their early birthdates and the fact that many of these earliest-generated cells are not present in the adult, whereas in studies of rodent subplate great care must be taken to ensure that only the earliest-generated cells (rather than the permanent cells of layer 6) are being examined. For example, in studies of neurogenesis in the mouse cortex, Wood et al (1992) pinpointed accurately the first day of neurogenesis of subplate and marginal zone cells as E12 by demonstrating (a) that a single injection of label on E11 produced no heavily labeled cells and (b) that few if any cells labeled at E12 come to reside in the cortical plate, but instead the vast majority were located in the subplate and marginal zones. In addition, the majority of cells labeled at E12 had disappeared by P21 (see Table 1). Woo et al (1991) reached similar conclusions in their studies of the hamster cortex, in which they carefully followed the fates of only the heavily-labeled cells following a single injection of tritiated thymidine at E9 or E10 (see Table 1). In the rat, Bayer & Altman (1990) used an indirect subtraction method involving injection of thymidine on two consecutive days to calculate the percentage of subplate or marginal zone cells generated on a given day. They concluded that the period of neurogenesis of subplate and marginal zone cells began at E13 (= E12, using E0 as day of insemination) and was rather prolonged. However, they did not examine any animals receiving thymidine injections earlier than E13, so the exact time of onset of subplate neurogenesis was not determined. In addition, they did not follow their thymidine-labeled cells into late postnatal life, so they also could not determine whether these cells disappear in the adult. König & Marty (1981) also birthdated cortical neurons in the rat and showed that no cells could be heavily labeled on E11, which suggests that E12 is the first day on which postmitotic neurons are generated. Valverde et al (1989) have confirmed and extended these results by showing that the majority of the neurons generated on E12 come to reside in layer 1 and layer 6b, which most likely corresponds to the subplate in the rodent. Al-Ghoul & Miller (1989) gave a single thymidine injection at E12 (see Table 1) to study rat subplate neurons. In their studies, heavily-labeled cells were exclusively restricted to a zone at the base of the cortical plate at birth, but the majority had disappeared by P20, indicating that cells generated

Table 1 Birth and death of subplate neurons in various species

	Mouse	Rat	Hamster	Ferret	Cat	Sheep	Monkey	Human
Birthdate[a]	E12–13	E12–?	E9–10	E20–24	E24–30	E26–31	E38–48	
Waiting period[b] (thalamic)		E16–17		E27–P5[c]	E36–50		E78–124[d]	
Death[e]	≥80% by P21	yes	50–80% between P4 & adult		90% by 4 mos.	yes	yes	yes
SP/CP ratio[f]	1 : 2	1 : 2			1 : 1		3 : 1	4 : 1

[a] Angevine & Sidman 1961, Kostovic & Rakic 1980, König & Marty 1981, Luskin & Shatz 1985a, Al-Ghoul & Miller 1989, Jackson et al 1989, Bayer & Altman 1990, Woo et al 1991, Saunders et al 1992, Wood et al 1992. Based on ^{3}H-thymidine or BrdU labeling.

[b] Rakic 1974, Wise & Jones 1978, Catalano et al 1991, Erzurumlu & Jhaveri 1992, Ghosh & Shatz 1992b. Based on direct labeling of thalamic afferents.

[c] E27: Johnson & Casagrande 1993; P5: K Herrmann & CJ Shatz, unpublished observations. Estimates based on preliminary DiI labeling.

[d] Rakic 1977. By E78, fibers have already reached visual subplate, by E124, many have grown in to layer 4. Times earlier than E78 or between E78 and E124 have not been reported.

[e] Al-Ghoul & Miller 1989, Chun & Shatz 1989b, Kostovic & Rakic 1990, Woo et al 1991, Saunders et al 1992, Wood et al 1992. Where quantitation is available, numbers have been included.

[f] Kostovic & Rakic 1990. Calculated on the basis of area measurements.

at E12 in the rat cortex are not only the earliest generated, but are also transient. Thus, a consensus is emerging concerning the birthdates of the earliest-generated cells in several species of rodents as well (see Table 1).

Interestingly, in the reeler mouse (Caviness & Sidman 1973, Caviness & Rakic 1978, Goffinet 1992), the preplate layer is not split into two tiers (Ogawa et al 1992); instead, the cortical plate neurons continue to accumulate in an outside-in fashion, such that the cortex forms beneath the early-generated cells with layer 6 distal and layer 2 proximal to the ventricular zone (Caviness & Rakic 1978, Caviness & Frost 1983). Ogawa et al (1992) performed molecular studies of the subplate and marginal zone regions of reeler mice to determine whether interactions between preplate and cortical plate neurons were essential for normal laminar formation. By immunizing reeler mice with homogenates of wild-type fetal cortices, they generated a monoclonal antibody, CR-50, that recognizes a transient cell-surface molecule expressed on the marginal zone cells of wild-type but not reeler embryos. When wild-type embryos are treated with injections of the CR-50 antibody at E11, the cortex at E13 and E16 resembles that of reeler embryos: the preplate is not split by the cortical plate neurons, suggesting that the antigen recognized by CR-50 is instrumental in directing this early split of the preplate into subplate and marginal zone. This observation strengthens the hypothesis that the cellular framework consisting of the early-generated marginal zone and subplate cells plays a role in patterning the subsequent development of the cortical layers.

Subplate Neurons Mature Early and Participate in Functional Neural Circuits

Not only are subplate neurons the earliest-generated neurons of the cortex, they are also the earliest to mature, differentiate, and participate in complex neural circuits during fetal life. In the cat, cells labeled with ^{3}H-thymidine at E24 and then immunostained have been shown to express MAP-2 (Chun & Shatz 1989a) and peptide neurotransmitters (summarized in Table 2) well before the neurons of the cortical plate do (Chun et al 1987; Chun & Shatz 1989a,b; reviewed in Shatz et al 1988, 1990), demonstrating not only that they are neurons, but also that they acquire their adult neuronal characteristics very early.

In the adult, two classes of cortical plate neurons, interneurons and projection neurons, exhibit different transmitter phenotypes: interneurons are immunoreactive for GABA and frequently colocalize a neuropeptide such as somatostatin, NPY, or CCK (Hendry et al 1984), whereas at least some projection neurons are likely to use excitatory amino acids (Baughman & Gilbert 1980, Giuffrida & Rustioni 1989). To investigate whether subplate neurons exhibit similar phenotypes, Antonini & Shatz (1990) injected retrograde tracers into the subplate, cortical plate, or distant targets such as

Table 2 Phenotypes of subplate and marginal zone cells[a]

Projection sites	Localization	Species (reference)
Thalamus	SPN	Cat (Gilbert & Kelly 1975, McConnell et al 1989)
	SPN	Ferret (Antonini & Shatz 1990)
	SPN	Rat (De Carlos & O'Leary 1992)
Contralateral hemisphere	SPN	Cat (Chun et al 1987)
	SPN	Ferret (Antonini & Shatz 1990)
Superior colliculus	SPN	Cat (McConnell et al 1989)
Cortical plate	SPN	Cat (Chun et al 1987, Wahle & Meyer 1987, Antonini & Shatz 1990, Friauf et al 1990)
	SPN	Ferret (Antonini & Shatz 1990)
Local connections within subplate	SPN	Cat (Chun et al 1987, Antonini & Shatz 1990)
	SPN	Ferret (Antonini & Shatz 1990)
Classical neurotransmitters/receptors		
GABA	SPZ/?	Monkey (Huntley et al 1988, Meinecke & Rakic 1992)
	SPN/MZ	Cat (Chun & Shatz 1989a, 1989b)
	SPN/?	Ferret (Antonini & Shatz 1990)
	SPZ/?	Rat (Lauder et al 1986)
—GABA-A receptor	SPZ/?	Monkey (Meinecke & Rakic 1992)
EAA uptake mechanisms	SPN/?	Cat (Antonini & Shatz 1990)
	SPN/?	Ferret (Antonini & Shatz 1990)
—glutamate receptors	SPN/?	Ferret (Herrmann & Shatz 1992)
Calcium binding protein	SPN/?	Cat (Antonini & Shatz 1990)
(CaBP, calbindin)	SPN/?	Ferret (Antonini & Shatz 1990)
	SPN/MZ	Rat (Liu & Graybiel 1992, Sánchez et al 1992)
MAP2	SPN/MZ	Human (Sims et al 1988, Honig et al 1990)
	SPN/MZ	Monkey (Mehra & Hendrickson 1993)
	SPN/MZ	Cat (Chun et al 1987, Chun & Shatz 1989a)
	SPN/MZ	Ferret (Allendoerfer et al 1990)
	SPN/MZ	Mouse (Crandall et al 1986)
CR-50	-/MZ	Mouse (Ogawa et al 1992)
Neuropeptides		
Neuropeptide Y (NPY)	SPN/MZ	Monkey (Huntley et al 1988, Mehra & Hendrickson 1993)
	SPN/-	Cat (Chun et al 1987, Wahle & Meyer 1987, Chun & Shatz 1989a)
	SPN/?	Ferret (Antonini & Shatz 1990)
Cholecystokinin (CCK)	SPN/MZ	Cat (Chun & Shatz 1989a)
Somatostatin (SRIF)	SPZ/?	Monkey (Huntley et al 1988)
	SPN/-	Cat (Chun et al 1987, Chun & Shatz 1989a)
	SPN/?	Ferret (Antonini & Shatz 1990)
	SPZ/?	Rat (Feldman et al 1990)
—SRIF receptor	SPZ/?	Rat (Gonzalez et al 1989)
Substance P (or SP-like)	SPN/-	Monkey (Mehra & Hendrickson 1993)
	SPZ/-	Rat (Del Rio et al 1991)

Table 2 *(continued)*

Projection sites	Localization	Species (reference)
Putative degeneration/death markers		
Alz-50	SPZ/-	Human (Wolozin et al 1988)
	SPN/-	Cat (Valverde et al 1990)
	SPN/-	Rat (Al-Ghoul & Miller 1989)
Subplate-1	SPN/-	Cat (Naegele et al 1991)
Putative growth factors/receptors		
Neurotrophin family-like	SPZ/MZ	Cat (KL Allendoerfer, A Hohn & CJ Shatz, unpublished observations)
	SPZ/MZ	Ferret (Allendoerfer & Shatz 1991)
AC3	SPN/-	Rat (Yu et al 1992)
Estrogen-R	SPZ/MZ	Rat (Miranda & Toran-Allerand 1992)
P75-NGFR	SPZ/-	Human (Kordower & Mufson 1992)
	SPZ/MZ	Monkey (Meinecke & Rakic 1993)
	SPN/-	Cat (Allendoerfer et al 1990)
	SPN/-	Ferret (Allendoerfer et al 1990)
	SPZ/-	Rat (Koh & Higgins 1991)
	SPZ/?	Mouse (Wayne et al 1991)
trkB	SPZ/?	Ferret (Allendoerfer et al 1994)
trkC	SPZ/?	Ferret (Allendoerfer et al 1994)
Adhesion molecules		
Chondroitin sulfate proteoglycan (CSPG)	SPZ/MZ	Mouse (Bicknese et al 1991, Sheppard et al 1991, Miller et al 1992)
Fibronectin	SPN/MZ	Cat (Chun & Shatz 1988b)
	SPN/MZ	Mouse (Pearlman 1987, Sheppard et al 1991, Pearlman et al 1992)
Laminin	SPZ/?	Rat (Hunter et al 1992)
L1	SPZ/?	Mouse (Godfraind et al 1988, Chung et al 1991)
NCAM	SPZ/?	Mouse (Godfraind et al 1988, Chung et al 1991), Human (Terkelsen et al 1992)
Plasma proteins		
Albumin	SPZ/MZ	Human (Møllgård & Jacobsen 1984)
Alpha-fetoprotein	SPZ/MZ	Human (Møllgård & Jacobsen 1984)
Fetuin/AHSG	SPZ/MZ	Human (Dziegielewska et al 1987)
	SPN/MZ	Sheep (Saunders et al 1992, Dziegielewska et al, 1993)
Ig-like immunoreactivity	SPZ/MZ	Rat (Fairén et al 1992)
Prealbumin	SPZ/MZ	Human (Møllgård & Jacobsen 1984)
Transferrin	SPZ/MZ	Human (Møllgård & Jacobsen 1984)

[a] Abbreviations:

SPN: subplate neurons (marker is associated with subplate neurons, as proven by ^{3}H-thymidine birth-dating studies, colocalization with another known marker, excitotoxic deletion of subplate neurons, or retrograde labeling)

SPZ: subplate zone (marker is located within the subplate zone, but is not proven to be associated with subplate neurons)

MZ: marginal zone

-: not found in marginal zone

?: the question of whether the marker was located in the marginal zone is not clear or was not addressed in the study

EAA: excitatory amino acids

the thalamus or opposite hemisphere and then immunostained sections for NPY or somatostatin. Results indicated that at least some of the local circuit subplate neurons are peptide or calbindin immunoreactive, whereas subplate neurons with long projections are not. Instead, subplate neurons with thalamic or interhemispheric projections could be retrogradely labeled with ^{3}H-aspartate, indicating that they may use an excitatory amino acid as a transmitter. These observations suggest that the subplate exhibits basic features of cortical organization later echoed in the layers of the adult cerebral cortex.

Subplate neurons also apparently participate in early functional circuits. Some of the earliest synapses in the telencephalon are found in the subplate and marginal zone, at times well before the onset of synaptogenesis in the cortical plate (Molliver et al 1973; Kostovic & Rakic 1980; König & Marty 1981; Blue & Parnavelas 1983a,b; Chun & Shatz 1988a). To investigate whether subplate neurons are capable of firing action potentials and whether they receive functional synaptic inputs, intracellular microelectrode recordings or current source density recordings were made from subplate neurons in acute slices of fetal and neonatal cat visual cortex (Friauf et al 1990, Friauf & Shatz 1991). As early as E50, subplate neurons received synaptic inputs and fired action potentials in response to electrical stimulation of the optic radiations, indicating that some of the synapses seen in the electron microscope are indeed capable of functional synaptic transmission.

To determine the presynaptic origin of at least some of these synapses, Herrmann et al (1991) injected the anterograde tracer PHAL into the thalamus of neonatal ferrets, and showed with electron microscopy that labeled thalamic axons made synaptic contacts onto subplate neurons. This finding is entirely consistent with many previous light microscope observations showing that in cat, ferret, and primate thalamic axons accumulate and "wait" in the subplate in large numbers before they grow into the cortical plate. (We will consider the subject of the "waiting" period in more detail below.) Subplate neurons are also likely to make synaptic contacts with each other in view of the fact that they extend local axon collaterals within the subplate, as revealed by immunohistochemistry (Wahle & Meyer 1987, Chun & Shatz 1989a) and intracellular injections of biocytin (Friauf et al 1990).

Subplate neurons not only have descending axons and collateral branches within the subplate, but many send axons into the cortical plate. Intracellular injections of biocytin into subplate neurons have shown that their axons can terminate within the marginal zone and cortical layer 4, particularly at neonatal ages (Figure 2) (Friauf et al 1990). It is not clear, however, whether subplate neurons that project to distant targets such as the thalamus also have collaterals within the cortical plate, or whether instead some or all of the local circuit subplate neurons are responsible. Subplate neurons receive synaptic inputs from waiting thalamic axons and in turn make axonal projections into the

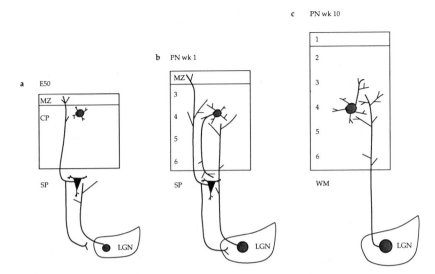

Figure 2   Subplate neurons may participate in a transient synaptic circuit that also includes thalamic axons and neurons of cortical layer 4. (*a*) E50 cat: While LGN axons accumulate in the subplate (SP) and make synaptic contacts with subplate neurons (Ghosh & Shatz 1992b, Herrmann & Shatz 1992), subplate neurons send axons up into the cortical plate (CP) and marginal zone (MZ) (Friauf et al 1990). They also send a projection back to LGN (McConnell et al 1989). It is not known whether a single subplate neuron can make both of these axonal projections simultaneously. (*b*) By postnatal week 1 in the cat, LGN axons have invaded layer 4 (Ghosh & Shatz 1992b), and subplate neurons elaborate collaterals within layer 4 (Wahle & Meyer 1987, Friauf et al 1990). In turn, layer 4 neurons send a transient projection into the subplate (Callaway & Katz 1992). Subplate neurons may also project to the LGN, but this has not been confirmed directly. Thus a recurrent pathway between layer 4 and subplate exists. (*c*) By postnatal week 10, the majority of subplate neurons have disappeared (Chun & Shatz 1989b) and the mature pattern of connectivity between the LGN and layer 4 is present.

cortical plate (Figure 2*a*), raising the possibility that they may function as a cellular scaffold that forms a crucial but transient link between developing thalamic axons and their ultimate target cells in cortical layer 4 (Figure 2*b,c*).

ROLE OF SUBPLATE NEURONS

Subplate Neurons Pioneer the Intracortical Pathway to Thalamus

During the development of the brain, growing axons must traverse considerable distances to find their targets. The pathway from LGN to cortical layer 4, or the reciprocal pathway from cortical layer 6 to LGN, is many tens of millimeters by adulthood, and even early in development the sets of neurons that must ultimately interconnect are many cell body diameters distant from each other. A solution to this difficult problem may be found in the early outgrowth of axons from pioneer neurons (Bate 1976, Kuwada 1986, Klose & Bentley 1989; reviewed in Goodman & Shatz 1993) when distances between neurons and their targets are small and local environmental cues appear sufficient to point growth cones in the right direction. To examine the earliest efferent projections from the cerebral cortex, several investigators (McConnell et al 1989, Blakemore & Molnár 1990, De Carlos & O'Leary 1992, Erzurumlu & Jhaveri 1992; SK McConnell, A Ghosh & CJ Shatz, submitted) have used the lipophilic fluroescent tracer 1,1-dioctadecyl-3,3,3',3'-tetramethyl in-docarbocyanine perchlorate (DiI) to trace connections in fetal brains. McConnell et al (1989) found that when DiI was placed in the internal capsule at E30 in the cat, the injection retrogradely labeled exclusively neurons in the preplate, which at this early age is composed only of subplate and marginal zone neurons. At later ages, after the preplate has split into subplate and marginal zone, similar injections labeled only subplate, but not marginal zone, neurons. Hence, it is likely that subplate but few if any marginal zone neurons supply the early axons to the internal capsule. DiI injections into the preplate itself at E30 labeled axons traveling through the intermediate zone and entering the internal capsule (McConnell et al 1989), showing that some subplate neurons had extended descending axons very early, before many of the neurons of layers 5 and 6, which form the descending projection in the adult, had even become postmitotic and begun to migrate (Luskin & Shatz 1985a). Moreover, at E26, DiI injections into the preplate never retrogradely labeled any neurons outside of cortex. Placement of DiI directly into the thalamus starting at E30 retrogradely labeled subplate neurons in temporal cortex, and then, by E36, also the subplate neurons underlying visual cortex. Thus, these observations collectively indicate that the axons of subplate neurons are the first to traverse the pathway from cortex through the internal capsule in the cat—that is, subplate axons "pioneer" these pathways in the classic sense of the word (Bate 1976).

In rodents, subplate neurons perform a similar pioneering function in vivo for the corticothalamic projection (De Carlos & O'Leary 1992, Erzurumlu & Jhaveri 1992). Unlike McConnell et al's (1989) findings for carnivores, however, De Carlos & O'Leary (1992) did not find any evidence that subplate neurons extended axons beyond the thalamus to the superior colliculus. In addition, in rodents, subplate neurons apparently do not send axons into the spinal cord or even into the midbrain superior peduncle (De Carlos & O'Leary 1992). The observation that subplate axons do project to superior colliculus in carnivores but do not in rodents could be a result of a species difference; the pathway taken by corticotectal axons in the cat is a dorsal one, extending posteriorly past the LGN and through the pretectum (SK McConnell, A Ghosh

& CJ Shatz, submitted), whereas in the rat, the colliculus is innervated by collaterals that branch dorsally from ventral spinally-projecting axons (O'Leary et al 1990). Thus the mechanism for layer 5 axon pathfinding toward the superior colliculus may differ between rodents and carnivores. On the other hand, given the sensitivity of this type of experiment to the exact placement of the DiI injection, it is possible that the DiI was placed in such a way in the rodent as to miss the small population of subplate axons that project to the superior colliculus; alternately, dye leakage could conceivably have labeled subplate neurons in the cat that project only to the posterior thalamus. Future experiments are necessary to resolve this question definitively.

In contrast, ascending axons from subcortical structures such as the brainstem do not invade the cortex until much later, after the reciprocal connections between cortex and thalamus have already begun to form in the rodent (De Carlos & O'Leary 1992). Immunohistochemical studies of the monoaminergic and catecholaminergic pathways to cortex (Schlumpf et al 1980) have indicated that these axons traverse the intermediate zone at E16 at the earliest. A similar conclusion has been reached by Erzurumlu & Jhaveri (1992), in a study of rat cortical development in which the first brainstem neurons were not retrogradely labeled with DiI injections into the cortex until E17, well after subplate neurons could be labeled with an injection in the internal capsule. All of these considerations suggest that the first axon pathway laid down within the intermediate zone of the cerebral cortex derives from the subplate neurons. If so, this raises the possibility that as in invertebrates the later-growing efferent axon systems, such as those arising from cortical layers 5 and 6, travel along the pathway originally laid down by the subplate axons, and that this early pathway is necessary for target selection.

A good way to test directly whether the subplate neurons play an essential role in subsequent pathfinding and target selection by the axons of cortical layers 5 and 6 would be to ablate subplate neurons at very early times in development, before their axons have reached the internal capsule, and then to examine the consequences for the corticothalamic or corticotectal projections. At present, no methods are available that permit the selective ablation of subplate neurons at these ages. However, in the cat, by E37 kainic acid injections selectively remove subplate neurons, leaving the neurons of the cortical plate essentially intact (Chun & Shatz 1988b, Ghosh et al 1990; SK McConnell, A Ghosh & CJ Shatz, submitted). At this time, although subplate axons have traversed the internal capsule and invaded the thalamus, cortical plate axons have just begun to elongate within the intermediate zone and have not yet reached the internal capsule (SK McConnell, A Ghosh & CJ Shatz, submitted). The effects of subplate ablation on cortical axogenesis were assessed by allowing lesioned animals to develop until around birth, when

[3]H-leucine was injected into the visual cortex to trace the pattern of the descending projection. In half of the lesioned animals, descending projections formed normally. In the other half, however, the labeled axons traversed the internal capsule and then gathered at the antero-medial borders of the LGN (in the region of the perigeniculate nucleus), where they failed to grow into the body of the LGN. This situation contrasts markedly to that in normal animals at comparable ages, in which cortical axons have already grown well into the LGN and are concentrated in interlaminar regions.

In these same animals where the LGN projection was abnormal, cortical projections to and within other subcortical regions (e.g. the superior colliculus) were also missing, although intracortical pathways such as the projections to the claustrum and across the corpus callosum appeared normal. These observations imply that subplate neurons are involved in the process of target invasion by cortical axons, but that there may be additional cues available as well. Similar conclusions have been drawn from experiments in lower vertebrates (Kuwada 1986) and invertebrates (Bastiani et al 1985) when pioneer neurons have been ablated genetically or with a laser.

Taken together, these observations imply a critical role for subplate pioneers in the normal innervation of subcortical targets by cortical plate neurons. Surprisingly, however, in the absence of subplate neurons, cortical plate neurons are apparently able to navigate more or less correctly through the internal capsule and grow toward the appropriate thalamic target, although they do not innervate it. A simple hypothesis is that cortical plate axons employ subplate axons as a substrate for growth through the internal capsule and into the subcortical targets. At the earliest times of their outgrowth from the cortical plate, these axons would have easy access to subplate axons and could grow along them. The subplate neurons were ablated in this study at the earliest at E37, when some cortical plate outgrowth has begun. These early interactions between subplate and cortical plate axons may be sufficient to start the cortical plate axons down the correct path. It is also possible that once the pioneer pathway has been laid down by the initial subplate neuron outgrowth, some residual molecular cues remain in the internal capsule which serve to guide the cortical axons in a crude manner; and it is likely that additional environmental cues are present in the extracellular matrix, independent of subplate neurons. In vitro studies have shown that axons can employ more than one extracellular cue for growth, and that elimination of any single cue can be insufficient to block outgrowth (Neugebauer et al 1988, Tomaselli et al 1988). The failure of cortical plate neurons to innervate their subcortical targets in half of the ablated animals implies that target recognition involves two distinct steps, pathfinding and innervation, and suggests an additional role for the subplate axons in the second. The absence of a projection to the superior colliculus in these animals implies that in the cat, at least, there is indeed some role for the subplate neurons in pioneering not only the corticothalamic pathway but also the corticotectal pathway. It would also be of great interest to perform similar experiments in other species to examine the generality of these observations.

Several groups have cultured slices of rat cortex either alone or in conjunction with a target slice, such as LGN or another slice of cortex (Yamamoto et al 1989, 1992; Bolz et al 1990; Molnár & Blakemore 1991) with the goal of further investigating pathfinding mechanisms. Bolz et al (1990) and Yamamoto et al (1992) did not comment on the projections of the subplate neurons in their cortical slices to the thalamic or other cortical explants in their cultures; nonetheless, the thalamic slices were innervated by the appropriate projection neurons (in the deep layers) of cortical explants, whereas the cortical slices were appropriately innervated by neurons in more superficial layers. Bolz et al (1990) therefore concluded that the projections of subplate neurons are not necessary for the formation of corticothalamic projections in vitro. However, the presence or absence of subplate neurons was not confirmed directly in these slices by labeling with ^{3}H-thymidine or other markers (see Table 2). In addition, the cocultures involved slices of cortex taken at P1, after connections between thalamus and cortex have already been established in vivo, raising the possibility that the growing cortical axons are responding to cues laid down much earlier. Finally, it may be that the subplate neurons are required for the targeting of cortical axons to the correct thalamic nucleus, rather than for the generic ingrowth of cortical axons into thalamus, which is being assessed in the coculture studies. Further experiments are required to resolve these issues and to establish the appropriateness of this in vitro approach for elucidating mechanisms of pathfinding as opposed to target selection. For example, it would be of great interest to repeat these coculture experiments with "naive" cortex and thalamus taken from fetal rats, particularly because the subplate deletion experiments considered above were performed in cat much earlier in development than these in vitro studies.

Are Subplate Neurons Required for Thalamocortical Pathfinding?

As described above, the axons of subplate neurons may create or themselves serve as a scaffold on which the descending corticothalamic projection is built. This subplate neuron scaffold may also serve an additional purpose in the establishment of the ascending thalamocortical projections. Thalamus and cortex must specifically connect with each other in two different ways: they must first find their their appropriate modality-specific cortical area, and then must grow into their appropriate laminar targets, cortical layers 4 and 6 [for a recent review, see also O'Leary & Koester (1993)]. This specificity of connections could come about by several mechanisms, including selective fasciculation, timing of axon outgrowth, extracellular matrix or cell-surface cues within the cortical plate and subplate, chemotropic guidance mechanisms, or even competition between ingrowing thalamocortical axons for common target neurons located in the subplate or cortical plate.

In considering mechanisms by which thalamic axons might grow to their appropriate cortical target areas, one possibility is that thalamic axons require subplate axons. This idea arises from observations, based on DiI labeling of subplate and thalamus, that the first growth cones of LGN neurons and visual subplate neurons are seen within the internal capsule at the same time (E30 in cat: McConnell et al 1989, Ghosh & Shatz 1992b; E14–15 in rat: Blakemore & Molnár 1990, De Carlos & O'Leary 1992), and likewise for axons from somatosensory thalamus and corresponding subplate in rat that were labeled with DiI and DiA and visualized simultaneously at E15 (Erzurumlu & Jhaveri 1992). These observations raise the possibility that thalamic and subplate axons from corresponding regions might meet in the internal capsule and subsequently fasciculate on each other as they make their final traverse to the appropriate targets within thalamus or cortex, and that this fasciculation could form the basis for the establishment of specific connections between thalamus and cortex (Blakemore & Molnár 1990, Ghosh & Shatz 1993). Such selective fasciculation between anterior- and posterior-growing axons has been observed in the development of longitudinal pathways in the grasshopper embryo. Moreover, in these embryos, laser ablation of two specific posterior-growing axons prevents the anterior-growing axon from growing normally and reaching its target (Bastiani et al 1986).

However, an apparent contradiction with the suggestion that selective fasciculation underlies the process of pathfinding by thalamocortical and corticothalamic axons is that at later times, DiI-labeled axon bundles from the cortex and from the thalamus apparently run in separate fascicles, with the afferent thalamocortical axons traveling more superficially in the subplate (Bicknese et al 1991, Miller et al 1991a), and the efferent corticothalamic axons deep in the intermediate zone (Shatz & Rakic 1981, Bicknese et al 1991, Miller et al 1991a; SK McConnell, A Ghosh & CJ Shatz, submitted; in all of these studies, the location of subplate axons in relation to the afferent and efferent pathways was not determined). It may be important that these two pathways are also different with regard to cell-surface molecules. Two chondroitin sulfate protoglycan (CSPG) core proteins are apparently restricted to the afferent thalamocortical pathway, but not the corticothalamic pathway (Bicknese et al 1991, Sheppard et al 1991, Miller et al 1992) (see Table 2), suggesting that specific cortical pathways can be marked with different molecules. Even though it appears that at later times there are spatially distinct pathways to and from cortex, it may be that the earliest thalamic axons and subplate axons do fasciculate with each other, but the later-growing thalamic or cortical axons prefer to grow on like axons. This would result in the formation of two separate pathways, but with the axons of subplate neurons and the earliest growing LGN neurons located at the interface. Some support for this suggestion comes from preliminary studies in which one fluorescent dye was used to label the corticothalamic axons and another to label the thalamocortical axons, and the two axons pathways were seen to be apposed to each other in the intermediate zone (Blakemore & Molnár 1990) or internal capsule (Bicknese & Pearlman 1992). More definitive evidence is obviously required here, including: (a) examining these pathways at very early ages, (b) showing at the ultrastructural level that thalamic axons fasciculate on identified subplate axons, and (c) demonstrating that deleting subplate neurons before thalamic axons traverse their intracortical pathways prevents them from arriving at their correct cortical targets.

Subplate Neurons are Required for Target Selection and Ingrowth by Thalamic Axons

Once axons from the thalamus have successfully navigated through the internal capsule and intermediate zone, they must stop at the appropriate cortical target and grow into the cortical plate. Early studies of the development of connections between geniculocortical axons and their ultimate targets, neurons of cortical layer 4, used relatively low resolution tract tracing techniques such as transneuronal labeling with ^{3}H-proline (Rakic 1977, Shatz & Luskin 1986) or thalamic lesions followed by examination of degenerating axon terminals (Lund & Mustari 1977). These studies showed that label representing thalamic axons accumulated within the subplate for an extended period of time, from three days in rats (Lund & Mustari 1977) to several weeks in cats and primates (Rakic 1977, Shatz & Luskin 1986), before invading the cortical plate. From these observations developed the concept of a "waiting period" (Rakic 1977), in which axons accumulated and paused in the subplate for a period of time before growing into layer 4. In addition to the geniculocortical axons from visual thalamus, other afferent axonal systems, including thalamocortical projections from somatosensory thalamus (Wise & Jones 1978) and interhemispheric connections via the corpus callosum (Wise & Jones 1976, 1978; Innocenti 1981), appear to "wait" in the subplate before invading the cortical plate, suggesting that waiting periods may be a general feature of the development of corticopetal axonal systems.

More recent studies that reexamined the waiting period with DiI tracing techniques provide further evidence for potential interactions between the "waiting" axons and cellular elements in the subplate. In the cat, Ghosh & Shatz (1992b) found that the geniculocortical axons arrived earlier in the subplate (E36 rather than E40) and invaded the cortical plate earlier (E50 rather than E55) than expected based on data from ^{3}H-proline autoradiography (Shatz & Luskin 1986). During the intervening time between E36 and E50, the majority of LGN axons accumulate within the subplate beneath the visual cortex, forming extensive terminal branches confined to the visual subplate. In addition, as axons grow through the optic radiations on their way to visual cortex, they send out interstitial collaterals into the subplate of inappropriate areas (e.g. the auditory subplate) as if the axons were "sampling" a variety of cortical areas. Naegele et al (1988), studying geniculocortical development in the hamster by using HRP tracing, also showed that between P3 and P5 multiple short collaterals with no terminal arbors were extended into the subplate and deeper portions of the cortical plate. In both species, these collaterals appear to be a transient feature of the developing geniculocortical pathway, since they are absent by P7 (Ghosh & Shatz 1992b) in cat and during the second postnatal week in the hamster (Naegele et al 1988). The existence of collaterals and terminal branches within the subplate during the waiting period contrasts with the simple morphology of thalamocortical axons—consisting of a single parent axon tipped with a growth cone—at earlier times in development. The existence of these collaterals in a zone also filled with subplate neurons and many synapses implies that thalamocortical axons may be involved in ongoing dynamic interactions both as they grow towards their appropriate area and while they are "waiting" within the subplate appropriate for their thalamic nucleus of origin.

Although there is no dispute concerning the existence of a long waiting period in cats and primates, where the pace of development is comparatively slow and it is possible to define a clear period in which thalamocortical axons are confined to the subplate, in rodents the very rapid pace of development has led to some controversy about whether there is a waiting period and if so, how long (Wise & Jones 1978, Catalano et al 1991). This controversy is based on more recent experiments in which DiI has been used to label somatosensory thalamocortical axons in the rat, and then the timecourse of their ingrowth into the cortical plate has been assesed by inspecting fluorescence micrographs (Catalano et al 1991). Such experiments suggest that thalamic axons proceed directly from the subplate into the cortical plate without a significant delay (e.g. days to weeks). However, as mentioned above, Naegele et al (1988) have shown that thalamic axons make modest branches in the rodent subplate, suggesting that interactions, albeit much more limited in extent and duration than in higher mammals, may still take place. Clarification of this point would be aided if the precise border between the subplate and the cortical plate could be pinpointed with more accuracy in rodents, perhaps by combining axon labeling studies with ^{3}H-thymidine birthdating, in order to determine exactly when thalamic axons leave the subplate.

The role of the interactions between subplate neurons and thalamocortical axons was tested by deleting them during the waiting period (Ghosh et al 1990; Ghosh & Shatz 1992b, 1993). Visual subplate neurons in the cat were deleted with an injection of kainic acid at E42, and the histology of the subplate zone and the morphology of the LGN axons were examined at E60. In subplate-ablated animals, the entire zone that normally contains the subplate neurons had collapsed, and the LGN axons no longer arborized below the cortical plate. Instead of fanning out and branching in the subplate, the axons grew past the visual cortex in a tight fascicle. These observations suggest that subplate neurons are necessary for LGN axons to stop and arborize below their normal cortical target before they grow into the cortical plate. If the cortex is observed later, at P5, the axons still have not grown in, even though normally they would have done so more than two weeks earlier; in fact, they grow past visual cortex and into the white matter underlying the adjacent cingulate gyrus. This phenomenon is not specific for visual subplate. If a deletion is made in the subplate beneath the auditory cortex and DiI is injected into the MGN, it can also be seen that MGN axons grow past auditory cortex (Ghosh & Shatz 1993). It remains to be seen whether at later ages these axons are then able to innervate layer 4 in other, non-auditory areas, which still contain subplate neurons. Such findings could shed light on the issue of whether intrinsic differences among subplate neurons specify the identity of overlying cortical areas, or whether instead there are competitive interactions within the subplate, coupled with timing of arrival of axons at their correct cortical target areas. Whatever the case, these experiments imply that the cortical plate alone does not contain sufficient information to promote axon ingrowth; interactions in the subplate are necessary.

Although the nature of these interactions is unknown, several lines of evidence suggest that thalamocortical axons exhibit an affinity for the earliest-generated subplate neurons. For example, in the reeler mouse, where the preplate fails to split (Caviness & Rakic 1978, Ogawa et al 1992) and consequently subplate is located above the cortical plate as a kind of "superplate", the geniculocortical axons still initially grow into the superplate at E15 (Molnár & Blakemore 1992). They traverse the cortical plate in diagonal fascicles toward the early-generated superplate cells, wait on top of the developing cortical plate (Molnár & Blakemore 1992), and finally loop down to innervate cortical layer 4 (Frost & Caviness 1980, Caviness & Frost 1983, Molnár & Blakemore 1992). Thus, thalamocortical axons may select subplate neurons regardless of the radial position of these neurons within cortex, again implying close interactions.

It is possible that the affinity of thalamic axons for subplate neurons could be mediated by adhesive interactions. An array of adhesion and extracellular matrix molecules have been shown to be localized to the subplate and marginal zone, but not the cortical plate, early in development (see Table 2). For example, L1, a cell surface glycoprotein, and J1, a secreted glycoprotein, were detected in the subplate and marginal zone of developing mouse embryos (Godfraind et al 1988). Additionally, fibronectin is found in these zones in both cat and rodent (Chun & Shatz 1988b, Pearlman et al 1992), and at least some of it is thought to be associated with the subplate neurons themselves (Chun & Shatz 1988b, Pearlman et al 1992). Moreover, deletion of subplate neurons abolishes the fibronectin immunoreactivity, implicating a possible role for this matrix molecule during the period in which axons normally wait in the subplate. Not only might there be permissive influences on axon growth, but preliminary evidence from coculture experiments suggests that the cortical plate itself may be nonpermissive for thalamic axon ingrowth. Götz et al (1992) demonstrated that when thalamic explants were cocultured with E16 cortex, rather than P1 cortex, there was no innervation of the cortical slice. In a separate in vitro assay, it was then shown that membranes from E16 but not P7 cortex cause growth cone collapse (Hubener et al 1992), suggesting that repulsive influences from embryonic cortex might also contribute to the waiting period. If so, the relative affinities of the thalamic axons for subplate and cortical plate would be expected to change at the end of the waiting period and the onset of invasion of the cortical plate.

Do Subplate Neurons Play a Role in the Specification of Cortical Areas?

Much discussion has addressed the issue of how cortical areas are specified during development. One view is that cortical areas have intrinsic differences that are mapped out very early in development, perhaps even in the ventricular zone (Rakic 1988). Alternatively, cortical areas may emerge gradually from an undifferentiated "protocortex" through epigenetic influences such as interactions with afferent inputs (O'Leary 1989, Shatz 1992). The results of Ghosh et al (1990) and Ghosh & Shatz (1993) indicate that subplate neurons stand as a crucial link in cortical target recognition, whichever strategy is used. The observation that LGN axons grow past the visual cortex in the absence of subplate neurons indicates that the cortical plate alone has insufficient information to allow the ingrowth of appropriate axons.

Several pieces of evidence suggest, however, that thalamic axons are not guided to their appropriate target areas solely by molecular "labels" present on subplate neurons in different cortical areas. In coculture experiments, axons from LGN explants appear to terminate appropriately in layer 4 of visual cortical explants and to form functional connections, as revealed by DiI labeling (Molnár & Blakemore 1991, Yamamoto et al 1992) and current source density analysis (Yamamoto et al 1989, 1992). However, when LGN explants are challenged to make a decision between cortical explants derived from both appropriate and inappropriate regions (i.e. visual vs frontal), the thalamic explants are able to innervate both cortical explants with appropriate laminar specificity, regardless of the area of origin (Molnár & Blakemore 1991, Yamamoto et al 1992). In vivo experiments also suggest that different areas of cortex have multiple developmental potentials and their ultimate fate is dependent on the afferents they receive. Schlaggar & O'Leary (1991) transplanted occipital (visual) cortex to the presumptive barrel field of parietal (somatosensory) cortex, and found that barrel-like morphologies developed in the occipital-derived transplant. Therefore, the ability to form barrels is not unique to "somatosensory" cortex, the region that forms barrels normally, but is also possessed by the embryonic "visual" cortex.

These experiments would seem to suggest that the subplate does not possess intrinsic positional information that marks an area as appropriate for ingrowth, or thalamic explants should have exhibited a preference in vitro, and/or might not innervate a transplant from another cortical area that contained inappropriate subplate neurons. However, in the case of the transplant studies, the transplants were made in newborn rats, when VB axons had already begun to grow into the somatosensory cortex in vivo. The axons were thus cut during the transplantation experiment and presumably regrew from the thalamic radiations into the transplant. They did not have to renavigate the entire pathway from thalamus to cortex. Thus the cues needed may have been present in the subplate early on when the VB axons were first growing towards cortex but were perhaps no longer relevant, or even present, after birth.

Visual Subplate Neurons Are Involved in the Formation of Ocular Dominance Columns

In the visual cortex of higher mammals, thalamic axons from the LGN are segregated in cortical layer 4 according to eye preference into alternating patches of input that represent the anatomical basis for the system of ocular dominance columns (Hubel & Wiesel 1977). The segregation of LGN axons into the ocular dominance columns within cortical layer 4 represents the end-point of their developmental history, and it occurs shortly after LGN axons have left the subplate and invaded layer 4. Axons representing both eyes are initially intermixed with each other in layer 4, and then through a process of remodeling and selective growth, segregate to form eye-specific patches. In the cat and monkey visual system, segregation is complete at about six weeks postnatal (LeVay et al 1978, 1980); in cat, the process begins at about three weeks postnatal, whereas in the monkey this process begins prenatally (Rakic 1977).

Several lines of evidence suggest that subplate neurons may play a role in this final stage of thalamocortical development, at least in the visual system. As shown in the schematic diagram of Figure 2b, intracellular injections of biocytin into subplate neurons during the first postnatal week in the cat visual system, a time when LGN axons have already invaded layer 4 but before the onset of segregation, indicate that some subplate neurons can send extensive axonal collaterals within layer 4 of the cortical plate (Friauf et al 1990). At the same time, similar intracellular injection techniques have revealed that many layer 4 neurons send a transient axon collateral that traverses layers 5 and 6 to branch within the subplate (Callaway & Katz 1992). These observations suggest that a transient and reciprocal synaptic circuit between the subplate and layer 4 may be present at the onset of ocular dominance column formation. Although future physiological experiments and ultrastructural studies are necessary to confirm the presence of this proposed circuit, electron microscopic studies following PHAL injections into the cortical plate have shown that in neonatal cats some descending cortical axons can make synaptic contacts within the subplate (Lowenstein & Shering 1992).

More direct evidence for a role for subplate neurons in the segregation of LGN axons comes from an experiment in which kainic acid was injected into the subplate during the first postnatal week in the cat to ablate subplate neurons, and the consequences were examined six to eight weeks later, when ocular dominance columns in layer 4 should have normally formed. Results obtained by using the technique of transneuronal transport following intraocular injection of ^{3}H-amino acids into one eye showed that in the region of visual cortex corresponding to the subplate ablation, LGN axons had failed to segregate into eye-specific patches within layer 4, whereas elsewhere columns had formed normally (Ghosh & Shatz 1992a). This observation suggests that subplate neurons play a role not only at early times in the formation of thalamocortical connections, but also in the final patterning of the geniculocortical projection within cortical layer 4. Although this type of ablation experiment cannot shed light on the underlying cellular interactions that operate normally to produce ocular dominance columns, it is worthwhile to consider whether a similar manpulation in rodent somatosensory cortex might disrupt barrel formation. If so, this would indicate a broad role for subplate neurons in the final patterning of thalamocortical projections within layer 4.

DEATH OF SUBPLATE NEURONS

In the adult, it has long been known that the white matter and layer 1 contain only a few scattered neurons. These are the interstitial cells in the white matter (Ramón y Cajal 1911) and the Cajal-Retzius cells of layer 1 (Marin-Padilla 1971, 1988). Birth-dating studies have proven that many if not all of these neurons are derived from the subplate and marginal zone neurons, because they can only be labeled by injections of ^{3}H-thymidine at the earliest ages of cortical neurogenesis (Chun & Shatz 1989b, Woo et al 1991). The scarcity of neurons in the adult in what was a dense neuropil during development suggests that many of the subplate and marginal zone cells are eliminated by cell death. Several lines of evidence strongly support this possibility. First, degenerating neurons with the appropriate interstitial or Cajal-Retzius cell morphology have been observed in the subplate (Kostovic & Rakic 1980, Valverde & Facal-Valverde 1988) and marginal zone (Shoukimas & Hinds 1978, Derer & Derer 1990) of neonates by using electron microscopy. These cells exhibit a swelling of the endoplasmic reticulum and Golgi followed by a progressive darkening of the cytoplasm, a type of degeneration classified as "cytoplasmic" (Pilar & Landmesser 1976, Cunningham 1982) and thought to be characteristic of dying mature neurons. Wahle & Meyer (1987) also observed degenerating NPY immunoreactive "axonal loop" cells in the neonatal cat white matter.

Moreover, the density of subplate neurons immunostained for neuropeptides or for MAP2 in the cat decreases dramatically during the first four postnatal weeks until adult levels are reached (Chun & Shatz 1989b). For example, the density of MAP2-immunoreactive neurons remaining in the white matter of the cat lateral gyrus at 25 weeks postnatal is only about 10–20% of the density at birth. In contrast, the area of the white matter increases by about twofold at most, indicating that growth alone cannot account for this decrease in density. In two rodent studies (Woo et al 1991, Wood et al 1992) a more detailed statistical analysis was performed to account for apparent dilution of subplate neurons by growth of the cortex. In the hamster (Woo et al 1991), a ^{3}H-thymidine labeled subplate population was compared with a ^{3}H-thymidine labeled "reference" population in layer 6 directly adjacent to the subplate. Woo et al (1991) showed that the ratio of subplate neurons to layer 6 neurons decreased approximately fourfold between P4 and adulthood. A similar analysis was also performed by comparing the subplate neurons with a reference population of LGN neurons, with nearly identical results. In the mouse, Wood et al (1992) estimated that a greater than 75-fold decrease occurred in labeled subplate neurons cells per unit volume between birth and P21, whereas the upper limit for increase in white matter volume was only 15-fold. They found no E12-labeled cells remaining in the marginal zone at P21, although some cells labeled at E12 were present in the marginal zone at E17. Thus, again it is not possible to account for the fall in subplate neuron number entirely on the basis of growth of the white matter: a major portion of the subplate and marginal zone neurons must be eliminated by cell death during early postnatal life. Although developmentally regulated cell death appears to be a universal feature of subplate neurons in all species where it has been examined, the extent of subplate neuron death is almost certainly not the same in all species, nor is it likely to be uniform between cortical areas or among the different subplate neuron phenotypes in a given species (Chun & Shatz 1989b, Mehra & Hendrickson 1993).

Molecular Changes Accompany Subplate Neuron Death

The mechanism by which subplate neurons die remains an open question. Immunohistochemical studies have provided evidence that subplate neurons undergo molecular changes during the period of cell death. Naegele et al (1991) isolated an antibody, SP-1, that recognizes a 56-kDa polypeptide expressed transiently and exclusively within subplate neurons (but not marginal zone cells) just after birth in the cat, a time that corresponds to the peak of cell death. Many of the neurons expressing the SP-1 antigen possess an inverted pyramidal morphology, project to thalamus, and are glutamatergic; only a few SP-1$^+$ cells express GABA or peptides. The identity of this antigen is unknown, but it is thought to be intracellular.

Staining with Alz-50, a monoclonal antibody isolated from a screen directed against neurofibrillary tangles in the brains of Alzheimer's patients (Wolozin et al 1986), also labels subplate neurons in rat (Al-Ghoul & Miller 1989), cat (Valverde et al 1990), and human (Wolozin et al 1988) during a period of active cell death. The Alz-50 epitope is now known to correspond to an abnormally phosphorylated form of the microtubule-associated structural protein tau (Ueda et al 1990, Lee et al 1991). The SP-1 antigen appears to be distinct from Alz-50 in immunoblots, and exhibits an overlapping, but more restricted, pattern of cellular localization (Naegele et al 1991). In addition to its expression in Alzheimer's disease, Down's syndrome, and developmentally regulated cell death, Alz-50 can be induced in neurons by environmental insults such as axotomy (Miller et al 1991b) or by the elevation of protein kinase C (Mattson 1991), and is thought to be a general marker for dying neurons.

The expression of antigens or proteins unique to the death process suggests that the subplate neurons may actively turn on a program of "death genes." This view of neuron death as an active process that requires transcription and protein synthesis derives from studies of cultured sympathetic neurons (Martin et al 1988) or PC12 cells (DiBenedetto et al 1992) deprived of NGF, where the death of these NGF-deprived neurons is inhibited by transcription and protein-synthesis inhibitors. Another possibility, not mutually exclusive, is that certain genes are turned off during the death process. For example, the proto-oncogene *bcl-2* is expressed developmentally in lymphocytes that do not die (Korsmeyer 1992) and, in the immune system, is thought of as a repressor of cell death. It would be interesting to know whether the interstitial neurons and Cajal-Retzius cells that survive into adulthood express *bcl-2* or other relevant proteins.

Neurotrophic Factors May Control Survival and Death of Subplate Neurons

It is also possible that neurotrophic factors may play a critical role in subplate neuron survival. Subplate neurons express the low-affinity NGF receptor ($p75^{LNGFR}$) soon after they become postmitotic (Allendoerfer et al 1990). Expression of $p75^{LNGFR}$ remains high during maturation of the subplate neurons and the waiting period for thalamic axons (Allendoerfer et al 1990, Koh & Higgins 1991, Wayne et al 1991, Kordower & Mufson 1992, Meinecke & Rakic 1993). Subplate neurons are also immunoreactive for a member of the neurotrophin family, based on immunostaining with a pan-neurotrophin antibody (Allendoerfer & Shatz 1991) at the same time that the $p75^{LNGFR}$ staining is strong. This staining for neurotrophin, also observed in basal forebrain neurons that are known to retrogradely transport NGF in the adult (Seiler & Schwab 1984) and during development (Allendoerfer & Shatz 1991), suggests that the subplate neurons could also internalize and transport NGF or a related neurotrophin (Allendoerfer & Shatz 1991). Just before the subplate neurons begin to die, they lose $p75^{LNGFR}$ immunoreactivity, suggesting that they die because they are no longer able to respond to the ligand (Allendoerfer et al 1990). Several days later, the neurotrophin immunoreactivity also disappears from the subplate (Allendoerfer & Shatz 1991).

Although death resulting from trophic factor deprivation is an intruiging hypothesis, the identity of such a factor for subplate neurons has so far been elusive. The $p75^{LNGFR}$ was identified first on neurons that retrogradely transport ^{125}I-NGF from their target tissues, such as basal forebrain neurons (Seiler & Schwab 1984, Yan & Johnson 1989, Ferguson et al 1991); however, retrograde transport of ^{125}I-NGF to subplate neurons is not observed when labeled factor is injected into any subplate neuron targets, such as thalamus or internal capsule, nor when it is injected directly into the subplate itself (Allendoerfer & Shatz 1991). Moreover, the addition of NGF to E30 ferret subplate neurons in culture does not enhance their survival (KL Allendoerfer & CJ Shatz, unpublished observations). Since $p75^{LNGFR}$ is able to bind equally well to other members of the neurotrophin family, such as BDNF and NT-3 (Rodriguez-Tébar et al 1990, 1992), the possibility remains that a neurotrophin distinct from NGF supports subplate neurons. High-affinity receptors for different neurotrophins, including trkB, truncated trkB, and trkC, are also present in the subplate zone, both early, when the subplate neurons first become postmitotic, and later, as they are dying (Allendoerfer et al 1994). Furthermore, the elegant model of neurotrophin action developed for sympathetic and sensory neurons in which a population of neurons projects to a single target, from which it derives trophic support, may not apply to a heterogeneous population of neurons such as the subplate, which has a number of different transmitter phenotypes and projection targets both local and distant—subplate neurons project into the cortical plate, across the corpus callosum, back to thalamus, and locally, to one another (see Table 2). The neurotrophin immunoreactivity within the subplate may not result from retrogradely transported neurotrophin but rather from neurotrophin synthesized locally within the subplate that could act in an autocrine fashion, as has been hypothesized for the peripheral nervous system during development (Schecterson & Bothwell 1992).

In addition to neurotrophins, other putative growth factors have also been shown either to be present in the subplate neurons or to have receptors there. For example, Yu & Bottenstein (1991) generated an antibody (AC3) to an uncharacterized growth factor that increases O-2A progenitor cell proliferation; this antibody stains neurons in the subplate zone of fetal and early postnatal rat and has been shown to colocalize with $p75^{NGFR}$ staining in both the subplate and septum (Yu et al 1992). Miranda & Toran-Allerand (1992) have shown that estrogen receptor mRNA and ^{125}I-estrogen binding sites are also present in the subplate and marginal zone as these neurons mature. Thus, estrogen may be important in the maturation of the subplate zone and/or subplate neurons, either alone or in concert with one or more growth factors.

Additional Mechanisms May Play a Role in Subplate Cell Death

It is also possible that growth factors serve a maturation and/or differentiation function, but not a survival function within the subplate, and that neuron death takes place through an excitotoxic mechanism. As the subplate neurons mature, they develop a sensitivity to kainate toxicity, and as mentioned earlier this early sensitivity has been used in experimental manipulations to delete subplate neurons selectively (Chun & Shatz 1988b, Ghosh et al 1990; McConnell et al, submitted). In addition, subplate neurons also undergo increases in intracellular Ca^{2+} in response to application of glutamate or agonists in acute slice preparations (Herrmann & Shatz 1992), indicating that developing subplate neurons also possess an assortment of glutamate receptors, including kainate, NMDA, quisqualate, and metabotropic (Herrmann and Shatz 1992; K Herrmann, unpublished observations). Thus the subplate neurons are likely to be vulnerable to glutamate neurotoxicity (Choi 1992) during the cell death period. Molnár et al (1991) proposed that glutamate released by the waiting thalamic axons normally plays a role in the death of subplate neurons. They labeled rat subplate neurons with ^{3}H-thymidine at E12, and then mechanically lesioned the internal capsule at E16, thereby depriving the subplate of all innervation. When they counted ^{3}H-thymidine–labeled cells postnatally, they found that the number on the lesioned side was greater than that on the control side. They suggested that the increased subplate neuron survival results from the absence of "excitotoxic input" during the period when subplate neurons would normally receive synaptic contacts from the waiting thalamic axons. Unfortunately, this interpretation is somewhat at odds with the presence of a prolonged waiting period observed in "higher" mammals such as cats and monkeys, where the thalamic axons make synaptic contacts with the subplate neurons for weeks or months, without any signs of excitoxic damage. On the other hand, the sensitivity of subplate neurons to glutamate may continue to increase throughout the waiting period, only becoming sufficient for excitotoxic damage very near the end. It is also possible that it is actually the departure of the thalamic axons from the subplate or their ingrowth into layer 4 that triggers subplate neuron death. The afferents may induce dependence in the subplate cells on a trophic factor derived from LGN neurons, which is withdrawn as they leave the subplate. Obviously, many additional experiments are required to resolve these issues or reveal other possible mechanisms governing survival and death of subplate neurons.

CONCLUSIONS

Many lines of evidence suggest that the subplate neurons play a functional role in setting up connections between cortex and thalamus during development. Experiments have revealed that subplate axons are early pioneers of the corticothalamic pathway, that subplate neurons may play a role in the subsequent waiting period and ingrowth of thalamocortical axons into the cortical plate, and that subplate neurons may even be important for the formation of ocular dominance columns in the visual cortex. However, many of the molecular and mechanistic questions still remain to be addressed. An important tool for investigations of subplate neuron function has been the ability to delete them selectively with kainic acid. The ability to perform such lesions even earlier, before either the thalamocortical or corticothalamic axons have even begun to grow out, would permit a true test of the pioneer hypothesis. In both the cortex and the thalamus, the subplate neurons appear at least in vivo to be required for innervation of appropriate targets. Do the local interactions between subplate neurons and either the waiting axons or postsynaptic cells of the cortical plate create a permissive environment for ingrowth? And if so, why are growth cones of subplate neurons themselves immune to this need for a permissive environment? It is likely that molecular interactions between the subplate axon growth cones and their target cells and/or extracellular matrix will be involved.

For this and other reasons, further molecular characterization of subplate neurons will be essential. Neocortical subplate neurons may not possess unique area-specific molecular labels that direct ingrowing afferents to the appropriate cytoarchitectonic area. How do specific areal subdivisions then emerge if subplate neurons are not intrinsically "marked"? It may be that again, competitive interactions between different thalamic axons will enable inputs to sort out into specific patterns of innervation. In addition, further quantitation of the molecular and anatomical phenotypes that are already known to exist would be useful for understanding which cells are responsible for the diverse roles attributed to subplate neurons; for example, what percentage of subplate neurons project to the thalamus, superior colliculus, or cortical plate, and what percentage are local circuit neurons? What percentage die with respect to transmitter phenotype and projection site, and is there anything unique in a molecular or projection field sense about the subplate neurons that survive into adulthood?

The question of why subplate neurons die could have implications for both developmental brain defects and neurodegenerative disease. Subplate neurons, which are present throughout the neocortex, reach a high degree of functional and morphological maturity early in life and may be uniquely susceptible at this point to the effects of trophic factor deprivation or excitotoxicity. The timely elimination of subplate neurons is a normal developmental process, but if certain aspects of this process are reactivated abnormally in other parts of the brain during aging, an understanding of subplate neuron death may ultimately lead to an understanding of neuron death at any age. In addition, subplate neurons may be uniquely sensitive to prenatal or perinatal trauma owing to their high degree of maturity; even a small subplate lesion during fetal life could conceivably lead to large axon targeting errors during subsequent development.

Finally, the number of subplate neurons and the size of the subplate zone relative to the cortical plate increases dramatically as one ascends the phylogenetic scale (see Table 1); in rodents, the ratio of subplate area to cortical plate area is 1:2, whereas in human it reaches 4:1 (Mrzljak et al 1988, Kostovic & Rakic 1990). Not only is the subplate larger in extent in primates and cats than in rodents, but it persists for a much longer developmental period as well. Thus interactions that occur within the subplate during development may be a basis for the increased complexity in the radial and tangential organization present in the neocortex of higher mammals.

ACKNOWLEDGMENTS

The authors thank Dr. Susan McConnell and Dr. Dennis O'Leary for their critical reading of the manuscript. Research from the authors' laboratory was supported by NIH R37 EY02858 and the Alzheimer's and Related Disorders Association to C.J.S. and NIH NS07158 to K.L.A.

Literature Cited

Al-Ghoul WM, Miller MW. 1989. Transient expression of Alz-50 immunoreactivity in developing rat neocortex: a marker for naturally occurring neuronal death? *Brain Res.* 481:361–67

Allendoerfer KL, Cabelli RJ, Escandón E, Nikolics K, Shatz CJ. 1994. Regulation of neurotrophin receptors during the maturation of the mammalian visual system. *J. Neurosci.* In press

Allendoerfer KL, Shatz CJ. 1991. Neurotrophic factor family immunoreactivity is present in developing mammalian cerebral cortex. *Soc. Neurosci. Abstr.* 17(1):220 (Abstr.)

Allendoerfer KL, Shelton DL, Shooter EM, Shatz CJ. 1990. Nerve growth factor receptor immunoreactivity is transiently associated with the subplate neurons of the mammalian telencephalon. *Proc. Natl. Acad. Sci. USA* 87:187–90

Angevine JB Jr, Sidman RL. 1961. Autoradiographic study of cell migration during histogenesis of cerebral cortex in the mouse. *Nature* 192:766–68

Antonini A, Shatz CJ. 1990. Relation between putative transmitter phenotypes and connectivity of subplate neurons during cerebral cortical development. *Eur. J. Neurosci.* 2: 744–61

Bastiani MJ, Doe CQ, Helfand SL, Goodman CS. 1985. Neuronal specificity and growth cone guidance in grasshopper and *Drosophila* embryos. *Trends Neurosci.* 8:257–66

Bastiani MJ, du Lac S, Goodman CS. 1986. Guidance of neuronal growth cones in the grasshopper embryo. I. Recognition of a specific axonal pathway by the pCC neuron. *J. Neurosci.* 6(12):3518–31

Bate CM. 1976. Pioneer neurones in an insect embryo. *Nature* 260:54–56

Baughman RW, Gilbert CD. 1980. Aspartate and glutamate as possible neurotransmitters of cells in layer 6 of the visual cortex. *Nature* 287:848–49

Bayer SA, Altman J. 1990. Development of layer I and the subplate in rat neocortex. *Exp. Neurol.* 107:48–62

Bicknese AR, Pearlman AL. 1992. Growing corticothalamic and thalamocortical axons interdigitate in a restricted portion of the forming internal capsule. *Soc. Neurosci. Abstr.* 18(1):778 (Abstr.)

Bicknese AR, Sheppard AM, O'Leary DDM, Pearlman AL. 1991. Thalamocortical axons preferentially extend along a chondroitin sulfate proteoglycan enriched pathway coincident with the neocortical subplate and distinct from the efferent path. *Soc. Neurosci. Abstr.* 17(1):764 (Abstr.)

Blakemore C, Molnár Z. 1990. Factors involved in the establishment of specific interconnections between thalamus and cerebral cortex. *Cold Spring Harbor Symp. Quant. Biol.* 55:491–504

Blue ME, Parnavelas JG. 1983a. The formation and maturation of synapses in the visual cortex of the rat. I. Qualitative analysis. *J. Neurocytol.* 12(4):599–616

Blue ME, Parnavelas JG. 1983b. The formation and maturation of synapses in the visual cortex of the rat. I. Quantitative analysis. *J. Neurocytol.* 12(4):697–712

Bolz J, Novak N, Götz M, Bonhoeffer T. 1990. Formation of target-specific neuronal projections in organotypic slice cultures from rat visual cortex. *Nature* 346(6282):359–62

Boulder Committee. 1970. Embryonic vertebrate central nervous system: revised terminology. *Anat. Rec.* 166:257–62

Brodmann K. 1909. *Vergleichenole Lokalisationslehre der Grosshirinde.* Leipzig: J. A. Barth. In German

Callaway EM, Katz LC. 1992. Development of axonal arbors of layer 4 spiny neurons in cat striate cortex. *J. Neurosci.* 12(2):570–82

Catalano SM, Robertson RT, Killackey HP. 1991. Early ingrowth of thalamocortical afferents to the neocortex of the prenatal rat. *Proc. Natl. Acad. Sci. USA* 88:2999–3003

Caviness VS Jr, Frost DO. 1983. Thalamocortical projections in the reeler mutant mouse. *J. Comp. Neurol.* 219:182–202

Caviness VS Jr, Rakic P. 1978. Mechanisms of cortical development: a view from mutations in mice. *Annu. Rev. Neurosci.* 1:297–326

Caviness VS Jr, Sidman RL. 1973. Time of origin of corresponding cell classes in the cerebral cortex of normal and reeler mutant mice: an autoradiographic analysis. *J. Comp. Neurol.* 148(2):141–52

Choi DW. 1992. Excitotoxic cell death. *J. Neurobiol.* 23(9):1261–76

Chun JJM, Nakamura MJ, Shatz CJ. 1987. Transient cells of the developing mammalian telencephalon are peptide immunoreactive neurons. *Nature* 325:617–20

Chun JJM, Shatz CJ. 1988a. Distribution of synaptic vesicle antigens is correlated with the disappearance of a transient synaptic zone in the developing cerebral cortex. *Neuron* 1:297–310

Chun JJM, Shatz CJ. 1988b. A fibronectin-like molecule is present in the developing cat cerebral cortex and is correlated with subplate neurons. *J. Cell Biol.* 106:857–72

Chun JJM, Shatz CJ. 1989a. The earliest-generated neurons of the cat cerebral cortex: characterization by MAP2 and neurotransmitter immunohistochemistry during fetal life. *J. Neurosci.* 9(5):1648–67

Chun JJM, Shatz CJ. 1989b. Interstitial cells of the adult neocortical white matter are the remnant of the early generated subplate population. *J. Comp. Neurol.* 282:555–69

Chung WW, Lagenaur CF, Yan YM, Lund JS. 1991. Developmental expression of neural cell adhesion molecules in the mouse neocortex and olfactory bulb. *J. Comp. Neurol.* 314(2):290–305

Crandall JE, Jacobson M, Kosik K. 1986. Ontogenesis of microtubule-associated protein 2 (MAP-2) in embryonic mouse cortex. *Dev. Brain Res.* 28:127–33

Cunningham TJ. 1982. Naturally occurring neuron death and its regulation by developing neural pathways. *Int. Rev. Cytol.* 74: 163–86

De Carlos JA, O'Leary DDM. 1992. Growth and targeting of subplate axons and establishment of major cortical pathways. *J. Neurosci.* 12(4):1194–1211

Del Rio JA, Soriano E, Ferrer I. 1991. A transitory population of substance P-like immunoreactive neurones in the developing cerebral cortex of the mouse. *Dev. Brain Res.* 64(1–2):205–11

Derer P, Derer M. 1990. Cajal-Retzius cell ontogenesis and death in mouse brain visualized with horseradish peroxidase and electron microscopy. *Neuroscience* 36(3):839–56

DiBenedetto AJ, Wang S, Pittman RN. 1992. Isolation of genes up- or down-regulated in cells undergoing programmed neuronal cell death. *Soc. Neurosci. Abstr.* 18(1):43 (Abstr.)

Dziegielewska KM, Møllgård K, Reynolds ML, Saunders NR. 1987. A fetuin-related glycoprotein (α2HS) in human embryonic and fetal development. *Cell Tissue Res.* 248:33–41

Dziegielewska KM, Reader M, Matthews N, Brown WM, Møllgård K, Saunders NR. 1993. Synthesis of the foetal protein fetuin by early developing neurons in the immature neocortex. *J Neurocytol.* 22(4):266–72

Erzurumlu RS, Jhaveri S. 1992. Emergence of connectivity in the embryonic rat parietal cortex. *Cereb. Cortex* 2(4):336–52

Fairén A, Smith-Fernández A, Martí E, DeDiego I, de la Rosa EJ. 1992. A transient immunoglobulin-like reactivity in the developing cerebral cortex of rodents. *Neuro Report* 3:881–84

Feldman SC, Harris MR, Laemle LK. 1990. The maturation of the somatostatin systems in the rat visual cortex. *Peptides* 11(6):1055–64

Ferguson IA, Schweitzer JB, Bartlett PF, Johnson EM Jr. 1991. Receptor-mediated retrograde transport in CNS neurons after intraventricular administration of NGF and growth factors. *J. Comp. Neurol.* 313(4): 680–92

Friauf E, McConnell SK, Shatz CJ. 1990. Functional synaptic circuits in the subplate during fetal and early postnatal development of cat visual cortex. *J. Neurosci.* 10(8): 2601–13

Friauf E, Shatz CJ. 1991. Changing patterns of synaptic input to subplate and cortical plate during development of visual cortex. *J. Neurophysiol.* 66(6):2059–71

Frost DD, Caviness VS Jr. 1980. Radial organization of thalamic projections to the neocortex in the mouse. *J. Comp. Neurol.* 194:369–93

Ghosh A, Antonini A, McConnell SK, Shatz CJ. 1990. Requirement for subplate neurons in the formation of thalamocortical connections. *Nature* 347:179–81

Ghosh A, Shatz CJ. 1992a. Involvement of subplate neurons in the formation of ocular dominance columns. *Science* 255:1441–43

Ghosh A, Shatz CJ. 1992b. Pathfinding and target selection by developing geniculocortical axons. *J. Neurosci.* 12(1):39–55

Ghosh A, Shatz CJ. 1993. A role for subplate neurons in the patterning of connections from thalamus to neocortex. *Development* 117:1047–47

Gilbert CD, Kelly JD. 1975. The projections of cells in different layers of the cat's visual cortex. *J. Comp. Neurol.* 183:81–106

Giuffrida R, Rustioni A. 1989. Glutamate and aspartate immunoreactivity in cortico-cortical neurons of the sensory motor cortex of rats. *Exp. Brain Res.* 74:41–46

Godfraind C, Schachner M, Goffinet AM. 1988. Immunohistological localization of cell adhesion molecules L1, J1, N-CAM and their common carbohydrate L2 in the embryonic cortex of normal and reeler mice. *Brain Res.* 470(1):99–111

Goffinet AM. 1992. The reeler gene: a clue to brain development and evolution. *Int. J. Dev. Biol.* 36(1):101–7

Gonzalez BJ, Leroux P, Bodenant C, et al. 1989. Ontogeny of somatostatin receptors in the rat brain: biochemical and autoradiographic study. *Neuroscience* 29(3):629–44

Goodman CS, Shatz CJ. 1993. Developmental mechanisms that generate precise patterns of neuronal connectivity. *Cell* 72 and *Neuron* 10:77–98

Götz M, Novak N, Bastmeyer M, Bolz J. 1992. Membrane-bound molecules in rat cerebral cortex regulate thalamic innervation. *Development* 116(3):507–19

Hendry SHC, Jones EG, DeFelipe J, et al. 1984. Neuropeptide-containing neurons of the cerebral cortex are also GABAergic. *Proc. Natl. Acad. Sci. USA* 81:6526–30

Herrmann K, Antonini A, Shatz CJ. 1991. Thalamic axons make synaptic contacts with subplate neurons in cortical development. *Soc. Neurosci. Abstr.* 17(1):899 (Abstr.)

Herrmann K, Shatz CJ. 1992. Glutamate-induced calcium responses of developing subplate cells. *Soc. Neurosci. Abstr.* 18(1):924 (Abstr.)

Honig LS, Herrmann K, Shatz CJ. 1990. An immunohistochemical study of developing human visual cortex. *Soc. Neurosci. Abstr.* 16(1):493 (Abstr.)

Hubel DH, Wiesel TN. 1977. Ferrier lecture. Functional architecture of macaque monkey visual cortex. *Proc. R. Soc. London* 198 (1130):1–59

Hubener M, Götz M, Klostermann S, Bolz J. 1992. Growth behavior of thalamic axons on cortical membranes studied with time-lapse video microscopy. *Soc. Neurosci. Abstr.* 18(1):924 (Abstr.)

Hunter DD, Llinas R, Ard M, et al. 1992. Expression of s-laminin and laminin in the developing rat central nervous system. *J. Comp. Neurol.* 323(2):238–51

Huntley GH, Hendry SHC, Killakey HP, et al. 1988. Temporal sequence of neurotransmitter expression by developing neurons of fetal monkey visual cortex. *Dev. Brain Res.* 43: 69–96

Innocenti GM. 1981. Growth and reshaping of axons in the establishment of visual callosal connections. *Science* 212:824–27

Jackson CA, Peduzzi JD, Hickey TL. 1989. Visual cortex development in the ferret. I. Genesis and migration of visual cortical neurons. *J. Neurosci.* 9(4):1242–53

Johnson JK, Casagrande VA. 1993. Prenatal development of axon outgrowth and connectivity in the ferret visual system. *Vis. Neurosci.* 10:117–30

Klose M, Bentley D. 1989. Transient pioneer neurons are essential for formation of an embryonic peripheral nerve. *Science* 245: 982–83

Koh S, Higgins GA. 1991. Differential regulation of the low-affinity nerve growth factor receptor during postnatal development of the rat brain. *J. Comp. Neurol.* 313(3):494–508

König N, Marty R. 1981. Early neurogenesis and synaptogenesis in cerebral cortex. *Bibl. Anat.* 19:152–60

Kordower JH, Mufson EJ. 1992. Nerve growth factor receptor-immunoreactive neurons within the developing human cortex. *J. Comp. Neurol.* 323:25–41

Korsmeyer SJ. 1992. Bcl-2: a repressor of lymphocyte death. *Immunol. Today* 13(8): 285–88

Kostovic I, Molliver ME. 1974. A new interpretation of the laminar development of cerebral cortex: synaptogenesis in different layers of neopallium in the human fetus. *Anat. Rec.* 178:395

Kostovic I, Rakic P. 1980. Cytology and time of origin of interstitial neurons in the white matter in infant and adult human and monkey telencephalon. *J. Neurocytol.* 9:219–42

Kostovic I, Rakic P. 1990. Developmental history of the transient subplate zone in the visual and somatosensory cortex of the macaque monkey and human brain. *J. Comp. Neurol.* 297:441–70

Kuwada J. 1986. Cell recognition by neuronal growth cones in a simple vertebrate embryo. *Science* 233:740–46

Lauder JM, Han VKM, Henderson P, et al. 1986. Prenatal ontogeny of the GABAergic system in the rat brain: an immunocytochemical study. *Neuroscience* 19:465–93

Lee VM, Balin BJ, Otvos L Jr, Trojanowski JQ. 1991. A68: a major subunit of paired helical filaments and derivatized forms of normal tau. *Science* 251(4994):675–78

LeVay S, Stryker MP, Shatz CJ. 1978. Ocular dominance columns and their development in layer IV of the cat's visual cortex. *J. Comp. Neurol.* 179:223–44

LeVay S, Wiesel TN, Hubel DH. 1980. The development of ocular dominance columns in normal and visually deprived monkeys. *J. Comp. Neurol.* 191:1–51

Liu F-C, Graybiel AM. 1992. Transient calbindin-D28K-positive systems in the telencephalon: ganglionic eminence, developing striatum and cerebral cortex. *J. Neurosci.* 12(2):694–90

Lowenstein PR, Shering AF. 1992. Synaptic input to subplate neurons in the cat: infragranular neurons provide synaptic input to underlying subplate cells during early postnatal neocortical development. A light and electron microscopical study. *J. Anat.* 180: 383 (Abstr.)

Lund RD, Mustari MJ. 1977. Development of the geniculocortical pathway in rats. *J. Comp. Neurol.* 173:289–306

Luskin MB, Shatz CJ. 1985a. Neurogenesis of the cat's primary visual cortex. *J. Comp. Neurol.* 242:611–31

Luskin MB, Shatz CJ. 1985b. Studies of the earliest generated cells of the cat's visual cortex: cogeneration of subplate and marginal zones. *J. Neurosci.* 5(4):1062–75

Marin-Padilla M. 1971. Early prenatal ontogenesis of the cerebral cortex (neocortex) of the cat (*Felis Domestica*). A golgi study. *Z. Anat. Entwicklungsgesch.* 134:117–45

Marin-Padilla M. 1988. Early ontogenesis of the human cerebral cortex. In *Cerebral Cortex*, ed. A Peters & EG Jones, 7:1–34. New York/London: Plenum.518 pp.

Martin DP, Schmidt RE, DiStefano PS, et al. 1988. Inhibitors of protein synthesis and

RNA synthesis prevent neuronal death caused by nerve growth factor deprivation. *J. Cell Biol.* 106(3):829–44

Mattson MP. 1991. Evidence for the involvement of protein kinase C in neurodegenerative changes in cultured human cortical neurons. *Exp. Neurol.* 112(1):95–103

McConnell SK, Ghosh A, Shatz CJ. 1989. Subplate neurons pioneer the first axon pathway from the cerebral cortex. *Science* 245: 978–82

Mehra RD, Hendrickson AE. 1993. A comparison of the development of neuropeptide and MAP2 immunocytochemical labeling in the macaque visual cortex during prenatal and postnatal development. *J. Neurobiol.* 24(1): 101–24

Meinecke DL, Rakic P. 1992. Expression of GABA and GABA-A receptors by neurons of the subplate zone in developing primate occipital cortex: evidence for transient local circuits. *J. Comp. Neurol.* 317:91–101

Meinecke DL, Rakic P. 1993. Low-affinity p75 nerve growth factor receptor expression in the embryonic monkey telencephalon: timing and localization in diverse cellular elements. *Neuroscience* 54(1):105–16

Miller B, Chou L, Finlay BL. 1991a. The early development of visual corticothalamic and thalamocortical projections in the golden hamster. *Soc. Neurosci. Abstr.* 17(1):764 (Abstr.)

Miller B, Sheppard AM, Pearlman AL. 1992. Expression of two chondroitin sulfate proteoglycan core proteins in the subplate pathway of early cortical afferents. *Soc. Neurosci. Abstr.* 18(1):778 (Abstr.)

Miller MW, Al-Ghoul WM, Murtaugh M. 1991b. Expression of Alz-50 immunoreactivity in the developing principal sensory nucleus of the trigeminal nerve: effect of transecting the infraorbital nerve. *Brain Res.* 560(1–2):132–38

Miranda RC, Toran-Allerand CD. 1992. Developmental expression of estrogen receptor mRNA in the rat cerebral cortex: a nonisotopic in situ hybridization histochemistry study. *Cereb. Cortex* 2(1):1–15

Møllgård K, Jacobsen M. 1984. Immunohistochemical identification of some plasma proteins in human embryonic and fetal forebrain with particular reference to the development of the neocortex. *Dev. Brain Res.* 13:49–63

Molliver ME, Kostovic I, Van der Loos H. 1973. The development of synapses in cerebral cortex of the human fetus. *Brain Res.* 50:403–7

Molnár Z, Blakemore C. 1991. Lack of regional specificity for connections formed between thalamus and cortex in coculture. *Nature* 351:475–77

Molnár Z, Blakemore C. 1992. How are thalamocortical axons guided in the reeler mouse? *Soc. Neurosci. Abstr.* 18(1):778 (Abstr.)

Molnár Z, Yee K, Lund R, Blakemore C. 1991. Development of rat thalamus and cerebral cortex after embryonic interruption of their connections. *Soc. Neurosci. Abstr.* 17(1):764 (Abstr.)

Mrzljak L, Uylings HBM, Kostovic I, Van Eden CG. 1988. Prenatal development of neurons in the human prefrontal cortex: I. A qualitative Golgi study. *J. Comp. Neurol.* 271:355–86

Naegele JR, Barnstable CJ, Wahle PR. 1991. Expression of a unique 56-kDa polypeptide by neurons in the subplate zone of the developing cerebral cortex. *Proc. Natl. Acad. Sci. USA* 88:330–34

Naegele JR, Jhaveri S, Schneider GE. 1988. Sharpening of topographical projections and maturation of geniculocortical axon arbors in the hamster. *J. Comp. Neurol.* 277(4): 593–607

Neugebauer KM, Tomaselli KJ, Lillien J, Reichardt LF. 1988. N-cadherin, NCAM, and integrins promote retinal neurite outgrowth on astrocytes in vitro. *J. Cell Biol.* 107:1177–87

O'Leary DDM. 1989. Do cortical areas emerge from a protocortex? *Trends Neurosci.* 12(10):400–6

O'Leary DDM, Bicknese AR, DeCarlos JA, et al. 1990. Target selection by cortical axons: alternative mechanisms to establish axonal connections in the developing brain. *Cold Spring Harbor Symp. Quant. Biol.* 55:453–68

O'Leary DDM, Koester SE. 1993. Development of axonal pathways and patterned connections of the mammalian cortex. *Neuron* 10:991–1006

Ogawa M, Miyata T, Nakajima K, Mikoshiba K. 1992. *Antibodies bind to Cajal-Retzius neurons.* Presented at the Ann. Meet. Physiol. Soc. Jpn., 69th (Abstr.)

Pearlman AM, Broekelmann TJ, McDonald JA, Sheppard AM. 1992. Stage-specific production of fibronectin in neocortical development: evidence for synthesis by both neurons and glia. *Soc. Neurosci. Abstr.* 18(1):777 (Abstr.)

Pilar G, Landmesser L. 1976. Ultrastructural differences during embryonic cell death in normal and peripherally derived ciliary ganglia. *J. Cell Biol.* 68(2):339–56

Rakic P. 1974. Neurons in rhesus monkey visual cortex: systematic relation between time of origin and eventual disposition. *Science* 183:425–27

Rakic P. 1977. Prenatal development of the visual system in rhesus monkey. *Philos. Trans. R. Soc. London Ser. B* 278:245–60

Rakic P. 1988. Specification of cerebral cortical areas. *Science* 241(4862):170–76

Ramón y Cajal S. 1911. *Histologie du systeme nerveux de l'homme et des vertebres.* Paris: Maloine. In French

Rickmann M, Chronwall BM, Wolff JR. 1977. On the development of non-pyramidal neurons and axons outside the cortical plate: The early marginal zone as a pallial anlage. *Anat. Embryol.* 151:285–307

Rodriguez-Tébar A, Dechant G, Barde Y-A. 1990. Binding of brain-derived neurotrophic factor to the nerve growth factor receptor. *Neuron* 4(4):487–92

Rodriguez-Tébar A, Dechant G, Gotz R, Barde Y-A. 1992. Binding of neurotrophin-3 to its neuronal receptors and interactions with nerve growth factor and brain-derived neurotrophic factor. *EMBO J.* 11(3):917–22

Sánchez MP, Frassoni C, Álvarez-Bolardo G, Spreafico R, Fairén A. 1992. Distribution of calbindin and paralbumin in the developing somatosensory cortex and its primordium in the rat: an immunohistochemical study. *J. Neurocytol.* 21:717–36

Saunders NR, Habgood MD, Ward RA, Reynolds ML. 1992. Origin and fate of fetuin-containing neurons in the developing neocortex of the fetal sheep. *Anat. Embryol.* 186:477–86

Schecterson LC, Bothwell M. 1992. Novel roles for neurotrophins are suggested by BDNF and NT-3 mRNA expression in developing neurons. *Neuron* 9(3):449–63

Schlaggar BL, O'Leary DDM. 1991. Potential of visual cortex to develop an array of functional units unique to somatosensory cortex. *Science* 252:1556–60

Schlumpf M, Shoemaker WJ, Bloom FE. 1980. Innervation of embryonic rat cerebral cortex by catecholamine-containing fibers. *J. Comp. Neurol.* 192(2):361–76

Seiler M, Schwab ME. 1984. Specific retrograde transport of nerve growth factor (NGF) from neocortex to nucleus basalis in the rat. *Brain Res.* 300:33–39

Shatz CJ. 1992. Dividing Up the Neocortex. *Science* 258:237–38

Shatz CJ, Chun JJM, Luskin MB. 1988. The role of the subplate in the development of the mammalian telencephalon. In *Cerebral Cortex,* ed. A Peters & EG Jones, 7:35–58. New York/London: Plenum. 518 pp.

Shatz CJ, Ghosh A, McConnell SK, et al. 1990. Pioneer neurons and target selection in cerebral cortical development. *Cold Spring Harbor Symp. Quant. Biol.* 55:469–80

Shatz CJ, Luskin MB. 1986. The relationship between the geniculocortical afferents and their cortical target cells during development of the cat's primary visual cortex. *J. Neurosci.* 6(12):3655–68

Shatz CJ, Rakic P. 1981. The genesis of efferent connections from the visual cortex of the fetal rhesus monkey. *J. Comp. Neurol.* 196:287–307

Sheppard AM, Hamilton SK, Pearlman AL. 1991. Changes in the distribution of extracellular matrix components accompany early morphogenetic events of mammalian cortical development. *J. Neurosci.* 11(12): 3928–42

Shoukimas G, Hinds JW. 1978. The development of the cerebral cortex in the embryonic mouse: an electron microscopic serial section analysis. *J. Comp. Neurol.* 179:795–830

Sims KB, Crandall JE, Kosik KS, Williams RS. 1988. Microtubule-associated protein 2 (MAP2) immunoreactivity in human fetal neocortex. *Brain Res.* 449:192–200

Stewart GR, Pearlman AL. 1987. Fibronectin-like immunoreactivity in the developing cerebral cortex. *J. Neurosci.* 7:3325–33

Terkelsen OB, Stagaard Janas M, Bock E, Møllgård K. 1992. NCAM as a differentiation marker of postmigratory immature neurons in the developing human servous system. *Int. J. Dev. Neurosci.* 10(6):505–16

Tomaselli KJ, Neugebauer KM, Bixby JL, et al. 1988. N-cadherin and integrins: two receptor systems that mediate neuronal process outgrowth on astrocyte surfaces. *Neuron* 1:33–43

Ueda K, Masliah E, Saitoh T, et al. 1990. Alz-50 recognizes a phosphorylated epitope of tau protein. *J. Neurosci.* 10(10):3295–3304

Valverde F, Facal-Valverde MV. 1988. Postnatal development of interstitial (subplate) cells in the white matter of the temporal cortex of kittens: a correlated golgi and electron microscopic study. *J. Comp. Neurol.* 269:168–92

Valverde F, Facal-Valverde MV, Santacana M, Heredia M. 1989. Development and differentiation of early generated calls of sublayer VIb in the somatosensory cortex of the rat: a correlated golgi and autoradiographic study. *J. Comp. Neurol.* 290: 118–40

Valverde F, Lopez-Mascaraque L, De Carlos JA. 1990. Distribution and morphology of Alz-50-immunoreactive cells in the developing visual cortex of kittens. *J. Neurocytol.* 19:662–71

Wahle P, Meyer G. 1987. Morphology and quantitative changes of transient NPY-ir neuronal populations during early postnatal development of the cat visual cortex. *J. Comp. Neurol.* 261:165–92

Wayne DB, Zupan AA, Pearlman AL. 1991. An antiserum to the low affinity nerve growth factor receptor labels subplate cells of developing neocortex in the mouse and a subset of neurons and glia in tissue culture. *Soc. Neurosci. Abstr.* 17(2):1304 (Abstr.)

Wise SP, Jones EG. 1976. The organization and postnatal development of the commissural projection of the rat somatic sensory cortex. *J. Comp. Neurol.* 168(3):313–43

Wise SP, Jones EG. 1978. Developmental studies of thalamocortical and commissural connections in the rat somatic sensory cortex. *J. Comp. Neurol.* 178:187–208

Wolozin BL, Pruchnicki A, Dickson DW, Davies P. 1986. A neuronal antigen in the brains of Alzheimer patients. *Science* 232:648–50

Wolozin BL, Scicutella A, Davies P. 1988. Reexpression of a developmentally regulated antigen in Down syndrome and Alzheimer disease. *Proc. Natl. Acad. Sci. USA* 85:6202–6

Woo TU, Beale JM, Finlay BL. 1991. Dual fate of subplate neurons in a rodent. *Cereb. Cortex* 1(5):433–43

Wood JG, Martin S, Price DJ. 1992. Evidence that the earliest generated cells of the murine cerebral cortex form a transient population in the subplate and marginal zone. *Dev. Brain Res.* 66:137–40

Woolsey TA, van der Loos H. 1970. The structural organization of layer IV in the somatosensory region (SI) of mouse cerebral cortex. The description of a cortical field composed of discrete cytoarchitectonic units. *Brain Res.* 17:205–42

Yamamoto N, Kurotani T, Toyama K. 1989. Neural connections between the lateral geniculate nucleus and visual cortex in vitro. *Science* 245:192–94

Yamamoto N, Yamada K, Kurotani T, Toyama K. 1992. Laminar specificity of extrinsic cortical connections studied in coculture preparations. *Neuron* 9:217–28

Yan Q, Johnson EM Jr. 1989. Immunohistochemical localization and biochemical characterization of nerve growth factor receptor in adult rat brain. *J. Comp. Neurol.* 290(4):585–98

Yu Z-Y, Anderson DE, Bottenstein JE. 1992. Colocalization of nerve growth factor receptor- and AC3-immunoreactivity in embryonic rat cortical subplate and basal forebrain neurons. *Soc. Neurosci. Abstr.* 18(1):53 (Abstr.)

Yu Z-Y, Bottenstein JE. 1991. A novel marker for cortical subplate neurons and some subventricular zone cells in early postnatal rat brain. *Soc. Neurosci. Abstr.* 17(2):1308 (Abstr.)

Annu. Rev. Neurosci. 1994. 17:219–46

SYNAPTIC VESICLES AND EXOCYTOSIS

R. Jahn

Howard Hughes Medical Institute and Departments of Pharmacology and Cell Biology, Yale University School of Medicine, New Haven, Connecticut 06510

T. C. Südhof

Howard Hughes Medical Institute and Department of Molecular Genetics, Southwestern Medical Center, University of Texas, Dallas, Texas 75235

KEY WORDS: neurotransmitter release, synaptic membrane proteins, nerve terminal function, intracellular membrane traffic, membrane fusion

INTRODUCTION

Neurons transmit information by releasing neurotransmitters from presynaptic nerve endings. In the resting stage, transmitters are stored in small organelles of uniform size and shape, the synaptic vesicles. When an action potential arrives in the nerve terminal, the membrane depolarizes and voltage-gated Ca^{2+} channels open. The resulting Ca^{2+} influx triggers exocytosis of synaptic vesicles, resulting in the release of neurotransmitter. The synaptic vesicle membrane is rapidly retrieved by endocytosis and reutilized for the formation of synaptic vesicles. These vesicles are reloaded with transmitter for another round of exocytosis. This cycle is repeated many times and can be studied in nerve terminals that have no connection with the neuronal cell body. Thus, the presynaptic compartment consists of an autonomous unit that contains all components required for repetitive exocytosis and membrane recycling. In contrast, release of neuropeptides involves large dense core vesicles that follow a different route; they will not be considered in this discussion (for review, see Thureson-Klein & Klein 1990).

Although the principles of synaptic vesicle membrane traffic were worked out many years ago (reviewed by Ceccarelli & Hurlbut 1980), it remains a major challenge to define the molecular events underlying the individual

219

trafficking steps, especially exocytosis. In recent years, rapid progress has been made in the molecular characterization of presynaptic proteins. In particular, the membrane composition of synaptic vesicles is at present better understood than the composition of any other organelle. Although the emerging picture is still sketchy, it is now possible to develop working models for the function of individual proteins that can be tested experimentally. In addition, it has become evident that the synaptic vesicle cycle shares basic properties with other intracellular trafficking pathways currently being studied in simpler systems. Here, we briefly review the current knowledge of synaptic vesicle proteins. We then focus on neuronal exocytosis in an attempt to integrate the information obtained from a variety of biological exocytotic and fusion events into a coherent picture. We refer the reader to several recent reviews that discuss aspects of vesicular membrane traffic not covered in this chapter (De Camilli & Jahn 1990, Südhof & Jahn 1991, Trimble et al 1991, Linstedt & Kelly 1992).

THE SYNAPTIC VESICLE

Synaptic vesicles are small and uniform organelles. In the mammalian CNS, their diameter averages approximately 40–50 nm. Although direct measurements are lacking, theoretical considerations indicate that an individual vesicle contains 8,000–10,000 phospholipid molecules and proteins with a combined molecular weight of no more than $3–5 \times 10^6$ (Jahn & Südhof 1993). This limits the number of proteins that fit onto a vesicle and indicates that only a restricted set of proteins are required for all synaptic vesicle functions. In addition, the high degree of curvature of the vesicle membrane requires asymmetric packing of both phospholipids and proteins. Liposomes of this size are generally less stable and more fusogenic than larger vesicles (reviewed by Blumenthal 1991), a factor that may play a role in synaptic vesicle exocytosis.

Synaptic vesicle membrane proteins can be broadly classified, according to function, into proteins involved in neurotransmitter uptake and storage and proteins involved in membrane trafficking (Südhof & Jahn 1991). The proteins that have been characterized at the molecular level are listed in Table 1.

Proteins Involved in Neurotransmitter Uptake and Storage

Three elements are involved in neurotransmitter uptake by synaptic vesicles: (a) an electrogenic proton ATPase of the vacuolar type that creates a proton electronmotive force which drives transmitter uptake against a concentration gradient, (b) ion channels, and in some cases, electron transporters that allow for charge compensation or provide reduction equivalents, and (c) carriers specific for individual neurotransmitters.

Table 1 Molecularly characterized synaptic vesicle proteins[a]

Synaptic vesicle proteins involved in neurotransmitter uptake and storage
Vacuolar proton pump[d]

Protein family	Synonyms	Number of isoforms[b]	Cloned from	Structural properties[c]	Evolutionary conservation	Function	References
116-kDa subunit	TJG-immuno-suppressor, VPH1	1 2 in yeast	Rat, mouse, C. ele-gans, yeast	M_r = 96,267 (rat). Transmembrane glycoprotein with 6–8 (?) TMRs	High over entire sequence (2 long stretches, 71% identical between rat and C. elegans)	Coupling of hydrolysis to proton pumping? In yeast essential for assembly	Adachi et al 1990 Perin et al 1991b Manolson et al 1992 Sulston et al 1992
70-kDa subunit	A-subunit TFP1 VMA1	1	Bovine, fungi, yeast, plants	M_r = 68,339 (bovine). Peripheral protein	High; 60% identity between bovine and yeast or Neurospora	Essential for catalysis, ATPase?	Bowman et al 1988 Zimniak et al 1988 Hirata et al 1990 Puopolo et al 1991
58-kDa subunit	B-subunit VMA2 VAT2 VATP-B	>2 (human kidney)	Human insects, yeast, plants, fungi	M_r = 56,661 (human kidney). Peripheral protein	High	Essential for catalysis	Manolson et al 1988 Nelson et al 1989, 1992 Südhof et al 1989b Taiz et al 1990 Gill & Ross 1991 Puopolo et al 1992
40-kDa subunit	C-subunit VMA5	1	Human, bovine, yeast	M_r = 43,989 (bovine). Peripheral protein, very hydrophilic	Low; 37% identity between bovine and yeast	Modulatory or essential?	Xie & Stone 1988 Stone et al 1989 Nelson et al 1990 Ho et al 1993

Table 1 (continued)

Protein family	Synonyms	Number of isoforms[b]	Cloned from	Structural properties[c]	Evolutionary conservation	Function	References
38-kDa subunit	D-subunit	1?	Bovine	$M_r = 31,495$. Hydrophilic, contains hydrophobic stretch of 25 aa not considered as TMR	?	?	Wang et al 1988
33-kDa subunit	E-subunit	1	Bovine	$M_r = 26,139$. No TMR, but hydrophobic	?	?	Hirsch et al 1988
17-kDa subunit	mediatophore VATPc, VMA3 VATP-Pl	1; >4 in plants	Bovine, mouse, Torpedo, Drosophila, yeast, plants	$M_r = 15,849$ (bovine). Transmembrane protein with 4 TMR, very hydrophobic, proteolipid	Very high; 80% identity between mammals and plants	Proton channel	Sun et al 1987 Mandel et al 1988 Gogarten et al 1989 Nelson & Nelson 1989 Birman et al 1990 Umemoto et al 1990 Zhang et al 1990 Hanada et al 1991 Lai et al 1991
Monoamine transporter		2	Rat	$M_4 = 55,931$. Transmembrane glycoprotein with 12 TMR	?	Transporter	Liu et al 1992
SV2		2	Rat, bovine	$M_r = 82,700$. Transmembrane glycoprotein with 12 TMR		Transporter?	Bajjalieh et al 1992 Feany et al 1992 Gingrich et al 1992
Cytochrome b_{561}		1	Bovine	$M_r = 30,061$. Transmembrane protein with 6 TMR and prosthetic heme groups	?	Electron conductor	Perin et al 1988

Synaptic vesicle proteins presumably involved in membrane trafficking

Integral membrane proteins

Name		No.	Species	Structure	Conservation	Function	References
Synaptotagmin	p65	4	Human, rat, Torpedo, Drosophila	M_r = 47,476 (human, isoform 1). Transmembrane glycoprotein with 1 TMR, contains C_2-domains homologous to protein kinase C	High; 57% identity between Drosophila and humans, 74% in cytoplasmic tail	Ca^{2+} receptor for exocytosis? Docking protein?	Perin et al 1990, 1991a,c; Geppert et al 1991; Tugal et al 1991; Wendland et al 1991; Brose et al 1992
VAT-1		1	Torpedo	M_r = 41,572. Integral membrane glycoprotein, number of TMR unclear	? (probably not conserved)	Specific for electric organ, function presumably tissue-specific	Linial et al 1989
Synaptophysin	p38; Synaptoporin, HL-5 (isoforms only)	3	Human, rat bovine, Torpedo	M_r = 33,312 (rat). Integral membrane glycoprotein with 4 TMR	Low (high in mammals); 63% identity between Torpedo and mammals	Fusion pore? Docking protein? Structural protein? Channel?	Buckley et al 1987; Leube et al 1987; Südhof et al 1987; Johnston et al 1989; Cowan et al 1990; Knaus et al 1990
SCAMP		1	Rat	M_r = 37,900. Integral membrane protein with 4 TMR and putative metal binding domain	?	Fusion protein?	Brand et al 1991; Brand & Castle, in press
HPC-1	syntaxin; epimorphin (3rd isoform)	3	Rat, mouse	M_r = 33,989 (rat). Integral membrane protein with 1 TMR at the C-terminal end	?	Docking protein? Ca^{2+} channel binding protein?	Bennett et al 1992; Hirai et al 1992; Inoue & Akagawa 1992
Synaptobrevin	VAMP; cellubrevin (ubiquitous isoform only)	3	Human, rat, bovine, Torpedo, Drosophila	M_r = 12,650 (bovine SBI). Integral membrane protein with 1 TMR at the C-terminal end	High in conserved region of 84 aa (2/3 of the protein, 79% identity between all species and forms)	Fusion protein? Interaction with small GTP-binding proteins?	Trimble et al 1988; Elferink et al 1989; Südhof et al 1989c; Archer et al 1990; McMahon et al 1993

Table 1 (continued)

Protein family	Synonyms	Number of isoforms[b]	Cloned from	Structural properties[c]	Evolutionary conservation	Function	References
Peripheral membrane proteins							
Amphiphysin		1	Rat	$M_r = 75,204$. Hydrophilic; contains hydrophobic stretch of 21 aa not considered as TMR	?	?	Lichte et al 1992
Synapsin	Protein Ia,b, protein IIIa,b	4 (syn. Ia, Ib, IIa, IIb)	Human, rat, bovine	$M_r = 73,996$ (rat Ia). Isoforms range from 74–52 kDa. Peripheral protein with amphiphilic domains, globular head and elongated tail region	High in mammals	Linkage of vesicles to cytoskeleton? Regulated by multiple phosphorylations	Südhof et al 1989a Valtorta et al 1992
rab3[e]	smg p25	4 (a, b, c, d)	Human, rat, bovine, Discopyge, Drosophila	$M_r = 24,954$ (rat 3a). Modified by geranyl-geranyl-, palmitoyl-, and carboxymethyl groups	High; 78% identity between human and Drosophila	GTPase, regulation of membrane traffic?	Matsui et al 1988 Zahraoui et al 1988, 1989 Johnston et al 1991 Volknandt et al 1991 Baldini et al 1992

[a] Abbreviations: TMR, transmembrane region; aa, amino acids. Additional proteins were found on purified mammalian synaptic vesicles, such as p29 (Baumert et al 1990), SVAPP 120 (Bähler et al 1991), pp60[c-src] (Barnekow et al 1990, Linstedt et al 1992), Ca²⁺-calmodulin-dependent protein kinase II (Benfenati et al 1992b), ceramide kinase (Bajjalieh et al 1989), calmodulin (Hooper & Kelly 1984; probably by means of binding to CaM-kinase II), kinesin (Sato-Yoshitake et al 1992), dynein (Lacey & Haimo 1992), aldolase (Bähler et al 1991), glutaraldehyde 3-phosphate dehydrogenase (M Baumert & R Jahn, unpublished data), 30- and 36-kDa membrane proteins (Obata et al 1987). The proteins involved in forming the clathrin coat during endocytosis [clathrin chains and assembly proteins (adaptors) are not included (see text)]. See also footnote e.

[b] As of December 1992. The term "isoform" indicates structural homology and not necessarily functional identity.

[c] All M_r predicted from cDNA or gene structure (predominant isoform only unless indicated otherwise), not including posttranslational modifications.

[d] The vacuolar proton ATPase of brain synaptic/coated vesicles contains additional subunits including those of apparent M_r of 34, 31, and 19 kDa, respectively, that are not yet structurally characterized (Forgac 1989, Stone et al 1989, Adachi et al 1990).

[e] Purified synaptic vesicles contain several additional small GTP-binding proteins of the rab family (Ngsee et al 1990), including rab5 and rab7 (G Fischer von Mollard, TC Südhof & R Jahn, unpublished data).

The proton ATPase is a very complex enzyme consisting of multiple subunits, most of which are cloned (Table 1). It is present in all tissues on all acidifying organelles (including lysosomes), but some subunits exist in multiple isoforms that display tissue-specific expression (reviewed by Nelson 1992). In the brain, most studies have dealt with a proton ATPase purified from clathrin-coated vesicles (Forgac 1989) that are essentially derived from synaptic vesicles (Maycox et al 1992). There is still no agreement about the precise subunit composition of the vacuolar proton ATPase, which may vary between different species or even between different organelles in a single cell. However, at least four of the subunits (116-, 70-, and 58-kDa subunits and the 17-kDa proton channel) are highly conserved through evolution; the latter three exhibit significant homologies to corresponding subunits of proton ATPases in archebacteria, mitochondria, and chloroplasts. Although the functions of most of the subunits are still unclear, both in vitro reconstitution experiments and genetic studies in yeast show that at least the 70-kDa subunit, the 58-kDa subunit, and the proton channel are essential for activity (Table 1 and references therein).

Synaptic vesicles contain at least four biochemically distinct transport activities for neurotransmitters that are differentially distributed in the CNS: one for monoamines and serotonin (Johnson 1988), one for acetylcholine (Marshall & Parsons 1987), one for glutamate (Maycox et al 1990), and one for GABA and glycine (Fykse & Fonnum 1988, Hell et al 1988, Burger et al 1991). All transporters use the electrochemical proton gradient as the driving force, but they differ in the transport mechanism and the relative dependence on the membrane potential vs the proton concentration gradient (reviewed by Johnson 1988, Maycox et al 1990). Recently, genes for a vesicular monoamine transporter were cloned from PC12 cells and brain (Liu et al 1992). They encode proteins with overall structural similarities to plasma membrane transporters, such as the glucose transporters [12 predicted transmembrane regions (TMR), with cytoplasmic C- and N-termini; see Wright et al (1992) for review]. Although no significant homologies to other eukaryotic transport proteins were found, the monoamine transporters appear to be structurally related to a class of bacterial drug resistance transporters (Liu et al 1992). This suggests that the monoamine transporters may be the founding members of a new family of eukaryotic transport proteins. In addition, the cloning of SV2, a well characterized abundant synaptic vesicle protein (Buckley & Kelly 1985), revealed that this protein's structure also resembles that of a transporter (Bajjalieh et al 1992, Feany et al 1992, Gingrich et al 1992). Like the monoamine transporters, SV2 has 12 TMRs with cytoplasmic amino and carboxyl termini and is homologous to a family of bacterial transporters. However, SV2 and the monoamine transporters show homology to different classes of bacterial transporters that are not homologous to each other. In addition, SV2 and the monoamine transporters are not homologous to each other. These findings suggest that SV2 may also represent the first member of a new class of eukaryotic transporters, although its substrate has not yet been identified.

Furthermore, monoamine and peptide-containing vesicles possess an electron transporter, cytochrome b561, which provides redox equivalents for biosynthetic and processing enzymes such as dopamine-β-hydroxylase and α-amidating enzyme. The protein uses ascorbate as the electron donor/acceptor on both sides of the membrane (Beers et al 1986).

Much less is known about ion channels of synaptic vesicles. Biochemical and electrophysiological experiments have demonstrated the presence of channels for monovalent cations and for chloride (Rahamimoff et al 1988). A chloride channel recently purified from brain coated vesicles (Xie et al 1989) and from kidney microsomes (Redhead et al 1992) may be related to the vesicular chloride channel, but structural information is not yet available. In addition, the vesicle protein synaptophysin was shown to possess channel activity under certain experimental conditions (see below).

Proteins with Putative Trafficking Functions

Intensive study of synaptic vesicles over the past years has led to the structural characterization of major components of the synaptic vesicle membrane. Several novel protein families were discovered that are not related to previously characterized proteins; at least one member of every family is present on every synaptic vesicle, regardless of the neurotransmitter phenotype. The functions of most of these proteins are not understood at present. However, many of them are highly conserved in evolution and possess unique properties that suggest a role in protein-protein interactions or protein-membrane interactions during vesicular membrane traffic.

As outlined in Table 1, the putative trafficking proteins can be grouped into integral membrane proteins with one or more TMR, and peripheral proteins that are anchored to the vesicles by means of amphiphilic domains or posttranslational hydrophobic modifications. Integral membrane proteins are permanent residents of the vesicle membrane that remain on the vesicle during its entire cycle. Some of the peripheral proteins may be associated with the membrane only during certain parts of the membrane cycle and may be involved in the control of individual segments of the vesicle cycle. Examples of the latter group include small GTP-binding proteins of the rab protein family and proteins involved in the formation of clathrin coats (see below).

INTEGRAL MEMBRANE PROTEINS   The common integral membrane proteins described so far are members of small and structurally diverse protein families:

the synaptotagmins, synaptophysins, synaptobrevins (VAMPs), HPC-1/syn-taxins, and SCAMP. None of these proteins possesses a cleavable signal sequence. Aside from this, there is no similarity or sequence homology between the protein families. The synaptophysins and probably also SCAMP are polytopic membrane proteins with four TMRs and both the C- and N-terminus facing the cytoplasm. The synaptobrevins and HPC-1/syntaxins are small, largely cytoplasmic proteins with a single membrane spanning domain at the C-terminus. The synaptotagmins are type I transmembrane glycoproteins with one TMR and the N-terminus facing the vesicle lumen (see Table 1 for references).

Synaptophysin was the first of these proteins to be studied in detail. The protein forms high molecular weight complexes that contain multiple syn-aptophysin monomers and additional proteins of low molecular weight that have not yet been characterized (Johnston & Südhof 1990). Although reported to bind Ca^{2+} (Rehm et al 1986), this interaction could not be confirmed in a later study (Brose et al 1992). Incorporation of purified synaptophysin into black lipid membranes led to the formation of voltage-gated ion channels (Thomas et al 1988). It was proposed that synaptophysin (and by analogy, its isoform synaptoporin) forms a fusion channel by means of interactions with a putative plasmalemma receptor, allowing neurotransmitter release without (or before) complete exocytosis (reviewed by Betz 1990). Thus, synaptophysin may be the molecule responsible for the fusion pore that forms at the onset of exocytosis (see below). However, the properties of the synaptophysin channel (voltage-dependency, conductance) are difficult to reconcile with the view of a pore that is open only when bound to its partner within the plasma membrane (Südhof & Jahn 1991). Alternatively, it has been suggested that synaptophysin may play a role in the structural organization of the synaptic vesicle as a small and highly curved organelle (Johnston & Südhof 1990). We were unable to reconstitute synaptophysin in an inside-out orientation by using a variety of liposome-forming techniques (in contrast to most other vesicle proteins; M Baumert & R Jahn, unpublished observations). This indicates that synaptophysin is sensitive to membrane curvature. Inter-estingly, a protein with similar overall structure (four TMRs, intramolecular disulfide bonds in the loops at the luminal side) but no sequence homology, named peripherin/rds, was recently found to be selectively localized at the rim of photoreceptor discs. Peripherin/rds was invoked to function in forming and maintaining the highly curved rim region of the discs (Connell & Molday 1990, Arikawa et al 1992).

The synaptotagmins have two large, highly conserved repeats in their cytoplasmic domain that are homologous to the C_2-domain of protein kinase C (Perin et al 1990). In addition, C_2-like domains were recently identified in cytoplasmic phospholipase A_2 (Clark et al 1991), phospholipase $C\gamma1$ (Stahl et al 1988), GTPase activating protein (Vogel et al 1988), and unc-13 (Maruyama & Brenner 1991). In *C. elegans,* mutations in the unc-13 gene cause diverse defects in the nervous system, suggestive of a possible role in neurotransmission.

The C_2-domains are thought to confer the Ca^{2+} and phospholipid binding properties of protein kinase C and phospholipase A_2 (Clark et al 1991, Nishizuka 1992). Purified synaptotagmin was shown to bind at least 4 mol Ca^{2+}/mol synaptotagmin monomer in a phospholipid-dependent manner (Brose et al 1992). The K_d was in the range of 10^{-6}–10^{-7} M, with the precise value depending on the proportion of acidic phospholipids. Thus, synaptotagmin is currently the most promising candidate for an exocytotic Ca^{2+} receptor (Brose et al 1992). However, it did not induce fusion of liposomes in these experiments. Synaptotagmin undergoes protein-protein interactions, discussed below. In addition, synaptotagmin was identified as a possible autoantigen of Lambert-Eaton myasthenic syndrome, an autoimmune disease associated with the degeneration of presynaptic structures (Leveque et al 1992), further supporting an essential role of this protein in synaptic function.

The synaptobrevins (VAMPs) have recently been identified as the specific target for the tetanus toxin and botulinum B toxin (Link et al 1992, Schiavo et al 1992; McMahon et al 1993). These neurotoxins completely block synaptic vesicle exocytosis without affecting synaptic structure, energy metabolism, or ion gradients (reviewed by Niemann 1991). Their light chains are Zn^{2+}-dependent proteases that selectively cleave synaptobrevin 2 and cellubrevin (the ubiquitously distributed isoform), but not rat synaptobrevin 1, between positions 76 and 77. These findings suggest that synaptobrevin is essential for exocytotic membrane fusion. Synaptobrevins are similar to the yeast gene products SNC1 (Gerst et al 1992) and SLY1 (Dascher et al 1991). These gene products interact genetically with the small GTP-binding proteins ras and ypt1, respectively, raising the possibility of interactions between synaptobrevins and rab proteins in the synaptic vesicle pathway. However, specific protein-protein interactions of the synaptobrevins have not yet been demonstrated.

SCAMP is a recently characterized protein whose structure has yet to be studied in detail (Brand et al 1991; Brand & Castle, In press). So far, only one form of SCAMP has been cloned, but immunoblotting data suggest heterogeneity of the antigen (Brand et al 1991). SCAMP appears to be a ubiquitous resident of trafficking organelles in neuronal and nonneuronal cells. The large N-terminal cytoplasmic domain of SCAMP possesses amphiphilic helices, a region capable of forming a leucine zipper, and at the very N-terminal end, a putative metal (Ca^{2+})-binding domain. These features make this protein an attractive candidate for protein-protein interactions and possibly for protein-phospholipid interactions (membrane fusion).

Finally, three members of a new protein family, HPC-1 (syntaxin), were recently characterized at the molecular level. Similar to the synaptobrevins, these proteins have a single putative TMR at the C-terminus; the majority of the protein is exposed to the cytoplasm. Originally, these proteins were found on the presynaptic plasma membrane. However, recent results from our laboratories indicate that these proteins are also components of synaptic vesicles. If significant amounts of HPC-1 are indeed present on both the synaptic vesicle and the presynaptic plasma membrane, this would constitute a major difference from all other characterized vesicle membrane proteins, which are only transiently incorporated into the plasma membrane during exocytosis.

PERIPHERAL MEMBRANE PROTEINS   Specific association of peripheral proteins with synaptic vesicles is more difficult to demonstrate and is subject to experimental artifacts. Thus, many proteins have been identified in purified synaptic vesicles that also (or mainly) act in other parts of the neuron (Table 1). However, the presence of a variety of peripheral proteins on synaptic vesicles has been definitely demonstrated. These proteins can be roughly classified into four categories (although some of the assignments are somewhat arbitrary):

"Bona-fide" synaptic vesicle proteins   This group comprises proteins that are exclusively localized to the synaptic vesicle membrane and whose function is not completely understood. The most abundant and intensely studied proteins of this group are the synapsins (reviewed by Valtorta et al 1992). The synapsins encompass a small protein family of four homologous isoforms (synapsin Ia and Ib, synapsin IIa and IIb) that are derived from two genes by alternative splicing. All four synapsins are phosphorylated by cAMP-dependent protein kinase and CaM kinase I close to the N-terminus. In addition, synapsins Ia and Ib, but not IIa and IIb, are phosphorylated by CaM kinase II and a proline-directed protein kinase at sites localized to the C-terminal domain. In synapsin I, major conformational changes were found to be associated with phosphorylation at the C-terminal sites. Synapsin I is a highly surface-active molecule. This probably results from extensive amphiphilic domains in the N-terminal region, which may be responsible for the ability to form unusually stable monolayers at phase boundaries. In agreement with these observations, binding of synapsin I to synaptic vesicles involves interactions not only with the membrane surface but also with the hydrophobic part of the bilayer. Furthermore, synapsin I is complexed with CaM kinase II on the vesicle surface. This association may allow rapid phosphorylation of synapsin I following a rise of intrasynaptic Ca^{2+} levels (Valtorta et al 1992 and references therein).

In vitro, synapsin I was demonstrated to interact with various cytoskeletal elements, including F-actin, microtubules, spectrin, and neurofilaments. The interactions with actin have been studied in detail and were found to be very complex, involving bundling and an increase in the initial rate of actin polymerization, the latter probably by means of an increase in the number of filaments. All interactions are diminished, if not abolished, by CaM kinase II phosphorylation of the C-terminal phosphorylation site. Furthermore, vesicle-bound dephospho-synapsin I is as effective as free synapsin I in nucleating F-actin formation and is capable of binding but not of bundling F-actin (Benfenati et al 1992a). Vesicle-bound synapsin I was recently identified as the protein responsible for spectrin-binding to purified synaptic vesicles (Sikorski et al 1992).

Greengard and coworkers have integrated many of these observations into a model for synapsin function (Valtorta et al 1992). According to their concept, synapsin I forms links between the vesicle membrane and a primarily actin-based cytomatrix, thus preventing synaptic vesicles from moving to the presynaptic plasma membrane. Increase of cytosolic Ca^{2+} levels upon activation causes phosphorylation of synapsin, which would result in the release of the vesicles from the cytoskeletal network, transferring the vesicles from a resting into an active pool (Valtorta et al 1992). However, several properties of synapsins have not been explained and thus remain unclear. These include the role of synapsin II that is not regulated by C-terminal phosphorylation (Südhof et al 1989a) and the function of spectrin binding by synapsin I. In addition, it has recently been questioned whether reserve pools of synaptic vesicles exist that are mobilized only upon high activity (Betz & Bewick 1992). Hence it is possible that synapsins have additional, as yet unidentified functions.

Amphiphysin is a hydrophilic protein that is specifically localized to synaptic vesicles but also occurs in an unbound form in the cytosol (Lichte et al 1992). Although the function of this protein is not known, autoantibodies against amphiphysin were identified in three patients with stiff-man syndrome and breast cancer (Folli et al 1993; P De Camilli, personal communucation), indicating that this protein may participate in the pathogenesis of these diseases.

Cycle-dependent proteins   This group contains proteins that are membrane-associated only during specific parts of the synaptic vesicle pathway. The best characterized proteins of this group include small GTP-binding proteins and proteins involved in the formation of clathrin-coated vesicles.

Since the original discovery in yeast (Gallwitz et al 1983), many ras-related small GTP-binding proteins were found to be required for defined steps of intracellular membrane traffic (reviewed by Pfeffer 1992). According to the current concept, these proteins exist in an active (GTP) and inactive (GDP) forms. Although our knowledge concerning the GTP-GDP cycle and the participating accessory proteins is rapidly growing, their mechanism of action remains unclear. The proteins are anchored to the membrane by hydrophobic posttranslational modifications (Table 1), but the GDP forms can be removed from the membrane by specific proteins (GDP-dissociation inhibitor), apparently without alteration of the hydrophobic moieties. Synaptic vesicles contain several different small GTP-binding proteins, the most abundant being rab3A (Fischer von Mollard et al 1990, Ngsee et al 1990). In synaptosomal preparations, rab3A dissociates from synaptic vesicles upon stimulation of exocytosis (Fischer von Mollard et al 1991). In addition, coated vesicles from nerve terminals lack rab3A (Maycox et al 1992). This suggests that rab3A acts on the exocytotic limb of the vesicle cycle.

The second group of state-specific vesicle proteins includes the proteins involved in the formation of the clathrin coat. The route by which synaptic vesicles recycle under normal conditions is still under debate. Most of the evidence, however, favors recycling via clathrin-coated vesicles (Ceccarelli & Hurlbut 1980, Heuser 1989). Coated vesicles increase in number upon stimulation of nerve terminals and are labeled by extracellular markers (Heuser 1989); their membrane composition is virtually identical to that of synaptic vesicles (with the exception of rab proteins; Maycox et al 1992; G Fischer von Mollard & R Jahn, unpublished observations).

The coat structure of clathrin-coated vesicles has recently received considerable attention (for more detailed reviews, see Pearse & Robinson 1990, Brodsky et al 1991, Schmid 1992). The coat is composed of clathrin triskelia containing clathrin heavy and light chains. Assembly of the clathrin coat is promoted by assembly proteins or adaptors that are constituents of the clathrin coat and presumably link the clathrin cage to the underlying membrane. Two classes of adaptors are distinguished, AP1 (HA1) and AP2 (HA2), which are present in most cell types, including neurons. The adaptors are composed of several partially homologous subunits, all of which have been cloned and sequenced. AP2 adaptors are confined to endocytotic coated pits and vesicles and are found in the nerve terminal (reviewed by Brodsky et al 1991, Pearse & Robinson 1990).

Interestingly, neurons contain not only specific splicing variants of the clathrin light chains and some of the adaptors, but also have additional, apparently monomeric clathrin-binding proteins, namely AP180 (identical with AP3, NP-185, and F1-20) and auxilin (Morris et al 1992 and references therein, Zhou et al 1992). Furthermore, a new component (p140) was recently isolated from brain coated vesicles, but its tissue distribution remains to be established (Lindner & Ungewickell 1992). The adaptor proteins AP180 and auxilin, as well as αa_1 and αc_1 adaptin, are enriched in nerve terminals

(Maycox et al 1992). In addition, AP180 was shown by immunocytochemistry to be selectively localized in synapses (Su et al 1991, Sousa et al 1992). This protein was recently shown to be four times more active in promoting clathrin assembly than AP2 or any of the other neuronal adaptors (Lindner & Ungewickell 1992). These findings suggest that at least some of the neuronal adaptor proteins function exclusively in the recycling of synaptic vesicles. This may reflect a functional specialization of the coated vesicle pathway in nerve terminals that may allow for an increase in turnover rates and assembly efficiency.

Cytoskeletal and motor proteins   Movements of synaptic vesicles to and from nerve terminals as well as within the nerve terminal require interaction with cytoskeletal elements and with molecular motors. Dynein was shown to bind to purified synaptic vesicles with high affinity in a saturable manner, indicating that this protein interacts directly with synaptic vesicles during retrograde axonal transport (Lacey & Haimo 1992). Similarly, saturable binding of kinesin to synaptic vesicles that is reduced upon phosphorylation by cAMP-dependent protein kinase has been reported (Sato-Yoshitake et al 1992). In the nematode *C. elegans,* a mutation in a kinesin-related gene (unc-104) results in the selective reduction of synapses and synaptic vesicles, whereas axonogenesis is normal, suggesting that a specific member of the kinesin family is responsible for axonal transport of synaptic vesicles (Hall & Hedgecock 1991). However, the motor proteins responsible for vesicle movement within the nerve terminal are not known. In addition, purified synaptic vesicles contain substantial amounts of actin and tubulin.

Enzymes   Several enzymes were reported to be tightly associated with synaptic vesicles. It remains unclear why purified synaptic vesicle preparations contain high amounts of glycolytic enzymes such as aldolase and glutaralde-hyde 3-phosphate dehydrogenase (Bähler et al 1991, and unpublished observations). However, enzymes that modify vesicle proteins or lipids may play a regulatory role in synaptic vesicle membrane traffic. The tyrosine kinase pp60[c-src] is highly enriched in synaptic vesicles (Barnekow et al 1990, Linstedt et al 1992), where its major substrate appears to be synaptophysin (Pang et al 1988, Barnekow et al 1990). Similarly, CaM-kinase II is tightly associated with synaptic vesicles, where it appears to be complexed with its substrate synapsin I (Benfenati et al 1992b).

EXOCYTOSIS

Membrane fusion is a basic feature of every cell. Whether the steps of membrane fusion are similar in different intracellular fusion events or whether eukaryotic cells have evolved independent fusion mechanisms for different events is unclear. In phospholipid vesicles, aggregation and fusion can easily be induced by dehydration and/or local perturbation of lipid contact zones involving agents as diverse as Ca^{2+} ions, polyethylene glycol, or lysophosphatides in conjunction with lipid-binding proteins, or by introduction of energy, such as with alternating electric fields (reviewed byBurger & Verkleij 1990, Blumenthal 1991). Such studies allow the definition of basic biophysical parameters under which phospholipid bilayers are able to fuse. However, there is general agreement that biological fusion is a highly organized process that is catalyzed by specific proteins and is strictly regulated by a vast array of control mechanisms.

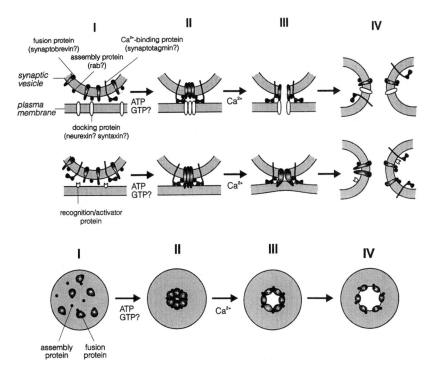

Figure 1 Model describing the hypothetical steps involved in exocytotic membrane fusion in nerve terminals. (A) Side view (cross-section) of the fusion complex. The upper panel shows a model that requires a docking protein in the plasma membrane for the assembly of the fusion complex and the formation of the fusion pore. In the model shown in the lower panel, assembly requires interactions with a recognition molecule of the plasma membrane, which does not participate in the formation of the fusion pore. Note that the numbers of molecules depicted in the models are arbitrary. (A) En face view of the protein arrangements in the synaptic vesicle membrane. This view is applicable to both models shown in Figure 1a. Cytosolic factors are not included. See text for details.

Fusion between two biological membranes occurs in three consecutive steps (see also Figure 1): (*i*) docking, i.e. establishing a contact between the two membranes; (*ii*) engagement, a process in which the proteins involved in the fusion event are aligned and activated; and (*iii*) the actual fusion event, during which phospholipids are rearranged and a hydrophilic continuum across the two fusing membranes is generated. Finally, the fusion complex disassembles (*iv*).

Docking

One of the most conspicuous specializations of fast synapses is the tight association of subpopulations of synaptic vesicles with specialized areas of the presynaptic plasma membrane. These "active zones" are organized structures in which vesicles are lined up in rows or arrays with regular intervals, separated by electron-dense, probably proteinaceous thickenings (Peters et al 1991). Such preassembly of the fusion machine may be one of the factors that allow exocytosis of synaptic vesicles to proceed orders of magnitude faster (200- to 300-μs delay time) than in other secretory systems.

What are the proteins involved in vesicle binding to the presynaptic plasma membrane? Recent data suggest that some synaptic vesicle proteins may bind directly to putative components of the presynaptic membrane. In detergent extracts, synaptotagmin was shown to bind to the receptor for α-latrotoxin, the active ingredient of black widow spider venom (Petrenko et al 1991). α-Latrotoxin causes massive, Ca^{2+}-independent exocytosis in the neuromuscular junction (Rosenthal & Meldolesi 1989). Cloning of the high molecular weight component of the receptor, based on amino acid sequences covering more than 20% of the entire protein, led to the discovery of the neurexins, a family of polymorphic neuron-specific cell surface receptors. The neurexins are derived from at least three genes, each of which contains two independent promoters and multiple splicing sites, potentially resulting in several hundred variants (Ushkaryov et al 1992; TC Südhof, unpublished observations). Neurexin Iα probably constitutes the high molecular weight component of the α-latrotoxin receptor and is primarily localized to nerve terminals. A physiological interaction between the α-latrotoxin receptor and synaptotagmin would have to occur by means of the cytoplasmic domains of these proteins. The three neurexins are highly homologous to each other, especially in their short cytoplasmic carboxyl-terminal domains, suggesting that all three neurexins should interact with synaptotagmin. Indeed, recent experiments demonstrate that the carboxyl termini of all neurexins bind to synaptotagmin and that this interaction is dependent on a 40–amino acid sequence in the neurexin tails (Hata et al 1993). These data support a role for neurexins in the targeting or docking of synaptic vesicles in the nerve terminal. In addition, this interaction suggests a potential mechanism of action for α-latrotoxin, but no data are yet available from physiological responding preparations to support this model.

Synaptic vesicles may also bind directly to ω-conotoxin-sensitive presynaptic Ca^{2+} channels (Bennett et al 1992, Yoshida et al 1992). This suggestion was based upon the observation that a complex consisting of synaptotagmin, HPC-1, and ω-conotoxin-binding sites cosedimented in immunprecipitation experiments. Clustering of Ca^{2+} channels at active zones has been suggested from physiological experiments (reviewed by Smith & Augustine 1988); the opening of these channels may lead to localized Ca^{2+} concentrations as high as 200–300 μM (Llinas et al 1992). However, despite the rapid progress in the molecular characterization of Ca^{2+} channels, it is still debated which types of Ca^{2+} channels are involved in synaptic transmission in the mammalian CNS. Synaptic Ca^{2+} currents are caused by channels that are partially sensitive to ω-Aga-IVA toxin (P-channels) but apparently insensitive to ω-conotoxin (Llinas et al 1992, Turner et al 1992). This indicates that most central synapses in mammals do not utilize ω-conotoxin-sensitive Ca^{2+} channels for neurotransmitter release.

In addition to synaptotagmin, synaptophysin was suggested to function in vesicle docking at the plasma membrane. Thomas & Betz (1990) have identified a putative binding protein for synaptophysin (physophilin) that is enriched in synaptic membrane fractions but has to be further characterized.

Engagement and Fusion

The mechanism of membrane fusion is still unclear. With the exception of viral fusion proteins, proteins that catalyze fusion are unknown. However, our knowledge of factors relevant to membrane fusion has advanced, primarily owing to studies in the following areas: (*a*) biochemical and genetic analysis of viral fusion proteins (reviewed by Stegmann et al 1989, White 1992), (*b*) genetic analysis of the secretory pathway in yeast (reviewed by Pryer et al 1992), (*c*) reconstitution of intracellular membrane fusion in permeabilized cells and, more importantly, cell-free assay systems (Pryer et al 1992, Burgoyne & Morgan 1993), and (*d*) detailed analysis of the fusion event by using capacitance measurements (Penner & Neher 1989, Almers & Tse 1990, Monck & Fernandez 1992). These studies have led to the formulation of fusion models that, while tenuous, accommodate many of the experimental observations.

Below, we attempt to outline a model for exocytotic membrane fusion in the nerve terminal that explains some of the discrepancies between observations made with different experimental systems and approaches. Obviously, this model is highly speculative and necessarily vague on several counts, but we hope that it offers a conceptual framework for future experiments. Our model builds on the concept of a proteinaceous fusion pore with defined dimensions as the initial event in fusion (Almers & Tse 1990, Monk & Fernandez 1992). Such pores were observed not only in secretory cells but also in fusion events mediated by viral fusion proteins (reviewed by White 1992), suggesting that they are general features of protein-mediated membrane fusion. However, our model contains important modifications in order to address some of the biochemical problems associated with the original models.

In our model, the central element in exocytotic membrane fusion is a ring-like aggregate of fusion proteins within the vesicle membrane that may encircle a small patch of phospholipids. This aggregate is generated from individual monomers (or oligomers) that assemble only upon contact with the plasma membrane (Figure 1). In this respect, the model differs from that proposed by Betz (1990), who envisions the fusion channel as a permanent structure that bears similarities with an ion channel and disintegrates only during fusion. Possibly, corresponding partner proteins are required in the plasma membrane that coassemble with the vesicular fusion proteins (Figure 1A, *upper panel*). Viral fusion proteins, however, do not require partner proteins (see Figure 1A, *lower panel*).

Assembly of the fusion proteins probably requires interactions with recruitment proteins (or protein complexes) that bind to the fusion proteins and regulate the formation of the prefusion complex. Such regulation of assembly may be an important control point for membrane fusion. Recruitment proteins would exist in active and inactive states, which could be controlled by recognition molecules within the plasma (target) membrane. Candidates for this role are GTP-binding proteins that are generally regarded to function as molecular switches (Bourne et al 1991, Pfeffer 1992). Both ras-related monomeric as well as heterotrimeric G-proteins are required for intracellular membrane fusion and exocytosis.

Although in our model reversible aggregation of the fusion proteins is necessary for fusion to occur, we believe that additional activation is required in order to unfold the fusogenic sequences. Without such a control step, fusion proteins would be in an active conformation even in the resting (non-aggregated) state, and a vesicle contacting into a nontarget membrane would have some probability of fusion. Activation of the fusion proteins may be catalyzed by cytoplasmic molecules that recognize the fusion aggregate. Candidates for this role are proteins such as NEM-sensitive factor (NSF) and SNAP, both soluble proteins that bind to a still unidentified membrane protein (reviewed by Mellman & Simons 1992, Rothman & Orci 1992). These proteins are essential for several intracellular fusion events in cell-free assays. In addition, two of them (NSF and α-SNAP) correspond to two yeast genes required for secretion (SEC18 and SEC17, respectively). In our model, activation of the fusion proteins may occur some time before the actual fusion event and may involve metabolic energy (priming; see Figure 1). Evidence for ATP-dependent priming that must precede Ca^{2+}-dependent exocytosis was recently obtained in chromaffin cells (Bittner & Holz 1992) and PC12 cells (Hay & Martin 1992).

The steps discussed so far are probably elements of most intracellular fusion events. According to this scenario, fusion would proceed as soon as the fusion complex is assembled and activated. However, in regulated exocytosis additional control proteins must be involved that are capable of inhibiting the function of the fusion complex, either by preventing assembly and activation or by interfering with the engaged complex. These proteins are inactivated by Ca^{2+} ions, either directly or through intermediate proteins, thus releasing inhibition and allowing fusion to proceed in response to a Ca^{2+} stimulus.

Fusion itself is initiated when fusogenic domains of the fusion proteins engage the target phospholipids, possibly involving amphiphilic helices analogous to those of many viral fusion proteins. This leads to the generation of a hydrophilic pore in the middle of the membrane, requiring either the opening of a hydrophilic protein-lined fusion channel in the middle of the aggregate or a rearrangement of the phospholipid patch in the middle of the aggregate, for instance by guiding the phospholipids out of the planes of the two membranes along the hydrophobic parts of the fusion helices (White 1992, Monck & Fernandez 1992). In any case, the enlargement of the fusion pore requires that phospholipids invade the space between the fusion protein units, leading to the dilation of the pore and to the disassembly of the fusion aggregate. During the onset of this invasion, phospholipids would have to curve around the lining of the neck of the fusion channel, a configuration that is, thermodynamically, highly unfavorable. It is possible that the phospholipids are assisted by specialized helper proteins in this process. Such proteins could consist of globular phospholipid-binding proteins that are bound to the cytoplasmic side of the contact sites and that allow the lipids to curve around them. Candidates for this role are certain members of the annexin protein family. Annexins comprise a group of related, abundant cytosolic proteins that easily form oligomers and bind to phospholipids in a Ca^{2+}-dependent manner (Creutz 1992; see also above). However, annexins are not enriched in secretory or nervous tissues and are not present at high concentrations in nerve terminals, as would be expected for such a role. During fusion or after completion of fusion, the disassembled fusion proteins are inactivated and the various cytosolic factors dissociate from the membrane.

Clearly, all functional assignments of proteins here are hypothetical. Additional or alternative proteins may be involved, such as a 145-kD protein shown to be essential for exocytosis in permeabilized PC12 cells (Walent et al 1992). Despite these uncertainties, the model makes the following predictions:

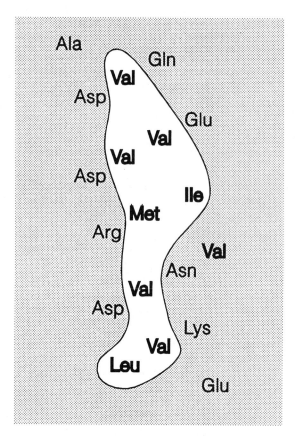

Figure 2   α-Helical projection of an amphiphilic domain of synaptobrevin that is shared by all isoforms and whose characteristics resemble fusion peptides of viral fusion proteins. Bulky hydrophobic residues (hydrophobic index > 0.64) are in bold type. A circle surrounds the contiguous area of bulky hydrophobic amino acids.

1. The fusion protein is an integral membrane protein of the vesicle membrane. It is unlikely that soluble cytoplasmic proteins carry out the actual fusion event, but an important role of proteins in the plasma membrane cannot be excluded (Figure 1A, *lower panel*). Pore formation requires assembly of multiple subunits. Thus, the fusion protein must be relatively abundant, and the odds are high that it is identical with one of the already characterized vesicle proteins (see Table 1). The prime candidate for a fusion protein is synaptobrevin. The striking block of exocytosis by botulinum and tetanus toxin that is caused by synaptobrevin proteolysis occurs at a very late step of exocytosis. All synaptobrevins possess a clearly defined amphiphilic helix in

their cytoplasmic domain (Figure 2), which is highly reminiscent of fusion peptides contained in viral fusion proteins. However, other candidates should be considered, such as SCAMP or HPC-1/syntaxin, both of which contain domains capable of forming amphiphilic helices.

2. The model predicts that the Ca^{2+} trigger molecule is not identical with the fusion protein(s) and that it acts as an inhibitor rather than an activator of fusion. Such an inhibitor would arrest fusion at a late state in the assembly and engagement of the fusion complex, thus greatly reducing the number of steps between arrival of the Ca^{2+} trigger and exocytosis. Supporting this view are studies of vesicular membrane traffic early in neuronal development, i.e. before mature synapses have formed. These studies suggest that synaptic vesicle precursors are fusion-competent and recycle actively in a Ca^{2+}-independent fashion, clearly demonstrating the independence of synaptic vesicle fusion from the Ca^{2+} trigger. Apparently, recycling is reduced upon vesicle clustering and docking during synaptogenesis (Young & Poo 1983, Zoran et al 1991, Matteoli et al 1992), and Ca^{2+}-dependent control is established in certain invertebrate neurons within seconds after contact formation (Haydon & Zoran 1989). Spontaneous fusion events, as evidenced by miniature endplate potentials, may reflect vesicles that have escaped the control of the Ca^{2+} regulated inhibitor. It cannot be ruled out, however, that the Ca^{2+}-binding proteins not only disinhibit the fusion complex but also play an active role in promoting fusion. Our prime candidate for this role is synaptotagmin, whose properties and localization are in agreement with the expected Ca^{2+} receptor. Recently, however, a PC12 cell line has been identified that is apparently devoid of synaptotagmin but responds to Ca^{2+} stimulation with a higher rate of secretion than the parental cell line's. How can these findings be reconciled with synaptotagmin being the Ca^{2+} receptor for exocytosis? If synaptotagmin is indeed an inhibitor of fusion in the Ca^{2+}-free form, loss of synaptotagmin would enhance secretion, as observed in these experiments, leaving the control to other, less efficient Ca^{2+}-dependent proteins involved in the steps preceding fusion. Such proteins may involve, for instance, Ca^{2+}-dependent actin severing proteins that dissolve actin barriers between vesicles and the plasma membrane (reviewed by Burgoyne & Morgan 1993) or Ca^{2+}-dependent proteins in the priming event (Bittner & Holz 1992). In neurons, the interaction between synaptotagmin and the α-latrotoxin receptor may suggest that α-latrotoxin induces the Ca^{2+}-bound conformation of synaptotagmin even in the absence of Ca^{2+} ions, resulting in disinhibition of the fusion complex and massive, constitutive exocytosis. Such a role of synaptotagmin is in agreement with recent data suggesting that in PC12 cells interference with synaptotagmin function impairs exocytosis (Elferink et al 1993).

3. The model provides a possible mechanistic explanation for the dualism

of nonhydrolyzable GTP-analogues and Ca^{2+} ions as triggers for exocytosis that has been found in many secretory systems (reviewed by Gomperts 1990). In neuronal exocytosis, there is little doubt that Ca^{2+} has ultimate control over membrane fusion. In other systems, however, assembly of the fusion complex and its control by means of a Ca^{2+}-binding protein may be less tightly coupled. If activated GTP-binding proteins are indeed involved in the assembly of the fusion machine, then nonhydrolyzable GTP-analogues will drive assembly, probably resulting in a permanently assembled complex, and promote fusion, overriding partially (or fully) the Ca^{2+}-sensitive inhibitor.

In summary, the model described above introduces several new elements into the previously described models. These include the requirements for regulated assembly proteins and for cytosolic activators and the proposal that the Ca^{2+} receptor acts as an inhibitor blocking a fusion complex that otherwise would proceed to fusion.

CONCLUSION

The field of intracellular membrane traffic is rapidly advancing, and the specialized area of nerve terminal membrane traffic is no exception. Although it is still too early to draw any general conclusions, it appears that the synaptic vesicle pathway uses mechanisms that are similar to other trafficking pathways in eukaryotic cells, in particular the recycling of plasma membrane receptors (Cameron et al 1991). This similarity probably extends to the molecular components catalyzing individual steps of the cycle, as documented by the recent identification of nonneuronal homologues for major synaptic vesicle proteins. However, the high degree of specialization for speed in the synaptic vesicle pathway certainly requires additional components that may act in concert with, or occasionally even replace, the general trafficking proteins. The complete characterization of the protein components of the synaptic vesicle membrane is within reach. Synaptic vesicles may become the first trafficking organelle whose complete structure will be known. Furthermore, the restricted localization and tight regulation of synaptic vesicle membrane traffic offer unique opportunities to study the mechanisms involved. As is already evident, the conclusions from these studies not only will be relevant for understanding neurotransmission but also will offer general insights into the nature and mechanisms of membrane trafficking events in eukaryotic cells.

ACKNOWLEDGMENTS

We are indebted to Drs. P De Camilli, A Helenius, and J White and the members of our laboratories for many helpful discussions; Drs. RD Burgoyne and D Castle for manuscript preprints; and Drs. P De Camilli, R Chow and

ER Chapman for critical reading of the manuscript. Work in the authors' laboratories was supported by grants from the Deutsche Forschungsgemeinschaft (RJ), the Perot Family Foundation, and the National Institute of Health (TCS).

Literature Cited

Adachi I, Puopolo K, Marquez-Sterling N, et al. 1990. Dissociation, cross-linking, and glycosylation of the coated vesicle proton pump. *J. Biol. Chem.* 265:967–73

Almers W, Tse FW. 1990. Transmitter release from synapses: Does a preassembled fusion pore initiate exocytosis? *Neuron* 4:813–18

Archer BT, Ozcelik T, Jahn R, et al. 1990. Structures and chromosomal localizations of two human genes encoding synaptobrevins 1 and 2. *J. Biol. Chem.* 265:17267–73

Arikawa K, Molday LL, Molday RS, Williams DS. 1992. Localization of peripherin/rds in the disk membranes of cone and rod photoreceptors: relationship to disk membrane morphogenesis and retinal degeneration. *J. Cell Biol.* 116:659–67

Bähler M, Klein RL, Wang JKT, et al. 1991. A novel synaptic vesicle-associated phosphoprotein: SVAPP-120. *J. Neurochem.* 57:423–30

Bajjalieh SM, Martin TFJ, Floor E. 1989. Synaptic vesicle ceramide kinase. A calcium-stimulated lipid kinase that co-purifies with brain synaptic vesicles. *J. Biol. Chem.* 264:14354–60

Bajjalieh SM, Peterson K, Shinghal R, Scheller RH. 1992. SV2, a brain synaptic vesicle protein homologous to bacterial transporters. *Science* 257:1271–73

Baldini G, Hohl T, Lin HY, Lodish HF. 1992. Cloning of a rab3 isotype predominately expressed in adipocytes. *Proc. Natl. Acad. Sci. USA* 89:5049–52

Barnekow A, Jahn R, Schartl M. 1990. Synaptophysin: A substrate for the protein tyrosine kinase pp60 *c-src* in intact synaptic vesicles. *Oncogene* 5:1019–24

Baumert M, Takei K, Hartinger J, et al. 1990. P29, a novel tyrosine-phosphorylated membrane protein present in small clear vesicles of neurons and endocrine cells. *J. Cell Biol.* 110:1285–94

Beers MF, Johnson RG, Scarpa A. 1986. Evidence for an ascorbate shuttle for the transfer of reducing equivalents across chromaffin granule membranes. *J. Biol. Chem.* 261:2529–35

Benfenati F, Valtorta F, Chieregatti E, Greengard P. 1992a. Interaction of free and synaptic vesicle-bound synapsin I with F-actin. *Neuron* 8:377–86

Benfenati F, Valtorta F, Rubenstein JL, et al. 1992b. Synaptic vesicle-associated $Ca^{2+}/$ calmodulin-dependent protein kinase II is a binding protein for synapsin I. *Nature* 359: 417–20

Bennett MK, Calakos N, Scheller RH. 1992. Syntaxin: a synaptic protein implicated in docking of synaptic vesicles at presynaptic active zones. *Science* 257:255–59

Betz H. 1990. Homology and analogy in transmembrane channel design: Lessons from synaptic membrane proteins. *Biochemistry* 29:3591–99

Betz WJ, Bewick GS. 1992. Optical analysis of synaptic vesicle recycling at the frog neuromuscular endplate. *Science* 255:200–3

Birman S, Meunier FM, Lesbats B, et al. 1990. A 15 kDa proteolipid found in mediatophore preparations from Torpedo electric organ presents high sequence homology with the bovine chromaffin granule protonophore. *FEBS Lett.* 261:303–6

Bittner MA, Holz RW. 1992. Kinetic analysis of secretion from permeabilized adrenal chromaffin cells reveals distinct components. *J. Biol. Chem.* 267:16219–25

Blumenthal, 1991. Membrane fusion. *Curr. Top. Membr. Transp.* 29:203–54

Bourne HR, Sanders DA, McCormick F. 1991. The GTPase superfamily: conserved structure and molecular mechanism. *Nature* 349: 117–27

Bowman EJ, Tenney K, Bowman BJ. 1988. Isolation of genes encoding the Neurospora vacuolar ATPase. Analysis of vma-1 encoding the 67-kDa subunit reveals homology to other ATPases. *J. Biol. Chem.* 263:13994–4001

Brand S, Castle JD. Is SCAMP 37 a prototypic fusion protein common to exocytic and endocytic pathways? *EMBO J.* In press

Brand SH, Laurie SM, Mixon MB, Castle JD. 1991. Secretory carrier membrane proteins 31–35 define a common protein composition among secretory carrier membranes. *J. Biol. Chem.* 266:18949–57

Brodsky FM, Hill BL, Acton SL, et al. 1991. Clathrin light chains: arrays of protein motifs that regulate coated-vesicle dynamics. *TIBS* 16:208–13

Brose N, Petrenko AG, Südhof TC, Jahn R. 1992. Synaptotagmin: a Ca^{2+} sensor on the synaptic vesicle surface. *Science* 256:1021–25

Buckley K, Kelly RB. 1985. Identification of a transmembrane glycoprotein specific for secretory vesicles of neural and endocrine cells. *J. Cell Biol.* 100:1284–94

Buckley KM, Floor E, Kelly RB. 1987. Cloning and sequence analysis of cDNA encoding p38, a major synaptic vesicle protein. *J. Cell Biol.* 105:2447–56

Burger KNJ, Verkleij AJ. 1990. Membrane fusion. *Experientia* 46:631–44

Burger PM, Hell J, Mehl E, et al. 1991. GABA and glycine in synaptic vesicles: Storage and transport characteristics. *Neuron* 7:287–93

Burgoyne RD, Morgan A. 1993. Regulated exocytosis. *Biochem. J.* 293:305–16

Cameron PL, Südhof TC, Jahn R, De Camilli P. 1991. Co-localization of of synaptophysin with transferrin receptors: implications for synaptic vesicle biogenesis. *J. Cell Biol.* 115:151–64

Ceccarelli B, Hurlbut WP. 1980. Vesicle hypothesis of the release of quanta of acetylcholine. *Physiol. Rev.* 60:396–441

Clark JD, Lin LL, Kriz RW, et al. 1991. A novel arachidonic acid-selective cytosolic PLA$_2$ contains a Ca^{2+}-dependent translocation domain with homology to PKC and GAP. *Cell* 65:1043–51

Connell GJ, Molday RS. 1990. Molecular cloning, primary structure and orientation of the vertebrate photoreceptor cell protein peripherin in the rod outer segment disk membrane. *Biochemistry* 29:4691–98

Cowan D, Linial M, Scheller RH. 1990. Torpedo synaptophysin: Evolution of a synaptic vesicle protein. *Brain Res.* 509:1–7

Creutz CE. 1992. The annexins and exocytosis. *Science* 258:924–31

Dascher C, Ossig R, Gallwitz D, Schmitt HD. 1991. Identification and structure of four yeast genes (sly) that are able to suppress the functional loss of ypt1, a member of the ras superfamily. *Mol. Cell. Biol.* 11:872–85

De Camilli P, Jahn R. 1990. Pathways to regulated exocytosis in neurons. *Annu. Rev. Physiol.* 52:625–645

Elferink LA, Peterson MR, Scheller RH. 1993. A role for synaptotagmin (p65) in regulated exocytosis. *Cell* 72:153–59

Elferink LA, Trimble WS, Scheller RH. 1989. Two vesicle-associated membrane protein genes are differentially expressed in the rat central nervous system. *J. Biol. Chem.* 264:11061–64

Feany MB, Lee S, Edwards RH, Buckley KM. 1992. The synaptic vesicle protein SV2 is a novel type of transmembrane transporter. *Cell* 70:861–67

Fischer von Mollard G, Mignery GA, Baumert M, et al. 1990. Rab3 is a small GTP-binding protein exclusively localized to synaptic vesicles. *Proc. Natl. Acad. Sci. USA* 87:198892

Fischer von Mollard G, Südhof TC, Jahn R. 1991. A small GTP-binding protein (rab3A) dissociates from synaptic vesicles during exocytosis. *Nature* 349:79–81

Folli F, Solimena M, Cofiell R, et al. 1993. Autoantibodies to a 128 kd synaptic protein in three women with the stiff-man syndrome and breast cancer. *N. Engl. J. Med.* In press

Forgac M. 1989. Structure and function of vacuolar class of ATP-driven proton pumps. *Physiol. Rev.* 69:765–96

Fykse E, Fonnum F. 1988. Uptake of γ-aminobutyric acid by a synaptic vesicle fraction isolated from rat brain. *J. Neurochem.* 50:1237–42

Gallwitz D, Donath C, Sander C. 1983. A yeast gene encoding a protein homologous to the human c-has/bas proto-oncogene product. *Nature* 306:704–7

Geppert M, Archer III BT, Südhof TC. 1991. Synaptotagmin II. A novel differentially distributed form of synaptotagmin. *J. Biol. Chem.* 266:13548–52

Gerst JE, Rodgers L, Riggs M, Wigler M. 1992. SNC1, a yeast homolog of the synaptic vesicle-associated membrane protein/synaptobrevin gene family: Genetic interactions with the RAS and CAP genes. *Proc. Natl. Acad. Sci. USA* 89:4338–42

Gill SS, Ross LS. 1991. Molecular cloning and characterization of the B subunit of a vacuolar H$^+$-ATPase from the midgut and Malpighian tubules of *Helicoverpa virescens*. *Arch. Biochem. Biophys.* 291:92–99

Gingrich JA, Andersen PH, Tiberi M, et al. 1992. Identification, characterization, and molecular cloning of a novel transporter-like protein localized to the central nervous system. *FEBS Lett.* 312:115–22

Gogarten JP, Rausch T, Bernasconi P, et al. 1989. Molecular evolution of H$^+$-ATPases. 1. *Methanococcus* and *Sulfolobus* are monophyletic with respect to eukaryotes and eubacteria. *Z. Naturforsch.* 44:641–50

Gomperts BD. 1990. G$_E$: A GTP-binding protein mediating exocytosis. *Annu. Rev. Physiol.* 52:591–606

Hall DH, Hedgecock EM. 1991. Kinesin-related gene unc-104 is required for axonal transport of synaptic vesicles in *C. elegans*. *Cell* 65:837–47

Hanada H, Hasebe M, Moriyama Y, et al. 1991. Molecular cloning of a cDNA encoding the 16 kDa subunit of vacuolar H$^+$-ATPase from mouse cerebellum. *Biochem. Biophys. Res. Commun.* 176: 1062–67

Hata Y, Davletov B, Petrenko AG, Jahn R, Südhof TC. 1993. Interaction of synaptotagmin with the cytoplasmic domains of neurexins. *Neuron* 10:307–15

Hay JC, Martin TFJ. 1992. Resolution of regulated secretion into sequential MgATP-dependent and calcium-dependent stages mediated by distinct cytosolic proteins. *J. Cell Biol.* 119:139–51

Haydon PG, Zoran MJ. 1989. Formation and modulation of chemical connections: Evoked acetylcholine release from growth cones and neurites of specific identified neurons. *Neuron* 2:1483–90

Hell JW, Maycox PR, Stadler H, Jahn R. 1988. Uptake of GABA by rat brain synaptic vesicles isolated by a new procedure. *EMBO J.* 7:3023–29

Heuser J. 1989. The role of coated vesicles in recycling of synaptic vesicle membrane. *Cell Biol. Int. Rep.* 13:1063–76

Hirai Y, Takebe K, Takashina M, et al. 1992. Epimorphin: A mesenchymal protein essential for epithelial morphogenesis. *Cell* 69: 471–81

Hirata R, Ohsumi Y, Nakano A, et al. 1990. Molecular structure of a gene, *VMA1*, endocing the catalytic subunit of H^+-translocating adenosine triphosphatase from vacuolar membranes of saccharomyces cerevisiae. *J. Biol. Chem.* 265:6726–33

Hirsch S, Strauss A, Masood K, et al. 1988. Isolation and sequence of a cDNA clone encoding the 31 kDa subunit of bovine kidney vacuolar H^+-ATPase. *Proc. Natl. Acad. Sci. USA* 85:3004–8

Ho MN, Hill KJ, Lindorfer MA, Stevens TH. 1993. Isolation of vacuolar membrane H^+-ATPase-deficient yeast mutants: the *VMA5* and *VMA4* genes are essential for assembly and activity of the vacuolar H^+-ATPase. *J. Biol. Chem.* 268:221–27

Hooper JE, Kelly RB. 1984. Calmodulin is tightly associated with synaptic vesicles independent of calcium. *J. Biol. Chem.* 259: 148–53

Inoue A, Akagawa K. 1992. Cloning and sequence analysis of cDNA for a neuronal cell membrane antigen, HPC-1. *J. Biol. Chem.* 267:10613–19

Jahn R, Südhof TC. 1993. Synaptic vesicle traffic: Rush hour in the nerve terminal. *J. Neurochem.* 61:12–21

Johnson RG. 1988. Accumulation of biological amines into chromaffin granules: A model for hormone and neurotransmitter transport. *Physiol. Rev.* 68:232–307

Johnston PA, Archer BT, Robinson K, et al. 1991. Rab3A attachment to the synaptic vesicle membrane mediated by a conserved polyisoprenylated carboxy-terminal sequence. *Neuron* 7:101–9

Johnston PA, Jahn R, Südhof TC. 1989. Transmembrane topography and evolutionary conservation of synaptophysin. *J. Biol. Chem.* 264:1268–73

Johnston PA, Südhof TC. 1990. The multi-

subunit structure of synaptophysin. Relationship between disulfide bonding and homo-oligomerization. *J. Biol. Chem.* 265: 8869–73

Kanner BI, Schuldiner S. 1987. Mechanism of transport and storage of neurotransmitters. *Crit. Rev. Biochem.* 22:1–38

Knaus P, Marqueze-Pouey B, Scherer H, Betz H. 1990. Synaptoporin, a novel putative channel protein of synaptic vesicles. *Neuron* 5:453–62

Lacey ML, Haimo LT. 1992. Cytoplasmic dynein is a vesicle protein. *J. Biol. Chem.* 267:4793–98

Lai SP, Watson JC, Hanson JN, Sze H. 1991. Molecular cloning and sequencing of cDNAs encoding the proteolipid subunit of the vacuolar H^+-ATPase from a higher plant. *J. Biol. Chem.* 266:16078–84

Leube RE, Kaiser P, Seiter A, et al. 1987. Synaptophysin: molecular organization and mRNA expression as determined from cloned cDNA. *EMBO J.* 6:3261–68

Leveque C, Hoshino T, David P, et al. 1992. The synaptic vesicle protein synaptotagmin associates with calcium channels and is a putative Lambert-Eaton myasthenic syndrome antigen. *Proc. Natl. Acad. Sci. USA* 89:3625–29

Lichte B, Veh RW, Meyer HE, Kilimann MW. 1992. Amphiphysin, a novel protein associated with synaptic vesicles. *EMBO J.* 11: 2521–30

Lindner R, Ungewickell E. 1992. Clathrin-associated proteins of bovine brain coated vesicles. An analysis of their number and assembly-promoting activity. *J. Biol. Chem.* 267:16567–73

Linial M, Miller K, Scheller RH. 1989. VAT-1: An abundant membrane protein from Torpedo cholinergic synaptic vesicles. *Neuron* 2:1265–73

Link E, Edelmann L, Chou JH, et al. 1992. Tetanus toxin action: Inhibition of neurotransmitter release linked to synaptobrevin proteolysis. *Biochem. Biophys. Res. Commun.* 189:1017–23

Linstedt A, Kelly RB. 1992. The molecular architecture of the nerve terminal. *Curr. Opin. Neurobiol.* 1:382–387

Linstedt AD, Vetter ML, Bishop JM, Kelly RB. 1992. Specific association of the proto-oncogene product pp60$^{c\text{-}src}$ with an intracellular organelle, the PC12 synaptic vesicle. *J. Cell Biol.* 117:1077–84

Liu Y, Peter D, Roghani A, et al. 1992. A cDNA that suppresses MPP$^+$ toxicity encodes a vesicular amine transporter. *Cell* 70:539–51

Llinas R, Sugimori M, Silver RB. 1992. Microdomains of high calcium concentration in a presynaptic terminal. *Science* 256:677–79

Mandel M, Moriyama Y, Hulmes JD, et al.

1988. cDNA sequence encoding the 16-kDa proteolipid of chromaffin granules implies gene duplication in the evolution. *Proc. Natl. Acad. Sci. USA* 85:5521–24

Manolson MF, Ouellette BFF, Filion M, Poole RJ. 1988. cDNA sequence and homologies of the "57-kDa" nucleotide-binding subunit of the vacuolar ATPase from Arabidopsis. *J. Biol. Chem.* 263:17987–94

Manolson MF, Proteau D, Preston RA, et al. 1992. The *VPH1* gene encodes a 95-kDa integral membrane polypeptide required for in vivo assembly and activity of the yeast vacuolar H^+-ATPase. *J. Biol. Chem.* 267: 14294–303

Marshall IG, Parsons SM. 1987. The vesicular acetylcholine transport system. *TINS* 10: 174–77

Maruyama IN, Brenner S. 1991. A phorbol ester/diacylglycerol-binding protein encoded by the unc-13 gene of *Caenorhabditis elegans*. *Proc. Natl. Acad. Sci. USA* 88:5729–33

Matsui Y, Kikuchi A, Kondo J, et al. 1988. Nucleotide and deduced amino acid sequences of a GTP-binding protein family with molecular weights of 25,000 from bovine brain. *J. Biol. Chem.* 263:11071–74

Matteoli M, Takei K, Perin MS, et al. 1992. Exo-endocytotic recycling of synaptic vesicles in developing processes of cultured hippocampal neurons. *J. Cell Biol.* 117:849–61

Maycox PR, Hell JW, Jahn R. 1990. Amino acid neurotransmission: Spotlight on synaptic vesicles. *TINS* 13:83–87

Maycox PR, Link E, Reetz A, et al. 1992. Clathrin-coated vesicles in nervous tissue are involved primarily in synaptic vesicle recycling. *J. Cell Biol.* 118:1379–88

McMahon HT, Ushkaryov YA, Edelmann L, Link E, Binz T et al. 1993. Cellubrevin is a ubiquitous tetanus-toxin substrate homologous to a putative synaptic vesicle fusion protein. *Nature* 364:346–49

Mellman I, Simons K. 1992. The Golgi complex: in vitro veritas? *Cell* 68:829–40

Monck JR, Fernandez JM. 1992. The exocytotic fusion pore. *J. Cell Biol.* 119:1395–404

Morris SA, Schröder S, Plessmann U, et al. 1993. Clathrin assembly protein AP180: Primary structure, domain organization and identification of a clathrin binding site. *EMBO J.* 12:667–75

Nakata T, Sobue K, Hirokawa N. 1990. Conformational change and localization of calpactin I complex involved in exocytosis as revealed by quick-freeze, deep-etch electron microscopy and immunocytochemistry. *J. Cell Biol.* 110:13–25

Nelson H, Mandiyan S, Nelson N. 1989. A conserved gene encoding the 57-kDa subunit of the yeast vacuolar H^+-ATPase. *J. Biol. Chem.* 264:1775–78

Nelson H, Mandiyan S, Noumi T, et al. 1990. Molecular cloning of cDNA encoding the C subunit of H^+-ATPase from bovine chromaffin granules. *J. Biol. Chem.* 265:20390–93

Nelson H, Nelson N. 1989. The progenitor of ATP synthases was closely related to the current vacuolar H^+-ATPase. *FEBS Lett.* 247:147–53

Nelson N. 1992. Organellar proton-ATPases. *Curr. Opin. Cell Biol.* 4:654–60

Nelson RD, Guo XL, Masood K, et al. 1992. Selectively amplified expression of an isoform of the vacuolar H^+-ATPase 56-kilodalton subunit in renal intercalated cells. *Proc. Natl. Acad. Sci. USA* 89:3541–45

Ngsee JK, Miller K, Wendland B, Scheller RH. 1990. Multiple GTP-binding proteins from cholinergic synaptic vesicles. *J. Neurosci.* 10:317–22

Niemann H. 1991. Molecular biology of clostridial neurotoxins. In *Sourcebook of Bacterial Toxins*, eds. JE Alour, JH Freer, pp. 303–348. New York: Academic

Nishizuka Y. 1992. Intracellular signaling by hydrolysis of phospholipids and activation of protein kinase C. *Science* 258:607–614

Obata K, Kojima N, Nishiye H, et al. 1987. Four synaptic vesicle-specific proteins: identification by monoclonal antibodies and distribution in the nervous tissue and the adrenal medulla. *Brain Res.* 404:169–79

Pang DT, Wang J, Valtorta F, et al. 1988. Protein tyrosine phosphorylation in synaptic vesicles. *Proc. Natl. Acad. Sci. USA* 85: 762–66

Pearse BMF, Robinson MS. 1990. Clathrin, adaptors and sorting. *Annu. Rev. Cell Biol.* 6:151–71

Penner R, Neher E. 1989. The patch-clamp technique in the study of secretion. *TINS* 12:159–63

Perin MS, Brose N, Jahn R, Südhof TC. 1991a. Domain structure of synaptotagmin (p65). *J. Biol. Chem.* 266:623–29

Perin MS, Fried VA, Slaughter CA, Südhof TC. 1988. The structure of cytochrome b561: a secretory vesicle-specific electron transport protein. *EMBO J.* 7:2697–703

Perin MS, Fried VA, Mignery GA, et al. 1990. Phospholipid binding by a synaptic vesicle protein homologous to the regulatory region of protein kinase C. *Nature* 345:260–63

Perin MS, Fried VA, Stone DK, et al. 1991b. Structure of the 116-kDa polypeptide of the clathrin-coated vesicle/synaptic vesicle proton pump. *J. Biol. Chem.* 266:3877–81

Perin MS, Johnston PA, Ozcelik T, et al. 1991c. Structural and functional conservation of synaptotagmin (p65) in *Drosophila* and humans. *J. Biol. Chem.* 266:615–22

Peters A, Palay SL, Webster HD. 1991. *The Fine Structure of the Nervous System.* Oxford Univ. Press. 494 pp.

Petrenko AG, Perin MS, Davletov BA, et al. 1991. Binding of synaptotagmin to the α-latrotoxin receptor implicates both in synaptic vesicle exocytosis. *Nature* 353:65–68

Pfeffer SR. 1992. GTP-binding proteins in intracellular transport. *TICS* 2:41–46

Pryer NK, Wuestehube LJ, Schekman R. 1992. Vesicle-mediated protein sorting. *Annu. Rev. Biochem.* 61:471–516

Puopolo K, Kumamoto C, Adachi I, Forgac M. 1991. A single gene encodes the catalytic "A" subunit of the bovine vacuolar H^+-ATPase. *J. Biol. Chem.* 266:24564–72

Puopolo K, Kumamoto C, Adachi I, et al. 1992. Differential expression of the "B" subunit of the vacuolar H^+-ATPase in bovine tissues. *J. Biol. Chem.* 267:3696–706

Rahamimoff R, DeRiemer SA, Sakmann B, et al. 1988. Ion channels in synaptic vesicles from Torpedo electric organ. *Proc. Natl. Acad. Sci. USA* 85:5310–14

Redhead CR, Edelman AE, Brown D, et al. 1992. A ubiquitous 64-kDa protein is a component of a chloride channel of plasma and intracellular membranes. *Proc. Natl. Acad. Sci. USA* 89:3716–20

Rehm H, Wiedenmann B, Betz H. 1986. Molecular characterization of synaptophysin, a major calcium-binding protein of the synaptic vesicle membrane. *EMBO J.* 5:535–41

Rosenthal L, Meldolesi J. 1989. α-Latrotoxin and related toxins. *Pharmacol. Ther.* 42:115–34

Rothman JE, Orci L. 1992. Molecular dissection of the secretory pathway. *Nature* 355:409–15

Sato-Yoshitake R, Yorifuji H, Inagaki M, Hirokawa N. 1992. The phosphorylation of kinesin regulates its binding to synaptic vesicles. *J. Biol. Chem.* 267:23930–36

Schiavo G, Benfenati F, Poulain B, et al. 1992. Tetanus and botulinum-B neurotoxins block neurotransmitter release by proteolytic cleavage of synaptobrevin. *Nature* 359:832–35

Schmid S. 1992. The mechanism of receptor-mediated endocytosis: More questions than answers. *BioEssays* 14:589–96

Sikorski AF, Terlecki G, Zagon IS, Goodman SR. 1992. Synapsin I-mediated interaction of brain spectrin with synaptic vesicles. *J. Cell Biol.* 114:313–18

Smith SJ, Augustine GJ. 1988. Calcium ions, active zones and synaptic transmitter release. *TINS* 11:458–64

Sousa R, Tannery NH, Zhou S, Lafer EM. 1992. Characterization of a novel synapse-specific protein. I. Developmental expression and cellular localization of the F1–20

protein and mRNA. *J. Neurosci.* 12:2130–43

Stahl ML, Ferenz CR, Kelleher KL, et al. 1988. Sequence similarity of phospholipase C with the non-catalytic region of src. *Nature* 332:269–272

Stegmann T, Doms RW, Helenius A. 1989. Protein-mediated membrane fusion. *Annu. Rev. Biophys. Biophys. Chem.* 18:187–211

Stone DK, Crider BP, Südhof TC, Xie XS. 1989. Vacuolar proton pumps. *J. Bioenerget. Biomembr.* 21:605–20

Su B, Hanson V, Perry D, Puszkin S. 1991. Neuronal specific protein NP185 is enriched in nerve endings: binding characteristics for clathrin light chains, synaptic vesicles and synaptosomal plasma membrane. *J. Neurosci. Res.* 29:461–73

Südhof TC, Baumert M, Perin MS, Jahn R. 1989c. A synaptic vesicle membrane protein is conserved from mammals to *Drosophila*. *Neuron* 2:1475–81

Südhof TC, Czernik AJ, Kao HT, et al. 1989a. Synapsins: Mosaics of shared and individual domains in a family of synaptic vesicle phosphoproteins. *Science* 245:1474–80

Südhof TC, Fried VA, Stone DK, et al. 1989b. Human endomembrane H^+ pump strongly resembles the ATP-synthetase of Archaebacteria. *Proc. Natl. Acad. Sci. USA* 86:6067–71

Südhof TC, Jahn R. 1991. Proteins of synaptic vesicles involved in exocytosis and membrane recycling. *Neuron* 6:665–77

Südhof TC, Lottspeich F, Greengard P, et al. 1987. A synaptic vesicle protein with a novel cytoplasmic domain and four transmembrane regions. *Science* 238:1142–44

Sulston J, Du Z, Thomas K, et al. 1992. The *C. elegans* genome sequencing project: a beginning. *Nature* 356:37–40

Sun SZ, Xie XS, Stone DK. 1987. Isolation and reconstitution of the dicyclohexylcarbodiimide-sensitive proton pore of the clathrin-coated vesicle proton translocating complex. *J. Biol. Chem.* 262:14790–94

Taiz L, Bernascon P, Rausch T, et al. 1990. An mRNA from human brain encodes an isoform of the B subunit of the vacuolar H^+-ATPase. *J. Biol. Chem.* 265:17428–31

Thomas L, Betz H. 1990. Synaptophysin binds to physophilin, a putative synaptic plasma membrane protein. *J. Cell Biol.* 111:2041–52

Thomas L, Hartung K, Langosch D, et al. 1988. Identification of synaptophysin as a hexameric channel protein of the synaptic vesicle membrane. *Science* 242:1050–53

Thureson-Klein AK, Klein RL. 1990. Exocytosis from neuronal large dense-cored vesicles. *Int. Rev. Cytol.* 121:67–126

Trimble WS, Cowan DM, Scheller RH. 1988. VAMP-1: A synaptic vesicle-associated integral membrane protein. *Proc. Natl. Acad. Sci. USA* 85:4538–42

Trimble WS, Linial M, Scheller RH. 1991. Cellular and molecular biology of the presynaptic nerve terminal. *Annu. Rev. Neurosci.* 14:93–122

Tugal HB, van Leeuwen F, Apps DK, et al. 1991. Glycosylation and transmembrane topography of bovine chromaffin granule p65. *Biochem. J.* 279:699–703

Turner TJ, Adams ME, Dunlap K. 1992. Calcium channels coupled to glutamate release identified by ω-Aga-IVA. *Science* 258:310–13

Umemoto N, Yoshihisa T, Hirata R, Anraku Y. 1990. Roles of the VMA3 gene product, subunit c of the vacuolar membrane H^+-ATPase on vacuolar acidification and protein transport. A study with *VMA3*-disrupted mutants of *Saccharomyces cerevisiae*. *J. Biol. Chem.* 265:18447–53

Ushkaryov YA, Petrenko AG, Geppert M, Südhof TC. 1992. Neurexins: Synaptic cell surface proteins related to the α-latrotoxin receptor and laminin. *Science* 257:50–56

Valtorta F, Benfenati F, Greengard P. 1992. Structure and function of the synapsins. *J. Biol. Chem.* 267:7195–98

Vogel US, Dixon RAF, Schaber MD, et al. 1988. Cloning of bovine GAP and its interaction with oncogenic ras p21. *Nature* 335:90–93

Volknandt W, Pevsner J, Elferink LA, et al. 1991. A synaptic vesicle specific GTP-binding protein from ray electric organ. *Mol. Brain Res.* 11:283–90

Walent JH, Porter BW, Martin TFJ. 1992. A novel 145 kd brain cytosolic protein reconstitutes Ca^{2+}-regulated secretion in permeable neuroendocrine cells. *Cell* 70:765–75

Wang SY, Moriyama Y, Mandel M, et al. 1988. Cloning of cDNA encoding a 32-kDa protein. An accessory polypeptide of the H^+-ATPase from chromaffin granules. *J. Biol. Chem.* 263:17638–42

Wendland B, Miller KG, Schilling J, Scheller RH. 1991. Differential expression of the p65 gene family. *Neuron* 6:993–1007

White JM. 1992. Membrane fusion. *Science* 258:917–23

Wright EM, Hager KM, Turk E. 1992. Sodium cotransport proteins. *Curr. Opin. Cell Biol.* 4:696–702

Xie XS, Crider BP, Stone DK. 1989. Isolation and reconstitution of the chloride transporter of clathrin-coated vesicles. *J. Biol. Chem.* 264:18870–73

Xie XS, Stone DK. 1988. Partial resolution and reconstitution of the subunits of the clathrin-coated vesicle proton ATPase responsible for Ca^{2+}-activated ATP hydrolysis. *J. Biol. Chem.* 263:9859–67

Yoshida A, Oho C, Kuwahara R, et al. 1992. HPC-1 is associated with synaptotagmin and ω-conotoxin receptor. *J. Biol. Chem.* 2267:24925–28

Young SH, Poo MM. 1983. Spontaneous release of transmitter from growth cones of embryonic neurons. *Nature* 305:634–37

Zahraoui A, Touchot N, Chardin P, Tavitian A. 1988. Complete coding sequences of the ras related rab 3 and 4 cDNAs. *Nucleic Acids Res.* 16:1204

Zahraoui A, Touchot N, Chardin P, Tavitian A. 1989. The human rab genes encode a family of GTP-binding proteins related to yeast YPT1 and SEC4 products involved in secretion. *J. Biol. Chem.* 264:12394–401

Zhang J, Myers M, Forgac M. 1990. Characterization of the V_0 domain of the coated vesicle (H^+)-ATPase. *J. Biol. Chem.* 267:9773–78

Zhou S, Sousa R, Tannery NH, Lafer EM. 1992. Characterization of a novel synapse-specific protein. II. cDNA cloning and sequence analysis of the F1–20 protein. *J. Neurosci.* 12:2144–55

Zimniak L, Dittrich P, Gogarten JP, et al. 1988. The cDNA sequence of the 69 kDa subunit of the carrot vacuolar H^+-ATPase. Homology to the beta chain of F_0F_1-ATPases. *J. Biol. Chem.* 263:9102

Zoran MJ, Doyle RT, Haydon PG. 1991. Target contact regulates the calcium responsiveness of the secretory machinery during synaptogenesis. *Neuron* 6:145–51

Annu. Rev. Neurosci. 1994. 17:247–65

THE EPIGENETICS OF MULTIPLE SCLEROSIS: Clues to Etiology and a Rationale for Immune Therapy

Lawrence Steinman

Department of Neurology and Neurological Sciences, Stanford University, Stanford, California 94305

Ariel Miller

Department of Cell Biology, Weizman Institute of Science, Rehovot, Israel

Claude CA Bernard

Laboratory of Neuroimmunology, LaTrobe University, Victoria 3083, Australia

Jorge R. Oksenberg

Department of Neurology and Neurological Sciences, Stanford University, Stanford, California 94305

KEY WORDS: multiple sclerosis, T cell receptor, HLA, autoimmunity

Introduction

Multiple sclerosis (MS) is a chronic disease involving an inflammatory reaction within the white matter of the central nervous system, mediated by T cells, B cells, and macrophages. The target of the inflammatory response in MS has been elusive. In this review, we summarize the significant advances made in the past few years, aimed at defining the nature of the immune response within the MS lesion. Surprisingly, analysis of the T cell receptor gene rearrangements in the MS lesion has indicated that at least one of the major immune responses is directed to myelin basic protein (Oksenberg et al 1993). The characteristics of this immune response, how it is initiated, and

247

how it may be suppressed using novel therapies, now in clinical trials, are described.

Studies on T Cell Receptor Rearrangements in the MS Lesion

Multiple sclerosis is an inflammatory demyelinating disease of the central nervous system. T lymphocytes may play a critical role in the pathogenesis of the disease (McFarlin & McFarland 1982, Traugott et al 1982). Although the antigen triggering MS is unknown and may be an infectious agent, a component of self, or even a pathogen mimicking self (Fujinami & Oldstone 1985), an analysis of the two polymorphic components of the trimolecular complex involved in T cell recognition of antigen, specifically the T cell receptor (TCR) and the HLA molecule, may further the understanding of the autoimmune reaction. Indeed, some evidence suggests that certain TCR genes (Beall et al 1989, Oksenberg et al 1989, Seboun et al 1989) and HLA class II genes linked to HLA-DR2,Dw2 (Oksenberg & Steinman 1990, Terasaki et al 1976, Vartdal et al 1989) are associated with susceptibility to MS.

A restricted humoral and cellular immune response within the central nervous system may be characteristic of certain stages in the pathogenesis of MS. Approximately 90% of patients with MS have oligoclonal IgG in their cerebrospinal fluid (CSF). Moreover IgG of restricted clonality with reactivity to myelin basic protein can be eluted from MS brain (Bernard et al 1981, Mattson et al 1980, Walsh & Tourtellotte 1986). B cells producing antibody to myelin basic protein can be found in the cerebrospinal fluid of 57% of MS patients (Olsson et al 1990, Martino et al 1991). At the cellular level, oligoclonality of T cells has been described in the CSF of MS patients (Lee et al 1991).

In 1990 we described the use of the polymerase chain reaction (PCR) to amplify TCR Vα sequences from transcripts derived from MS brain lesions in three patients (Oksenberg et al 1990). In each of the three MS brains, only two to four rearranged TCR Vα transcripts were detected. No Vα sequences could be found in three control brains without inflammation. Sequence analysis of 25 cDNA clones from MS white matter plaques from one patient with Vα 12.1-Jα-Cα rearrangements demonstrated only two Jα sequences, reinforcing further the notion of a limited heterogeneity of TCR transcripts in MS brain (Oksenberg et al 1990).

We recently extended these observations by examining an additional 13 MS brains, another 8 control brains without inflammation, 2 brains with viral encephalitis, and 1 brain specimen with progressive multifocal leuco-encephalopathy (Oksenberg et al 1993). All patients studied had chronic progressive MS, and no correlations with TCR V gene usage or HLA molecular phenotype could be made based on sex, duration of disease, or the region of brain sampled. TCR Vα-Jα-Cα and Vβ-Dβ-Jβ-Cβ rearrangements were confirmed by Southern blotting and hybridization of the PCR product with C-region specific probes, following amplification with one of 18 Vα or one of 21 Vβ specific oligonucleotide primers (Panzara et al 1992). The primers amplify rearrangements of all Vα and Vβ families surveyed in PHA stimulated peripheral blood lymphocytes from a pool of donors (Oksenberg et al 1990) and in lymph nodes taken at biopsy. Specificity of each primer pair was confirmed by identifying single rearrangements in antigen-specific T cell clones for *B. bergdorferi*, pertussis toxin, peptide epitopes of myelin basic protein, and acetylcholine receptor (Panzara et al 1992). The panel of Vα and Vβ primers used in this survey is broad, detecting the specific TCR α and β gene rearrangements in nearly 90% of over 150 T cell clones with a single Vα-Jα or Vβ-Dβ-Jβ rearrangement (Panzara et al 1992).

A limited number of TCR Vα gene rearrangements was seen in 15 of the 16 MS specimens. In one specimen no Vα genes of the 18 families we surveyed were rearranged, although this patient had two Vβ rearrangements. Use of a consensus primer (Broeren et al 1991) failed to reveal a Vα rearrangement not surveyed with our set of primers. The number of different TCR Vα genes transcribed ranged from 0 to 9 per brain, with a mean of 4.4 $\pm$ 2.8 ($\pm$ 1 SD). TCR Vβ rearrangements were more diverse, with a range of 2 to 13 per brain, with a mean of 7.0 $\pm$ 3.4 ($\pm$ 1 SD). TCR Vα or Vβ transcripts were detected in one of the 11 brains of individuals who died of non-neurologic diseases. In all specimens, as a positive control for the integrity of the cDNA, actin was amplified.

Because (i) T cells recognize antigen bound to molecules of the major histocompatibility complex (MHC), (ii) MHC genes influence the peripheral T cell receptor repertoire (Kappler et al 1987, Gulwani-Akolkar 1991), and (iii) susceptibility to MS has been associated with certain MHC class II genes on the HLA-DR2,Dw2 haplotype (Oksenberg & Steinman 1990, Terasaki et al 1976), we analyzed the DNA-defined alleles of the patients' HLA-DR, DQ, and DP loci. All of the 16 MS patients were typed for the HLA class II loci DRB1, DQA1, DQB1, and DPB1 using PCR and sequence specific oligonucleotide (SSO) probes (Bugawan et al 1990, Bugawan & Erlich 1991, Helmuth et al 1990, Scharf et al 1991). Eight of 16 patients were DRB1*1501, DQA1* 0102, DQB1* 0602 and either DPB1* 0401 or 0402. This molecular HLA-DR:DQ haplotype, which corresponds to the cellular type HLA-DR2,Dw2, is associated with increased susceptibility to MS in certain caucasoid populations (Terasaki et al 1976, Vartdal et al 1989).

HLA DRB1*1501, DQA1*0102, DQB1*0602 are in strong linkage disequilibrium, while DPB1*0401 is in weaker linkage disequilibrium with DRB1*1501, DQA1*0102, DQB1*0602. DPB1*0402 is not in linkage disequilibrium with this DR2 haplotype (Begovich et al 1992). We do not know whether the DPB1 allele and the DRB1,DQA1 and DQB1 alleles are all in a *cis* configuration on a single chromosome. Therefore we refer to this constellation of alleles as a molecular phenotype, avoiding the designation of "haplotype" which can be discerned only from segregation analysis in families. Patients who were DRB1*1501, DQA1*0102, DQB1*0602, and either DPB1*0401 or 0402, hereinafter called molecular phenotype "1," showed an increased frequency of certain TCR rearrangements.

Given the increased frequency of Vβ 5 and Vβ 6 rearrangements in HLA-DR2 restricted T cell clones reactive to MBP (Kotzin et al 1991), we asked whether Vβ 5 and 6 were transcribed more frequently in MS brains of patients with phenotype 1, compared to MS patients who were not phenotype 1 or non-MS controls. Thus, of 8 patients with phenotype "1," 7 had rearrangements of Vβ 5.2, and all 8 MS patients rearranged either Vβ 5.1 or 5.2, or both, compared to 2 of 7 MS brains from patients who were not phenotype 1. Vβ 6 was transcribed in 6 of 8 MS brains with phenotype "1," compared to 4 of 7 MS brains of patients who were not phenotype "1," and 1 of 7 control brains which were not phenotype "1." Other frequent rearrangements seen in patients with phenotype "1 were Vβ 7 and Vβ 8 in 6/8, and Vβ 12 in 4/8.

Among patients who were phenotype "1," four patients had multiple plaques sampled. Patient KL had Vβ 5.2 rearranged in all three plaques and Vβ 6 in two of three plaques. Patient LJ transcribed Vβ 5.2 in one of five plaques and Vβ 5.1 in three of five plaques. Patient LJ was the only individual of phenotype 1 who had plaques (2/5) without Vβ 5 or 6. Patient PM transcribed Vβ 5.2 and 6 in both plaques sampled. Patient GL transcribed Vβ 6 in all three plaques and Vβ 5.2 in two of three plaques. In summary, Vβ 5.1 and Vβ 5.2 were present in ten of thirteen plaques.

In the HLA-DR2 patients analyzed by Kotzin and colleagues, of the 30 clones specific for MBP, rearrangements of Vα 1, 2, 7, 8, and 10 were well represented. These Vα TCRs were rearranged in all seven phenotype I patients and were present in 13 of 17 plaques analyzed from these patients. In general, when multiple areas were sampled, there was a common Vα or Vβ gene rearranged in two or more of the regions studied.

We sequenced Vβ 5.2 rearrangements from two plaques from different anatomic regions in the brain of MS patient (KL), who was HLA phenotype "1." We also sequenced Vβ 5.2 rearrangements from another brain (LJ) of an MS patient who was also HLA phenotype "1." We found five predominant Vβ5.2 motifs in brain KL (Table 1): (Q)LR or LR (5 /35 sequences), LGG (4/35), LVAG (4/35), LDG (3/35), and (Q)PT (3/35). The Vβ5.2 LR, LVAG, and PT motifs were seen in both anatomic regions sampled from KL. Moreover they were also seen in brain LJ (Table 1). None of these sequences was seen in Vβ 5.2 sequences from a muscle biopsy of an individual with Duchenne's muscular dystrophy who was phenotype "1," or from Vβ 5.2

Table 1 CDR3 sequences of TCR rearrangements amplified from MS brains and controls

Vβ5.2/3	N-D-N-J		Cβ	
KL-1				
LCASS	LPGTP	YGYFGSGTRLTVV	(Jβ 1.2)	EDLKN
LCASS	LPGTP	YGYTFGSGTRLTVV	(Jβ 1.2)	EDLKN
LCASS	LRLAN	SPLHFGNGTRLTVT	(Jβ 1.6)	EDLKN
LCASS	LDRL	YNSPLHFGNGTRLTVT	(Jβ 1.6)	EDLKN
LCAS	QLRLA	NSPLHFGNGTRLTVT	(Jβ 1.6)	EDLKN
LCASS	QLRLA	NSPLHFGNGTRLTVT	(Jβ 1.6)	EDLKN
LCASS	FLG	YNSPLHFGNGTRLTVT	(Jβ 1.6)	EDLKN
LCASS	QPTV	YNNEQFFGQRTRLLVL	(Jβ 2.1)	EDLKN
LCASS	SDGRM	STQYFGPGTRLLVL	(Jβ 2.3)	EDLKN
LCASS	LVAG	SIYEQYFGPGTRLTVT	(Jβ 2.7)	EDLKN
LCASS	SEREG	RAQYFGQGTRLTVL	(Jβ ?)	EDLKN
LCASS	GGEG	RAQYFGQGTRLTVL	(Jβ ?)	EDLKN
KL-3				
LCASS	LDGVP	YGYTFGSGTGLTVV	(Jβ 1.2)	EDLKN
LCASS	LDGVP	YGYTFGSGTRLTVV	(Jβ 1.2)	EDLKN
LCASS	LDGV	NYGYTFGSGTRLTVV	(Jβ 1.2)	EDLKN
LCASS	LVGRGP	YGYTFGSGTRLTVV	(Jβ 1.2)	EDLKN
LCASS	LGGVP	YGYTFGSGTGLTVV	(Jβ 1.2)	EDLKN
LCASS	LRGTP	YGYTFGSGTRLTVV	(Jβ 1.2)	EDLKN
LCASS	QPAV	YNEQFFGPGTRLTVL	(Jβ 2.1)	EDLKN
LCASS	LELAG	YNEQFFGPGTRLTVL	(Jβ 2.1)	EDLKN
LCASS	LGGSEE	DTQYFGPGTRLTVL	(Jβ 2.3)	EDLKN
LCASS	LGGSE	ETQYFGPGTRLLVL	(Jβ 2.5)	EDLKN
LCASS	LGGSV	ETQYFGPGTRLLVL	(Jβ 2.5)	EDLKN
LCASS	LGSGTL	QETQYFGPGTRLLVL	(Jβ 2.5)	EDLKN
LCASS	LASGTL	QETQYFGPGTRLLVL	(Jβ 2.5)	EDLKN
LCASS	LASGTL	QETQYFGPGTRLLVL	(Jβ 2.5)	EDLKN
LCASS	PT	GANVLTFGAGSRLTVL	(Jβ 2.6)	EDLKN
LCASS	PT	GANVLTFGAGSRLTVL	(Jβ 2.6)	EDLKN
LCASS	QGS	TFGAGSRLTVL	(Jβ 2.6)	EDLKN
LCASS		SGANVLTFGAGSRLTVL	(Jβ 2.6)	EDLKN
LCASS	L	GANVLTFGAGSRLTVL	(Jβ 2.6)	EDLKN
LCASS	LR	GANVLTFGAGSRLTVL	(Jβ 2.6)	EDLKN
LCASS	LVAG	SIYEQYFGPGTRLTVT	(Jβ 2.7)	EDLKN
LCASS	LVAG	SIYEQYFGPGTRLTVT	(Jβ 2.7)	EDLKN
LCASS	LVAG	SIYEQYFGPGTRLTVT	(Jβ 2.7)	EDLKN
LJ 1				
LCAS	TLRL	GNSPLHFGNGTRLTVT	(Jβ 1.6)	EDLNK
LCASS	DSS	ETQYFGPGTRLLVL	(Jβ 2.5)	EDLKN
LCASS	LR	GANVLTFGAGSRLTVL	(Jβ 2.6)	EDLKN
LCASS	LR	GANVLTFGAGSRLTVL	(Jβ 2.6)	EDLKN

Table 1 (*continued*)

Vβ5.2/3	N-D-N-J		Cβ	
LCASS	PT	GANVLTFGAGSRLTVL	(Jβ 2.6)	EDLKN
LCASS	LVAG	IYEQYFGPGTRLTVT	(Jβ 2.7)	EDLKN
LCASS	LVAG	SIYEQYFGPSTRLTVT	(Jβ 2.7)	EDLKN
LCASS	LVAG	SIYEQYFGPSTRLTVT	(Jβ 2.7)	EDLKN
SE (Viral encephalitis)				
LCASS	PCLFNR	GYGYTFGSGTRLTVV	(Jβ 1.2)	EDLNK
LCASS	GRSGT	NYGYTFGSGTRLTVV	(Jβ 1.2)	EDLNK
LCASS	NQGH	YNSPLHFGNGTRLTVT	(Jβ 1.6)	EDLNK
LCASS	QATG	YNSPLHFGNGTRLTVT	(Jβ 1.6)	EDLNK
LCASS	RLDRGQGE	EQFFGPGTRLTVL	(Jβ 2.1)	EDLKN
LCASS	ITCGLAGARD	EQFFGPGTRLTVL	(Jβ 2.1)	EDLKN
LCASS	WGLAGARD	EQFFGPGTRLTVL	(Jβ 2.1)	EDLKN
LCASS	TSGQP	GELFFGEGSRLTVL	(Jβ 2.2)	EDLKN
LCASS	LDLGKA	DTQYFGPGTRLTVL	(Jβ 2.3)	EDLKN
LCASS	LVAG	KNIQYFGAGTRLSVLK	(Jβ 2.4)	EDLKN
LCASS	GTL	VLTFGAGSRLTVL	(Jβ 2.6)	EDLKN
LCASS	YGTSGI	YEQYVGPGTRLTVT	(Jβ 2.7)	EDLKN
LCASS	FIWG	YEQYFGPGTRLTVT	(Jβ 2.7)	EDLKN
LCASS	TAGT	YEQYFGPGTRLTVT	(Jβ 2.7)	EDLKN
Muscle infiltrating lymphocytes				
LCASS	LGSPGYR	TNEKLFFGSGTQLSVL	(Jβ 1.4)	EDLNK
LCASS	FTGAY	YNEQFFGPGTRLTVL	(Jβ 2.1)	EDLKN
LCASS	RRTSGFVH	DTQYFGPGTRLTVL	(Jβ 2.3)	EDLKN
LCAS	ARRTSGFV	TDTQYFGPGTRLTVL	(Jβ 2.3)	EDLKN
LCAS	TARRTSGFV	TDTQYFGPGTRLTVL	(Jβ 2.3)	EDLKN
LCA	TARRTSGFV	TDTQYFGPGTRLTVL	(Jβ 2.3)	EDLKN
LCA	TARRTSGFV	TDTQYFGPGTRLTVL	(Jβ 2.3)	EDLKN
LCA	TARRTSGFV	TDTQYFGPGTRLTVL	(Jβ 2.3)	EDLKN
LCA	TARRTSGFV	TDTQYFGPGTRLTVL	(Jβ 2.3)	EDLKN
LCAS	RQGART	GANVLTFGAGSRLTVL	(Jβ 2.6)	EDLKN
JO (PBLs)				
LCASS	VALQDR	YGYTFGSGTGLTVV	(Jβ 1.2)	EDLNK
LCASS	TVRGS	QPQHFGDGTRLSIL	(Jβ 1.5)	EDLNK
LCASS	PGM	KNIQYFGAGTRLSVL	(Jβ 2.4)	EDLKN
LCASS	DSPSG	QETQYFGPGTRLTVL	(Jβ 2.5)	EDLKN
LCASS	RPGNIR	ETQYFGPGTRLSVL	(Jβ 2.5)	EDLNK
LCASS	RSQGART	GANVLTFGAGSRLTVL	(Jβ 2.6)	EDLKN
BM (PBLs)				
LCASS	DAG	YNSPLHFGNGTRLTVT	(Jβ 1.6)	EDLNK
LCASS	YRTQL	NSPLHFGNGTRLTVT	(Jβ 1.6)	EDLNK
LCASS	LEHRPT	AKNIQYFGAGTRLSVL	(Jβ 2.4)	EDLKN
LCASS	PER	GANVLTFGAGSRLTVL	(Jβ 2.6)	EDLKN
LCASS	QEA	SYEQYFGPGTRLTVT	(Jβ 2.7)	EDLKN
LCAS	RLVRDL	SHEQYFGPSTRLTVT	(Jβ 2.7)	EDLKN

sequences from peripheral blood lymphocytes of healthy individuals (JO, HLA DR3DRw8; BM, phenotype "1).

Rearrangements at the site of inflammation in the brains of three patients with encephalitis were studied. KW (HLA DRB1*0404,0901, DQB1*0302, 0305, DPB1*0401,0402), a patient with viral encephalitis, rearranged Vα 5, 8, 10, 12, 14 and 16 and Vβ 1, 2, 3, 6, 7, 11, 12, 13 and 15. SE (HLA DRB1*1201,1501 DQB1*0301,0602, DPB1*0401,0501), another patient with viral encephalitis, rearranged Vα 2, 3, 6, 8, 10, 12, 13, 14, 16, 17, and 18, and all Vβ genes other than Vβ 20. SP (HLA DRB1*1303,11, DQB1* 0301,0180, DPB1*0201,0401), with progressive multifocal leucoencephalopathy, rearranged Vα 1 through 10, 12, 13, 16, 17 and 18 and Vβ 2, 3, 9, 10 and 14 through 20. TCR V region transcripts were in general more diverse in viral encephalitis compared to MS. Since patient SE was HLA phenotype "1," we sequenced Vβ5.2 rearrangements from inflammatory lesions. There were no repeated sequences. We did not see the LR motif, though we did find one example of the LVAG sequence (Table 1).

One of the repeated motifs found in MS brains KL and LJ contained the sequence Vβ 5.2 LCASSLRGA at the V-D-J junction, which is identical to that found in a T cell clone recognizing MBP peptide 87-106 in an MS patient (Martin et al 1991). We determined that the HLA type of this patient is phenotype "1." This CDR3 sequence is also seen in T cell clones isolated from the spinal cords of Lewis rats with EAE. These clones are reactive to MBP peptide 87-99, and cause paralysis when adoptively transferred (Gold et al 1992). The nucleotide sequences encoding the Vβ 5.2 LR motif in the KL and LJ brain specimens reveal strong selection for these amino acids. It is unlikely that these sequences represent a cDNA cloning or PCR artifact (Table 2). It is remarkable that the same CDR3 sequence, Vβ5.2LRGA, is selected in different MS patients' brains, though different codons are used, and that these codons were in turn different from those encoding Vβ5.2LRGA in a T cell clone recognizing MBP 87-106 (Martin et al 1991).

The LR motif was described in other CDR3 sequences. A CDR3 sequence LRAG was found in a cDNA library derived from human tonsillar tissue (Tillinghast et al 1986). A CDR3 sequence LCASSLLRS was seen in a human T cell clone reactive to MBP peptide 139-153 plus HLA DR2a, while a CDR3 sequence, LCASSREFSS, was observed in a human T cell clone recognizing MBP peptide 80-99 and HLA DRB1*1501 (Giegerich et al 1992). Another sequence, LRDF, is seen in a murine cytotoxic T cell clone specific for influenza A (Morahan et al 1989). MBP and influenza share amino acid sequences (Jahnke et al 1985). It may only be necessary for an infectious pathogen to share a few amino acids within an epitope to provoke a pathogenic cross-reaction. EAE could be induced with only four core amino acids out of an eleven-mer peptide of MBP (Gautam et al 1992).

Table 2 Nucleotide sequence homology in the use of *LeuArgGly*, *LeuGlyGlyGlu*, *and LeuValAlaGly*

Sample		N-D-N-J			
KL3	AGCAGC	**CTA CGC GGG** GCC AAC	SS	*LRGAN*	(Vβ5.2/Jβ2.6)
	AGCAGC	**TTA CGC GGG** ACA CCC	SS	*LRGTP*	(Vβ5.2/Jβ1.2)
KL1	AGCAGC	**TTG CGC** TTG GCT AAT	SS	*LRLAN*	(Vβ5.2/Jβ1.6)
	AGC	CAG**TTG CGC** TTG GCT AAT	S	*QLRLA*	(Vβ5.2/Jβ1.6)
	AGCAGC	CAG**TTG CGC** TTG GCT AAT	SS	*QLRLA*	(Vβ5.2/Jβ1.6)
	AGCAGC	**TTG** GAT **CGC** TTG TAT AAT	SS	*LDRLA*	(Vβ5.2/Jβ1.6)
LJ1	AGC	ACG**TTG CGC** TTG **GGT**	S	*TLRLG*	(Vβ5.2/Jβ1.6)
	AGCAGC	**CTA CGG GGG** GCC AAC	SS	*LRGAN*	(Vβ5.2/Jβ2.6)
	AGCAGC	**CTA CGG GGG** GCC AAC	SS	*LRGAN*	(Vβ5.2/Jβ2.6)
MS18[a]	ACGACG	**TTG AGG GGG** GCG CTA	SS	*LRGAL*	(Vβ5.2/Jβ2.4)
BF1[b]	AGCAGC	**CTC AGG GGG**	SS	*LRG*	(Vβ6/Jβ1.6)
E[b]	AGCAGC	**ATA AGG GGA** AGC	SS	*IRGS*	(Vβ6/Jβ2.7)
BD3[b]	AGCAGC	**ATC** GTC **AGG GGA** TCG	SS	*IVRGS*	(Vβ6/Jβ2.7)
ph 11[c]	AGCAGT	**TTA AGG** GCG **GGA**	SS	*LRAG*	(Vβ8/Jβ1.1)
12H6[d]	AGCAGC	**CTC CGG** GAC TTT	SS	*LRDF*	(Vβ13/Jβ2.1)
KL3	AGCAGC	**TTG GGA GGG** GTA CCC TAT	SS	*LGGVPY*	(Vβ5.2/Jβ1.2)
	AGCAGC	**TTG GGA GGG** TCC **GAA GAG**	SS	*LGGSEE*	(Vβ5.2/Jβ2.3)
	AGCAGC	**TTG GGA GGG** TCC **GAA GAG**	SS	*LGGSEE*	(Vβ5.2/Jβ2.5)
	AGCAGC	**TTG GGA GGG** TCC GTT **GAG**	SS	*LGGSVE*	(Vβ5.2/Jβ2.5)
4[e]	AGCAGC	**CTG GGG GGC GAA**	SS	*LGGE*	(Vβ8.2/Jβ2.5)
KL3	AGCAGC	**TTA GTG GCG GGA** TCT ATC	SS	*LVAGSI*	(Vβ5.2/Jβ2.7)
	AGCAGC	**TTA GTG GCG GGA** TCT ATC	SS	*LVAGSI*	(Vβ5.2/Jβ2.7)
	AGCAGC	**TTG GTG GCG GGA** TCT ATC	SS	*LVAGSI*	(Vβ5.2/Jβ2.7)
KL1	AGCAGC	**TTA GTG GCG GGA** TCT ATC	SS	*LVAGSI*	(Vβ5.2/Jβ2.7)
LJ1	AGCAGC	**TTA GTG GCG GGA** ATC	SS	*LVAGI*	(Vβ5.2/Jβ2.7)
	AGCAGC	**TTA GTG GCG GGA** TCT ATC	SS	*LVAGSI*	(Vβ5.2/Jβ2.7)[f]
	AGCAGC	**TTA GTG GCG GGA** TCT ATC	SS	*LVAGSI*	(Vβ5.2/Jβ2.7)[f]
C*	AGCAGC	ATA **GCT GGC** GGT	SS	*IAGG*	(Vβ6/Jβ2.3)

[a] CDR3 usage in human MBP 88-99 specific T cell line (Martin et al 1991)
[b] CDR3 usage in rat spinal cord derived T cell clones specific for BP 85-99 (Gold et al 1992)
[c] Clone derived from a human tonsil cDNA library (Tillinghast et al 1986)
[d] Noncytolytic mouse T cell clone specific for the influenza virus strain A/PR8/34 (Morahan et al 1989)
[e] CDR3 usage in rat lymph node derived T cell clone specific for BP 85-99 (Gold et al 1992)
[f] G- > S substitution in Jβ2.7

Another motif found in plaque KL-3 was LCASSLGGSEET (Table 1). In the Lewis rat some T cell clones, isolated from lymph nodes and specific for MBP peptide 87-99, have a CDR3 sequence of LCASSLGGEET (Gold et al 1992). In plaque KL-1 a CDR3 sequence LCASSGGEG was seen (Table 1). Another motif LVAG was also found in the CDR3 region (IAG) of T cell clones specific for MBP peptide 87-99 and that cause EAE (Table 1). Taken together the presence and frequency (16/40 unique sequences, not counting any repeated sequences which might represent sister clones) of these three CDR3 motifs, LCAS(S)LRG, LCASSLVAG and LCASSLGG(S)E, suggest that an immune response to an epitope within MBP peptide 87-106 may be occurring in MS lesions. The frequency of these common CDR3 motifs is described in Table 3. This immune response may be deleterious, given that T cells expressing the same CDR3 motifs and recognizing this epitope of MBP cause paralysis in the EAE model (Gold et al 1992).

Although both α and β TCR chains contribute to and define the peptide-MHC specificity of a T cell (Jorgensen et al 1992), identification of TCR β CDR3 sequences identical to those found in bona fide clones with specificity

Table 3 Summary of restricted CDR3 motifs in Vβ5.2 TCR transcripts from MS brain lesions

Patient KL		
Vβ5.2 LCASS	(Q)LR	5/35
Vβ5.2 LCASS	LGG	4/35
Vβ5.2 LCASS	LVAG	4/35
Vβ5.2 LCASS	LDG	3/35
Vβ5.2 LCASS	(Q)PT	3/35
Patient LJ		
Vβ5.2 LCASS	LR (GA)	3/8
Vβ5.2 LCASS	LVAG	3/8
Vβ5.2 LCASS	PT	1/8

CDR3 in human MBP 88-99 specific T cell line (Martin et al 1991)
    Vβ5.2 LCASS   LRGA
CDR3 in rat spinal cord derived T cell clones specific for BP 85-99 (Gold et al 1992)
    Vβ6CASS   LRG
CDR3 in rat lymph node derived T cell clone specific for BP 85-99 (Gold et al 1992)
    Vβ6 LCASS   LGG
CDR3 in rat lymph node derived T cell clone specific for BP 85-99 (Gold et al 1992)
    Vβ6 LCASS   IAG

LCASSLR, LCASSLVAG, AND **LCASSLGG,** occurred in 44% (19/43) of the TCR Vβ5.2 transcripts from MS brain plaques of two different DRB1*1501, DQB1*0602 patients.

for MBP 87-106 is highly suggestive that T cells of such antigenic specificity are present in MS brain lesions, and that they represent a substantial fraction of the cells expressing Vβ5.2 TCR. An immune response directed to other antigens such as proteolipid protein, myelin-oligodendroglial glycoprotein, and heat shock proteins may also be critical in MS. As sequences for TCRs specific for these antigens become available, attribution of the specificity of some of the other rearrangements seen within lesions may be possible.

The Immune Response to Myelin Basic Protein in Multiple Sclerosis

Multiple sclerosis may arise from pathogenic T cells that somehow evaded mechanisms promoting self-tolerance. These pathogenic T cells may rearrange a restricted number of TCR Vα and Vβ genes. This conclusion stems mainly from animal studies analyzing T cell clones that recognize MBP causing EAE (Acha-Orbea et al 1988, Burns et al 1989, Urban et al 1988), and it has recently been confirmed in some studies of T cell clones from the peripheral blood in humans that recognize MBP (Ota et al 1990, Wucherpfennig et al 1990, Kotzin et al 1991). In helper T cell clones recognizing MBP in MS patients who were HLA-DR2, the restricted usage of Vβ 5.2 and Vβ 6 is seen. Of 41 MBP-specific T cell clones from these HLA-DR2 MS patients, 27 rearranged Vβ 5.2 and 10 rearranged Vβ 6.1 (Kotzin et al 1991).

Further evidence for restricted heterogeneity of the TCR response to the self-antigen myelin basic protein comes from studies on the junctional region of the TCR, particularly the third complementarity determining region (CDR3). Modeling the trimolecular interaction of the TCR, MHC, and bound peptide suggest that the CDR1 and CDR2 interact with the alpha helical regions of the MHC, while the CDR3 region, (N)Jα and (N)Dβ(N)Jβ, interacts with peptide bound in the MHC cleft (Davis & Bjorkman 1988, Chothia et al 1988). For MBP-specific T cell clones in the mouse and rat, selection of CDR3 amino acids correlates well with the specificity of the TCR for the bound MBP peptide. Thus, the TCR CDR3 involved in the recognition of MBP Ac1-9 in B10.PL and PL/J mice (both H-2u) are highly conserved (Acha-Orbea et al 1988, Urban et al 1988). For recognition of MBP peptide 68-88, the major encephalitogenic epitope in the Lewis rat, there is strong selection for the first two amino acids of the junctional CDR3 regardless of the associated Jβ (Gold et al 1991). Further functional evidence for the signficance of the CDR3 region in the recognition by peptide bound to MHC comes from recent studies with TCR transgenic mice. Charge substitutions in the amino acid sequence of the bound peptide elicited reciprocal changes in the CDR3 sequences of the TCR Vα or Vβ chains (Jorgensen et al 1992). It is highly likely that the CDR3 region recognizes peptide bound in the HLA cleft.

These studies on the functional importance of CDR3 support the interpretation that the Vβ5.2 LR motif found in MS lesions of DRB1*1501, DQB1*0602 patients might indicate T cell recognition of MBP in lesions, since Vβ5.2 LR is found in a cytotoxic T cell clone reactive to MBP 89-106 in a DRB1*1501,DQB1*0602 patient (Martin et al 1991). This CDR3 motif, either LCASSLRG or LCASSIRG, was also seen in T cell clones reactive to MBP peptide 87-99 derived from the spinal cord of Lewis rats with EAE (Gold et al 1992). MBP peptide 87-99 induces paralysis and EAE in the Lewis rat (Gold et al 1992) and in the SJL/J mouse (Sakai et al 1988a,b). The observation that this motif occurred in 21% (9/43, counting some repeated sequences) of the TCR Vβ5.2 transcripts from MS brain plaques of two different DRB1*1501, DQB1*0602 patients that we sequenced may also imply that it is a rather common specificity for T cells in the MS lesion. These sequences, which could be specific for MBP p89-106, are consistent with an MBP-specific T cell response in these lesions.

Two other motifs involving LCASSLGGSEE or LCASSGGE and LCASSLVAG were seen in plaques from brain KL. This motif was seen in T cell clones reactive to MBP peptide 87-99 derived from lymph nodes of Lewis rats (Gold et al 1992). Altogether we saw three CDR3 motifs, LR, (L)GG(S)E, and LVAG, that have been identified in T cell clones known to recognize a stretch of MBP from amino acids 87-106, in 44% (19/43, counting some repeats) of rearranged TCR sequences involving Vβ5.2 from two MS brains (Table 3)

Earlier work on MBP-specific T cell clones was devoted to studies on peripheral blood lymphocyte-derived MBP T cells (Ota et al 1990, Martin et al 1991, Pette et al 1990, Ben Nun et al 1991, Allegretta et al 1990). These studies suggested a potential role for a T cell immune response to MBP in MS. The B cell response to MBP in MS has also been studied extensively. Nearly two thirds of MS patients have Ig reactive to MBP in the CNS (Olson et al 1990). Taken together these data indicate that a cellular and humoral immune response to MBP may be critical in the pathogenesis of MS.

How might an immune response to MBP arise in MS? Two strong possibilities emerge, which might be called (i) "the molecular mimicry model," or (ii) "the innocent bystander model." Molecular mimicry refers to structural homologies between a self-protein and a protein in a viral or bacterial pathogen. For example, MBP shares extensive homologies at the amino acid level with a number of common pathogens including measles, influenza virus, and adenovirus. Residues 91-101 of MBP share stretches of four to six amino acids with adenovirus for instance (Jahnke et al 1985). One can induce EAE by immunizing with a peptide sequence from a pathogen with homology to MBP (Fujinami & Oldstone 1985). Homology may be necessary at only 4 out of 10 amino acids comprising a T cell epitope. Conservation of the native amino acid sequence at only 4 of 11 amino acids of an MBP epitope is sufficient to induce EAE (Gautam et al 1992). Besides molecular mimicry, tolerance to MBP can be broken by viral infection of the brain, "the innocent bystander model." Thus anti-MBP responses are seen in measles encephalitis in humans (Johnson et al 1984). In rats, infection of the brain with corona virus permits a breakdown in tolerance to MBP; MBP-reactive T cells, capable of transferring EAE, can be isolated from brain infected with corona virus (Watanabe et al 1983). Thus, a neurotrophic virus infecting brain induces an immune response to the perturbed brain tissue. These mechanisms might be operative in MS.

The Specificity and Diversity of TCR in a Cellular Infiltrate

What is the likelihood of detecting TCR rearrangements associated with pathogenic T cells in a cellular infiltrate in inflamed brain? The homing of T cells to the CNS in experimental demyelinating disease has an antigen-specific and a nonspecific phase (Yednock et al 1992, Karin et al 1993). We have recently measured the frequency of TCR Vβ genes rearranged in the brains of rats with EAE, at various times after the injection of an encephalitogenic T cell line. The pathogenic T cells express Vβ8.2. Within a few hours after the intravenous injection of the pathogenic T cell line, until day 4, the brain lesions are comprised almost entirely of Vβ8.2 T cells. In contrast, during active clinical disease with paralysis, the TCR genes rearranged in the brain are quite diverse, including not only Vβ8.2, but nearly the entire repertoire of Vβ genes. After the rats recover from the acute attack, around day 15 post-infusion, the TCR repertoire in the lesions is again quite restricted (Karin et al 1993). In MS in acute lesions, TCR Vβ gene transcripts are quite diverse, whereas in chronic lesions they are more restricted (Wucherpfennig et al 1992).

Based on these observations, our material, which contained mainly chronic MS lesions, with fewer lymphocytes and more macrophages than acute lesions, may have allowed us to detect a few common motifs among the TCR genes that were rearranged in the lesions. Some of the motifs may be related to pathogenic responses. In our study of chronic lesions, the critical signals among the TCRs rearranged in the cellular infiltrate could be deciphered from the noise. This may not have been feasible if more acute lesions had been examined.

The restriction of certain TCR Vβ gene rearrangements in brain may reflect their prominence on myelin reactive T cells in peripheral blood and their subsequent trapping in brain (McFarlin & McFarland 1982, Traugott, et al 1982). Vβ 5.2 bearing T cells constitute less than 5% of the T cells in the peripheral blood of MS patients (Oksenberg et al 1993, Kotzin et al 1991). No evidence could be found for distortions in the levels of any particular T cell expressing Vβ5.2 or Vβ6, or other TCR Vβ genes (including Vβ8 and Vβ12) in the peripheral blood of MS patients. Though superantigens might play a role in MS, as they appear to have in rheumatoid arthritis (Paliard et al 1991), recent studies do not provide any evidence for the influence of a superantigen in MS. A sequential study of TCR in MS patients before, during, or after a relapse must be undertaken. In EAE, relapses can be triggered by superantigen (Brocke et al, In press).

HLA and TCR Genes and Susceptibility to MS

Studies of germ line polymorphisms of TCR Vα and Vβ genes have indicated correlations between TCR haplotype and susceptibility to MS (Beall et al 1989, 1991, Oksenberg et al 1989, Seboun et al 1989, Sherritt et al 1992), although other work indicates no such correlations (Hillert et al 1992). In one study, a TCR Vβ haplotype defined by an RFLP linked to Vβ 8 was prevalent in MS patients who were HLA-DR2,Dw2 (Beall et al 1989). More detailed mapping of this polymorphism indicates that susceptibility for MS is determined by inheritance of a 175 kb region of the TCR-β chain locus between Vβ8.1 and Vβ11 particularly in patients who are HLA DR2 (Beall et al 1991). In our study (Oksenberg et al 1993), Vβ8 was seen in 6 of 8 brains of haplotype 1 (5 of 13 plaques) and in 5 of 7 brains of MS patients who were not haplotype 1. Thus, Vβ8 may be another critical TCR V gene involved in the autoimmune reaction underlying MS.

Genetic susceptibility is associated with the HLA-DR2 haplotype in Caucasoid populations (Oksenberg & Steinman 1990, Terasaki et al 1976, Vartdal et al 1989). The results of the study by Oksenberg and colleagues (Oksenberg et al 1993) support the notion that an HLA-DR2 phenotype, HLA-DRB1*1501, DQA1*0102, DQB1*0602, may be strongly associated with susceptibility to MS.

Though HLA class II is not constitutively expressed in brain, HLA-DR, DQ, and DP are expressed in MS brain. The aberrant expression of HLA-DR, DQ, and DP in MS brain may play a central role in the pathogenesis of MS (Bottazzo et al 1985, Traugott et al 1982), especially in allowing for migration of lymphocytes into brain (Steinman et al 1983). These results indicate that in studies analyzing TCR usage in MS lesions, strong associations may become apparent only after complete molecular genotyping of HLA class II molecules. At present a picture of genetic susceptibility to MS is emerging where particular rearranged TCR V genes are found in regions of demyelination, and these particular V genes are expressed in the brains of individuals with particular HLA class II genes. Some of these TCR V genes such as Vβ 5.2 and 6 are also transcribed in T cell clones recognizing MBP from MS patients who are HLA DR2 (Kotzin et al 1991, Martin et al 1991).

Direct cloning and sequencing of TCR rearrangements from MS brain material have indicated that some of these rearrangements in MS brain plaques are encoding TCR CDR3 regions identical to those found in T cells recognizing MBP. Testing T cells from brain for a functional response to MBP is technically and logistically difficult. Brain biopsy in MS is rarely performed, given that the diagnosis can be made by history, exam, cerebro-spinal fluid examination, and magnetic resonance images. As sequences become available from other T cell clones reactive to other myelin proteins like PLP, MAG, and MOG, as well as to other antigens like heat shock proteins (Selmaj et al 1991), more correlations may emerge between T cell clones with defined specificities and particular T cell receptor rearrangements in diseased areas of MS brain. The three other common motifs, which we have thus far not found exemplified in any known antigen-specific T cell clone, may encode CDR3 regions recognizing other myelin proteins or heat shock proteins for example. In addition other TCR Vβ genes found in the lesions, particularly Vβ6 (Kotzin et al 1991), Vβ8 (Beall et al 1989), and Vβ12 (Allegretta et al 1990 and in preparation), may also encode CDR3 regions involved in disease pathogenesis. Given the correlations that have emerged thus far, it may not be unreasonable to suggest that at least some of the other rearranged TCRs involving other Vα and Vβ families may also encode restricted CDR3 regions.

These findings may lead to an understanding of the pathogenesis of MS and may have therapeutic implications, given the success of reversing or preventing EAE with reagents targeting the products of TCR V genes, including the CDR3 region, involved in the recognition of MBP (Acha-Orbea et al 1988, Burns et al 1989, Howell et al 1989, Urban et al 1988, Vandenbark et al 1989), the MBP molecule itself (Miller et al 1991, Bornstein et al 1987, Teitelbaum et al 1988) or MHC class II molecules (Steinman et al 1981, Wraith et al 1989a,b, Gautam et al 1992).

Selective Immunotherapy Targeting CD4, T Cell Receptor Vβ 5.2, and Myelin Basic Protein: Initial Clinical Trials In Multiple Sclerosis

The success of selective immunotherapy in the EAE system aimed at targeting the CD4 molecule, the T cell receptor, myelin basic protein, and MHC class II molecules has been the subject of extensive and frequent reviews by us in the past three years (Miller et al 1991a, Steinman 1991, Bell & Steinman 1991, Zamvil & Steinman 1992). The reader is kindly referred to these recent reviews for an extensive discussion of the development of selective immune therapy in the EAE model. We focus here on a summary of the results of clinical trials utilizing chimeric antibodies to CD4 (Hodgkinson et al 1992), Vβ5.2 peptides (Vandenbark et al 1992), Cop-1, a synthetic copolymer which may inhibit T cell recognition of myelin basic protein (Bornstein et al 1987) and oral administration of myelin (Weiner et al 1993.

Clinical Trials with Cop-1 and Orally Administered MBP

Suppression of EAE with oral administration of myelin basic protein and its peptide fragments (Higgins & Weiner 1988, Bitar & Whitacre 1988) provided preclinical support for a clinical trial with oral MBP in MS. Recent experiments have elucidated the physiologic mechanisms underlying oral tolerance induction. Oral tolerance can be adoptively transferred by CD8+ T cells that are generated following oral administration of antigens (Lider et al 1989) and that release TGFβ after being triggered by the specific antigen (Miller et al 1991a, Miller et al 1992). TGFβ suppresses immune responses in the microenvironment, creating a form of bystander suppression. The phenomenon of "antigen driven bystander suppresson" following oral administration of antigens (Miller et al 1991b) may explain suppression induced by "tolerogenic epitopes" distinct from the immunogenic epitopes of the autoantigen involved in the pathogenesis of the autoimmune process.

In the first clinical trial fifteen patients with relapsing-remitting MS (RRMS) were treated with daily administration of bovine myelin by mouth, while another fifteen patients with RRMS served as controls. Over a year 6 of the 15 patients fed bovine myelin protein experienced relapses, while 12 of the 15 given placebo had relapses. Men who were HLA DR2-negative received the most benefit. The frequency of T cells reacting to MBP, including those T cells responding to MBP epitope 87-106, was reduced in the group fed myelin (Weiner et al 1993). These results are indeed encouraging and are likely to be followed by larger scale trials. The results are similar to that obtained with Cop-1, a synthetic random copolymer of tyrosine, glutamate, alanine, and lysine, which blocks T cell recognition of MBP (Teitelbaum et al 1988). In 1987, Bornstein et al showed that relapses were reduced by daily administration of Cop-1 to MS patients with relapsing-remitting disease (Bornstein et al, 1987). Cop-1 has been less effective in treatment of chronic, progressive MS.

Clinical Trials with Chimeric Anti-CD4 Antibody

Patients with chronic progressive MS have been treated with a chimeric anti-CD4 antibody with a humanized Fc region. Initial results of a phase I trial will be reported in detail in 1993. Preliminary results indicate that toxicity is low; opportunistic infections occur only rarely and respond to conventional antibiotics (Hodgkinson et al 1992). The antibody can be used repeatedly without eliciting significant immunogenicity. The antibody markedly decreased the number of circulating CD4+ cells, with CD4 counts remaining low for more than a month. Some reduction of activity on gadolinium-enhanced MR scans was seen following infusion of the anti-CD4 antibody (Hodgkinson et al 1992).

Clinical Trials with Vβ5.2 Peptides

Patients with chronic progressive MS have been treated with a peptide from the CDR2 region of Vβ5.2 and Vβ6.1. Preliminary results indicate that the Vβ peptides elicit an immune response, with anti-TCR specific antibodies and DTH responses to the respective TCR CDR2 regions (Vandenbark et al 1992). T cells specific for myelin basic protein were decreased in frequency. Detailed results of these trials should be described in 1993.

As a clearer understanding of the pathogenesis of MS emerges, rational therapies will be designed, aimed at countering the pathophysiology of the disease. Other approaches not reviewed here, including targeting molecules involved in lymphocyte homing to MS lesions (Yednock et al 1992), or cytokines like TNFα (Sharief & Hentges 1991, Powell et al 1990, Brosnan et al 1988, Ruddle et al 1990), are likely to be tested in clinical trials in the near future. The ultimate goal of selective immune therapy for MS, which leaves the immune system intact while reversing established disease and preventing further progression, may be possible. By the end of the decade combinations of these selective therapies may be employed by neurologists, just as oncologists use various modalities of therapy to treat cancer.

ACKNOWLEDGMENTS

This work was supported by the National Institutes of Health, the National Multiple Sclerosis Societies of the United States and Australia, the Phil N. Allen Fund, Tocor Inc., and a grant from Mr. Hyman Abadi and the Fahnstock family.

Literature Cited

Acha-Orbea, H, Mitchell DJ, Timmermann L, et al. 1988. Limited heterogeneity of T cell receptors from lymphocytes mediating autoimmune encephalomyelitis allows specific immune intervention. Cell 54:263–73

Allegretta M, Nicklas J, Sriram S, Albertini RJ. 1990. T cells responsive to myelin basic protien in patients with multiple sclerosis. Science 47:718–21

Beall SS, Concannon P, Charmley P, et al. 1989. The germline repertoire of T cell receptor β-chain genes in patients with chronic progressive multiple sclerosis. J. Neuroimmunol. 21:59–66

Beall SS, McFarlin D, McFarland H, et al. 1991. Susceptibility for MS is determined by inheritance of a 175 kb region of the TcR-β

chain locus in conjunction with HLA class II genes. Am. J. Hum. Genetics 49S:21

Begovich AB, McClure GR, Suraj VC, et al. 1992. Polymorphism, recombination and linkage disequilibrium within the HLA class II region. J. Immunol. 148:249–58

Bell RB, Steinman L. 1992. Specific immunotherapeutic strategies: Lessons from myelin basic protein induced EAE. In Treatment of MS: Trial Design, Results and Future Perspectives, ed. RA Rudick, DE Goodkin, pp. 282–99. London: Springer Verlag

Ben-Nun A, Liblau R, Cohen L, et al. 1991. Restricted TCR Vβ gene usage by myelin basic protein-specific T cell clones in multiple sclerosis: predominant genes vary in individuals. *Proc. Natl. Acad. Sci. USA* 88:2466–70

Bernard CCA, Randell VB, Horvath L, et al. 1981. Antibody to myelin basic protein in extracts of multiple sclerosis brain. *Immunology* 43:447–57

Bitar D, Whitacre CC. 1988. Suppression of EAE by the oral administration of myelin basic protein. *Cell. Immunol.* 112:364–70

Bornstein MB, Miller A, Slagle S, et al. 1987. A pilot trial of Cop 1 in exacerbating-remitting multiple sclerosis. *N. Engl. J. Med.* 317:408–14

Bottazzo GF, Dean BM, McNally JM, et al. 1985. *In situ* characterization of autoimmune phenomena and expression of HLA molecules in the pancreas in diabetic insulin. *N. Engl. J. Med.* 313:353–57

Brocke S, Gaur A, Piercy C, Gautam A, Gijhelsk K, et al. Induction of relapsing paralysis in EAE by bacterial superantigen. *Nature* In press

Broeren CPM, Verjans GM, van Eden W, et al. 1991. Conserved nucleotide sequences at the 5′ end of T cell receptor variable genes facilitate polymerase chain reaction amplification. *Eur. J. Immunol.* 21:569–75

Brosnan C, Selmaj K, Raine CS. 1988. Hypothesis: a role for tumor necrosis factor in immune-mediated demyelination and its relevance to multiple sclerosis. *J. Neuroimmunol.* 18:87–94

Bugawan TL, Begovich AB, Erlich HA. 1990. Rapid HLA-DPB typing using enzymatically amplified DNA and nonradioactive sequence-specific oligonucleotide probes. *Immunogenetics* 32:231–41

Bugawan TL, Erlich HA. 1991. Rapid typing of HLA-DQ1 DNA polymorphism using nonradioactive oligonucleotide probes and amplified DNA. *Immunogenetics* 33:163–70

Burns FR, Li X, Shen N, et al. 1989. Both rat and mouse T cell receptors specific for the encephalitogenic determinants of myelin basic protein use similar Vα and Vβ chain genes. *J. Exp. Med.* 169:27–39

Chothia C, Boswell DR, Lesk AM. 1988. The outline structure of the T cell alpha-beta receptor. *EMBO J.* 7:3745–55

Davis M, Bjorkman PJ. 1988. T cell antigen receptor genes and T-cell recognition. *Nature* 334:395–402

Fujinami RS, Oldstone MBA. 1985. Amino acid homology between the encephalitogenic site of myelin basic protein and virus: Mechanism for autoimmunity. *Science* 230:1043–45

Gautam AM, Pearson C, Smilek D, et al. 1992. A polyalanine peptide containing only five native myelin basic protein residues induces autoimmune encephalomyelitis. *J. Exp. Med.* 176:605–9

Giegerich G, Pette M, Meinl E, et al. 1992. Diversity of T cell receptor alpha and beta chain genes expressed by human T cells specific for similar myelin basic protein peptide/major histocompatibility complexes. *Eur. J. Immunol.* 22:753–58

Gold DP, Offner H, Sun D, et al. 1991. Analysis of TCR β chains in Lewis rats with experimental allergic encephalomyelitis: conserved complementarity determining region 3. *J. Exp. Med.* 174:1467–76

Gold DP, Vainiene M, Celnik B, et al. 1992. Characterization of the immune response to a secondary encephalitogenic epitope of basic protein in Lewis rats. II. Biased TCR Vβ expression predominates in spinal cord infiltrating T cells. *J. Immunol.* 148:1712–17

Gulwani-Akolkar B, Posnett DN, Janson CH, et al. 1991. T cell receptor V-segment frequencies in peripheral blood T cells correlate with human leukocyte antigen type. *J. Exp. Med.* 174:1139–46

Helmuth R, Fildes N, Blake E, et al. 1990. HLA-DQ alpha allele and genotype frequencies in various human populations, determined by using enzymatic amplification and oligonucleotide probes. *Am. J. Hum. Genet.* 47:515–23

Higgins PJ, Weiner HL. 1988. Suppression of EAE by oral administration of myelin basic protein and its fragments. *J Immunol.* 140:440–5

Hillert J, Leng C, Olerup O. 1992. T-cell receptor α chain germline gene polymorphisms in multiple sclerosis. *Neurology* 42:80–84

Hodgkinson S, Lindsey JW, Allegretta M, et al. 1992. Phase I study of chimeric anti-CD4 monoclonal antibody in MS. *Neurology* 42:209 (Abstr)

Howell MD, Winters ST, Olee T, et al. 1989. Vaccination against experimental allergic encephalomyelitis with T cell receptor peptides. *Science* 246:668–70

Jahnke U, Fischer E, Alvord E. 1985. Sequence homology between certain viral proteins and proteins related to encephalomyelitis and neuritis. *Science* 229:282–84

Johnson RT, Griffin RE, Hirsch RL, et al. 1984. Measles encephalitis—clinical and immunological studies. *N. Engl. J. Med.* 310:137–41

Jorgensen J, Esser U, Fazekas de St. Groth B, et al. 1992. Mapping TCR/peptide contacts by variant peptide immunization of single-chain TCR transgenics. *Nature* 355:224–30

Kappler JW, Roehm N, Marrack P. 1987. T cell tolerance by clonal elimination in the thymus. *Cell* 49:273–80

Karin N, Szafer F, Mitchell D, et al. 1993. Selective and non-selective stages homing of T lymphocytes to the central nervous system during experimental allergic enephalo-myelins. *J. Immunol.* 150:4116–24

Kotzin B, Karaturi S, Chou Y, et al. 1991. Preferential T cell receptor V beta usage in myelin basic protein reactive to T cell clones from patients with multiple sclerosis. *Proc. Natl. Acad. Sci. USA* 88:9161–65

Lee SJ, Wucherpfennig KW, Brod SA, et al. 1991. Common T cell receptor V beta usage in oligoclonal T lymphocytes derived from cerebrospinal fluid and blood of patients with multiple sclerosis. *Ann. Neurol.* 29:33–40

Lider O, Santos LMB, Lee CSY. 1989. Suppression of experimental allergic encephalomyelitis by oral administration of myelin basic protein. II. Suppression of disease and in vitro immune responses is mediated by antigen-specific CD8+ T lymphocytes. *J. Immunol.* 142:748–52

Martin R, Howell MD, Jaraquemada D, et al. 1991. A myelin basic protein peptide is recognized in the context of four HLA-DR types associated with multiple sclerosis. *J. Exp. Med.* 173:19–24

Martino G, Olsson T, Fredrikson S, et al. 1991. Cells producing antibodies specific for myelin basic protein region 70–89 are predominant in cerbrospinal fluid from patients with MS. *Eur. J. Immunol.* 21:2971–6

Mattson DH, Roos RP, Arnason BGW. 1980. Isoelectric focusing of IgG eluted from multiple sclerosis and subacute sclerosing panenecephalitis brains. *Nature* 287:335–37

McFarlin DE, McFarland HF. 1982. Multiple sclerosis. *N. Engl. J. Med.* 307:1183–88

Miller A, Hafler DA, Weiner HL. 1991. Tolerance and suppressor mechanisms in EAE: Implications for immunotherapy of human autoimmune diseases. *FASEB J.* 5:2560–6

Miller A, Lider O, Roberts B, et al. 1992. Suppressor T cells generated by oral tolerization to myelin basic protein supress both in vitro and in vivo immune responses by the release of TGFβ after antigen-specific triggering. *Proc. Natl. Acad. Sci. USA* 89:421–5

Miller A, Lider O, Weiner HL. 1991. Antigen-driven bystander suppression following oral administration of antigens. *J. Exp. Med.* 174:791

Morahan G, Allison J, Peterson MG, Malcolm L. 1989. Sequence of the Vβ13 gene used by an influenza-specific T cell. *Immunogenetics* 30:311–13

Oksenberg JR, Panzara MA, Begovich AB, et al. 1993. Selection for T cell receptor Vβ-Dβ-Jβ gene rearrangements with specificity for myelin basic protein peptide 87–106 in brain lesions of multiple sclerosis. *Nature.* 362:68–70

Oksenberg JR, Sherritt M, Begovich AB. 1989. T-cell receptor V alpha and C alpha

alleles associated with multiple sclerosis and myasthenia gravis. *Proc. Natl. Acad. Sci. USA* 86:988–92

Oksenberg JR, Steinman, L. 1990. The role of the MHC and T cell receptor in susceptibility to multiple sclerosis. *Curr. Opinion Immunol.* 2:619–21

Oksenberg JR, Stuart S, Begovich AB, et al. 1990. Limited heterogeneity of rearranged T-cell receptor V alpha transcripts in brains of multiple sclerosis patients. *Nature* 345:344–46

Olsson T, Baig S, Hojeberg B, Link H. 1990. Antimyelin basic protein and antimyelin antibody-producing cells in MS. *Ann. Neurol.* 27:132–6

Ota K, Matsui M, Milford E, et al. 1990. T cell recognition of an immunodominant myelin basic protein epitope in multiple sclerosis. *Nature* 346:183–87

Paliard X, West S, Lafferty J, et al. 1991. Evidence for the effects of a superantigen in rheumatoid arthritis. *Science* 253:325–29

Panzara MA, Gussoni E, Steinman L, Oksenberg J. 1992. Analysis of the T cell repertoire using the PCR and specific oligonucleotide primers. *Biotechniques* 12(5): 728–735

Pette M, Fujita K, Kitze B, et al. 1990. Myelin basic protein specific T cell lines from MS patients and healthy individuals. *Neurology* 40:1770–76

Powell MB, Mitchell D, Lederman J, et al 1990. Lymphotoxin and tumor necrosis factor-alpha production by myelin basic protein specific T cell clones correlates with encephalitogenicity. *Int. Immunol.* 2:539–44

Ruddle NH, Bergman CM, McGrath KM. 1990. An antibody to lymphotoxin and TNF prevents transfer of EAE. *J. Exp. Med.* 172:1193–200

Sakai K, Sinha A, Mitchell DJ. 1988a. Involvement of distinct T cell receptors in the autoimmune encephalitogenic response to nested epitopes of myelin basic protein. *Proc. Natl. Acad. Sci. USA* 85:8608–12

Sakai K, Zamvil SS, Mitchell DJ, et al. 1988b. Characterization of a major encephalitogenic T cell epitope in SJL/J mice with synthetic oligopeptides of myelin basic protein. *J. Neuroimmunol.* 19:21–32

Scharf SJ, Griffith R, Erlich HA. 1991. Rapid typing of DNA sequence polymorphism at the HLA-DRB1 locus using the polymerase chain reaction and nonradioactive oligonucleotide probes. *Hum. Immunol.* 30:190–201

Seboun E, Robinson MA, Doolittle TH, et al. 1989. A susceptibility locus for multiple sclerosis is linked to the T cell receptor beta chain complex. *Cell* 57:1095–100

Selmaj K, Brosnan CF, Raine CS. 1991. Colocalization of TCR γδ lymphocytes and hsp65+ oligodendrocytes in multiple sclerosis. *Proc. Natl. Acad. Sci. USA* 88:6452–56

Sharief MK, Hentges R. 1991. Association between TNFa and disease progression in patients with MS. *N. Engl. J. Med.* 325:467–72

Sherritt MA, Oksenberg J, Kerlero de Rosbo N, Bernard CCA. 1992. Influence of HLA-DR2, HLA-DPw4, T cell receptor α chain genes on the susceptibility to multiple sclerosis. *Int. Immunol.* 4(2):177–181

Steinman L. 1991. The development of rational strategies for selective immunotherapy against demyelinating disease. *Adv. Immunol.* 49:357–79

Steinman L, Rosenbaum JT, Sriram, S, McDevitt HO. 1981. *In vivo* effects of antibodies to immune response gene products: prevention of experimental allergic encephalitis. *Proc. Natl. Acad. Sci. USA* 78: 7111–14

Steinman L, Solomon D, Lim M, et al. 1983. Prevention of experimental allergic encephalitis with in vivo administration of anti I-A antibody. Decreased accumulation of radiolabelled lymph node cells in the central nervous system. *J. Neuroimmunol.* 5:91–97

Teitelbaum D, Aharoni R, Arnon R, Sela M. 1988. Specific inhibition of the T-cell response to myelin basic protein by the synthetic copolymer Cop 1. *Proc. Natl. Acad. Sci. USA* 85:9724–8

Terasaki PI, Park MS, Opelz G, Ting A. 1976. Multiple sclerosis and high incidence of a B-lymphocyte antigen. *Science* 193:1245–47

Tillinghast JP, Behlke MA, Loh D. 1986. Structure and diversity of the human TCR beta chain variable region genes. *Science* 233:879–83

Traugott U, Reinherz E, Raine CS. 1982. Multiple sclerosis: Distribution of T cell subsets within active chronic lesions. *Science* 219:308–10

Urban JL, Kumar V, Kono DH, et al. 1988. Restricted use of T cell receptor V genes in murine autoimmune encephalomyelitis raises possibilities for antibody therapy. *Cell* 54:577–92

Vandenbark AA, Chou Y, Bourdette D, et al.

1992. TCR peptides induce autoregulation in MS. *Neurology* 42:186 (abstract)

Vandenbark AA, Hashim G, Offner H. 1989. Immunization with a synthetic T-cell receptor V-region peptide protects against experimental autoimmune encephalomyelitis. *Nature* 341:541–44

Vartdal F, Sollid L, Vandvik B. 1989. Patients with multiple sclerosis carry DQβ1 genes which encode shared polymorphic amino acid sequences. *Hum. Immunol.* 25:103–10

Walsh MJ, Tourtellotte WW. 1986. Temporal invariance and clonal uniformity of brain and cerebrospinal IgG, IgA and IgM in multiple sclerosis. *J. Exp. Med.* 163:41–53

Watanabe R, Wege H, ter Meujlen V. 1983. Adoptive transfer of EAE-like lesions from rats with corona virus-induced demyelination and encephalomyelitis. *Nature* 305:50–3

Weiner HL, Mackin GA, Matsui M, et al. 1993. Double blind pilot trial of oral polerization with myelin antigens in multiple sclerosis. *Science* 259:1321–24

Wraith DC, McDevitt HO, Steinman L, Acha-Orbea H. 1989a. T cell recognition as the target for immune intervention in autoimmune disease. *Cell* 57:709–15

Wraith DC, Smilek DE, Mitchell DJ, et al. 1989b. Antigen recognition in autoimmune encephalomyelitis and the potential for peptide-mediated immunotherapy. *Cell* 59:247–55

Wucherpfennig K, Newcombe J, Li H, et al. 1992. Polyclonal TCR Vα-Vβ repertoire in active multiple sclerosis lesions. *J. Exp. Med.* 175:993–1002

Wucherpfennig K, Ota K, Endo N. 1990. Shared human T cell receptor Vβ usage to immunodominant regions of myelin basic protein. *Science* 248:1016–19

Yednock TA, Cannon C, Fritz L, et al. 1992. Prevention of experimental autoimmune encephalomyelitis by antibodies against α4β1 integrin. *Nature* 356:63–66

Zamvil SS, Steinman L. 1992. The pathogenesis of demyelinating disease in the central nervous system and the development of selective immunotherapy. *Curr. Neurol.* 12: 42–69

Annu. Rev. Neurosci. 1994. 17:267–310

NEURONAL POLARITY

Ann Marie Craig and Gary Banker

Department of Neuroscience, University of Virginia School of Medicine, Charlottesville, Virginia 22908

KEY WORDS: axons, dendrites, protein sorting, neuronal development, nerve cell culture

THE CONCEPT OF NEURONAL POLARITY

The cell body is continuous with a variable number of processes which frequently branch....These ultimately become extremely thin and disappear in the spongy ground substance....These processes which, even in their ultimate, invariable branches, must not be considered to be the source of axis cylinders or to have a nerve fiber growing from them, will hereafter be called, for the sake of convenience, "protoplasmic processes". [The axis cylinder,] a prominent, single process which originates either in the body of the cell or in one of the largest protoplasmic processes, immediately at its origin from the cell is distinguishable from these [protoplasmic processes]. As soon as it leaves the cell, it appears at once as a rigid, hyaline mass, much more resistant to reagents and on the whole with a different reaction to them; and, from the start, it does not branch....This characteristic is not peculiar merely to the large motor cells [spinal motoneurons] in which Remak has already partially recognized it, but also to the sensory ones, to those of the olive, the pons, and on the whole to all which could so far be examined; indeed, if I am not mistaken, it is also peculiar to the cells of the cerebrum.

O Dieters (in Shepherd 1991)[1]

Some History

So wrote Otto Deiters 130 years ago. The terminology Deiters used sounds archaic now—it was not until the turn of the century that the modern terms "axon" and "dendrite" replaced the "axis cylinder" and "protoplasmic pro-

---

[1]Early studies on the form and function of nerve cells, particularly as they regard the development of the neuron doctrine, are the subject of a valuable new book by Gordon Shepherd (1991). This work includes translations of key passages from many early papers, including those mentioned here.

cesses" of Deiters' day. But Deiters' description, and the remarkably beautiful lithographs that accompany it, formulated the modern view of the morphology of neurons.

An appreciation of the significance of these morphological features required another thirty years and a spokesman with the insight and authority of Cajal: "The transmission of the nervous impulse is always from the dendritic branches and the cell body to the axon or functional process. Every neuron, then, possesses a receptor apparatus, the body and the dendritic prolongations, an apparatus of emission, the axon, and an apparatus of distribution, the terminal arborization of the nerve fiber" (Cajal 1989). This proposition, referred to as the Law of Dynamic Polarization, formed the basis for analysis of the circuits underlying brain function for the next 75 years.

The Law of Dynamic Polarization is, of course, incorrect. Since the discovery of dendro-dendritic synapses by Rall, Shepherd, Reese & Brightman (1966), it has become clear that neurons participate in a much more varied repertoire of synaptic interactions than Cajal and his contemporaries envisioned. Thus from a physiological standpoint it is incorrect to view neurons as having an overall polarity, with dendrites that conduct information toward the soma and axons away.

But are axons and dendrites fundamentally different, apart from the ambiguities of their physiological polarity? In the past decade, evidence from two lines of investigation has indicated that they are, and has returned the concept of neuronal polarity to fashion. The first line of investigation, marked by the publication of Matus et al (1981), was the immunocytochemical demonstration of dramatic and unexpected molecular differences between the axonal and dendritic cytoskeleton. One reason for the impact of these studies is that the differences they revealed between axons and dendrites seemed unambiguous—black and white, or in some cases, red and green—far more concrete and compelling than the differences in shape and structure revealed by microscopy. Moreover, molecular specialization of the axonal and dendritic cytoskeletons bore no obvious relationship to other structural specializations, such as synapses or nodes of Ranvier. Rather this specialization seemed to reveal an inherent difference between axons and dendrites as a whole.

The second of these developments was the demonstration that many of the morphological and molecular distinctions between axons and dendrites are reproduced when neurons develop processes in culture. Since this is our area of expertise, it merits a section of its own.

A Little Culture

Many experiments during the past decade have demonstrated that when embryonic neurons are dissociated and placed into culture, they develop a single axon and several dendrites that can be readily distinguished based on their morphology (Neale et al 1978, Kriegstein & Dichter 1983, Bartlett & Banker 1984a). Because of the advantages of cell culture for cell biological studies—advantages such as accessibility, control of the environment, and the ability to visualize living cells—cell culture has become the method of choice for many studies of neuronal polarity. Two culture systems have been particularly well characterized from this point of view, one derived from the embryonic rat hippocampus (Goslin & Banker 1991), the other from embryonic or neonatal rat sympathetic ganglia (Peng et al 1986, Higgins et al 1991). Both types of culture are relatively homogeneous. Under appropriate conditions, sympathetic cultures contain exclusively principal neurons and are free from contamination by ganglionic interneurons and nonneuronal cells. Our best estimates suggest that about 90% of the neurons in hippocampal cultures are pyramidal cells; the rest are interneurons of several types (Benson et al 1992, Craig et al 1993). Few nonneuronal cells are present. Cells from both sources can be maintained for a month or more in culture, enough time for the cells to become quite mature and to allay fears that the cells are terminally ill from the outset. In situ both of these cell types engage primarily in classical synaptic relationships; in culture their axons are consistently presynaptic and their dendrites postsynaptic (Bartlett & Banker 1984b, Furshpan et al 1986). In both cultures, axons and dendrites display their distinctive and characteristic morphologies, and the cells appropriately compartmentalize a variety of marker proteins. For example, Figure 1 shows the complementary distribution of MAP2 to dendrites and GAP-43 to axons of cultured rat hippocampal neurons. This polarized phenotype is equally characteristic of neurons that develop in low-density cultures and do not have an opportunity to contact other cells or form synapses (Bartlett & Banker 1984a, Caceres et al 1984). Such observations indicate that, from a cell biological point of view, neurons exhibit a fundamental polarity, independent of physiological considerations concerning information flow, and suggest that this organization may be largely governed by an endogenous program of development.

Several additional culture models are available for studies of neuronal polarity, including cultures derived from embryonic rat mesencephalon (Autillo-Touati et al 1988, Lafont et al 1992) and cerebellum (Ferreira & Caceres 1989, Caceres & Kosik 1990). The neuronal populations in these cultures have not been fully characterized, and the extent to which they develop a mature, fully polarized phenotype remains uncertain. Nevertheless, they have already provided novel insights into mechanisms that control the development of neuronal polarity (see below). Retinal photoreceptors, although hardly typical neurons, have also proved valuable for studies of polarity (Adler 1986). Given the diversity of neuronal form and function (see below), the development and characterization of additional culture models suitable for studies of neuronal polarity would be welcome.

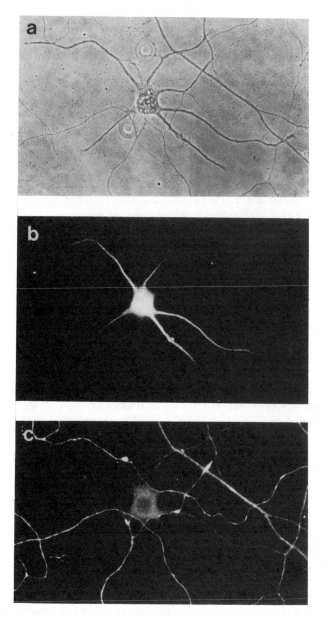

Figure 1   Complementary localization of MAP2 (*b*) and GAP43 (*c*) in a rat hippocampal neuron after eight days in culture. MAP2 immunoreactivity is present only in the cell body and the five dendrites, whereas GAP43 immunoreactivity is absent from the dendrites but present in the numerous axons traversing the field. (From Goslin et al 1988, with permission).

We are firmly convinced of the validity of studying neuronal polarity in culture—all of our own research hinges on this assumption—but we emphasize that there is no guarantee that neurons will become polarized in culture, even if they exhibit a well-polarized phenotype in situ. For example, under some culture conditions sympathetic neurons fail to develop dendrites and distribute normally dendritic markers to their axons, even though the cells survive for many months (Higgins et al 1988). Thus in attempting to characterize the polarity of cultured neurons it is critically important to be certain that markers, morphology, and synaptic polarity give a consistent picture.

Suitable continuous nerve cell lines would provide a particularly valuable complement to the use of primary cultures for studies of polarity. Cell lines can be used to obtain large quantities of genetically homogeneous cells, which would facilitate biochemical studies of neuronal polarity, and they can also be used to prepare stably transfected lines that overexpress normal proteins or that express mutant proteins. Unfortunately, most commonly used neural cell lines apparently are not polarized. For example, PC12 cells make an extensive fiber network when exposed to nerve-growth factor (NGF), but they do not form synapses or become polarized (Greene et al 1991). It is unclear whether their neurites represent axons, dendrites, or a hybrid that expresses some features of each. A neuroblastoma-glioma hybrid cell line, which does not ordinarily make synapses, can be induced to form presynaptic specializations through overexpression of synapsin IIb, but even under these circumstances it is not obvious that individual neurites display a consistent synaptic polarity (Han et al 1991).

Recently Pleasure et al (1992) described a promising method for obtaining a relatively pure population of neuronal cells from a human embryocarcinoma cell line and showed that some cells develop two types of processes that resemble axons and dendrites, as judged by their content of cytoskeletal antigens. The method involves prolonged treatment of the parent cell line (N-Tera 2/D1 cells) with retinoic acid, which causes many of the cells to acquire a neuronal phenotype. These cells are then dislodged mechanically and replated onto a substrate appropriate for studies of neuronal growth and morphological differentiation.

By using novel approaches, including retrovirus-mediated cell immortalization or cell hybridization, researchers are now generating new neural cell lines. It may prove worthwhile to screen some of these lines by using the standard immunological markers of neuronal polarity. For example, the morphology of MAH cells, a v-myc-immortalized line of sympathoadrenal progenitor cells that acquire many of the features of sympathetic neurons when exposed to fiboblast growth factor (FGF) and NGF, suggests that they might develop both axons and dendrites following differentiation by FGF and NGF (Birren & Anderson 1990).

The Size and Shape of Things

Although neurons obviously have a highly extended and complex shape compared with other types of cells and have axons that can attain great length, actual measurements of the absolute size of axonal and dendritic arbors are difficult to obtain. Table 1 provides estimates of the volume and surface area of the axonal and dendritic arbors of two types of neurons, based on published measurements of their dendritic arbors and estimates of their axonal size. Spinal motoneurons are representative of projection neurons, and granule neurons from the dentate gyrus are typical short-axon cells. These data make clear just how large neurons can become—the size of most nonneuronal cells is about that of a neuron's cell body—and illustrate the remarkable extent of variation among different neuronal types. This is true not just in absolute terms, but also in the relative size of their dendritic versus axonal arbors. The somatodendritic domain accounts for only 8% of the surface area of the motoneuron, but is more than 80% of the surface area of a dentate granule cell. It was surprising to find that the somatodendritic domain can account

Table 1   Neuronal dimensions

	Spinal motoneurons[a]	Dentate granule cells[b]
Surface Area (x 10^3 μm^2)		
Cell body	8.4	0.4
Dendrites	550	12
Axon	7,100	2.5
Total	7,700	15
Soma:Dendrites:Axon	1:65:845	1:28:6
Soma/Dendrites:Axon	1:13	5:1
Volume (x 10^3 μm^3)		
Soma	72	0.8
Dendrites	414	4.4
Axon	17,700	0.13
Total	18,200	5.3
Soma:Dendrites:Axon	1:6:245	6:34:1
Soma/Dendrites:Axon	1:35	40:1

[a] Cat lumbosacral α-motoneurons (Cullheim et al 1987; R Burke, personal communication). These cells have on average 11.7 dendrites, whose combined total length is 107 mm. Their axons were assumed to be 225 mm long and 10 μm in diameter.

[b] Rat dentate gyrus granule cells (Desmond & Levy 1982, 1984, and personal communication; Claiborne et al 1986). These cells have on average 2.2 dendrites, whose combined total length is 3.7 mm. Their axons were assumed to be 4 mm in total length (including collaterals) and 0.2 μm in diameter. The contribution of dendritic spines and axonal varicosities is not included. This could affect these estimates by as much as two-fold.

for so large a proportion of cell surface and volume, because most discussions of neuronal geometry focus on the extreme length and volume of the axon.

Axons and dendrites differ in caliber at their origin, and the differences in caliber are amplified during branching. At dendritic branch points, total cross-sectional area is conserved or, most frequently, decreases slightly following the so-called 3/2's power rule ($P^{3/2} = D_1^{\frac{3}{2}} + D_2^{\frac{3}{2}}$, where P and D represent the diameters of parent and daughter branches, respectively).[2] These branch power relationships are conserved among neurons with dendritic branching patterns as diverse as Purkinje cells, motoneurons, cerebellar granule cells, and cultured hippocampal neurons (Hillman 1979, Banker & Waxman 1988). Assuming that dendritic branches must attain some minimum diameter for structural stability (typically 0.5–1.0 μm), the diameter of each stem dendrite as it emerges from the cell body limits both the number of branches it can give rise to and, because dendrites usually branch at regular intervals, its total length. The decrease in diameter at branch points explains why dendrites resemble trees and how they got their name (*dendron* is *tree,* in Greek). Although the branching of individual axons has not been analyzed in detail, as that of dendrites has, the total cross-sectional area of axons clearly can increase with branching. For example, as motoneuron axons ramify to innervate different muscle fibers, their total cross-sectional area increases by a factor of 1.25–1.5 at each branch point (Zenker & Hohberg 1973, Pfeiffer & Friede 1985). Cullheim & Kellerth (1978) illustrate a motoneuron collateral in the spinal cord that is 2 μm in diameter and gives rise to 39 branches whose average diameter is 0.73 μm (a 14-fold increase in total cross-sectional area). Finally, axonal diameter can also increase independent of branching. The diameter of a motoneuron axon doubles between the initial segment and the axon proper (Cullheim & Kellerth 1978). Dendrites and axons also typically differ in branching angles (Hillman 1979, Lasek 1988). These differences in caliber and branching presumably reflect important, but as yet unknown, differences in the properties of their cytoskeletal elements.

The Facts

Table 2 lists some of the constituents reported to be differentially distributed in neurons. Complementary studies in situ and in culture provide the best confirmation for the distribution of a given protein. Data from studies in situ

---

[2]In reality, the exponent in the branch power equation varies among different classes of neurons between 3/2 and 2. If the 3/2's power ratio holds for a particular cell, the summed cross-sectional area of both daughter branches is about 80% that of the parent, whereas the summed circumference (proportional to membrane area) increases by about 126%. If the exponent in the branch power equation is 2, the summed cross-sectional area of the daughters is the same as the parent, and summed circumference increases about 141%.

Table 2 Compartmentation in neurons

	Localization and basis (1: in situ; 2: in culture)		References
Organelles:			
ribosomes and protein synthetic machinery	somatodendritic	1,2	Peters et al 1991, Bartlett & Banker 1984a,b
mRNAs	most restricted to soma; selected mRNAs are somatodendritic	1,2	Steward & Banker 1992
endoplasmic reticulum	throughout	1	Villa et al 1992, Satoh et al 1990, Walton et al 1992
Golgi complex	soma and proximal dendrites	1,2	Stieber et al 1987, Croul et al 1990, Peters et al 1991
synaptic vesicles	predominantly axonal (presynaptic)	1,2	De Camilli et al 1983, Navone et al 1986, Fletcher et al 1991
early endosomes	throughout	2	Parton et al 1992
late endosomes	soma and proximal dendrites	2	Parton et al 1992
mitochondria	throughout	1,2	Peters et al 1991, Bartlett & Banker 1984a
Cytoskeletal proteins:			
MAP1a (MAP1)	throughout	1	Huber & Matus 1984, Bernhardt et al 1985
MAP1b (MAP5)	throughout; phosphorylated in axons but not dendrites	1,2	Sato-Yoshitake et al 1989, Riederer et al 1990, Mansfield et al 1991
MAP2	somatodendritic	1,2	Bernhardt & Matus 1984, Caceres et al 1984
MAP 3/4	axonal	1	Huber et al 1985, Bernhardt et al 1985
tau	axonal (or differentially phosphorylated); not well segregated in culture	1,2	Binder et al 1985, Dotti et al 1987, Kosik & Finch 1987
kinesin	throughout	2	Ferreira et al 1992
RII subunit of cAMP kinase	somatodendritic (binds to MAP2)	1	Vallee et al 1981, De Camilli et al 1986
neurofilament subunits	throughout; highly phosphorylated in axons but not dendrites	1,2	Sternberger & Sternberger 1983, Lee et al 1987, Lein & Higgins 1989
Membrane proteins:			
Na^+, K^+-ATPase (α1 and α3 subunits)	throughout	1,2	Pietrini et al 1992
glycine receptors	somatodendritic (postsynaptic)	1	Triller et al 1985, Seitandou et al 1988
$GABA_A$ receptors	somatodendritic (postsynaptic)	1,2	Somogyi et al 1989, Killisch et al 1991, Craig & Banker, unpublished observations
glutamate receptors— gluR1-4; mGluR1α	somatodendritic (postsynaptic)	1,2	Rogers et al 1991, Petralia & Wenthold 1992, Martin et al 1992, 1993, Craig et al 1993

Table 2   (*continued*)

	Localization and basis (1: in situ; 2: in culture)		References
Na channels			
α subunit, Type I and III	somatodendritic	1	Westenbroek et al 1989, 1992b
α subunit, Type II	axonal	1	Westenbroek et al 1989, 1992b
K channels			
A-type, Kv1.4	axonal	1	Sheng et al 1992
A-type, Kv4.2	somatodendritic	1	Sheng et al 1992
delayed rectifier (drk1)	somatodendritic (clustered)	1	Trimmer 1991
transferrin receptor	somatodendritic	2	Cameron et al 1991
L1	axonal	1,2	Persohn & Schachner 1987, 1990; Esch & Banker, unpublished observations
N-CAM 180	somatodendritic (postsynaptic)	1	Persohn & Schachner 1990
integrin, α8 subunit	axonal	1	Bossy et al 1991
Thy-1 (GPI-linked)	conflicting reports	1,2	Morris et al 1985, Xue et al 1990, Dotti et al 1991
TAG-1 (GPI-linked)	axonal	1	Furley et al 1990, Yamamoto et al 1990
F3/F11/contactin (GPI-linked)	axonal in some neurons, throughout in others	1	Faivre-Sarrailh et al 1992
GAP-43 (lipid-linked)	axonal	1,2	Van Lookeren Campagne et al 1990, Goslin et al 1988, 1990
neurogranin (RC-3; lipid-linked?)	somatodendritic	1	Represa et al 1990, Watson et al 1992

obviously set the standard because they refer to "real" neurons, but studies in cell culture have proved to be a useful, sometimes essential complement. Cell culture allows direct visualization of large portions of individual cells, whereas it is sometimes difficult to construct an accurate view of a protein's localization from the incomplete picture available from a single tissue section. In addition, cell culture often enables the researcher to examine the distribution of constituents as intercellular interactions are manipulated. Finally, antibodies can be applied to living cells so that only extracellular antigens are accessible. Thus light microscopy of cell cultures is sometimes sufficient to draw inferences—for instance, whether a given receptor is pre- or postsynaptic or a given receptor epitope is intra- or extracellular—that would be difficult to draw from studies of tissue sections, even with electron microscopy.

Several conclusions are apparent from this compilation of differentially distributed neuronal constituents. First and most obvious is that many of the proteins that are most important in neuronal function have a highly specific localization. This observation is hardly novel, but it emphasizes once again the importance of understanding how neuronal constituents come to be differentially distributed.

A surprisingly large proportion of membrane proteins apparently have a polarized distribution. This could be coincidental, because the catalog of protein distributions is far from complete, but it could also be related to the mechanisms that govern the sorting of membrane proteins. In other types of polarized cells, most proteins are segregated in one domain or the other. It is also striking that different members of the same family often have different distributions. This is particularly obvious in the case of the voltage-gated Na^+ and K^+ channels, and it applies to calcium channels as well (Westenbroek et al 1992a). One reason for the remarkable diversity of channel types may be to allow channels with different functional characteristics to be targeted to different cellular domains and microdomains.

Synaptic constituents, including synaptic vesicle proteins and postsynaptic glutamate receptors, polarize to axonal or somatodendritic domains in uninnervated cultured hippocampal neurons and subsequently cluster upon innervation (Fletcher et al 1991, Craig et al 1993). This has led to the suggestion that the localization of synaptic constituents involves two distinct mechanisms: an endogenously programmed targeting either to axons or cell bodies and dendrites, followed by a clustering at synaptic sites that depends on cell-cell interactions. A similar two-step model might apply to the localization of proteins at other microdomains, such as nodes of Ranvier.

With regard to the cytoskeleton, polarity is reflected not in the partitioning of the basic constituents—both axons and dendrites are built on a central core that contains microtubules and neurofilaments surrounded by an actin-rich cortex—but rather in the organization of microtubules, the differential distribution of some cytoskeleton-associated proteins, and local differences in phosphorylation. Microtubules in the axon are all oriented plus-end distal, whereas dendrites contain microtubules of both polarity orientations (Baas et al 1988, Burton 1988). The high molecular weight form of the microtubule-associated protein MAP2 is present in the somatodendritic domain but is absent from axons. On the other hand, neurofilaments are present in both dendrites and axons (although their abundance varies), but their highly phosphorylated forms are restricted to axons. In some cases (e.g. tau and MAP 3/4) it is not yet clear if the differential immunoreactivity of axons and dendrites reflects the distribution of the protein itself or that of a specific posttranslational modification (Papasozomenos & Binder 1987).

The distribution of intracellular organelles presents a somewhat more complicated picture. Some organelles, such as late endosomes and the Golgi complex, extend only into proximal dendrites, indicating a partial partitioning between soma and dendrites. Others, such as ribosomes, are present throughout the somatodendritic domain but excluded from the axons, consistent with the conventional view of neuronal polarity. But an analysis of individual mRNAs suggests an underlying complexity. Many mRNAs are confined to the cell body, whereas some are present throughout the dendritic tree (Steward & Banker 1992). Still other organelles, such as mitochondria and endoplasmic reticulum, are present throughout the cell, but even here complications may emerge. A differential distribution of mitochondria has been observed in some circumstances (Hevner & Wong-Riley 1991), presumably reflecting the metabolic demands of different portions of the neuron.

Several important classes of molecules are conspicuously absent or underrepresented in the data summarized in Table 2. For example, different extracellular matrix (ECM) molecules exert profoundly different effects on axonal versus dendritic growth, yet virtually nothing is known about the distribution of the integrins and other potential receptors likely to mediate these effects. Likewise, surprisingly few systematic studies of the distribution of cell adhesion molecules have been carried out, especially in culture, where the axonal or dendritic distribution of cell adhesion molecules might be more obvious. Given the important role of these molecules in neuronal development generally, and the specific possibility that a differential distribution of adhesion molecules could account for the differential growth that marks the emergence of neuronal polarity, this constitutes an important gap in our knowledge.

Also missing from Table 2 are the proteins that make up the spectrin-based submembranous cytoskeleton. Neurons express multiple forms of spectrin, ankyrin, and a band 4.1 homologue (Riederer et al 1986, Zagon et al 1986, Bennett et al 1991), together with other, novel proteins thought to associate with the submembranous cytoskeleton (Hayes et al 1991, Shirao et al 1992). Many of these constituents are differentially expressed in different types of neurons and are differentially distributed within these neurons. For example, $ankyrin_{R0}$ is restricted to cell bodies and dendrites, whereas a different isoform, $ankyrin_R$, is highly concentrated at nodes of Ranvier (Kordeli & Bennett 1991). The available data are somewhat difficult to interpret in the context of neuronal polarity, but based on observations in other cell types there is every reason to suspect that these proteins play an important role in the establishment and maintenance of neuronal polarity (see below).

All the Facts

Just as the Law of Dynamic Polarization is an oversimplification, so "the facts" just presented paint a distorted picture. In order to make a case for the concept of neuronal polarity, researchers have focused almost entirely on types of neurons that fit the classical view. Perhaps it is now safe to admit that this concept is oversimplified and to begin to come to grips with the diversity of the neuronal phenotype. Some neurons, such as the granule cells of the olfactory bulb, lack axons (Shepherd 1979). Some classes of amacrine cells have as many as six axons that arise from different dendrites (Famiglietti 1992). Still others, including certain classes of horizontal and amacrine cells, have shapes so unusual that it is uncertain if the terms axons and dendrites apply at all (Fisher & Boycott 1974, Bloomfield & Miller 1986). It should be instructive to examine some of these unusual cells with accepted markers of polarity.

Cells such as these are sufficiently rare that they might be regarded as curiosities, but neurons whose axons and dendrites are simultaneously pre- and postsynaptic are not uncommon (Shepherd 1979). Thus the cellular mechanisms that in hippocampal neurons ensure that synaptic vesicles and postsynaptic receptors are selectively segregated to axonal and dendritic domains must permit the same constituents to assume a different distribution in other types of nerve cells.

And what of invertebrate neurons? In the nerve nets of coelenterates, neurons are apparently unpolarized. Cells give rise to two or three apparently identical neurites, which can conduct impulses both toward and away from the cell body and which form bidirectional, chemical synapses where they contact other cells (Bullock & Horridge 1965, Anderson 1985). In higher invertebrates a typical cell gives rise to a single process that has the properties of an axon, and this process gives off groups of dendritic branches that arborize within the neuropil and are postsynaptic (Bullock & Horridge 1965, Cohen 1970; also see illustrations in Young 1989). In some cases, clusters of dendritic branches are given off in different ganglia (Kennedy & Mellon 1964). Such neurons exhibit a clear localization of function, which presumably must be reflected in a differential distribution of membrane receptors and channels, but the organization of their axonal and dendritic domains is sufficiently different from that in vertebrate neurons to suggest that the details of molecular sorting may also differ.

THE CELLULAR BASIS OF NEURONAL POLARITY

We must admit from the outset that, in a factual sense, we know almost nothing about the cellular mechanisms responsible for compartmentation in neurons. If we confined ourselves to the facts, this discussion would be quite short. But even from the little we know, it is possible to make reasonable conjectures about the mechanisms that might account for the polarity of membrane proteins. Although the segregation of intracellular organelles, cytoskeletal proteins, and RNA is undoubtedly of equal importance, we know much less about their cellular bases.

We propose that the segregation of membrane proteins involves the following steps (Figure 2): (a) the sorting of axonal and dendritic proteins

1. SORTING AND VESICLE FORMATION
  -in the TGN?
  -receptor-mediated?
  -lipid-mediated?
  -different classes of vesicles?
  -default pathway?

2. TRANSPORT
  -microtubule-based?
  -dependent on microtubule polarity?
  -directed by domain-specific motors?

3. PLASMA MEMBRANE ADDITION
  -at specific sites?
  -at different sites than final localization?
  -mediated by docking proteins?
  -involves domain-specific rabs?

4. ANCHORING
  -to cytoskeleton via ankyrins, spectrins...?
  -limits diffusion between domains?
  -generates microdomains?
  -other barriers to diffusion?

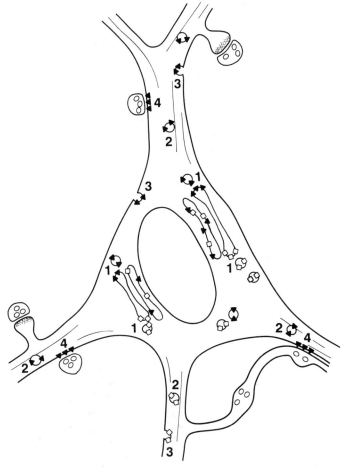

Figure 2   Proposed targeting mechanisms for axonal and somatodendritic membrane proteins. The neuron illustrated has three dendrites and a single axon, which emerges from the base of the cell body. Axonal proteins are represented by open squares, and somatodendritic proteins by closed triangles.

into distinct carrier vesicles, (*b*) directed transport to the appropriate domain, (*c*) insertion into the membrane at specific sites, and (*d*) anchoring to limit diffusion within the membrane. Local protein synthesis, localized post-translational modification, and domain-specific degradation may also contribute. Some of these mechanisms are likely to contribute to the polarized distribution of other neuronal constituents as well.

Much of the current discussion concerning the targeting of membrane proteins in neurons derives from possible analogies with polarized epithelial cells, whose sorting mechanisms are better understood. Thus before discussing mechanisms in detail, we will examine the parallels between polarity in neurons and polarity in epithelial cells.

Are Neurons Just Long, Skinny Epithelial Cells?

Epithelial cells form sheets that line the gastrointestinal tract and the luminal surfaces of organs such as the kidneys, liver, and lungs. The epithelial cell surface is divided by a circumferential band of tight junctions into two functionally and molecularly distinct domains: the basolateral domain, which is in contact with the basement membrane and adjacent epithelial cells, and the apical domain, which faces the lumen. The idea that neurons and epithelial cells may share common mechanisms for segregating cellular constituents was first explicitly proposed by Dotti & Simons (1990), who infected cultured hippocampal neurons with viruses and examined the targeting of viral glycoproteins. This method had been used with great success to elucidate the mechanisms of protein sorting in epithelial cells. Glycoproteins encoded by the viral DNA are routed through the ER and Golgi complex of infected cells, presumably by normal cellular mechanisms, and are inserted into the apical or basolateral plasma membrane, where they direct the budding of the nucleocapsid from the cell surface. In Madin-Darby canine kidney (MDCK) cells, a polarized epithelial cell line, the vesicular stomatitis virus (VSV) G protein, is delivered to the basolateral surface, whereas the influenza fowl plague virus (FPV) hemagglutinin (HA) is delivered apically (Rodriguez-Boulan & Pendergast 1980). Dotti & Simons (1990) found that VSV G protein was preferentially targeted to the somatodendritic domain of infected hippocampal neurons, and FPV HA to the axonal domain. These results, taken in the context of the then-current model of epithelial polarity (Simons & Wandinger-Ness 1990), led to the proposal that the axonal and somatodendritic domains of neurons correspond respectively to the apical and basolateral domains of epithelial cells, and that the mechanisms for targeting membrane proteins to the corresponding domains may be similar in the two cell types.

One simple means to evaluate this hypothesis is to compare the distribution of membrane proteins in epithelial cells and in neurons. Few proteins are endogenous to both cell types, so such comparisons usually require the introduction of exogenous proteins, either through viral infection or through

DNA transfection. Table 3 compares the distribution of membrane proteins whose localizations have been determined in both hippocampal neurons and MDCK cells. In support of the neuron/epithelial hypothesis, Dotti et al (1991) reported that endogenously expressed Thy-1 is preferentially distributed to the axons of cultured hippocampal neurons; it has an apical distribution when expressed in MDCK cells (Powell et al 1991b). Although the distribution of several other proteins is consistent with the epithelial/neuronal hypothesis, apparent contradictions have also come to light. For example, three proteins that have a polarized distribution in MDCK cells—the $\alpha 1$ and $\alpha 3$ subunits of Na^+,K^+-ATPase and CD-8, a lymphocyte cell surface protein—are not obviously polarized in hippocampal neurons. A second possible exception is Thy-1. Although Dotti et al (1991) reported that Thy-1 is present only in axons, other workers have detected this protein in both dendrites and axons of hippocampal neurons and other types of neurons in situ (Morris et al 1985, Xue et al 1990).

Table 3 Comparison of the distribution of membrane proteins in hippocampal neurons and MDCK cells

Protein	Localization in neurons[a]	Localization in MDCK cells[a]	Parallel
FPV-HA	axonal[b] (I)	apical[c] (I)	yes
GABA transporter	axonal[d] (E)	apical[d] (I)	yes
Thy-1	axonal[e]; uniform[f] (E)	apical[g] (I)	?
VSV-G	somatodendritic[b] (I)	basolateral[b] (I)	yes
N-CAM 180	somatodendritic[h] (E)	basolateral[i] (I)	yes
transferrin receptor	somatodendritic[j] (E)	basolateral[k] (E)	yes
GABA$_A$ receptor, $\alpha 1$	somatodendritic[l] (E)	basolateral[m] (I)	yes
SFV E protein	somatodendritic[n] (I)	basolateral[o] (I)	yes
Na^+K^+ATPase, $\alpha 1$	not polarized[p] (E)	basolateral[q] (E)	no
Na^+K^+ATPase, $\alpha 3$	not polarized[p] (E)	basolateral[t] (I)	no
CD-8	not polarized[r] (I)	apical[s] (I)	no

[a] (E), endogenous; (I), introduced by transfection or viral infection.
[b] Dotti & Simons (1990).
[c] Rodriguez-Boulan & Pendergast (1980).
[d] Pietrini et al (1993).
[e] Dotti et al (1991).
[f] Xue et al (1990).
[g] Powell et al (1991b).
[h] Persohn & Schachner (1990).
[i] Powell et al (1991a).
[j] Cameron et al (1991).
[k] Fuller & Simons (1986).
[l] Killisch et al (1991).
[m] Perez-Velasquez & Angelides (1993).
[n] Dotti et al (1993).
[o] Roman & Garoff (1986).
[p] Pietrini et al (1992).
[q] Caplan et al (1986).
[r] A. M. Craig, R. Wyborski, D. Gottlieb & G. Banker, unpublished research.
[s] Migliaccio et al 1990.

Recent studies of epithelial cells that indicate a diversity of protein distribution among different cell types are also difficult to reconcile with the epithelial/neuronal hypothesis as originally proposed. For example, Na^+K^+-ATPase is distributed to the basolateral domain in MDCK cells and to the apical domain in choroid plexus epithelium, and is also uniformly distributed in cultured retinal pigment epithelial (RPE-J) cells (Caplan et al 1986, Rodriguez-Boulan & Powell 1992, Marrs & Nelson 1993). GPI-linked proteins, which are exclusively apical in MDCK cells, are sorted to the basolateral domain in Fisher rat thyroid (FRT) cells (Zurzolo et al 1993). Observations such as these have led to the concept of a flexible epithelial phenotype, in which the mechanisms of protein targeting permit identical proteins to be distributed differently in different cell types (Rodriguez-Boulan & Powell 1992). Given the diversity of neuronal cell types (as discussed above), it would hardly be surprising if neurons exhibited a similar flexibility in the expression of polarity. Indeed, recent evidence indicates that the cell adhesion molecule F3/F11, which is restricted to axons of cerebellar granule cells, is uniformly distributed in cerebellar Golgi cells (Faivre-Sarrailh et al 1992).

In summary, the idea that the axonal and dendritic domains of neurons are strictly equivalent to the apical and basolateral domains of epithelial cells requires modification. But, in a broader sense, the possibility that neurons and epithelial cells use common targeting mechanisms remains valid and has considerable intuitive appeal. Equally important, the comparison with epithelial cells has provided a valuable paradigm for analyzing the cellular mechanisms underlying polarity in neurons.

The Sorting Problem

In epithelial cells, apical and basolateral proteins are both synthesized in the ER and traverse the Golgi complex together. Segregation of apical and basolateral proteins first occurs in the trans-Golgi network (TGN), where proteins destined for different domains are packaged into different transport vesicles (Rodriguez-Boulan & Nelson 1989). In addition, some apical proteins are initially transported to the basolateral membrane, then endocytosed, sorted within endosomes, and transported to the apical domain. In neurons, sorting of peptides into different classes of dense core vesicles occurs in the TGN (reviewed by Huttner & Dotti 1991, Jung & Scheller 1991). Sorting of axonal and somatodendritic membrane proteins probably also occurs in the TGN but direct evidence is lacking. Synaptic vesicle proteins may be sorted in the TGN or after arriving at the nerve terminal, using the same mechanisms that underlie local recycling (reviewed by Kelly & Grote 1993).

Independent of where sorting occurs, the sorting process is thought to involve (*a*) signals contained within the proteins to be sorted and (*b*) a mechanism for their segregation and selective incorporation into distinct populations of transport vesicles. The use of deletion mutants and protein chimeras has led to remarkable progress in the identification of apical and basolateral sorting signals in epithelial cells. The first sorting signal identified with this approach was the glycosyl phosphatidylinositol (GPI) anchor. GPI-anchored proteins contain a conventional N-terminal signal sequence that mediates entry into the ER and a hydrophobic C-terminal sequence that directs cotranslational cleavage of a C-terminal peptide and the addition of a GPI anchor (reviewed by Lisanti & Rodriguez-Boulan 1990). In their final form, such proteins consist of a single ectodomain that is anchored to the exoplasmic leaflet of the membrane via the GPI link. In MDCK cells all endogenous GPI-anchored proteins are targeted apically (Lisanti et al 1990), and the addition of a GPI anchor to the ectodomains of basolateral proteins is sufficient to redirect them to the apical domain (Brown et al 1989, Lisanti et al 1989).

By using a similar approach, researchers have identified targeting signals in the endoplasmic domains of several basolateral proteins (Brewer & Roth 1991, Casanova et al 1991, Hunziker et al 1991, Le Bivic et al 1991, Mostov et al 1992). For example, Casanova et al (1991) identified a 14–amino acid region of the polymeric immunoglobulin receptor that acts as an autonomous basolateral sorting signal. The basolateral signals that have been identified do not contain common sequences, but they are thought to share a common structural motif, possibly related to the beta-turn motif that marks proteins for endocytosis via clathrin-coated pits (reviewed by Mostov et al 1992). An apical sorting signal also appears to be present in the ectodomains of some proteins (Vogel et al 1993).

Do similar sorting signals govern the targeting of neuronal proteins? Following from the suggestion of Dotti et al (1991) that GPI-anchored proteins are selectively distributed to the axonal domain, Lowenstein et al (1993 and personal communication) expressed a chimeric protein composed of an exogenous secretory protein (tissue inhibitor of metalloproteinases) linked to the C-terminal region of Thy-1 (which contains the signal for GPI anchoring) in cultured neocortical neurons. The chimeric protein, which was inserted into the neuronal plasma membrane via a GPI anchor, appeared first in the dendrites and later in the axons. Thus, in this case the GPI anchor was not sufficient to target a foreign protein to a single domain in cultured cortical neurons.

This approach, which is now being applied to the search for axonal and somatodendritic targeting signals in neurons, is likely to yield the first real information about sorting signals and sorting pathways in neurons. But it is not without pitfalls. First, mutations can interfere with protein oligomerization and exit from the ER, a phenomenon that confounded many of the initial studies of sorting signals in epithelia (discussed by Mostov et al 1992). Second, steady-state distributions are influenced by protein instability, selective endocytosis, and transcytosis, in addition to the selectivity of protein delivery. In epithelial cells, protein delivery can be measured directly, but at present this is not possible in neurons. The behavior of chimeric proteins may be complicated if they contain distinct, contradictory signals (Roth et al 1987, Brown et al 1989). Finally, assembly of subunits into different multisubunit complexes can alter the distribution of individual subunits (Perez-Velazquez & Angelides 1993).

Do all membrane proteins in polarized epithelial cells have sorting signals? This has proved to be a surprisingly difficult question to answer. Although it was once thought that all basolateral proteins did not require sorting signals but were targeted by default (Simons & Wandinger-Ness 1990), this is now regarded as unlikely (Matlin 1992, Mostov et al 1992). Whether or not a signalless protein would be selectively targeted remains anyone's guess. If default pathways do exist, either in epithelial cells or in neurons, the interpretation of experiments to identify sorting signals becomes that much more complicated.

The mechanisms that underlie the sorting of membrane proteins in polarized cells are not known, either in epithelial cells or neurons, but two possibilities have been suggested. One mechanism is based on the association of proteins with specific lipids. This suggestion follows from the observation that in model membranes, glycosphingolipids aggregate to form discrete domains, and that the apical membranes of most epithelial cells are enriched in glycosphingolipids (Simons & van Meer 1988). Thus it has been suggested that glycosphingolipids, which are synthesized in the Golgi, may aggregate into patches that entrap apical proteins, either by direct binding or by binding to a sorting receptor that binds to epical proteins. Consistent with this possibility, biochemical evidence suggests that GPI-linked proteins become associated with sphingolipid domains shortly after their synthesis (Brown & Rose 1992, LeBivic et al 1993). In addition, in FRT cells glycosphingolipids and GPI-anchored proteins are both targeted basolaterally, whereas most transmembrane proteins have the same distribution as they do in MDCK cells (Zurzolo et al 1993).

A second mechanism is receptor-mediated sorting, as exemplified by the sorting of soluble lysosomal enzymes that bind to the mannose-6-phosphate receptor in the TGN and are thereby targeted to the lysosomal pathway (Kornfeld & Mellman 1989). A similar mechanism underlies the receptor-mediated endocytosis of extracellular proteins such as transferrin and low density lipoprotein (Pearse & Robinson 1990). Cytosolic adaptor proteins bind to the receptor-ligand complex, which causes them to cluster in the membrane. Interaction between adaptor proteins and other coat proteins, such as clathrin, induces vesicle budding. Sorting of axonal and somatodendritic proteins in the TGN could occur in an analogous fashion. Novel proteins mediating intra-Golgi traffic via nonclathrin coated vesicles have recently been identified (Duden et al 1991, Serafini et al 1991).

Both trimeric G proteins and small G proteins of the Rab family are also thought to be involved in vesicle budding from and fusion to membrane compartments; hydrolysis of bound GTP to GDP is thought to mediate directionality (reviewed by Gruenberg & Clague 1992). Different Rab proteins are associated with different vesicular compartments. Two members of the Rab family are of particular interest to neurobiologists. Rab3A associates with synaptic vesicles distal to the Golgi complex and dissociates during exocytosis, suggesting a role in targeting vesicles to active zones (Fischer von Mollard et al 1990, 1991; Matteoli et al 1991). Rab8, a novel member of the Rab family cloned from MDCK cells, is specifically associated with TGN-derived basolateral transport vesicles and with the basolateral plasma membrane (Chavrier et al 1990, Huber et al 1992). Rab8 is localized to cell bodies and dendrites of cultured hippocampal neurons (Huber et al 1992), and may prove to be the first identified component of the somatodendritic targeting machinery.

Getting to the Right Place...

Compared with other cell types, the extended geometry of neurons imposes special demands on the control of vesicle traffic. The microtubule-based transport systems present in all cells have evolved in neurons into highly efficient means for transporting vesicles over long distances, and intuitively it seems that there must be labels to ensure that vesicles interact only with microtubules destined for the appropriate domains. One possibility is that axonal and dendritic transport vesicles have different affinities for the motors that direct plus-end and minus-end transport. Because axons contain solely plus-end distal microtubules but dendrites contain both plus- and minus-end distal microtubules (Baas et al 1988), selective transport into dendrites could be mediated by minus-end directed motors. Axonal transport vesicles must associate with a plus-end directed motor, but that by itself would not prevent their entry into dendrites. Furthermore, during development many constituents become selectively localized to axons before these differences in microtubule organization arise. Thus directed transport likely involves more than a differential affinity for microtubule motor proteins.

The large variation in the relative surface areas of axonal versus dendritic arbors among different neuronal cell types, as discussed above, has interesting implications for transport vesicle traffic. For example, assuming that the turnover rate of membrane constituents in axons and dendrites is similar, and based on relative surface areas (see Table 1), a dentate granule cell with two dendrites would have about about twice as many vesicles entering each of its dendrites as enter its axon. However, similar calculations for a spinal motoneuron suggest that the number of vesicles entering the axon will be about 200 times the number entering each of its 12 dendrites.

Just as there must be mechanisms to prevent the fusion of vesicles with inappropriate intracellular compartments, so there are likely to be plasma membrane docking proteins that govern the fusion of exocytic transport vesicles. Such proteins could interact with Rabs or coat proteins on the vesicle surface, but the identity of the domain-specific docking proteins themselves is unknown. Proteins destined for specific membrane microdomains, such as postsynaptic receptor clusters, need not be specifically inserted at these sites, but may accumulate there by diffusion within the membrane and selective trapping. Given the diversity of microdomains within the neuronal membrane, it is difficult to imagine that proteins are sorted into separate vesicle populations that are selectively targeted to each specialized site.

In our discussion so far we have assumed that axonal and somatodendritic proteins are targeted directly to the appropriate domains, but this need not be the exclusive pathway for the polarized addition of membrane proteins. Whereas epithelial cells transport some membrane proteins directly from the TGN to the appropriate membrane surface, other membrane proteins reach their final destination indirectly, through transcytosis. For example, in hepatocytes most apical membrane proteins are initially transported to the basolateral domain. There they are endocytosed, sorted from authentic basolateral proteins in endosomes, and transported to the apical surface (reviewed by Simons & Wandinger-Ness 1990). The possibility of transcytosis in neurons remains unexplored. Considering the much more extended geometry of neurons compared with epithelial cells, transcytosis seems unlikely to be the predominant mechanism for the polarized delivery of neuronal membrane proteins.

...And Staying There

In epithelial cells, circumferential tight junctions act as fences to block the diffusion of membrane proteins and lipids across the apical-basolateral boundary. Neurons lack tight junctions or any other obvious physical barrier at the junction between the somatodendritic and axonal domains. A characteristic dense membrane undercoating is present at the axon initial segment of most neurons in situ (Peters et al 1991), but this cannot be essential for polarity because it is not present in cultured neurons, which are nonetheless polarized (Bartlett & Banker 1984a). Might there be a functional barrier to diffusion at the axon hillock? Kobayashi et al (1992) recently addressed this question by fusing liposomes carrying fluorescently labeled lipids and a hemagglutinin receptor glycolipid to the surface membrane of cultured hippocampal neurons that had been infected with fowl plague virus to induce expression of hemagglutinin on the axonal surface. Thus the labeled lipids were introduced exclusively into the axonal membrane. After 30–60 minutes, labeled lipids were not detected in cell bodies. This was taken as evidence for a barrier to diffusion between somata and axons, although direct measurements of the diffusion coefficients of the labeled lipids were not made. However, based on lipid diffusion coefficients measured in other neurons, Futerman et al (1993) have argued that movement of the label into the cell body would have been difficult to detect, even after one hour.

One mechanism known to limit the diffusion of membrane proteins is anchoring to the membrane-associated cytoskeleton. Spectrins and ankyrins link membrane proteins to the cortical actin network (reviewed by Goodman et al 1988, Coleman et al 1989, Baines 1990). Spectrin also interacts with microtubules and intermediate filaments. The clustering and immobilization of voltage-dependent sodium channels at nodes of Ranvier and the initial segment is thought to result from their association with ankyrin and spectrin (Angelides et al 1988, Srinivasan et al 1988). Interactions with spectrin may also contribute to the concentration of glutamate receptors and of N-CAM 180 at postsynaptic sites (Siman et al 1985, Pollerberg et al 1987). It is tempting to suggest that the differential localization of neuronal ankyrin and spectrin isotypes contributes to the generation or maintenance of micro-domains within the neuronal membrane.

In most epithelial cells, spectrin and ankyrin are selectively localized to the basolateral domain. Proteins that interact with them, such as Na^+, K^+, ATPase, are not subject to endocytosis, and hence they are stabilized in the membrane (Nelson & Veshnock 1987). In one MDCK cell isolate, Na^+, K^+-ATPase is delivered at equal rates to both membrane surfaces, but becomes concentrated in the basolateral domain by association with the spectrin/ankyrin network (Hammerton et al 1991). Stabilization by association with domain-specific isoforms of spectrin or ankyrin may also contribute to the polarized distribution of some neuronal membrane proteins.

What's Bred in the Bone

As it applies to the cytoskeleton, neuronal polarity involves two distinct issues. The first concerns the organization of the cytoskeletal polymers, particularly the organization of microtubules; the second concerns molecular differences between the axonal and dendritic cytoskeleton.

Microtubule organization in neurons, and the special cell biological problems it poses, can best be appreciated by comparing neurons with other types of cells (Bershadsky & Vasiliev 1988). In the typical interphase cell, microtubules are anchored at their minus-ends to the microtubule-organizing center (usually associated with the centriole pair), and their plus-ends are directed toward the cell periphery. The centrosomal organizing center is thought to nucleate the assembly of new microtubules (e.g. following mitosis) and to stabilize their minus-ends against dissociation. Microtubules that are not attached to the microtubule organizing center do not ordinarily form, and when they are induced to form by a microtubule stabilizing agent such as taxol, they rapidly disassemble when the drug is removed (De Brabander et al 1982). Plus-ends undergo alternating phases of polymerization and depoly-merization, a process referred to as dynamic instability (Mitchison & Kirschner 1988, Caplow 1992).

Neurons contain a centrosome in their somata, but most microtubules are not attached to it. The total increase in axonal cross-sectional area that occurs with branching would suggest that some microtubules must begin within the axon (assuming no great decrease in microtubule density). In unbranched axonal segments, measurements have shown that microtubules are typically 100–800 µm long (reviewed in Lasek 1988), far too short to originate from the cell body. Nonetheless, axonal microtubules are uniformly oriented with plus-ends distal to the soma. The plus-ends of axonal microtubules exhibit dynamic instability, but minus-ends appear to be stabilized (Okabe & Hirokawa 1988, Baas & Black 1990). Attempts to discover possible minus-end stabilizers in axons, such as known constituents of microtubule organizing centers, have so far been unsuccessful (Baas & Joshi 1992).

How is the microtubule organization in axons established and maintained? One possible explanation is that microtubule nucleation occurs exclusively at an organizing center in the cell body, and that assembled microtubules (or short segments of microtubules) are transported into and along the axon (Baas & Joshi 1992). This would suggest that the uniform polarity of axonal microtubules might be a function of the motors that govern their translocation. This model is consistent with metabolic labeling studies, which demonstrate a coherent proximodistal transport of tubulin (as well as of neurofilament proteins), but attempts to visualize the transport of microtubule polymers have yielded conflicting results (Reinsch et al 1991, Okabe & Hirokawa 1992). An alternative possibility is that tubulin subunits are transported into the growing axon and assembled at its tip (Bamburg et al 1986). If this is the case, it is not clear how new microtubules might be nucleated or what might determine their polarity orientation. Microtubule-associated proteins (MAPs) might play a role in the latter process. The expression of tau in Sf9 cells induces the formation of microtubule bundles that are of uniform polarity (Baas et al 1991).

In contrast to axons, dendrites are short enough so that individual microtubules could extend from soma to tip. If this were the case, it could help to explain the rules for conservation of dendritic diameter at branch points. The only two attempts to estimate the length of dendritic microtubules have given diametrically opposite results (Sasaki et al 1983, Hillman 1988). The presence in dendrites of microtubules of opposite polarity orientation is especially problematic. The explanation could be that dendrites, unlike axons, contain sites for microtubule nucleation. Alternatively, a different population of motors might permit the somatodendritic transport of minus-end–out microtubules.

Among the molecular differences between the axonal and dendritic cytoskeletons, perhaps the best known is the restriction of the high molecular weight form of MAP2 to the somatodendritic domain. Possible mechanisms that might restrict the distribution of such a soluble protein include the local synthesis of MAP2 in dendrites (Garner et al 1988, Bruckenstein et al 1990, Kleiman et al 1990) and the local degradation of any MAP2 that enters in axons (Okabe & Hirokawa 1989).

Differential phosphorylation of cytoskeletal proteins may also contribute to the generation or maintenance of polarity. For example, axonal, but not somatodendritic, neurofilaments are highly phosphorylated (Nixon & Sihag 1991), and some MAPs also exhibit a differential phosphorylation (Sato-Yoshitake et al 1989, Mansfield et al 1991). Phosphorylation regulates the ability of some MAPs to bind to microtubules in vitro and might regulate the spacing between neurofilaments (Carden et al 1987), but how these processes might contribute to the differences in shape between axons and dendrites is not well understood. Identification and localization of the kinases and phosphatases that control the phosphorylation of these cytoskeletal elements could also help to explain how local differences in phosphorylation are generated in the first place.

Getting the Message Through

Polyribosomes are excluded from the axons of mammalian neurons but extend far into the dendrites. Likewise, when cultured hippocampal neurons are labeled with ^{3}H-uridine, newly synthesized RNA is transported into dendrites but is largely excluded from the axon (Davis et al 1987). Most mRNAs are detectable only in neuronal cell bodies and initial segments of proximal dendrites, but a few, including those that encode high molecular weight MAP2 and the alpha subunit of calcium/calmodulin-dependent protein kinase II, extend far into dendrites (Garner et al 1988, Burgin et al 1990). The mechanisms that govern the differential distribution of RNA in neurons are not understood, but have nevertheless been discussed at length in a recent review (Steward & Banker 1992). Studies of other cell types suggest that the differential localization of many mRNAs depends on signals in the 3' untranslated regions (reviewed by Kislauskis & Singer 1992), which are thought to mediate interaction with the cytoskeleton (Hesketh & Pryme 1991, Singer 1992). We expect that the identification of dendritic RNA targeting signals by expression of chimeric mRNAs will enable researchers to identify putative RNA binding proteins and other components of the RNA targeting machinery.

THE DEVELOPMENT AND PLASTICITY OF NEURONAL POLARITY

It should be clear that researchers generally agree about the cellular mechanisms that must form the basis for neuronal polarity, even though we know very little about how things really work. As for the development of polarity, a cursory survey of the literature suggests that even the relevant questions are in doubt. Different workers use different experimental systems and focus on different aspects of the problem. Some concentrate on the intracellular mechanisms that underlie the emergence of polarity, others on extracellular signals that influence the development of polarity. Of necessity, nearly all work relies on culture models, and it is difficult to know how closely the details of development in culture parallel those in situ. But having come this far in our review, we will not be deterred by such uncertainties.

Because polarity is so fundamental to the organization and function of neurons, one might expect that, once established, polarity would be rigidly maintained. In fact, it displays a surprising plasticity. For example, cutting a cell's axon can lead to a profound respecification of the identity of its processes, and the response to axotomy observed in culture bears a surprising (and satisfying) similarity to changes in neuronal organization that occur following axotomy in situ. Even in mature neurons, polarity is not to be taken for granted.

Being Five Ages

The developmental events that lead to the establishment of neuronal polarity have been most thoroughly characterized in cultures of rat hippocampal neurons. Dotti et al (1988) have identified five developmental stages, based on observations of individual cells over a seven-day period combined with immunostaining for appropriate marker proteins (Figure 3). Stage 1, which occurs immediately after attachment to the substrate, is characterized by the formation of lamellipodia around the circumference of the cell. Stage 2 is marked by the growth of four to five short neurites, typically 15–25 μm in length. We refer to these as minor processes, because at this point their features are not obviously axonal or dendritic. Some 12–24 hours later, one of the minor processes begins to extend rapidly until it becomes many times longer than the remaining neurites. This process is the cell's axon, and its selective elongation is the first morphological evidence of neuronal polarity. Stage 4, which begins after three to four days in culture, is characterized by the elongation of the remaining minor processes and their acquisition of the morphological features that mark them as dendrites. Stage 5 does not involve a change in polarity, but refers to the continued maturation of both axonal and dendritic arbors, including dendritic branching, synaptogenesis, and the formation of dendritic spines (Bartlett & Banker 1984b, Banker & Waxman 1988, Fletcher et al 1991, Craig et al 1993). These latter events, unlike those that occur in stages 1 to 4, appear to be strongly influenced by cell-cell interactions. Time-lapse video microscopy of hypothalamic neurons in culture indicates they undergo much the same sequence of developmental events (Diaz et al 1992). The morphology of neurons in cultures of the cerebellum suggests that they may also undergo a similar sequence of morphological changes, although their development has not been followed at the single-cell level (Ferreira & Caceres 1989, Caceres et al 1991).

How are these morphological events related to the development of molecular differences between axons and dendrites? First, none of the markers thus far examined exhibit a polarized distribution in hippocampal neurons at developmental stage 2. This finding forms the starting point for most models of the initial development of polarity (see below). By stage 3, several axonal markers (GAP-43, synaptic vesicle antigens, and L1) are preferentially concentrated in the axon, although a low level of expression persists in other processes (Goslin et al 1990, Fletcher et al 1991, Jareb et al 1993). These proteins apparently become preferentially distributed as soon as the presumptive axon can be identified by its greater length. In contrast, several dendritic proteins (including MAP2 and several GABA$_A$ and glutamate receptor subunits) appear to be unpolarized at this stage (Caceres et al 1986, Killisch et al 1991, Craig et al 1993). They become restricted to dendrites by the end of stage 4, when residual dendritic expression of axonal proteins has also declined. Thus the polarized distribution of membrane proteins appears to arise in two distinct phases that parallel the morphological stages of develop-

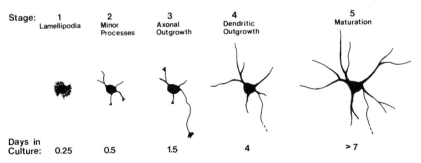

Stage:	1	2	3	4	5
	Lamellipodia	Minor Processes	Axonal Outgrowth	Dendritic Outgrowth	Maturation

| Days in Culture: | 0.25 | 0.5 | 1.5 | 4 | > 7 |

Figure 3   Morphological stages in the development of hippocampal neurons in culture and the approximate times when cells enter each of the stages. (From Dotti et al 1988, with permission).

ment. If these results can be generalized, they would suggest that axonal proteins become selectively distributed before those of the somatodendritic domain. Polyribosomes are an interesting exception. They are preferentially excluded from axons early in stage 3 (Deitch & Banker 1993).

Why is it that some membrane proteins acquire a polarized distribution before others? One possible explanation is that neurons become capable of sorting membrane proteins into different vesicle populations early in development, perhaps even before they are placed into culture, but the mechanisms for the selective delivery of vesicles to axons or dendrites develop later, and at different times. Selective transport to dendrites might require the presence of minus-end distal microtubules, which arise during stage 4 of development, well after the outgrowth of the axon.

An even more basic question is why the axon grows out and begins to differentiate before dendrites emerge. This appears to be true not only in culture but also in situ. The invariance of this sequence suggests that some aspect of axonal outgrowth is necessary to allow dendrites to mature. Using tau antisense oligonucleotides to selectively inhibit axonal growth in cerebellar cultures, Caceres et al (1991) found that retraction of an existing axon allowed continued growth of other neurites "on schedule," whereas a delay in initial axonal outgrowth resulted in a delay in the growth of other neurites as well. These authors suggested that initial axon outgrowth is a prerequisite for dendritic development, regardless of the subsequent fate of the elongated axon.

In the Beginning

In cultured neurons, the first step in the development of polarity is marked by the emergence of the axon. In the case of hippocampal neurons, we have so far been unable to identify any features that indicate which of the several minor processes present at stage 2 of development will become the axon. We have found no significant difference among minor processes in light microscopic morphology or behavior, in ultrastructural characteristics, in microtubule density or polarity orientation, or in their content of markers that distinguish axons and dendrites in mature neurons (Dotti et al 1988, Baas et al 1989, Goslin & Banker 1990, Deitch & Banker 1993). How is it that one of several, apparently identical processes acquires axonal properties? What determines which minor process becomes the axon?

Although the site of emergence of the axon might be governed by intrinsic determinants, two lines of evidence argue against this. First, when the axons of stage 3 cells are transected near the cell body, there is no tendency for the new axons to reform in their original locations (Dotti & Banker 1987). Second, the location of the two obvious candidate organizers, the Golgi complex and the microtubule organizing center, is not correlated with the site where the axon emerges (Dotti & Banker 1991). Despite these negative results, a more sensitive assay, such as a comparison of the morphologies of daughters of a single dividing cell (Solomon 1979), might reveal the existence of intrinsic determinants that bias the site of origin of the axon. Mattson et al (1989) report a tendency for the daughter neurons of hippocampal cells that divide in culture to develop a similar number of minor processes, but they do not comment on possible symmetries in the site of origin of the axon.

Observation of cellular behavior during the transition from stage 2 to 3 has led us to suggest that in culture the site from which the axon emerges is determined by chance. During stage 2 several minor processes often undergo alternating periods of elongation and retraction, a competition that appears to end only when one grows significantly longer than the others and becomes the axon (Goslin & Banker 1990). Analysis of the response of neurons to axotomy at different sites along the axon also indicates that the longest remaining process has the best chance to become the new axon (Goslin & Banker 1989). Thus, we have proposed that small differences in length that occur by chance may be enhanced through a positive feedback mechanism until some threshold is passed and one process becomes the axon. In addition, negative feedback would probably be necessary to prevent additional minor processes from also becoming axons.

Any of a number of mechanisms could underlie a positive feedback between length and growth rate. Positive feedback could occur if the accumulation of some protein in a process acted to stimulate elongation, and if its accumulation increased in proportion to process length. Another possibility is that local activation of a kinase or phosphatase could lead to a change in phosphorylation of some protein whose altered activity results in process outgrowth. A localized cascade involving changes in phosphorylation could be perpetuated by activation of a self-regulatory kinase or phosphatase. The putative growth regulatory protein could be a cell adhesion molecule, a cytoskeletal protein, a channel protein that causes a change in growth cone calcium concentration, a receptor that governs the concentration of some second messenger—any of a wide variety of molecules could have such an effect. And therein lies the principal difficulty with this model. Although it is possible to assess changes in the accumulation and in some cases the phosphorylation state of specific molecules in one or another minor process (e.g. Goslin & Banker 1990), one might expect that selective delivery and posttranslational modification of many axonal proteins would correlate with the emergence of the axon. Indeed, as discussed above, this seems to be the case. To test the possible involvement of each in the specification of the axon is a daunting task indeed.

An alternative approach might begin with an analysis of the mechanisms that are necessary for the rapid and selective growth of a single process. These mechanisms must involve the selective elongation of microtubules in a single process as well as the selective delivery of the cytoskeletal and membrane components needed to support its growth. Because MAPs are known to regulate the stability of microtubules, Caceres & Kosik and their collaborators have investigated the role of MAPs in the development of polarity in cultures of cerebellar macroneurons (Caceres & Kosik 1990; Caceres et al 1991, 1992). They have found that inhibition of tau expression by the addition of antisense oligonucleotides at the time of plating dramatically inhibited axonal outgrowth during the first 24 hours (Figure 4). There was no effect on the number or length of minor neurites, indicating a selective inhibition of axonal outgrowth. When addition of tau antisense oligonucleotides was delayed until cells had already polarized, the axons that had formed retracted, but the remaining neurites continued to grow. Inhibition of MAP2 expression beginning at the time of plating blocks the outgrowth of all processes.

Kosik & Caceres (1991) propose that tau selectively binds to and stabilizes microtubules in the emergent axon, thus permitting its elongation. Although tau is present in all of the neurites of cerebellar macroneurons, a local posttranslational modification, such as dephosphorylation, might alter its affinity for microtubules. Consistent with this possibility, tau in the axon is preferentially associated with the cytoskeleton (Ferreira et al 1989), and axons, but not minor neurites, contain posttranslationally modified forms of tubulin indicative of stable microtubules (Ferreira & Caceres 1989, Arregui et al 1991). As Kosik & Caceres point out (1991), these observations could be incorporated into a stochastic model of axonal determination if the distribution or activity of a phosphatase that dephosphorylates tau (or of some protein that regulates its activity) was governed by chance. When the level of dephosphorylated tau exceeded some threshold value in one neurite, its microtubules would be selectively stabilized and it would become the axon.

Although stable microtubules are probably necessary for axonal growth, regional differences in microtubule stability may play a less important role in governing the selective outgrowth of the axon in other cell types. In the case of neocortical and hippocampal neurons, stable microtubules are present in all processes, not just in the axon (Dotti & Banker 1991, Mansfield & Gordon-Weeks 1991).

The relationship between axonal outgrowth and the transport of the membrane constituents needed to support it has so far received little attention. Treatment of developing hippocampal neurons with brefeldin A, a drug that disrupts the Golgi complex and interferes with the delivery of Golgi-derived exocytic vesicles, inhibits the growth of axons and causes their selective retraction (Jareb & Banker 1991). Inhibition of glycosphingolipid synthesis also inhibits axonal growth (Harel & Futerman 1993). On the other hand, suppression of the expression of kinesin, one of the motors thought to mediate anterograde vesicle transport, does not prevent polarization (Ferreira et al

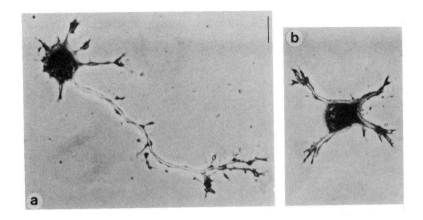

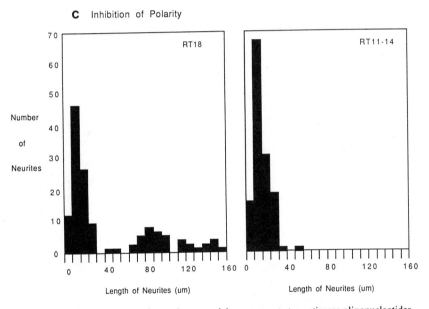

C   Inhibition of Polarity

Figure 4   Selective inhibition of axonal outgrowth by exposure to tau antisense oligonucleotides. Cerebellar neurons were cultured for one day in the presence of (*a*) control sense oligonucleotide, or (*b*) a tau antisense oligonucleotide. Tau antisense oligonucleotides specifically inhibited outgrowth of axons but not of other neurites, as shown also in the bar chart (*c*). (RT18, control sense oligonucleotide; RT11-14, tau antisense oligonucleotide.) (From Caceres & Kosik 1990, with permission).

1992). Instead, the rate of elongation of both axons and minor processes is slowed to an equal extent, as might be predicted from the presence of plus-end out microtubules in both types of processes.

The Outside World

Evidently, in situ the development of neuronal polarity must be regulated by extrinsic influences, and several groups have used cell culture to examine these extrinsic factors. The role of ECM proteins has been most thoroughly examined. In a series of studies, Higgins and his collaborators have investigated the influences of ECM proteins on the development of neurons from rat superior cervical ganglia (Bruckenstein & Higgins 1988a,b; Tropea et al 1988; Johnson et al 1989; Lein & Higgins 1989; Lein et al 1991; Osterhout et al 1992). When these neurons are cultured on a polylysine substrate in the absence of glia or serum, they develop an axon but not dendrites (Figure 5a). Growth on a laminin substrate increases the number of axons per cell and the rate of axonal growth (Figure 5b). Thrombospondin and collagen IV also stimulate the rate of axon growth but do not affect the number of axons that each cell forms. Matrigel, a basement membrane extract that contains a complex mixture of ECM proteins, induces the development of dendrites

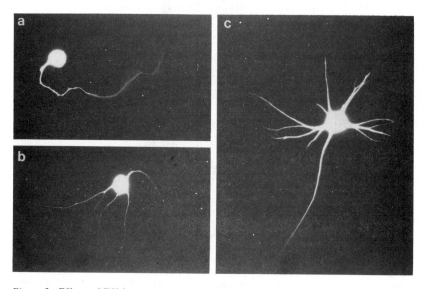

Figure 5 Effects of ECM molecules on the morphology of cultured rat sympathetic neurons. (a) Neurons grown on polylysine alone usually had only one long axon. (b) Growth on laminin typically induced supernumerary axons. (c) Exposure to a basement membrane extract resulted in outgrowth of a single axon and multiple dendrites. Cell morphology was visualized by injection with Lucifer yellow. (From Lein & Higgins 1989, with permission).

(Figure 5c). Culture at high density or in the presence of Schwann cells also promotes dendritic development.

ECM molecules also influence the number and character of the processes extended by embryonic rat mesencephalic and striatal neurons (Denis-Donini et al 1984; Chamak et al 1987; Autillo-Touati et al 1988; Rousselet et al 1988, 1990; Chamak & Prochiantz 1989; Lafont et al 1992). Growth of dendrite-like neurites is enhanced by dermatan sulfate, as well as by culture on a monolayer of homotypic astrocytes and by the addition of a soluble fraction from astrocyte-conditioned media. Growth of axon-like neurites is promoted by proteoglycans, heparan sulfate, chondroitin sulfate, soluble laminin, bound fibronectin, and an insoluble fraction from astrocyte-conditioned media. Based on a further analysis of their results, Prochiantz et al (1990) have proposed that dendritic growth is dependent on tight adhesion to the substrate, but axonal growth is not.

The effects of ECM molecules are cell-type specific. Whereas laminin increases the number of axons extended by sympathetic neurons and decreases the number of dendrite-like processes extended by mesencephalic neurons, it has no effect on the number of axons or dendrites extended by hippocampal neurons (Chamak & Prochiantz 1989, Lein & Higgins 1989, Lein et al 1992, Osterhout et al 1992). This is not because hippocampal neurons cannot respond to laminin. As with other cell types, laminin selectively increases the rate of axonal growth by hippocampal neurons. Chamak & Prochiantz (1989) also note that conditions that stimulate axonal growth in mesencephalic neurons tend to inhibit their dendritic growth. In contrast, laminin-induced stimulation of axonal growth in hippocampal neurons neither enhances nor retards dendritic growth (Lein et al 1992).

The effects of ECM molecules also depend on the cell's maturity. Whereas under some conditions superior cervical ganglion neurons from prenatal rats extend only axons, postnatal neurons extend both axons and dendrites under identical conditions (Bruckenstein et al 1989). If neurons from prenatal animals are cultured in the presence of nonneuronal cells for three weeks (a condition that permits dendritic outgrowth), they gain the ability to extend both axons and dendrites when dissociated and replated in the absence of glia. Conversely, sensory neurons from the neonatal nodose ganglion, which do not form dendrites in situ, can be induced to form dendrites when cultured in the presence of NGF and the absence of satellite cells (DeKoninck et al 1993). Under the same conditions, nodose neurons from 14-day-old rats form only axons.

These results could be explained if ECM molecules influence neuronal gene transcription. It may be that extracellular factors regulate the expression of specific genes that control the ability of neurons to form axons and dendrites. It is also possible these changes reflect an influence of ECM molecules on the specification of cell identity: ECM molecules might convert cells from a type that does not ordinarily make dendrites to a type that does, or vice versa. A variety of experiments have demonstrated that other environmental signals play a key role in determining the phenotype of neural crest derivatives (Landis 1990, Nawa et al 1990). Many of the effects of ECM molecules, particularly their enhancement of the rate of neurite elongation, are unlikely to involve gene regulation because they occur rapidly and are not blocked by inhibitors of protein synthesis (Lein & Higgins 1991, Rivas et al 1992). As Lein et al (1991) propose, "a testable hypothesis is that receptors for ECM molecules with axon-specific effects are expressed on axons but not dendrites." Work to identify the relevant receptors has just begun (Lein et al 1991, Tomaselli 1991), and the distribution of these receptors is not yet known. Evidence from in situ studies does suggest that at least one integrin subunit, $\alpha 8$, is concentrated in axons (Bossy et al 1991).

In addition to ECM molecules, many other factors selectively influence axonal or dendritic outgrowth in culture. For example, glia-derived nexin, a serine protease inhibitor, selectively enhances axonal growth by hippocampal neurons (Farmer et al 1990), whereas glutamate selectively suppresses their dendritic outgrowth (Mattson 1988, Mattson et al 1988, Mattson & Kater 1989). These effects may be important for selectively shaping axonal or dendritic growth, but it seems unlikely that they contribute to the initial specification of polarity.

The Real World

What bearing, if any, do these studies of cultured cells have on the question of how neurons develop in situ? In particular, do real neurons develop the equivalent of minor processes that compete to become the axon? Based on the available data, these questions are difficult to answer. In the embryonic hippocampus and neocortex, cells have been described that bear a surprising resemblance to cultured neurons at stage 2 of development (Shoukimas & Hinds 1978, Nowakowski & Rakic 1979), but there is no way of knowing if the "minor processes" of these cells actually give rise to definitive axons and dendrites, as the equivalent processes do in culture. Even the identity of these cells, which were originally described as recently generated neurons in the process of migrating to the cortex, is now in doubt (McConnell 1988). New video techniques for following the development of individual cells in slices or explants may finally resolve these questions (O'Rourke et al 1992). In a preliminary report, Brittis & Silver (1992) have used such an approach and found that retinal ganglion cells initially extend and retract several "exploratory" processes before one becomes the definitive axon.

There are also many examples of cells that do not follow this pattern of development. Zebrafish motoneurons, whose development can be followed in living embryos after dye injection, initially put out only a single process, the axon (Myers et al 1986). Dendrites develop at a later stage (Westerfield et al 1986). A similar pattern of development is followed by GABAergic and glycinergic neurons in *Xenopus* spinal cord (Roberts et al 1987, 1988). On the other hand, all of these cells are capable of forming several neurites but develop only a single axon. It seems reasonable to assume that the cellular mechanisms that ensure formation of a single axon are similar, even if they do not always manifest themselves in the same sequence of developmental events. Whatever the mechanisms that underlie the development of polarity in situ, it seems evident that they must be regulated by extrinsic factors. For example, in most cases the axon consistently emerges at one pole of the cell body, usually oriented toward its immediate destination. In some instances this regularity in axonal orientation may be an illusion, reflecting the specific details of a cell's developmental history rather than a direct environmental influence on axonal outgrowth. When the axon emerges during cell migration, it ultimately comes to lie at the trailing pole of the cell, regardless of its initial orientation (Puelles & Privat 1977, Shoukimas & Hinds 1978). But in most cases axonal orientation is likely to be directly determined by extrinsic factors.

How might the environment influence the development of neuronal polarity? In culture, we assume that the molecular asymmetries that underlie axonal outgrowth arise independent of the environment, but these same molecular mechanisms could permit oriented axonal outgrowth in response to asymmetries in the environment. For example, if in culture cell adhesion molecules or matrix receptors become concentrated on a single process, this could initiate a series of signalling events that would permit that process to grow more rapidly than the rest and become the axon. On the other hand, if these same adhesion molecules or receptors were homogeneously distributed on the cell surface but exposed to a gradient of their appropriate ligands, this could initiate an identical series of signalling events. In fact, a gradient of proteoglycans is believed to govern the orientation of axon outgrowth by retinal ganglion cells, although in this case the matrix molecules are inhibitory rather than stimulatory (Brittis & Silver 1992, Brittis et al 1992). It seems likely that environmental signals might also accelerate the process by which the axon is specified. In culture, in the absence of spatially organized environmental cues, this phase of neuronal development might be accentuated and prolonged.

As the World Turns

Once established, neuronal polarity is not immutable. In hippocampal cultures, axonal transection at stage 3 of development can lead to a respecification of polarity (Dotti & Banker 1987). Often a new axon arises from one of the minor processes, which would otherwise have become a dendrite, and the stump of the transected axon becomes a dendrite. The response to axotomy depends on the site of transection (Goslin & Banker 1989). If the axon is transected at some distance from the cell body, so that it remains significantly longer than the minor processes, then the axon simply regenerates and begins to regrow almost immediately. If the transection is made near the cell body, so that the axonal stump is about the same length as the minor processes, the cell goes through a period when its minor processes alternately extend and retract, just as if the cell had been returned to a previous stage of development. Ultimately, and apparently by chance, one of the minor processes becomes the new axon. Preliminary observations suggest that when axotomy occurs at stage 4 of development, multiple axons may arise (K. Goslin & G. Banker, unpublished observation).

Surprisingly, similar disruptions of polarity also occur following axotomy of vertebrate neurons in situ, even in the case of neurons that are fully mature. It has long been known that transecting the axon of a spinal motoneuron in a ventral root or peripheral nerve leads to predictable changes in the cell body, which are usually followed by regeneration of the axon. Only comparatively recently has it been appreciated that a very different result occurs when motoneuron axons are transected within the spinal cord, near the cell body (Lindå et al 1985, 1992; Havton & Kellerth 1987). In some cases a new axon forms from the axonal stump, but often up to three new sprouts arise from the distal dendrites. These sprouts resemble axons in morphology, and they form presynaptic specializations on nearby cells (Havton & Kellerth 1987). The formation of multiple axonal sprouts from dendrites has also been observed after close axotomy of rat retinal ganglion cells (Cho & So 1989) and of lamprey reticulospinal neurons (Hall & Cohen 1988; Hall et al 1989, 1991). Axotomy also causes abnormalities in the remaining dendrites. They often partially retract and sometimes become ensheathed by myelin (Sumner & Watson 1971, Lindå et al 1992). Thus close axotomy commonly leads to profound changes in neuronal polarity. Deafferentation can have similar consequences. Membranes that normally are exclusively postsynaptic can form presynaptic specializations (Hamori 1990), and the expression and localization of MAPs can be disrupted (Caceres & Dotti 1985, Caceres et al 1988, Shaw et al 1988). These results suggest that feedback mechanisms such as those thought to be involved in specification of the axon during development may continue to operate throughout the life of the cell.

Alterations in neuronal polarity may also contribute to the pathology of some neurological diseases. Abnormally phosphorylated tau isoforms that fail to bind microtubules are at the core of paired helical filaments that comprise the neurofibrillary tangles characteristic of Alzheimer's disease (Kosik 1990, Lee et al 1991, Goedert et al 1992). Abnormal phosphorylation appears to be an early step that leads to accumulation of tau in the cell body prior to incorporation into paired helical filaments (Bancher et al 1991). In addition, sprouts that contain both MAP2 and fibrillar tau have been observed to arise from proximal dendrites (McKee et al 1989). Thus, reorganization of the microtubule system appears to be a key pathological feature of Alzheimer's disease. A recent report describes deficits in MAP2 and MAP5 in dendrites of the subiculum and entorhinal cortex in schizophrenia, suggesting that distortions in the polarized geometry of neurons may also occur in this disease (Arnold et al 1991). It would not be surprising if defects in neurite specification or protein targeting were associated with other diseases as well.

ACKNOWLEDGMENTS

Our thanks to Bob Burke for his critical comments concerning measurements of neuronal shape.

Our work and ideas are a product of the many people who have worked with us on this project, lab members past and present, and many generous collaborators. We are indebted as well to many whom we do not know—to the reviewers of our manuscripts and grants and to those anonymous members of the audience who have raised such thought-provoking questions when this work has been presented. We thank you all.

Our research is supported by NIH grants NS17112 and NS23094, and by NRSA NS09248.

Literature Cited

Adler R. 1986. Developmental predetermination of the structural and molecular polarization of photoreceptor cells. *Dev. Biol.* 117: 520–27

Anderson PAV. 1985. Physiology of a bidirectional, excitatory, chemical synapse. *J. Neurophysiol.* 53:821–35

Angelides KJ, Elmer LW, Loftus D, Elson E. 1988. Distribution and lateral mobility of voltage-dependent sodium channels in neurons. *J. Cell Biol.* 106:1911–25

Arnold SE, Lee VMY, Gur RE, Trojanowski JQ. 1991. Abnormal expresson of two microtubule-associated proteins (MAP2 and MAP5) in specific subfields of the hippocampal formation in schizophrenia. *Proc. Natl. Acad. Sci. USA* 88:10850–54

Arregui C, Busciglio J, Caceres A, Barra HS. 1991. Tyrosinated and detyrosinated microtubules in axonal processes of cerebellar macroneurons grown in culture. *J. Neurosci. Res.* 28:171–81

Autillo-Touati A, Chamak B, Araud D, et al. 1988. Region-specific neuro-astroglial interactions: ultrastructural study of the in vitro expression of neuronal polarity. *J. Neurosci. Res.* 19:326–42

Baas PW, Black MM. 1990. Individual microtubules in the axon consist of domains that differ in both composition and stability. *J. Cell Biol.* 111:495–509

Baas PW, Black MM, Banker GA. 1989. Changes in microtubule polarity orientation during the development of hippocampal neurons in culture. *J. Cell Biol.* 109:3085–94

Baas PW, Deitch JS, Black MM. 1988. Polarity orientation of microtubules in hippocampal neurons: uniformity in the axon and nonuniformity in the dendrite. *Proc. Natl. Acad. Sci. USA* 85:8335–39

Baas PW, Joshi HC. 1992. Gamma-tubulin distribution in the neuron: implications for the origins of neuritic microtubules. *J. Cell Biol.* 119:171–78

Baas PW, Pienkowski TP, Kosik KS. 1991. Processes induced by tau expression in Sf9 cells have an axon-like microtubule organization. *J. Cell Biol.* 115:1333–44

Baines AJ. 1990. Ankryin and the node of Ranvier. *Trends Neurosci.* 13:119–21

Bamburg JR, Bray D, Chapman K. 1986. Assembly of microtubules at the tip of growing axons. *Nature* 321:788–90

Bancher C, Grundke-Iqbal I, Iqbal K, et al.

1991. Abnormal phosphorylation of tau precedes ubiquitination in neurofibrillary pathology of Alzheimer disease. *Brain Res.* 539:11–18

Banker GA, Waxman AB. 1988. Hippocampal neurons generate natural shapes in cell culture. In *Intrinsic Determinants of Neuronal Form and Function*, ed. R. J. Lasek MM Black, pp. 61–82. New York: Liss

Bartlett WP, Banker GA. 1984a. An electron microscopic study of the development of axons and dendrites by hippocampal neurons in culture. I. Cells which develop without intercellular contacts. *J. Neurosci.* 4:1944–53

Bartlett WP, Banker GA. 1984b. An electron microscopic study of the development of axons and dendrites by hippocampal neurons in culture. II. Synaptic relationships. *J. Neurosci.* 4:1954–65

Bennett V, Otto E, Kunimoto M, et al. 1991. Diversity of ankyrins in the brain. *Biochem. Soc. Trans.* 19:1034–39

Benson DL, Watkins FH, Steward O, Banker G. 1992. Characterization of GABA-ergic neurons in hippocampal cultures. *Soc. Neurosci. Abstr.* 18:34 (Abstr.)

Bernhardt R, Huber G, Matus A. 1985. Differences in the developmental patterns of three microtubule-associated proteins in the rat cerebellum. *J. Neurosci.* 5:977–91

Bernhardt R, Matus A. 1984. Light and electron microscopic studies of the distribution of microtubule-associated protein 2 in rat brain: a difference between dendritic and axonal cytoskeletons. *J. Comp. Neurol.* 226:203–21

Bershadsky AD, Vasiliev JM. 1988. *Cytoskeleton*, pp. 1–298. New York: Plenum

Binder LI, Frankfurter A, Rebhun LI. 1985. The distribution of tau in the mammalian central nervous system. *J. Cell Biol.* 101:1371–78

Birren SJ, Anderson DJ. 1990. A v-myc-immortalized sympathoadrenal progenitor cell line in which neuronal differentiation is initiated by FGF but not NGF. *Neuron* 4:189–201

Bloomfield SA, Miller RF. 1986. A functional organization of ON and OFF pathways in the rabbit retina. *J. Neurosci.* 6:1–13

Bossy B, Bossy-Wetzel E, Reichardt LF. 1991. Characterization of the integrin alpha8 subunit: a new integrin beta 1-associated subunit, which is prominently expressed on axons and on cells in contact with basal laminae in chick embryos. *EMBO J.* 10:2375–85

Brewer CB, Roth MG. 1991. A single amino acid change in the cytoplasmic domain alters the polarized delivery of influenza virus hemagglutinin. *J. Cell Biol.* 114:413–21

Brittis PA, Canning DR, Silver J. 1992. Chondroitin sulfate as a regulator of neuronal patterning in the retina. *Science* 255:733–36

Brittis PA, Silver J. 1992. Glycosaminoglycans as mediators of retinal ganglion cell body and axon polarity. *Soc. Neurosci. Abstr.* 18:1461 (Abstr.)

Brown DA, Crise B, Rose JK. 1989. Mechanism of membrane anchoring affects polarized expression of two proteins in MDCK cells. *Science* 245:1499–501

Brown DA, Rose JK. 1992. Sorting of GPI-anchored proteins to glycolipid-enriched membrane subdomains during transport to the apical cell surface. *Cell* 68:533–44

Bruckenstein D, Johnson MI, Higgins D. 1989. Age-dependent changes in the capacity of rat sympathetic neurons to form dendrites in tissue culture. *Dev. Brain Res.* 46:21–32

Bruckenstein DA, Higgins D. 1988a. Morphological differentiation of embryonic rat sympathetic neurons in tissue culture. I. Conditions under which neurons form axons but not dendrites. *Dev. Biol.* 128:324–36

Bruckenstein DA, Higgins D. 1988b. Morphological differentiation of embryonic rat sympathetic neurons in tissue culture. II. Serum promotes dendritic growth. *Dev. Biol.* 128:337–48

Bruckenstein DA, Lein PJ, Higgins D, Fremeau RT Jr. 1990. Distinct spatial localization of specific mRNAs in cultured sympathetic neurons. *Neuron* 5:809–19

Bullock TH, Horridge GA. 1965. *Structure and Function in the Nervous Systems of Invertebrates*, pp. 1–798. San Francisco & London: Freeman

Burgin KE, Waxham MN, Rickling S, et al. 1990. In situ hybridization histochemistry of Ca2+/calmodulin-dependent protein kinase in developing rat brain. *J. Neurosci.* 10:1788–98

Burton PR. 1988. Dendrites of mitral cell neurons contain microtubules of opposite polarity. *Brain Res.* 473:107–15

Caceres A, Banker G, Steward O, et al. 1984. MAP2 is localized to the dendrites of hippocampal neurons which develop in culture. *Dev. Brain Res.* 13:314–18

Caceres A, Banker GA, Binder L. 1986. Immunocytochemical localization of tubulin and microtubule-associated protein 2 during the development of hippocampal neurons in culture. *J. Neurosci.* 6:714–22

Caceres A, Busciglio J, Ferreira A, Steward O. 1988. An immunocytochemical and biochemical study of the microtubule-associated protein MAP-2 during post-lesion dendritic remodeling in the central nervous system of adult rats. *Brain Res.* 427:233–46

Caceres A, Dotti C. 1985. Immunocytochemical localization of tubulin and the high molecular weight microtubule-associated protein 2 in Purkinje cell dendrites deprived of climbing fibers. *Neuroscience* 16:133–50

Caceres A, Kosik KS. 1990. Inhibition of neurite polarity by tau antisense oligonucleotides in primary cerebellar neurons. *Nature* 343:461–63

Caceres A, Mautino J, Kosik KS. 1992. Suppression of MAP2 in cultured cerebeller macroneurons inhibits minor neurite formation. *Neuron* 9:607–18

Caceres A, Potrebic S, Kosik KS. 1991. The effect of tau antisense oligonucleotides on neurite formation of cultured cerebellar macroneurons. *J. Neurosci.* 11:1515–23

Cajal SR. 1989. *Recollections of My Life,* pp. 1–638. Cambridge: MIT

Cameron PL, Sudhof TC, Jahn R, De Camilli P. 1991. Colocalization of synaptophysin with transferrin receptors: implicatons for synaptic vesicle biogenesis. *J. Cell Biol.* 115:151–64

Caplan MJ, Anderson HC, Palade GE, Jamieson JD. 1986. Intracellular sorting and polarized cell surface delivery of (Na+, K+)ATPase an endogenous component of MDCK cell basolateral plasma membranes. *Cell* 46:623–31

Caplow M. 1992. Microtubule dynamics. *Curr. Opin. Cell Biol.* 4:58–65

Carden MJ, Trojanowski JQ, Schlaepfer WW, Lee VMY. 1987. Monoclonal antibodies distinguish several differentially phosphorylated states of the two largest rat neurofilament subunits (NF-H and NF-M) and demonstrate their existence in the normal nervous system of adult rats. *J. Neurosci.* 7:3489–504

Casanova JE, Apodaca G, Mostov KE. 1991. An autonomous signal for basolateral sorting in the cytoplasmic domain of the polymeric immunoglobulin receptor. *Cell* 66:65–75

Chamak B, Fellous A, Glowinski J, Prochiantz A. 1987. MAP2 expression and neuritic outgrowth and branching are coregulated through region-specific neuro-astroglial interactions. *J. Neurosci.* 7:3163–70

Chamak B, Prochiantz A. 1989. Influence of extracellular matrix proteins on the expression of neuronal polarity. *Development* 106: 483–91

Chavrier P, Vingron M, Sander C, et al. 1990. Molecular cloning of YPT1/SEC4-related cDNAs from an epithelial cell line. *Mol. Cell. Biol.* 10:6578–85

Cho EY, So KF. 1989. De novo formation of axon-like processes from axotomized retinal ganglion cells which exhibit long distance growth in a peripheral nerve graft in adult hamsters. *Brain Res.* 484:371–77

Claiborne BJ, Amaral DG, Cowan WM. 1986. A light and electron microscopic analysis of the mossy fibers of the rat dentate gyrus. *J. Comp. Neurol.* 246:435–58

Cohen MJ. 1970. A comparison of invertebrate and vertebrate central neurons. In *The Neurosciences: Second Study Program,* ed. F. O. Schmitt GC Quarton T Melnechuk G Adelman, pp. 798–812. New York: Rockefeller

Coleman TR, Fishkind DJ, Mooseker MS, Morrow JS. 1989. Functional diversity among spectrin isoforms. *Cell Motil. Cytoskelet.* 12:225–47

Craig AM, Blackstone CD, Huganir RL, Banker G. 1993. The distribution of glutamate receptors in cultured rat hippocampal neurons: postsynaptic clustering of AMPA-selective subunits. *Neuron* 10:1055–68

Croul S, Mezitis SG, Stieber A, et al. 1990. Immunocytochemical visualization of the Golgi apparatus in several species, including human, and tissues with an antiserum against MG-160, a sialoglycoprotein of rat Golgi apparatus. *J. Histochem. Cytochem.* 7:957–63

Cullheim S, Fleshman JW, Glenn LL, Burke RE. 1987. Membrane area and dendritic structure in type-identified triceps surae alpha motoneurons. *J. Comp. Neurol.* 255: 68–81

Cullheim S, Kellerth JO. 1978. A morphological study of the axons and recurrent axon collaterals of cat sciatic alpha-motoneurons after intracellular staining with horseradish peroxidase. *J. Comp. Neurol.* 178:537–58

Davis L, Banker GA, Steward O. 1987. Selective dendritic transport of RNA in hippocampal neurons in culture. *Nature* 330:477–79

De Brabander M, Geuens G, Nuydens R, et al. 1982. Microtubule stability and assembly in living cells: the influence of metabolic inhibitors, taxol and pH. *Cold Spring Harbor Symp. Quant. Biol.* 46:227–40

De Camilli P, Cameron R, Greengard P. 1983. Synapsin I (Protein I), a nerve cell-specific phosphoprotein. I. Its general distribution in synapses of the central and peripheral nervous system demonstrated by immunofluorescence in frozen and plastic sections. *J. Cell Biol.* 96:1337–54

De Camilli P, Moretti M, Denis Donini S, et al. 1986. Heterogeneous distribution of the cAMP receptor protein RII in the nervous system: evidence for its intracellular accumulation on microtubules, microtubule-organizing centers, and in the area of the Golgi complex. *J. Cell Biol.* 103:189–203

De Koninck P, Carbonetto S, Cooper E. 1993. NGF induces neonatal rat sensory neurons to extend dendrites in culture after removal of satellite cells. *J. Neurosci.* 13:577–85

Deitch JS, Banker GA. 1993. An electron microscopic analysis of hippocampal neurons developing in culture: early stages in the emergence of polarity. *J. Neurosci.* In press

Denis-Donini S, Glowinski J, Prochiantz A. 1984. Glial heterogeneity may define the three-dimensional shape of mouse mesencephalic dopaminergic neurons. *Nature* 307: 641–43

Desmond NL, Levy WB. 1982. A quantitative anatomical study of the granule cell dendritic fields of the rat dentate gyrus using a novel probabilistic method. *J. Comp. Neurol.* 212: 131–45

Desmond NL, Levy WB. 1984. Dendritic caliber and the 3/2 power relationship of dentate granule cells. *J. Comp. Neurol.* 227:589–96

Diaz H, Lorenzo A, Carrer HF, Caceres A. 1992. Time lapse study of neurite growth in hypothalamic dissociated neurons in culture: sex differences and estrogen effects. *J. Neurosci. Res.* 33:266–81

Dotti CG, Banker G. 1991. Intracellular organization of hippocampal neurons during the development of neuronal polarity. *J. Cell Sci. Suppl.* 15:75–84

Dotti CG, Banker GA. 1987. Experimentally induced alteration in the polarity of developing neurons. *Nature* 330:254–56

Dotti CG, Banker GA, Binder LI. 1987. The expression and distribution of the microtubule-associated proteins tau and microtubule-associated protein 2 in hippocampal neurons in situ and in cell culture. *Neuroscience* 23:121–30

Dotti CG, Kartenbeck J, Simons K. 1993. Polarized distribution of the viral glycoproteins of vesicular stomatitis, fowl plague, and Semliki forest viruses in hippocampal neurons in culture: a light and electron microscopy study. *Brain Res.* 610:141–47

Dotti CG, Parton RG, Simons K. 1991. Polarized sorting of glypiated proteins in hippocampal neurons. *Nature* 349:158–61

Dotti CG, Simons K. 1990. Polarized sorting of viral glycoproteins to the axon and dendrites of hippocampal neurons in culture. *Cell* 62:63–72

Dotti CG, Sullivan CA, Banker GA. 1988. The establishment of polarity by hippocampal neurons in culture. *J. Neurosci.* 8:1454–68

Duden R, Griffiths G, Frank R, et al. 1991. Beta-COP, a 110 kd protein associated with non-clathrin-coated vesicles and the golgi complex, shows homology to beta-adaptin. *Cell* 64:649–65

Faivre-Sarrailh C, Gennarini G, Goridis C, Rougon G. 1992. F3/F11 cell surface molecule expression in the developing mouse cerebellum is polarized at synaptic sites and within granule cells. *J. Neurosci.* 12:257–67

Famiglietti EV. 1992. Polyaxonal amacrine cells of rabbit retina: morphology and stratification of PA1 cells. *J. Comp. Neurol.* 316:391–405

Farmer L, Sommer J, Monard D. 1990. Glia-derived nexin potentiates neurite extension in hippocampal pyramidal cells in vitro. *Dev. Neurosci.* 12:73–80

Ferreira A, Busciglio J, Caceres A. 1989. Microtubule formation and neurite growth in cerebellar macroneurons which develop in vitro: evidence for the involvement of the microtubule-associated proteins MAP-1a HMW-MAP2 and Tau. *Dev. Brain Res.* 49:215–28

Ferreira A, Caceres A. 1989. The expression of acetylated microtubules during axonal and dendritic growth in cerebellar macroneurons which develop in vitro. *Dev. Brain Res.* 49:205–13

Ferreira A, Niclas J, Vale R, et al. 1992. Suppression of kinesin expression in cultured hippocampal neurons using antisense oligonucleotides. *J. Cell Biol.* 117:595–606

Fischer von Mollard G, Mignery GA, Baumert M, et al. 1990. Rab3 is a small GTP-binding protein exclusively localized to synaptic vesicles. *Proc. Natl. Acad. Sci. USA* 87:1988–92

Fischer von Mollard G, Sudhof TC, Jahn R. 1991. A small GTP-binding protein dissociates from synaptic vesicles during exocytosis. *Nature* 349:79–81

Fisher S, Boycott B. 1974. Synaptic connections made by horizontal cells within the outer plexiform layer of the retina of the cat and the rabbit. *Proc. Natl. Acad. Sci. USA* 186:317–31

Fletcher TL, Cameron P, De Camilli P, Banker GA. 1991. The distribution of synapsin I and synaptophysin in hippocampal neurons developing in culture. *J. Neurosci.* 11:1617–26

Fuller SD, Simons K. 1986. Transferrin receptor polarity and recycling accuracy in "tight" and "leaky" strains of Madin-Darby canine kidney cells. *J. Cell Biol.* 103:1767–79

Furley AJ, Morton SB, Manalo D, et al. 1990. The axonal glycoprotein TAG-1 is an immunoglobulin superfamily member with neurite-promoting activity. *Cell* 61:157–70

Furshpan EJ, Landis SC, Matsumoto SG, Potter DD. 1986. Synaptic functions in rat sympathetic neurons in microcultures. I. Secretion of norepinephrine and acetylcholine. *J. Neurosci.* 6:1061–79

Futerman AH, Khanin R, Segel LA. 1993. Lipid diffusion in neurons. *Nature* 362:119

Garner CC, Tucker RP, Matus A. 1988. Selective localization of messenger RNA for cytoskeletal protein MAP2 in dendrites. *Nature* 336:674–77

Goedert M, Spillantini MG, Cairns NJ, Crowther RA. 1992. Tau proteins of Alzheimer paired helical filaments: Abnormal phosphorylation of all six brain isoforms. *Neuron* 8:159–68

Goodman SR, Krebs KE, Whitfield CF, et al.

1988. Sprectrin and related molecules. *CRC Crit. Rev. Biochem.* 23:197–234

Goslin K, Banker G. 1989. Experimental observations of the development of polarity by hippocampal neurons in culture. *J. Cell Biol.* 108:1507–16

Goslin K, Banker G. 1990. Rapid changes in the distribution of GAP-43 correlate with the expression of neuronal polarity during normal development and under experimental conditions. *J. Cell Biol.* 110:1319–31

Goslin K, Banker G. 1991. Rat hippocampal neurons in low-density culture. In *Culturing Nerve Cells,* ed. G. Banker, K Goslin, pp. 251–81. Cambridge: MIT

Goslin K, Schreyer DJ, Skene JHP, Banker G. 1988. Development of neuronal polarity: GAP-43 distinguishes axonal from dendritic growth cones. *Nature* 336:672–74

Goslin K, Schreyer DJ, Skene JHP, Banker G. 1990. Changes in the distribution of GAP-43 during the development of neuronal polarity. *J. Neurosci.* 10:588–602

Greene LA, Sobeih MM, Teng KK. 1991. Methodologies for the culture and experimental use of the PC12 rat pheochromocytoma cell line. In *Culturing Nerve Cells,* ed. G. Banker K Goslin, pp.207–26. Cambridge: MIT

Gruenberg J, Clague MJ. 1992. Regulation of intracellular membrane transport. *Curr. Opin. Cell Biol.* 4:593–99

Hall GF, Cohen MJ. 1988. The pattern of dendritic sprouting and retraction induced by axotomy of lamprey central neurons. *J. Neurosci.* 8:3584–97

Hall GF, Lee VMY, Kosik KS. 1991. Microtubule destabilization and neurofilament phosphorylation precede dendritic sprouting after close axotomy of lamprey central neurons. *Proc. Natl. Acad. Sci. USA* 88:5016–20

Hall GF, Poulos A, Cohen J. 1989. Sprouts emerging from the dendrites of axotomized lamprey central neurons have axonlike ultrastructure. *J. Neurosci.* 9:588–99

Hammerton RW, Krzeminski KA, Mays RW, et al. 1991. Mechanism for regulating cell surface distribution of Na+, K+-ATPase in polarized epithelial cells. *Science* 254:847–50

Hamori J. 1990. Morphological plasticity of postsynaptic neurones in reactive synaptogenesis. *J. Exp. Biol.* 153:251–60

Han HQ, Nichols RA, Rubin MR, et al. 1991. Induction of formation of presynaptic terminals in neuroblastoma cells by synapsin IIb. *Nature* 349:697–700

Harel R, Futerman AH. 1993. Inhibition of sphingolipid synthesis affects axonal outgrowth in cultured hippocampal neurons. *J. Biol. Chem.* In press

Havton L, Kellerth J-O. 1987. Regeneration by supernumerary axons with synaptic terminals in spinal motoneurons of cats. *Nature* 325:711–14

Hayes NVL, Rayner DA, Woods A, Baines AJ. 1991. p103 and A60: novel proteins of the neuronal membrane-associated cytoskeleton. *Biochem. Soc. Trans.* 1042–48

Hesketh JE, Pryme IF. 1991. Interaction between mRNA, ribosomes and the cytoskeleton. *Biochem. J.* 277:1–10

Hevner RF, Wong-Riley MTT. 1991. Neuronal expression of nuclear and mitochondrial genes for cytochrome oxidase (CO) subunits analyzed by in situ hybridization: comparison with CO activity and protein. *J. Neurosci.* 11:1942–58

Higgins D, Lein PJ, Osterhout DJ, Johnson MI. 1991. Tissue culture of mammalian autonomic neurons. In *Culturing Nerve Cells,* ed. G. Banker K Goslin, pp. 177–205. Cambridge: MIT

Higgins D, Waxman A, Banker GA. 1988. The distribution of microtubule-associated protein 2 changes when dendritic growth is induced in rat sympathetic neurons in vitro. *Neuroscience* 24:583–92

Hillman DE. 1979. Neuronal shape parameters and substructures as a basis of neuronal form. In *The Neuroscience: Fourth Study Program,* ed. F. O. Schmitt FG Worden, pp. 477–97. Cambridge: MIT

Hillman DE. 1988. Parameters of dendritic shape and substructure: intrinsic and extrinsic determination. In *Intrinsic Determinants of Neuronal Form and Function,* ed. R. J. Lasek MM Black, pp. 83–113. New York: Liss

Huber G, Alaimo-Beuret D, Matus A. 1985. MAP3: characterization of a novel microtubule-associated protein. *J. Cell Biol.* 100: 496–507

Huber G, Matus A. 1984. Immunocytochemical localization of microtubule-associated protein 1 in rat cerebellum using monoclonal antibodies. *J. Cell Biol.* 98:777–81

Huber LA, Dotti C, Parton R, et al. 1992. GTP-binding proteins involved in post Golgi transport. *Eur. J. Cell Biol.* 57:35 (Abstr.)

Hunziker W, Harter C, Matter K, Mellman I. 1991. Basolateral sorting in MDCK cells requires a distinct cytoplasmic domain determinant. *Cell* 66:907–20

Huttner WB, Dotti CG. 1991. Exocytotic and endocytotic membrane traffic in neurons. *Curr. Opin. Neurobiol.* 1:388–92

Jareb M, Banker G. 1991. Inhibition of axonal growth by brefeldin A in cultures of embryonic rat hippocampal neurons. *Soc. Neurosci. Abstr.* 17:739 (Abstr.)

Jareb M, Esch T, Craig AM, Banker G. 1993. The development of polarity by hippocampal neurons in culture. *J. Cell. Biochem.* 17B:267 (Abstr.)

Johnson MI, Higgins D, Ard MD. 1989. Astrocytes induce dendritic development in cultured sympathetic neurons. *Dev. Brain Res.* 47:289–92

Jung LJ, Scheller RH. 1991. Peptide processing and targeting in the neuronal secretory pathway. *Science* 251:1330–35

Kelly RB, Grote E. 1993. Protein targeting in the neuron. *Annu. Rev. Neurosci.* 16:95–127

Kennedy D, Mellon D. 1964. Synaptic activation and receptive fields in crayfish interneurons. *Comp. Biochem. Physiol.* 13:275–300

Killisch I, Dotti CG, Laurie DJ, et al. 1991. Expression patterns of GABA-A receptor subtypes in developing hippocampal neurons. *Neuron* 7:927–36

Kislauskis EH, Singer RH. 1992. Determinants of mRNA localization. *Curr. Opin. Cell Biol.* 4:975–78

Kleiman R, Banker G, Steward O. 1990. Differential subcellular localization of particular mRNAs in hippocampal neurons in culture. *Neuron* 5:821–30

Kobayashi T, Storrie B, Simons K, Dotti CG. 1992. A functional barrier to movement of lipids in polarized neurons. *Nature* 359:647–50

Kordeli E, Bennett V. 1991. Distinct ankyrin isoforms at neuron cell bodies and nodes of Ranvier resolved using erythrocyte ankyrin-deficient mice. *J. Cell Biol.* 114:1243–59

Kornfeld S, Mellman I. 1989. The biogenesis of lysosomes. *Annu. Rev. Cell Biol.* 5:483–525

Kosik KS. 1990. Tau protein and neurodegeneration. *Mol. Neurobiol.* 4:171–79

Kosik KS, Caceres A. 1991. Tau protein and the establishment of an axonal morphology. *J. Cell Sci.* 15:69–74

Kosik KS, Finch EA. 1987. MAP2 and Tau segregate into dendritic and axonal domains after the elaboration of morphologically distinct neurites: an immunocytochemical study of cultured rat cerebrum. *J. Neurosci.* 7:3142–53

Kriegstein AR, Dichter MA. 1983. Morphological classification of rat cortical neurons in cell culture. *J. Neurosci.* 3:1634–47

Lafont F, Rouget M, Triller A, et al. 1992. In vitro control of neuronal polarity by glycosaminoglycans. *Development* 114:17–29

Landis SC. 1990. Target regulation of neurotransmitter phenotype. *Trends Neurosci.* 13:344–50

Lasek RJ. 1988. Studying the intrinsic determinants of neuronal form and function. In *Intrinsic Determinants of Neuronal Form and Function*, ed. R. J. Lasek MM Black, pp. 3–58. New York: Liss

Le Bivic A, Garcia M, Quaroni A, Mirre C. 1993. Mechanisms of sorting in an intestinal epithelial cell line Caco-2. *J. Cell. Biochem.* 17B:275 (Abstr.)

Le Bivic A, Sambuy Y, Patzak A, et al. 1991. An internal deletion in the cytoplasmic tail reverses the apical localization of human NGF receptor in transfected MDCK cells. *J. Cell Biol.* 115:607–18

Lee VMY, Balin BJ, Otvos L Jr, Trojanowski JQ. 1991. A68: A major subunit of paired helical filaments and derivatized forms of normal tau. *Science* 251:675–78

Lee VMY, Carden MJ, Schlaepfer WW, Trojanowski JQ. 1987. Monoclonal antibodies distinguish several differentially phosphorylated states of the two largest rat neurofilament subunits (NF-H and NF-M) and demonstrate their existence in the normal nervous system of adult rats. *J. Neurosci.* 7:3474–88

Lein P, Banker G, Higgins D. 1992. Laminin selectively enhances axonal growth and accelerates the development of polarity by hippocampal neurons in culture. *Dev. Brain Res.* 69:191–97

Lein PJ, Higgins D. 1989. Laminin and a basement membrane extract have different effects on axonal and dendritic outgrowth from embryonic rat sympathetic neruons in vitro. *Develop. Biol.* 136:330–45

Lein PJ, Higgins D. 1991. Protein synthesis is required for the initiation of dendritic growth in embryonic sympathetic neurons in vitro. *Dev. Brain Res.* 60:187–96

Lein PJ, Higgins D, Turner DC, et al. 1991. The NC1 domain of type IV collagen promotes axonal growth in sympathetic neurons through interaction with the $\alpha 1\beta 1$ integrin. *J. Cell Biol.* 113:417–28

Lindá H, Cullheim S, Risling M. 1992. A light and electron microscopic study of intracellularly HRP-labeled lumbar motoneurons after intramedullary axotomy in the adult cat. *J. Comp. Neurol.* 318:188–208

Lindá H, Risling M, Cullheim S. 1985. "Dendraxons" in regenerating motoneurones in the cat: do dendrites generate new axons after central axotomy. *Brain Res.* 358:329–33

Lisanti MP, Caras IW, Davitz MA, Rodriguez-Boulan E. 1989. A glycophospholipid membrane anchor acts as an apical targeting signal in polarized epithelial cells. *J. Cell Biol.* 109:2145–56

Lisanti MP, Le Bivic A, Saltiel AR, Rodriguez-Boulan E. 1990. Preferred apical distribution of glycosyl-phosphatidylinositol (GPI) anchored proteins: A highly conserved feature of the polarized epithelial cell phenotype. *J. Membr. Biol.* 113:155–67

Lisanti MP, Rodriguez-Boulan E. 1990. Glycophospholipid membrane anchoring provides clues to the mechanism of protein sorting in polarized epithelial cells. *Trends Biochem. Sci.* 15:113–18

Lowenstein PR, Morrison E, Douglas P, et al.

1993. Neocortical neuronal polarity: polarized distribution of a marker of the trans-Golgi network (TGN 38) and targeting of a foreign protein linked to a glycosyl-phosphatidylinositol (GPI) anchor in postmitotic neurons. *Biochem. Soc. Trans.* In press

Mansfield SG, Diaz-Nido J, Gordon-Weeks PR, Avila J. 1991. The distribution and phosphorylation of the microtubule-associated protein MAP 1B in growth cones. *J. Neurocytol.* 20:1007–22

Mansfield SG, Gordon-Weeks PR. 1991. Dynamic post-translational modification of tubulin in rat cerebral cortical neurons extending neurites in culture. *J. Neurocytol.* 20:654–66

Marrs JA, Nelson WJ. 1993. Membrane-cytoskeleton, B-Cadherin and apical Na$^+$, K$^+$-ATPase polarity in the choroid plexus epithelium. *J. Cell. Biochem.* 17B:271 (Abstr.)

Martin LJ, Blackstone CD, Huganir RL, Price DL. 1992. Cellular localization of a metabotropic glutamate receptor in rat brain. *Neuron* 9:259–70

Martin LJ, Blackstone CD, Levey AI, et al. 1993. AMPA glutamate receptor subunits are differentially distributed in rat brain. *Neuroscience* 53:327–58

Matlin KS. 1992. W(h)ither default? Sorting and polarization in epithelial cells. *Curr. Opin. Cell Biol.* 4:623–28

Matteoli M, Takei K, Cameron R, et al. 1991. Association of Rab3A with synaptic vesicles at late stages of the secretory pathway. *J. Cell Biol.* 115:625–33

Mattson MP. 1988. Neurotransmitters in the regulation of neuronal cytoarchitecture. *Brain Res. Rev.* 13:179–212

Mattson MP, Dou P, Kater SB. 1988. Outgrowth-regulating actions of glutamate in isolated hippocampal pyramidal neurons. *J. Neurosci.* 8:2087–100

Mattson MP, Guthrie PB, Hayes BC, Kater SB. 1989. Roles for mitotic history in the generation and degeneration of hippocampal neuroarchitecture. *J. Neurosci.* 9:1223–32

Mattson MP, Kater SB. 1989. Development and selective neurodegeneration in cell cultures from different hippocampal regions. *Brain Res.* 490:110–25

Matus A, Bernhardt R, Hugh-Jones T. 1981. High molecular weight microtubule-associated proteins are preferentially associated wtih dendritic microtubules in brain. *Proc. Natl. Acad. Sci. USA* 780:3010–14

McConnell SK. 1988. Development and decision-making in the mammalian cerebral cortex. *Brain Res.* 427:1–23

McKee AC, Kowall NW, Kosik KS. 1989. Microtubular reorganization and dendritic growth response in Alzheimer's disease. *Ann. Neurol.* 26:652–59

Migliaccio G, Zurzolo C, Nitsch L, et al. 1990. Human CD8a glycoprotein is expressed at the apical plasma membrane domain in permanently transformed MDCK II clones. *Eur. J. Cell Biol.* 52:291–96

Mitchison T, Kirschner M. 1988. Cytoskeletal dynamics and nerve growth. *Neuron* 1:761–72

Morris RJ, Beech JN, Barber PC, Raisman G. 1985. Early stage of Purkinje cell maturation demonstrated by Thy-1 immunohistochemistry on postnatal rat cerebellum. *J. Neurocytol.* 14:427–52

Mostov K, Apodaca G, Aroeti B, Okamoto C. 1992. Plasma membrane protein sorting in polarized epithelial cells. *J. Cell Biol.* 116:577–83

Myers PZ, Eisen JS, Westerfield M. 1986. Development and axonal outgrowth of identified motoneurons in the zebrafish. *J. Neurosci.* 6:2278–89

Navone F, Jahn R, Di Gioia D, et al. 1986. Protein p38: an integral membrane protein specific for small clear vesicles of neurons and neuroendocrine cells. *J. Cell Biol.* 103:2511–27

Nawa H, Yamamori T, Le T, Patterson PH. 1990. Generation of neuronal diversity: analogies and homologies with hematopoiesis. *Cold Spring Harbor Symp. Quant. Biol.* 55:247–53

Neale EA, MacDonald RL, Nelson PG. 1978. Intracellular horseradish peroxidase injection for correlation of light and electron microscopic anatomy with synaptic physiology of cultured mouse spinal cord neurons. *Brain Res.* 152:265–82

Nelson WJ, Veshnock PJ. 1987. Ankyrin binding to Na$^+$,K$^+$,ATPase and implications for the organization of membrane domains in polarized cells. *Nature* 328:533–36

Nixon RA, Sihag RK. 1991. Neurofilament phosphorylation: a new look at regulation and function. *Trends Neurosci.* 14:501–6

Nowakowski RS, Rakic P. 1979. The mode of migration of neurons to the hippocampus: a Golgi and electron microscopic analysis in foetal rhesus monkey. *J. Neurocytol.* 8:697–718

O'Rourke NA, Dailey ME, Smith SJ, McConnell SK. 1992. Diverse migratory pathways in the developing cerebral cortex. *Science* 258:299–302

Okabe S, Hirokawa N. 1988. Microtubule dynamics in nerve cells: analysis using microinjection of biotinylated tubulin into PC12 cells. *J. Cell Biol.* 107:651–64

Okabe S, Hirokawa N. 1989. Rapid turnover of microtubule-associated protein MAP2 in the axon revealed by microinjection of biotinylated MAP2 into cultured neurons. *Proc. Natl. Acad. Sci. USA* 86:4127–31

Okabe S, Hirokawa N. 1992. Differential behavior of photoactivated microtubules in growing axons of mouse and frog neurons. *J. Cell Biol.* 117:105–20

Osterhout DJ, Frazier WA, Higgins D. 1992. Thrombospondin promotes process outgrowth in neurons from the peripheral and central nervous systems. *Develop. Biol.* 150:256–65

Papasozomenos SC, Binder LI. 1987. Phosphorylation determines two distinct species of Tau in the central nervous system. *Cell Motil. Cytoskelet.* 8:210–26

Parton RG, Simons K, Dotti CG. 1992. Axonal and dendritic endocytic pathways in cultured neurons. *J. Cell Biol.* 119:123–37

Pearse BMF, Robinson MS. 1990. Clathrin, adaptors, and sorting. *Annu. Rev. Cell Biol.* 6:151–71

Peng I, Binder LI, Black MM. 1986. Biochemical and immunological analyses of cytoskeletal domains of neurons. *J. Cell Biol.* 102:252–62

Perez-Velazquez JL, Angelides KJ. 1993. Assembly of GABA-A receptor subunits determines sorting and localization in polarized cells. *Nature* 361:457–60

Persohn E, Schachner M. 1987. Immunoelectron-microscopic localization of the neural cell adhesion molecules L1 and N-CAM during postnatal development of the mouse cerebellum. *J. Cell Biol.* 105:569–76

Persohn E, Schachner M. 1990. Immunohistological localization of the neural adhesion molecules L1 and N-CAM in the developing hippocampus of the mouse. *J. Neurocytol.* 19:807–19

Peters A, Palay SL, Webster HdeF. 1991. *The Fine Structure of the Nervous System.* New York: Oxford

Petralia RS, Wenthold RJ. 1992. Light and electron immunocytochemical localization of AMPA-selective glutamate receptors in the rat brain. *J. Comp. Neurol.* 318:329–54

Pfeiffer G, Friede RL. 1985. The axon tree of rat motor fibres: morphometry and fine structure. *J. Neurocytol.* 14:809–24

Pietrini G, Matteoli M, Banker G, Caplan MJ. 1992. Immunolocalization of two Na,K-ATPase isoforms in hippocampal neurons in culture. *Proc. Natl. Acad. Sci. USA* 89:8414–18

Pietrini G, Suh YJ, Edelmann L, et al. 1993. Neuronal and epithelial polarity—the GABA and betaine transporters are sorted to opposite membranes of polarized epithelial cells. *J. Cell. Biochem.* 17B:280 (Abstr.)

Pleasure SJ, Page C, Lee VMY. 1992. Pure, postmitotic, polarized human neurons derived from NTera 2 cells provide a system for expressing exogenous proteins in terminally differentiated neurons. *J. Neurosci.* 12:1802–15

Pollerberg GE, Burridge K, Krebs KE, et al.

1987. The 180-kD component of the neural cell adhesion molecule N-CAM is involved in cell-cell contacts and cytoskeleton-membrane interactions. *Cell Tissue Res.* 250:227–36

Powell SK, Cunningham BA, Edelman GM, Rodriguez-Boulan E. 1991a. Targeting of transmembrane and GPI-anchored forms of N-CAM to opposite domains of a polarized epithelial cell. *Nature* 353:76–77

Powell SK, Lisanti MP, Rodriguez-Boulan EJ. 1991b. Thy-1 expresses two signals for apical localization in epithelial cells. *Am. J. Physiol.* 260:715–20

Prochiantz A, Rousselet A, Chamak B. 1990. Adhesion and the in vitro development of axons and dendrites. *Prog. Brain Res.* 86:331–36

Puelles L, Privat A. 1977. Do oculomotor neuroblasts migrate across the midline in the fetal rat brain?. *Anat. Embryol.* 150:187–206

Rall W, Shepherd GM, Reese TS, Brightman MW. 1966. Dendro-dendritic synaptic pathway for inhibition in the olfactory bulb. *Exp. Neurol.* 14:44–56

Reinsch SS, Mitchison TJ, Kirschner MW. 1991. Microtubule polymer assembly and transport during axonal elongation. *J. Cell Biol.* 115:365–79

Represa A, Deloulme JC, Sesenbrenner M, et al. 1990. Neurogranin: immunocytochemical localization of a brain-specific protein kinase C substrate. *J. Neurosci.* 10:3782–92

Riederer BM, Guadano-Ferraz A, Innocenti GM. 1990. Difference in distribution of microtubule-associated proteins 5a and 5b during the development of cerebral cortex and corpus callosum in cats: dependence on phosphorylation. *Dev. Brain Res.* 56:235–43

Riederer BM, Zagon IS, Goodman SR. 1986. Brain spectrin (240/235) and brain spectrin (240/235E): Two distinct spectrin subtypes with different locations within mammalian neural cells. *J. Cell Biol.* 102:2088–97

Rivas RJ, Burmeister DW, Goldberg DJ. 1992. Rapid effects of laminin on the growth cone. *Neuron* 8:107–15

Roberts A, Dale N, Ottersen OP, Storm-Mathisen J. 1987. The early development of neurons with GABA immunoreactivity in the CNS of Xenopus laevis embryos. *J. Comp. Neurol.* 261:435–49

Roberts A, Dale N, Ottersen OP, Storm-Mathisen J. 1988. Development and characterization of commissural interneurons in the spinal cord of *Xenopus laevis* embryos revealed by antibodies to glycine. *Development* 103:447–61

Rodriguez-Boulan E, Nelson WJ. 1989. Morphogenesis of the polarized epithelial cell phenotype. *Science* 245:718–25

Rodriguez-Boulan E, Pendergast M. 1980. Polarized distribution of viral envelope proteins in the plasma membrane of infected epithelial cells. *Cell* 20:45–54

Rodriguez-Boulan E, Powell SK. 1992. Polarity of epithelial and neuronal cells. *Annu. Rev. Cell Biol.* 8:395–427

Rogers SW, Hughes TE, Hollmann M, et al. 1991. The characterization and localization of the glutamate receptor subunit GluR1 in the rat brain. *J. Neurosci.* 11:2713–24

Roman LM, Garoff H. 1986. Alteration of the cytoplasmic domain of the membrane-spanning glycoprotein p62 of Semliki forest virus does not affect its polar distribution in established lines of Madin-Darby canine kidney cells. *J. Cell Biol.* 103:2607–18

Roth MG, Gundersen D, Patil N, Rodriguez-Boulan E. 1987. The large external domain is sufficient for the correct sorting of secreted or chimeric influenza virus hemagglutinins in polarized monkey kidney cells. *J. Cell Biol.* 104:769–82

Rousselet A, Autillo-Touati A, Araud D, Prochiantz A. 1990. In vitro regulation of neuronal morphogenesis and polarity by astrocyte-derived factors. *Dev. Biol.* 137: 33–45

Rousselet A, Fetler L, Chamak B, Prochaintz A. 1988. Rat mesencephalic neurons in culture exhibit different morphological traits in the presence of media conditioned on mesencephalic or striatal astroglia Paris. *Dev. Biol.* 129:495–504

Sasaki S, Jacobs JR, Stevens JK. 1983. Intracellular control of axial shape in non-uniform neurites: a serial electron microscopic analysis of organelles and microtubules in AI and AII amacrine neurites. *J. Cell Biol.* 98: 1279–90

Sato-Yoshitake R, Shiomura Y, Miyasaka H, Hirokawa N. 1989. Microtubule-associated protein 1B: molecular structure, localization, and phosphorylation-dependent expression in developing neurons. *Neuron* 3:229–38

Satoh T, Ross CA, Villa A, et al. 1990. The inositol 1,4,5-triphosphate receptor in cerebellar Purkinje cells: quantitative immunogold labeling reveals concentration in an ER subcompartment. *J. Cell Biol.* 111:615–24

Seitandou T, Triller A, Korn H. 1988. Distribution of glycine receptors on the membrane of a central neuron: an immunoelectron microscopy study. *J. Neurosci.* 8:4319–33

Serafini T, Stenbeck G, Brecht A, et al. 1991. A coat subunit of Golgi-derived non-clathrin-coated vesicles with homology to the clathrin-coated vesicle coat protein beta-adaptin. *Nature* 349:215–20

Shaw G, Winialski D, Reier P. 1988. The effect of axotomy and deafferentation on phosphorylation dependent antigenicity of neurofilaments in rat superior cervical ganglion neurons. *Brain Res.* 460:227–34

Sheng M, Tsaur ML, Jan YN, Jan LY. 1992.

Subcellular segregation of two A-type K^+ channel proteins in rat central neurons. *Neuron* 9:271–84

Shepherd GM. 1979. *The Synaptic Organization of the Brain*, pp. 1–436. New York: Oxford

Shepherd GM. 1991. *Foundations of the Neuron Doctrine*, pp. 1–338. New York & Oxford: Oxford

Shirao T, Kojima N, Obata K. 1992. Cloning of drebrin A and induction of neurite-like processes in drebrin-transfected cells. *Dev. Neurosci.* 3:109–12

Shoukimas GM, Hinds JW. 1978. The development of the cerebral cortex in the embryonic mouse: an electron microscopic serial section analysis. *J. Comp. Neurol.* 179:795–830

Siman R, Baudry M, Lynch G. 1985. Regulation of glutamate receptor binding by the cytoskeletal protein fodrin. *Nature* 313:225–28

Simons K, van Meer G. 1988. Lipid sorting in epithelial cells. *Biochemistry* 27:6197–202

Simons K, Wandinger-Ness A. 1990. Polarized sorting in epithelia. *Cell* 62:207–10

Singer RH. 1992. The cytoskeleton and mRNA localization. *Curr. Opin. Cell Biol.* 4:15–19

Solomon F. 1979. Detailed neurite morphologies of sister neuroblastoma cells are related. *Cell* 16:161–65

Somogyi P, Takagi H, Richards JG, Mohler H. 1989. Subcellular localization of benzodiazepine/GABA-A receptors in the cerebellum of rat, cat, and monkey using monoclonal antibodies. *J. Neurosci.* 9:2197–209

Srinivasan Y, Elmer L, Davis J, et al. 1988. Ankyrin and spectrin associate with voltage-dependent sodium channels in brain. *Nature* 333:177–80

Sternberger LA, Sternberger NH. 1983. Monoclonal antibodies distinguish phosphorylated and nonphosphorylated forms of neurofilaments in situ. *Proc. Natl. Acad. Sci. USA* 80:6126–30

Steward O, Banker G. 1992. Getting the message from the gene to the synapse: sorting and intracellular transport of RNA in neurons and other spatially complex cells. *Trends Neurosci.* 15:180–86

Stieber A, Gonatas JO, Gonatas NK, Louvard D. 1987. The Golgi apparatus-complex of neurons and astrocytes studied with an anti-organelle antibody. *Brain Res.* 408:13–21

Sumner BEH, Watson WE. 1971. Retraction and expansion of the dendritic tree of motor neurones of adult rats induced in vivo. *Nature* 233:273–75

Tomaselli KJ. 1991. Beta1-integrin-mediated neuronal responses to extracellular matrix proteins. *Ann. N. Y. Acad. Sci.* 633:100–4

Triller A, Cluzeaud F, Pfeiffer F, et al. 1985. Distribution of glycine receptors at central synapses: an immunoelectron microscopy study. *J. Cell Biol.* 101:683–88

Trimmer JS. 1991. Immunological identification and characterization of a delayed rectifier K+ channel polypeptide in rat brain. *Proc. Natl. Acad. Sci. USA* 88:10764–68

Tropea M, Johnson MI, Higgins D. 1988. Glial cells promote dendritic development in rat sympathetic neurons in vitro. *Glia* 1:380–92

Vallee RB, DiBartolomeis MJ, Theurkauf WE. 1981. A protein kinase bound to the projection portion of MAP2 (microtubule-associated protein 2). *J. Cell Biol.* 90:568–76

Van Lookeren Campagne M, Oestreicher AB, Van Bergen en Henegouwen PMP, Gispen WH. 1990. Ultrastructural double localization of B-50/GAP43 and synaptophysin (p38) in neonatal and adult rat hippocampus. *J. Neurocytol.* 19:948–61

Villa A, Podini P, Clegg DO, et al. 1992. Intracellular Ca^{2+} stores in chicken Purkinje neurons: differential distribution of the low affinity-high capacity Ca^{2+} binding protein, calsequestrin, of Ca^{2+} ATPase and of the ER lumenal protein Bip. *J. Cell Biol.* 113:779–91

Vogel LK, Spiess M, Noren O, Sjostrom H. 1993. Aminopeptidase N carries a signal for apical sorting on its catalytic head group. *J. Cell. Biochem.* 17B:278 (Abstr.)

Walton PD, Airey JA, Sutko JL, et al. 1992. Ryanodine and inositol triphosphate receptors coexist in avian cerebellar Purkinje neurons. *J. Cell Biol.* 113:1145–57

Watson JB, Sutcliffe JG, Fisher RS. 1992. Localization of the protein kinase C phosphorylation/calmodulin-binding substrate RC3 in dendritic spines of neostriatal neurons. *Proc. Natl. Acad. Sci. USA* 89:8581–85

Westenbroek RE, Hell JW, Warner C, et al. 1992a. Biochemical properties and subcellular distribution of an N-type calcium channel alpha1 subunit. *Neuron* 9:1099–115

Westenbroek RE, Merrick DK, Catterall WA. 1989. Differential subcellular localization of the RI and RII Na^{+} channel subtypes in central neurons. *Neuron* 3:695–704

Westenbroek RE, Noebels JL, Catterall WA. 1992b. Elevated expression of type II Na^{+} channels in hypomyelinated axons of shiverer mouse brain. *J. Neurosci.* 12:2259–67

Westerfield M, McMurray JV, Eisen JS. 1986. Identified motoneurons and their innervation of axial muscles in the zebrafish. *J. Neurosci.* 6:2267–77

Xue GP, Calvert RA, Morris RJ. 1990. Expression of the neuronal cell surface glycoprotein Thy-1 is under post-transcriptional control and is spatially regulated in the developing olfactory system. *Development* 109:851–64

Yamamoto M, Hassinger L, Crandall JE. 1990. Ultrastructural localization of stage-specific neurite-associated proteins in the developing rat cerebral and cerebellar cortices. *J. Neurocytol.* 19:619–27

Young D. 1989. *Nerve Cells and Animal Behavior.* Cambridge MA: Cambridge Univ. Pr.

Zagon IS, Higbee R, Riederer BM, Goodman SR. 1986. Spectrin subtypes in mammalian brain: an immunoelectron microscopic study. *J. Neurosci.* 6:2977–86

Zenker W, Hohberg E. 1973. Alpha-motorische nervenfaser: axonquerschnittsflache von stammfaser und endasten. *Z. Anat. Entwicklungsgesch.* 139:163–72

Zurzolo C, Van't Hof W, Lisanti M, et al. 1993. Basolateral targeting of GPI-anchored proteins and glycosphingolipids in a polarized thyroid epithelial cell line. *J. Cell. Biochem.* 17B:278 (Abstr.)

Annu. Rev. Neurosci. 1994. 17:311–39

PRION DISEASES AND NEURODEGENERATION

Stanley B. Prusiner and Stephen J. DeArmond

Departments of Neurology, Biochemistry and Biophysics, and Pathology, University of California, San Francisco, California 94143

KEY WORDS: transgenic mice, amyloid plaques, genetic linkage, conformational changes, strains of prions

INTRODUCTION

Prion diseases are neurodegenerative disorders of humans and animals. In humans, the three forms of the prion disease are manifest as infectious, sporadic, and inherited disorders (Table 1) (Prusiner 1991). Prions cause six transmissible neurodegenerative diseases of animals (Table 2) and four of humans. Prions are composed largely, if not entirely, of an abnormal isoform of the prion protein (PrP) designated PrP^{Sc}. Genetic linkage studies of people suffering from prion diseases as well as investigations with transgenic (Tg) mice expressing mutant PrP genes indicate that mutations in human PrP genes cause neurodegeneration.

Studies of Tg mice expressing Syrian hamster (SHa) PrP or chimeric SHa/mouse (Mo) PrP suggest that an interaction between PrP^{Sc} and homologous PrP^{C} is important in the replication of prions (Prusiner 1991). By using these Tg mice, researchers have found the pattern of PrP^{Sc} accumulation in brain to be specific for a particular prion isolate. The existence of distinct isolates or "strains" of prions poses a conundrum, since a wealth of experimental data indicates that prions are devoid of nucleic acid, yet distinct isolates each possess a unique set of properties characterized by the length of the incubation time, distribution of neuropathologic lesions, and pattern of PrP^{Sc} accumulation (Dickinson et al 1968, Fraser & Dickinson 1973, Hecker et al 1992, DeArmond et al 1993). The unique patterns of PrP^{Sc} deposition suggest that each prion isolate possesses a cell-specific trophism that in turn leads to the conversion of PrP^{C} into PrP^{Sc} within a restricted population of cells. The synthesis of PrP^{Sc} is a post-translational process (Borchelt et al

Table 1 The human prion diseases

Suggested nomenclature	Traditional name[a]
Sporadic prion disease	CJD
Inherited prion diseases	
Inherited prion disease (PrP-P102L)	GSS
Inherited prion disease (PrP-A117V)	GSS
Inherited prion disease (PrP-D178N)	Familial CJD, FFI
Inherited prion disease (PrP-F198S)	GSS-nft
Inherited prion disease (PrP-E200K)	Familial CJD
Inherited prion disease (PrP-V2101)	Familial CJD
Inherited prion disease (PrP-Q217R)	GSS-nft
Inherited prion disease (PrP octarepeat insert)	Familial CJD
Infectious prion diseases	Kuru
	Iatrogenic CJD

[a] Creutzfeldt-Jakob disease (CJD), Gerstmann-Sträussler-Scheinker syndrome (GSS), GSS with neurofibrillary tangles (GSS-nft), fatal familial insomnia (FFI).

Table 2 The animal prion diseases

Suggested nomenclature	Traditional name
Sporadic prion disease	Natural scrapie
Inherited prion disease	Tg(GSSMoPrP) mice
	Natural scrapie?
Infectious prion disease	Bovine spongiform encephalopathy
	Feline spongiform encephalopathy
	Transmissible mink encephalopathy
	Chronic wasting disease
	Exotic ungulate encephalopathy
	Experimental scrapie
	Natural scrapie?

1990, 1992; Caughey & Raymond 1991) that is thought to involve the refolding of α-helical regions of PrP^C into β-sheets (Bazan et al 1987; Caughey et al 1991b; Gasset et al 1992, 1993). No evidence has been found to suggest that PrP^{Sc} synthesis involves a covalent change (Stahl et al 1993). Indeed, prion diseases seem to be disorders of protein conformation.

Among the human prion diseases, the most common is sporadic Creutzfeldt-Jakob disease (CJD). Sporadic CJD cases are defined by the failure to find either an infectious or genetic etiology. The infectious cases result from transmission of prions through either ritualistic cannibalism of brains from relatives of New Guinea highlanders dying of kuru (Alpers 1968, Gajdusek 1977) or medical accidents as in iatrogenic CJD cases caused by prion-contaminated growth hormone derived from human pituitaries (Brown et al 1992). Inherited prion diseases are caused by mutations of the PRNP gene that result in nonconservative amino acid substitutions (Hsiao et al 1989a, Prusiner 1991). An unusual characteristic of all prion diseases, whether sporadic, infectious, or familial, is that they are frequently transmissible to experimental animals (Gajdusek et al 1966, Gibbs et al 1968, Masters et al 1981a).

Prions Are Not Viruses

The transmission of prion diseases to experimental animals after prolonged incubation times fostered the notion that these disorders are caused by "slow viruses." Over the past four decades, scientists have sought to identify the putative "scrapie virus" and to isolate its nucleic acid genome. No immunologic, morphologic, or molecular biological evidence for the "scrapie virus" has been found. Instead, a wealth of experimental data now indicates that the infectious prion particle causing scrapie is not a virus, but is composed largely, if not entirely, of PrP^{Sc} molecules (Prusiner et al 1982, 1990; McKinley et al 1983; Oesch et al 1985). The term "prion" was introduced to designate this class of infectious agents and to distinguish them from viruses and viroids (Prusiner 1982).

Terminology

Numerous novel findings in studies of the prion diseases have necessitated the introduction of a new terminology to describe these phenomena (Prusiner et al 1993). The prion protein (PrP) in purified scrapie prion preparations was designated PrP^{Sc} to distinguish it from the normal cellular isoform designated PrP^C (Oesch et al 1985). PrP^{Sc} can be distinguished from PrP^C by its resistance to protease digestion, insolubility after detergent extraction, deposition in secondary lysosomes, posttranslational synthesis, and enrichment during copurification of prion infectivity. Limited proteolysis of PrP^{Sc} removes the N-terminal ~ 67 amino acid residues to produce PrP 27–30 without loss of infectivity, whereas PrP^C is completely digested under the same conditions. The term PrP^{Sc} is used to designate the abnormal, scrapie isoform of PrP^C in animals as well as humans; however, some investigators prefer PrP^{CJD} for the abnormal isoform in humans. When the disease is transmitted to experimental animals by brain extracts from patients dying of CJD, the abnormal PrP isoform is produced by conversion of host encoded

PrP^C into PrP^{Sc}. This contrasts with viral proteins, which are encoded by the viral genome.

The human PrP gene maps to the short arm of chromosome 20 and is designated PRNP; the mouse PrP gene maps to the homologous region of chromosome 2 and is designated *Prn-p* (Sparkes et al 1986).

Transgenic mice expressing foreign or mutant PrP genes are labeled with the transgene in parentheses. For example, transgenic mice harboring SHaPrP genes are designated Tg(SHaPrP) mice. Transgenic mice with both PrP alleles ablated are denoted as $Prn-p^{0/0}$ mice.

Several hypothetical mechanisms of prion replication have been proposed; these include the possibility that an as yet unidentified nucleic acid genome replicates and serves as a trigger for the synthesis of PrP^{Sc}. Against this hypothesis is the lack of evidence for a scrapie-specific nucleic acid. Alternatively, prions may be devoid of nucleic acid, and replication of infectivity requires only the conversion of PrP^C to PrP^{Sc}. This hypothesis is consistent with all of the currently available data (Prusiner 1991).

Because diversity among strains of bacteria, viruses, and viroids resides in the sequence of the nucleic acid genomes for each of these infectious pathogens, we have chosen to avoid using strains to describe distinct prion inocula that have been passaged through the same animal species. We use the term "isolate" in order to avoid prejudging the mechanism by which distinct prion inocula produce scrapie with different clinical and neuropathologic features. We name specific isolates according to the designation given them when they were first isolated, preceded by the hosts in which they were passaged. For example, Sc237 prions passaged in Syrian hamsters are labeled SHa(Sc237) prions, whereas those passaged in Tg mice are denoted TgSHaPrPSHa(Sc237) prions.

STRUCTURE OF THE INFECTIOUS PRION PARTICLE

Copurification of Infectious Prions and Protein

Development of an incubation time assay for scrapie infectivity accelerated scrapie research by nearly a factor of 100 (Prusiner et al 1980). This improvement in the bioassay permitted the development of effective purification protocols. Enriching fractions from brains of scrapie-infected Syrian hamsters for infectivity led to the discovery of PrP 27–30 (Prusiner et al 1982), which was subsequently shown to be derived by limited proteolysis from a larger protein, designated PrP^{Sc} (Meyer et al 1986, Oesch et al 1985). Copurification of infectious prions and PrP^{Sc} was demonstrated first with biochemical procedures commonly used in protein purification (Prusiner et al

1982, 1983) and later with immunoaffinity chromatography using monoclonal antibodies raised against denatured PrP 27–30 (Gabizon et al 1987, 1988). Some investigators have argued that scrapie infectivity and PrP^{Sc} can be dissociated, but these results have not been reproducible. On the contrary, we and others have found an excellent correlation between prion titers and PrP^{Sc} levels (Jendroska et al 1991). In Tg(SHaPrP) mice, scrapie incubation times were found to be a function of the level of SHaPrP expression when SHa prions were inoculated (Prusiner et al 1990).

Search for a Scrapie-Specific Nucleic Acid

In contrast to the discovery of PrP^{Sc}, the search for the other "half of the putative scrapie virus" has been unrewarding (Meyer et al 1991, Kellings et al 1992). Despite great effort, no candidate scrapie-specific nucleic acid has been identified. Although physical analyses of nucleic acids in highly purified fractions have failed to demonstrate a viral-like polynucleotide, small nucleic acid fragments of < 100 nucleotides could not be eliminated as candidates for a scrapie genome (Meyer et al 1991). Whether these small nucleic acids are contaminants or are essential for infectivity remains uncertain, but there is no evidence to favor such a postulate. Indeed, an impressive array of studies have shown that procedures that hydrolyze or modify proteins inactivate prions, whereas those that hydrolyze or modify nucleic acids do not (Alper et al 1967, 1978; McKinley et al 1983; Bellinger-Kawahara et al 1987a,b).

It has been proposed that an accessory cellular RNA called a coprion can modify the properties of PrP^{Sc} and that each prion "strain" carries a coprion with a unique sequence (Weissmann 1991). Although the coprion hypothesis presents an interesting proposal that could explain the conundrum posed by the isolation of multiple prion "strains," there is no physical or chemical evidence demonstrating the existence of a coprion.

NEW APPROACHES TO PRION DISEASES

Five new approaches to the study of prion diseases have dramatically enhanced our understanding of these disorders. First, the finding that mutations in the PrP gene result in nonconservative substitutions has helped us to understand how these diseases can present as inherited, sporadic, and infectious disorders. Second, Tg mouse studies have illuminated many facets of the prion problem, including control of incubation times, prion synthesis, genetic modeling of neurodegeneration, neuropathology, and prion isolates. Third, combining Tg mice with a new technique for in situ detection of PrP^{Sc} has allowed substantial progress in understanding the biogenesis of prion isolates of "strains." Fourth, some synthetic PrP peptides with predicted α-helical structure possess a high b-sheet content, providing a model for the conversion of PrP^C into PrP^{Sc}. Fifth, chronically infected mouse and Syrian hamster cultured cells have helped clarify the posttranslational synthesis of PrP^{Sc}.

HUMAN PRION DISEASES

Genetic Linkage

Studies of PrP genes (*Prn-p*) in mice with short and long incubation times have demonstrated genetic linkage between a *Prn-p* restriction fragment length polymorphism and a gene modulating incubation times (*Prn-i*) (Carlson et al 1986). Other investigators have confirmed the genetic linkage, and one group has shown that the incubation time gene *Sinc* is also linked to PrP (Hunter et al 1987, Race et al 1990). *Sinc* was first described by Dickinson and colleagues more than 20 years ago (Dickinson et al 1968); whether the genes for PrP, *Prn-i*, and *Sinc* are all congruent remains to be established. The PrP sequences of NZW (*Prn-p^a*) and I/Ln (*Prn-p^b*) mice with short and long scrapie incubation times, respectively, differ at codons 108 (L→F) and 189 (T→V) (Westaway et al 1987). Although these amino acid substitutions argue for the congruency of *Prn-p* and *Prn-i*, experiments with *Prn-p^a* mice expressing *Prn-p^b* transgenes demonstrated a paradoxical shortening of incubation times (Westaway et al 1991) instead of a prolongation as predicted from (*Prn-p^a* × *Prn-p^b*) F1 mice, which exhibit long incubation times that are dominant (Dickinson et al 1968, Carlson et al 1986). Whether this paradoxical shortening of scrapie incubation times in Tg(*Prn-p^b*) mice results from high levels of PrP^C-B expression remains to be established (Westaway et al 1991).

Inherited Prion Diseases

Based on the genetic linkage studies in mice described above, we searched for PrP gene mutations in patients suffering from familial CJD (Neugut et al 1979, Masters et al 1981b) and Gerstmann-Sträussler-Scheinker syndrome (GSS) (Gerstmann et al 1936). To date, 13 mutations of PRNP have been found to segregate with the inherited prion diseases (Figure 1). Wherever the families were of sufficient size and DNA samples were available, significant genetic linkage between the PrP gene mutation and the development of disease has been established. The finding of mutations in the PRNP gene that segregate with the inherited human prion diseases is most compatible with the hypothesis that prions are composed only of PrP^{Sc} molecules. Besides familial CJD and GSS, a new inherited human prion disease designated fatal familial insomnia (FFI) has recently been identified (Goldfarb et al 1992b, Medori et al 1992b) (Table 1). Mutant PrP^C molecules

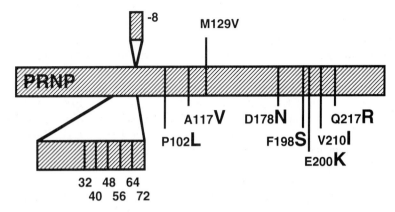

Figure 1   Human prion protein gene (PRNP). The open reading frame (ORF) is denoted by the large gray rectangle. Human PRNP wild-type polymorphisms are shown above the rectangle, whereas mutations that segregate with the inherited prion diseases are depicted below. The wild-type human PrP gene contains five octarepeats [P(Q/H)GGG(G/-)WGQ] from codons 51 to 91 (Kretzschmar et al 1986). Deletion of a single octarepeat at codon 81 or 82 is not associated with prion disease (Laplanche et al 1990, Puckett et al 1991, Vnencak-Jones & Phillips 1992); whether this deletion alters the phenotypic characteristics of a prion disease is unknown. There are common polymorphisms at codons 117 (Ala→Ala) and 129 (Met→Val); homozygosity for Met or Val at codon 129 appears to increase susceptibility to sporadic CJD (Palmer et al 1991). Octarepeat inserts of 32, 40, 48, 56, 64, and 72 amino acids at codons 67, 75, or 83 are designated by small rectangles below the ORF. These inserts segregate with familial CJD, and significant genetic linkage has been demonstrated where sufficient specimens from family members are available (Collinge et al 1989, 1990; Owen et al 1989, 1990; Crow et al 1990; Goldfarb et al 1990c, 1991a; Collinge & Palmer, 1992 unpublished data). Point mutations are designated by the wild-type amino acid, the codon number, and the mutant residue, e.g. P102L. These point mutations segregate with the inherited prion diseases, and significant genetic linkage has been demonstrated where sufficient specimens from family members are available. Mutations at codons 102 (Pro→Leu), 117 (Ala→Val), 198 (Phe→Ser), and 217 (Gln→Arg) are found in patients with GSS (Doh-ura et al 1989; Goldgaber et al 1989; Hsiao et al 1989a,b, 1991b; Goldfarb et al 1990a,c,d; Hsiao & Prusiner 1990; Tateishi et al 1990). Point mutations at codons 178 (Asp→Asn), 200 (Glu→Lys), and 210 (Val→Iso) are found in patients with familial CJD (Goldfarb et al 1990b, 1991c; Gabizon et al 1991; Hsiao et al 1991a; Ripoll et al 1993). Point mutations at codons 198 (Phe→Ser) and 217 (Gln→Arg) are found in patients with GSS who have PrP amyloid plaques and neurofibrillary tangles (Dlouhy et al 1992, Hsiao et al 1992). Single letter code for amino acids are as follows: A, Ala; D, Asp; E, Glu; F, Phe; 1, Iso; K, Lys; L, Leu; M, Met; N, Asn; P, Pro; Q, Gin; R, Arg; S, Ser; T, Thr; and V, Val.

apparently undergo spontaneous conversion into PrP^{Sc}, the accumulation of which causes CNS degeneration. Whether all of the inherited human prion diseases involve PrP^{Sc} formation is unknown; the possibility that mutant PrP^{C} can also cause disease must also be considered.

GSS in humans has been genetically linked to a mutation at codon 102 in the PrP gene that results in the substitution of leucine for proline (Figure 1)

(Hsiao et al 1989a,b; Kretzschmar et al 1991). Six additional point mutations and six different octarepeat insertions have been found to segregate with the inherited human prion diseases. The telencephalic form of GSS has been found to segregate with a mutation at codon 117 in the PrP gene (Doh-ura et al 1989, Hsiao et al 1991b). Mutations at codons 200 and 210 have been identified in some familial CJD pedigrees (Goldfarb et al 1990b,d, 1991b; Gabizon et al 1991; Hsiao et al 1991a; Ripoll et al 1993). Inserts of four, five, six, seven, eight, and nine additional octarepeats have been reported in families with CJD (Goldfarb et al 1991a; Owen et al 1989, 1991). Mutations at codons 198 and 217 (Dlouhy et al 1992, Hsiao et al, in preparation) have been found in a unique form of GSS in which neuritic plaques characteristic of Alzheimer's disease were found to contain PrP instead of the β-A4 peptide (Ghetti et al 1989, Giaccone et al 1990, Tagliavini et al 1991). A codon 178 mutation (Goldfarb et al 1992a; Medori et al 1992a,b) has been found in families with fatal familial insomnia, a unique disorder in which nerve cell loss is highly localized to the mediodorsal and anterior ventral nuclei of the thalamus (Manetto et al 1992). Many familial cases of CJD have also been shown to segregate with the codon 178 mutation. Consistent with failures to demonstrate a scrapie-specific nucleic acid are Tg(GSSMoPrP) mice carrying the codon 102 PrP gene mutation that spontaneously develop neurodegeneration (see Hsiao & Prusiner 1990, Prusiner 1991 for reviews).

Sporadic Prion Diseases

Sporadic cases of CJD (Creutzfeldt 1920, Jakob 1921, Masters & Richardson 1978) comprise $\sim$ 85% of human prion diseases (Table 1). Sporadic CJD occurs with an incidence of $\sim 1/10^6$ across the earth, and all clusters of CJD, once thought to result from the communicable spread of CJD prions, have been shown to be inherited forms of the prion diseases caused by PrP gene mutations. Whether sporadic CJD results from the rare spontaneous conversion of wild-type PrP^C into PrP^{Sc}, or a somatic mutation of the PrP gene produces mutant PrP^C, which is transformed into PrP^{Sc}, remains to be established. An alternative proposal argues that prions contain a scrapie-specific nucleic acid that is ubiquitous; people with wild-type PrP genes are resistant to infection whereas those with PrP gene mutations are highly susceptible (Kimberlin 1990, Chesebro 1992). This proposal must explain how a foreign polynucleotide recruits PrP^C and stimulates its conversion into PrP^{Sc}.

Infectious Prion Diseases

Prion replication can be initiated by inoculation of prions. The results of studies with Tg mice expressing SHaPrP indicate that prion replication involves the formation of a complex between inoculated PrP^{Sc} and PrP^C

synthesized by the host (Prusiner et al 1990). These findings are most compatible with the hypothesis that prions are composed only of PrP^{Sc} molecules. Kuru and iatrogenic CJD illustrate the infectious forms of the human prion diseases (Alpers 1979, 1987; Gajdusek 1977). Kuru was transmitted among New Guinea aborigines by ritualistic cannibalism, whereas prion-contaminated corneas, growth hormone, dura mater grafts, and surgical instruments have all been implicated in iatrogenic CJD (Brown et al 1992). Both familial and infectious forms of human prion disease are rare, accounting for less than 15% of cases.

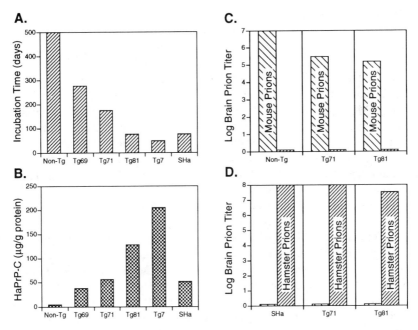

Figure 2 Transgenic mice (Tg) expressing Syrian hamster (SHa) prion protein (PrP) exhibit species-specific scrapie incubation times, infectious prion synthesis and neuropathology (Prusiner et al 1990). (*A*) Scrapie incubation times in nontransgenic mice (Non-Tg) and four lines of Tg mice expressing SHaPrP and Syrian hamsters inoculated intracerebrally with ~ 10^7 ID$_{50}$ units of Sc237 prions serially passaged in Syrian hamsters. The four lines of Tg mice have different numbers of transgene copies: Tg69 and 71 mice have two to four copies of the SHaPrP transgene, whereas Tg81 have 30 to 50 and Tg7 mice have > 60. Incubation times are number of days from inoculation to onset of neurologic dysfunction. (*B*) Brain SHaPrP^C in Tg mice and hamsters. SHaPrPC levels were quantitated by an enzyme-linked immunoassay. (*C*) Prion titers in brains of clinically ill animals after inoculation with Mo prions. Brain extracts from Non-Tg, Tg71, and Tg81 mice were bioassayed for prions in mice (*left*) and hamsters (*right*). (*D*) Prion titers in brains of clinically ill animals after inoculation with SHa prions. Brain extracts from Syrian hamsters as well as Tg71 and Tg81 mice were bioassayed for prions in mice (*left*) and hamsters (*right*). Reprinted, with permission, from Prusiner (1992).

TRANSGENIC MICE EXPRESSING FOREIGN AND MUTANT PrP GENES

In constructions of Tg mice, foreign or mutant PrP transgenes have been shown to profoundly influence experimental scrapie in mice. Investigations with these Tg mice have shown that prion synthesis, scrapie incubation times, and neuropathology are governed by a highly selective interaction of the inoculated prion isolates with the substrate PrP^C that is synthesized by the host (Scott et al 1989, Prusiner et al 1990).

Species Barrier for Prion Transmission

Not only do amino acid substitutions in PrP produce inherited prion diseases, but they are also responsible for the species barrier. Pattison and colleagues were the first to describe greatly prolonged incubation times upon passage of prions from one species to another. Subsequently, we reported that after crossing the species barrier, the PrP^{Sc} molecules synthesized were encoded by the host and did not reflect the PrP^{Sc} in the inoculum (Bockman et al 1987). More persuasive evidence was obtained when Tg mice expressing SHaPrP were constructed. In these Tg(SHaPrP) mice, the species barrier for SHa(Sc237) prions was abolished. In the four different Tg(SHaPrP) mouse lines, scrapie incubation times varied between 50 and 280 days and were found to be a function of the level of $SHaPrP^C$ expression (Scott et al 1989, Prusiner et al 1990) (Figure 2).

The neuropathological changes in the Tg(SHaPrP) mouse lines that developed in response to SHa(Sc237) prions were indistinguishable from those caused by SHa(Sc237) prions in Syrian hamsters (Prusiner et al 1990). These changes included spongiform degeneration confined to grey matter and numerous subependymal, subpial, and subcallosal amyloid plaques composed of SHaPrP based on immunohistochemical staining with a SHaPrP monoclonal antibody (13A5). In contrast, when the Tg(SHaPrP) mice were inoculated with the Mo(RML) prions, scrapie incubation times were similar to those in the parent mouse strain inoculated with Mo(RML) prions, i.e. 160–200 days. The neuropathology resembled that found in C57BL mice with spongiform degeneration in both grey and white matter and no amyloid plaques. Also noteworthy was the finding that Tg(SHaPrP) mice infected with SHa(Sc237) or SHa(139H) prions produced nascent prions with properties indistinguishable from those synthesized in hamsters.

Interactions Between PrP^C and PrP^{Sc} During Prion Replication

Although the search for a scrapie-specific nucleic acid continues to be unrewarding, some investigators have continued work along this line. If prions are found to contain a scrapie-specific nucleic acid, then such a molecule would be expected to direct scrapie agent replication by using a strategy similar to that employed by viruses. In the absence of any chemical or physical evidence for a scrapie-specific polynucleotide (Aiken & Marsh 1990, Prusiner 1991, Kellings et al 1992), it seems reasonable to consider some alternative mechanisms that might feature in prion biosynthesis.

The propagation of prion infectivity is an exponential process in which the posttranslational conversion of PrP^C or a precursor to PrP^{Sc} appears to be obligatory (Borchelt et al 1990). In the simplest model, one PrP^{Sc} molecule combines with one PrP^C molecule to form a heterodimeric intermediate that is transformed into two molecules of PrP^{Sc}. In the next cycle, each of the two PrP^{Sc} molecules combines with a PrP^C molecule, giving rise to four PrP^{Sc} molecules. In the third cycle, each of the four PrP^{Sc} molecules combines with a PrP^C molecule, giving rise to eight PrP^{Sc} molecules, thus creating an exponential process (Prusiner 1991). Assuming prion biosynthesis simply involves amplification of posttranslationally altered PrP molecules, we might expect Tg(SHaPrP) mice to produce both SHa and Mo prions after inoculation with either prion, since these mice synthesize both SHa and $MoPrP^C$. Yet Tg(SHaPrP) mice produce only those prions present in the inoculum (Figures 2C and D). We have interpreted these results as indicating that the incoming PrP^{Sc} molecules interact with the homologous PrP^C substrate to replicate more of the same prions (Prusiner et al 1990).

Additional evidence in support of the proposed model for prion replication comes from Tg(Mo/SHaPrP) mice expressing chimeric $Mo/SHaPrP^C$ (M Scott, D Groth, M Torchia, D Foster & SB Prusiner, in preparation). The chimeric Mo/SHaPrP gene was constructed by substituting the SHaPrP sequence for MoPrP from codon 94 to 188; within this domain, there are five amino acid substitutions which distinguish Mo from SHaPrP. When inoculated with either Mo or SHa prions, these Tg(Mo/SHaPrP) mice develop scrapie after ~ 140 days. The chimeric Tg mice produce $Mo/SHaPrP^{Sc}$ and Mo/SHa prions after inoculation with SHa prions and probably Mo prions as well. Evidence for chimeric Mo/SHa prions comes from the development of scrapie in Tg(Mo/SHaPrP) mice at ~ 70 days after inoculation with brain extracts from Tg(Mo/SHaPrP) mice containing the chimeric prions.

Ablation of the PrP Gene

Ablation of the PrP gene in Tg ($Prn-p^{0/0}$) mice has, unexpectedly, not affected the development of these animals (Büeler et al 1992). In fact, they are healthy at almost two years of age. $Prn-p^{0/0}$ mice are resistant to prions and do not propagate scrapie infectivity (Büeler et al 1993). Furthermore, since the absence of PrP^C expression does not provoke disease, we can conclude that scrapie and other prion diseases are a consequence of PrP^{Sc} accumulation rather than the inhibition of PrP^C function. To date, the function of PrP^C remains unknown.

SYNTHESIS OF PrP^{Sc} IS A POSTTRANSLATIONAL PROCESS

Metabolic labeling studies of scrapie-infected cultured cells have shown that PrP^C is synthesized and degraded rapidly whereas PrP^{Sc} is synthesized slowly by an as yet undefined posttranslational process (Figure 1) (Caughey et al 1989; Borchelt et al 1990, 1992; Caughey & Raymond 1991). These observations are consistent with earlier findings showing that PrP^{Sc} accumulates in the brains of scrapie-infected animals whereas PrP mRNA levels remain unchanged (Oesch et al 1985). Furthermore, the structure and organization of the PrP gene make it likely that PrP^{Sc} is formed during a posttranslational event (Basler et al 1986).

Both PrP isoforms appear to transit through the Golgi apparatus, where their Asn-linked oligosaccharides are modified and sialylated (Bolton et al 1985, Manuelidis et al 1985, Endo et al 1989, Haraguchi et al 1989, Rogers et al 1990). PrP^C is presumably transported within secretory vesicles to the external cell surface, where it is anchored by a glycosyl phosphatidylinositol (GPI) moiety (Figure 3; Stahl et al 1987, 1992; Safar et al 1990). In contrast, PrP^{Sc} accumulates primarily within cells, where it is deposited in cytoplasmic vesicles, many of which appear to be secondary lysosomes (Taraboulos et al 1990b, 1992b; Caughey et al 1991a; McKinley et al 1991b; Borchelt et al 1992).

Whether PrP^C is the substrate for PrP^{Sc} formation or a restricted subset of PrP molecules are precursors for PrP^{Sc} remains to be established. Several experimental results suggest that PrP molecules destined to become PrP^{Sc} exit to the cell surface, as does PrP^C (Stahl et al 1987), prior to their conversion into PrP^{Sc} (Caughey & Raymond 1991, Borchelt et al 1992, Taraboulos et al 1992b). Interestingly, the GPI anchors of both PrP^C and PrP^{Sc}, which presumably are important in directing the subcellular trafficking of these molecules, are sialylated (Stahl et al 1992). It is unknown whether sialylation of the GPI anchor participates in some aspect of PrP^{Sc} formation.

Although most of the difference between the mass of PrP 27–30 predicted from the amino acid sequence and that observed after posttranslational modification results from complex-type oligosaccharides, these sugar chains are not required for PrP^{Sc} synthesis in scrapie-infected cultured cells, based on experiments with the Asn-linked glycosylation inhibitor tunicamycin and site-directed mutagenesis studies (Taraboulos et al 1990a).

Cell-free translation studies have demonstrated two forms of PrP: a transmembrane form, which spans the bilayer twice at the transmembrane

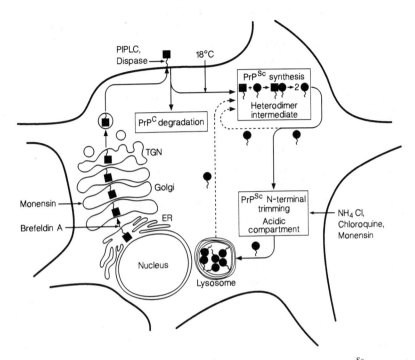

Figure 3 Pathways of prion protein synthesis and degradation in cultured cells. PrP^{Sc} is denoted by circles; squares designate PrP^{C} and the PrP^{Sc} precursor, which may be indistinguishable. Rectangular boxes denote as yet unidentified subcellular compartments. Prior to becoming protease-resistant, the PrP^{Sc} precursor transits through the plasma membrane and is sensitive to dispase or phosphatidylinositol-specific phospholipase C (PIPLC) added to the medium. PrP^{Sc} synthesis probably occurs in a compartment accessible from the plasma membrane, such as caveolae or endosomes; PrP^{Sc} formation is blocked at 18°C. PrP^{Sc} synthesis probably occurs through the interaction of PrP^{Sc} precursor with existing PrP^{Sc}; the dotted lines denote possible feedback pathways for the reflection of PrP^{Sc} in the active site. Acidic pH within vesicles is not obligatory for PrP^{Sc} synthesis. One to two hours after PrP^{Sc} formation, it is N-terminally trimmed by an acidic protease; PrP^{Sc} then accumulates primarily in secondary lysosomes. The inhibition of PrP^{Sc} synthesis by Brefeldin A demonstrates that the endoplasmic reitculum (ER)-Golgi is not competent for its synthesis and that transport of PrP down the secretory pathway is required for the formation of PrP^{Sc}. Reprinted, with permission, from Taraboulos et al (1992b).

(TM) and amphipathic helix domains, and a secretory form (Lopez et al 1990, Yost et al 1990). The stop transfer effector (STE) domain controls the topogenesis of PrP. The fact that PrP contains both a TM domain and a GPI anchor poses a topologic conundrum. It seems likely that membrane-dependent events are important in the synthesis of PrP^{Sc}, especially since Brefeldin A, which selectively destroys the Golgi stacks (Doms et al 1989), prevents PrP^{Sc} synthesis in scrapie-infected cultured cells (Taraboulos et al 1992b).

For many years, the association of scrapie infectivity with membrane fractions has been appreciated (Gibbons & Hunter 1967); indeed, hydrophobic interactions are thought to account for many of the physical properties displayed by infectious prion particles (Prusiner et al 1980, Gabizon & Prusiner 1990).

CONFORMATIONAL CHANGES IN PRION PROPAGATION

Structure Prediction and Synthetic PrP Peptides

In the absence of any candidate posttranslational chemical modifications (Stahl et al 1993) that differentiate PrP^C from PrP^{Sc}, we are forced to consider the possibility that conformation distinguishes these isoforms. In a comparison of the amino acid sequences of 11 mammalian and one avian prion proteins, structural analyses predicted four α-helical regions (J-M Gabriel, F Cohen, RA Fletterick & SB Prusiner, in preparation; Cohen et al 1986). Peptides corresponding to these regions of the SHaPrP were synthesized and, contrary to predictions, three of the four spontaneously formed amyloids as shown by electron microscopy and Congo red staining (Gasset et al 1992). With infrared spectroscopy, these amyloid peptides were found to exhibit a secondary structure comprised largely of β-sheets. The first of the predicted helices is the 14-residue peptide corresponding to SHaPrP codons 109–122; this peptide and the overlapping 15-residue sequence 113–127 both form amyloid. The most highly amyloidogenic peptide is the sequence AGAAAAGA corresponding to PrP codons 113–120. This peptide is in a region of PrP that is conserved across all known species. Two other predicted α-helices corresponding to SHaPrP codons 178–191 and 202–218 form amyloids and exhibit considerable β-sheet structure when synthesized as peptides. These findings suggest the possibility that the conversion of PrP^C to PrP^{Sc} involves the transition of one or more putative PrP α-helices into β-sheets. Infrared spectroscopy of PrP 27–30 has shown a high β-sheet content (Caughey et al 1991b), which decreased when PrP 27–30 was denatured; scrapie infectivity diminished concomitantly (Gasset et al 1993).

These structural investigations of synthetic PrP peptides and the correlations between PrP 27–30 secondary structure and scrapie infectivity offer a structural model for the conversion of PrP^C to PrP^{Sc} as well as the replication of infectious prion particles involving a transition from α-helices to β-sheets in PrP. Whether any of these synthetic PrP peptides can induce brain degeneration, PrP^{Sc} formation, or prion infectivity is currently being investigated. If additional data can be obtained to support the hypothesis set forth here, then it may be useful to examine other degenerative diseases with respect to proteins undergoing similar structural changes.

In humans carrying point mutations or inserts in their PrP genes, mutant PrP^C molecules might spontaneously convert into PrP^{Sc}. Although the initial stochastic event may be inefficient, once it happens the process becomes autocatalytic. The proposed mechanism is consistent with individuals harboring germline mutations who do not develop CNS dysfunction for decades, and with studies on Tg(GSSMoPrP) mice that spontaneously develop CNS degeneration (Hsiao et al 1990). Whether all GSS and familial CJD cases contain infectious prions or some represent inborn errors of PrP metabolism in which neither PrP^{Sc} nor prion infectivity accumulates is unknown. If the latter is found to be the case, then presumably, mutant PrP^C molecules alone can produce CNS degeneration.

PrP Amyloid is Not Necessary for Prion Propagation

Some investigators have suggested that scrapie agent multiplication proceeds through a crystallization process involving PrP amyloid formation (Gajdusek 1988, 1990; Gajdusek & Gibbs 1990). Against this hypothesis is the absence or rarity of amyloid plaques in many prion diseases, as well as the inability to identify any amyloid-like polymers in cultured cells chronically synthesizing prions (McKinley et al 1991a, Prusiner et al 1990). Purified infectious preparations isolated from scrapie-infected hamster brains exist as amorphous aggregates; only if PrP^{Sc} is exposed to detergents and limited proteolysis does it then polymerize into prion rods exhibiting the ultrastructural and tinctorial features of amyloid (McKinley et al 1991a). Furthermore, dispersion of prion rods into detergent-lipid-protein complexes results in a 10- to 100-fold increase in scrapie titer, and no rods could be identified in these fractions by electron microscopy (Gabizon et al 1987).

IN SITU MEASUREMENTS OF PrP^{Sc}

Histoblotting of PrP^{Sc}

A new highly sensitive and specific technique termed "histoblotting" has been developed to localize and quantify PrP^{Sc} (Taraboulos et al 1992a). By using the histoblotting procedure, persuasive evidence has been acquired indicating that the deposition of PrP^{Sc} is responsible for the neuropathologic changes found in the prion diseases. Histoblotting overcomes two obstacles that plagued PrP^{Sc} detection in brain by standard immunohistochemical techniques: the presence of PrP^C and weak antigenicity of PrP^{Sc} (DeArmond et al 1987). The histoblot is made by pressing 10 μm thick cryostat sections of fresh frozen brain tissue to nitrocellulose paper. To localize protease-resistant PrP^{Sc} in brain, the histoblot is digested with proteinase K to eliminate PrP^C, then the undigested PrP^{Sc} is denatured with guanidinium thiocyante (GdnSCN) or

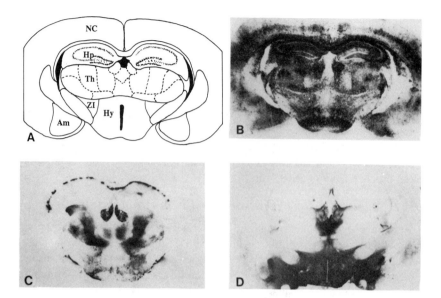

Figure 4   Distribution of PrP^{Sc} as a function of prion isolate at the level of the hippocampus and thalamus revealed by histoblots. In each case, the animal was in the terminal stages of scrapie. For the Sc237 and 139H prion isolates, the animals developed clinical signs between 49 and 55 days postinoculation. For the Me7H isolate, animals developed clinical signs between 180 and 200 days. Am, amygdala; Hp, hippocampus; Hy, hypothalamus; NC, neocortex; Th, thalamus; ZI, zona incerta.

NaOH to enhance binding of PrP antibodies. Immunohistochemical staining of the resulting histoblot yields a far more intense, specific, and reproducible PrP signal than was ever achieved by immunohistochemistry on standard tissue sections. Furthermore, the intensity of immunostaining in each brain region correlates well with neurochemical estimates of PrP^{Sc} concentration in homogenates of dissected brain regions. PrP^{C} can also be localized in histoblots of normal brains by eliminating the proteinase K digestion step (Figure 4).

Distinct Isolates of Prions

In 1961, Pattison & Millson demonstrated that the different clinical syndromes in scrapie-infected sheep could be transferred experimentally to and maintained through serial passages in goats. They concluded that "...the type of clinical syndrome produced by experimental inoculation will resemble the syndrome exhibited by the animal from which the material for inoculation was obtained" (Pattison & Millson 1961). Subsequently, multiple scrapie isolates were identified by using rodents (Dickinson et al 1968, Bruce & Dickinson 1987); each isolate was defined by (*a*) scrapie incubation times, (*b*) distribution and intensity of spongiform degeneration (Fraser & Dickinson 1973), and (*c*) cerebral amyloid plaques (Fraser & Bruce 1973, Bruce et al 1976). These characteristics were preserved during sequential passages within a single inbred mouse strain or hamster species but varied markedly or even failed to appear when a particular scrapie isolate was transferred to a different inbred mouse or animal species.

The existence of distinct prion isolates or "strains" has been frequently cited as requiring a scrapie-specific nucleic acid to explain prion diversity, yet none has been found. Because multiple prion isolates can be prepared from a single host animal, each capable of transmitting a different clinical-pathological syndrome, the challenge is to explain how this can be accomplished if the only functional component of the prion is PrP^{Sc}. Specifically, one has to postulate that PrP^{Sc} can exist in multiple, stable configurations. There is considerable evidence that the synthesis of PrP^{Sc} involves the posttranslational conversion of PrP^{C} into PrP^{Sc} (Borchelt et al 1990, 1992; Caughey & Raymond 1991) and that PrP^{Sc} recognizes homologous PrP^{C} molecules as they form a transient complex. The propagation of distinct prion isolates, each of which carries a particular set of characteristics that give rise to a specific scrapie incubation time and neuropathology, raises the questions: how can infectious prions be composed of PrP^{Sc} molecules alone? If prions are devoid of nucleic acid, how can biological information be retained as they are passaged from one host to another? With other infectious pathogens, such as bacteria, viruses, and viroids, biological characteristics are clearly specified by a nucleic acid genome.

The high specificity and sensitivity of the histoblotting technique has played a major role in the evolution of our recent thinking about prion isolates. Unexpectedly, we found that the pattern of PrP^{Sc} distribution in the brain is characteristic of and unique to each prion isolate (Hecker et al 1992, DeArmond et al 1993). This finding has led to a new hypothesis that each distinct prion isolate is propagated in a different set of cells. The biological properties exhibited by each isolate may arise from different modifications of PrP^{Sc} that are unique to a particular set of cells in which a given isolate is propagated.

Three Prion Isolates Produce Different Patterns of PrP^{Sc} Deposition

We have studied three prion isolates, all passaged in Syrian hamsters. Each has a unique incubation time and a unique lesion profile. We asked if each of the three isolates produces a unique pattern of PrP^{Sc} accumulation (Hecker et al 1992, DeArmond et al 1993). To minimize the time required for these studies, we used Tg(SHaPrP)7 mice instead of Syrian hamsters. Tg(SHaPrP)7 mice were inoculated intracerebrally with each of the three isolates designated Sc237, 139H, and Me7H. The mean scrapie incubation time for both Sc237 and 139H was ~ 50 days and for Me7H was ~ 144 days in Tg(SHaPrP)7 mice. As shown in Figure 4, each isolate produced a unique pattern of PrP^{Sc} deposition.

What is the Molecular Basis of Prion Diversity?

The molecular mechanism responsible for distinct isolates or "strains" of prions remains enigmatic. As noted above, there is no evidence for a scrapie-specific nucleic acid encoding this observed diversity. Recently, two new proposals have been set forth. First, it has been proposed that isolate-specific information might be encoded within the tertiary and quaternary structure of PrP^{Sc} (Prusiner 1991). Since prion propagation apparently involves formation of a replication intermediate, i.e. a PrP^{C}/PrP^{Sc} complex, it was suggested that isolate-specific information resides in the tertiary structure of PrP^{Sc}. But such a hypothesis must account for the fact that the number of different conformations that PrP^{Sc} might assume is likely to be quite limited (Bessen & Marsh 1992, Prusiner 1992).

An alternative hypothesis suggests that the diversity of prions may reside in PrP^{Sc} glycoforms. This hypothesis is attractive because the diversity of PrP^{Sc} glycoforms seems to be sufficient to account for a large number of prion isolates, and the propagation of a particular prion isolate might result from the synthesis of nascent PrP^{Sc} molecules within a restricted subset of cells (Hecker et al 1992). If this explanation proves to have merit, then receptors on the surface of this subset of cells would recognize the N-linked CHOs of PrP^{Sc} that are homologous to those that are attached to the PrP^{C} synthesized within the cells. In other words, these cells express a repetoire of glycosyltransferases necessary to synthesize PrP^{C} molecules with the same CHOs as those found attached to PrP^{Sc} molecules in the inoculum. This hypothesis not only is supported by a number of experimental observations but it also explains how each prion isolate exhibits a specific scrapie incubation time, neuropathologic lesion profile, and pattern of PrP^{Sc} accumulation. Moreover, the propagation of a given prion isolate is a consequence of the particular population of cells in which PrP^{Sc} is synthesized. The postulate set forth here does not preclude PrP^{Sc} interacting with receptor proteins on the surface of cells that might restrict its entry or even PrP^{C}. The validity of these hypotheses remains to be determined. Although PrP^{Sc} formation has been shown to occur in the absence of Asn-linked glycosylation (Taraboulos et al 1990a), it is unknown whether infectious prions can be formed with unglycosylated PrP^{Sc}.

PrP^Sc-Deposition Precedes Spongiform Degeneration

Although immunohistochemistry of PrP^{Sc} suggested that deposition of PrP^{Sc} preceded and thus might cause spongiform degeneration of neurons and reactive astrocytic gliosis found in rodents with scrapie, the technical problems cited above prevented any firm conclusion (DeArmond et al 1987). With the advent of the histoblotting procedure, an excellent correlaton between PrP^{Sc} deposition and neuropathologic lesions was demonstrated for three different isolates of scrapie prions propagated in Syrian hamsters and designated Sc237, 139H, and Me7H (Figures 5 and 6; DeArmond et al 1993, Hecker et al 1992). There is also a temporal correlation between the accumulation of PrP^{Sc} and the development of neuropathology (Jendroska et al 1991). With Sc237 prions, spongiform degeneration and reactive astrocytic gliosis appear to follow the accumulation of PrP^{Sc} in Syrian hamster brains. Finally, PrP amyloid plaques in Tg(SHaPrP) mice were found when SHa prions were inoculated but not when Mo prions were injected. When found, PrP amyloid plaques are diagnostic of the human prion diseases (DeArmond et al 1985, 1987; Prusiner & DeArmond 1987). Additional evidence for the role of PrP in causing spongiform degeneration of grey matter and reactive astrocytic gliosis comes from studies with Tg(GSSMoPrP) mice that spontaneously develop neurodegeneration (Hsiao et al 1990). In this animal model of an inherited prion disease, overexpression of a mutant transgene triggers the development of neurodegeneration.

Targeting and Spread of Prions to Specific Neuronal Populations

The application of the histoblot technique for localizing and quantifying PrP^{Sc} in the brains of scrapie-infected Syrian hamsters and Tg mice revealed the site-specific formation and accumulation of PrP^{Sc}. The pattern and rate of PrP^{Sc} accumulation in Syrian hamsters inoculated in the thalamus with SHa(Sc237) prions has been examined in detail with both neurochemical measurements of PrP^{Sc} in dissected brain regions (DeArmond et al 1987, Jendroska et al 1991) and with the histoblot method (Hecker et al 1992, Taraboulos et al 1992a). De novo synthesis of PrP^{Sc} began at the site of inoculation of SHa(Sc237) prions in the thalamus, where its accumulation became detectable 14–21 days postinoculation by the neurochemical method and unilaterally in the thalamus at 7 days by the histoblot technique. The second site to accumulate PrP^{Sc} was the septum at about 28 days. This was puzzling because it had been assumed that scrapie spread from region to region in the brain along neuroanatomical pathways (Kimberlin et al 1987, Scott et al 1989) and because the septum is poorly interconnected with the thalamus.

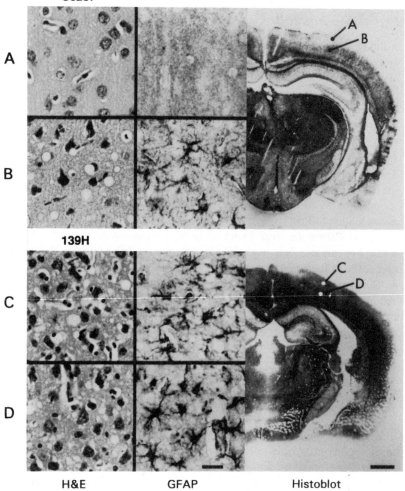

Figure 5 Spongiform degeneration of grey matter colocalizes with PrP^{Sc} deposition in Syrian hamsters and Tg(SHaPrP) mice inoculated with Sc237, 139H, or Me7H prion isolates. *Left panels:* Spongiform degeneration assessed on brain sections stained with hematoxylin and eosin. *Middle panels:* Reactive astrocytic gliosis demonstrated by immunostaining with antibodies to glial fibrillary acidic protein (GFAP). With Sc237 prions, spongiform degeneration and reactive astrocytic gliosis were detectable in layer IV (*upper panels*). In contrast with 139H prions, spongiform degeneration and reactive astrocytic gliosis were present in layers II and IV (*lower panels*). Areas of PrP^{Sc} deposition in the brain of a hamster inoculated with Sc237 prions (*right upper panel*) differed from those with 139H prions (*right lower panel*). The differences are most striking in the cerebral cortex. With Sc237 prions, PrP^{Sc} accumulated in layers III–VI (see histoblot), whereas with 139H prions, PrP^{Sc} was deposited throughout full thickness of the cerebral cortex (layers I–VI). Approximate locations of the sections stained with hematoxylin and eosin or immunostained for GFAP are indicated by the dots labeled with letters (*A–D*) in the cerebral cortex in the histoblot from layer II to the outer portion of layer IV.

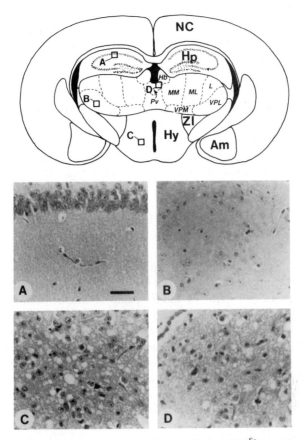

Figure 6   In Tg(SHaPrP) mice, vacuolation also colocalized with PrP^Sc deposition. Compare the neuropathology in this figure with the distribution of PrP^Sc in Me7-inoculated mice shown in Figure 4D. Little or no PrP^Sc was deposited in the CA1 region of the hippocampus (*A*) or the ventral posterior lateral (VPL) nucleus of the thalamus (*B*), and neither contained vacuoles, whereas two regions with intense PrP^Sc deposition, the hypothalamus (*C*) and the paraventricular nucleus of the thalamus (*D*), were severely vacuolated. Hematoxylin- and eosin-stained histological sections. Bar in *A* is 50 μm and applies to all photomicrographs.

The combination of high sensitivity and anatomical localization of PrP^Sc by histoblots provided an explanation. The histoblots indicated that the septum was uniquely infected via the cerebrospinal fluid (CSF). Within the first two weeks following intrathalamic inoculation, PrP^Sc was identified along the ventricular lining of the septum and the medial surface of the caudate nucleus (Taraboulos et al 1992a). During the next two weeks, the PrP^Sc signal increased in thickness and intensity, suggesting that it was derived from newly synthesized PrP^{Sc} in the thalamus and not merely residual PrP^{Sc} from the original inoculum. By 28 days, the time when PrP^{Sc} was first detected in the septum by neurochemical methods, the histoblot revealed PrP^{Sc} accumulation in the medial septal nucleus and diagonal band of Broca. Other brain regions, including the caudate nucleus, apparently became infected via neuroanatomical pathways, based on the timing and pattern of spread from the thalamus. For example, disease spread to the cerebral cortex two to three weeks after involvement of the thalamus. That cortical involvement was the result of anterograde transport along thalamocortical pathways was supported by the accumulation of PrP^{Sc} mainly in neocortical layers 4 and 6, the regions where the thalamic projection systems terminate. The caudate nucleus, which is highly interconnected with the cerebral cortex, accumulated PrP^{Sc} two weeks later.

The finding that the disease spread from the thalamus to the septum but not to the caudate nucleus via the CSF was the first to indicate that prions target specific nerve cell populations. The second finding supporting this conclusion was the pattern of PrP^{Sc} spread within the thalamus itself. The entire thalamus appeared to accumulate PrP^{Sc} by 65 days, when clinical signs appeared; this did not occur by radial spread of disease from the site of inoculation; rather, the spread must have occurred by a discontinuous process. At 28 and 35 days, new accumulations of PrP^{Sc} were found in the contralateral thalamic nuclei. The two halves of the thalamus are not directly interconnected by neuroanatomical pathways, suggesting that as in the septum, newly formed PrP^{Sc} was released into the CNS extracellular space from thalamic nuclei on the side of inoculation and was selectively targeted to the identical neuronal populations on the opposite side of the thalamus.

Formation of PrP Amyloid Plaques

Corroborating evidence that PrP^{Sc} is released into the extracellular space comes from the presence of symmetrically localized PrP amyloid plaques, particularly in the subcallosal region. PrP amyloid plaques are composed of extracellular deposits of polymerized, protease-resistant PrP (DeArmond et al 1985). The location of amyloid plaques can be duplicated precisely by unilateral injection of dilute India ink suspension into the brain. Diffusion of India ink within the CNS extracellular space, into the CSF, and to the subcallosal region in particular occurs rapidly, within one hour of injection. This experiment suggests that the subcallosal amyloid plaques represent PrP^{Sc} trapped in the potential space between the hippocampus and corpus callosum, which is continuous with the walls of the lateral ventricles. The similarity between the distribution of PrP amyloid plaques and that of India ink injected into the brain also indicates that PrP^{Sc} originates within the brain and not in sites outside the CNS, such as the spleen.

The observations of SHa(Sc237) prions in Syrian hamsters indicate that there is a selective vulnerability of some neuron populations to infection by prions formed early after inoculation. The medial septal and diagonal band nuclei were more vulnerable than the caudate nucleus, and the lateral thalamic nuclei were more vulnerable than the medial nuclei.

CONCLUDING REMARKS

The study of prions has taken several unexpected directions over the past few years. The discovery that prion diseases in humans are uniquely both genetic and infectious has greatly strengthened and extended the prion concept. To date, 13 different mutations in the human PrP gene, all resulting in nonconservative substitutions, have been found to segregate with the inherited prion diseases, and for many mutations genetic linkage has been established. Yet the transmissible prion particle is composed largely, if not entirely, of an abnormal isoform of the prion protein designated PrP^{Sc} (Prusiner 1991). These findings suggest that prion diseases should be considered pseudoinfections, since the particles transmitting disease appear to be devoid of a foreign nucleic acid and thus differ from all known microorganisms as well as viruses and viroids. Because much information, especially about scrapie of rodents, has been derived by using experimental protocols adapted from virology, we continue to use terms such as infection, incubation period, transmissibility, and endpoint titration in studies of prion diseases.

Although relatively little is known about the replication of prions, Tg mice expressing foreign or mutant PrP genes now permit virtually all facets of prion diseases to be studied and have provided a framework for future investigations. Furthermore, the structure and organization of the PrP gene suggested that PrP^{Sc} is derived from PrP^C or a precursor by a posttranslational process. Studies with scrapie-infected cultured cells have provided evidence that the conversion of PrP^C to PrP^{Sc} is a posttranslational process that probably occurs in the endocytic pathway. The molecular basis of the PrP^{Sc} synthetic process remains to be elucidated, but extensive protein chemical studies suggest that this process is likely to involve a conformational change.

It seems likely that the principles learned from the study of prion diseases will be applicable to elucidating the causes of more common neurodegenerative diseases. Such disorders include Alzheimer's disease, amyotrophic lateral sclerosis, and Parkinson's disease. Since people at risk for inherited prion diseases can now be identified decades before neurologic dysfunction is evident, the development of an effective therapy is imperative. If PrP^C can be diminished in humans without deleterious effects, as is the case for $Prn-p^{0/0}$ mice (Büeler et al 1992), then reducing the level of PrP mRNA with antisense oligonucleotides might prove an effective therapeutic maneuver in delaying the onset of CNS symptoms and signs.

The study of prion biology and diseases seems to be a new and emerging area of biomedical investigation. While prion biology has its roots in virology, neurology, and neuropathology, its relationships with the disciplines of molecular and cell biology as well as protein chemistry have become evident only recently. A better understanding of how prions multiply and cause disease will surely open new vistas in biochemistry and genetics.

ACKNOWLEDGMENTS

The authors thank M. Baldwin, D. Borchelt, G. Carlson, F. Cohen, C. Cooper, R. Fletterick, D. Foster, J.-M. Gabriel, M. Gasset, R. Gabizon, D. Groth, L. Hood, K. Hsiao, V. Lingappa, M. McKinley, W. Mobley, B. Oesch, D. Riesner, M. Scott, A. Serban, N. Stahl, A. Taraboulos, M. Torchia, C. Weissmann, and D. Westaway for their help in these studies. Special thanks to Lorraine Gallagher who assembled this manuscript. This research is supported by grants from the National Institutes of Health (NS14069, AG08967, AG02132, and NS22786) and the American Health Assistance Foundation, as well as by gifts from Sherman Fairchild Foundation, Bernard Osher Foundation, and National Medical Enterprises.

Literature Cited

Aiken JM, Marsh RF. 1990. The search for scrapie agent nucleic acid. *Microbiol. Rev.* 54:242–46

Alper T, Cramp WA, Haig DA, Clarke MC. 1967. Does the agent of scrapie replicate without nucleic acid? *Nature* 214:764–66

Alper T, Haig DA, Clarke MC. 1978. The scrapie agent: evidence against its dependence for replication on intrinsic nucleic acid. *J. Gen. Virol.* 41:503–16

Alpers M. 1987. Epidemiology and clinical aspects of kuru. In *Prions—Novel Infectious Pathogens Causing Scrapie and Creutzfeldt-Jakob Disease*, eds. SB Prusiner, MP McKinley, pp. 451–65. Orlando: Academic Press

Alpers MP. 1968. Kuru: implications of its transmissibility for interpretation of its changing epidemiological pattern. In *The Central Nervous System, Some Experimental Models of Neurological Diseases*, eds. OT Bailey, DE Smith, pp. 234–51. Baltimore: Williams and Wilkins

Alpers MP. 1979. Epidemiology and ecology of kuru. In *Slow Transmissible Diseases of the Nervous System*, eds. SB Prusiner, WJ Hadlow, 1:67–90. New York: Academic Press

Basler K, Oesch B, Scott M, et al. 1986. Scrapie and cellular PrP isoforms are encoded by the same chromosomal gene. *Cell* 46:417–28

Bazan JF, Fletterick RJ, McKinley MP, Prusiner SB. 1987. Predicted secondary structure and membrane topology of the scrapie prion protein. *Protein Eng.* 1:125–35

Bellinger-Kawahara C, Cleaver JE, Diener TO, Prusiner SB. 1987a. Purified scrapie prions resist inactivation by UV irradiation. *J. Virol.* 61:159–66

Bellinger-Kawahara C, Diener TO, McKinley MP, et al. 1987b. Purified scrapie prions resist inactivation by procedures that hydrolyze, modify, or shear nucleic acids. *Virology* 160:271–74

Bessen RA, Marsh RF. 1992. Biochemical and physical properties of the prion protein from two strains of the transmissible mink encephalopathy agent. *J. Virol.* 66:2096–2101

Bockman JM, Prusiner SB, Tateishi J, Kingsbury DT. 1987. Immunoblotting of Creutzfeldt-Jakob disease prion proteins: host species-specific epitopes. *Ann. Neurol.* 21:589–95

Bolton DC, Meyer RK, Prusiner SB. 1985.

Scrapie PrP 27–30 is a sialoglycoprotein. *J. Virol.* 53:596–606

Borchelt DR, Scott M, Taraboulos A, et al. 1990. Scrapie and cellular prion proteins differ in their kinetics of synthesis and topology in cultured cells. *J. Cell Biol.* 110: 743–52

Borchelt DR, Taraboulos A, Prusiner SB. 1992. Evidence for synthesis of scrapie prion proteins in the endocytic pathway. *J. Biol. Chem.* 267:6188–99

Brown P, Preece MA, Will RG. 1992. "Friendly fire" in medicine: hormones, homografts, and Creutzfeldt-Jakob disease. *Lancet* 340:24–27

Bruce ME, Dickinson AG. 1987. Biological evidence that the scrapie agent has an independent genome. *J. Gen. Virol.* 68:79–89

Bruce ME, Dickinson AG, Fraser H. 1976. Cerebral amyloidosis in scrapie in the mouse: effect of agent strain and mouse genotype. *Neuropathol. Appl. Neurobiol.* 2:471–78

Büeler H, Aguzzi A, Sailer A, et al. 1993. Mice devoid of PrP are resistant to scrapie. *Cell* 73:1139–47

Büeler H, Fischer M, Lang Y, et al. 1992. The neuronal cell surface protein PrP is not essential for normal development and behavior of the mouse. *Nature* 356:577–82

Carlson GA, Kingsbury DT, Goodman PA, et al 1986. Linkage of prion protein and scrapie incubation time genes. *Cell* 46:503–11

Caughey B, Race RE, Ernst D, et al. 1989. Prion protein biosynthesis in scrapie-infected and uninfected neuroblastoma cells. *J. Virol.* 63:175–81

Caughey B, Raymond GJ. 1991. The scrapie-associated form of PrP is made from a cell surface precursor that is both protease- and phospholipase-sensitive. *J. Biol. Chem.* 266: 18217–23

Caughey B, Raymond GJ, Ernst D, Race RE. 1991a. N-terminal truncation of the scrapie-associated form of PrP by lysosomal protease(s): implications regarding the site of conversion of PrP to the protease-resistant state. *J. Virol.* 65:6597–6603

Caughey BW, Dong A, Bhat KS, et al. 1991b. Secondary structure analysis of the scrapie-associated protein PrP 27–30 in water by infrared spectroscopy. *Biochemistry* 30: 7672–80

Chesebro B. 1992. PrP and the scrapie agent. *Nature* 356:560

Cohen FE, Abarbanel RM, Kuntz ID, Fletterick RJ. 1986. Turn prediction in proteins using a pattern-matching approach. *Biochemistry* 25:266–75

Collinge J, Harding AE, Owen F, et al. 1989. Diagnosis of Gerstmann-Straussler syndrome in familial dementia with prion protein gene analysis. *Lancet* 2:15–17

Collinge J, Owen F, Poulter H, et al. 1990. Prion dementia without characteristic pathology. *Lancet* 336:7–9

Creutzfeldt HG. 1920. Über eine eigenartige herdförmige Erkrankung des Zentrainervensystems. *Z. Gesamte Neurol. Psychiatr.* 57: 1–18. In German

Crow TJ, Collinge J, Ridley RM, et al. 1990. Mutations in the prion gene in human transmissible dementia. *Seminar on Molecular Approaches to Research in Spongiform Encephalopathies in Man*. London: Medical Research Council (Abstr.)

DeArmond SJ, McKinley MP, Barry RA, et al. 1985. Identification of prion amyloid filaments in scrapie-infected brain. *Cell* 41: 221–35

DeArmond SJ, Mobley WC, DeMott DL, et al. 1987. Changes in the localization of brain prion proteins during scrapie infection. *Neurology* 37:1271–80

DeArmond SJ, Yang S-L, Lee A, et al. 1993. Three scrapie prion isolates exhibit different accumulation patterns of the prion protein scrapie isoform. *Proc. Natl. Acad. Sci. USA* 90:6449–53

Dickinson AG, Meikle VMH, Fraser H. 1968. Identification of a gene which controls the incubation period of some strains of scrapie agent in mice. *J. Comp. Pathol.* 78:293–99

Dlouhy SR, Hsiao K, Farlow MR, et al. 1992. Linkage of the Indiana kindred of Gerstmann-Sträussler-Scheinker disease to the prion protein gene. *Nature Genetics* 1:64–67

Doh-ura K, Tateishi J, Sasaki H, et al. 1989. Pro→Leu change at position 102 of prion protein is the most common but not the sole mutation related to Gerstmann-Sträussler syndrome. *Biochem. Biophys. Res. Commun.* 163:974–79

Doms RW, Russ G, Yewdell JW. 1989. Brefeldin A redistributes resident and itinerant Golgi proteins to the endoplasmic reticulum. *J. Cell. Biol.* 109:61–72

Endo T, Groth D, Prusiner SB, Kobata A. 1989. Diversity of oligosaccharide structures linked to asparagines of the scrapie prion protein. *Biochemistry* 28:8380–88

Fraser H, Bruce ME. 1973. Argyrophilic plaques in mice inoculated with scrapie from particular sources. *Lancet* 1:617

Fraser H, Dickinson AG. 1973. Scrapie in mice. Agent-strain differences in the distribution and intensity of grey matter vacuolation. *J. Comp. Pathol.* 83:29–40

Gabizon R, McKinley MP, Groth DF, Prusiner SB. 1988. Immunoaffinity purification and neutralization of scrapie prion infectivity. *Proc. Natl. Acad. Sci. USA* 85:6617–21

Gabizon R, McKinley MP, Prusiner SB. 1987. Purified prion proteins and scrapie infectivity copartition into liposomes. *Proc. Natl. Acad. Sci. USA* 84:4017–21

Gabizon R, Meiner Z, Cass C, et al. 1991. Prion protein gene mutation in Libyan Jews with Creutzfeldt-Jakob disease. *Neurology* 41:160 (Abstr.)

Gabizon R, Prusiner SB. 1990. Prion liposomes. *Biochem. J.* 266:1–14

Gajdusek DC. 1977. Unconventional viruses and the origin and disappearance of kuru. *Science* 197:943–60

Gajdusek DC. 1988. Transmissible and nontransmissible amyloidoses: autocatalytic post-translational conversion of host precursor proteins to β-pleated sheet configurations. *J. Neuroimmunol.* 20:95–110

Gajdusek DC. 1990. Subacute spongiform encephalopathies: transmissible cerebral amyloidoses caused by unconventional viruses. In *Virology*, 2nd ed., eds. BN Fields, DM Knipe, RM Chanock, MS Hirsch, JL Melnick, TP Monath, B Roizman, pp. 2289–2324. New York: Raven Press

Gajdusek DC, Gibbs CJ Jr. 1990. Brain amyloidoses-precursor proteins and the amyloids of transmissible and nontransmissible dementias: scrapie-kuru-CJD viruses as infectious polypeptides or amyloid enhancing vactor. In *Biomedical Advances in Aging*, ed. A Goldstein, pp. 3–24. New York: Plenum Press

Gajdusek DC, Gibbs CJ Jr, Alpers M. 1966. Experimental transmission of a kuru-like syndrome to chimpanzees. *Nature* 209:794–96

Gasset M, Baldwin MA, Fletterick RJ, Prusiner SB. 1993. Perturbation of the secondary structure of the scrapie prion protein under conditions associated with changes in infectivity. *Proc. Natl. Acad. Sci. USA* 90:1–5

Gasset M, Baldwin MA, Lloyd D, et al. 1992. Predicted α-helical regions of the prion protein when synthesized as peptides form amyloid. *Proc. Natl. Acad. Sci. USA* 89:10940–44

Gerstmann J, Sträussler E, Scheinker I. 1936. Über eine eigenartige hereditär-familiäre erkrankung des zentralnervensystems zugleich ein beitrag zur frage des vorzeitigen lokalen alterns. *Z. Neurol.* 154:736–62. In German

Ghetti B, Tagliavini F, Masters CL, et al. 1989. Gerstmann-Straussier-Scheinker disease. II. Neurofibrillary tangles and plaques with PrP-amyloid coexist in an affected family. *Neurology* 39:1453–61

Giaccone G, Tagliavini F, Verga L, et al. 1990. Neurofibrillary tangles of the Indiana kindred of Gerstmann-Sträussler-Scheinker disease share antigenic determinants with those of Alzheimer disease. *Brain Res.* 530:325–29

Gibbons RA, Hunter GD. 1967. Nature of the scrapie agent. *Nature* 215:1041–43

Gibbs CJ Jr, Gajdusek DC, Asher DM, et al. 1968. Creutzfeldt-Jakob disease (spongiform encephalopathy): transmission to the chimpanzee. *Science* 161:388–89

Goldfarb L, Brown P, Goldgaber D, et al. 1990a. Identical mutation in unrelated patients with Creutzfeldt-Jakob disease. *Lancet* 336:174–75

Goldfarb L, Korczyn A, Brown P, et al. 1990b. Mutation in codon 200 of scrapie amyloid precursor gene linked to Creutzfeldt-Jakob disease in Sephardic Jews of Libyan and non-Libyan origin. *Lancet* 336:637–38

Goldfarb LG, Brown P, Goldgaber D, et al. 1990c. Creutzfeldt-Jakob disease and kuru patients lack a mutation consistently found in the Gerstmann-Sträussler-Scheinker syndrome. *Exp. Neurol.* 108:247–50

Goldfarb LG, Brown P, Haltia M, et al. 1992a. Creutzfeldt-Jakob disease cosegregates with the codon 178^Asn *PRNP* mutation in families of European origin. *Ann. Neurol.* 31:274–81

Goldfarb LG, Brown P, McCombie WR, et al. 1991a. Transmissible familial Creutzfeldt-Jakob disease associated with five, seven, and eight extra octapeptide coding repeats in the *PRNP* gene. *Proc. Natl. Acad. Sci. USA* 88:10926–30

Goldfarb LG, Brown P, Mitrova E, et al. 1991b. Creutzfeldt-Jacob disease associated with the PRNP codon 200^Lys mutation: an analysis of 45 families. *Eur. J. Epidemiol.* 7:477–86

Goldfarb LG, Haltia M, Brown P, et al. 1991c. New mutation in scrapie amyloid precursor gene (at codon 178) in Finnish Creutzfeldt-Jakob kindred. *Lancet* 337:425

Goldfarb LG, Mitrova E, Brown P, et al. 1990d. Mutation in codon 200 of scrapie amyloid protein gene in two clusters of Creutzfeldt-Jakob disease in Slovakia. *Lancet* 336:514–15

Goldfarb LG, Petersen RB, Tabaton M, et al. 1992b. Fatal familial insomnia and familial Creutzfeldt-Jakob disease: disease phenotype determined by a DNA polymorphism. *Science* 258:806–8

Goldgaber D, Goldfarb LG, Brown P, et al. 1989. Mutations in familial Creutzfeldt-Jakob disease and Gerstmann-Sträussler-Scheinker's syndrome. *Exp. Neurol.* 106:204–6

Haraguchi T, Fisher S, Olofsson S, et al. 1989. Asparagine-linked glycosylation of the scrapie and cellular prion proteins. *Arch. Biochem. Biophys.* 274:1–13

Hecker R, Taraboulos A, Scott M, et al. 1992. Replication of distinct prion isolates is region specific in brains of transgenic mice and hamsters. *Genes Dev.* 6:1213–28

Hsiao K, Baker HF, Crow TJ, et al. 1989a. Linkage of a prion protein missense variant to Gerstmann-Sträussler syndrome. *Nature* 338:342–45

Hsiao K, Dloughy S, Ghetti B, et al. 1992. Mutant prion proteins in Gerstmann-Sträussler-Scheinker disease with neurofibrillary tangles. *Nature Genetics* 1:68–71

Hsiao K, Meiner Z, Kahana E, et al. 1991a. Mutation of the prion protein in Libyan Jews with Creutzfeldt-Jakob disease. *N. Engl. J. Med.* 324:1091–97

Hsiao K, Prusiner SB. 1990. Inherited human prion diseases. *Neurology* 40:1820–27

Hsiao KK, Cass C, Schellenberg GD, et al. 1991b. A prion protein variant in a family with the telencephalic form of Gerstmann-Sträussler-Scheinker syndrome. *Neurology* 41:681–84

Hsiao KK, Doh-ura K, Kitamoto T, et al. 1989b. A prion protein amino acid substitution in ataxic Gerstmann-Sträussler syndrome. *Ann. Neurol.* 26:137

Hsiao KK, Scott M, Foster D, et al. 1990. Spontaneous neurodegeneration in transgenic mice with mutant prion protein of Gerstmann-Sträussler syndrome. *Science* 250:1587–90

Hunter N, Hope J, McConnell I, Dickinson AG. 1987. Linkage of the scrapie-associated fibril protein (PrP) gene and Sinc using congenic mice and restriction fragment length polymorphism analysis. *J. Gen. Virol.* 68:2711–16

Jakob A. 1921. Über eigenartige Erkrankungen des Zentralnervensystems mit bemerkenswertem anatomischen Befunde (spastische Pseudosklerose-Encephalomyelopathie mit disseminierten Degenerationsherden). *Z. Gesamte Neurol. Psychiatr.* 64:147–228. In German

Jendroska K, Heinzel FP, Torchia M, et al. 1991. Proteinase-resistant prion protein accumulation in Syrian hamster brain correlates with regional pathology and scrapie infectivity. *Neurology* 41:1482–90

Kellings K, Meyer N, Mirenda C, et al. 1992. Further analysis of nucleic acids in purified scrapie prion preparations by improved return refocussing gel electrophoresis (RRGE). *J. Gen. Virol.* 73:1025–29

Kimberlin RH. 1990. Scrapie and possible relationships with viroids. *Semin. Virol.* 1: 153–62

Kimberlin RH, Cole S, Walker CA. 1987. Temporary and permanent modifications to a single strain of mouse scrapie on transmission to rats and hamsters. *J. Gen. Virol.* 68:1875–81

Kretzschmar HA, Kufer P, Riethmuller G, et al. 1991. Prion protein mutation at codon 102 in an Italian family with Gerstmann-Sträussler-Scheinker syndrome. *Neurology* 42:809–10

Kretzschmar HA, Stowring LE, Westaway D, et al. 1986. Molecular cloning of a human prion protein cDNA. *DNA* 5:315–24

Laplanche J-L, Chatelain J, Launay J-M, et al. 1990. Deletion in prion protein gene in a Moroccan family. *Nucleic Acids. Res.* 18: 6745

Lopez CD, Yost CS, Prusiner SB, et al. 1990. Unusual topogenic sequence directs prion protein biogenesis. *Science* 248:226–29

Manetto V, Medori R, Cortelli P, et al. 1992. Fatal familial insomnia: clinical and pathological study of five new cases. *Neurology* 42:312–19

Manuelidis L, Valley S, Manuelidis EE. 1985. Specific proteins associated with Creutzfeldt-Jakob disease and scrapie share antigenic and carbohydrate determinants. *Proc. Natl. Acad. Sci. USA* 82:4263–67

Masters CL, Gajdusek DC, Gibbs CJ Jr. 1981a. Creutzfeldt-Jakob disease virus isolations from the Gerstmann-Sträussler syndrome. *Brain* 104:559–88

Masters CL, Gajdusek DC, Gibbs CJ Jr. 1981b. The familial occurrence of Creutzfeldt-Jakob disease and Alzheimer's disease. *Brain* 104:535–58

Masters CL, Richardson EP Jr. 1978. Subacute spongiform encephalopathy Creutzfeldt-Jakob disease—the nature and progression of spongiform change. *Brain* 101:333–44

McKinley MP, Masiarz FR, Isaacs ST, et al. 1983. Resistance of the scrapie agent to inactivation by psoralens. *Photochem. Photobiol.* 37:539–45

McKinley MP, Meyer R, Kenaga L, et al. 1991a. Scrapie prion rod formation *in vitro* requires both detergent extraction and limited proteolysis. *J. Virol.* 65:1440–49

McKinley MP, Taraboulos A, Kenaga L, et al. 1991b. Ultrastructural localization of scrapie prion proteins in cytoplasmic vesicles of infected cultured cells. *Lab. Invest.* 65:622–30

Medori R, Montagna P, Tritschler HJ, et al. 1992a. Fatal familial insomnia: a second kindred with mutation of prion protein gene at codon 178. *Neurology* 42:669–70

Medori R, Tritschler H-J, LeBlanc A, et al. 1992b. Fatal familial insomnia, a prion disease with a mutation at codon 178 of the prion protein gene. *N. Engl. J. Med.* 326: 444–49

Meyer N, Rosenbaum V, Schmidt B, et al. 1991. Search for a putative scrapie genome in purified prion fractions reveals a paucity of nucleic acids. *J. Gen. Virol.* 72:37–49

Meyer RK, McKinley MP, Bowman KA, et al. 1986. Separation and properties of cellular and scrapie prion proteins. *Proc. Natl. Acad. Sci. USA* 83:2310–14

Neugut RH, Neugut AI, Kahana E, et al. 1979. Creutzfeldt-Jakob disease: familial clustering among Libyan-born Israelis. *Neurology* 29:225–31

Oesch B, Westaway D, Wälchli M, et al. 1985. A cellular gene encodes scrapie PrP 27–30 protein. *Cell* 40:735–46

Owen F, Poulter M, Collinge J, et al. 1991. Insertions in the prion protein gene in atypical dementias. *Exp. Neurol.* 112: 240–42

Owen F, Poulter M, Lofthouse R, et al. 1989. Insertion in prion protein gene in familial Creutzfeldt-Jakob disease. *Lancet* 1:51–52

Owen F, Poulter M, Shah T, et al. 1990. An in-frame insertion in the prion protein gene in familial Creutzfeldt-Jakob disease. *Mol. Brain Res.* 7:273–76

Palmer MS, Dryden AJ, Hughes JT, Collinge J. 1991. Homozygous prion protein genotype predisposes to sporadic Creutzfeldt-Jakob disease. *Nature* 352:340–42

Pattison IH, Millson GC. 1961. Scrapie produced experimentally in goats with special reference to the clinical syndrome. *J. Comp. Pathol.* 71:101–8

Prusiner SB. 1982. Novel proteinaceous infectious particles cause scrapie. *Science* 216: 136–44

Prusiner SB. 1991. Molecular biology of prion diseases. *Science* 252:1515–22

Prusiner SB. 1992. Chemistry and biology of prions. *Biochemistry* 31:12278–88

Prusiner SB, Baldwin M, Collinge J, et al. 1993. Classification and nomenclature of viruses: Prions. *Arch. Virol.* In press

Prusiner SB, Bolton DC, Groth DF, et al. 1982. Further purification and characterization of scrapie prions. *Biochemistry* 21: 6942–50

Prusiner SB, DeArmond SJ. 1987. Biology of disease: Prions causing nervous system degeneration. *Lab. Invest.* 56:349–63

Prusiner SB, Groth DF, Cochran SP, et al. 1980. Molecular properties, partial purification, and assay by incubation period measurements of the hamster scrapie agent. *Biochemistry* 19:4883–91

Prusiner SB, McKinley MP, Bowman KA, et al. 1983. Scrapie prions aggregate to form amyloid-like birefringent rods. *Cell* 35:349–58

Prusiner SB, Scott M, Foster D, et al. 1990. Transgenetic studies implicate interactions between homologous PrP isoforms in scrapie prion replication. *Cell* 63:673–86

Puckett C, Concannon P, Casey C, Hood L. 1991. Genomic structure of the human prion protein gene. *Am. J. Hum. Genet.* 49:320–29

Race RE, Graham K, Ernst D, et al. 1990. Analysis of linkage between scrapie incubation period and the prion protein gene in mice. *J. Gen. Virol* 71:493–97

Ripoll L, Laplanche J-L, Salzmann M, et al. 1993. A new point mutation in the prion protein gene at codon 210 in Creutzfeldt-Jakob disease. *Neurology* In press

Rogers M, Taraboulos A, Scott M, et al. 1990. Intracellular accumulation of the cellular prion protein after mutagenesis of its Asn-linked glycosylation sites. *Glycobiology* 1: 101–9

Safar J, Ceroni M, Piccardo P, et al. 1990. Subcellular distribution and physicochemical properties of scrapie associated precursor protein and relationship with scrapie agent. *Neurology* 40:503–8

Scott M, Foster D, Mirenda C, et al. 1989. Transgenic mice expressing hamster prion protein produce species-specific scrapie infectivity and amyloid plaques. *Cell* 59:847–57

Sparkes RS, Simon M, Cohn VH, et al. 1986. Assignment of the human and mouse prion protein genes to homologous chromosomes. *Proc. Natl. Acad. Sci. USA* 83:7358–62

Stahl N, Baldwin MA, Hecker R, et al. 1992. Glycosylinositol phospholipid anchors of the scrapie and cellular prion proteins contain sialic acid. *Biochemistry* 31:5043–53

Stahl N, Baldwin MA, Teplow DB, et al. 1993. Structural analysis of the scrapie prion protein using mass spectrometry and amino acid sequencing. *Biochemistry* 32:1991–2002

Stahl N, Borchelt DR, Hsiao K, Prusiner SB. 1987. Scrapie prion protein contains a phosphatidylinositol glycolipid. *Cell* 51:229–40

Tagliavini F, Prelli F, Ghisto J, et al. 1991. Amyloid protein of Gerstmann-Sträussler-Scheinker disease (Indiana kindred) is an 11-kd fragment of prion protein with an N-terminal glycine at codon 58. *EMBO J.* 10:513–19

Taraboulos A, Jendroska K, Serban D, et al. 1992a. Regional mapping of prion proteins in brains. *Proc. Natl. Acad. Sci. USA* 89: 7620–24

Taraboulos A, Raeber AJ, Borchelt DR, et al. 1992b. Synthesis and trafficking of prion proteins in cultured cells. *Mol. Biol. Cell* 3:851–63

Taraboulos A, Rogers M, Borchelt DR, et al. 1990a. Acquisition of protease resistance by prion proteins in scrapie-infected cells does not require asparagine-linked glycosylation. *Proc. Natl. Acad. Sci. USA* 87:8262–66

Taraboulos A, Serban D, Prusiner SB. 1990b. Scrapie prion proteins accumulate in the cytoplasm of persistently-infected cultured cells. *J. Cell Biol.* 110:2117–32

Tateishi J, Kitamoto T, Doh-ura K, et al. 1990. Immunochemical, molecular genetic, and transmission studies on a case of Gerstmann-

Sträussler-Scheinker syndrome. *Neurology* 40:1578–81

Vnencak-Jones CL, Phillips JA. 1992. Identification of heterogeneous PrP gene deletions in controls by detection of allele-specific heteroduplexes (DASH). *Am. J. Hum. Genet.* 50:871–72

Weissmann C. 1991. A "unified theory" of prion propagation. *Nature* 352:679–83

Westaway D, Goodman PA, Mirenda CA, et al. 1987. Distinct prion proteins in short and long scrapie incubation period mice. *Cell* 51:651–62

Westaway D, Mirenda CA, Foster D, et al. 1991. Paradoxical shortening of scrapie incubation times by expression of prion protein transgenes derived from long incubation period mice. *Neuron* 7:59–68

Yost CS, Lopez CD, Prusiner SB, et al. 1990. A non-hydrophobic extracytoplasmic determinant of stop transfer in the prion protein. *Nature* 343:669–72

Annu. Rev. Neurosci. 1994. 17:341–71

DENDRITIC SPINES: CELLULAR SPECIALIZATIONS IMPARTING BOTH STABILITY AND FLEXIBILITY TO SYNAPTIC FUNCTION

Kristen M. Harris

Department of Neurology and Program in Neuroscience, Children's Hospital and Harvard Medical School, Boston, Massachusetts 02115

S. B. Kater

Program in Neuronal Growth and Development, Department of Anatomy and Neurobiology, Colorado State University, Fort Collins, Colorado 80523

KEY WORDS: serial electron microscopy, long-term potentiation, hippocampus, synapses, ultrastructure, learning, memory, modeling

INTRODUCTION

Dendritic spines, the tiny protrusions that stud the surface of many neurons, are the location of over 90% of all excitatory synapses that occur in the CNS. Their small size has, in large part, made them refractory to conventional experimental approaches. Yet their widespread occurrence and likely involvement in learning and memory has motivated extensive efforts to obtain quantitative descriptions of spines in both steady state and dynamic conditions. Since the seminal mathematical analyses of D'Arcy Thompson (1992), the power of quantitatively establishing key parameters of structure has become recognized as a foundation of successful biological inquiry. For dendritic spines, highly precise determinations of structure and its variation are again proving to be essential for establishing a valid concept of function. The recent conjunction of high quality information about the structure, function, and theoretical implications of dendritic spines has, in fact, produced a flurry of

341

new considerations of their role in synaptic transmission (Rall 1970, 1974; Diamond et al 1970; Kawato & Tsukahara 1983; Horwitz 1984; Wilson 1984; Perkel & Perkel 1985; Shepherd et al 1985; Gamble & Koch 1987; Shepherd & Brayton 1987; Wickens 1988; Segev & Rall 1988; Brown et al 1988; Qian & Sejnowski 1989; Holmes 1990; Zador et al 1990; Baer & Rinzel 1991; Larson & Lynch 1991; Koch et al 1992; Koch & Zador 1993).

A powerful working hypothesis is that the structure of dendritic spines sets the boundaries within which synaptic function can be modulated. Theory defines the limits of what can happen within these boundaries and experimental manipulation defines what actually happens to spine morphology. As measurements of spine dimensions and organelle and molecular composition have become more precise, so too have the theoretical models of spine function improved. In the following discussion we evaluate some of the morphological, theoretical, and experimental evidence indicating that dendritic spine structure and composition can influence synaptic efficacy. In this context, we consider how spines might serve the cellular mechanisms that establish specific and enduring memories.

INTEGRATION OF SYNAPTIC INPUT ON SPINY NEURONS

The intrigue of understanding the cellular mechanisms involved in learning and memory has provided a strong impetus for investigating many aspects of neural organization. Several discrete levels of integration, ranging from changes in cellular ensembles to changes in individual molecules, have been proposed as key loci for learning and memory. Neurons are, however, more than globes filled with talented molecules, and the complexity of neuronal structure sets neurons apart from all other cells. Highly branched dendrites receive and integrate input from hundreds, even thousands, of other neurons. Neurons with different functions can be classified according to the shape of their dendritic arbor and the density of spines occurring along their dendrites. The degree of dendritic branching, the length of individual dendritic branches, and the frequency of dendritic spines are all modified by experience and probably represent the growth of new synapses (Greenough & Bailey 1988).

Spiny neurons tend to be the principal input/output cells of a given brain region (Shepherd 1990). As such, they integrate diverse excitatory and inhibitory input, from different regions and from different cell types within a brain region. For example, a spiny pyramidal cell is the principal cell of hippocampal area CA1 (Figure 1). The spines that are located on the proximal two thirds of CA1 pyramidal cell dendrites receive excitatory synapses from both the ipsilateral and the contralateral hippocampus. Distal spines of the same pyramidal cells receive excitatory synapses from the entorhinal cortex.

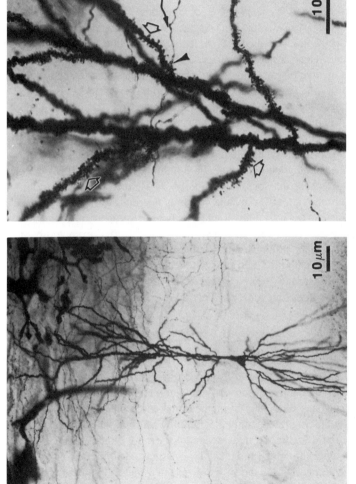

Figure 1   Light micrographs of a Golgi-impregnated spiny pyramidal cell in area CA1 of the rat hippocampus. (*A*) From the apex of the pyramidal cell body emerges an apical dendrite with multiple lateral branches. From the base emerge several basilar dendrites. (*B*) At higher magnification, the dendritic spines (*open arrows*) that stud the surface of these dendrites become just visible, along with the presynaptic axons (*wavy arrow*), which occasionally come into close apposition with a spine (*arrowhead*). For unknown reasons, the Golgi method usually impregnates less than 1% of the cells and axons that are present in the neuropil.

In contrast, nonspiny neurons tend to be the local interneurons of a brain region (Shepherd 1990). Their dendrites are usually spine-free or sparsely spiny and have large swellings or varicosities along their lengths. Both excitatory and inhibitory synapses occur directly onto the dendritic shafts, often with a higher frequency at the varicosities (KM Harris, personal observation). The axons of the nonspiny cells usually remain within a brain region to form inhibitory or modulatory synapses on the dendritic shafts between the spines and on the somata of the spiny pyramidal cells. The excitatory synapses typically have an asymmetric appearance, featuring a thickened postsynaptic density adjacent to a presynaptic axonal bouton containing round, clear vesicles (Peters et al 1991). The inhibitory synapses tend to have a symmetric appearance, owing to the near equal thickening of the pre- and postsynaptic membranes, with both round and flattened vesicles in the presynaptic bouton. Glutamate and aspartate are the predominant excitatory neurotransmitters, whereas GABA is the primary inhibitory neuro-transmitter functioning at synapses on spiny neurons throughout the CNS (Shepherd 1990). The presynaptic axons of symmetric synapses also contain several substances that either modulate the inhibitory influence of GABA or alter the excitability of the spiny cell directly. Thus, the dendritic, axonal, and synaptic morphologies of spiny and nonspiny neurons differ dramatically, along with their electrophysiological and biochemical properties, indicating that the roles played by spiny neurons are likely to differ from those of nonspiny neurons in an ensemble that involves both. When integration of the diverse excitatory, inhibitory, and modulatory actions drives the spiny cells past threshold, they usually send a signal to the next brain region, where its neurons undergo similar integrative activities.

STRUCTURE OF DENDRITIC SPINES

As the first postsynaptic element encountered by the excitatory neurotrans-mitter, dendritic spines are uniquely situated to be a fundamental integrative unit. The synaptic strength at different spines determines the pattern of activity of individual cells and ultimately of the neuronal ensemble. Dendritic spines are so small and intermingled within the complex neuropil that contemporary electrophysiological and biochemical techniques cannot directly evaluate the activity and composition of individual living spines within this neuropil. However, recent confocal microscopy has revealed that individual dendritic spines do persist over periods of several hours in hippocampal slices in vitro (Hosokawa et al 1992). This observation establishes a necessary prerequisite for spines as fundamental integrative units, namely that once formed they are relatively persistent structures.

Most of what we suspect about spine function is based on computer

Table 1 Range in spine dimensions and their association(s) with presynaptic axons in six brain regions[a]

Brain region	Total length (μm)	Neck diameter (μm)	Neck length (μm³)	Total volume (μm³)	Total surface (μm²)	PSD area (μm²)	PSD: head surface area	Maximum number of boutons per spine	Maximum number of branches per spine
Cerebellum	0.7–3	0.1–0.3	0.1–2	0.06–0.2	0.7–2	0.04–0.4	0.17 ± 0.09	2	5
Hippocampal CA1	0.2–2	0.04–0.5	0.1–2	0.004–0.6	0.1–4	0.01–0.5	0.12 ± 0.06	3	3
Visual cortex	0.5–3	0.07–0.5	—	0.02–0.8	0.5–5	0.02–0.7	0.10 ± 0.04	2	2
Neostriatum	—	0.1–0.3	0.6–2	0.04–0.3	0.6–3	0.02–0.3	0.13	2	2
Hippocampal CA3	0.6–6.5	0.2–1	0.1–1	0.1–2	1–3	0.01–0.6	0.09 ± 0.04	3	16
Hippocampal Dentata	1 ± 0.6	0.2 ± 0.1	0.8	—	—	—	—	—	3

[a] For the first five brain regions all of the data are from three-dimensional analyses of serial EM reconstructions. Cerebellar spines are from the Purkinje spiny branchlets (Harris & Stevens 1988), comparable to Spacek & Hartman (1983). Spines of hippocampal CA1 pyramidal cells are from Harris & Stevens (1989); for visual cortex from pyramidal cells, from Spacek & Hartman (1983). Spines in the neostriatum are from Wilson et al (1983), and spines in hippocampal area CA3 are from the proximal portion of the pyramidal cells (Chicurel & Harris 1992). The data for hippocampal area dentata are from three-dimensional analyses of Golgi-impregnated cells viewed with high-voltage EM and measured on stereo pair images of Hama et al (1989). The volume and surface area of the spine head is typically about 90% of the total. Only branched spines contacted more than one excitatory axonal bouton; all unbranched spines synapsed with just one bouton. In neostriatum and visual cortex, a second inhibitory synapse occurs on the neck of about 8% of the spines Difiglia et al 1982, Wilson et al 1983, De Zeeuw et al 1990). In most brain regions less than 10% of the spines are branched; in contrast, about 90% of the proximal CA3 spines are branched. "Maximum" refers to the maximum number viewed to date.

simulations that vary the structural dimensions and the locations of active molecules in simulated spines. For simplicity, the theoretical models have used ideal geometries, such as spheres with variable dimensions for the heads, connected to cylinders with variable lengths and widths for the necks. For some spines, these descriptions are adequate, and the theoretical conclusions are generally interpretable (e.g. Wilson et al 1983, Harris & Stevens 1988, Brown et al 1988).

Physiological evidence readily shows that different excitatory synapses can have very different efficacies (Manabe et al 1992, reviewed in Lisman & Harris 1993). If dendritic spine structure participates in defining the differences in synaptic efficacy, then the heterogeneity of synaptic strength should

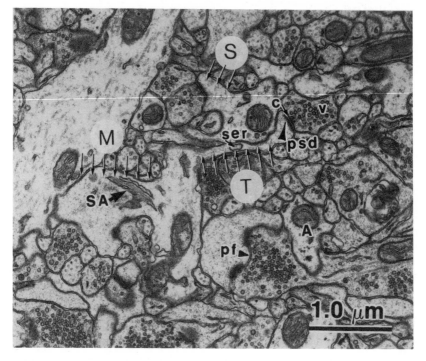

Figure 2 Electron micrograph of a section through dendritic spines in stratum radiatum of hippocampal area CA1. In this fortuitous section, three spines were sectioned parallel to their longitudinal axis, revealing spines of the stubby (*S*), mushroom (*M*), and thin (*T*) morphologies. The postsynaptic density (*psd*) occurs on the spine head (see *T*) immediately adjacent to the synaptic cleft (*c*) and to the presynaptic axonal bouton that is filled with round vesicles (*v*). This *T* spine contains a small tube of smooth endoplasmic reticulum (*ser*) in its neck. In the *M* spine, a spine apparatus (*SA*) is visible. A perforated postsynaptic density (*pf*) is evident on the head of another mushroom spine. Near to this spine is a large astrocytic process (*A*), identified by the black glycogen granules and clear cytoplasm.

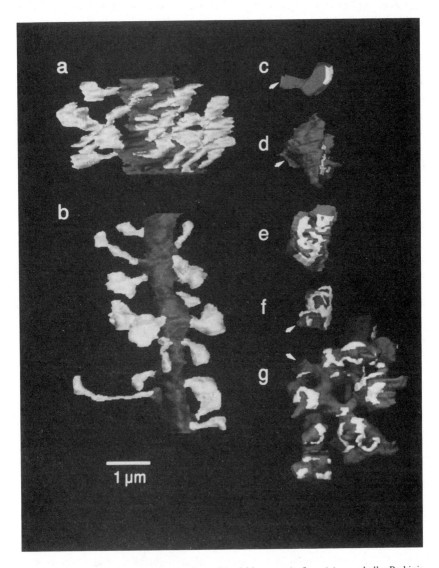

Figure 3   Three-dimensional reconstructions of dendritic segments from (*a*) a cerebellar Purkinje spiny branchlet and (*b*) a hippocampal CA1 pyramidal cell. The dendritic shaft is illustrated in gray, and the individual spines are illustrated in white. In *c–g* the spines are gray, the PSDs are white, and the small white arrows indicate where the spines joined their parent dendrites. (*c*) Profile of a typical cerebellar spine with a macular PSD. (*d*) A side view and (*e*) a top view of a short mushroom-shaped CA1 spine. This spine has a complex PSD with several segments and perforations. (*f*) A single headed spine in hippocampal area CA3, also with a complex PSD, and (*g*) a highly branched spine from hippocampal area CA3 which has multiple PSDs.

be evident in spine structure. Indeed, there are dramatic differences in spine and synaptic shape (Figure 2). Spines necks can be long or short, fat or thin, straight or bent, cylindrical or irregular, and branched or unbranched in all combinations. Spine heads can be small or large, and spherical, oval, or irregular in shape. This heterogeneity in spine structure occurs both among spines on a single dendrite and across different cell types. Figure 3 reveals large differences in the three-dimensional shape of dendritic spines, and Table 1 summarizes the variability in spine dimensions across several brain regions. The greater-than-tenfold differences in spine dimensions shown in this summary easily provide sufficient heterogeneity in spine structure to account for the large heterogeneity in synaptic strengths. Despite these gross differences in spine structure, it is possible to construct rather distinct categories of dendritic spine shapes (e.g. thin, mushroom, stubby, branched) both within and across brain regions (Jones & Powell 1969, Peters & Kaiserman-Abramof 1970, Harris et al 1992).

The diversity in spine morphology may reflect dynamic states during the life history of individual spines and/or different synaptic efficacies occurring along a single dendrite at a particular time. The distinct categories also might well represent specific spine functions or the stages through which individual spines must pass to achieve a "mature" state.

COMPOSITION OF DENDRITIC SPINES AND THEIR SYNAPTIC COMPLEX

Dendritic spines must be considered within the context of the overall synaptic complex, which includes the spine, the postsynaptic density, the synaptic cleft, the presynaptic axonal bouton and its vesicles, and the neighboring astrocytic processes. Morphological and biochemical evidence shows that multiple organelles and molecules are localized within dendritic spines. The specific composition of spines and their synapses may result in further discrimination in the functions of spines. Here we summarize the composition of dendritic spines and their synapses and refer the reader to other articles and reviews for more detail.

Postsynaptic Density (PSD)

One of the most conspicuous ultrastructural features in the CNS is the postsynaptic density (PSD) (Peters et al 1991). The PSD is a structure about 50 nm thick that is apposed to the cytoplasmic side of the postsynaptic membrane. It is found at virtually all excitatory synapses, including those occurring on the heads of dendritic spines (Figure 2). Three-dimensional reconstructions have shown PSDs to be either disc (macular) shaped or highly irregular in shape with perforations, which are electron lucent regions within the PSD (Cohen & Siekevitz 1978, Spacek 1985a, Harris & Stevens 1989). Some PSDs on a single spine head are segmented into discrete zones (Geinisman et al 1992; Figures 2 *pf* and 3*e,f*). In all brain regions, spine dimensions are proportional to the total area of the PSD or segments of the PSD added together (Westrum & Blackstad 1962; Peters & Kaiserman-Abramof 1970; Wilson et al 1983; Harris & Stevens 1988, 1989; Harris et al 1992; Chicurel & Harris 1992). In freeze-fracture preparations, the extracellular half of the synaptic membrane of dendritic spines has an aggregate of particles, ranging in size from 6–17 nm, with mean densities of about 2800 particles/μm^2 on hippocampal dendritic spines and 3600 particles/μm^2 on cerebellar dendritic spines (Harris & Landis 1986). It has been proposed that these particles are anatomical representations of the molecules involved in synaptic function.

More than 30 proteins that are highly enriched in PSDs have been identified in subcellular fractions from the brain (Kelly & Cotman 1978; Carlin et al 1980, 1981, 1983; Siekevitz 1985; Wu et al 1986; Wu & Siekevitz 1988; Kennedy et al 1990; Walsh & Kuruc 1992). These proteins have been grouped into five classes: (*a*) neuroreceptor glycoproteins (e.g. binding sites for excitatory amino acids and GABA, ion channels for Ca^{2+} and K$^+$ and others), (*b*) protein kinases [calcium/calmodulin-dependent protein kinase type II (CaM-kinase II), protein kinase C (PKC), and the associated regulatory protein calmodulin], (*c*) structural and mechanochemical proteins (tubulin, actin, brain spectrin/fodrin, myosin, dynamin, mapII, adducin, dystrophin, microtubule associated protein 2 (MAP2), neurofilament proteins), (*d*) proteins involved in endocytosis (elongation factor 1 alpha; N-ethylmaleimide sensitive factor; BiP, a resident protein of the ER), and (*e*) proteins involved in the glycolytic pathway (e.g. possibly glyceraldehyde-3-phosphate dehydrogenase and pyruvate kinase). These constituents must be regarded with caution, as the PSD-enriched preparations are known to have some contamination from mitochondrial and other membranes as well as from polyribosomes. Under some conditions, 50% of the total PSD fraction contains what has been referred to as the major PSD protein (Goldenring et al 1984), which is CaM-kinase II (Carlin et al 1981, Kennedy et al 1983, Kelly et al 1984). The CaM-kinase II molecule has aroused considerable interest because it can switch from a calcium/calmodulin-dependent state to a calcium/calmodulin-independent, autophosphorylating state after a brief exposure to calcium and calmodulin. Lisman & Goldring (1988) have postulated that this switch in the state of the CaM-kinase II could mediate short-term changes in synaptic efficacy through phosphorylation of certain proteins (e.g. MAP2 or tubulin) in the PSD, though many of the specific substrates for CaM-kinase II in the PSD remain to be identified (Kennedy 1992). It has long been thought that changes in the structure of the PSD reflect alterations in synaptic efficacy (Cohen & Siekevitz

1978, Nieto-Sampedro et al 1982, S kevitz 1985). The molecular composition of the PSD certainly provides many candidate molecules that could work in consort to mediate the plasticity of synaptic structure and electrophysiology.

Organelles

All spines contain smooth endoplasmic reticulum (SER in Figure 1C) (Peters et al 1991; Harris & Stevens 1988, 1989; Spacek 1985a,b), an organelle known to be involved in membrane synthesis (Hall 1992) and to store calcium. The volume of the SER is proportional to spine volume and PSD area and occupies about 10–20% of the total spine volume (Harris & Stevens 1988). The more complex spines contain sacs of SER laminated with dense-staining material into a structure known as the spine apparatus (Gray 1959; Figure 2 sa). The spine apparatus appears to be similar to the Golgi apparatus in both its overall structure and its intimate association with the SER, though the Golgi apparatus is typically restricted to the soma and proximal dendrites. Whether the spine apparatus performs similar functions to those of the Golgi apparatus, e.g. modification of proteins to form proteoglycans and vesicle formation (Hall 1992), is not known.

The SER is also thought to be involved in the sequestration and intracellular release of calcium, like the sarcoplasmic reticulum of muscle cells (e.g. Hall 1992). X-ray microanalysis of cerebellar dendritic spines has revealed a preferential localization of calcium in the spine SER (Andrews et al 1988), and precipitates of calcium-oxalate occur in the SER of hippocampal and cortical dendritic spines (Burgoyne et al 1983, Fifkova et al 1983). The inositol triphosphate (IP3) receptor has been identified on the SER in spines and dendrites (Mignery et al 1989, Walton et al 1991). Since the IP3 receptor is activated by calcium in the cytoplasm, release of the stored calcium could be triggered by a brief rise in intracellular calcium, as discussed below.

Polyribosomes have been revealed through three-dimensional reconstructions in more than three quarters of visual cortical spines (Spacek 1985b) and in at least one head of nearly all the highly branched CA3 dendritic spines (Chicurel & Harris 1992). In addition, polyribosomes have been detected both within spines and at the base of spines in the dendrites of hippocampal area dentata and area CA1 neurons (Steward & Levy 1982, Steward & Reeves 1988). The frequency of polyribosomes in the vicinity of dendritic spines increases during synaptogenesis (McWilliams & Lynch 1978, Steward 1983, Steward & Falk 1985) and with rearing of rats in an enriched environment (Greenough et al 1985). The mRNAs that encode for MAP2 and CaM-kinase II, and the brain cytoplasmic mRNA (BC1) are prominent in dendritic laminae throughout the CNS, suggesting that these two proteins (and probably others) are locally synthesized within dendrites (Garner et al 1988, Burgin et al 1990, Tiedge et al 1991, reviewed in Steward & Banker 1992). The preferential positioning of polyribosomes near to or within dendritic spines indicates that spines and their synapses may be recipients of proteins that are synthesized locally in the dendrites or spines and reinforces the view of spines as autonomous components. This local synthesis of proteins may provide a cellular mechanism whereby new proteins can be specifically targeted in response to synaptic activation (Steward & Banker 1992).

Mitochondria rarely occur in dendritic spines and are typically restricted to the very complex or very large dendritic spines such as those found in the cerebral cortex (Ebner & Colonnier 1975, 1978; Westrum et al 1980), in the branched spines of hippocampal area CA3 (Hamlyn 1962, Amaral & Dent 1981, Chicurel & Harris 1992), or in spines of the olfactory bulb that have both pre- and postsynaptic functions (Cameron et al 1991). Similarly, multivesicular bodies are restricted to large spines (Chicurel & Harris 1992) and the base of dendritic spines (KM Harris, personal observation). The function of the multivesicular bodies has not been clarified for spines; however, studies in other neuronal systems (Rosenbluth & Wissig 1964, Schmied & Holtman 1987, Bailey et al 1992) support their role in the endolysosomal system and involvement in synaptic turnover and plasticity. Coated vesicles are occasionally found in dendritic spines of the adult brain; their frequency also increases with synaptogenesis, and it has been proposed that they may facilitate the formation of new synapses (McWilliams & Lynch 1981).

Cytoskeleton and Cytoplasm

The cytoskeleton of dendritic spines is characterized by a loose network of filaments (Gray 1959). It is distinguished from the dendritic cytoskeleton by the near absence of microtubules, except for an occasional microtubule in the largest and most complex spines (Westrum et al 1980, Chicurel & Harris 1992). The filamentous network of spines is comprised of actin and actin-regulating proteins (Landis & Reese 1983, Fifkova 1985, Cohen et al 1985). The actin filaments of the spine neck are longitudinally situated, whereas those in the head are organized into a lattice surrounding the SER or spine apparatus. This organization of the actin filaments suggests that they provide the scaffolding for the basic spine structure. Other molecules found in the spine cytoplasm that may interact with the actin cytoskeleton, usually in a calcium-dependent manner, include calmodulin, myosin, brain spectrin (fodrin), and MAP2. The organization of the actin filaments within spines does not seem to differ dramatically across the brain regions studied to date. However, some of the actin-associated proteins are heterogeneously distributed and together with local calcium concentrations may contribute to the diversity in spine structure described above. Surprisingly, the growth-associated protein GAP-43, normally thought to be involved in growth cones, neurotransmitter release, and the function of presynaptic axons (Benowitz & Perrone-Bizzozero 1991), has occasionally been found in dendritic spines or appendages of neostriatal neurons (DiFiglia et al 1990).

Synaptic Plasma Membrane and Cleft Material

The plasma membrane of dendritic spines is similar in appearance to the membrane surrounding the rest of the neuron and the presynaptic axonal bouton and vesicles. It is characterized by a lipid bilayer, which when cross-sectioned can be readily discerned in osmium-stained material (Peters et al 1991). Between the pre- and postsynaptic membranes is the synaptic cleft, a region where the extracellular space widens slightly to about 10–20 nm and is filled with a dense-staining material. The plasma membrane also contains many integral proteins, of which some are specific to the synapse and others are generally found throughout the neuron. For example, two G proteins (G_i & G_o) that are involved in the opening of Ca^{2+} and K^+ channels are found in the synaptic plasma membrane fraction (Wu et al 1992). Although the composition of the synaptic cleft has not yet been delineated, it is likely comprised of cell surface molecules involved in cell-cell adhesion (McDonald 1989, Akiyama et al 1990). Emerging evidence suggests that the synaptic plasma membrane fraction contains integrin-type adhesion receptors (Bahr & Lynch 1992) and neural cell-adhesion molecules (NCAMS) (Persohn et al 1989). In addition, peptides that block a subclass of the integrins disrupt the stabilization of synaptic potentiation (Staubli et al 1990, Xiao et al 1991), suggesting an important role in structural plasticity. Several lines of evidence have led to the hypothesis that the basal lamina protein agrin may be responsible for the aggregation of synaptic proteins on the surface of muscle fibers (Ferns & Hall 1992). Isoforms of this protein are produced throughout the CNS, where they may perform similar synaptic functions (McMahan et al 1992). Whether similar proteins are specifically found in the dense material of the CNS synaptic cleft remains to be determined.

Presynaptic Vesicles

The boutons associated with dendritic spines have numerous round clear vesicles (Figure 2) which contain glutamate (Storm-Mathisen et al 1983; Otterson et al 1990a,b; Clements et al 1990). On the presynaptic membrane is the presynaptic grid (Aghajanian & Bloom 1967, Vrensen & Cardozo 1981), which is characterized by dense projections on the cytoplasmic side of the membrane and which may be the equivalent of the actin-like filaments (Landis 1988) thought to be the "vesicle docking" sites (Schwartz 1992).

The full composition of the vesicles and the presynaptic bouton is very complex and beyond the scope of this review (Maycox et al 1990, Verhage et al 1991). However, it is noteworthy that certain kinds of structural data can be brought to bear on physiological issues (e.g. Clements et al 1992, Larkman et al 1992). For example, the dimensions of the presynaptic and postsynaptic elements are tightly linked. The total number of vesicles is closely correlated with spine volume, SER volume, and the area of the PSD on dendritic spines (Harris & Stevens 1988, 1989). These correlations hold for a large range in vesicle number, from 38–1234 and 3–1606 for cerebellar and CA1 spine synapses, respectively. These data suggest that a coordinating process coregulates the dimensions of these pre- and postsynaptic structures (Lisman & Harris 1993).

Astrocytes

Astrocytes are identified by the presence of dark glycogen granules and astrocytic fibrils in the cytoplasm, which is typically light in electron micrographs (Peters et al 1991). In some brain regions, such as the cerebellum, the astrocytic processes have been found through EM reconstruction to surround the synaptic complex, involving dendritic spines and their presynaptic axonal boutons (Spacek 1985c). In other brain regions, such as the hippocampus and neocortex, the tiny astrocytic processes that occur in the vicinity of the spines do not surround the entire complex, though their presence becomes obvious through immunolabeling and EM reconstruction (Aoki 1992; KM Harris, unpublished observation).

Astrocytes perform many important functions for the synapses involving the regulation of the extracellular milieu and uptake of potassium and glutamate (Kuffler 1967, Barres 1991). In cultures of dissociated cerebral cortex, astrocytes and astrocytic processes surround a thick layer of neuropil that is full of synapses on dendritic spines and shafts; in contrast, the neuropil of astrocyte-poor cortical cultures is very thin, and few synapses form (Harris & Rosenberg 1993). Neurons in the astrocyte-poor cultures are 100-fold more sensitive to glutamate-induced toxicity (Rosenberg & Aizenman 1989, Rosenberg et al 1992); in fact, cell death occurs at glutamate concentrations that normally occur in the extracellular fluid of a healthy brain. It was proposed that the astrocytes provide a physical buffer in vivo like that seen in vitro, allowing the astrocytes to clear the extracellular fluid of glutamate in the immediate vicinity of the synapses. Astrocytes in the vicinity of hippocampal dendritic spines reportedly proliferate during synaptic plasticity, suggesting an increased need for glutamate regulation at the larger synapses (Sirevaag & Greenough 1987, Wenzel et al 1991). Astrocytes also may regulate calcium in response to stimulation by glutamate (Cornell-Bell et al 1990a,b). Finally, it has been shown that growth of cerebellar dendritic spines is induced by an astrocyte-secreted factor even in the absence of presynaptic axons (Seil et al 1992). Together, these observations suggest an elaborate functional relationship between dendritic spines and their astrocytic partners.

FUNCTIONS OF DENDRITIC SPINES

As Postsynaptic Targets

Ultrastructural evaluation of dendritic spines reveals them to be the major site of excitatory synaptic input. Occasionally, inhibitory/modulatory synapses form on the heads, on the necks, or at the bases of dendritic spines (Figure 4a; Colonnier 1968, DiFiglia et al 1982, De Zeeuw et al 1990, Dehay et al 1991, Fifkova et al 1992), which could act to "veto" or modify the strength of the excitatory input (Qian & Sejnowski 1990). Because most dendritic spines have a single excitatory synapse on their head, more spines means more synapses and accordingly more point-to-point connections in a neuronal ensemble involving spiny neurons. Thus, one function of the spine is to preserve the individuality of inputs.

Ramon y Cajal originally postulated that spines could increase the surface area available for new synapses to form. Most of the dendritic shaft between spines, however, does not have synapses, and ample room is available for more synapses to occur even in the absence of more dendritic spines (Gray 1959, Harris & Stevens 1988). Spines allow dendrites to reach multiple axons as they weave through the neuropil (Figure 4b; Swindale 1981). For nonspiny dendrites to attain the same radius of access to the axons, they must be thicker than spiny dendrites (which typically they are) and must occupy a significantly greater volume of the neuropil (Figure 4c). Spiny dendrites thus allow more synaptic connections to be compacted into a limited brain volume, and hence they can be considered the microscopic parallel to sulci and gyri in the brain. Since their discovery around the turn of the century, however, it has been suspected that dendritic spines do more than simply connect neurons. In fact, both Ramon y Cajal (1893) and Tanzi (1893) suggested that changes in dendritic spine number and/or morphology could provide a cellular basis for learning and memory.

Spines and LTP

A major driving force for establishing a functional description of spine morphology has been the desire to understand the morphological substrate for the profound synaptic plasticities seen in the hippocampus and cortex. One of the most extensively investigated has been long-term potentiation (LTP), a long-lasting enhancement of the post-synaptic response resulting from repetitive or appropriately patterned activation of the neurons (reviewed in Madison et al 1991, Bliss & Collingridge 1993). LTP is widely considered to be a cellular mechanism of at least some forms of learning and memory. In spite of continuing controversy over the exact sequence of events, a growing consensus holds that changes in the properties of both the pre- and postsynaptic elements are involved (Kullman & Nicoll 1992, Larkman et al 1992, Bliss &

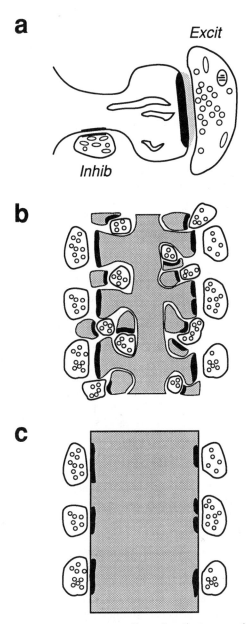

Figure 4 Functions of dendritic spines. (*a*) Sites of excitatory synaptic input (*excit*) and occasionally inhibitory/modulatory (*inhib*) synaptic input. (*b*) Longitudinal section through a spiny dendrite illustrating its reach to many axonal boutons and the interdigitation of other processes between the spines. (*c*) Nonspiny dendrite with the same "axonal reach" as the spiny dendrite in *b*, but no other processes can occupy the space between the synapses.

Collingridge 1993, Lisman & Harris 1993). In the next four sections we consider how the structure of dendritic spines could contribute to the cellular mechanisms that mediate the induction, associativity, specificity, and endurance of LTP. We propose that if spines serve these roles for LTP, they could similarly facilitate learning and memory.

ROLE SPINES MAY SERVE IN THE INDUCTION OF LTP   Induction of LTP requires entry of calcium into the postsynaptic cell (Madison et al 1991). To achieve this calcium entry at most of the synapses where LTP is induced, glutamate must be released from the presynaptic terminal at (or near) the same time that the postsynaptic element is depolarized. The postsynaptic depolarization is necessary to relieve a magnesium block in the calcium channel that is associated with the N-methyl-D-aspartate (NMDA) receptor (Madison et al 1991). The constriction in dendritic spine necks, if it poses a resistive barrier, results in an amplification of the depolarization attained in the immediate vicinity of the synapse, relative to that which would be generated if the synapse occurred directly on the dendritic shaft (Figure 5a; Perkel 1982, Turner 1984, Coss & Perkel 1985, Brown et al 1988). Thus, spine neck constriction could facilitate induction of LTP by allowing the voltage-dependent channels to open in response to a lower synaptic activation than would be required to depolarize synapses on nonspiny dendrites.

Results from ontogenetic studies on LTP, the NMDA receptors, and dendritic spines in the rat hippocampus lend support to this hypothesis. At birth, no potentiation is elicited from tetanic stimulation in area CA1, but by postnatal days 3–4 posttetanic potentiation, lasting less than a minute, can be induced (Harris & Teyler 1984). By days 5–7 a more enduring potentiation can be induced, but it lasts for only about 45 minutes post tetanus (Harris & Teyler 1984, Bekenstein & Lothman 1991). By days 10–11 the potentiation endures for 2.5 hours, and by day 15 some animals show persistent LTP (for at least 9 hours in vitro). These findings cannot be explained simply by the development of NMDA receptors: In area CA1, the NMDA receptors are present at about 75% of adult values from birth through day 7 (Insel et al 1990, McDonald & Johnston 1990, McDonald et al 1990). The development of a minimum number of dendritic spines may be required for the induction of LTP, as spines are first present at days 5–7, when a nonpersistent form of LTP is first induced (Minkwitz 1976; Pokorny & Yamamoto 1981a,b; Harris et al 1989). Notably, with maturation more spines have constricted necks, and LTP can be induced at lower stimulus intensities than those required at the younger ages (Harris & Teyler 1984, Bekenstein & Lothman 1991). Spines may similarly facilitate the effectiveness of the maturing NMDA receptors and ontogeny of LTP in the cortex (e.g. Mates & Lund 1983; Wilson &

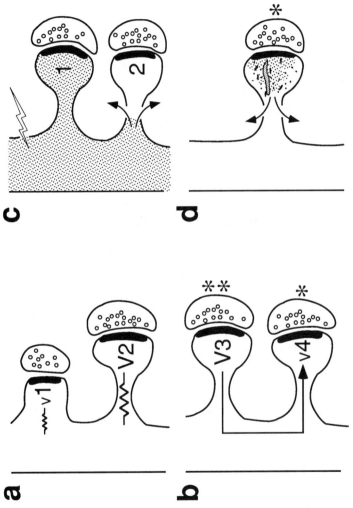

Figure 5 Functions of dendritic spines, continued. (*a*) Amplifying potential (*V*) in spine head. This amplification could contribute to the induction of LTP. (*b*) Sharing postsynaptic potential among neighboring spines. This coactivation could facilitate the associativity of LTP; **, strongly activated synapse; *, weakly activated synapse. (*c,d*) Biochemical compartmentation. This compartmentation could contribute to the specificity and endurance of LTP. (*d*) illustrates a tube of SER as an intracellular store of calcium that can be released upon synaptic activation. The small dots represent calcium concentration and the larger black shapes represent stimulus-activated molecules. (See text.)

Racine 1983; Kleinschmidt et al 1987; Komatsu & Toyama 1988, 1989; Perkins & Teyler 1988; Tsumoto et al 1989; Insel et al 1990; Tsumoto 1992).

SPINES PERMIT ASSOCIATIVITY IN LTP   It has been shown that weak tetanic stimulation is insufficient to induce LTP, but that when the weak stimulation at one set of synapses is coupled with strong stimulation at another set of synapses on the same cell, LTP is induced at both sets of synapses (McNaughton et al 1978, Levy & Steward 1979, Barrionuevo & Brown 1983, Kelso & Brown 1986, Kelso et al 1986, Larson & Lynch 1986, Sastry et al 1986, Brown et al 1991). The weak and strong stimulation must occur within 100 ms of one another to potentiate the synapses at the weak site, and for this reason it is thought that the "associative messenger" is the polarization state of the postsynaptic membrane (Madison et al 1991).

A longstanding hypothesis has been that the narrow dimensions of the spine neck attenuate current flow between the spine head and the dendrite (Rall 1970, 1974; Coss & Perkel 1985). Morphological evidence suggests, however, that most spine necks are not thin and long enough to significantly reduce the charge transferred to the parent dendrite, if the conductance changes at the synapse are less than 5 nS (Wilson et al 1983; Wilson 1984; Brown et al 1988; Harris & Stevens 1988, 1989). Recent electrophysiological evidence from hippocampal CA1 cells suggests that the mean synaptic conductance for a minimal evoked response is 0.21 ± 0.12 nS, such that the current generated by the release of 10–20 quanta would likely be fully transmitted to the postsynaptic dendrite (Bekkers et al 1990). Thus, spines should permit the addition of voltage changes among coactivated synapses via the dendrites connecting them (Figure 5b), thereby allowing the associativity observed with LTP. Other models endow the spine with active membrane (Miller et al 1985, Perkel & Perkel 1985, Shepherd et al 1985, Rall & Segev 1988, Segev & Rall 1988). These models suggest that the excitable membrane might facilitate communication between spines; such facilitated communication would enhance the associativity of the postsynaptic potential among spines.

POSSIBLE EFFECT OF SPINES ON THE SPECIFICITY OF LTP   LTP has long been known to be specific to the inputs that are activated during tetanic stimulation (Bliss & Lomo 1973; reviewed in Wigstrom & Gustafsson 1988). In the early experiments, input pathways from different brain regions were tested. LTP was subsequently shown also to be restricted to the stimulated axons within a single input pathway. In these experiments, two subsets of axons were first shown to converge on the same CA1 dendrites, but at different synapses. Then one set of axons was tetanized and LTP was induced. LTP was restricted to those axons that were tetanized, and LTP was not evoked in the nontetanized subset of axons. This specificity is partly explained by the requirement for glutamate to be released from the presynaptic axon at the same time that the postsynaptic area of the synapse is sufficiently depolarized (i.e. during the tetanus). However, if the molecules relevant to LTP at the tetanized synapses were to diffuse rapidly to the neighboring synapses, then they might also modify those synapses, resulting in a nonspecific spread of the potentiation. The subcellular localization of specific molecules and their regulating organelles within spines may be important factors in establishing the specificity of LTP.

Several modeling studies have predicted that changes in the concentration of calcium and other molecules occurring in the spine will not necessarily transfer to the dendritic shaft, and vice versa (Gamble & Koch 1987, Brown et al 1988, Wickens 1988, Holmes 1990). Recently, visualization of events within living spines in vitro has provided direct confirmation of these predictions. Two sets of experiments examined whether calcium diffuses freely between the dendrites and the spines (Guthrie et al 1991, Muller & Connor 1991). Guthrie et al (1991) exploited the ability of cobalt to quench fura 2 fluorescence to test whether specificity is accomplished by a physical diffusion barrier between spines and their dendritic shaft. As cobalt diffused along a dendritic shaft from a distant region of locally induced entry (*lightning bolt* in Figure 5c), the loss of fluorescence occurred virtually simultaneously in the dendrite and the adjacent spines (Figure 5c, *spine 1*). That is, there was essentially no absolute physical barrier to diffusion of small ions from the dendritic shaft into the spine. Nonetheless, when a parallel experiment was done with calcium as the diffusing ion, large rises in calcium in the shaft did not occur in many of the adjacent spines (24/74), indicating that calcium in the dendrite, in contrast to cobalt, is indeed isolated from some spines (Figure 5c, *spine 2*). Muller & Connor (1991) employed synaptic activation to demonstrate that stimulation of spine synapses results in sustained elevation of spine calcium that long outlasts changes in the dendritic shaft (Figure 5d). Were stimulation-activated release of calcium to occur from the SER in the spine, the calcium concentration in the spine head would be amplified. Thus, the local concentration of postsynaptic calcium has emerged as a candidate for the mechanism by which specificity could be achieved.

Mathematical modeling provides plausible insights into how such calcium compartmentation could occur in the absence of an absolute barrier to diffusion between spines and dendrites (Zador et al 1990). Three conditions could achieve localization of this second messenger: (*a*) the spine neck could provide a narrow diffusion path that limits calcium ion flux into or out of some spine heads; (*b*) even a small rise in spine calcium could cause a controlled release from intracellular calcium stores, thereby amplifying the calcium signal; and finally (c) only a very few calcium pumps would be required to extrude the few calcium ions that might diffuse through the limited volume of the spine neck from even micromolar concentrations in the dendritic shaft (arrows in Figure 5c) or from the spine head to the dendrite (arrows in Figure 5d). Not all spine morphologies would be expected to restrict diffusion. Similarly, it is possible that the distribution of intracellular calcium stores and pumps is not the same on all spines. For example, only the large, mushroom-shaped dendritic spines have laminated spine apparatuses, whereas the smaller, thin spines have a thin tube of SER (see Figure 2 above). Perhaps a subset of spines, or alternatively all spines, but only at a restricted time during their developmental history, achieve the compartmentalization required to confer this specificity.

ROLE OF SPINES IN THE ENDURANCE OF LTP   The hallmark of LTP in mature animals is its longevity; it can last for hours to days to weeks depending on the exact experimental conditions (Bliss & Gardner-Medwin 1973, Barnes 1979, Racine et al 1983, Staubli & Lynch 1987). Considerable attention has been devoted to understanding the cellular mechanisms mediating this endurance. From the postsynaptic perspective, the local compartments that spines create in the vicinity of the synapses may allow the concentration of calcium and other molecules relevant to LTP (Figure 5d) to remain high enough for sufficient time to stabilize changes in the synaptic machinery leading to persistent LTP.

Study of the ontogeny of LTP also supports this hypothesis. During development, LTP does not persist longer than 2.5 hours until postnatal day 15 (Harris & Teyler 1984, Jackson et al 1993). Different 15-day-old animals express one of two patterns of potentiation: some animals show enduring potentiation like that seen in adults, whereas others show an elevated response for 2.5 hours, which then decays to baseline by 4 hours posttetanus. These findings suggest that day 15 may be a threshold age for expressing persistent LTP. At day 15 about half of the synapses that will be found in the adults have been formed (Harris et al 1992). Spines of the thin, mushroom, and stubby shapes are all present at about equal frequencies. By the time the animals are young adults (days 48–60), however, the majority of synapses are on small, thin spines. These observations suggest that a sufficient number of spines with constricted necks may be required for persistent LTP. They also illustrate the importance of potentially dynamic changes in spine structure during development.

CHANGES IN DENDRITIC SPINE STRUCTURE WITH LTP   Average spine and synaptic dimensions have been compared in preparations that have undergone plasticity with those in preparations that have not. Several other reviews have considered the changes in spine morphology that accompany synaptogenesis during development, behavioral changes associated with learning and memory, and pathological changes associated with neural dysfunction (Scheibel & Scheibel 1968, Huttenlocher 1975, Coss & Perkel 1985, Greenough & Bailey 1988, Calverly & Jones 1990). Considerable accumulated evidence suggests that changes in spine and synaptic structure occur during LTP (for review see Wallace et al 1991). Where tested, the reported changes in spine and synaptic morphology have been specific to the tetanized input (Van Harreveld & Fifkova 1975, Fifkova & Van Harreveld 1977, Desmond & Levy 1988a). Controversy remains as to whether new spines and synapses form or if the geometries of existing spines and synapses change (Wallace et al 1991, Harris et al 1992). For example, in hippocampal area dentata, some results suggest that dendritic spines swell during LTP (Van Harreveld & Fifkova 1975, Fifkova & Van Harreveld 1977), and others suggest a change in the morphology of existing PSDs during LTP (Desmond & Levy 1986, 1988b, 1990). In contrast, results from a study utilizing serial EM reconstructions suggest that during LTP the total spine number doubles and the number of branched spines and spines with wide necks increases (Andersen et al 1987a,b; Trommald et al 1990). In hippocampal area CA1, no significant changes in overall spine density have been detected, although there is evidence for spine "rounding" and an increase in the frequency of stubby dendritic spines (Lee et al 1980, Chang & Greenough 1984). Except for extremely fortuitous sections (such as the one shown in Figure 2 above), the morphology of most spines cannot be identified on a single section, and therefore no data exist in these studies on the fate of the predominant thin, mushroom, and branched dendritic spines during LTP (Harris et al 1992).

As implied by the description of dendritic spine composition above, several molecular mechanisms exist that could mediate rapid short-term and long-term changes in spine and synaptic morphology. For example, glutamate and its analogues activate proteolysis of brain spectrin (fodrin) by the neuron-specific protease, calpain I (Siman & Noszek 1988). Degradation of fodrin, a structural protein of the (spine) cytoskeleton (Perlmutter et al 1988), could allow the spine to undergo shape changes (Siman et al 1990), possibly in response to growth of the synapse. The state of actin polymerization is regulated by calcium concentration and determines the viscosity of the spine cytoplasm (Fifkova 1985). The actin filaments are transient structures that can change rapidly in response to the calcium-activated second messenger systems involving stimulation of phosphorylation by calmodulin. Actin could serve to stabilize spine structure through its binding to the subplasmalemmal cytoskeleton or to alter spine structure through "contraction" (Crick 1982, Katsumaru et al 1982, Eccles 1983).

Spines Might Prevent Neuronal Pathology During Normal Synaptic Transmission and Plasticity, Such as LTP

Spines are nearly absent or have gross distortions in the cerebral cortex and hippocampus of individuals suffering from severe mental retardation (Marin-Padilla 1972, 1974, 1976; Purpura 1974, 1975a,b; Huttenlocher 1975, Williams et al 1990), epileptic seizures (Scheibel et al 1974), and the neuropathology associated with hypoxia, ischemia, and stroke (Fischer et al 1974, 1980; Rothman & Olney 1986; von Bossanyi & Dietzmann 1990). Under normal conditions, the compartmentation of calcium in dendritic spines could allow its concentration to achieve levels that can activate the second messenger systems. Such intracellular calcium levels could be toxic in the dendrite, but the spines contain a sufficiently small volume that the endoplasmic reticulum and cytoplasmic calcium buffers can return the calcium concentration to basal levels shortly after synaptic activation.

In experimental animal models of seizures and hypoxia/ischemia, a characteristic sequence of ultrastructural lesions occurs (Olney et al 1979, 1983; Evans et al 1983; von Lubitz & Diemer 1983; Sloviter & Dempster 1985; Siman & Card 1988; Allen et al 1989; Remis et al 1989; Yamamoto et al 1990). Notably, spines are lost in one of the first steps during this process, and the dendrites and their organelles become grossly swollen during these early stages (i.e. the rapid toxicity on a time scale of minutes to hours). Subsequently the animals or cells are returned to normal conditions, and over a prolonged time the neurons die (i.e. the delayed toxicity). Excitotoxicity induced by seizures or hypoxia/ischemia may be forms of neuronal pathology that result from excessive use of the same synaptic mechanisms that are normally used in LTP and learning and memory. A compelling hypothesis is that the cellular changes involving dendritic and cellular swelling during the phase of rapid toxicity are mediated through the influx of large amounts of sodium, chloride, and water following excessive activation of the glutamatergic receptors. The delayed toxicity may well result from a prolonged elevation in calcium throughout the neuron, which can cause hyperexcitability, proteolysis of neurofilaments, irreversible mitochondrial damage, and breakdown of membrane phospholipids with release of arachidonic acid and oxygen-free radicals (Rothman & Olney 1986, Choi 1988, Meyer 1989, Meldrum & Garthwaite 1990). These findings support the speculation that dendritic spines promote synaptic stability and plasticity under normal conditions and additionally protect the dendrites and postsynaptic cells from changes in molecular composition that might otherwise be pathological.

PROSPECTUS

Here we have described how the structure of dendritic spines could facilitate not only the stability and reliability of excitatory synaptic transmission but also the functioning of cellular mechanisms that mediate the induction, associativity, specificity, and endurance of LTP. We have also discussed how spines may serve to prevent cytotoxicity during normal synaptic transmission and plasticity. Furthermore, evidence suggests that the morphology of spines and their synaptic complexes changes with both LTP and cytotoxicity. This evidence is based largely on the comparison of static images from experimental and control neurons. Though powerful in its own right, such a statistical or population approach cannot follow dynamic changes in individual spines.

Foreseeable advances in light microscope technology will undoubtedly allow the life histories of individual spines to be followed at a gross level. The formation of new spines or the resorption of existing spines is clearly in the realm of this level of resolution. Gross changes in the volume of the spine head will also be detectable. However, the following key features of spine and synaptic structure are all below the resolution of light and require ultrastructural analysis: (*a*) spine neck diameter, which could modulate ionic flux; (*b*) irregularities in spine surface area, which could affect channel number and capacitance; (*c*) PSD area, which predicts the availability of several molecules involved in synaptic transmission; (*d*) SER volume, which could regulate the ionic and other molecular composition of the cytoplasm; and (*e*) presynaptic vesicles, the size and distribution of which may predict availability of neurotransmitter for release. Since theory predicts that even subtle changes in spine structure can influence synaptic transmission and plasticity, it will be important to establish the degree to which these alterations occur. Obviously, significant progress in spine research will best be made when single experiments can make use of the full range of temporal and spatial resolution of both light and electron microscopy.

ACKNOWLEDGMENTS

This work was supported by NIH grant number NS21184 (KMH) and the Alzheimer's Association (SBK).

Literature Cited

Aghajanian GK, Bloom FE. 1967. The formation of synaptic junctions in developing rat brain: a quantitative electron microscopic study. *Brain Res.* 6:716–27

Akiyama SK, Nagata K, Yamada KM. 1990. Cell surface receptors for extracellular matrix components. *Biochim. Biophys. Acta* 1031:91–110

Allen A, Yanushka J, Fitzpatrick JH, et al. 1989. Acute ultrastructural response of hypoxic hypoxia with relative ischemia in the isolated brain. *Acta Neuropathol.* 78:637–48

Amaral DG, Dent JA. 1981. Development of the mossy fibers of the dentate gyrus: I. A light and electron microscopic study of the mossy fibers and their expansions. *J. Comp. Neurol.* 195:51–86

Andersen P, Blackstad T, Hulleberg G, et al. 1987a. Dimensions of dendritic spines of rat dentate granule cells during long-term potentiation (LTP). *J. Physiol. (London)* 390:264

Andersen P, Blackstad T, Hulleberg G, et al. 1987b. Changes in spine morphology associated with LTP in rat dentate granule cells. *Proc. Physiol. Soc.* PC50:288P

Andrews SB, Leapman RD, Landis DMD, Reese TS. 1988. Activity-dependent accumulation of calcium in Purkinje cell dendritic spines. *Proc. Natl. Acad. Sci. USA.* 85:1682–85

Aoki C. 1992. β-Adrenergic receptors: astrocytic localization in the adult visual cortex and their relationship to catecholamine axon terminals as revealed by electron microscopic immunocytochemistry. *J. Neurosci.* 12:781–92

Baer SM, Rinzel J. 1991. Propagation of dendritic spikes mediated by excitable spines: a continuum theory. *J. Neurophysiol.* 65:874–90

Bahr BA, Lynch G. 1992. Purification of an Arg-Gly-Asp selective matrix receptor from brain synaptic plasma membrane. *Biochem. J.* 281:137–42

Bailey CH, Chen M, Keller F, Kandel ER. 1992. Serotonin-mediated endocytosis of apCAM: An early step of learning-related synaptic growth in Aplysia. *Science* 256:645–50

Barnes CA. 1979. Memory deficits associated with senescence: A neurophysiological and behavioral study in the rat. *J. Comp. Physiol. Psychol.* 93:74–104

Barres BA. 1991. New roles for glia. *J. Neurosci.* 11:3685–94

Barrionuevo G, Brown TH. 1983. Associative long-term potentiation in hippocampal slices. *Proc. Natl. Acad. Sci. USA* 80:7347–51

Bekenstein JW, Lothman EW. 1991. An in vivo study of the ontogeny of long-term potentiation (LTP) in the CA1 region and in the dentate gyrus of the rat hippocampal formation. *Dev. Brain Res.* 63:245–51

Bekkers JM, Richerson GB, Stevens CF. 1990. Origin of variability in quantal size in cultured hippocampal neurons and hippocampal slices. *Proc. Natl. Acad. Sci. USA* 87:5359–62

Benowitz LI, Perrone-Bizzozero NI. 1991. The relationship of GAP-43 to the development and plasticity of synaptic connections. *Ann. NY Acad. Sci.* 627:58–74

Bliss TVP, Collingridge GL. 1993. A synaptic model of memory: long-term potentiation in the hippocampus. *Nature* 361:31–39

Bliss TVP, Gardner-Medwin AR. 1973. Long-lasting potentiation of synaptic transmission in the dentate area of the unanaesthetized rabbit following stimulation of the perforant path. *J. Physiol. (London)* 232:357–74

Bliss TVP, Lomo T. 1973. Long-lasting potentiation of synaptic transmission in the dentate area of the anaesthetized rabbit following stimulation of the perforant path. *J. Physiol. (London)* 232:331–56

Brown TH, Chang VC, Ganong AH, et al. 1988. Biophysical properties of dendrites and spines that may control the induction and expression of long-term synaptic potentiation. *Neurol. Neurobiol.* 35:201–64

Brown TH, Zador AM, Mainen ZF, Claiborne BJ. 1991. Hebbian modifications in hippocampal neurons. In *Long-term Potentiation: A Debate of Current Issues*, ed. M Baudry, JL Davis, pp. 357–89. Cambridge, MA: MIT Press

Burgin KE, Waxham MN, Rickling S, et al. 1990. In situ hybridization histochemistry of Ca^{2+}/calmodulin-dependent protein kinase in developing rat brain. *J. Neurosci.* 10:1788–98

Burgoyne RD, Gray EG, Barron J. 1983. Cytochemical localization of calcium in the dendritic spine apparatus of the cerebral cortex and at synaptic sites in the cerebellar cortex. *J. Anat.* 136:634–35

Calverly RKS, Jones DG. 1990. Contributions of dendritic spines and perforated synapses to synaptic plasticity. *Brain Res. Rev.* 15:215–49

Cameron HA, Kaliszewski CK, Greer CA. 1991. Organization of mitochondria in olfactory bulb granule cell dendritic spines. *Synapse* 8:107–18

Carlin RK, Bartelt DC, Siekevitz P. 1983. Identification of fodrin as a major calmodulin-binding protein in postsynaptic density preparations. *J. Cell Biol.* 96:443–48

Carlin RK, Grab DJ, Cohen RS, Siekevitz P. 1980. Isolation and characterization of postsynaptic densities from various brain regions: Enrichment of different types of postsynaptic densities. *J. Cell Biol.* 86:831–43

Carlin RK, Grab DJ, Siekevitz P. 1981. Function of a calmodulin in postsynaptic densities. III. Calmodulin-binding proteins of the postsynaptic density. *J. Cell Biol.* 89:449–55

Chang FF, Greenough WT. 1984. Transient and enduring morphological correlates of synaptic activity and efficacy change in the rat hippocampal slice. *Brain Res.* 309:35–46

Chicurel ME, Harris KM. 1992. Three-dimensional analysis of the structure and composition of CA3 branched dendritic spines and their synaptic relationships with mossy fiber boutons in the rat hippocampus. *J. Comp. Neurol.* 325:169–82

Choi DW. 1988. Glutamate neurotoxicity and diseases of the nervous system. *Neuron* 1:623–34

Clements JD, Lester RA, Tong G, et al. 1992. The time course of glutamate in the synaptic cleft. *Science* 258:1498–501

Clements JR, Magnusson KR, Beitz AJ. 1990. Ultrastructural description of glutamate-, aspartate-, taurine-, and glycine-like immunoreactive terminals from five rat brain regions. *J. Electron Microsc. Tech.* 15:49–66

Cohen RS, Chung SK, Pfaff DW. 1985. Immunocytochemical localization of actin in dendritic spines of the cerebral cortex using colloidal gold as a probe. *Cell. Mol. Neurobiol.* 5:271–84

Cohen RS, Siekevitz P. 1978. Form of the postsynaptic density. A serial section study. *J. Cell Biol.* 78:36–46

Colonnier M. 1968. Synaptic patterns on different cell types in the different laminae of the cat visual cortex. An electron microscope study. *Brain Res.* 9:268–87

Cornell-Bell AH, Finkbeiner SM, Cooper MS, Smith SJ. 1990a. Glutamate induces calcium waves in cultured astrocytes: Long-range glial signaling. *Science* 247:470–73

Cornell-Bell AH, Thomas PG, Smith SJ. 1990b. The excitatory neurotransmitter glutamate causes filopodia formation in cultured hippocampal astrocytes. *Glia* 3:322–34

Coss RG, Perkel DH. 1985. The function of dendritic spines: A review of theoretical issues. *Behav. Neural Biol.* 44:151–85

Crick F. 1982. Do dendritic spines twitch. *Trends Neurosci.* 5:44–46

De Zeeuw CI, Ruigrok TJH, Holstege JC, et al. 1990. Intracellular labeling of neurons in medial accessory olive of the cat: II. Ultrastructure of dendritic spines and their GABAergic innervation. *J. Comp. Neurol.* 300:478–94

Dehay C, Douglas RJ, Martin KAC, Nelson C. 1991. Excitation by geniculocortical synapses is not 'vetoed' at the level of dendritic spines in cat visual cortex. *J. Physiol. (London)* 440:723–34

Desmond NL, Levy WB. 1986. Changes in the postsynaptic density with long-term potentiation in the dentate gyrus. *J. Comp. Neurol.* 253:476–82

Desmond NL, Levy WB. 1988a. Anatomy of associative long-term synaptic modification. In *Long-term Potentiation: From Biophysics to Behavior*, ed. V Chan-Palay, C Kohlu, pp. 265–305. New York: Liss

Desmond NL, Levy WB. 1988b. Synaptic interface surface area increases with long-term potentiation in the hippocampal dentate gyrus. *Brain Res.* 453:308–14

Desmond NL, Levy WB. 1990. Morphological correlates of long-term potentiation imply the modification of existing synapses, not synaptogenesis, in the hippocampal dentate gyrus. *Synapse* 5:139–43

Diamond J, Gray EG, Yasargil GM. 1970. The function of the dendritic spine: an hypothesis. In *Excitatory Synaptic Mechanisms*, ed. P Andersen, JKS Jensen, pp. 213–222. Oslo: Universitets Forlaget

DiFiglia M, Aronin N, Martin JB. 1982. Light and electron microscopic localization of immunoreactive leu-enkephalin in the monkey basal ganglia. *J. Neurosci.* 2:303–20

DiFiglia M, Roberts RC, Benowitz LI. 1990. Immunoreactive GAP-43 in the neuropil of adult rat neostriatum: Localization in unmyelinated fibers, axon terminals, and dendritic spines. *J. Comp. Neurol.* 302:992–1001

Ebner FF, Colonnier M. 1975. Synaptic patterns in the visual cortex of turtle: An electron microscopic study. *J. Comp. Neurol.* 160:51–80

Ebner FF, Colonnier M. 1978. A quantitative study of synaptic patterns in turtle visual cortex. *J. Comp. Neurol.* 179:263–76

Eccles JC. 1983. Calcium in long-term potentiation as a model for memory. *Neuroscience* 10:1071–81

Evans M, Griffiths T, Meldrum B. 1983. Early changes in the rat hippocampus following seizures induced by Bicuculline or L-Allylglycine: A light and electron microscopic study. *Neuropathol. Appl. Neurobiol.* 9:39–52

Ferns MJ, Hall ZW. 1992. How many agrins does it take to make a synapse? *Cell* 70:1–3

Fifkova E. 1985. Actin in the nervous-system. *Brain Res. Rev.* 9:187–215

Fifkova E, Eason H, Schaner P. 1992. Inhibitory contacts on dendritic spines of the dentate fascia. *Brain Res.* 577:331–36

Fifkova E, Markham JA, Delay RJ. 1983. Calcium in the spine apparatus of dendritic spines in the dentate molecular layer. *Brain Res.* 266:163–68

Fifkova E, Van Harreveld A. 1977. Long-lasting morphological changes in dendritic spines of dentate granular cells following stimulation of the entorhinal area. *J. Neurocytol.* 6:211–30

Fischer J, Jilek L, Trojan S. 1974. Qualitative and quantitative neurohistological changes produced in the rat brain by prolonged aerogenic hypoxia in early ontogeny. *Physiol. Bohemoslov.* 23:211–19

Fischer J, Langmeier M, Trojan S. 1980. Changes in the length and width of the postsynaptic density and the synaptic cleft in the cerebral cortex synapses of rats exposed to prolonged aerogenic hypoxia during early ontogenesis. An electron microscopic morphometric study. *Physiol. Bohemoslov.* 29: 561–67

Gamble E, Koch C. 1987. The dynamics of free calcium in dendritic spines in response to repetitive synaptic input. *Science* 236: 1311–15

Garner CC, Tucker RP, Matus A. 1988. Selective localization of messenger RNA for cytoskeletal protein MAP2 in dendrites. *Nature* 336:674–77

Geinisman Y, Morrell F, de Toledo-Morrell L. 1992. Increase in the number of axospinous synapses with segmented postsynaptic densities following hippocampal kindling. *Brain Res.* 569:341–47

Goldenring JR, McGuire JS, DeLorenzo RJ. 1984. Identification of the major postsynaptic density protein as homologous with the major calmodulin-binding subunit of a calmodulin-dependent protein kinase. *J. Neurochem.* 42:1077–84

Gray EG. 1959. Axo-somatic and axo-dendritic synapses of the cerebral cortex: An electron microscopic study. *J. Anat.* 83:420–33

Greenough WT, Bailey CH. 1988. The anatomy of a memory: Convergence of results across a diversity of tests. *Trends Neurosci.* 11:142–47

Greenough WT, Hwang H-MF, Gorman C. 1985. Evidence for active synapse formation or altered postsynaptic metabolism in visual cortex of rats reared in complex environments. *Proc. Natl. Acad. Sci. USA* 82:4549–52

Guthrie PB, Segal M, Kater SB. 1991. Independent regulation of calcium revealed by imaging dendritic spines. *Nature* 354:76–80

Hall ZW. 1992. *An Introduction to Molecular neurobiology.* Sunderland, MA: Sinauer

Hama K, Arii T, Kosaka T. 1989. Three-dimensional morphometrical study of dendritic spines of the granule cell in the rat dentate gyrus with HVEM stereo images. *J. Electron Microsc. Tech.* 12:80–87

Hamlyn LH. 1962. The fine structure of the mossy fibre endings in the hippocampus of the rabbit. *J. Anat.* 96:112–20

Harris KM, Jensen FE, Tsao B. 1989. Ultrastructure, development, and plasticity of dendritic spine synapses in area CA1 of the rat hippocampus: Extending our vision with serial electron microscopy and three-dimensional analyses. In *The Hippocampus—New Vistas, Neurology and Neurobiology,* eds. V Chan-Palay, C Kohler, 52:33–52. New York: Liss

Harris KM, Jensen FE, Tsao B. 1992. Three-dimensional structure of dendritic spines and synapses in rat hippocampus (CA1) at postnatal day 15 and adult ages: Implications for the maturation of synaptic physiology and long-term potentiation. *J. Neurosci.* 12: 2685–705

Harris KM, Landis DM. 1986. Membrane structure at synaptic junctions in area CA1 of the rat hippocampus. *Neuroscience* 19: 857–72

Harris KM, Rosenberg PA. 1993. Localization of synapses in rat cortical cultures. *Neuroscience* 53:495–508

Harris KM, Stevens JK. 1988. Dendritic spines of rat cerebellar Purkinje cells: Serial electron microscopy with reference to their biophysical characteristics. *J. Neurosci.* 8: 4455–69

Harris KM, Stevens JK. 1989. Dendritic spines of CA1 pyramidal cells in the rat hippocampus: serial electron microscopy with reference to their biophysical characteristics. *J. Neurosci.* 9:2982–97

Harris KM, Teyler TJ. 1984. Developmental onset of long-term potentiation in area CA1 of the rat hippocampus. *J. Physiol. (London)* 346:27–48

Holmes WR. 1990. Is the function of dendritic spines to concentrate calcium? *Brain Res.* 519:338–42

Horwitz B. 1984. Electrophoretic migration due to postsynaptic potential gradients: Theory and application to autonomic ganglion neurons and to dendritic spines. *Neuroscience* 12:887–905

Hosokawa T, Bliss TVP, Fine A. 1992. Persistence of individual spines in living brain slices. *NeuroReport* 3:477–80

Huttenlocher PR. 1975. Synaptic and dendritic development and mental defect. In *Brain Mechanisms in Mental Retardation,* ed. N Buchwald, MA Brazier, pp. 123–140. NY: Academic

Insel TR, Miller LP, Gelhard RE. 1990. The ontogeny of excitatory amino acid receptors in rat forebrain. I. N-methyl-D-aspartate and quisqualate receptors. *Neuroscience* 35:31–43

Jackson PJ, Suppes T, Harris KM. 1993. Stereotypical changes in the pattern and duration of long-term potentiation expressed at postnatal days 11 and 15 in the rat hippocampus. *J. Neurophysiol.* In Press

Jones EG, Powell TPS. 1969. Morphological variations in the dendritic spines of the neocortex. *J. Cell Sci.* 5:509–29

Katsumaru H, Murakami F, Tsukahara N. 1982. Actin filaments in dendritic spines of red nucleus neurons demonstrated by immunoferritin localization and heavy meromyosin binding. *Biomed. Res.* 3:337–40

Kawato M, Tsukahara N. 1983. Theoretical study on electrical properties of dendritic spines. *J. Theor. Biol.* 103:507–22

Kelly PT, Cotman CW. 1978. Synaptic proteins. Characterization of tubulin and actin and identification of a distinct postsynaptic density polypeptide. *J. Cell Biol.* 79:173–83

Kelly PT, McGuiness TL, Greengard P. 1984.

Evidence that the major postsynaptic protein is a component of a Ca^{2+}/calmodulin-dependent protein kinase. *Proc. Natl. Acad. Sci. USA* 81:945–49

Kelso SR, Brown TH. 1986. Differential conditioning of associative synaptic enhancement in hippocampal brain-slices. *Science* 232:85–87

Kelso SR, Ganong AH, Brown TH. 1986. Hebbian synapses in the hippocampus. *Proc. Natl. Acad. Sci. USA* 83:5326–31

Kennedy MB. 1992. Second messengers and neuronal function. In *An Introduction to Molecular Neurobiology*, ed. ZW Hall, pp. 207–246. Sunderland, MA: Sinauer

Kennedy MB, Bennett MK, Bulliet RF, et al. 1990. Structure and regulation of type II calcium/calmodulin-dependent protein kinase in central nervous system neurons. *Cold Spring Harbor Symp. Quant. Biol.* 55:101–10

Kennedy MB, Bennett MK, Erondu NE. 1983. Biochemical and immunochemical evidence that the "major postsynaptic density protein" is a subunit of a calmodulin-dependent protein kinase. *Proc. Natl. Acad. Sci. USA* 80:7357–61

Kleinschmidt A, Bear MF, Singer W. 1987. Blockade of "NMDA" receptors disrupts experience-dependent plasticity of kitten striate cortex. *Science* 238:355–58

Koch C, Zador A. 1993. The function of dendritic spines: devices subserving biochemical rather than electrical compartmentalization. *J. Neurosci.* 13:413–22

Koch C, Zador A, Brown TH. 1992. Dendritic spines: convergence of theory and experiment. *Science* 256:973–74

Komatsu Y, Toyama K. 1988. Relevance of NMDA receptors to the long-term potentiation in kitten visual cortex. *Biomed. Res.* 2:39–41

Komatsu Y, Toyama K. 1989. Long-term potentiation of excitatory synaptic transmission in kitten visual cortex. *Biomed. Res.* 10:57–59

Kuffler SW. 1967. Neuroglial cells: physiological properties and a potassium mediated effect of neuronal activity on the glial membrane potential. *Proc. R. Soc. London* 168:1–21

Kullman DM, Nicoll RA. 1992. Long-term potentiation is associated with increases in quantal content and quantal amplitude. *Nature* 357:240–44

Landis DM, Reese TS. 1983. Cytoplasmic organization in cerebellar dendritic spines. *J. Cell Biol.* 97:1169–78

Landis DMD. 1988. Membrane and cytoplasmic structure at synaptic junctions in the mammalian central nervous system. *J. Electron Microsc. Tech.* 10:129–51

Larkman A, Hannay T, Stratford K, Jack J.

1992. Presynaptic release probability influences the locus of long-term potentiation. *Nature* 360:70–73

Larson J, Lynch G. 1986. Induction of synaptic potentiation in hippocampus by patterned stimulation involves two events. *Science* 232:985–88

Larson J, Lynch G. 1991. A test of the spine resistance hypothesis for LTP expression. *Brain Res.* 538:347–50

Lee KS, Schottler F, Oliver M, Lynch G. 1980. Brief bursts of high-frequency stimulation produce two types of structural change in rat hippocampus. *J. Neurophysiol.* 44:247–58

Levy WB, Steward O. 1979. Synapses as associative memory elements in the hippocampal formation. *Brain Res.* 175:233–45

Lisman J, Goldring MA. 1988. Feasibility of long-term storage of graded information by the Ca^{2+}/calmodulin-dependent protein kinase molecules of the postsynaptic density. *Proc. Natl. Acad. Sci. USA* 85:5320–24

Lisman J, Harris KM. 1993. Quantal analysis and synaptic anatomy—integrating two views of hippocampal plasticity. *Trends Neurosci.* 16:141–47

Madison DM, Malenka RC, Nicoll RA. 1991. Mechanisms underlying long-term potentiation of synaptic transmission. *Annu. Rev. Neurosci.* 14:379–97

Manabe T, Renner P, Nicoll RA. 1992. Postsynaptic contribution to long-term potentiation revealed by the analysis of miniature synaptic currents. *Nature* 355:50–55

Marin-Padilla M. 1972. Structural abnormalities of the cerebral cortex in human chromosomal aberrations: a Golgi study. *Brain Res.* 44:625–29

Marin-Padilla M. 1974. Structural organization of the cerebral cortex (motor area) in human chromosomal aberrations. A golgi study. I. D1 (13–15 Trisomy, Patau syndrome. *Brain Res.* 66:375–91

Marin-Padilla M. 1976. Pyramidal cell abnormalities in the motor cortex of a child with Down's syndrome. A Golgi study. *J. Comp. Neurol.* 167:63–82

Mates SL, Lund JS. 1983. Spine formation and maturation of type 1 synapses on spiny stellate neurons in primate visual cortex. *J. Comp. Neurol.* 221:91–97

Maycox PR, Hell JW, Jahn R. 1990. Amino acid neurotransmission: Spotlight on synaptic vesicles. *Trends Neurosci.* 13:83–87

McDonald JA. 1989. Receptors for extracellular matrix components. *Am. J. Physiol.* 257:L331–L337

McDonald JW, Johnston MV. 1990. Physiological and pathophysiological roles of excitatory amino acids during central nervous system development. *Brain Res. Rev.* 15:41–70

McDonald JW, Johnston MV, Young AB.

1990. Differential ontogenic development of three receptors comprising the NMDA receptor/channel complex in the rat hippocampus. *Exp. Neurol.* 110:237–47

McMahan UJ, Horton SE, Werle MJ, et al. 1992. Agrin isoforms and their role in synaptogenesis. *Cell Biol.* 4:869–74

McNaughton BL, Douglas RM, Goddard GV. 1978. Synaptic enhancement in fascia dentata: cooperativity among coactive afferents. *Brain Res.* 157:277–93

McWilliams JR, Lynch G. 1981. Sprouting in the hippocampus is accompanied by an increase in coated vesicles. *Brain Res.* 211: 158–64

McWilliams R, Lynch G. 1978. Terminal proliferation and synaptogenesis following partial deafferentation: The reinnervation of the inner molecular layer of the dentate gyrus following removal of its commissural afferents. *J. Comp. Neurol.* 180:581–616

Meldrum B, Garthwaite J. 1990. Excitatory amino acid neurotoxicity and neurogenerative disease. *Trends Pharm. Sci.* 11:379–87

Meyer FB. 1989. Calcium, neuronal hyperexcitability and ischemic injury. *Brain Res. Rev.* 14:227–43

Mignery GA, Sudhof TC, Takei K, DeCamilli P. 1989. Putative receptor for inositol 1,4,5-trisphosphate similar to ryanodine receptor. *Nature* 342:192–95

Miller JP, Rall W, Rinzel J. 1985. Synaptic amplification by active membrane in dendritic spines. *Brain Res.* 325:325–30

Minkwitz HG. 1976. Zur Entwicklung der Neuronenstruktur des Hippocampus wahrend der praund postnatalen Ontogenese der Albinoratte. III. Mitteilung: Morphometrische Erfassung der ontogenetischen Veranderungen in Dendritenstruktur und Spinebesatz an Pyramidenneuronen (CA1) des Hippocampus. *J. Hirnforsch.* 17:255–75

Muller W, Connor JA. 1991. Dendritic spines as individual neuronal compartments for synaptic Ca^{2+} responses. *Nature* 354:73–76

Nieto-Sampedro M, Hoff SF, Cotman CW. 1982. Perforated postsynaptic densities: Probable intermediates in synapse turnover. *Proc. Natl. Acad. Sci. USA* 79:5718–22

Olney JW, de Gubareff T, Sloviter RS. 1983. "Epileptic" Brain damage in rats induced by sustained electrical stimulation of the perforant path. II. Ultrastructural analysis of acute hippocampal pathology. *Brain Res. Bull.* 10:699–712

Olney JW, Fuller T, de Gubareff T. 1979. Acute dendrotoxic changes in the hippocampus of kainate treated rats. *Brain Res.* 176: 91–100

Otterson OP, Storm-Mathisen J, Bramham C. 1990a. A quantitative electron microscopic immunocytochemical study of the distribution and synaptic handling of glutamate in rat hippocampus. *Prog. Brain Res.* 83:99–114

Ottersen OP, Storm-Mathisen J, Bramham C, et al. 1990b. A quantitative electron microscopic immunocytochemical study of the distribution and synaptic handling of glutamate in rat hippocampus. *Prog. Brain Res.* 83:99–114

Perkel DH. 1982. Functional role of dendritic spines. *J. Physiol. (Paris)* 78:695–99

Perkel DH, Perkel DJ. 1985. Dendritic spines: role of active membrane in modulating synaptic efficacy. *Brain Res.* 325:331–35

Perkins AT, Teyler TJ. 1988. A critical period for long-term potentiation in the developing rat visual cortex. *Brain Res.* 439:222–29

Perlmutter LS, Siman R, Gall C, et al. 1988. The ultrastructural localization of calcium-activated protease "calpain" in rat brain. *Synapse* 2:79–88

Persohn E, Pollerberg GE, Schachner M. 1989. Immunoelectron-microscopic localization of the 180 kD component of the neural cell adhesion molecule N-CAM in postsynaptic membranes. *J. Comp. Neurol.* 288:92–100

Peters A, Kaiserman-Abramof IR. 1970. The small pyramidal neuron of the rat cerebral cortex. The perikaryon, dendrites and spines. *J. Anat.* 127:321–56

Peters A, Palay SL, Webster Hdef. 1991. *The Fine Structure of the Nervous System: The Neurons and Supporting Cells.* Philadelphia: Saunders. 13th ed.

Pokorny J, Yamamoto T. 1981a. Postnatal ontogenesis of hippocampal CA1 area in rats. I. Development of dendritic arborization in pyramidal neurons. *Brain Res. Bull.* 7:113–20

Pokorny J, Yamamoto T. 1981b. Postnatal ontogenesis of hippocampal Ca1 area in rats. II. Development of ultrastructure in stratum lacunosum and moleculare. *Brain Res. Bull.* 7:121–30

Purpura DP. 1974. Dendritic spine "dysgenesis" and mental retardation. *Science* 186: 1126–28

Purpura DP. 1975a. Normal and aberrant neuronal development in the cerebral cortex of human fetus and young infant. In *Brain Mechanisms in Mental Retardation,* ed. NA Buchwald, MAB Brazier, pp. 141–170. New York: Academic

Purpura DP. 1975b. Dendritic differentiation in human cerebral cortex: Normal and aberrant developmental patterns. In *Physiology and Pathology of Dendrites,* ed. GW Kreutzberg, pp. 91–116. New York: Raven

Qian N, Sejnowski TJ. 1989. An electro-diffusion model for computing membrane potentials and ionic concentrations in branching dendrites, spines and axons. *Biol. Cybern.* 62:1–15

Qian N, Sejnowski TJ. 1990. When is an

inhibitory synapse effective? *Proc. Natl. Acad. Sci. USA* 87:8145–49

Racine RJ, Milgram NW, Hafner S. 1983. Longterm potentiation phenomena in the rat limbic forebrain. *Brain Res.* 260:823–35

Rall W. 1970. Cable properties of dendrites and effects of synaptic location. In *Excitatory Synaptic Mechanisms,* ed. P Andersen, JKS Jensen, pp. 175–187. Oslo: Universitets Forlaget

Rall W. 1974. Dendritic spines, synaptic potency, and neuronal plasticity. In *Cellular Mechanisms Subserving Changes in Neuronal Activity,* ed. C Woody, K Brown, T Crow, J Knispel, pp. 13–21. Los Angeles: Brain Information Service

Rall W, Segev I. 1988. Synaptic integration and excitable dendritic spine clusters: structure/function. *Neurol. Neurobiol.* 37:263–82

Ramon y Cajal S. 1893. *Neue Darstellung vom Histologischen Bau des Centralnervensystems.* Archiv fur Anatomie und Entwickelungsgeschichte. Anatomische abtheilung des archives fur anatome und physiologie. 319–428. In German

Remis T, Benuska J, Masarova M. 1989. The ultrastructural picture of cerebral cortex of rat after hypoxia I. Findings after inhalation of 5% O₂ and 95% N₂. *Z. Mikrosk.-Anat. Forsch.* 103:297–308

Rosenberg PA, Aizenman E. 1989. Hundredfold increase in neuronal vulnerability to glutamate toxicity in astrocyte-poor cultures of rat cerebral cortex. *Neurosci. Lett.* 103:162–68

Rosenberg PA, Amin S, Leitner M. 1992. Glutamate uptake disguises neurotoxic potency of glutamate agonists in cerebral cortex in dissociated cell culture. *J. Neurosci.* 12:56–61

Rosenbluth J, Wissig SL. 1964. The distribution of exogenous ferritin in toad spinal ganglia and the mechanism of its uptake by neurons. *J. Cell Biol.* 23:307–25

Rothman SM, Olney JW. 1986. Glutamate and pathophysiology of hypoxic-ischemic brain damage. *Ann. Neurol.* 19:105–11

Sastry BR, Goh JW, Auyeung A. 1986. Associative induction of posttetanic and long-term potentiation in CA1 neurons of rat hippocampus. *Science* 232:988–90

Scheibel ME, Crandall PH, Scheibel AB. 1974. The hippocampal-dentate complex in temporal lobe epilepsy. *Epilepsia* 15:55–80

Scheibel ME, Scheibel AB. 1968. On the nature of dendritic spines: report of a workshop. *Commun. Behav. Biol.* A1:231–65

Schmied R, Holtman E. 1987. A phosphatase activity and synaptic vesicle antigen in multivesicular bodies of frog retinal photoreceptor terminals. *J. Neurocytol.* 16:627–37

Schwartz JH. 1992. Synaptic vesicles. In *Principles of Neural Science,* pp. 225–34. New York: Elsevier. 3rd ed.

Segev I, Rall W. 1988. Computational study of an excitable dendritic spine. *J. Neurophysiol.* 60:499–523

Seil FJ, Eckenstein FP, Reier PJ. 1992. Induction of dendritic spine proliferation by an astrocyte secreted factor. *Exp. Neurol.* 117:85–89

Sheperd GM, Brayton RK. 1987. Logic operations are properties of computer-simulated interactions between excitable dendritic spines. *Neuroscience* 21:151–65

Shepherd GM. 1990. *The Synaptic Organization of the Brain.* New York, NY: Oxford Univ. Press. 3rd ed.

Shepherd GM, Brayton RK, Miller JP, et al. 1985. Signal enhancement in distal cortical dendrites by means of interactions between active dendritic spines. *Proc. Natl. Acad. Sci. USA* 82:2192–95

Siekevitz P. 1985. The postsynaptic density: a possible role in long-lasting effect in the central nervous system. *Proc. Natl. Acad. Sci. USA* 82:3494–98

Siman R, Baudry M, Lynch GS. 1990. Calcium-activated proteases as possible mediators of synaptic plasticity. In *Synaptic Function,* ed. GM Edelman, WE Gall, WM Cowan. New York: Wiley & Sons

Siman R, Card JP. 1988. Excitatory amino acid neurotoxicity in the hippocampal slice preparation. *Neuroscience* 26:433–47

Siman R, Noszek JC. 1988. Excitatory amino acids activate calpain I and induce structural protein breakdown in vivo. *Neuron* 1:279–87

Sirevaag AM, Greenough WT. 1987. Differential rearing effects on rat visual cortex synapses. III. Neuronal and glial nuclei, boutons, dendrites, and capillaries. *Brain Res.* 424:320–32

Sloviter RS, Dempster DW. 1985. "Epileptic" Brain damage is replicated qualitatively in the rat hippocampus by central injection of glutamate or aspartate but not by GABA or Acetylcholine. *Brain Res. Bull.* 15:39–60

Spacek J. 1985a. Relationships between synaptic junctions, puncta adhaerentia and the spine apparatus at neocortical axo-spinous synapses. *Anat. Embryol.* 173:129–35

Spacek J. 1985b. Three-dimensional analysis of dendritic spines. II. Spine apparatus and other cytoplasmic components. *Anat. Embryol.* 171:235–43

Spacek J. 1985c. Three-dimensional analysis of dendritic spines III. Glial sheath. *Anat. Embryol.* 171:245–52

Staubli U, Lynch G. 1987. Stable hippocampal long-term potentiation elicited by 'theta' pattern stimulation. *Brain Res.* 435:227–34

Staubli U, Vanderklish P, Lynch G. 1990. An inhibitor of integrin receptors blocks long-term potentiation. *Behav. Neural Biol.* 53:1–5

Steward O. 1983. Alterations in polyribosomes associated with dendritic spines during the reinnervation of the dentate gyrus of the adult rat. *J. Neurosci.* 3:177–88

Steward O, Banker GA. 1992. Getting the message from the gene to the synapse: sorting and intracellular transport of RNA in neurons. *Trends Neurosci.* 15:180–86

Steward O, Falk PM. 1985. Polyribosomes under developing spine synapses: Growth specializations of dendrites at sites of synaptogenesis. *J. Neurosci. Res.* 13:75–88

Steward O, Levy WB. 1982. Preferential localization of polyribosomes under the base of dendritic spines in granule cells of the dentate gyrus. *J. Neurosci.* 2:284–91

Steward O, Reeves TM. 1988. Protein-synthetic machinery beneath postsynaptic sites on CNS neurons: association between polyribosomes and other organelles at the synaptic site. *J. Neurosci.* 8:176–84

Storm-Mathisen J, Leknes Ak, Bore AT, et al. 1983. First visualization of glutamate and GABA in neurones by immunocytochemistry. *Nature* 301:517–20

Swindale NV. 1981. Dendritic spines only connect. *Trends Neurosci.* 4:240–41

Tanzi E. 1893. I Fatti i le induzioni nell'odierna istologia del sistema nervoso. *Riv Sperim Freniatria Med Leg* 19:419–72. In Italian

Thompson DW. 1992. *On Growth and Form.* New York: Dover. 2nd ed.

Tiedge H, Fremeau RT Jr, Weinstock PH, et al. 1991. Dendritic location of neural BC1 RNA. *Proc. Natl. Acad. Sci. USA* 88:2093–97

Trommald M, Vaaland JL, Blackstad TW, Andersen P. 1990. Dendritic spine changes in rat dentate granule cells associated with long-term potentiation. In *Neurotoxicity of Excitatory Amino Acids,* ed. A Guidotti, E Costa, pp. 163–74. New York: Raven

Tsumoto T. 1992. Long-term potentiation and long-term depression in the neocortex. *Prog. Neurobiol.* 39:209–28

Tsumoto T, Kimura F, Nishigori A, Shirokawa T. 1989. Long-term potentiation and NMDA receptors in the developing visual cortex. *Biomed. Res.* 10:61–66

Turner DA. 1984. Conductance transients onto dendritic spines in segmental cable model of hippocampal neurons. *Biophys. J.* 46:85–96

Van Harreveld A, Fifkova E. 1975. Swelling of dendritic spines in the fascia dentata after stimulation of the perforant fibers as a mechanism of post-tetanic potentiation. *Exp. Neurol.* 49:736–49

Verhage M, McMahon HT, Ghijsen WEJM, et al. 1991. Differential release of amino acids, Neuropeptides, and Catecholamines from isolated nerve terminals. *Neuron* 6:517–24

von Bossanyi P, Dietzmann K. 1990. Infantile brain damage due to hypoxia with special reference to formation of dendritic spines. *Neuropatol. Pol.* 28:225–32

von Lubitz DKJE, Diemer NH. 1983. Cerebral ischemia in the rat: Ultrastructural and morphometric analysis of synapses in stratum radiatum of the hippocampal CA-1 region. *Acta Neuropathol. (Berlin)* 61:52–60

Vrensen G, Cardozo JN. 1981. Changes in size and shape of synaptic connections after visual training: an ultrastructural approach of synaptic plasticity. *Brain Res.* 218:79–97

Wallace C, Hawrylak N, Greenough WT. 1991. Studies of synaptic structural modifications after long-term potentiation and kindling: Context for a molecular morphology. In *Long-Term Potentiation: A Debate of Current Issues,* ed. M Baudry, JL Davis. Cambridge, MA: MIT Press

Walsh MJ, Kuruc N. 1992. The postsynaptic density: constituent and associated proteins characterized by electrophoresis, immunoblotting, and peptide sequencing. *J. Neurochem.* 59:667–78

Walton PD, Airey JA, Sutko JL, et al. 1991. Ryanodine and inositol trisphosphate receptors coexist in avian cerebellar Purkinje Neurons. *J. Cell Biol.* 113:1145–57

Wenzel J, Lammert G, Meyer U, Krug M. 1991. The influence of long-term potentiation on the spatial relationship between astrocyte precesses and potentiated synapses in the dentate gyrus neuropil of rat brain. *Brain Res.* 560:122–31

Westrum LE, Blackstad T. 1962. An electron microscopic study of the stratum radiatum of the rat hippocampus (regio superior, CA1) with particular emphasis on synaptology. *J. Comp. Neurol.* 119:281–309

Westrum LE, Jones DH, Gray EG, Barron J. 1980. Microtubules, dendritic spines and spine apparatuses. *Cell Tissue Res.* 208:171–81

Wickens J. 1988. Electrically coupled but chemically isolated synapses: dendritic spines and calcium in a rule for synaptic modification. *Prog. Neurobiol.* 31:507–28

Wigstrom H, Gustafsson B. 1988. Presynaptic and postsynaptic interactions in the control of hippocampal long-term potentiation. In *Long-Term Potentiation: From Biophysics to Behavior,* ed. PW Landfield, SA Deadwyler, pp. 73–108. New York: Liss

Williams RS, Hauser SL, Purpura DP, et al. 1990. Autism and mental retardation: Neuropathological studies performed in four retarded persons with autistic behavior. *Arch. Neurol. (Chicago)* 37:749–53

Wilson CJ. 1984. Passive cable properties of dendritic spines and spiny neurons. *J. Neurosci.* 4:281–97

Wilson CJ, Groves PM, Kitai ST, Linder JC.

1983. Three dimensional structure of dendritic spines in rat striatum. *J. Neurosci.* 3:383–98

Wilson DA, Racine RJ. 1983. The postnatal development of post-activation potentiation in the rat neocortex. *Dev. Brain Res.* 7:271–726

Wu K, Nigam SK, LeDoux M, et al. 1992. Occurrence of α subunits of G proteins in cerebral cortex synaptic modulation of ADP-ribosylation by Ca^{2+}/calmodulin. *Proc. Natl. Acad. Sci. USA* 89:8686–90

Wu K, Sachs L, Carlin RK, Siekevitz P. 1986. Characteristics of a Ca^{2+}/calmodulin-dependent binding of the Ca^{2+} channel antagonist, nitrendipine, to a postsynaptic density fraction isolated from canine cerebral cortex. *Brain Res.* 387:167–84

Wu K, Siekevitz P. 1988. Neurochemical characteristics of a postsynaptic density fraction isolated from adult canine hippocampus. *Brain Res.* 457:98–112

Xiao P, Bahr BA, Staubli U, et al. 1991. Evidence that matrix recognition contributes to stabilization but not induction of LTP. *NeuroReport* 2:461–64. Published erratum appears in 1991. *NeuroReport* 2(9):546 et seq.

Yamamoto K, Hayakawa T, Mogami H, et al. 1990. Ultrastructural investigation of the CA1 region of the hippocampus after transient cerebral ischemia in gerbils. *Acta Neuropathol. (Berlin)* 80:487–92

Zador A, Koch C, Brown TH. 1990. Biophysical model of a Hebbian synapse. *Proc. Natl. Acad. Sci. USA* 87:6718–22

Annu. Rev. Neurosci. 1994. 17:373–97

DETERMINATION OF NEURONAL CELL FATE: LESSONS FROM THE R7 NEURON OF *DROSOPHILA*

S. Lawrence Zipursky

Howard Hughes Medical Institute, Department of Biological Chemistry, University of California at Los Angeles, School of Medicine, Los Angeles, California 90024

Gerald M. Rubin

Howard Hughes Medical Institute, Department of Molecular and Cell Biology, University of California at Berkeley, Berkeley, California 94720

KEY WORDS: signal transduction, induction, receptor tyrosine kinase, *D. melanogaster* eye

INTRODUCTION

How is the remarkable cellular diversity in the nervous system established during development? Do cells assume specific fates as a consequence of intrinsic factors, or are they acquired through cellular interactions? Cell lineage studies provide a way to distinguish between these two extreme mechanistic alternatives. For instance, lineage studies in the vertebrate central nervous system (CNS) reveal that at late stages of development pluripotent cells exist within the neuroepithelium that give rise to different neuronal and glial cell types (e.g. Turner & Cepko 1987). Qualitatively similar results were obtained from genetic mosaic studies carried out some 20 years ago, which showed that there were no strict lineage relationships between different classes of neurons and support cells in the compound eye of *Drosophila melanogaster* (Ready et al 1976, Lawrence & Green 1979). In the absence of lineage, a prominent role for cellular interactions in regulating the development of both the fly eye and the vertebrate CNS has been proposed. In recent years considerable progress has been made in understanding the mechanisms by

373

which pattern formation and cell fate are regulated in the developing compound eye of *D. melanogaster*. (For general reviews of *D. melanogaster* eye development, see Tomlinson 1988, Ready 1989, Cagan & Zipursky 1992). We anticipate that similar mechanisms will be shown to play important roles in regulating cell fate determination in the vertebrate CNS.

In this review we focus on the development of one particular cell type in the compound eye, the R7 photoreceptor neuron. Through a multifaceted approach utilizing tools of cell biology, genetics, and biochemistry, important insights have been gained into the molecular strategies of cell fate determination. We discuss in detail the molecular nature of the cellular interactions leading to the induction of the R7 cell, the mechanisms that restrict induction to a subclass of cells in the developing eye, and the mechanisms by which the signal is transduced in the responding cells.

Several features of compound eye development make it a favorable system for molecular and genetic studies. First, the compound eye has a relatively simple modular structure. It comprises some 800 identical eyes called ommatidia (Figure 1), each containing 8 photoreceptor neurons (R cells), which minimally fall into 5 classes. There is one R7 cell within each ommatidium; it is a distinct neuronal cell type defined by its position, morphology, action spectrum, and synaptic connections. In addition to the R cells, each ommatidium contains nonneuronal support cells, including the cone cells, which secrete the simple positive lens capping each ommatidium, and the pigment cells, which ensheath the R cells, thereby optically isolating them from R cells in adjacent ommatidia. Second, the cellular dynamics of pattern formation have been described with single cell resolution (Ready et al 1976, Tomlinson 1985, Tomlinson & Ready 1987b, Cagan & Ready 1989, Wolff & Ready 1991). This description has provided a critical foundation of knowledge for interpreting developmental defects in mutants. Third, the eye is a dispensible structure; it is straightforward to identify mutations affecting the developing eye. As we shall see, the reiterative nature of the compound eye has proven to be an important feature in the design of highly successful genetic screens for interacting genes defining the R7 pathway. Recently developed genetic gadgetry facilitates the identification of mutations in genes that are essential for development to proceed through embryogenesis and that are also required for the development of postembryonic structures, such as the compound eye (Xu & Rubin 1993). Hence, in principle, all the genes necessary for compound eye development can be isolated. And finally, it is straightforward to clone genes from *D. melanogaster* and to reintroduce them into the germline under either their own promoter or one that drives their expression in different subsets of cells in the developing eye (reviewed in Rubin 1988).

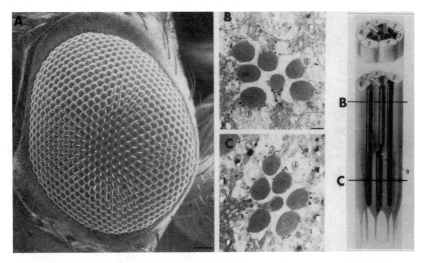

Figure 1 The structure of the compound eye of *Drosophila melanogaster*. (A) Scanning electron micrograph of the compound eye. The compound eye is a reiterated pattern of hexagonally arranged units called ommatidia. Each ommatidium is comprised of an invariant number of cells including eight photoreceptor neurons (R cells) and additional nonneuronal cells. (B) and (C) Transmission electron micrographs of sections in the distal and proximal regions of the ommatidium, respectively. The large, electron-dense structures are the rhabdomeres (Rh), the photosensitive organelles of the R cells. The panel on the right is a schematic representation of the cluster of R cells in an adult ommatidium. Scale bar is 10 μm in A and 1 μm in B and C. Panels B and C and the schematic representations are from Reinke & Zipursky (1988); reprinted by permission of publisher.

INDUCTION IN THE COMPOUND EYE

Assembly of ommatidia begins in the third larval instar in a columnar epithelium called the eye imaginal disc (Figure 2). Ommatidial development does not occur synchronously throughout the disc, but instead begins at the posterior edge and progresses anteriorly. Eye discs removed from larvae just prior to pupariation show a smoothly graded series of ommatidia at different stages of development, covering just over half of the disc (Ready et al 1976; see Figure 2B). Examination of individual cells in the forming ommatidia has shown that the photoreceptors differentiate in a fixed sequence, beginning with the central R8 photoreceptor and proceeding pairwise with R2 and R5, R3 and R4, R1 and R6, and finally R7 (Tomlinson & Ready 1987b; see Figure 2B). Although photoreceptor differentiation occurs in this fixed sequence, genetic mosaic analysis has failed to detect any role for cell-lineage relation-

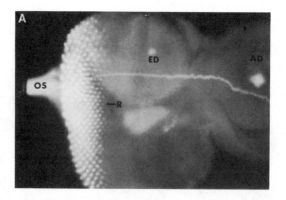

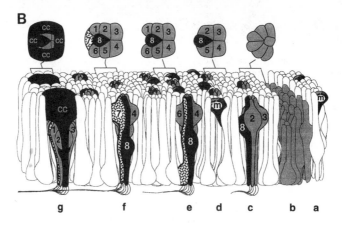

Figure 2 The development of the compound eye. The primordium of the compound eye, the eye imaginal disc, is set aside in early development as an invagination of the embryonic ectoderm. These cells proliferate during early larval development. Pattern formation and differentiation commence in the third and final stage of larval development. The retinal primordium is referred to as the eye imaginal disc.

Panel *A* shows a differentiating eye primordium, referred to as the eye imaginal disc, stained with a monoclonal antibody to chaoptin, a membraneglycoprotein expressed on the surface of developing photoreceptor neurons. Pattern formation in the eye commences in the posterior region and progresses anteriorly as a wave. (Posterior is to the left.) The leading edge of this wave is called the morphogenetic furrow. The following structures are shown: *AD*, antennal disc; *ED*, eye disc; *OS*, optic stalk; *R*, R cell cluster. Scale bar, 50 μm.

Panel *B* is a schematic representation of development in the eye imaginal disc. Because development proceeds as a wave across the disc, each disc represents a series of older steps in ommatidial assembly as one proceeds from the morphogenetic furrow (*b*) towards the posterior (*left*). (*a*) Anterior to the furrow, cells are actively dividing and unpatterned. Although there are no overt signs of cellular differentiation or pattern formation in this region of the disc, a few molecular markers are expressed in this region. (*b*) As cells enter the morphogenetic furrow (MF) they form an array of clusters containing 10–15 cells. It is thought that the MF reflects a series

ships in determining cell fates in the developing fly retina (Ready et al 1976, Lawrence & Green 1979). These genetic and morphological observations led to the proposal that retinal cell fate is governed by the specific combination of signals received by a cell from its immediate neighbors (Tomlinson & Ready 1987b). Such a mechanism would likely require the precise temporal and spatial regulation of the genes directly involved in generating, receiving, and interpreting these signals.

Recruitment of the R7 Cell

The R7 photoreceptor is the last of the eight photoreceptors to differentiate. Mutation of the *sevenless* (*sev,* Tomlinson & Ready 1986, 1987a) or *bride of sevenless* (*boss;* Reinke & Zipursky 1988) genes leads to a transformation of the R7 precursor cell—the cell that finds itself in the pocket created by the differentiating R8, R1, and R6 cells—into a nonneuronal, lens-secreting cone cell (Figures 2 and 3). Thus, the R7 precursor cell appears to face a simple choice between two alternative cell fates: it will develop into an R7 photoreceptor if it receives appropriate instructions, or it will adopt a nonneuronal cone cell fate if these instructions are disrupted by mutations in *sev* or *boss.*

Lack of the *sev* or *boss* functions might be expected to block R7 development, either because the inducing cells are unable to provide a signal or because the receiving cells are unable to perceive or implement the signal. It has been possible to distinguish between these alternatives by determining which cells require a wild-type allele of the *sev* or *boss* genes for development to proceed normally. Mosaic individuals in which some somatic cells have lost their wild-type copy of a gene of interest can be generated through mitotic recombination (Figure 4). In this way, it has been demonstrated that the *sev* gene product is required only in the presumptive R7 cell itself (Campos-Ortega et al 1979, Tomlinson & Ready 1987a), implying that the gene has a role in receiving or implementing a signal required for R7 cell development. In

---

of coordinated cell movements. Some cells in the initial cluster dissociate, giving rise to clusters containing six or seven cells. (*c*) The five-cell precluster contains the cells that will give rise to R2, R3, R4, R5, and R8. (*d*) The cells that are not part of the five-cell precluster undergo an additional round of cell division (m). (*e*) The R1 and R6 cells then join the cluster. (*f*) They are followed by R7. When the nuclei of R1 and R6 rise into an apical location, the nucleus of the R7 precursor is found basally. The nucleus of the R7 precursor rises apically. As it begins to differentiate, its nucleus shifts more basally. (*g*) The cone cells then surround this core of neuronal cells. The upper panel shows a cross-sectional view of the developing cluster at the apical region of the disc epithelium. The lower panel shows the profile of these cells along the apical-basal axis. A single bundle of axons extends from the basal region of each cluster of photoreceptor neurons. Although the nuclei of the developing cells move up and down in the disc epithelium, there is very little lateral movement of cells.

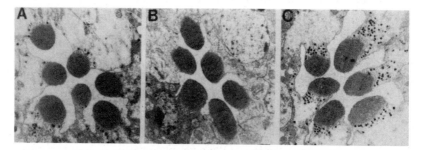

Figure 3   The R7 cell is missing in both *sev* and *boss* mutant ommatidia. (*A*) Wild type: In a distal section of an ommatidium the cellular profiles of seven of the eight R cells are seen. The rhabdomere of the R7 cell in the center of the ommatidium is smaller than the rhabdomeres of the surrounding R1–R6 cells. The R7 neuron has a unique projection pattern, cellular morphology, and spectral sensitivity. The R7 cell is missing in both *sev* (panel *B*) and *boss* mutants (panel *C*). Scale bar, 1 μm. From Reinke & Zipursky 1988; reprinted by permission of publisher.

contrast, *boss* is required in, and only in, the R8 cell for successful development of the R7 cell in the same ommatidium (Reinke & Zipursky 1988). These results indicate that *boss* is required to generate a product in R8 that is necessary, not for the development of R8, but for the development of the R7 cell that it contacts. As described below, the structures and biochemical activities of the Sev and Boss proteins fit their inferred roles quite well.

SEVENLESS IS A RECEPTOR PROTEIN TYROSINE KINASE

The sequence of the *seven less* gene revealed a striking similarity to the primary sequence of receptor tyrosine kinases (RTK) (Hafen et al 1987). Biochemical studies demonstrated that (Sev) is synthesized as a 280-kDa glycoprotein precursor that is subsequently cleaved into two subunits of 220 kDa (N-terminal) and 58 kDa (C-terminal) that remain associated by noncovalent interactions (Simon et al 1989). Sev differs from other RTKs in its unusually large extracellular domain ($\sim$ 2000 amino acids) and in the noncovalent nature of the association between its subunits. Gel filtration experiments suggest that Sev may exist as an $\alpha_2\,\beta_2$ heterotetramer. The carboxy-terminal subunit contains the transmembrane domain and a cytoplasmic domain that is highly homologous to known tyrosine kinases. In addition, a lower level of homology extends into the extracellular domain with the vertebrate RTK, the c-ros protein (Chen et al 1991a; Narayana & Nagarajan 1992); however, it is unclear if there is any functional homology between these two receptors. The Sev protein has been shown to have protein tyrosine kinase activity in vitro (Simon

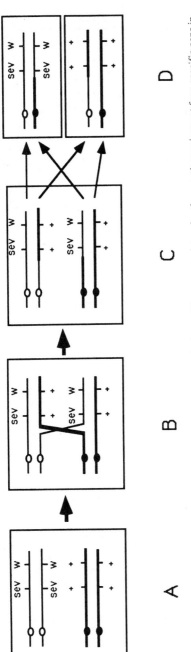

Figure 4 Genetic mosaic analysis of *sev* function. Genetic mosaic analysis provides a way to determine the genetic requirement for a specific gene in any of the pigmented cells in the eye. In this example we describe the method used to determine the cellular requirement for the *sev* gene. The goal of this experiment is to induce clones of mutant cells in an otherwise wild-type eye. Owing to the nonclonal mechanism of ommatidial assembly, ommatidia containing both mutant and wild-type cells will be found. The genotype of each cell at the *sev* locus can be assessed by determining the presence of pigment granules controlled by the white (*w*) gene: pigmented cells carry the wild-type allele of *sev*, and conversely unpigmented cells are homozygous for the mutant *sev* allele. The cellular requirement for *sev* can be determined by correlating the phenotype of each ommatidium (i.e. the presence of the R7 cell) and the genotype of each cell. Genetic mosaic analysis of *sev* showed that it was strictly required in R7; no unpigmented R7 cells were found. In contrast, a similar genetic mosaic study with the *boss* gene showed that the R7 cell could form regardless of its *boss* genotype. An examination of the genotype of other cells in mosaic ommatidia showed that the presence of an R7 cell was strictly dependent on the *boss* genotype of the R8 cell. In the example above, females heterozygous for both *sev* and *w* are subjected to X rays in late first instar of larval development. X rays stimulate mitotic recombination. Mitotic recombination occurs subsequent to chromosomal duplication (*A*, *B*, and *C*). Mitotic recombination between *sev* and the centromere can give rise to two daughter cells that are genotypically distinct (*D*). The daughter cell that is homozygous for the mutant *sev* and *w* will give rise to an unpigmented patch of tissue in the adult eye. Its sister cell will give rise to a red patch of tissue that is indistinguishable from the heterozygous parental cell. In practice, only a fraction of the irradiated larvae give rise to mosaic eyes. Not surprisingly, an eye rarely contains more than one patch.

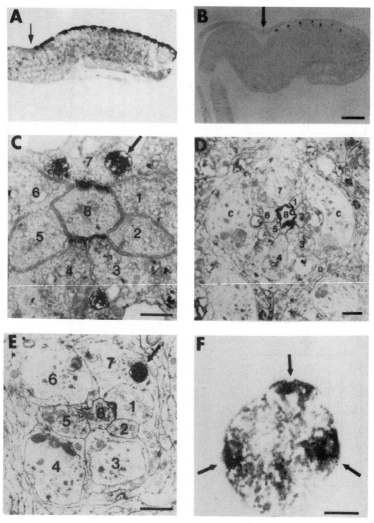

Figure 5   Immunolocalization of Sev and Boss in the developing eye imaginal disc. (A) and (B) The distribution of Sev and Boss proteins, respectively, along the apical basal axis in the eye imaginal discs. The arrows indicate the position of the morphogenetic furrow. Scale bar, 25 μm.

(C) Electron micrograph of a developing cluster stained with an antibody to Sev showing expression in R3, R4, and R7. Sev is also expressed in R1 and R6 and in cone cell precursors that surround the R cell cluster and cells between clusters (not shown in this micrograph). In this micrograph the Sev protein is seen capping on the R8 cell. In addition, Sev immunoreactivity is seen in a multivesicular body (MVB) in the R7 cell (*arrow*). Scale bar, 0.5 μm.

(D) Expression of Boss on the surface of the R8 cell. Scale bar, 1 μm.

(E) Accumulation of Boss immunoreactivity in an MVB in R7. Various lines of evidence indicate that the accumulation of Boss in R7 is the result of its transfer from R8 via a Sev-receptor mediated process. Scale bar, 1 μm.

et al 1989), and this activity appears to be essential for Sev function in vivo (Basler & Hafen 1988).

Expression of Sevenless

Sev is expressed in each developing ommatidium prior to differentiation of the R7 cell and is found transiently at high levels in at least nine cells (Banerjee et al 1987, Tomlinson et al 1987). In all Sev-expressing cells, the heaviest staining occurred in the microvilli at the apical surface of the epithelium (Figure 5A). In addition, staining was found in large multivesicular bodies (MVBs) in all cells in which Sev was detected at the cell surface (Tomlinson et al 1987). A striking staining of plasma membranes occurred at the level of the adherens-type junctions in the first few microns of the tissue below the microvilli. Here, accumulation of stain within R3, R4, and R7 was seen precisely at the position where they oppose R8, but not where they abut other cells or below the level of the adherens junctions (Figure 5C). This led Tomlinson et al (1987) to propose that the ligand for the Sev RTK was expressed on the surface of the R8 cell.

Expression of Sev is first detected prior to the cell division that generates the R7 precursor cell. The onset of expression in R3, R4, and other cells of the ommatidium closely follows the sequence of, but precedes, the maturation of these cells, as revealed by their expression of neural antigens (Tomlinson & Ready 1987b). In particular, Sev staining is seen at high levels in the R7 precursor cell some eight hours prior to overt differentiation (Tomlinson et al 1987). However, unlike the staining revealed with antibodies to neuronal markers, the expression of Sev is transient, and only some photoreceptors express detectable levels, with R2, R5, and R8 being the exceptions (Tomlinson et al 1987). This complex pattern of expression appears to be controlled strictly at the level of transcription (Basler et al 1989; Bowtell et al 1989a, 1991).

The appearance of Sev prior to the specification of R7 and its presence in many cells are consistent with a role for Sev in receiving a signal required to induce the R7 developmental pathway. The observation that the Sev expression pattern was much more complex and dynamic than those observed for all other RTKs suggested that its distribution might be important in determining which cells respond to the proposed inductive signal. However, this specificity was not compromised when Sev was expressed in all cells of the eye disc under the control of a heat-shock promoter (Basler & Hafen 1989,

---

(F) High magnification of view of an MVB in R7 stained with the Boss antibody. Scale bar, 0.3 μm.

Panels A and C are from Tomlinson et al (1987); reprinted by permission of publisher. Panels B and D–F are from Krämer et al (1991); reprinted by permission of publisher.

Bowtell et al 1989b). Thus, the complex spatial distribution of Sev is not a crucial part of the positional information specifying R7 cell fate. Consistent with a role in receiving a transient inductive signal, expression of Sev was found to be required only during a brief period of ommatidial development; Sev activity is required for the initiation of R7 cell development but not for its subsequent differentiation, maintenance, or function (Basler & Hafen 1989, Bowtell et al 1989b). Experiments with temperature-sensitive alleles of Sev indicate that the Sev-mediated signal must be maintained for several hours for R7 induction to occur (Mullins & Rubin 1991).

BRIDE OF SEVENLESS IS A LIGAND FOR THE SEV RTK

Boss is a Multiple Membrane Spanning Protein

The requirement of *boss* in the R8 cell led to the parsimonious model that *boss* encodes the inductive ligand for Sev (Reinke & Zipursky 1988). As a first step toward testing this model, *boss* was cloned and its primary structure was deduced from the cDNA sequence (Hart et al 1990). The *boss* gene encodes a polypeptide of 896 amino acids. Hydropathy analysis led to the prediction that Boss is a transmembrane protein with an extracellular N-terminal domain of 498 amino acids, 7 segments of sufficient length and hydropathy to span the membrane, and a C-terminal cytoplasmic tail of 115 amino acids. Biochemical and immunohistological studies confirmed that the N- and C-termini are oriented toward the outside and inside of the cell, respectively (Cagan et al 1992). Although the topology of Boss is remarkably similar to that of the superfamily of G protein-linked receptors, there is no primary sequence homology.

Boss is Expressed in the Apical Region of the R8 Cell

Boss was shown to be expressed at a time and place consistent with its proposed role as the R7 inductive ligand. As described above, Sev was expressed in the apical-most region of the R7 precursor cell, as well as other cells in the disc, and was required in a precisely defined window of developmental time. Immunohistology at the light and electron microscope level also showed that Boss was localized to the most apical region of the disc, in this case specifically on the membranes of the R8 cell (Figure 5B, D; Krämer et al 1991). Expression of Boss commences prior to R7 induction and remains in the R8 cell throughout the period required for *sev* activity, as defined with heat-shock promoter-driven constructs (hs sev; Basler & Hafen 1989, Bowtell et al 1989b; see below) and temperature-sensitive alleles of *sev* (Mullins & Rubin 1991).

Although Boss is also expressed in a large number of developing sensory neurons, it does not appear to be required for their development or the development of associated sensory structures. In principle, Boss could play a more subtle developmental role or a physiological role, or its expression in these cells may simply be gratuitous. It is important to note, however, that Boss and Sev are not coexpressed in tissues outside of the developing eye; apparently the Boss-Sev pathway is exquisitely specific for regulating the development of only one cell type in the entire organism.

Evidence for Interaction between Boss and Sev from In Vitro Studies

A tissue culture approach was used to test whether Boss and Sev directly interact (Krämer et al 1991). Two derivatives of the *D. melanogaster* S2 cell line were constructed, one expressing Sev (SL2-sev) (Simon et al 1989) and the other expressing Boss (S2:boss). These two cell lines formed single cell suspensions when incubated alone but aggregated upon mixing. Aggregates contained approximately equal numbers of S2:boss and SL2-sev cells. Aggregation was inhibited by antibodies to Boss and Sev and was calcium-dependent.

The ability of Boss to stimulate Sev tyrosine kinase activity was shown in two ways (Hart et al 1993). First, incubation of S2:boss cells with SL2-sev cells led to a rapid and transient increase in tyrosine phosphorylation on Sev, as assessed on protein immunoblots probed with antibodies to phosphotyrosine. Similar results were obtained with membrane preparations from S2:boss cells. An increase in tyrosine phosphorylation was not seen when an S2 cell line expressing a catalytically inactive Sev mutant was incubated with S2:boss cells. Histological studies in culture provided additional evidence that Sev was locally activated at the site of interaction between the S2:boss and the SL2-sev cells. Aggregates of S2:boss and SL2-sev cells were triple-stained with antibodies to Sev, Boss, and phosphotyrosine. The Sev protein and phosphotyrosine immunoreactivity were concentrated at the site of contact between the two cells; there was no apparent clustering of Boss. In contrast, although the catalytically inactive Sev protein capped on the Boss-expressing cell, a local increase in tyrosine phosphorylation was not observed. These data indicate that the increase in tyrosine phosphorylation likely results from activation of the Sev tyrosine kinase itself.

Internalization of Boss into the R7 Precursor Cell

The studies described in the previous sections indicate a direct interaction between Boss and Sev in vitro. Does this happen in vivo? Detailed histological studies provided evidence for such an interaction. By using immuno-electronmicroscopy, Krämer et al (1991) have found Boss within a multivesicular body (MVB) in the R7 precursor cell (Figure 5*E, F*). This appears to be the same structure in which Sev immunoreactivity was also detected, as described earlier in this review. MVBs are morphologically similar to late endosomal compartments in mammalian cells within which internalized receptor-ligand complexes dissociate; the ligands are then sorted to the lysosome for degradation, and the receptor, depending on the specific receptor in question, is either recycled to the cell surface or degraded.

Several lines of evidence indicated that Boss was transferred from the R8 cell to the R7 precursor cell through its interaction with Sev. First, Boss immunoreactivity was observed on the rough endoplasmic reticulum (RER) of the R8 cell, but not on the RER of any other cells in the developing eye disc. Hence, Boss is only translated in R8. Second, internalization of Boss by the R7 cell was strictly dependent upon Sev. Internalization of Boss is not sufficient for R7 development; an R7 precursor cell, expressing a kinase-negative Sev mutant protein that binds Boss in vitro, internalizes Boss in vivo even though the cell is transformed into a cone cell. Moreover, internalization is not a specialized property of the R7 precursor cell, as it is observed in aggregates of S2:boss and SL2-sev cells.

To assess which regions of Boss were internalized into Sev-expressing cells both in culture and in the developing eye disc, the Boss distribution was determined in studies that used antibodies to several different extracellular and intracellular epitopes. These studies showed that the entire Boss protein was internalized into the SL2-sev cells. Although the mechanism by which this occurs is not known, it seems likely that it proceeds by a variation of receptor-mediated endocytosis recently described in Aplysia; Bailey et al (1992) showed that double-membrane clathrin-coated pits and double-membrane clathrin-coated vesicles formed during active periods of membrane remodeling of cell contacts in culture. Interestingly, internalization of Boss requires the *shibire* (*shi*) gene product, a homologue of rat dynamin (Chen et al 1991b, Krämer et al 1991, Van der Bliek & Meyerowitz 1991). In temperature-sensitive alleles of *shi* incubated at nonpermissive temperatures, endocytic intermediates accumulate in which deep invaginations of the plasma membrane form, but vesicles fail to pinch off (Kosaka & Ikeda 1983, Masur et al 1990). In addition, very brief incubation at the nonpermissive temperature results in marked loss of clathrin-coated vesicles and pits.

Although Boss internalization provides strong evidence for a direct interaction between Boss and Sev in vivo, internalization does not appear to be an obligatory requirement for R7 induction. A constitutively activated form of Sev (Basler et al 1991; see below) promotes Boss-independent R7 development. It remains possible, however, that internalization of an activated receptor is a necessary requirement for induction and that during normal development the receptor is only active when complexed to Boss.

The Seven Transmembrane Domains of Boss are Required for its Function

Although other transmembrane ligands have been identified (e.g. Steel Factor; see Witte 1990), Boss is unique in containing seven putative transmembrane segments. Is the seven-transmembrane region of Boss important for its function? Several mutant forms of Boss were tested for their function both in vivo and in vitro (Hart et al 1993). The extracellular domain (designated EXboss) comprising 480 amino acids was purified to apparent homogeneity by lectin-affinity chromatography. Radiolabeled EXboss did not bind with high affinity to Sev. However, it did inhibit aggregation of S2:boss with the SL2-sev cells and Boss activation of the Sev tyrosine kinase. Expression of EXboss in the developing eye disc was shown to inhibit R7 development in a genetically sensitized system in which the expression of wild-type Boss was limiting. The inability to function as an agonist did not result from a requirement for membrane anchoring, as two transmembrane forms, one containing TM 7 and the other TM 1, 6, and 7, functioned neither as agonists nor antagonists of Sev. These data suggest that the extracellular domain has a low affinity for Sev and that the region containing the transmembrane segments is required for Boss function. The transmembrane domain may be required for the correct folding of the extracellular domain, or it may interact directly with Sev. Given the topological similarity of Boss to members of the seven-transmembrane superfamily of G-protein linked receptors, it is intriguing to consider the notion that Sev activates a G-protein cascade in the R8 cell. If such a cascade is activated, it is unclear what role it would play; development of the R8 cell is normal in *sev, boss,* or a *sev;boss* double mutant background.

A GENETIC APPROACH TO DISSECTING AN RTK SIGNAL TRANSDUCTION CASCADE

Activation of the Sev tyrosine kinase must result in intracellular changes in the R7 precursor cell that cause it to adopt the R7 cell fate rather than that of a cone cell. The mechanisms by which RTKs effect changes in cell physiology are still poorly understood. Biochemical studies with mammalian RTKs have led to the identification of proteins that bind to or are phosphorylated by them (see Cantley et al 1991). Although some of these interactions suggest potential mechanisms of signal transmission, their role in vivo is still unclear.

The Sev pathway is ideal for genetic approaches to understanding signaling by RTKs. First, both the R7 cell and the Sev protein are dispensable for viability and fertility. Second, the functioning of this signaling pathway can be inferred from the presence of the R7 cell in a live, anesthetized fly. Screens for mutations that alter the strength of Sev signaling can be used to identify other elements in the pathway.

Simon et al (1991) carried out a systematic genetic screen for dominant enhancers of *sev* that would presumably decrease the effectiveness of signaling by Sev. By adjusting the temperature at which flies carrying a temperature-sensitive allele of *sev* were grown, it was possible to adjust Sev activity to a level barely above the threshold necessary for R7 cell formation. Under these conditions, some 80% of the ommatidia had an R7 cell. Small reductions in the abundance or activity of other elements of the pathway might then be expected to lower signal strength sufficiently to cause a *sev* phenotype in most of the ommatidia. This sensitivity allowed Simon et al (1991) to identify seven genes encoding putative downstream elements of the pathway, by screening for genes in which inactivation of only one copy of the gene—which would be expected to reduce the level of gene product by half—resulted in the absence of the R7 cell. As the other copy of the gene remained functional, they were able to identify these loci even though their functions were essential for viability.

To determine whether these genes defined components of the Sev pathway specifically or RTK systems more generally, Simon et al (1991) assessed the ability of these mutations to suppress the *Ellipse* (Elp) mutation (Baker & Rubin 1989, 1992), a dominant allele of the *D. melanogaster* EGF receptor. The *Elp* mutation gives rise to a disorganized eye with a reduced number of ommatidia. Four *sev* enhancers were suppressors of *Elp*, suggesting that they encode proteins that participate in the transduction of signals from at least two different RTKs. One of the four loci corresponds to the *Ras1* gene, which encodes a p21*ras* protein (Simon et al 1991, Neumann-Silberberg et al 1984) and is the *D. melanogaster* homologue of human H-ras, Ki-ras, and N-ras. Studies in vertebrates also implicate ras in early steps in signaling cascades downstream from RTKs (Smith et al 1986).

Activation of Ras1 is a Key Consequence of Sev Activation

To determine whether Ras1 activation alone is sufficient for Sev-mediated signaling, Fortini et al (1992) used *sev* gene regulatory sequences to express dominant-activating Ras1 alleles in those cells of the eye imaginal disc that normally express Sev; the dominantly active Ras1 mutations were constructed in vitro based on studies of mammalian Ras proteins (Bourne et al 1990a,b). Previous studies by Basler et al (1991) demonstrated that expression of a truncated, constitutively active Sev protein under Sev gene control results in *boss*-independent rescue of the normal R7 cell and transformation of cone cell precursors into supernumerary R7 cells (Basler et al 1991; see below). Identical results were obtained with activated Ras1 (Ras1^{val12}), but not wild-type Ras, indicating that Ras1 activation can substitute for all of the Sev-mediated signal (Fortini et al 1992). However, these results cannot exclude the possibility that endogenous Ras1 protein is not activated to the levels achieved by Ras1^{val12} or that normal Sev-mediated signaling requires additional pathways operating in parallel to the Ras1 pathway.

Suppression of the *sev* phenotype by dominant activated *ras1*^{val12} is strikingly similar to genetic interactions in *Caenorhabditis elegans* vulval development. Vulva formation requires *ras* (reviewed in Sternberg & Horvitz 1991). Dominant activating mutations in the Ras protein bypass the requirement for receptor and produce a multivulva phenotype. The central role of Ras activation in signaling by these two structurally dissimilar RTKs is surprising, as mammalian RTKs are thought to act on a variety of downstream targets. Moreover, Ras activation alone cannot explain how different physiological responses are elicited in a single cell type by stimulation of different RTKs (Ullrich & Schlessinger 1990, Cross & Dexter 1991; see below). Nevertheless, the genetic results obtained in *D. melanogaster* and *C. elegans* suggest that Ras activation may be a primary means by which RTKs exert their effects in many signal transduction pathways.

How is Ras1 Activated?

How does the activation of Sev result in the activation of Ras1? An outline of the answer is emerging from the analysis of the other mutations isolated in genetic screens, such as those described above. The activity of Ras proteins is regulated by bound guanine nucleotides (reviewed by Bourne et al 1990a,b); the GTP-bound state is active, whereas the GDP-bound state is inactive. The ratio of GTP:Ras to GDP:Ras is determined by two antagonistic reactions (see Figure 6). An active GTP:Ras molecule is inactivated by the intrinsic GTPase activity of the Ras protein, a process that is greatly stimulated by RasGAP. An inactive GDP:Ras molecule is activated by the exchange of the bound GDP molecule for a GTP molecule, a reaction that is stimulated by guanine nucleotide exchange proteins (GNEPs).

The product of the *Son of sevenless* (*Sos*) locus is required for R7 development. Loss of function mutations at the *Sos* locus were isolated in the Simon et al (1991) screen, and a gain-of-function mutation of *Sos* was identified as a suppressor of a *sev* allele that retained a low level of activity (Rogge et al 1991). The Sos protein shows sequence similarity to the *Saccharomyces cerevisiae* CDC25 protein, a known GNEP (Simon et al 1991, Bonfini et al 1992), providing further support for a model in which the stimulation of Ras activity is a key element in signaling by Sev. Furthermore, it suggests that Sev stimulation may be achieved by activating the exchange of GDP for GTP by Ras proteins. In this context, the activity of another protein encoded by the E(sev)2B gene originally identified in the Simon et al

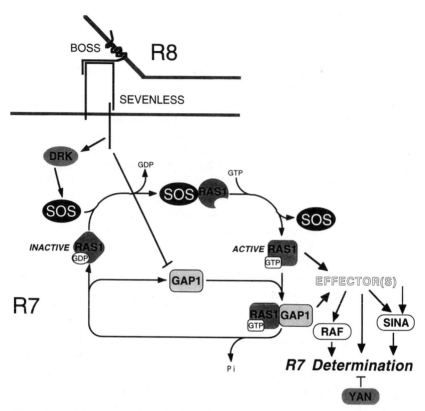

Figure 6   Model for the signal transduction pathway that acts downstream of Sevenless. See text for details.

(1991) screen is intriguing. The gene encodes a protein of the structure SH3-SH2-SH3 (Olivier et al 1993, Simon et al 1993), similar to the *C. elegans* SEM 5 (Clark et al 1992) and human GRB2 proteins (Lowenstein et al 1992). This gene, which has been renamed *downstream of receptor kinases* (*drk*), encodes a protein (Drk) that is required for activation of p21[Ras1], but not for any subsequent event, and it can bind both to Sev and to Sos (Olivier et al 1993, Simon et al 1993). These results suggest that Drk may stimulate Sos to activate p21[Ras1] by linking Sev and Sos in a signaling complex.

A RasGAP, Gap1, that regulates Ras1 during eye development has also been identified. Gap1 encodes a protein similar to mammalian RasGAP (Gaul et al 1992). Loss-of-function mutations of Gap1 mimic constitutive activation of Ras, implicating Gap1 as a negative regulator of Ras1 and hence R7 determination (Buckles et al 1992, Gaul et al 1992, Rogge et al 1992). Gap1 presumably functions by inhibiting *D. melanogaster* Ras1 protein; signaling by Sev overcomes this inhibition. The available data show that the levels of both the Sos and Gap1 activities can be limiting steps in the decision by the R7 precursor cell to become an R7 cell. Inactivation of one copy of the Sos gene decreases the effectiveness of Sev signaling (Simon et al 1991), whereas inactivation of one copy of the Gap1 gene increases it (Gaul et al 1992). Although these results are consistent with the possibility that Sev regulates the activity of both Sos and Gap1, biochemical studies will be required to determine whether these proteins are directly regulated by Sev.

The expression of Gap1 is highly restricted, whereas that of Ras1 is not. For example, within the eye disc, Gap1 expression is found only in the region posterior to the morphogenetic furrow, but Ras1 expression is found throughout the entire disc (Segal & Shilo 1986). Furthermore, whereas *Ras1* and *Sos* loss-of-function mutations are organismal lethal, *Gap1* homozygotes are viable. This suggests that distinct GTPase activating proteins may regulate Ras1 in different developmental pathways.

Raf May Function Downstream from Ras1 in R7 Determination

Little is known about the molecular mechanisms downstream from ras. Several lines of evidence in vertebrate systems indicate that the product of the c-raf proto-oncogene functions downstream of Ras in at least some cells (Kolch et al 1991, Wood et al 1992). Studies of the Torso RTK system in the *D. melanogaster* embryo place the *D. melanogaster* raf homologue, the product of the *1(1)polehole* gene (simply referred to here as the *raf* locus), downstream of the Torso RTK (Ambrosio et al 1988). Does *raf* function in the Sev RTK system downstream of *ras*? Dickson et al (1992b) have recently addressed this question. They showed that although complete loss-of-function mutations of *raf* are recessive lethals, hemizygous males carrying a weak *raf* allele (*raf*[HM7]) do survive into adulthood. Sections of *raf*[HM7] eyes reveal that many of the ommatidia lack R7 cells; other R cells are also occasionally missing. Conversely, as seen with Sev and Ras, a dominantly activated form of Raf also leads to additional R7 cells. Activation of Raf was achieved by fusing the extracellular and transmembrane domains of a constitutively active torso RTK mutant to the Raf kinase domain.

Two lines of evidence suggest that raf may act downstream of ras. First, the massive increase in R7 cells that develop as a consequence of expressing the dominantly-activated Ras protein is reverted in a genetic background hemizygous for the *raf*[HM7] allele. Second, reducing the gene dosage of *ras*, which as we saw earlier enhances the phenotype of a hypomorphic *sev* allele, has no effect on the development on the phenotype of a dominant *raf* mutation. Genetic analysis in *C. elegans* has also placed raf downstream from ras in the signal transduction cascade regulating vulval development (Han et al 1993).

sina Encodes a Nuclear Protein Essential for R7 Development

The *sina* gene encodes a nuclear protein required for R7 determination; loss of *sina* activity has the same effect on the presumptive R7 cell as does loss of *sev* function (Carthew & Rubin 1990). However *sina,* unlike *sev,* also functions in other tissues and other cells in the eye disc. The precise role of *sina* within the R7 precursor cell remains unclear; however, *sina* appears to act downstream of activated Ras (Fortini et al 1992) and activated Raf (Dickson et al 1992b). Modification of *sina* activity may be one end-point for the signal transduction pathway initiated by Sev, or alternatively, *sina* may function in parallel to it.

MULTIPLE MECHANISMS RESTRICT THE R7 PATHWAY OF DEVELOPMENT TO A SINGLE PRECURSOR CELL

Why does only one of the Sev-expressing cells in the developing ommatidium assume an R7 cell fate? In addition to the R7 precursor cell, Sev is expressed in the precursors to additional R cells (R1, R3, R4, and R6), the so-called mystery cells, the cone cells, and cells between developing clusters (Tomlinson et al 1987). Multiple mechanisms act in concert to restrict induction to the R7 precursor cell, including spatial localization of the ligand, commitment to alternative developmental pathways, and the existence of antagonistic, yet reversable, functions in uncommitted cells.

Spatial Localization of Boss Restricts R7 Induction

During normal development, the Sev-expressing cone cell precursors and cells between clusters do not contact the Boss-expressing R8 cell. To test whether the precise localization of Boss prevents these Sev-expressing cells from becoming R7 cells, Van Vactor et al (1991) examined the effects of expressing Boss in all cells in the eye disc under the control of the hsp 70 promoter. Cone cell precursors assume R7 cell fates in response to ectopic Boss expression (see Figure 7). Hence, during normal development the cone cell precursors fail to assume an R7 cell fate, owing to the spatial localization of the inductive ligand. Very similar results were obtained by expressing a constitutively active Sev RTK under the control of a *sev* enhancer element (Basler et al 1991, Dickson et al 1992a).

The restriction of the cone cell precursors is not exclusively controlled by the spatial localization of Boss. As described above, mutations at the *Gap1* locus lead to *sev*-independent transformation of cone cells into R7 cells

(Buckles et al 1992, Gaul et al 1992, Rogge et al 1992). Hence, during normal development the R7 signaling pathway is constitutively inhibited by Gap1, presumably through its inhibition of Ras. In addition to *Gap1,* the *yan* gene, a putative transcription factor, also represses R7 development, as well as the development of other R cells in the uncommitted cells surrounding early developing ommatidial clusters (Lai & Rubin 1992). These data indicate that both extrinsic and intrinsic mechanisms prevent R7 induction in cells that do not contact R8 (see Figure 8).

Commitment to Alternative Fates Prevents R7 Development in the R1-R6 Precursor Cells

The R1–R6 precursor cells do not assume an R7 cell fate in response to ectopic expression of Boss or in response to activated forms of Sev, Ras, or Raf. Several lines of evidence support the view that these cells are not able to respond because of their commitment to alternative pathways of development. Mutations at the *seven-up* locus, which encodes a member of the steroid receptor superfamily, lead to the transformation of R1, R3, R4, and R6 cells into R7 cells (Mlodzik et al 1990). Some of these cells, but not all, are transformed in a *sev*-dependent fashion. Similarly, mutations in the *rough* gene, which encodes another putative transcription factor containing a homeodomain (Saint et al 1988, Tomlinson et al 1988), also result in ectopic R7 cells (Heberlein et al 1991, Van Vactor et al 1991). Rough is required for the development of R2, R5, and, indirectly, additional R cells in the developing ommatidium (Tomlinson et al 1988). It is not clear which R1–R6 cells assume an R7 cell fate in a *rough* mutant background. The conclusion that additional R7 cells develop at the expense of R1–R6 cells is based largely on the observation of an increase in the number of R7 cells per ommatidium in the adult and a concomitant reduction in the number of R1–R6 cells.

Interestingly, the inability of the Sev-expressing R1–R6 precursor cells to respond to the inductive cue is correlated with the observation that, in contrast to the R7 precursor cell, these cells fail to internalize Boss. Under conditions in which these cells remain competent to respond to the inductive cue (e.g. in a *rough* mutant background), multiple cells, presumably precursors to the R1–R6 cell population, internalize Boss (Van Vactor et al 1991). This suggests that the molecular mechanism restricting the developmental potential of the R7 signaling pathway in the Sev-expressing R1–R6 cells is reflected at the level of the interaction between Sev and Boss; either Boss and Sev fail to bind to each other or the complex cannot be internalized. However, given that constitutively active Sev, Ras, and Raf cannot drive these cells to assume an R7 pathway, additional mechanisms must prevent activation of the R7 pathway in these cells.

The Control of R7 Identity

What is the nature of the Boss inductive cue? Is the R7 precursor cell multipotent, with one signal promoting R7 development, another cone cell development, and yet another R1–R6 development? Or is the R7 precursor cell already considerably restricted in its developmental potential, such that Boss activates an R7 cell fate, whereas in the absence of a signal the cell assumes a default pathway of a nonneuronal cone cell?

This has been a very complex question to address. Ideally one would like to assess the developmental consequences of transplanting the R7 precursor cell to ectopic locations in the developing disc or to regions of the disc representing different developmental times. These manipulations are not feasible. Evidence that the R7 precursor may be different from other cells in the developing disc, prior to the action of *sev*, is the finding that the H214 enhancer trap line is expressed specifically in the R7 precursor even in the absence of *sev* function (Mlodzik et al 1992). The information content of the developmental signal has been inferred from studies with transgenic animals. For instance, the results of ectopic expression of the *rough* gene in the R7 precursor cell suggest a rather nonspecific signal generated by Boss activation of Sev. Expression of *rough* in the R7 precursor cell leads to its transformation into an R1–R6-like cell in a *sev*-dependent fashion (Basler et al 1990, Kimmel et al 1990). This suggests that either *rough* alters the developmental potential prior to activation of Sev or, alternatively, the Rough protein itself is activated by the Sev signal transduction pathway to implement an R1–R6-like pathway of development. Given that Ras and Raf appear to function downstream of the *D. melanogaster* EGF, Torso, and Sev RTKs, and that activated forms of Ras and Raf drive *sev*-independent R7 development, it seems likely that the Sev signal is simply the last step in the sequential limitation of the developmental potential of the R7 precursor.

PERSPECTIVE

The Sev and Boss proteins comprise an unusual receptor/ligand pair. Both proteins have highly restricted distributions and functions, and activation of Sev apparently requires direct contact with Boss-expressing cells. It remains to be seen how many such specialized signaling systems are used during the development of complex organisms. The membrane-bound nature of the ligand makes it much more difficult to discover such signaling systems by using traditional biochemical approaches.

Recent genetic studies highlight the complementary nature of the information that has been, and can be, provided by molecular and genetic approaches to cell signaling. Screens for mutations have provided a way to identify genes that play an important role in eye development, and analysis of genetic mosaics has been used to determine in what cell the ommatidium requires a particular gene. In combination with detailed analyses of mutant phenotypes, these data indicate whether a gene product is involved in the sending or receiving of signals. For example, *sev* was unequivocally demonstrated to be required only in the R7 cell. Knowledge of the pattern of *sev* gene expression is strikingly uninformative in this regard; Sev is expressed in many cells, not just R7. On the other hand, the genetic analyses do not provide information on the nature of the gene product or its biochemical function. The genetic analyses are equally consistent with *sev*'s encoding a transcription factor, a receptor, or a component of the intracellular signaling machinery. Distinguishing among such alternatives has been the purview of molecular and biochemical studies.

Finally, we note the truly striking degree of evolutionary conservation seen in the components of the Sev-mediated signaling pathway. For example, the Sev tyrosine kinase domain is 47% identical to that of c-ros (Hafen et al 1987), *D. melanogaster* Ras1 is 77% identical to human H-ras (Simon et al

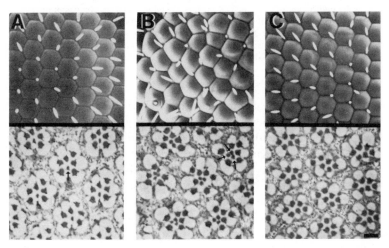

Figure 7   Ectopic activation of the Sev signal transduction pathway leads to the formation of additional R7 cells. Ectopic activation of the R7 pathway has been achieved by ubiquitous expression of Boss or by expression of activated forms of Sev, Ras, or Raf under the control of the *sev* enhancer (i.e. driving expression in all those cells expressing the Sev protein) or the hsp 70 promoter. In each case, additional R7 cells develop from a pool of competent cells (see text). Panel *A* shows a wild-type eye. The results of ubiquitous expression of Boss in a wild-type and *sev* mutant background are shown in panels *B* and *C*, respectively. Note the multiple R7 cells in each ommatidium (*arrow*) in panel *B* and the suppression of this effect in a *sev* mutant background. The upper part of each panel shows a scanning electron micrograph of the surface of the compound eye, and the lower panel shows a light micrograph of a plastic section of an eye of the same genotype. Scale bar in *C* is 5 μm.

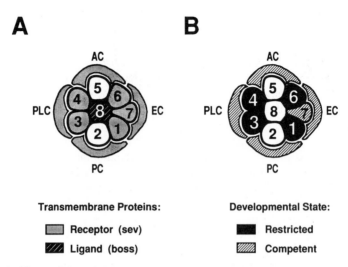

Figure 8 The restriction of the Boss inductive cue to a subset of cells in the developing ommatidium. (*A*) Protein expression in the developing ommatidium. (*B*) The developmental state of cells in the ommatidium. The cone cell precursors and R7 cells are competent to respond to the Boss inductive cue or constitutive activation of Sev, Raf, or Ras. In contrast, the R1, R3, R4, and R6 cells do not assume an R7 cell fate in response to constitutive activation of the R7 pathway, presumably owing to their commitment to alternative fates. Reprinted from Van Vactor et al (1991); reprinted by permission of publisher.

1991), Sos is 45% identical to its mouse homologue (Bowtell et al 1992), drk is 64% identical to the human GRB2 protein (Lowenstein et al 1992), and sina is 83% identical to its mouse counterpart (D. Bowtell, personal communication). This high degree of sequence conservation implies that the function of these proteins has changed little in the 500 million years that separate flies and man. In conclusion, we believe that the experimental advantages of being able to examine mutant phenotype at single cell resolution in a developing tissue that is dispensable for viability and fertility will ensure that new insights into the mechanisms used by neurons to acquire specific identities will continue to emerge from studies of the *D. melanogaster* eye.

Literature Cited

Ambrosia L, Mahowald AP, Perrimon N. 1989. Requirement of the *Drosophila* raf homologue for *torso* function. *Nature* 342: 288–91

Bailey CH, Chen M, Keller F, Kandel ER. 1992. Serotonin-mediated endocytosis of apCAM:An early step of learning-related synaptic growth in Aplysia. *Science* 256: 645–48

Baker NE, Rubin GM. 1989. Effect on eye development of dominant mutations in *Drosophila* homologue of the EGF receptor. *Nature* 340:150–53

Baker NE, Rubin GM. 1992. Ellipse mutations

in the *Drosophila* homologue of the EGF receptor affect pattern formation, cell division, and cell death in eye imaginal discs. *Dev. Biol.* 150:381–96

Banerjee U, Renfranz PJ, Pollock JA, Benzer S. 1987. Molecular characterization and expression of seveneless, a gene involved in neuronal pattern formation in the *Drosophila* eye. *Cell* 49:281–91

Basler K, Christen B, Hafen E. 1991. Ligand-independent activation of the sevenless receptor tyrosine kinase changes the fate of cells in the developing *Drosophila* eye. *Cell* 64:1069–81

Basler K, Hafen E. 1988. Control of photoreceptor cell fate by the sevenless protein requires a functional tyrosine kinase domain. *Cell* 54:299–311

Basler K, Hafen E. 1989. Ubiquitous expression of sevenless: Position-dependent specification of cell fate. *Science* 243:931–34

Basler K, Siegrist P, Hafen E. 1989. The spatial and temporal expression pattern of sevenless is exclusively controled by gene-internal elements. *EMBO J.* 8:2381–86

Basler K, Yen D, Tomlinson A, Hafen E. 1990. Reprogramming cell fate in the developing *Drosophila* retina: Transformation of R7 cells by ectopic expression of *rough*. *Genes Dev.* 4:728–39

Bonfini L, Karlovich CA, Dasgupta C, Banerjee U. 1992. The Son of *sevenless* gene product: a putative activator of ras. *Science* 255:603–5

Bourne HR, Sanders DA, McCormick F. 1990a. The GTPase superfamily: a conserved switch for diverse cell functions. *Nature* 348:125–32

Bourne HR, Sanders DA, McCormick F. 1990b. The GTPase superfamily: conserved structure and molecular mechanism. *Nature* 349:117–27

Bowtell D, Fu P, Simon M, Senior P. 1992. Identification of murine homologues of the *Drosophila* Son of sevenless gene: Potential activators of *ras*. *Proc. Natl. Acad. Sci. USA* 89:6511–15

Bowtell DDL, Kimmel BE, Simon MA, Rubin GM. 1989a. Regulation of the complex pattern of sevenless expression in the developing *Drosophila* eye. *Proc. Natl. Acad. Sci. USA* 86:6245–49

Bowtell DDL, Lila T, Michael WM, et al. 1991. Analysis of the enhancer element that controls expression of sevenless in the developing *Drosophila* eye. *Proc. Natl. Acad. Sci. USA* 88:6853–57

Bowtell DDL, Simon MA, Rubin GM. 1989b. Ommatidia in the developing *Drosophila* eye require and can respond to *sevenless* for only a restricted period. *Cell* 56:931–36

Buckles GR, Smith ZDJ, Katz FN. 1992. *mip* causes hyperinnervation of a retinotopic map

in *Drosophila* by excessive recruitment of R7 photoreceptor cells. *Neuron* 8:1015–29

Cagan RL, Krämer H, Hart AC, Zipursky SL. 1992. The bride of sevenless and sevenless interaction: Internalization of a transmembrane ligand. *Cell* 69:393–99

Cagan RL, Ready DF. 1989. The emergence of order in the *Drosophila* pupal retina. *Dev. Biol.* 136:346–62

Cagan RL, Zipursky SL. 1992. Cell choice and patterning in the *Drosophila* retina. In *Determinants of Neuronal Identity*, ed. M Shankland, ER Macagno, pp. 189–225. San Diego: Academic Press

Campos-Ortega JA, Jurgens G, Hofbauer A. 1979. Cell clones and pattern formation: studies on sevenless, a mutant of drosophila melanogaster. *Wilhelm Roux' Arch. Entwicklungsmech. Org.* 186:27–50

Cantley LC, Auger KR, Carpenter C, et al. 1991. Oncogenes and signal transduction. *Cell* 64:281–302

Carthew RW, Rubin GM. 1990. *seven in absentia*, a gene required for specification of R7 cell fate in the *Drosophila* eye. *Cell* 63:561–77

Chen J, Heller D, Poon B, et al. 1991a. The proto-oncogene *c-ros* codes for a transmembrane tyrosine protein kinase sharing sequence and structural homology with *sevenless* protein of *Drosophila melanogaster*. *Oncogene* 6:257–64

Chen MS, Obar RA, Schroeder CC, et al. 1991b. Multiple forms of dynamin are encoded by *shibire*, a *Drosophila* gene involved in endocytosis. *Nature* 351:583–86

Clark SG, Stern MJ, Horvitz HR. 1992. *C. elegans* cell-signalling gene *sem-5* encodes a protein with SH2 and SH3 domains. *Nature* 356:340–344

Cross M, Dexter TM. 1991. Growth factors in development, transformation, and tumorigenesis. *Cell* 64:271–80

Dickson B, Sprenger F, Hafen E. 1992a. Prepattern in the developing *Drosophila* eye revealed by an activated torso-sevenless chimeric receptor. *Genes Dev.* 6:2327–39

Dickson B, Sprenger F, Morrison D, Hafen E. 1992b. Raf functions downstream of ras1 in the sevenless signal transduction pathway. *Nature* 360:600–3

Fortini ME, Simon MA, Rubin GM. 1992. Signalling by the *sevenless* protein tyrosine kinase is mimicked by Ras1 activation. *Nature* 355:559–61

Gaul U, Mardon G, Rubin GM. 1992. A putative Ras GTPase activating protein acts as a negative regulator of signaling by the sevenless receptor tyrosine kinase. *Cell* 68:1007–19

Hafen E, Basler K, Edstroem JE, Rubin GM. 1987. Sevenless, a cell-specific homeotic gene of *Drosophila*, encodes a putative transmembrane receptor with a tyrosine kinase domain. *Science* 236:55–63

Han M, Golden A, Han Y, Sternberg PW. 1993. The *C. elegans lin-45 raf* gene participates in *let-60 ras*-stimulated vulval differentiation. *Nature* 363:133–40

Hart AC, Krämer H, Van Vactor DL Jr, et al. 1990. Induction of cell fate in the *Drosophila* retina: The bride of sevenless protein is predicted to contain a large extracellular domain and seven transmembrane segments. *Genes Dev.* 4:1835–47

Hart AC, Krämer H, Zipursky SL. 1993. Extracellular domain of the boss transmembrane ligand acts as an antagonist of the sev receptor. *Nature* 361:732–35

Kimmel BE, Heberlein U, Rubin GM. 1990. The homeo domain protein *rough* is expressed in a subset of cells in the developing *Drosophila* eye where it can specify photoreceptor cell subtype. *Genes Dev.* 4:712–27

Kolch W, Heidecker G, Lloyd P, Rapp UR. 1991. Raf-1 protein kinase is required for growth of induced NIH/3T3 cells. *Nature* 349:426–28

Kosaka T, Ikeda K. 1983. Reversible blockage of membrane retrieval and endocytosis in the garland cell of the temperature-sensitive mutant of *Drosophila melanogaster*, shibirets1. *J. Cell Biol.* 97:499–507

Krämer H, Cagan RL, Zipursky SL. 1991. Interaction of *bride of sevenless* membrane-bound ligand and the *sevenless* tyrosin-kinase receptor. *Nature* 352:207–12

Lai Z-C, Rubin GM. 1992. Negative control of photoreceptor development in *Drosophila* by the product of the *yan* gene, an ETS domain protein. *Cell* 70:609–20

Lawrence PA, Green SM. 1979. Cell lineage in the developing retina of *Drosophila*. *Dev. Biol.* 71:142–52

Lowenstein EJ, Daly RJ, Batzer AG, et al. 1992. The SH2 and SH3 domain-containing protein GRB2 links receptor tyrosine kinases to ras signaling. *Cell* 70:431–442

Masur SK, Kim Y-T, Wu C-F. 1990. Reversible inhibition of endocytosis in cultured neurons from the *Drosophila* temperature-sensitive mutant shibirets1. *J. Neurogenet.* 6:191–206

Mlodzik U, Mlodzik M, Rubin GM. 1991. Cell-fate determination in the developing *Drosophila* eye: Role of the *rough* gene. *Development* 112:703–12

Mlozkik M, Hiromi Y, Goodman CS, Rubin GM. 1992. The presumptive R7 cell of the developing *Drosophila* eye receives positional information independent of sevenless, boss and sina. *Mech. Dev.* 37:37–42

Mlokzik M, Hiromi Y, Weber U, et al. 1990. The *Drosophila seven-up* gene, a member of

the steroid receptor gene superfamily, controls photoreceptor cell fates. *Cell* 60:211–24

Mullins MC, Rubin GM. 1991. Isolation of temperature-sensitive mutations of the tyrosine kinase receptor sevenless (*sev*) in *Drosophila* and their use in determining its time of action. *Proc. Natl. Acad. Sci. USA* 88: 9387–91

Narayana L, Nagarajan L. 1992. A mouse c-ros genomic clone: Identification of a highly conserved 22-amino acid segment in the juxtamembrane domain. *Gene* 118: 297–98

Neumann-Silberberg FS, Schejter E, Hoffman FM, Shilo BZ. 1984. The *Drosophila* ras oncogene: Structure and nucleotide sequence. *Cell* 37:1027–33

Olivier JP, Raabe T, Henkemeyer M, et al. 1993. A *Drosophila* SH2/SH3 adaptor protein implicated in coupling the sevenless receptor tyrosine kinase to an activator of Ras guanine nucleotide exchange, Sos. *Cell* 73:179–91

Ready D. 1989. A multifaceted approach to neural development. *Trends Neurosci.* 12: 102–10

Ready DF, Hanson TE, Benzer S. 1976. Development of the *Drosophila* retina, a neurocrystalline lattice. *Dev. Biol.* 53:217–40

Reinke R, Zipursky SL. 1988. Cell-cell interaction in the *Drosophila* retina: The *bride of sevenless* gene is required in photoreceptor cell R8 for R7 cell development. *Cell* 55: 321–30

Rogge R, Cagan R, Majumdar A, et al. 1992. Neuronal development in the *Drosophila* retina: The sextra gene defines an inhibitory component in the developmental pathway of R7 photoreceptor cells. *Proc. Natl. Acad. Sci. USA* 89:5271–75

Rogge RD, Karlovich CA, Banerjee U. 1991. Genetic dissection of a neurodevelopmental pathway: Son of sevenless functions downstream of the *sevenless* and EGF receptor tyrosine kinases. *Cell* 64:39–48

Rubin GM. 1988. *Drosophila melanogaster* as an experimental organism. *Science* 240: 1453–59

Saint R, Kalionis B, Lockett TJ, Elizur A. 1988. Pattern formation in the developing eye of *Drosophila melanogaster* is regulated by the homoeo-box gene, rough. *Nature* 334:151–54

Segal D, Shilo BZ. 1986. Tissue localization of *Drosophila melanogaster* ras transcripts during development. *Mol. Cell Biol.* 6: 2241–48

Simon MA, Bowtell DDL, Dodson GS, et al. 1991. Ras1 and a putative guanine nucleotide exchange factor perform crucial steps in signaling by the sevenless protein tyrosine kinase. *Cell* 67:701–16

Simon MA, Bowtell DDL, Rubin GM. 1989. Structure and activity of the sevenless protein: A protein tyrosine kinase receptor required for photoreceptor development in *Drosophila*. *Proc. Natl. Acad. Sci. USA* 86:8333–37

Simon MA, Dodson GS, Rubin GM. 1993. An SH3-SH2-SH3 protein is required for p21*Ras1* activation and binds to sevenless and Sos proteins in vivo. *Cell* 73:169–77

Smith MR, DeGudicibus SJ, Stacey DW. 1986. Requirement for c-ras proteins during viral oncogene transformation. *Nature* 320:540–543

Sternberg PW, Horvitz HR. 1991. Signal transduction during *C. elegans* vulval induction. *Trends Genet.* 7:366–71

Tomlinson A. 1985. The cellular dynamics of pattern formation in the eye of *Drosophila*. *J. Embryol. Exp. Morphol.* 89:313–31

Tomlinson A. 1988. Cellular interactions in the developing *Drosophila* eye. *Development* 104:183–93

Tomlinson A, Bowtell DD, Hafen E, Rubin GM. 1987. Localization of the sevenless protein, a putative receptor for positional information, in the eye imaginal disc of *Drosophila*. *Cell* 51:143–5OX530

Tomlinson A, Kimmel BE, Rubin GM. 1988. *rough*, a *Drosophila* homeobox gene required in photoreceptors R2 and R5 for inductive interactions in the developing eye. *Cell* 55:771–84

Tomlinson A, Ready DF. 1986. Sevenless: A cell specific homeotic mutation of the *Drosophila* eye. *Science* 231:400–2

Tomlinson A, Ready DF. 1987a. Cell fate in the *Drosophila* ommatidium. *Dev. Biol.* 123:264–75

Tomlinson A, Ready DF. 1987b. Neuronal differentiation in the *Drosophila* ommatidium. *Dev. Biol.* 120:366–76

Turner DL, Cepko CL. 1987. A common progenitor for neurons and glia persists in rat retina late in development. *Nature* 238:131–36

Ullrich A, Schlessinger J. 1990. Signal transduction by receptors with tyrosine kinase activity. *Cell* 61:203–12

Van der Bliek AM, Meyerowitz EM. 1991. Dynamin-like protein encoded by the *Drosophila shibire* gene associated with vesicular traffic. *Nature* 351:411–14

Van Vactor DL Jr, Cagan RL, Krämer H, Zipursky SL. 1991. Induction in the developing compound eye of *Drosophila*: Multiple mechanisms restrict R7 induction to a single retinal precursor cell. *Cell* 67:1145–55

Witte ON. 1990. Steel locus defines new multipotent growth factor. *Cell* 63:5–6

Wolff T, Ready DF. 1991. The beginning of pattern formation in the *Drosophila* compound eye: The morphogenetic furrow and the second mitotic wave. *Development* 113:841–50

Wood KW, Sarneck C, Roberts TM, Blenis J. 1992. Ras mediates nerve growth factor receptor modulation of three signal-transducing protein kinases: MAP kinase, raf-1, and RSK. *Cell* 68:1041–50

Xu T, Rubin GM. 1993. Analysis of genetic mosaics in developing and adult *Drosophila* tissues. *Development* In press

Annu. Rev. Neurosci. 1994. 17:399–418

MOLECULAR BASIS FOR CA^{2+} CHANNEL DIVERSITY

F. Hofmann, M. Biel, V. Flockerzi

Institut für Pharmakologie und Toxikologie, TU München, Germany

KEY WORDS: voltage dependent ion channels, neurons, heart, dihydropyridines, protein
               phosphorylation

TYPES OF HIGH VOLTAGE–ACTIVATED CALCIUM CHANNELS

Neurotransmitter release, neurosecretion, neuronal excitation, survival of neurons, and many other neuronal functions are controlled by the cellular calcium concentration. Calcium entry across the plasma membrane in response to membrane depolarization or activation of neurotransmitter receptors represents a major pathway for the cellular control of calcium. The voltage-dependent calcium channels, activated and inactivated at a low or high membrane potential, are the best characterized plasmalemmal calcium entry pathway, primarily because powerful and specific channel blocking agents are available. The T (tiny)-type calcium channel is activated and inactivated at a low membrane potential and is present in a wide variety of excitable and non-excitable cells. T-type calcium channels are not considered in detail because the lack of specific blockers renders identification of cloned and expressed channels as T-type channels difficult.

In contrast, the high voltage activated and inactivated calcium channels have been subdivided into four distinct classes, using the organic calcium channel blockers (CaCB) (originally introduced by Fleckenstein and colleagues 1967) and several neurotoxins. They have been separated into the B (B stands for brain; B channels may include T-type channels)-, L (long lasting)-, N (neither L nor T channel)-, and P (Purkinje)-type calcium channels (Table 1). B-, L-, N- and P-type calcium channels are activated at a high membrane potential (around −30 mV), inactivate slowly (long lasting), and are expressed in neuronal and nonneuronal cells (Tsien et al 1991, Bertolini & Llinás 1992). N- and P-type calcium channels are blocked specifically by

Table 1 Cloned and expressed mammalian calcium channel subunit cDNAs[a]

Gene	Type	Numa nomenclature	Agreed[b] nomenclature	Source	Species	Functionally expressed	Sensitive to	Reference
α1 subunits								
CaCh1	L	Sk	S	Skeletal m	Rabbit	Yes	DHP	Tanabe et al 1987
CaCh2a	L	C	Ca	Heart	Rabbit	Yes	DHP	Mikami et al 1989
				Brain	Rat	—		Snutch et al 1991
				Heart	Rat	—		Diebold et al 1992
				Brain	Mouse	Yes		Ma et al 1992
CaCh2b	L	—	Cb	Lung, smooth m	Rabbit	Yes	DHP	Biel et al 1990
				Brain	Rat	—		Snutch et al 1991
				Aorta	Rat	Yes		Koch et al 1990
				Heart	Rat	—		Diebold et al 1992
				Fibroblast	Human	—		Soldatov 1992
				Brain	Mouse	Yes		Ma et al 1992
CaCh3	L	—	D	Brain	Human	Yes	DHP	Williams et al 1992b
				Pancreatic islet	Human	—		Seino et al 1992
CaCh4	P	BI	A	Brain	Rabbit	Yes	Spider venom ω-Aga IVA	Mori et al 1991
				Brain	Rat	—		Starr et al 1991
CaCh5	N	BIII	B	Brain	Human	Yes	ω-CTX GVIA	Williams et al 1992a
				Brain	Rat	—		Dubel et al 1992
				Brain	Rabbit	Yes	ω-CTX GVIA	Fujita et al 1993
CaCh6	B	BII	E	Brain	Rabbit	—	—	Niidome et al 1992
				Brain	Rat	Yes	—	Soong et al 1993

α₂/δ subunits

CaA₂1a	α2/δa	—	Skeletal m	Rabbit	Yes	—	Ellis et al 1988
CaA₂1b	α2/δb	—	Brain	Human	Yes	—	Williams et al 1992b
			Brain	Rat	—	—	Kim et al 1992

β subunits

CaB1*	β₁	—	Skeletal m	Rabbit	Yes	—	Ruth et al 1989
			Brain	Rat	—	—	Pragnell et al 1991
			Brain	Human	Yes	—	Williams et al 1992b
			Brain	Human	—	—	Powers et al 1992
CaB2*	β₂	—	Heart	Rabbit	Yes	—	Hullin et al 1992
			Brain	Rat	Yes	—	Perez-Reyes et al 1992
CaB3*	β₃	—	Heart	Rabbit	Yes	—	Hullin et al 1992
CaB4	β₄	—	Brain	Rat	Yes	—	Castellano et al 1993

γ subunit

CaG1	γ	—	Skeletal m	Rabbit	Yes	—	Bosse et al 1990
							Jay et al 1990

[a] Only full-length clones have been included in this table. The nomenclature for the α₁ subunit is adapted from Perez-Reyes et al (1990). The Numa nomenclature is used in his laboratory for the brain calcium channels. The agreed nomenclature is based on that of Snutch et al (1990). The references are to the first published sequence. In some cases functional expression of the particular clone has been reported in a different publication. —, not reported; DHP, dihydropyridine; ω-CTX GVIA, ω-conotoxin VIA; * at least three different variants (a – c) of the same gene have been identified.

[b] In August 1993 several laboratories agreed to use in the future following nomenclature for mammalian voltage-dependent calcium channels: α₁ subunits and their important splice variants are identified by the capital Roman letters A, B . . . (S for skeletal muscle) and the small Roman letters a, b, . . . respectively (example α1$_{Ca}$ = CaCh2a = α₁ subunit of the cardiac L-type calcium channel; α2/δ subunits are identified by α2/δ with, if necessary, small Roman letter a, b, . . . indicating the splice variant; β subunits are identified by Arabic numbers 1, 2, ... as index (example β₁); γ subunit is identified by γ; The complex of the skeletal muscle calcium channel will be α1sα2/δβ₁γ.

w-conotoxin GVIA and the funnel web spider toxin ω-Aga-IVA, respectively (Mintz et al 1992b). Both channels have been identified in neurons and neuroendocrine cells. P-type channels are also present in the distal convoluted tubule of rat kidney (Yu et al 1992).

L-type channels are readily blocked by the classical CaCB's nifedipine (a 1,4-dihydropyridine, DHP), verapamil (a phenylalkylamine, PAA), and diltiazem (a benzothiazepine) (Catterall et al 1988, Glossmann & Striessnig 1988, Hofmann et al 1990). L-type calcium channels are expressed in neuronal and endocrine cells, in cardiac, smooth and skeletal muscle, in fibroblasts and kidney. In skeletal muscle they are essential for excitation-contraction coupling, which does not require calcium influx through the channel (Rios et al 1992). In the normal heart they are necessary for the generation and propagation of electrical impulses and for the initiation of contraction in atrial and ventricular muscle. In smooth muscle they are involved in tension development, for which process they provide part of the necessary calcium. L-type channels apparently also control the intracellular calcium concentration in other cells. However, they are not involved in neurotransmitter secretion, which process is linked in many neuronal cells to N-type channels. B-type channels have been identified by cloning as a major neuronal calcium channel (Niidome et al 1992, Soong et al 1993). These cDNA clones could be responsible for a neuronal calcium current that is not blocked by a combination of blockers for L-, N- and P-type calcium channels (nimodipine, ω-conotoxin GVIA and ω-Aga-IVA) (Mintz et al 1992a).

COMPOSITION OF THE CALCIUM CHANNEL

The basic insight into the composition and the functional domains of calcium channels has been derived mainly from work carried out with the channel purified from skeletal muscle. However, caution is needed when the results obtained with the skeletal muscle channel are applied to the other channels because they may reflect specific properties of the skeletal muscle calcium channel that are not retained in the other channels. The L-type calcium channel as purified from skeletal muscle contains four proteins (Figure 1): the α_1 subunit (212 kDa), which contains the binding sites for all known CaCBs and the calcium conducting pore; the intracellularly located β subunit (57 kDa); the transmembrane γ subunit (25 kDa); and the α_2/δ subunit, a disulfide-linked dimer of 125 kDa (see Catterall et al 1988, Glossmann & Striessnig 1988, Hofmann et al 1990, and references cited there). Reconstitution of the purified complex into phospholipid bilayers results in functional calcium channels that are reversibly blocked by CaCBs and are modulated by cAMP-dependent phosphorylation (Flockerzi et al 1986, Hymel et al 1988, Nunoki et al 1989, Mundiña-Weilenmann et al 1991). The primary sequences of these proteins

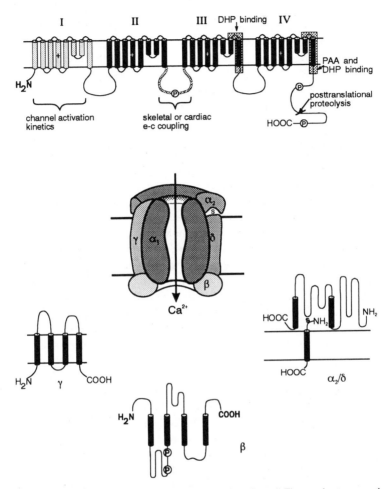

Figure 1 Proposed structure of the skeletal muscle calcium channel. The putative transmembrane configuration of individual subunits is taken from the hydropathicity analysis of the primary sequences. The suggested structure of the α_1 subunit is shown at the top. I, II, III, IV, proposed repeats of calcium channel α_1; +, proposed transmembrane amphipathic α helix, respectively the proposed voltage sensing helix of the channel; (P), sites phosphorylated in vitro by cAMP kinase; the SS1-SS2 region suggested to be part of the channel pore is shown by the short barrels; DHP and PAA, dihydropyridine and phenylalkylamine binding sites; e-c coupling, excitation contraction coupling. The brackets indicate parts of the protein that are responsible for the skeletal (CaCh1) or cardiac (CaCh2) properties of the channel. The dash at the carboxyterminal part indicates the area where the α_1 subunit is processed posttranslationally. The suggested structure of the γ, β, and α_2/δ subunit is shown on the bottom. s, disulfide bridge between the transmembrane δ and the extracellular α_2 subunit. The extracellular space is above the horizontal lines.

have been deduced by cloning their corresponding cDNAs from rabbit skeletal muscle (Tanabe et al 1987, Ellis et al 1988, Ruth et al 1989, Bosse et al 1990, Jay et al 1990). Using these cDNAs as probes, several laboratories have cloned different α_1 and β subunits from heart, smooth muscle, endocrine, and neuronal cells.

SUBUNITS OF THE CALCIUM CHANNEL

The α_1 Subunit

Complete cDNA clones of α_1 subunits that direct the expression of functional calcium channels in *Xenopus laevis* oocytes or cell culture cells have been isolated from skeletal, cardiac, smooth muscle, endocrine glands, and brain (Table 1). The primary sequences of these cDNAs are homologous to each other and encode proteins of predicted molecular masses of 212 to 273 kDa and exhibit homologies from 41% to 70%. Hydropathicity analysis of all α_1 subunits predicts a transmembrane topology similar to that of other voltage-dependent ion channels with four homologous repeats, each containing five hydrophobic putative transmembrane α helices and one amphiphathic segment (S4) (Figure 1). By functional expression of chimeras of the skeletal (CaCh1) and the cardiac (CaCh2a) muscle α_1 subunit, specific properties of the calcium channel were assigned to distinct parts of the ion conducting pore: Repeat I determines the activation time of the chimeric channel, i.e. slow activation upon membrane depolarization with the repeat from skeletal muscle and rapid activation with that from cardiac muscle (Tanabe et al 1991); the putative cytoplasmic loop between repeats II and III determines the type of excitation-contraction coupling. The loop from the skeletal muscle calcium channel α_1 subunit induces contraction in the absence of calcium influx, whereas the loop from the cardiac calcium channel α_1 subunit induces contraction only in the presence of calcium influx (Tanabe et al 1990). The "extracellular" loop between transmembrane helices 5 and 6 (SS1-SS2 region) is predicted to fold into the membrane, to form part of the pore of the channel (Guy & Conti 1990) and to take part in the control of ion selectivity (Heinemann et al 1992).

Photoaffinity labelling of skeletal muscle α_1 subunit followed by limited proteolysis and immunoprecipitation indicates that the DHP binding site is localized close to the SS1-SS2 region of repeat III (Striessnig et al 1991, Nakayama et al 1991) and to a sequence following the IVS6 segment (Regulla et al 1991), whereas the PAA binding site has been located directly after the IVS6 segment (Striessnig et al 1990) (Figure 1). Site directed mutagenesis of the α_1 subunit indicates that the loop between IVS5 and IVS6 is important for the inhibition of current by DHPs. Binding studies with radiolabelled DHPs demonstrate that the stably expressed α_1 subunits from skeletal and smooth muscle alone contain the allosterically coupled binding sites for the known CaCBs (Kim et al 1990, Bosse et al 1992). The cDNAs cloned so far for the α_1 subunits are encoded by the six different genes, CaCh1 to CaCh6 (Table 1).

The CaCh1 Gene

The mRNA of the CaCh1 gene was originally cloned from rabbit skeletal muscle (Tanabe et al 1987). Carp skeletal muscle expresses a homologous α_1 subunit (Grabner et al 1991). The product of the CaCh1 gene occurs in rabbit skeletal muscle in two isoforms: a minor form ($\sim$ 5%) of 212 kDa containing the complete amino acid sequence encoded by α_1 mRNA, and a major form ($\sim$ 95%) of 190 kDa which is derived from the full-length product by posttranslational proteolysis close to amino acid residue 1690 (De Jongh et al 1991). Presumably, the shorter form is involved functionally at the triad in E-C coupling of the skeletal muscle, because the expression of a skeletal muscle CaCh1 mRNA truncated at amino acid residue 1662 fully restores in dysgenic myotubes both excitation-contraction coupling and calcium current (Beam et al 1992). The short and the long form are phosphorylated rapidly in vitro by cAMP-dependent protein kinase at Ser-687 (Röhrkasten et al 1988), which is located at the cytosolic loop between repeat II and III, and Ser-1854 (Rotman et al 1992), which is present only in the larger untruncated form, and slowly at Ser-1617 (Röhrkasten et al 1988). Additional forms of the CaCh1 gene may be present in skeletal muscle. Malouf and coworkers (1992) cloned an α_1 subunit cDNA that lacks repeats II and III and has an apparent molecular weight of 155 kDa. This form might be expressed in the skeletal muscle of newborn rabbits. Using peptide specific antibodies, Brawley & Hosey (1992) detected an α_1 subunit–like protein in rabbit skeletal muscle which contained an immunological identical amino and carboxyterminus but lacked part of the loop between repeats II and III. This protein was smaller (160 kDa) than the mature α_1 subunit (165 kDa) and was not retained by a wheat germ agglutinin column. The α_1 subunit–like protein was enriched in fractions containing mostly the sarcoplasmic cisternal membranes. Neither the identity of the proteins described by Malouf et al (1992) and Brawley & Hosey (1992) nor their functional significance has been established. Partial clones of the CaCh1 gene mRNA have been detected in other tissues including kidney and brain by the PCR technique, which suggests that the CaCh1 gene may be functionally expressed in skeletal muscle and other tissues.

The CaCh2 Gene

The α_1 subunits from cardiac (CaCh2a) (Mikami et al 1989) and smooth (CaCh2b) muscle (Biel et al 1990) are splice products of the second gene (CaCh2). This gene is expressed in most excitable and many nonexcitable tissues including fibroblasts (see Table 1). The deduced amino acid sequence of CaCh2b is about 95% identical with that of CaCh2a. A major difference between the cardiac and the smooth muscle α_1 subunits is the use of alternative exons for the IVS3 segment (Perez-Reyes et al 1990). The alternative spliced primary transcripts of the CaCh2 gene have been identified in many tissues including kidney, heart, and brain (Perez-Reyes et al 1990, Biel et al 1991, Snutch et al 1991, Diebold et al 1992, Ma et al 1992, Yu et al 1992). Additional splice variations of the CaCh2 gene product have been noted (Perez-Reyes et al 1990, Biel et al 1990, Snutch et al 1991, Diebold et al 1992, Yu et al 1992, Schultz et al 1993). With one exception the functional significance of these splice variations is not clear. The two alternative splice variants CaCh2a and CaCh2b have been expressed transiently and stably in *Xenopus laevis* oocytes and CHO cells (Biel et al 1990, Biel et al 1991, Singer et al 1991, Bosse et al 1992, Lory et al 1993, Schultz et al 1993). No major differences have been observed in basic electrophysiological and pharmacological properties of the two isoforms including the amplitude of inward current, steady state activation and inactivation (Welling et al 1992a). However, recent experiments show that the DHP nisoldipine inhibits barium currents through the smooth muscle α_1 subunit (CaCh2b) at over 10-fold lower concentrations than that through the cardiac muscle α_1 subunit (CaCh2a) (Welling et al 1993b). The putative DHP binding sites, the loops between IIIS5 and IIIS6 and IVS5 and IVS6, are identical in both α_1 subunits and therefore not responsible for this remarkable difference. Northern blots and PCR analysis show that both splice variants are differentially expressed in heart and smooth muscle (Biel et al 1991) and during cardiac development (Diebold et al 1992). The vascular and cardiac muscle calcium channels exhibit in vivo a difference in their DHP sensitivity similar to that of the cloned and stably expressed channels, suggesting that the distinct pharmacology of the vascular and cardiac calcium channels is based on the expression of the alternatively spliced CaCh2 mRNA.

The CaCh3 and CaCh4 Genes

The cDNA of the third gene (CaCh3) was isolated from neural and endocrine tissues and represents a neuroendocrine specific L-type calcium channel (Williams et al 1992a, Seino et al 1992), whereas the gene products of the fourth and fifth gene (CaCh4 and CaCh5) were cloned from neuronal cDNA libraries. Calcium channels transiently expressed from cRNA of CaCh4 induce high voltage–activated calcium currents that are insensitive to nifedipine and ω-conotoxin but are inhibited by a mixture of toxins from the funnel web spider, thus characterizing this channel as a P-type calcium channel (Mori et al 1991). Transcripts of the CaCh3 and CaCh4 genes have been detected in both the brain and the kidney by PCR and Northern blots, suggesting again a widespread occurence of the channels. The P-type channel (CaCh4) mRNA was predominantly present in the distal convolute tubule of renal cortex (Yu et al 1992).

The CaCh5 and CaCh6 Genes

The gene product of the CaCh5 has been cloned exclusively from brain. Coexpression of the CaCh5 α_1 subunit in myotubes of dysgenic mice (Fujita et al 1993) or together with the α_2/δ and β subunit (see below) in *Xenopus laevis* oocytes induces a barium current that is inhibited by picomolar concentrations of ω-conotoxin GVIA (Williams et al 1992a). In addition, the cloned α_1 subunits bind ω-conotoxin GVIA with high affinity, (Dubel et al 1992) thus identifying the CaCh5 protein as a neuronal N-type calcium channel. A B-type calcium channel α_1 subunit (CaCh6) has been cloned from a brain library (Niidome et al 1992, Soong et al 1993). The mRNA of this channel is abundant in brain and has been expressed functionally in *Xenopus laevis* oocytes or other cells. The expressed channel has properties of a low voltage-activated T-type channel, i.e. activation occurs at membrane potentials below 0 mV; the inward current is not affected by DHPs or ω-conotoxin and only partially inhibited by ω-Aga IVA. In contrast to T-type channels, the expressed channel inactivated relatively slowly. Therefore, its identity with T-type channels remains to be established.

Other α_1 Subunits

Two additional α_1 (Doe 1 and Doe 4) have been cloned from the marine ray *Discopyge omnata* (Horne et al 1993). The Doe 4 sequence most closely resembles the mammalian CaCh5 gene product, a N-type channel. Doe 1 is related to the gene product of CaCh6. In contrast to the expressed CaCh6 cDNA (Soong et al 1993) expression of the Doe 1, cDNA yields a rapidly inactivating, high voltage activated channel, which is blocked by micromolar ω-conotoxin GVIA but not by DHPs or ω-Aga IVA (Ellinor et al 1993)

The α_2/δ Subunit

The skeletal muscle α_2/δ subunit (CaA1) is a glycosylated membrane protein of 125 kDa (Ellis et al 1988) which is apparently highly conserved in most tissues. In the skeletal muscle the primary protein product of the α_2/δ gene is processed posttranslationally by proteolysis, resulting in an α_2 protein containing amino acid 1 through 934 and a δ protein containing the amino acid 935 through 1080 (DeJongh et al 1990). The transmembrane δ subunit anchors the α_2 protein located extracellularly by disulfide bridges to the plasma membrane (Jay et al 1991). Two identical α_2/δ cDNAs isolated from human (Williams et al 1992b) and rat brain (Kim et al 1992) are splice variants of the skeletal muscle α_2/δ cDNA. The α_2/δ cDNA isolated from rat brain predicts an identical δ protein and a splice variant of the processed α_2 protein. Immunoblots (Norman et al 1987) and Northern blots (Ellis et al 1988, Biel et al 1991) show that similar or identical α_2/δ subunits exist in skeletal muscle, heart, brain, vascular, and intestinal smooth muscle, suggesting that the α_2/δ subunit is expressed together with the various α_1 subunits. However, this conclusion may be not valid for all cell types. Differentiating BC3H1 cells express DHP binding sites (Biel et al 1991) and skeletal muscle L-type calcium channel (Caffrey et al 1987). As expected, only the differentiating BC3H1 cells contain the mRNA for skeletal muscle α_1, β and γ subunits. They do not contain transcripts specific for the α_2/δ subunit (Biel et al 1991), suggesting that, at least in this cell line, the expression of the α_2/δ gene is not coordinated with that of the other calcium channel subunits.

The β Subunit

The skeletal β subunit (CaB1) is a membrane protein, located intracellularly, consisting of 524 amino acids (Ruth et al 1989). Its deduced amino acid sequence contains stretches of heptad repeat structure characteristic of cytoskeletal proteins. Transcripts of two other genes (CaB2 and CaB3) encoding β proteins different from the skeletal muscle β subunit have been isolated from a cardiac cDNA library (Hullin et al 1992). A fourth β subunit, CaB4, has been cloned from rat brain (Castellano et al 1993). The deduced amino acid sequence of CaB2 and CaB3 show an overall homology to CaB1 of 71% (CaB2) and 66.6% (CaB3). Differential splicing of the primary transcript of CaB1 results in at least three isoforms: CaB1a through CaB1c (Ruth et al 1989, Pragnell et al 1991, Williams et al 1992b, Powers et al 1992). CaB1a is expressed in skeletal muscle whereas two other isoforms are expressed in brain, heart, and spleen (Powers et al 1992). Four different splice variants have been characterized for the CaB2 gene (CaB2a through CaB2d); CaB2a and CaB2b have been isolated from a rabbit cardiac cDNA library whereas CaB2c and CaB2d have been cloned from rabbit and rat brain libraries (Hullin et al 1992, Perez-Reyes et al 1992). Like the CaB1 gene, the CaB2 and CaB3 genes are expressed tissue specifically with transcripts of CaB2 existing abundantly in heart and to a lower degree in aorta, trachea, and lung, whereas transcripts of CaB3 genes are found in brain and smooth muscle containing tissues such as aorta, trachea, and lung (Hullin et al 1992). This suggests that the CaB3 gene product may be expressed predominantly in neuronal and smooth muscle cells. The CaB4 gene is expressed in the brain, predominantly in the cerebellum and kidney (Castellano et al 1993).

The γ Subunit

The γ subunit (CaG1) consists of 222 amino acids and is an integral membrane protein (Bosse et al 1990, Jay et al 1990). Its deduced amino acid sequence contains four putative transmembrane domains and two glycosylation sites which are located at the extracellular side. Northern and PCR analysis have not identified the presence of γ subunit in other tissues; this fact suggests that the protein may be specific for skeletal muscle.

Other Subunits

The existence of additional subunits has been suggested from the copurification of unidentified peptides with the DHP-receptor protein (for example, see Kuniyasu et al 1992) without proving their functional necessity. In contrast, a novel subunit referred to as CCCS1 (candidate for calcium channel subunit) has been identified by suppression cloning in neuronal tissue of *Torpedo california* (Gundersen & Umbach 1992). The 21.7 kDa protein is required for the expression of the *Torpedo california* N-type calcium channel in *Xenopus* oocytes. The sequence of this protein is not related to any of the above described calcium channel subunits. It is similar to two cysteine rich proteins from *Drosophila* which were localized to nerve terminals. It is not known if mammalian N-type channels contain a related subunit, since small molecular weight proteins have not been detected in purified (McEnery et al 1991, Witcher et al 1993) nor in immunoprecipitated (Sakamoto & Campbell 1991, Ahlijanian et al 1991) rat brain ω-conotoxin binding sites.

FUNCTIONAL INTERACTION OF THE CALCIUM CHANNEL SUBUNITS

All cDNAs of Table 1 have been expressed singly or in combination with other subunits in *Xenopus laevis* oocytes or cell culture cells as functional ion channels. Transient expression in *Xenopus laevis* oocytes of CaCH2a cRNA (Mikami et al 1989) and CaCH2b cRNA (Biel et al 1990) induces DHP-sensitive currents with electrophysiological properties similar to those reported for cardiac and smooth muscle. Heterologous coexpression of the cardiac α_1 subunit together with the skeletal muscle β subunit and α_2/δ subunit enhanced consistently the inward current to amplitudes greater than 1 *mu*A/oocyte (Singer et al 1991). The α_2/δ or the β subunit alone or the combination of both decreased the activation time of the barium current twofold (Singer et al 1991, Wei et al 1991). Oocytes containing all four subunits (α_1, α_2/δ, β, γ subunit) had fast inactivating barium currents. The coexpression of the γ subunit shifted the steady state inactivation of I_{Ba} by 40 mV to negative membrane potentials (Singer et al 1991). Under each condition inward currents were increased severalfold by the calcium channel agonist BayK 8644. Homologous coexpression of the cardiac α_1 subunit with the cardiac β (CaB2) or the neuronal/smooth muscle β subunit (CaB3) with or without the α_2/δ

subunit results in an increase in the amplitude of I_{Ba} as well as in an acceleration of channel activation (Hullin et al 1992, Castellano et al 1993).

The four neuronal α_1 subunit cDNAs, i.e. the neuroendocrine L-type CaCh3 gene, the neuronal P-type CaCh4 gene, the neuronal CaCh5 gene, and the neuronal CaCh6 gene induce barium currents only when coexpressed with the α_2/δ and β subunit (Mori et al 1991, Williams et al 1992a,b; Fujita et al 1993, Soong et al 1993). The increase in current occurred always in the presence of the β subunit, most likely by an increased number of plasmalemmal calcium channel molecules (see also below). These results suggest that the skeletal muscle and other β subunits interact with different α_1 subunits by a common interaction site and mechanism.

Similar effects of the subunits were obtained by stable coexpression of the skeletal muscle α_1 and β subunit in mouse fibroblasts (L cells), which do not contain an endogenous calcium channel. The β subunit decreased the activation time of the expressed channel over 50-fold, and increased the number of DHP binding sites 2-fold (Lacerda et al 1991). In contrast to these results, coexpression of all four skeletal muscle subunits in L cells resulted in a decreased amplitude of the barium current and in a diminished response toward the calcium channel agonist BayK 8644 (Varadi et al 1991). In a later publication (Lory et al 1992) the same group reported that the decreased sensitivity toward the channel agonist BayK 8644 was caused by an overexpression of the β subunit in a 10:1 ratio over the α_1 subunit. The relative ratio of the α_1 and β subunit was estimated from Northern blots. Whether these Northern blots reflected the real mRNA and protein levels of the transfected α_1 and β subunits is unknown, making it difficult to evaluate the significance of this report. A similar phenomenon has not been observed by other groups using transient or permanent coexpression of other L-type α_1 subunits together with various β subunits in *Xenopus* oocytes, L or CHO cells.

The smooth muscle α_1 (CaCh2b) subunit expressed in CHO cells causes barium currents that are identical to those of native smooth muscle: the single channel conductance was 26 pSi in the presence of 80 mM Ba2+, the open probability increased with membrane depolarization, and the voltage-dependence of activation and inactivation was similar to that of the native smooth muscle channel (Bosse et al 1992). Stable expression of the CaCh2b with the skeletal muscle β gene (CaB1) increased in parallel the number of DHP binding sites and the amplitude of whole cell barium current; thus the amplitude of the inward current is probably directly related to the number of functional α_1 protein molecules (Welling et al 1993a). Coexpression of the α_1 subunit and the β subunit did not increase transcription or translation of the α_1 subunit gene, but apparently affected "maturation" of a functional channel (Nishimura et al 1993). The coexpression of the α_1 subunit with the CaB2 or CaB3 β subunit increased in parallel the inward current and the density of the DHP binding sites. The BayK 8644 sensitivity of the barium currents was retained in all cell lines, although different concentrations of β subunit protein were expressed (Welling et al 1993a). The coexpression of the CaB1 β subunit decreased the channel activation time two-fold and shifted the voltage dependence of steady-state inactivation by 18 mV to −13 mV. Coexpression of the skeletal muscle α_2/δ subunit together with the smooth muscle α_1 and skeletal muscle β subunit produced channels that inactivated faster when calcium was used as a charge carrier.

The expression of the cardiac α_1 subunit (CaCh2a) in the same cells induces currents indistinguishable from those induced by the smooth muscle α_1 subunit (Welling et al 1992a). This electrophysiological similarity is not surprising, since the primary sequence of both channels is 95% identical (Biel et al 1990). The only difference noted was a faster activation of the cardiac channel and a higher sensitivity of the smooth (CaCh2b) than cardiac (CaCh2a) muscle channel toward the block by the DHP nisoldipine (Welling et al 1993b).

HORMONAL MODULATION OF EXPRESSED CALCIUM CHANNELS

High voltage activated L- and N-type calcium channels are modulated by neurotransmitters and hormones through G proteins and protein kinases. The precise regulation of cloned neuronal channels is not known. In vitro experiments show that the α_1 subunit of an immunoprecipitated N-type channel is phosphorylated stoichiometrically by cAMP kinase and protein kinase C (Ahlijanian et al 1991). No detailed analyses are available for the hormonal modulation of expressed neuronal calcium channel, in contrast to results obtained with the cardiac L-type calcium channel.

In heart, β adrenergic stimulation increases the calcium current 3- to 7-fold either by cAMP-dependent phosphorylation of the channel (Osterrieder et al 1982, Kameyama et al 1985, Hartzell et al 1991, Hartzell & Fischmeister 1992) or by the activated α subunits of the trimeric GTP binding protein G_s (Yatani & Brown 1989) or a combination of the activated α subunit of the trimeric GTP binding protein G_s and the active cAMP kinase (Cavalié et al 1991). The L-type calcium current of isolated tracheal smooth muscle cells is also stimulated by activation of the β-adrenergic receptor (Welling et al 1992b). This β-adrenergic receptor effect is mediated directly by a G-protein and not by cAMP-kinase activation. These results suggest that the CaCh2 gene α_1 subunit may be regulated in vivo by the α subunit of a G protein and by cAMP-dependent phosphorylation.

Two subunits of the purified skeletal muscle calcium channel, the α_1 and β subunit, are substrates for cAMP-kinase in vitro (Jahn et al 1988, Ruth et al 1989, DeJongh et al 1989, Röhrkasten et al 1988, Rotman et al 1992).

Both splice variants of the CaCh2 calcium channel α_1 subunit contain an identical number of predicted cAMP kinase phosphorylation sites. One or two of these sites are phosphorylated by cAMP kinase in the expressed full length α_1 subunit of the CaCh2a gene (Yoshida et al 1992). The purified cardiac α_1 subunit is not phosphorylated by cAMP kinase (Schneider & Hofmann 1988, Chang & Hosey 1988). The differing result could be explained if the purified cardiac α_1 subunit protein lacks part of the carboxyterminal phosphorylation sites due to a posttranslational processing of the protein as described for the skeletal muscle α_1 subunit. The potential importance of phosphorylation sites was supported by experiments that showed (i) that rat cardiac poly(A+) RNA induced the expression of an L-type calcium current in *Xenopus laevis* oocytes, which was stimulated 1.34-fold by isoproterenol (Dascal et al 1986) or 2- to 3-fold by cAMP (Lory & Nargeot 1992); (ii) that in *Xenopus laevis* oocytes expressing the cardiac α_1 subunit, cAMP increased barium currents only in the presence of the skeletal muscle β subunit (Klöckner et al 1992), and (iii) that dibutyryl-cAMP stimulated two-fold barium currents in CHO cells expressing the CaCh2a gene (Yoshida et al 1992). Perfusion of a CHO cell expressing the CaCh2a gene with the catalytic subunit of cAMP kinase increased the inward current 1.6-fold, but did not affect the barium current in cells expressing the CaCh2b gene (G Mehrke & F Hofmann, unpublished results). However, the inward barium current was not significantly affected when the pipette solution contained 3 mM ATPγS, 3mM ATPγS and 10 or 100 μM 8Br-cAMP or 3 mM ATPγS and 10 μM GTPγS (Hofmann et al 1993). CHO cells contain a functional adenylyl cyclase, G_S and cAMP-dependent protein kinase. Isoproterenol failed to increase the barium current in a CHO cell line which expressed the β_2 adrenergic receptor at a concentration of 1 pmol/mg protein and the CaCh2 calcium channel. Perfusion of these CHO cells with up to 0.1 mM inhibitor peptide of the cAMP kinase inhibitor protein did not decrease the inward current (G Mehrke & F Hofmann, unpublished results). These negative findings could suggest that the α_1 subunit alone is not sufficient to restore the hormonal regulation of the native calcium channel.

Similar conclusions were made by Klöckner et al (1992) who injected the cardiac α_1 subunit alone or together with the skeletal muscle β subunit into *Xenopus laevis* oocyctes. These authors reported that cAMP increased barium currents only in *Xenopus laevis* oocytes expressing the cardiac α_1 and the skeletal muscle β subunit. However, the reported inward currents were small, and their sensitivity toward Bay K 8644 or a 1,4 dihydropyridine blocker was not tested. Therefore, the results reported by Dascal et al (1986), Lory & Nargeot (1992), and Klöckner et al (1992) do not exclude the possibility that the β subunit associated with the endogenous *Xenopus laevis* oocyte calcium channel (Singer et al 1991, Singer-Lahat et al 1992) and increased the inward barium current. The endogenous channel is insensitive to the 1,4 dihydropyridines but can be stimulated by cAMP in the presence of the skeletal muscle subunits (Dascal et al 1992). Furthermore, perfusion of CHO cells expressing the CaCh2 α_1 subunit and the skeletal muscle β subunit with cAMP or 8Br-cAMP had no effect on the size of the inward current; this observation suggests that, at least in CHO cells, the combination of these two subunits, which are not expressed in vivo in the same tissue, does not restore the hormonal control of the cardiac calcium channel.

These negative results are not due to a general inability of protein kinases to modulate the expressed channel, because different results were obtained with protein kinase C. The cardiac L-type calcium current is enhanced and subsequently inhibited by the activation of protein kinase C (Lacerda et al 1988). Similar results were obtained in *Xenopus laevis* oocytes after injection of rat heart mRNA (Bourinet et al 1992). The skeletal muscle α_1 subunit is rapidly phosphorylated by protein kinase C in vitro (Nastainczyk et al 1987). Currents through the CaCh2a α_1 subunit expressed in *Xenopus laevis* oocytes were modulated biphasic by the activation of protein kinase C (Singer-Lahat et al 1992). Initially the current increased, followed by a marked inhibition. The biphasic modulation was not modified significantly by the coexpression of the cardiac α_1 subunit with the α_2/δ, β and γ subunits from skeletal muscle suggesting that protein kinase C affected the current by phosphorylation of the α_1 subunit.

These results suggest that protein phosphorylation can modulate the properties of the expressed calcium channel. The failure to reconstitute the β adrenergic/cAMP kinase modulation of the cardiac calcium channel with the cloned CaCh2 gene probably indicates that an unphysiological combination of channel subunits has been used. The inability to see cAMP-dependent modulation of the channel expressed in *Xenopus laevis* oocytes could be caused by a constant phosphorylation of the α_1 subunit because oocytes have a high basal activity of the cAMP kinase. The primary sequences of cardiac and smooth muscle α_1 subunits are almost identical and contain identical potential phosphorylation sites. It is therefore conceivable that the cAMP-dependent stimulation of the cardiac calcium channel depends not solely on the phosphorylation of the α_1 subunit but also on the tissue-specific coexpression of other proteins, e.g. the β subunit. The deduced amino acid sequence of the skeletal muscle β subunit (CaB1) contains several phosphorylation sites. Two of these sites, Ser-182 and Thr-205, are phosphorylated in vitro by cAMP-dependent protein kinase (Ruth et al 1989, DeJongh et al 1990). The equivalent of Thr-205 is conserved in the "cardiac" β subunit (Thr-165 in CaB2a and Thr-191 in CaB2b) but is not present in the "smooth muscle" β subunit CaB3 (Hullin et al 1992). The sequence following this potential phosphorylation site is highly variable and determines several splice variants.

This variable region may be responsible for the tissue specific regulation calcium currents by hormones and neurotransmitters.

CONCLUSION

High voltage activated calcium channels are oligomeric complexes of four different subunits: α_1, α_2/δ, β, and γ. So far six different genes encoding α_1 subunits and four distinct genes encoding β subunits have been isolated from various tissues. The specific electrophysiological and pharmacological characteristics of calcium currents in different cells result from the expression of tissue-specific subunits of the calcium channel, leading to differences in functional interaction. This genetic polymorphism explains also the different regulatory mechanisms and possibly the different pharmacology of various calcium channels. An important observation appears to be that the expression not only of distinct α_1 subunits but also of small splice variations of one α_1 subunits changes significantly the sensitivity of the expressed channel against channel blockers. This nurtures the hope that the development of cell specific drugs is possible and may be of benefit in the treatment of disabling brain diseases.

ACKNOWLEDGMENT

The results obtained in the authors' laboratory were supported by grants from Deutsche Forschungsgemeinschaft and Fond der Chemie.

Literature Cited

Ahlijanian MK, Striessnig J, Catterall WA. 1991. Phosphorylation of an α_1-like subunit of an ω-conotoxin-sensitive brain calcium channel by cAMP-dependent protein kinase and protein kinase C. *J. Biol. Chem.* 266:20192–97

Beam KG, Adams BA, Niidome T, et al. 1992. Function of a truncated dihydropyridine receptor as both voltage sensor and calcium channel. *Nature* 360:169–71

Bertolini M, Llinás R. 1992. The central role of voltage-activated and receptor operated calcium channels in neuronal cells. *Annu. Rev. Pharmacol. Toxicol.* 32:399–421

Biel M, Hullin R, Freundner S, et al. 1991. Tissue-specific expression of high-voltage-activated dihydropyridine-sensitive L-type calcium channels. *Eur. J. Biochem.* 200:81–88

Biel M, Ruth P, Bosse E, Hullin R, Stühmer W, et al. 1990. Primary structure and functional expression of a high voltage activated calcium channel from rabbit lung. *FEBS Lett.* 269:409–12

Bosse E, Bottlender R, Kleppisch T, et al. 1992. Stable and functional expression of the calcium channel α_1 subunit from smooth muscle in somatic cell lines. *EMBO J.* 11:2033–38

Bosse E, Regulla S, Biel M, et al. 1990. The cDNA and deduced amino acid sequence of the γ subunit of the L-type calcium channel from rabbit skeletal muscle. *FEBS Lett.* 267:153–56

Bourinet E, Fournier F, Lory P, et al. 1992. Protein kinase C regulation of cardiac calcium channels expressed in *Xenopus* oocytes. *Pflügers Arch.* 421:247–55

Brawley RM, Hosey MM. 1992. Identification of two distinct proteins that are immunologically related to the α_1 subunit of the skeletal muscle dihydropyridine-sensitive calcium channel. *J. Biol. Chem.* 267:18218–23

Caffrey JM, Brown AM, Schneider MD. 1987. Mitogens and oncogenes can block the induction of specific voltage-gated ion channels. *Science* 236:570–73

Castellano A, Wei X, Birnbaumer L, Perez-Reyes E. 1993. Cloning and expression of a neuronal calcium channel β subunit. *J. Biol. Chem.* 268:12359–66

Catterall WA, Seagar MJ, Takahashi M. 1988. Molecular properties of dihydropyridine-sensitive calcium channels in skeletal muscle. *J. Biol. Chem.* 263:3533–38

Cavalié A, Allen TJA, Trautwein W. 1991. Role of the GTP-binding protein G_s in the β-adrenergic modulation of cardiac Ca channels. *Pflügers Arch.* 419:433–43

Chang FC, Hosey MM. 1988. Dihydropyridine and phenylalkylamine receptors associated with cardiac and skeletal muscle calcium channels are structurally different. *J. Biol. Chem.* 263:18929–37

Dascal N, Lotan I, Karni E, Gigi A. 1992. Calcium channel currents in *Xenopus* oocytes injected with rat skeletal muscle RNA *J. Physiol.* 450:469–90

Dascal N, Snutch TP, Lübbert H, et al. 1986. Expression and modulation of voltage-gated calcium channels after RNA injection in *Xenopus* oocytes. *Science* 231:1147–50

DeJongh KS, Merrick DK, Catterall WA. 1989. Subunits of purified calcium channels: A 212-kDa form of α1 and partial amino acid sequence of a phosphorylation site of an independent β subunit. *Proc. Natl. Acad. Sci. USA* 86:8585–89

DeJongh KS, Warner C, Catterall WA. 1990. Subunits of purified calcium channels; α2 and δ are encoded by the same gene. *J. Biol. Chem.* 265:14738–41

DeJongh KS, Warner C, Colvin AA, Catterall WA. 1991. Characterization of the two size forms of the α1 subunit of skeletal muscle L-type calcium channels. *Proc. Natl. Acad. Sci. USA* 88:10778–82

Diebold RJ, Koch WJ, Ellinor PT, et al. 1992. Mutually exclusive exon splicing of the cardiac calcium channel α1 subunit gene generates developmentally regulated isoforms in the rat heart. *Proc. Natl. Acad. Sci. USA* 89:1497–501

Dubel SJ, Starr TVB, Hell J, et al. 1992. Molecular cloning of the α-1 subunit of an ω-conotoxin-sensitive calcium channel. *Proc. Natl. Acad. Sci. USA* 89:5058–62

Ellinor PT, Zhang J-F, Randall AD, et al. 1993. Functional expression of a rapidly inactivating neuronal clacium channel. *Nature* 363:455–58

Ellis SB, Williams ME, Ways NR, et al. 1988. Sequence and expression of mRNAs encoding the α1 and α2 subunits of a DHP-sensitive calcium channel. *Science* 241:1661–64

Fleckenstein A, Kammermeier H, Döring HJ, Freund HJ. 1967. Zum Wirkungsmechanismus neuartiger Koronardilatatoren mit gleichzeitig Sauerstoff-einsparenden Myo-kard-Effekten. Prenylamin und Iproveratil. *Z. Kreislaufforsch.* 56:716–44

Flockerzi V, Oeken HJ, Hofmann F, et al. 1986. Purified dihydropyridine-binding site from skeletal muscle t-tubules is a functional calcium channel. *Nature* 323:66–68

Fujita Y, Mynlieff M, Dirksen RT, et al. 1993. Primary structure and functional expression of the ω-conotoxin-sensitive N-type calcium channel from rabbit brain. *Neuron* 10:585–

Glossmann H, Striessnig J. 1988. Calcium channels. *Vitam. Horm.* 44:155–328

Grabner M, Friedrich K, Knaus HG, et al. 1991. Calcium channels from Cyprinus carpio skeletal muscle. *Proc. Natl. Acad. Sci. USA* 88:727–31

Gundersen CB, Umbach JA. 1992. Suppression cloning of the cDNA for a candidate subunit of a presynaptic calcium channel. *Neuron* 9:527–37

Guy HR, Conti F. 1990. Pursuing the structure and function of voltage-gated channels. *Trends Neurosci.* 13:201–6

Hartzell HC, Fischmeister R. 1992. Direct regulation of cardiac Ca^{2+} channels by G proteins: neither proven nor necessary? *Trends Pharmacol. Sci.* 13:380–85

Hartzell HC, Mery P-F, Fischmeister R, Szabo G. 1991. Sympathetic regulation of cardiac calcium current is due exclusively to cAMP-dependent phosphorylation. *Nature* 351: 573–76

Heinemann SH, Terlau H, Stühmer W, et al. 1992. Calcium channel characteristics conferred on the sodium channel by single mutations. *Nature* 356:441–43

Hofmann F, Biel M, Bosse E, et al. 1993. Functional expression of cardiac and smooth muscle calcium channels. In *Ion Channels in the Cardiovascular Systems; Function and Dysfunction*, ed. AM Brown, WA Catterall, GJ Kaczorowski, PS Spooner, HC Strauss. Washington, DC: AAAS Press. In press

Hofmann F, Flockerzi V, Nastainczyk W, et al. 1990. The molecular structure and regulation of muscular calcium channels. *Curr. Top. Cell. Regul.* 31:223–39

Horne WA, Ellinor PT, Inman I, et al. 1993. Molecular diversity of Ca^{2+} channel α1 subunits from the marine ray *Discopygene ommata*. *Proc. Natl. Acad. Sci. USA* 90: 3787–91

Hullin R, Singer-Lahat D, Freichel M, et al. 1992. Calcium channel β subunit heterogeneity: functional expression of cloned cDNA from heart, aorta and brain. *EMBO J.* 11: 885–90

Hymel L, Striessnig J, Glossmann H, Schindler H. 1988. Purified skeletal muscle 1,4-dihydropyridine receptor forms phosphorylation-dependent oligomeric calcium channels in planar bilayers. *Proc. Natl. Acad. Sci. USA* 85:4290–94

Jahn H, Nastainczyk W, Röhrkasten A, et al. 1988. Site-specific phosphorylation of the purified receptor for calcium-channel blockers by cAMP- and cGMP-dependent protein kinases, protein kinase C, calmodulin-dependent protein kinase II and casein kinase II. *Eur. J. Biochem.* 178:535–42

Jay SD, Ellis SB, McCue AF, et al. 1990. Primary structure of the gamma-subunit of the DHP-sensitive calcium channel from skeletal muscle. *Science* 248:490–92

Jay SD, Sharp AH, Kahl St. D, et al. 1991. Structural characterization of the dihydropyridine-sensitive calcium channel α_2-subunit and the associated δ peptides. *J. Biol. Chem.* 266:3287–93

Kameyama M, Hofmann F, Trautwein W. 1985. On the mechanism of β-adrenergic regulation of the Ca channel in the guinea-pig heart. *Pflügers Arch.* 405:285–93

Kim HL, Kim H, Lee P, et al. 1992. Rat brain expresses an alternatively spliced form of the dihydropyridine-sensitive L-type calcium channel alpha-2 subunit. *Proc. Natl. Acad. Sci. USA* 89:3251–55

Kim HS, Wei XY, Ruth P, et al. 1990. Studies on the structural requirements for the activity of the skeletal muscle dihydropyridine receptor/slow Ca^{2+} channel. *J. Biol. Chem.* 265: 11858–63

Klöckner U, Itagaki K, Bodi I, Schwartz A. 1992. β-subunit expression is required for cAMP-dependent increase of cloned cardiac and vascular calcium channel currents. *Pflügers Arch.* 420:413–15

Koch WJ, Ellinor PT, Schwartz A. 1990. cDNA cloning of a dihydropyridine-sensitive calcium channel from rat aorta. *J. Biol. Chem.* 265:17786–91

Kuniyasu A, Oka K, Ide-Yamada T, et al. 1992. Structural characterization of the dihydropyridine receptor-linked calcium channel from porcine heart. *J. Biochem.* 112:235–42

Lacerda AE, Kim HS, Ruth P, et al. 1991. Normalization of current kinetics by interaction between the α_1 and β subunits of the skeletal muscle dihydropyridine-sensitive Ca^{2+} channel. *Nature* 352:527–30

Lacerda AE, Rampe D, Brown AM. 1988. Effects of protein kinase C activators on cardiac Ca^{2+} channels. *Nature* 335:249–51

Lory P, Nargeot J. 1992. Cyclic AMP-dependent modulation of cardiac Ca channels expressed in *Xenopus laevis* oocytes. *Biochem. Biophys. Res. Commun.* 182: 1059–65

Lory P, Varadi G, Schwartz A. 1992. The β subunit controls the gating and dihydropyridine sensitivity of the skeletal muscle Ca^{2+} channel. *Biophys. J.* 63:1421–24

Lory PL, Varadi G, Slish DF, et al. 1993. Characterization of β subunit modulation of a rabbit cardiac L-type Ca^{2+} channel α_1 subunit as expressed in mouse L cells. *FEBS Lett.* 315:167–72

Ma W-J, Holz RW, Uhler MD. 1992. Expression of a cDNA for a neuronal calcium channel α_1 subunit enhances secretion from adrenal chromaffin cells. *J. Biol. Chem.* 267:22728–32

Malouf NN, McMahon DK, Hainsworth CN, et al. 1992. A two-motif isoform of the major calcium channel subunit in skeletal muscle. *Neuron* 8:899–906

McEnery MW, Snowman AM, Sharp AH, et al. 1991. Purified ω-conotoxin GVIA receptor of rat brain resembles a dihydropyridine-sensitive L-type calcium channel. *Proc. Natl. Acad. Sci. USA* 88:11095–99

Mikami A, Imoto K, Tanabe T, et al. 1989. Primary structure and functional expression of the cardiac dihydropyridine-sensitive calcium channel. *Nature* 340:230–33

Mintz IM, Adams ME, Bean BP. 1992a. P-type calcium channels in rat central and peripheral neurons. *Neuron* 9:85–95

Mintz IM, Venema VJ, Swiderek KM, et al. 1992b. P-type calcium channels blocked by the spider toxin ω-Aga-IVA. *Nature* 355: 827–30

Mori Y, Friedrich T, Kim M-S, et al. 1991. Primary structure and functional expression from complementary DNA of a brain calcium channel. *Nature* 350:398–402

Mundina-Weilenmann C, Chang ChF, Gutierrez LM, Hosey MM. 1991. Demonstration of the phosphorylation of dihydropyridine-sensitive calcium channels in chick skeletal muscle and the resultant activation of the channels after reconstitution. *J. Biol. Chem.* 266:4067–73

Nakayama H, Taki M, Striessnig J, et al. 1991. Identification of 1,4-dihydropyridine binding regions within the α_1 subunit of skeletal muscle Ca^{2+} channels by photoaffinity labeling with diazipine. *Proc. Natl. Acad. Sci. USA* 88:9203–7

Nastainczyk W, Röhrkasten A, Sieber M, et al. 1987. Phosphorylation of the purified receptor for calcium channel blockers by cAMP kinase and protein kinase C. *Eur. J. Biochem.* 169:137–42

Niidome T, Kim MS, Friedrich T, Mori Y. 1992. Molecular cloning and characterization of a novel calcium channel from rabbit brain. *FEBS Lett.* 308:7–13

Nishimura S, Takeshima H, Hofmann F, et al. 1993. Requirement of the calcium channel β subunit for functional conformation. *FEBS Lett.* 324:283–86

Norman RI, Burgess AJ, Allen E, Harrison TM. 1987. Monoclonal antibodies against the 1,4-dihydropyridine receptor associated with voltage-sensitive Ca^{2+} channels detect similar polypeptides from a variety of tissues and species. *FEBS Lett.* 212:127–32

Nunoki K, Florio V, Catterall WA. 1989. Activation of purified calcium channels by stoichiometric protein phosphorylation. *Proc. Natl. Acad. Sci. USA* 86:6816–20

Osterrieder W, Brum G, Hescheler J, et al. 1982. Injection of subunits of cyclic AMP-dependent protein kinase into cardiac myocytes modulates Ca^{2+} current. *Nature* 298: 576–78

Perez-Reyes E, Castellano A, Kim HS, et al. 1992. Cloning and expression of cardiac/brain β subunit of the L-type calcium channel. *J. Biol. Chem.* 267:1792–97

Perez-Reyes E, Wei XY, Castellano A, Birnbaumer L. 1990. Molecular diversity of L-type calcium channel. Evidence for alternative splicing of the transcripts of three non-allelic genes. *J. Biol. Chem.* 265: 20430–36

Powers PA, Liu S, Hogan K, Gregg RG. 1992. Skeletal muscle and brain isoforms of the β-subunit of human voltage-dependent calcium channels are encoded by a single gene. *J. Biol. Chem.* 267:22967–72

Pragnell M, Sakamoto J, Jay SD, Campbell KP. 1991. Cloning and tissue-specific expression of the brain calcium channel β-subunit. *FEBS Lett.* 291:253–58

Regulla S, Schneider T, Nastainczyk W, et al. 1991. Identification of the site of interaction of the dihydropyridine channel blockers nitrendipine and azidopine with the calcium channel α1 subunit. *EMBO J.* 10:45–49

Rios E, Pizarro G, Stefani E. 1992. Charge movement and the nature of signal transduction in skeletal muscle excitation-contraction coupling. *Annu. Rev. Physiol.* 54:109–33

Röhrkasten A, Meyer HE, Nastainczyk W, Sieber M, Hofmann F. 1988. cAMP-dependent protein kinase rapidly phosphorylates serine-687 of the skeletal muscle receptor for calcium channel blockers. *J. Biol. Chem.* 263:15325–29

Rotman EI, DeJongh KS, Florio V, et al. 1992. Specific phosphorylation of a COOH-terminal site on the full-length form of the α1 subunit of the skeletal muscle calcium channel by cAMP-dependent protein kinase. *J. Biol. Chem.* 267:16100–5

Ruth P, Röhrkasten A, Biel M, et al. 1989. Primary structure of the β subunit of the DHP-sensitive calcium channel from skeletal muscle. *Science* 245:1115–18

Sakamoto J, Campbell KP. 1991. A monoclonal antibody to the β subunit of the skeletal muscle dihydropyridine receptor immunoprecipitates the brain x-conotoxin GVIA receptor. *J. Biol. Chem.* 266:18914–19

Schneider T, Hofmann F. 1988. The bovine cardiac receptor for calcium channel blockers is a 195-kDa protein. *Eur. J. Biochem.* 174:369–75

Schultz D, Mikala G, Yatani A, et al. 1993. Cloning, chromosomal localization, and functional expression of the α1 subunit of the L-type voltage-dependent calcium channel from normal human heart. *Proc. Natl. Acad. Sci. USA* 90:6228–32

Seino S, Chen L, Seino M, et al. 1992. Cloning of the α1 subunit of a voltage-dependent calcium channel expressed in pancreatic β cells. *Proc. Natl. Acad. Sci. USA* 89:584–88

Singer D, Biel M, Lotan I, et al. 1991. The roles of the subunits in the function of the calcium channel. *Science* 253:1553–57

Singer-Lahat D, Gershon E, Lotan H, et al. 1992. Modulation of cardiac Ca^{2+} channels in *Xenopus* oocytes by protein kinase C. *FEBS Lett.* 306:113–18

Snutch TP, Leonard JP, Gilbert MM, Lester HA, Davidson N. 1990. Rat brain expresses a heterogeneous family of calcium channels. *Proc. Natl. Acad. Sci. USA* 87:3391–95

Snutch TP, Tomlinson WJ, Leonard JP, Gilbert MM. 1991. Distinct calcium channels are generated by alternative splicing and are differentially expressed in the mammalian CNS. *Neuron* 7:45–57

Soldatov NM. 1992. Molecular diversity of L-type Ca^{2+} channel transcripts in human fibroblasts. *Proc. Natl. Acad. Sci. USA* 89:4628–32

Soong TW, Stea A, Hodson CD, et al. 1993. Structure and functional expression of a member of the low voltage-activated calcium channel family. *Science* 260:1133–36

Starr TVB, Prystay W, Snutch TP. 1991. Primary structure of a calcium channel that is highly expressed in the rat cerebellum. *Proc. Natl. Acad. Sci. USA* 88: 5621–25

Striessnig J, Glossmann H, Catterall WA. 1990. Identification of a phenylalkylamine binding region within the α1 subunit of skeletal muscle Ca^{2+} channels. *Proc. Natl. Acad. Sci. USA* 87:9108–12

Striessnig J, Murphy BJ, Catterall WA. 1991. Dihydropyridine receptor of L-type Ca^{2+} channels: Identification of binding domains for [3H](+)-PN200–110 and [3H]azidopine within the α1 subunit. *Proc. Natl. Acad. Sci. USA* 88:10769–73

Tanabe T, Adams BA, Numa S, Beam KG. 1991. Repeat I of the dihydropyridine receptor is critical in determining calcium channel activation kinetics. *Nature* 352:800–3

Tanabe T, Beam KG, Adams BA, et al. 1990. Regions of the skeletal muscle dihydropyridine receptor critical for excitation-contraction coupling. *Nature* 346:567–69

Tanabe T, Takeshima H, Mikami A, et al. 1987. Primary structure of the receptor for calcium channel blockers from skeletal muscle. *Nature* 328:313–18

Tsien RW, Ellinor PT, Horne WA. 1991. Molecular diversity of voltage-dependent Ca²⁺ channels. *Trends Pharmacol. Sci.* 12: 349–54

Varadi G, Lory P, Schultz D, et al. 1991. Acceleration of activation and inactivation by the β subunit of the skeletal muscle calcium channel. *Nature* 352:159–62

Wei XY, Perez-Reyes E, Lacerda AE, et al. 1991. Heterologous regulation of the cardiac Ca²⁺ channel α_1 subunit by skeletal muscle β and γ subunits. *J. Biol. Chem.* 266:21943–47

Welling A, Bosse E, Ruth P, et al. 1992a. Expression and regulation of cardiac and smooth muscle calcium channels. *J. Pharmacol.* 58(Suppl. II):258p-62p

Welling A, Bosse E, Bottlender R, et al. 1993a. Stable co-expression of calcium channel α_1, β, and α_2/δ subunits in a somatic cell line. *J. Physiol.* In press

Welling A, Kwan YW, Bosse E, et al. 1993b. Subunit-dependent modulation of recombinant L-type calcium channels: molecular basis for dihydropyridine tissue selectivity. *Circ. Res.* In press

Welling A, Felbel J, Peper K, Hofmann F. 1992b. Hormonal regulation of calcium current in freshly isolated airway smooth muscle cells. *Am. J. Physiol.* 262:L351–59 (Abstr.)

Williams ME, Brust PF, Feldman DH, et al. 1992a. Structure and functional expression of an ω-conotoxin-sensitive human N-type calcium channel. *Science* 257:389–95

Williams ME, Feldman DH, McCue AF, et al. 1992b. Structure and functional expression of α_1, α_2, and β subunits of a novel human neuronal calcium channel subtype. *Neuron* 8:71–84

Witcher DR, De Waard M, Sakamoto J, et al. 1993. Subunit identification and reconstitution of the N-type Ca²⁺ channel complex purified from brain. *Science* 261:486–89

Yatani A, Brown AM. 1989. Rapid β-adrenergic modulation of cardiac calcium channel currents by a fast G protein pathway. *Science* 245:71–74

Yoshida A, Takahashi M, Nishimura S, et al. 1992. Cyclic AMP-dependent phosphorylation and regulation of the cardiac dihydropyridine-sensitive Ca channel. *FEBS Lett.* 309:343–49

Yu ASL, Hebert SC, Brenner BM, Lytton J. 1992. Molecular characterization and nephron distribution of a family of transcripts encoding the pore-forming subunit of Ca²⁺ channels in the kidney. *Proc. Natl. Acad. Sci. USA* 89:10494–98

Annu. Rev. Neurosci. 1994. 17:419–39

SPECIFICATION OF NEOCORTICAL AREAS AND THALAMOCORTICAL CONNECTIONS

Dennis DM O'Leary, Bradley L Schlaggar, and Rebecca Tuttle

Molecular Neurobiology Laboratory, The Salk Institute, 10010 N. Torrey Pines Road, La Jolla, California 92037

KEY WORDS: axon targeting, transcription factors, subplate, preplate, protocortex, protomap, cortical plate, neuroepithelium, barrels, somatosensory cortex

INTRODUCTION

The mammalian neocortex has six main layers distinguished by neuronal morphology and density. Like the other distinct regions of the adult cerebral cortex, the neocortex is divided into functionally specialized areas that are anatomically distinguishable based on differences in connectional and architectural features (Brodmann 1909). These differences, as well as the modality of afferent input, dictate an area's functional identity. However, many features of the vertical organization of the neocortex, such as cell number per radial traverse (Rockel et al 1980, Hendry et al 1987), intrinsic circuitry (Gilbert 1983, Metin & Frost 1989, Sur et al 1988), and laminar organization of classes of input and output projections, (Jones 1984, McConnell 1988, 1989, O'Leary 1989) are consistent between areas of the adult neocortex, and even between mammalian species. An emerging view is that these common features of vertical organization are specified in the neuroepithelium (McConnell 1989, 1991, Katz & Callaway 1992). Even more prominent than the many similarities between neocortical areas are certain "area-specific" features, which include differences in connectivity and architecture with sharp borders delimiting these areas. Here we consider the relative contributions of genetic and epigenetic mechanisms in the emergence of area-specific features during neocortical development. In addition, because many studies indicate that

419

thalamocortical afferents play a fundamental role in cortical differentiation, we discuss recent findings on mechanisms that may contribute to the development of area-specific thalamocortical projections.

Neocortical cells are generated in the dorsal telencephalic neuroepithelium (Bayer & Altman 1991). The first postmitotic neurons leave the neuroepithelium and accumulate beneath the pial surface to form the preplate (Marin-Padilla 1971, 1972, Kostovic & Molliver 1974, Stewart & Pearlman 1987). Preplate neurons are distinct from cortical plate neurons, which are generated later and populate the definitive cellular layers 2 through 6 of the adult cortex (Luskin & Shatz 1985, Chun & Shatz 1989). Cortical plate neurons migrate superficially from the neuroepithelium along radial glia (Rakic 1978, 1981) and aggregate within the preplate, thereby splitting it into a superficial, neuron-sparse marginal zone (future layer 1) and a deep subplate (Luskin & Shatz 1985). The area-specific features in the adult neocortex are not evident in the immature cortical plate and emerge gradually over development.

Two models of cortical area differentiation, the protomap (Rakic 1988) and the protocortex (O'Leary 1989), have recently been put forward. The protomap model emphasizes the role of genetic events and applies the same mechanisms of development to areas throughout the cerebral cortex. Specifically, the protomap hypothesis (Rakic 1988) proposes a map of future cytoarchitectonic areas specified in the neuroepithelium and recapitulated in the developing cortical plate by a point-to-point migration of cortical neurons along a radial glial scaffolding. In this model, therefore, the neuroepithelium is programmed to generate area-specific cohorts of cortical plate neurons, and the relative positions and sizes of areas of the cerebral cortex are prespecified. The programmed emergence of discrete, cytoarchitectonic areas requires an interaction with afferents, but the capacity of developing cortical plate neurons to differentiate features normally associated with other areas is restricted by their commitment to specific area fates.

In contrast, the protocortex model (O'Leary 1989) emphasizes the role of epigenetic influences and addresses only the development of neocortical areas. This model proposes that the neocortical neuroepithelium is not programmed to generate cortical plate cells committed to a particular areal fate. Instead, neurons of the neocortical plate have the potential to develop the range of features associated with the diverse neocortical areas; their normal differentiation of area-specific connections and architecture requires instructions from afferents. An original tenet of the protocortex model is that molecular differences exist across the neocortex and contribute to this process of area differentiation, for example, by promoting the development of area-specific thalamocortical connections (O'Leary 1989). In summary, although these two models differ in their emphasis and scope, both recognize a contribution of genetic and epigenetic mechanisms in the differentiation of neocortical areas, as well as an important role for thalamocortical afferents in this process.

THALAMOCORTICAL AFFERENTS CONTROL DIFFERENTIATION OF AREA-SPECIFIC ARCHITECTURE

Functional Architecture Unique to the Somatosensory Cortical Area

Architectural differences between neocortical areas are composed in part by unique functional groupings—for example, the cytochrome oxidase–rich, color-sensitive blobs of primate visual cortex (Livingstone & Hubel 1984) and the vibrissae-related barrels of rodent primary somatosensory cortex (S1) (Woolsey & Van Der Loos 1970). Many studies relevant to the issue of neocortical areal differentiation have been done in the barrelfield of rodent S1. Barrels are discrete aggregates of layer 4 neurons (Woolsey & Van der Loos 1970) innervated by clusters of ventrobasal (VB) thalamic afferents (Killackey & Leshin 1975) and arranged in a pattern that parallels the pattern of vibrissae on the rodent's face (Woolsey & Van der Loos 1970). Barrels are also characterized by coincident or complementary distributions of certain enzymes and extracellular matrix (ECM) molecules (see Schlaggar & O'Leary 1993 for review). Sensory information is relayed from the vibrissae to S1 through a series of synaptic connections: The peripheral processes of trigeminal sensory neurons innervate vibrissae follicles, and their central processes pass this input to brainstem trigeminal nuclei, which in turn project to VB. As in S1, clustered afferents distributed in a periphery-related pattern innervate discrete groupings of neurons in both the brainstem and VB (reviewed in Woolsey 1990).

Developmental Plasticity in the Differentiation of S1

The differentiation of a normal barrel pattern depends upon information relayed from an intact sensory periphery via VB thalamocortical afferents during a critical period of development (for review see Woolsey 1990). For example, destruction of a row of vibrissae follicles in neonatal rodents results in a loss of barrel patterning in the corresponding row in S1, which is characterized by a fused band of VB afferents and layer 4 neurons; coincident with this effect, barrels in adjacent rows expand at the expense of the affected row (Woolsey & Warm 1976, Belford & Killackey 1980, Jeanmonod et al 1981, Crossin et al 1989, Cooper & Steindler 1989). This plasticity is mediated to some extent by an activity-dependent mechanism requiring activation of postsynaptic neurons by VB afferents (Schlaggar et al 1993b).

The afferent input also influences the body representation in S1, in terms of both its overall size and the relative sizes of portions of the somatotopic map. Supernumerary vibrissae that appear in some inbred strains of mice have a corresponding cortical barrel, but only if the vibrissal follicle is innervated by a suprathreshold number of trigeminal sensory fibers (Welker & Van der Loos 1986). When well innervated, extra vibrissal follicles transplanted onto the muzzles of newborn mice also have corresponding barrels in S1 (Andres 1990). Following bilateral enucleation in neonatal mice, some barrels grow larger (Rauschecker et al 1992, Bronchti et al 1992) and the overall size of the barrelfield expands (Bronchti et al 1992). These demonstrations of developmental plasticity in the S1 somatotopic map are not limited to the barrel representation. For example, forelimb removal in rat fetuses before E18 results in an increase in the territory in S1 devoted to the hindlimb (Killackey & Dawson 1989). Similarly, cutting the infraorbital nerve that carries processes of sensory cells in the trigeminal ganglion, in late embryonic rats not only massively disrupts the S1 barrelfield, but also results in an expansion of the lower jaw representation in S1 (Killackey et al 1993). Together, these studies demonstrate that the relative size of the representation of a given part of the body, or even whether it is represented at all, is not a fixed property of S1, but is dependent on the level of afferent input during development. Clearly, the sensory periphery, acting via thalamocortical afferents, has a prominent role in creating its cortical representation.

Thalamocortical Afferents Show the First Barrel-Patterning

Do VB thalamocortical afferents carry patterning information to a naive S1 or do they respond to patterning information intrinsic to S1? If barrels are prespecified in the neuroepithelium, then the first barrel-related patterning must be intrinsic to the cortex; it cannot, under such an assumption, be caused by extrinsic agents such as thalamocortical afferents. Considerable excitement was generated by reports showing that peanut agglutinin (PNA), a lectin marker for glycoconjugated molecules, revealed a barrel-related pattern in mouse (Cooper & Steindler 1986) and rat (McCandlish et al 1989) S1 earlier than any other cortical or afferent component noted at the time. The finding that lectin was binding to a cortical element suggested a molecular framework intrinsic to S1 that controlled barrel differentiation (Steindler et al 1989a,b, 1990, Waters et al 1990). However, a reevaluation of the development of VB thalamocortical afferents using more sensitive techniques, such as DiI, an axon tracer (Godement et al 1987, Honig & Hume 1989), or acetyl-cholinesterase (ACHE) histochemistry, which transiently marks developing principal sensory thalamic axons (Kristt 1979, Robertson 1987, Schlaggar et al 1993a), has revealed that VB afferents are organized in a periphery-related pattern well before the PNA revealed pattern, or any other barrel component yet described (Erzurumlu & Jhaveri 1990, Jhaveri et al 1991, Schlaggar & O'Leary 1993). In fact, the periphery-related molecular pattern revealed by PNA binding is molded by VB afferents (Waters et al 1990, Schlaggar & O'Leary 1991). These studies rule out initiation of barrel patterning by the PNA-ligand. However, other roles proposed for the glycoconjugate pattern, including acting as an intermediary in conferring patterning information from VB afferents to cortical cells, are consistent with current data (Steindler et al 1989a,b, 1990). Although these observations cannot exclude the existence of a still-covert molecular framework that might initiate barrel differentiation, they are consistent with the hypothesis that VB thalamocortical afferents contribute barrel-patterning information to the cortical plate of developing S1.

The Potential to Differentiate Barrels is Shared by Diverse Neocortical Areas

These development and plasticity studies do not address the questions of whether the S1 cortical plate is uniquely specified to differentiate barrels and their characteristic array, or whether other areas of the developing neocortex are competent to form these functional groupings. The capacity of VB afferents to direct barrel differentiation has been tested by heterotopically transplanting late embryonic visual (V1) cortex into the S1 barrelfield in newborn rats (Schlaggar & O'Leary 1991). This experiment tests two hypotheses: first, that information intrinsic to S1 is necessary to differentiate barrels; and second, that other neocortical areas are restricted in their fate such that they cannot differentiate the area-specific features unique to S1. V1, like S1, is characterized by a granular layer 4 receiving a dense innervation from a principal sensory thalamic nucleus—in this instance, the dorsal lateral geniculate nucleus (LG). In sharp contrast to the disjunctive distributions of VB afferents and layer 4 cells characteristic of S1, layer 4 neurons and LG thalamocortical afferents are more-or-less evenly distributed in V1. However, transplanted V1 develops the distinct features of normal S1 barrels: clusters of VB afferents coincident with aggregates of layer 4 neurons ringed by boundaries of PNA binding (Schlaggar & O'Leary 1991). Thus, VB afferents can organize appropriately in layer 4 of a foreign piece of neocortex. In response to these afferents, or possibly through an intermediary, layer 4 neurons distribute into aggregates. Since V1 has the capacity to develop functional groupings ordinarily unique to S1, late embryonic V1 must not be committed to develop the area-specific architecture and thalamocortical afferent distribution characteristic of mature V1. A corollary to this conclusion is that barrel architecture is not a prespecified feature of S1.

Thalamocortical Afferents Predict Areal Borders

Several other lines of evidence also suggest that thalamocortical afferents regulate cortical architecture. For example, in rodents, ACHE, which is transiently associated with developing VB afferents to S1 (Robertson 1987, Robertson et al 1991), not only provides the first indication of barrel patterning, but also delineates the location of S1 prior to its architectural definition (Schlaggar & O'Leary 1993). In primates, ACHE is transiently associated with developing pulvinar thalamic afferents to the second visual area, V2, and reveals the V1/V2 border before cytoarchitecture does (Kostovic & Rakic 1984). Therefore, the abrupt cytoarchitectural borders between neocortical areas in the adult become apparent only after the cortical plate is invaded by thalamocortical afferents. The evidence in the following section suggests that thalamocortical afferents create the architectural borders.

Area Borders Can Shift

Dramatic evidence for developmental plasticity in area-specific architecture and connectivity comes from the demonstration that mid-gestational bilateral eye removal in rhesus macaques results in a shift in the location of the border between V1 (area 17) and V2 (area 18) (Rakic 1988, Rakic et al 1991, Dehay et al 1989, 1991). In these animals, the LG, which provides the major thalamic input to V1, contains only 30–50% of the normal number of relay neurons; presumably a similar percentage reduction occurs in the number of geniculocortical axons projecting to V1. These studies find a corresponding reduction in total area of V1, but no significant deviations from normal in other parameters, including cortical depth and number of neurons in a radial traverse. Using the prostriate cortex, a nonneocortical region, as a fixed landmark, Dehay et al (1991) convincingly demonstrated that the reduction in the size of V1 is due, at least in part, to a border shift. The portion of neocortex surrounding the reduced V1 acquires not only the distinctive cytoarchitecture characteristic of the neighboring V2, but also its chemo-architecture and even its pattern of callosal connections. These findings, which again demonstrate that regions of the developing neocortex have the capacity to develop features normally associated with other areas, strongly implicate a controlling influence for thalamocortical afferents in this phenomenon.

As one might predict, peripheral manipulations done prior to thalamocortical ingrowth into the cortical plate result in more substantial plasticity than those done later. In monkeys, thalamocortical afferents invade the cortical plate of V1 around E80 (Kostovic & Rakic 1984); bilateral eye removals done around or after this age affect the size of V1 much less than those done earlier (Dehay et al 1991). In rats, VB afferents invade the cortical plate of S1 at E18 (Catalano et al 1991), at which age the critical period for intraareal compensation due to limb removal ends in S1 (Killackey & Dawson 1989, Killackey et al 1993). Thus, lesion-induced areal compensation seems to be related to the state of thalamocortical ingrowth into the cortical plate at the time of the lesion.

Thalamocortical Afferents Influence Architecture by Controlling Neuronal Number and Morphology

A striking feature of the primary sensory areas of the neocortex is the dense accumulation in layer 4 of small neurons with a stellate morphology, resulting in a granular appearance (Brodmann 1909). The high density of stellate neurons correlates with the prominent layer 4 terminations of thalamocortical afferents from the corresponding principal sensory thalamic nuclei. The granular appearance of primary sensory areas is in sharp contrast to agranular areas of the neocortex, such as the primary motor cortex, which has a poorly defined, sparsely populated layer 4. This difference is not apparent earlier in development; instead, the monkey primary motor cortex transiently exhibits a high density of small neurons in layer 4 which closely resembles that in S1 (Huntley & Jones 1991).

Two mechanisms, differential neuronal death and changes in neuronal morphology, each influenced by thalamocortical input, can account for much of the inter-area difference in the number and appearance of neurons found in specific layers. In rodents, about 30% of cortical plate neurons die (Finlay & Slattery 1983, Windrem & Finlay 1991). Most of this loss occurs in the superficial layers and to a lesser extent in layer 4. Recent evidence indicates that the number of cells in layer 4 is governed by the density of thalamic input. Windrem & Finlay (1991) have observed that regions with dense thalamocortical afferent input have fewer pyknotic cells in layer 4. Further, removal of dorsal thalamus in newborn hamsters produces large-scale neuronal degeneration in layer 4 and a loss of granularity. Therefore, thalamocortical afferents can influence cytoarchitecture by regulating cell death. Other evidence indicates that thalamocortical input seems to exert dramatic influences on the dendritic development of cortical neurons (Harris & Woolsey 1981, Peinado & Katz 1990). For example, layer 4 stellate neurons in rat S1 initially develop a pyramidal morphology characteristic of layer 3 neurons and only later take on a stellate appearance (Peinado & Katz 1990). This morphological change can be prevented by preventing the flow of sensory information from the periphery to S1 in neonatal rats.

Epigenetic Influences on the Development of Area-Specific Efferent Projections

In sharp contrast to their limited distributions in adults, most, if not all, classes of cortical projection neurons, including corticospinal, corticotectal, and callosal, are widely distributed in the tangential plane of the developing neocortex (for review see O'Leary 1989, 1992). These observations indicate that the area-specific patterns of cortical efferent projections do not reflect regional differences in the ability of the neocortical neuroepithelium to generate specific classes of cortical projection neurons. Rather, area-specific patterns of cortical outputs emerge as the consequence of an epigenetically regulated phenomenon, selective axon elimination (Innocenti 1981, O'Leary et al 1981, Ivy & Killackey 1982, Stanfield et al 1982, Stanfield & O'Leary 1985a,b, O'Leary & Stanfield 1989).

Manipulations resulting in the maintenance of normally transient cortical axons indicate that their removal is functionally determined. For example, the distribution of callosal neurons in V1 of adults is abnormally widespread in animals that are deprived visually during development (see O'Leary 1989, 1992 for review) or in cats with natural (Shatz 1977) or surgically induced strabismus (Lund et al 1978, Berman & Payne 1983). Visual deprivation also dramatically affects the normal refinement of horizontal intracortical projections in V1 (Katz & Callaway 1992). In all of these manipulations, visual thalamocortical input is altered in some manner. Similarly, when the somatosensory input is experimentally misrouted through LG to V1, cortical neurons in V1 retain their normally transient pyramidal tract axons to the spinal cord, a projection characteristic of the somatosensory cortex (O'Leary 1992). These experiments suggest that changes in thalamocortical input, or inputs being relayed through the thalamus, influence the process of selective axon elimination, and thus the output of a given area of the neocortex. This conclusion is consistent with the finding that heterotopic neocortical transplants, which become invaded by thalamic afferents appropriate for their new areal location (Chang et al 1986, O'Leary et al 1992), also establish the efferent projections appropriate for that area (Stanfield & O'Leary 1985b, O'Leary & Stanfield 1989).

RESTRICTION IN FATES BETWEEN REGIONS OF THE CEREBRAL CORTEX

Based on the studies described thus far, one may conclude that cortical plate cells in diverse areas of the developing neocortex have similar potentials to differentiate the range of connectional and architectural features seen in the adult neocortex. The features ultimately expressed by neocortical plate cells depend on where in the neocortex the cells develop, not where they were generated. Thus, the area-specific connectional and architectural features do not seem to be committed in the neuroepithelium, nor even at the time the neocortical plate is assembling. These conclusions are different from but not inconsistent with those derived from the heterotopic transplant studies of Barbe & Levitt (1991, 1992), who transplanted the perirhinal cortex, an area of limbic cortex, into the sensorimotor cortex, an area of neocortex, and vice versa, and later assayed the transplant for the source of thalamocortical input and the expression of limbic-associated membrane protein (LAMP) (Horton & Levitt 1988). LAMP is normally expressed on postmitotic cortical plate neurons in meso- and allocortical regions of the cerebral cortex, including the prefrontal, cingulate, perirhinal, and hippocampal cortices, but not by neocortical neurons (Barbe & Levitt 1991). Barbe & Levitt (1992) found that transplants taken before appreciable neurogenesis and placed in newborn rats regulate their phenotype to mimic that of their host cortical location, but those taken two days later, shortly after neurogenesis begins, do not change their fate. These findings suggest that this difference in molecular phenotype between limbic cortex and neocortex becomes committed early in corticogenesis. In conclusion, diverse areas of the neocortex have the capacity to differentiate area-specific architecture and connections normally associated with other neocortical areas even late in corticogenesis, whereas the ability of regions of the cerebral cortex to express features normally associated with other distinct regions of the cerebral cortex becomes limited much earlier in development.

DISPERSION OF CLONALLY RELATED CORTICAL NEURONS

Evidence presented here argues that at the time thalamocortical afferents invade the cortical plate of the neocortex, cortical plate neurons are not committed to a particular areal fate. Studies assessing spatial relationships between regions of the neocortical neuroepithelium and the overlying cortical plate suggest a similar conclusion. In this context, the most dramatic findings come from the work of Walsh & Cepko (1992, 1993), who have used a battery of recombinant retroviruses with unique tags to show that clonally related cortical plate neurons commonly disperse widely in the neocortex and often scatter over many areas. The dispersion they report could occur in a number of ways during the generation and migration of cortical plate neurons. For example, Fishell et al (1993) have shown that cells in the neocortical neuroepithelium make seemingly random tangential movements. Even if this mechanism were the sole contributor to dispersion, given the geometry of the developing cortex, small dispersions of clonally related cells in the neuroepithelium could result in large displacements by the time the cells reached the cortical plate. However, other mechanisms also contribute to dispersion, including tangential migration in the intermediate zone orthogonal to radial glia (O'Rourke et al 1992) and the shifting of migrating cells from one radial glial fascicle to another (Austin & Cepko 1990, Misson et al 1991). The large-scale dispersion of clonally related, cortical plate cells further questions whether they are committed to a particular areal fate when they leave the neuroepithelium, and has prompted Walsh & Cepko (1992) to conclude that the specification of neocortical areas must occur after neurogenesis.

EVIDENCE FOR MARKERS OF AREAL OR REGIONAL SPECIFICATION

Antigen Distribution

Any areal specification of the neocortical neuroepithelium and neocortical plate should be revealed by a parallel expression of molecules or genes in distinct patterns. The first evidence for molecular regionalization within the neocortical neuroepithelium came from immunocytochemical studies using antibodies against the peptides encoded by four proto-oncogenes (*sis*, *src*, *ras*, and *myc*) and the intermediate filament protein vimentin present in radial glia (Johnston & Van der Kooy 1988). In late embryonic rats, these antibodies stained discrete patches in the telencephalic neuroepithelium, including that of the neocortex. However, the patchy immunoreactivity for these peptides appears very late in neurogenesis and does not match any obvious areal organization. Hence, the antigen distributions are unlikely to reveal areal organization of the neuroepithelium.

Two other recently identified molecules also show patterns of distribution in the developing telencephalon relevant to this discussion. As described above, LAMP is present on neurons in cortical and subcortical components of the limbic system but is not expressed in the neocortex (Horton & Levitt 1988). Thus, LAMP reveals an organizational feature of the developing cerebral cortex at the regional level by distinguishing neocortex from limbic cortex. Likewise, the monoclonal antibody PC3.1 labels a subpopulation of neurons in ventrolateral cerebral cortex (Arimatsu et al 1992), including neocortical and nonneocortical regions. Most labeled neurons are in layer 6 and the subplate. As with LAMP, the commitment to PC3.1 occurs early in cortical neurogenesis, but unlike LAMP, it is not detectable until after corticogenesis is complete. The distribution of PC3.1-positive cells indicates some level of regionalization but does not match any known functional organization. Since PC3.1 recognizes a cytoplasmic antigen and does not appear until after cortical organizations are established, it is unlikely to play a direct role in this process.

Gene Expression

Another approach to understanding the mechanisms that control cortical differentiation has arisen from the pioneering work on *Drosophila* segmentation genes. In the process of identifying vertebrate homologs of these genes, a number of homeobox and other regulatory genes encoding putative transcription factors have been identified with temporal and spatial patterns of expression that suggest their involvement in regulating the regionalization of the CNS. A number of these genes are expressed in the rodent telencephalon. Two genes are reported to be restricted in their expression to the telencephalon: *Emx1* (Simeone et al 1992) and BF-1. BF-1, related to *Drosophila forkhead*, is specific to the cerebral cortex and caudate putamen (Tao & Lai 1992). However, both of these genes are expressed uniformly in the cerebral cortical neuroepithelium, as are several other genes whose expression is not restricted to the telencephalon, including the homeobox LH-2 (Xu et al 1993) and the POU domain genes, Bm-1, Bm-2, Tst-1/SCIP/Oct-6 (He et al 1989, Monuki et al 1990, Suzuki et al 1990, Frantz et al 1992). Others, such as *Dlx* (Price et al 1991) and *Tes-1* (Porteus et al 1991), both related to *Drosophila Distal-less*, and *Dbx* (Lu et al 1992), are expressed in ventral telencephalon. In summary, the expression patterns of these genes suggest that they may regulate the emergence of forebrain subdivisions during development—for example, the telencephalon from the forebrain, or the cerebral cortex from the rest of the telencephalon—but none is likely to be involved in the regionalization or arealization of the neocortex.

Three genes are more likely candidates for a role in the regionalization of the neocortex. *Pax-6*, a paired box gene, is expressed in a rostral to caudal gradient in the dorsal telencephalic neuroepithelium (Walther & Gruss 1991). At present, it is not clear whether the graded expression of *Pax-6* reflects position dependence or merely the rostrocaudal maturational gradient in developing cortex (Bayer & Altman 1991). The others, *Emx1* and *Emx2*, are closely related homologs of *Drosophila empty spiracles* (Simeone et al 1992). *Emx2* appears to have a caudal to rostral graded distribution in dorsal telencephalon. *Emx1* is expressed uniformly in the cerebral cortex, suggesting a nested expression with *Emx2* reminiscent of *Hox* gene expression in the hindbrain where the nested expression patterns are indicative of hierarchical control of segmentation (Lumsden & Keynes 1989). Although the expression patterns of these genes niether predict nor delineate specific neocortical areas, the position-dependent expression of *Emx2* in the developing neocortex suggests that it could contribute to the arealization of the developing neocortex.

DEVELOPMENT OF THALAMOCORTICAL PROJECTIONS

Introduction

From the preceding discussion it is evident that the developing neocortical plate depends on thalamocortical input to differentiate the area-specific features of the adult neocortex. At maturity, thalamocortical projections are organized such that specific thalamic nuclei project to specific neocortical areas (Hohl-Abrahao & Creutzfeldt 1991). The development of these projections is characterized by two distinct growth phases. The first is a targeting phase during which thalamocortical axons pass through the internal capsule to reach the neocortex and grow tangentially toward their target area along an intracortical pathway centered on the subplate (Bicknese et al 1991, Miller et al 1993, Ghosh & Shatz 1992). The second phase, the invasion of the cortical plate, is initiated, often after a delay (termed the "waiting period"), when thalamocortical afferents send collateral branches into the overlying cortical plate (Erzurumlu & Jhaveri 1990, Ghosh et al 1990, Ghosh & Shatz 1992, Naegle et al 1988, Reinoso & O'Leary 1988, 1990, Catalano et al 1991).

Development of Area-Specific Thalamocortical Connections

A crucial issue in neocortical differentiation is defining the mechanisms that control the development of area-specific thalamocortical projections. Heterotopic transplant and coculture experiments show that thalamic nuclei that normally innervate specific neocortical areas will readily innervate other, "inappropriate" areas of neocortex (Chang et al 1986, Schlaggar & O'Leary 1991, Molnar & Blakemore 1991, O'Leary et al 1992). It is unlikely that these results are due to respecification of a molecular phenotype of the neocortical transplants/explants because they were taken late in or even after neurogenesis. Therefore, by analogy to the commitment of the LAMP phenotype (Barbe & Levitt 1992), molecular phenotype in the transplants/explants should be committed. Findings from these transplant and coculture experiments should not be taken as evidence that molecular cues intrinsic to cortex do not normally provide targeting information for thalamocortical afferents. The transplants are placed in newborn host rats, several days after host thalamocortical afferents normally have begun to invade the overlying cortical plate. Thus, the thalamocortical axons already at the transplant site simply reextend and invade the transplanted cortex. Similarly, in the coculture experiments, individual thalamic axons were not provided with a choice of targets (for discussion see Tuttle & O'Leary 1993).

The targeting of developing thalamocortical axons from the principal sensory thalamic nuclei to their appropriate cortical area is remarkably accurate; thalamocortical axons rarely extend beyond their correct area or make gross directional errors (De Carlos et al 1992), nor do they extend collaterals into the cortical plate of inappropriate areas (Crandall & Caviness 1984, De Carlos et al 1992). These findings argue for the specification of position-dependent information in the developing neocortex—a specification that controls the process of area-specific thalamocortical matching. As noted above, there is limited evidence for position-dependent gene expression in the developing neocortex [e.g. *Emx2* (Simeone et al 1992), and possibly *Pax-6* (Walther & Gruss 1991)]. While these genes may play an important role in regulating cortical development, their products are far removed from the molecules that directly promote the area-specific targeting of thalamocortical axons. Transplant and in vitro experiments have implicated the cell surface glycoprotein LAMP in the development of thalamic connections to limbic cortex (Barbe & Levitt 1992), but LAMP is not a candidate molecule for controlling thalamocortical matching in the neocortex.

Shatz and colleagues (Ghosh et al 1990) have provided experimental evidence in embryonic cats that subplate neurons may be a source of targeting information for thalamocortical axons. They used the excitotoxin kainic acid to remove subplate neurons at a time when LG axons have grown beneath the cortical plate of V1 but have not invaded it. In these animals, LG axons did not invade regions of the V1 cortical plate overlying the depleted subplate; instead they grew beyond V1 aberrantly into the cingulate gyrus. Blakemore & Molnar (1990) have suggested that the axons of subplate neurons form a scaffolding that provides contact guidance for thalamocortical axons from the internal capsule to their appropriate cortical target areas. However, the intracortical pathways of cortical efferents and thalamocortical afferents are distinct in both adult (Woodward et al 1990) and fetal rats (De Carlos & O'Leary 1992, Miller et al 1993, Bicknese et al 1991). During development, subcortically projecting efferent axons from subplate and cortical plate neurons take an intracortical trajectory deep in the intermediate zone (De Carlos & O'Leary 1992). Thalamocortical axons, which arrive in the cortex after the extension of subplate axons, grow along a tangential intracortical path centered on the subplate layer, superficial to the path taken by the efferent subplate axons. This distinct intracortical layering of efferents and afferents makes it unlikely that thalamocortical axons are guided by subplate efferents. Furthermore, subplate axons cannot be responsible for providing directional cues for thalamocortical afferent growth from the thalamus into the internal capsule because subplate axons and thalamocortical axons enter the nascent internal capsule at about the same time (Blakemore & Molnar 1990, De Carlos & O'Leary 1992; Ghosh & Shatz 1992). The evidence, therefore, suggests that any influence of the subplate on the development of area-specific thalamocortical matching is likely limited to the subplate layer itself.

Establishment of Targeting Cues in the Subplate

Since thalamocortical axons travel in the subplate layer and make their targeting decisions within it, the cues that set up specific thalamocortical relationships are likely to operate there. To lay down these cues, neuroepithelial cells could impart positional information to their progeny that form the subplate, a mechanism postulated to operate widely in the developing nervous system (Jacobson 1991). This mechanism requires that the progeny of neighboring proliferative cells in the neuroepithelium maintain neighbor relationships in the subplate. If dispersion is substantial, targeting information likely has a source other than subplate cells or is specified by interactions that occur after the preplate is established. As described above, the use of recombinant retroviruses to mark clonally related cells shows that neurons generated at stages of cortical plate assembly disperse widely across diverse areas of the neocortex (Austin & Cepko 1990, Walsh & Cepko 1992). For technical reasons, the recombinant retroviral method has not been successfully applied to the study of neuronal dispersion during the generation of the preplate, the earliest stage of cortical neurogenesis. However, findings obtained by following the movements of fluorescently marked cells in the neocortical neuroepithelium suggest that the dispersion of neuroepithelial cells and their progeny is minimal during the period when the preplate is being established (O'Leary and Borngasser 1992). Preplate neurons rapidly form an intricate network of processes (De Carlos & O'Leary 1992) that presumably would restrict their movements relative to one another, thereby preserving their neighbor relationships as the cortex expands and the subplate emerges from the preplate. It would appear, then, that positional information imparted to preplate cells around the time they are generated could later be deployed in the subplate to control the area-specific targeting of thalamocortical axons.

Control of Thalamocortical "Waiting Periods" and Cortical Plate Invasion

Studies of developing thalamocortical projections in rats, cats and monkeys have reported that thalamic axons grow to the appropriate cortical region and then "wait" in the subplate for varying periods, depending on the species and the thalamocortical projection, before invading the overlying cortical plate (Lund & Mustari 1977, Rakic 1977, 1983, Shatz & Luskin 1986, Reinoso & O'Leary 1988, 1990, Catalano et al 1991, Ghosh & Shatz 1992). Little is known about the molecules that restrict the tangential growth of thalamocortical axons to the subplate or those that prompt their eventual invasion of the overlying cortical plate. One possibility is that until the end of the waiting period the cortical plate is too immature to attract or permit the ingrowth of thalamocortical axons. Indeed, at the time thalamic afferents reach their appropriate cortical area, layer 4 neurons, their principle target cells are still being generated and few if any have migrated into the cortical plate (Lund & Mustari 1977, Shatz & Luskin 1986, Reinoso & O'Leary 1990). Investigations of maturation-dependent changes in the distribution of extracellular matrix (ECM) and membrane-associated molecules in developing cortex have revealed a subset that exhibits pronounced spatiotemporal changes suggestive of a role in controlling thalamocortical afferent growth in cortex.

As thalamocortical axons grow intracortically to their target area, their path is centered on the subplate, which at this stage is enriched for chondroitin sulfate proteoglycans (CSPGs) (Bicknese et al 1991, Miller et al 1992). The growth of thalamocortical afferents into the overlying cortical plate coincides with a progressive deep to superficial increase in the amount of CSPGs in the cortical plate. In vitro, a substrate of cartilage CSPG affects the rate of CNS neurite growth, with high concentrations inhibiting neurite extension (Snow & Letourneau 1992). However, preparations of CSPGs from brain are reported to promote the growth of cortical neurites (Iijima et al 1991). CSPGs in cortex may, therefore, modulate thalamocortical axon extension, possibly via their core proteins (Grumet et al 1993). PNA lectin binding also correlates with thalamocortical afferent growth in cortex. In vitro studies suggest that PNA may bind growth promoting glycoconjugates since PNA decreases the number of thalamic neurite fascicles on postnatal day 6 cortical membranes (Gotz et al 1992). The PNA could be binding to CSPGs (Crossin et al 1989); it might also be binding other, as yet unidentified, glycoconjugated molecules that are upregulated in maturing cortical plate.

Other molecules that increase in the cortical plate during thalamocortical afferent invasion include the ECM glycoprotein, tenascin (also called cytotactin), and the membrane-associated cell adhesion molecule, L1. Tenascin immunoreactivity is initially most pronounced in the subplate and upper intermediate zone but becomes apparent in the cortical plate as it differentiates (Sheppard et al 1991). While soluble tenascin inhibits CNS neurite growth on a variety of ECM substrates, substrate-bound tenascin promotes neurite growth (Lochter et al 1991). L1 is present on postmigratory cortical plate neurons and their axons, as well as on thalamocortical axons (Fushiki & Schachner 1986, Chung et al 1991). In vitro, L1 promotes CNS neurite growth (Lagenaur & Lemmon 1987) and neuron-neuron adhesion (Keilhauer et al 1985). These data consistently claim that tenascin and L1 each play a role in promoting thalamocortical axon invasion of the cortical plate.

These immunocytochemical and histochemical studies describe maturation-dependent changes in molecules on the cell surface and in the ECM of developing neocortex. Presumably, thalamocortical afferents are responsive to these specific molecular changes in vivo, but this has yet to be directly demonstrated. It has, however, been shown that rat thalamic neurites can discriminate between embryonic and neonatal cortical plate membranes (Tuttle et al 1993). In this study, specific regions of the thalamus were explanted onto a substrate composed of alternating stripes of neonatal and embryonic cortical plate membranes. The embryonic cortical plate membranes were prepared at ages just prior to the invasion of thalamocortical afferents into the cortical plate. The thalamic neurites displayed a clear growth preference for the neonatal cortical plate membranes. This finding suggests that membrane-associated molecules in the cortical plate undergo maturation-dependent changes to which thalamic neurites are responsive; these changes might, therefore, promote thalamocortical afferent invasion of the cortical plate.

Once thalamocortical afferents have begun to invade the maturing cortical plate, how are the specific terminations in layer 4 established? Investigations of this issue using cocultures of embryonic dorsolateral thalamus with early postnatal neocortex show that thalamic neurites extend into the cortex and stop in the middle of layers 1–6 (Molnar & Blakemore 1991). This behavior is observed when the thalamic neurites enter the cortical explant from its ventricular surface, which approximates their entry in vivo, or from the pial surface, an aberrant entry (Bolz et al 1992, Molnar & Blakemore 1991). Birthdating data indicate that this region of termination corresponds to that containing layer 4 neurons (Bolz et al 1992, Molnar & Blakemore 1991). In long-term cocultures, thalamic neurites often exhibit a terminal arbor in or near layer 4 of the cortical explant, and at least some of these projections form functional connections (Bolz et al 1992, Yamamoto et al 1992, Yamamoto et al 1989). The distinct, layer-specific behaviors exhibited by these thalamocortical afferents suggest that these axons are responsive to layer-specific molecular cues. Molnar & Blakemore (1991) suggest that thalamocortical afferents might recognize a molecular stop signal associated with layer 4, which by itself or in association with other signals may bring about layer-specific axon arborization.

CONCLUSION

Converging lines of evidence suggest that at the time the neocortical plate is assembled, its cells are not committed to differentiate the area-specific architectural and connectional features that distinguish neocortical areas in the adult. Experiments that demonstrate developmental plasticity in the differentiation of area-specific features cannot rule out that neurons of the neocortical plate carry area-specific information, but they minimize the biological role that this information plays in areal differentiation. The considerable dispersion of clonally related cells forming the neocortical plate further suggests that they do not depend upon area-specific information that may be passed on to them by their progenitors. Numerous studies indicate a fundamental role of afferent input, especially thalamocortical afferents, in regulating the differentiation of area-specific features. The degree of specificity exhibited by developing thalamocortical afferents suggests that position-dependent molecular cues operating in the subplate control the development of area-specific thalamocortical projections. In contrast to the dispersion of cortical plate cells, the apparent minimal dispersion of subplate progenitors would allow for these cues to be effectively laid out in the subplate by the passing of positional information to subplate cells by their progenitors. Although the identity of these molecular cues is unknown, evidence for position-dependent expression of regulatory genes is consistent with this model.

Literature Cited

Andres FL. 1990. New barrels develop in the somatosensory cortex of mice after implantation at birth of the parts of vibrissal follicles containing large numbers of receptors. *Eur. Neurosci. Abstr.* 219:1990

Arimatsu Y, Miyamoto M, Nihonmatsu I, Hirata K, Uratani Y, et al. 1992. Early regional specification for a molecular phenotype in the rat neocortex. *Proc. Natl. Acad. Sci. USA* 89:8879–83

Austin CP, Cepko CL. 1990. Migration patterns in the developing mouse cortex. *Development* 110:713–32

Barbe MF, Levitt P. 1991. The early commitment of fetal neurons to the limbic cortex. *J. Neurosci.* 11:519–33

Barbe MF, Levitt P. 1992. Attraction of specific thalamic input by cerebral grafts depends on the molecular identity of the implant. *Proc. Natl. Acad. Sci. USA* 89: 3706–10

Bayer SA, Altman J. 1991. *Neocortical Development.* New York: Raven Press

Belford GR, Killackey HP. 1980. The sensitive period in the development of the trigeminal system in the neonatal rat. *J. Comp. Neurol.* 193:335–50

Berman NE, Payne BR. 1983. Alterations in connections of the corpus-callosum following convergent and divergent strabismus. *Brain Res.* 274:201–12

Bicknese AR, Sheppard AM, O'Leary DDM, Pearlman AL. 1991. Thalamocortical axons preferentially extend along a chondroitin sulfate proteoglycan enriched pathway coincident with the neocortical subplate and distinct from the efferent path. *Soc. Neurosci. Abstr.* 17:764

Blakemore C, Molnar Z. 1990. Factors involved in the establishment of specific interactions between thalamus and cerebral cortex. *Cold Spring Harbor Symp. Quant. Biol.* 55:491–504

Bolz J, Novak N, Staiger V. 1992. Formation of specific afferent connections in organotypic slice cultures from rat visual cortex cocultured with lateral geniculate nucleus. *J. Neurosci.* 12:3054–70

Brodmann K. 1909. Vergleichende Lokalisationslehre der Grosshirnrinde in ihren Prin-

zipien dargestellt auf Grund des Zellenbaues. Leipzig: Barth

Bronchti G, Schnenberger N, Welker E, Van der Loos H. 1992. Barrelfield expansion after neonatal eye removal in mice. *Neuro Report* 3:489–92

Catalano SM, Robertson RT, Killackey HP. 1991. Early ingrowth of thalamocortical afferents to the neocortex of the prenatal rat. *Proc. Natl. Acad. Sci. USA* 88:2999–3003

Chang F, Steedman JG, Lund RD. 1986. The lamination and connectivity of embryonic cerebral cortex transplanted to newborn rat cortex. *J. Comp. Neurol.* 244:401–11

Chun JJM, Shatz CJ. 1989. The earliest-generated neurons of the cat cerebral cortex: characterization by MAP-2 and neurotransmitter immunohistochemistry during fetal life. *J. Neurosci.* 9:1648–67

Chung WW, Lagenaur CF, Yan Y, Lund JS. 1991 Developmental expression of neural cell adhesion molecules in the mouse neocortex and olfactory bulb. *J. Comp. Neurol.* 314:290–305

Cooper NGF, Steindler DA. 1986. Lectins demarcate the barrel subfield in the somatosensory cortex of the early postnatal mouse. *J. Comp. Neurol.* 249:157–69

Cooper NGF, Steindler DA. 1989. Critical period-dependent alterations of the transient body image in the rodent cerebral cortex. *Dev. Brain Res.* 489:167–76

Crandall JE, Caviness VS. 1984. Thalamocortical connections in newborn mice. *J. Comp. Neurol.* 228:542–46

Crossin KL, Hoffman S, Tan S-S, Edelman GM. 1989. Cytotactin and its proteoglycan ligand mark structural and functional boundaries in somatosensory cortex of early postnatal mouse. *Dev. Biol.* 136:381–92

De Carlos JA, O'Leary DDM. 1992. Growth and targeting of subplate axons and establishment of major cortical pathways. *J. Neurosci.* 12:1194–211

De Carlos JA, Schlaggar BL, O'Leary DDM. 1992. Targeting specificity of primary sensory thalamocortical axons in developing rat neocortex. *Soc. Neuro. Abstr.* 18:57

Dehay C, Horsburgh G, Berland M, Killackey H, Kennedy H. 1989. Maturation and connectivity of the visual cortex in monkey is altered by prenatal removal of retinal input. *Nature* 337:265–67

Dehay C, Horsburgh G, Berland M, Killackey H, Kennedy H. 1991. The effects of bilateral enucleation in the primate fetus on the parcellation of visual cortex. *Dev. Brain Res.* 62:137–41

Erzurumlu RS, Jhaveri S. 1990. Thalamic axons confer a blueprint of the sensory periphery onto developing rat somatosensory cortex. *Dev. Brain Res.* 56:229–34

Finlay BL, Slattery M. 1983. Local differences in amount of early cell death in neocortex predict adult local specializations. *Science* 219:1349–51

Fishell G, Mason CA, Hatten ME. 1993. Dispersion of neural progenitors within the geminal zones of the forebrain. *Nature* 362: 636–38

Frantz GD, Burstein IM, Bohner AP, McConnell S. 1992. Developmental regulation of the POU-homeodomain gene SCIP /TST-1 during cerebral cortical development. *Soc. Neurosci. Abstr.* 18:957

Fushiki S, Schachner M. 1986. Immunocytological localization of cell adhesion molecules L1 and N-CAM and the shared carbohydrate epitope L2 during development of the mouse neocortex. *Dev. Brain Res.* 24:153–67

Ghosh A, Antonini A, McConnell SK, Shatz CJ. 1990. Requirement for subplate neurons in the formation of thalamocortical connections. *Nature* 347:179–81

Ghosh A, Shatz C. 1992. Pathfinding and target selection by developing geniculocortical axons. *J. Neurosci.* 12:39–55

Gilbert CD. 1983. Microcircuitry of the visual cortex. *Annu. Rev. Neurosci.* 6:217–48

Godement P, Vanselow J, Thanos S, Bonhoeffer F. 1987. A study in developing nervous systems with a new method of staining neurons and their processes in fixed tissue. *Development* 101:697–713

Gotz M, Novak N, Bastmeyer M, Bolz J. 1992. Membrane-bound molecules in rat cerebral cortex regulate thalamic innervation. *Development* 116:507–19

Grumet M, Flaccus A, Margolis RU. 1993. Functional characterization of chondroitin sulfate proteoglycans of brain: interactions with neurons and neural cell adhesion molecules. *J. Cell Biol.* 120:815–24

Harris RM, Woolsey TA. 1981. Dendritic plasticity in mouse barrel cortex following postnatal vibrissa follicle damage. *J. Comp. Neurol.* 220:63–79

He X, Treacy MN, Simmons DM, Ingraham HA, Swanson LW, Rosenfeld MG. 1989. Expression of a large family of POU-domain regulatory genes in mammalian brain development. *Science* 340:35–42

Hendry SHC, Schwark HD, Jones EG, Yan J. 1987. Numbers and proportions of GABA-immunoreactive neurons in different areas of monkey cerebral cortex. *J. Neurosci.* 7: 1503–19

Hohl-Abrahao JC, Creutzfeldt OD. 1991. Topographic mapping of the thalamocortical projections in rodents and comparison with that in primates. *Exp. Brain Res.* 87:283–94

Honig MG, Hume RI. 1989. DiI and DiO: versatile fluorescent dyes for neuronal labelling and pathway tracing. *Trends Neurosci.* 12(9):333–39

Horton HL, Levitt P. 1988. A unique membrane protein is expressed on early developing limbic system axons and cortical targets. *J. Neurosci.* 8:4653–61

Huntley GW, Jones EG. 1991. The emergence of architectonic field structure and areal borders in developing monkey sensorimotor cortex. *Neuroscience* 44:287–310

Iijima N, Oohira A, Mori T, Kitabatake K, Koshaka S. 1991. Core protein of chondroitin sulfate proteoglycan promotes neurite outgrowth from cultured neocortical neurons. *J. Neurochem.* 56:706–08

Innocenti GM. 1981. Growth and reshaping of axons in the establishment of visual callosal connections. *Science* 212:824–27

Ivy GO, Killackey HP. 1982. Ontogenetic changes in the projections of neocortical neurons. *J. Neurosci.* 2:735–43

Jacobson M. 1991. *Developmental Neurobiology.* New York: Plenum

Jeanmonod D, Rice FL, Van der Loos H. 1991. Mouse somatosensory cortex: alterations in the barrelfield following receptor injury at different early postnatal ages. *Neuroscience* 6:1503–35

Jhaveri S, Erzurumlu RS, Crossin K. 1991. Barrel construction in rodent neocortex: role of thalamocortical afferents versus extracellular matrix molecules. *Proc. Natl. Acad. Sci. USA* 88:4489–93

Johnston JG, Van der Kooy D. 1988. Protooncogene expression identifies a transient columnar organization of the forebrain within the late embryonic ventricular zone. *Proc. Natl. Acad. Sci. USA* 86:1066–70

Jones EG. 1984. Laminar distributions of cortical efferent cells. In *Cerebral Cortex,* ed. A Peters, EG Jones, 1:521–53. New York: Plenum

Katz LC, Callaway EM. 1992. Development of local circuits in mammalian visual cortex. *Annu. Rev. Neurosci.* 15:31–56

Keilhauer G, Faissner A, Schachner M. 1985. Differential inhibition of neurone-neurone, neurone-astrocyte and astrocyte-astrocyte adhesion by L1, L2 and N-CAM antibodies. *Nature* 316: 728–30

Killackey HP, Chiaia NL, Bennett-Clarke CA, Eck M, Rhoades RW. 1993. The mutability of somatotopic representations in the rat somatosensory cortex. *J. Neurosci.* In Press

Killackey HP, Dawson DR. 1989. Expansion of the central hindpaw representation following fetal forelimb removal in the rat. *Eur. J. Neurosci.* 1:210–21

Killackey HP, Leshin S. 1975. The organization of specific thalamocortical projections to the posteromedial barrel subfield of the rat somatic sensory cortex. *Brain Res.* 86: 469–72

Kostovic I, Molliver ME. 1974. A new interpretation of the laminar development of cerebral cortex: synaptogenesis in different layers of neopallium in the human fetus. *Anat. Rec.* 178:395

Kostovic I, Rakic P. 1984. Development of prestriate visual projections in the monkey and human fetal cerebrum revealed by transient cholinesterase staining. *J. Neurosci.* 4:25–42

Krist DA. 1979. Development of neocortical circuitry: histochemical localization of acetylcholinesterase in relation to the cell layers of rat somatosensory cortex. *J. Comp. Neurol.* 186:1–16

Lagenaur C, Lemmon V. 1987. An L1-like molecule, the 8D9 antigen, is a potent substrate for neurite extention. *Proc. Natl. Acad. Sci. USA* 84:7753–57

Livingstone MS, Hubel DH. 1984. Anatomy and physiology of a color system in the primate visual cortex. *J. Neurosci.* 4:309–56

Lochter A, Vaughan L, Kaplony A, Prochiantz A, Schachner M, Faissner A. 1991. J1/tenascin in substrate-bound and soluble form displays contrary effects on neurite outgrowth. *J. Cell Biol.* 113:1159–71

Lu S, Bogard LD, Murtha MT, Ruddle FH. 1992. Expression pattern of a murine homeobox gene, *Dbx*, displays extreme spatial restriction in embryonic forebrain and spinal cord. *Proc. Natl. Acad. Sci. USA* 89:8053–57

Lumsden A, Keynes R. 1989. Segmental patterns of neuronal development in the chick hindbrain. *Nature* 337:424–28

Lund RD, Mitchell DE, Henry GH. 1978. Squint-induced modification of callosal connections in cats. *Brain Res.* 144:169–72

Lund RD, Mustari MJ. 1977. Development of the geniculocortical pathway in rats. *J. Comp. Neurol.* 173:289–306

Luskin MB, Shatz CJ. 1985. Studies of the earliest generated cells of the cat's visual cortex: cogeneration of subplate and marginal zones. *J. Neurosci.* 5:1062–75

Marin-Padilla M. 1971. Early prenatal ontogenesis of the cerebral cortex (neocortex) of the cat (Felis domestica). A Golgi study. I. The primordial neocortical organization. *Z. Anat. Entwicklungsgesch.* 134:117–45

Marin-Padilla M. 1972. Prenatal ontogenetic story of the principal neurons of the neocortex of the cat (Felis domestica). Golgi study. II. Developmental differences and their significances. *Z. Anat. Entwicklungsgesch.* 136:125–42

McCandlish C, Waters RS, Cooper NGF. 1989. Early development of the representation of the body surface in S1 cortex barrelfield in neonatal rats as demonstrated with peanut agglutinin binding: evidence for differential development within the rattunculus. *Exp. Brain Res.* 77:425–31

McConnell SK. 1988. Development and decision-making in the mammalian cerebral cortex. *Brain Res. Rev.* 13:1–23

McConnell SK. 1989. The determination of neuronal fate in the cerebral cortex. *Trends Neurosci.* 12:342–49

McConnell SK, Kaznowski CE. 1991. Cell cycle dependence of laminar determination in developing cortex. *Science* 254:282–85

Metin C, Frost DO. 1989. Visual responses of neurons in somatosensory cortex of hamsters with experimentally induced retinal projections to somatosensory thalamus. *Proc. Natl. Acad. Sci. USA* 86:357–61

Miller B, Chou L, Finlay BL. 1993. The early development of thalamocortical and corticothalamic projections. *J. Comp. Neurol.* 335: 16–41

Miller B, Sheppard AM, Pearlman AL. 1992. Expression of two chondroitin sulfate proteoglycan core proteins in the subplate pathway of early cortical afferents. *Soc. Neurosci. Abstr.* 18:330.3

Misson J-P, Austin CP, Takahashi T, Cepko CL, Caviness VS, Jr. 1991. The alignment of migrating neural cells in relation to the murine neopallial radial glial fiber system. *Cerebral Cortex* 1:221–29

Molnar Z, Blakemore C. 1991. Lack of regional specificity for connections formed between thalamus and cortex in coculture. *Nature* 351:475–77

Monuki ES, Kuhn R, Weinmaster G, Trapp BD, Lemke G. 1990. Expression and activity of the POU transcription factor SCIP. *Science* 249:1300–3

Naegle JR, Jhaveri S, Schneider GE. 1988. Sharpening of topographical projections and maturation of geniculocortical axon arbors in the hamster. *J. Comp. Neurol.* 277:593–607

O'Leary DDM. 1989. Do cortical areas emerge from a proto-cortex? *Trends Neurosci.* 12: 400–6

O'Leary DDM. 1992. Development of connectional diversity and specificity in the mammalian brain by the pruning of collateral projections. *Curr. Op. Neurosci.* 2:70–77

O'Leary DDM, Borngasser D. 1992. Minimal dispersion of neuroepithelial cells and their progeny during generation of the cortical preplate. *Soc. Neurosci. Abstr.* 18:925

O'Leary DDM, Schlaggar BL, Stanfield BB. 1992. The specification of sensory cortex: lessons from cortical transplantation. *Exp. Neurol.* 115:121–26

O'Leary DDM, Stanfield BB. 1989. Selective elimination of axons extended by developing cortical neurons is dependent on regional locale: experiments utilizing fetal cortical transplants. *J. Neurosci.* 9:2230–46

O'Leary DDM, Stanfield BB, Cowan WM. 1981. Evidence that the early postnatal restriction of the cells of origin of the callosal projection is due to the elimination of axonal collaterals rather than to the death of neurons. *Dev. Brain Res.* 1:607–17

O'Rouke NA, Dailey ME, Smith SJ, McConnell SK. 1992. Diverse migratory pathways in the developing cerebral cortex. *Science* 258:299–302

Peinado A, Katz LC. 1990. Development of cortical spiny stellate cells: retraction of a transient apical dendrite. *Soc. Neurosci. Abstr.* 16:1127

Porteus MH, Bulfone A, Ciaranello RD, Rubenstein JLR. 1991. Isolation and characterization of a novel cDNA clone encoding a homeodomain that is developmentally regulated in the ventral forebrain. *Neuron* 7: 221–29

Price M, Lemaistre M, Pischetola M, Di Lauro R, Duboule D. 1991. A mouse gene related to *Distal-less* shows a restricted expression in the developing forebrain. *Nature* 351: 748–51

Rakic P. 1977. Prenatal development of the visual system in rhesus monkey. *Philos. Trans. R. Soc. London Ser. B* 278:245–60

Rakic P. 1978. Neuronal migration and contact guidance in primate telencephalon. *Postgrad. Med. J.* 54:25–40

Rakic P. 1981. Developmental events leading to laminar and areal organization of the neocortex. In *The Organization of the Cerebral Cortex*, ed. FO Schmitt, FG Worden, FG Adelman, SG. Dennis, pp. 74–128. Cambridge: MIT Press

Rakic P. 1983. Geniculo-cortical connections in primates: normal and experimentally altered development. *Prog. Brain Res.* 58: 393–404

Rakic P. 1988. Specification of cerebral cortical areas. *Science* 241:170–76

Rakic P, Suner I, Williams RW. 1991. A novel cytoarchitectonic area induced experimentally within the primate visual cortex. *Proc. Natl. Acad. Sci. USA* 88:2083–87

Rauschecker JP, Tian B, Korte M, Egert U. 1992. Crossmodal changes in the somatosensory vibrissa/barrel system of visually deprived animals. *Proc. Natl. Acad. Sci. USA* 89:5063–67

Reinoso BS, O'Leary DDM. 1988. Develop-ment of visual thalamocortical projections in the fetal rat. *Soc. Neurosci. Abstr.* 14:1113

Reinoso BS, O'Leary DDM. 1990. Correlation of geniculocortical growth into the cortical plate with the migration of their layer 4 and 6 target cells. *Soc. Neurosci. Abstr.* 16:439

Robertson RT. 1987. A morphogenic role for transiently expressed acetylcholinesterase in developing thalamocortical systems? *Neurosci. Lett.* 75:259–64

Robertson RT, Mostamand F, Kageyama GH, Gallardo KA, Yu J. 1991. Primary auditory cortex in the rat: transient expression of acetylcholinestrase activity in developing geniculocortical projections. *Dev. Brain Res.* 58:81–95

Rockel AJ, Hiorns RW, Powell TPS. 1980. The basic uniformity in the structure of the neocortex. *Brain* 103:221–44

Schlaggar BL, De Carlos JA, O'Leary DDM. 1993a. Acetylcholinesterase as an early marker of the differentiation of dorsal thalamus in embryonic rats. *Dev. Brain Res.* 75:19–30

Schlaggar BL, Fox K, O'Leary DDM. 1993b. Postsynaptic control of plasticity in developing somatosensory cortex. *Nature* 364:623–26

Schlaggar BL, O'Leary DDM. 1991. Potential of visual cortex to develop arrays of functional units unique to somatosensory cortex. *Science* 252:1556–60

Schlaggar BL, O'Leary DDM. 1993. Patterning of the barrel field in somatosensory cortex with implications for the specification of neocortical areas. *Perspect. Dev. Neurobiol.* 1:81–91

Shatz CJ. 1977. Anatomy of interhemispheric connections in the visual system of Boston Siamese and ordinary cats. *J. Comp. Neurol.* 173:497–518

Shatz CJ, Luskin MB. 1986. Relationship between the geniculocortical afferents and their cortical target cells during development of the cat's primary visual cortex. *J. Neurosci.* 6:3655

Sheppard AM, Hamilton SK, Pearlman AL. 1991. Changes in the distribution of extracellular matrix components accompany early morphogenetic events of mammalian cortical development. *J. Neurosci.* 11:3928–42

Simeone A, Gulisano M, Acampora D, Stornaiuolo A, Rambaldi M, Boncinelli E. 1992. Two vertebrate homeobox genes related to the Drosophila empty spiracles gene are expressed in the embryonic cerebral cortex. *EMBO J.* 11:2541–50

Snow DM, Letourneau PC. 1992. Neurite outgrowth on a step gradient of chondroitin sulfate proteoglycan (CS-PG). *J. Neurobiol.* 23:322–36

Stanfield BB, O'Leary DDM. 1985a. The transient corticospinal projection from the occipital cortex during the postnatal development of the rat. *J. Comp. Neurol.* 238: 236–48

Stanfield BB, O'Leary DDM. 1985b. Fetal occipital cortical neurons transplanted to the rostral cortex can extend and maintain a pyramidal tract axon. *Nature* 313: 135–37

Stanfield BB, O'Leary DDM, Fricks C. 1982. Selective collateral elimination in early postnatal development restricts cortical distribution of rat pyramidal tract axons. *Nature* 298:371–73

Steindler DA, Cooper NGF, Faissner A, Schachner M. 1989a. Boundaries defined by adhesion molecules during development of the cerebral cortex: the JI/tenascin glycoprotein in the mouse somatosensory cortical barrel field. *Dev. Biol.* 131:243–60

Steindler DA, Faissner A, Schachner M. 1989b. Brain "cordones": transient boundaries of glia and adhesion molecules that define developing functional units. *Comments Dev. Neurobiol.* 1:29–60

Steindler DA, O'Brien TF, Laywell E, Harrington K, Faissner A, Schachner M. 1990. Boundaries during normal and abnormal brain development: in vivo and in vitro studies of glia and glycoconjugates. *Exp. Neurol.* 109:35–56

Stewart GR, Pearlman AL. 1987. Fibronectin-like immunoreactivity in the developing cerebral cortex. *J. Neurosci.* 7:3325–33

Sur M, Garraghty PE, Roe AW. 1988. Experimentally induced visual projections into auditory thalamus and cortex. *Science* 242: 1437–41

Suzuki N, Rohdewohld H, Neuman T, Gruss P, Scholar HR. 1990. Oct-6: a POU transcription factor expressed in embryonal stem cells and in the developing brain. *EMBO J.* 9:3723–32

Tao W, Lai E. 1992. Telencephalon-restricted expression of BF-1, a new member of the HNK-3 *forkhead* gene family, in the developing rat brain. *Neuron* 8:957–66

Tuttle R, O'Leary DDM. 1993. Cortical connections in cocultures. *Curr. Biol.* 3: 70–72

Tuttle R, Schlaggar BL, O'Leary DDM. 1993. Maturational changes in membrane associated molecules in the cortical plate may control development of thalamocortical and intracortical connections. *Soc. Neurosci. Abstr.* 19:1088

Walsh C, Cepko CL. 1992. Widespread dispersion of neuronal clones across functional regions of the cerebral cortex. *Science* 255: 434–40

Walsh C, Cepko CL. 1993. Clonal dispersion in proliferative layers of developing cerebral cortex. *Nature* 362:632–35

Walther C, Gruss P. 1991. *Pax-6*, a murine paired box gene, is expressed in the developing CNS. *Development* 113:1435–49

Waters RS, McCandlish CA, Cooper NGF. 1990. Early development of S1 cortical barrel subfield representation of forelimb in normal and deafferented neonatal rat as delineated by peroxidase conjugated lectin, peanut agglutinin (PNA). *Exp. Brain Res.* 81:234–40

Welker E, Van Der Loos H. 1986. Quantitative correlation between barrelfield size and the sensory innervation of the whiskerpad: a comparative study in six strains of mice bred for different patterns of mystacial vibrissae. *J. Neurosci.* 6:3355–73

Windrem MS, Finlay BL. 1991. Thalamic ablations and neocortical development: effects on cortical cytoarchitecture and cell number. *Cereb. Cortex* 1:220–40

Woodward WR, Chiaia N, Teyler TJ, Leong L, Coull BM. 1990. Organization of cortical afferent and efferent pathways in the white matter of the rat visual system. *Neuroscience* 36:393–401

Woolsey TA. 1990. Peripheral alteration and somatosensory development. In *Development of Sensory Systems in Mammals*, ed. Coleman, pp. 461–516. New York: Wiley

Woolsey TA, Van Der Loos H. 1970. The structural organization of layer IV in the somatosensory regions (S1) of mouse cerebral cortex: the description of a cortical field composed of discrete cytoarchitectonic units. *Brain Res.* 17:205–42

Woolsey TA, Wann JR. 1976. Areal changes in mouse cortical barrel following vibrissal damage at different postnatal ages. *J. Comp. Neurol.* 170:53–66

Xu Y, Baldassare M, Fisher P, Rathbun G, Oltz EM, et al. 1993. LH-2: a LIM/homeodomain gene expressed in developing lymphocytes and neural cells. *Proc. Natl. Acad. Sci. USA* 90:227–31

Yamamoto N, Kurotani T, Toyama K. 1989. Neural connections between the lateral geniculate nucleus and visual cortex in vitro. *Science* 245:192–94

Yamamoto N, Yamada K, Kurotani T, Toyama K. 1992. Laminar specificity of extrinsic cortical connections studied in coculture preparations. *Neuron* 9:217–28

Annu. Rev. Neurosci. 1994. 17:441–64

DIRECT G PROTEIN ACTIVATION OF ION CHANNELS?

David E. Clapham

Mayo Foundation, Guggenheim 7, Rochester, Minnesota 55905

KEY WORDS: ion channels, patch clamp, muscarinic gated potassium channel, $G_{\beta\gamma}$, G_α, GTP, calcium channels, acetylcholine

INTRODUCTION

Cells are protected from their external environment by a lipid bilayer that can exclude molecules as complex as proteins or as simple as ions. In order to communicate with the outside world, cells have developed signal transduction mechanisms in which molecules bind to specific surface receptors and activate membrane-associated second messengers. These messengers translate the surface-bound message into a cellular response. Vision, smell, taste, hormones, and neurotransmitters all work through a similar mechanism involving binding of a molecule to the extracellular surface of a transmembrane receptor that catalyzes the transfer of guanosine triphosphate (GTP) onto a heterotrimeric protein known as a GTP-binding protein, or G protein. These systems are common and extremely important. Including olfactory receptors, there are over 1000 types of G protein-linked receptors. Of the heterotrimeric G proteins, there are at least 21 G_α (17 gene products), 4 G_β, and 7 G_γ subunits. Ion channels also comprise a diverse array of proteins; hundreds of channels with distinct characteristics have been identified.

This review focuses on the evidence for direct activation of ion channels by G proteins, not on the involvement of G protein–linked receptors in ion channel activation through other more indirect routes, such as phosphorylation. I do not repeat the content of several recent comprehensive reviews on G proteins (Bourne et al 1991, Clapham & Neer 1993, Hepler & Gilman 1992, Kaziro et al 1991, Neer & Clapham 1988, Simon et al 1991, Spiegel et al 1991, 1992) and G protein gating of ion channels (Armstrong & White

1992, Breitwieser 1991, Brown 1991, Dunlap et al 1987, Hille 1992, Houslay 1987, Kurachi et al 1992, Rosenthal et al 1988, Sternweis & Pang 1990). Instead I review basic information needed to understand the issues and then focus on the key question addressed in the title. As we shall see, there is currently little convincing evidence for direct gating of ion channels by G protein subunits. This is not to say that direct gating does not occur, even commonly, in nature; it only points out the dearth of convincing experiments that address this point.

RECENT DEVELOPMENTS IN G PROTEINS

G protein–linked receptors have evolved so that the receptors are activated by agents such as photons, proteases, peptides, and small molecules. As they are homologous with bacteriorhodopsin, they likely form seven transmembrane-spanning helices with an extracellular molecular morphology adapted to ligand binding and an intracellular structure adapted to G protein binding. (Although the seven-transmembrane motif is most common, we now have examples of other receptor structures that may interact with the heterotrimer.) Upon binding agonist, the receptor undergoes a conformational change that allows it to activate the $G_{\alpha\beta\gamma}$ heterotrimer (Figure 1). (Pertussis toxin disrupts $G_{\alpha i}$, $G_{\alpha o}$, and $G_{\alpha t}$ interaction with the receptor by catalyzing the transfer of ADP ribose to the carboxy terminus of these G_α subunits.) The G_α subunit has a higher affinity for $G_{\beta\gamma}$ in the GDP-liganded state, but upon association of the heterotrimer with the activated receptor in the presence of magnesium, GTP replaces GDP on the G_α subunit, and the G_α and $G_{\beta\gamma}$, subunits presumably dissociate. This reaction scheme shows the key importance of G_α because G_α is itself a GTPase, and thus it times the separation of G_α and $G_{\beta\gamma}$. As described below, G_α and $G_{\beta\gamma}$ both activate numerous effectors in in vitro assays. Cells contain ~100 μM GTP and ~10 μM GDP with K_ds for

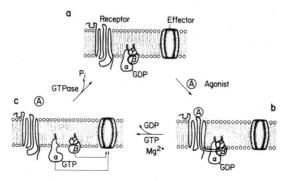

Figure 1   Conceptual model of the G protein cycle.

association with G_α in the nanomolar range. Thus, GTP and GDP are at saturating concentrations and probably do not limit exchange rates (Bourne et al 1991).

The separation of G_α and $G_{\beta\gamma}$ is the key activation step. The amounts of active G_α and $G_{\beta\gamma}$ depend directly on the rate of GTP exchange for GDP ($k_{GTP\text{-}GDP \text{ exchange}}$) and inversely on the rate of GTP hydrolysis (k_{cat}). Unlike the small G proteins, the heterotrimeric G_α subunits contain their own GTPase activity, which may be enhanced by interaction with some effectors. Excess GDP does not drive the pathway in the reverse direction (GDP replacing GTP on G_α before GTPase activity can be completed) for four reasons. First, the cell has an excess of GTP over GDP. Second, the G_α-GDP subunit preferentially binds $G_{\beta\gamma}$ so that the G_α-GDP-$G_{\beta\gamma}$ heterotrimer can bind to the receptor. Third, $G_{\beta\gamma}$ has a higher affinity for G_α-GDP than for G_α-GTP. Finally, the G_α-GTP rapidly dissociates from $G_{\beta\gamma}$. Thus, the relative amounts of active subunits, G_α-GTP and $G_{\beta\gamma}$, are set by the G_α subunit GTP hydrolysis rate, receptor catalysis of GTP exchange for GDP, and GTPase activating properties (GAP) of the effector. Once GTP binds G_α, the two active subunits are free to associate with effectors until GTP is hydrolyzed, presumably terminating the reaction by formation of the heterotrimer. The fate of individual subunits is far from understood, and swapping of $G_{\beta\gamma}$-G_α associations after effector binding has not been ruled out. Specific G_β subtypes associate preferentially with specific G_γ combinations. For example $G_{\beta1}$ can associate with both $G_{\gamma1}$ and $G_{\gamma2}$; $G_{\beta2}$ can associate with $G_{\gamma2}$ but not with $G_{\gamma1}$,

Table 1   G protein subunits

G_α	G_β	$G\gamma$	$G_{\beta\gamma}$
$G_{\alpha s}$	$G_{\beta1}$	$G_{\gamma1}$	28 unique $G_{\beta\gamma}$ combinations are
$G_{\alpha s1,2}$	$G_{\beta2}$	$G_{\gamma2}$	possible. However, there are
$G_{\alpha olf}$ (olfactory)	$G_{\beta3}$	$G_{\gamma3}$	preferred associations between
	$G_{\beta4}$	$G_{\gamma4}$	β and γ and thus not all combina-
$G_{\alpha i}$		$G_{\gamma5}$	tions are made[a]
$G_{\alpha i1,2,3}$		$G_{\gamma6}$	
$G_{\alpha oA,B}$		$G_{\gamma7}$	
$G_{\alpha Z}$			
$G_{\alpha+1,+2}$			
(transducins)			
$G_{\alpha g}$ (gustducin)			
$G_{\alpha q}$			
$G_{\alpha 15/16}$			
$G_{\alpha q,11,14}$			
$G_{\alpha 12}$			
$G_{\alpha 12,13}$			

[a] See Clapham & Neer (1993).

and $G_{\beta 3}$ cannot associate with either $G_{\gamma 1}$ or $G_{\gamma 2}$ (Clapham & Neer 1993). The G_γ subunit apparently is vital in determining the $G_{\beta\gamma}$ subunit's specific function. Specific $G_{\beta\gamma}$ and G_α subtypes are probably preassociated within receptor-effector geographical domains, limiting the number of possible outcomes for receptor–G protein–effector interactions.

Table 1 shows the current status of identified G_α and $G_{\beta\gamma}$ subunits. Sequence homology indicates there are four families of G_α subunits. Heterotrimeric G_α subunits are 39–46 kDa in size. They contain ~400 amino acids, bind and hydrolyze GTP, and are required for G protein interaction with a receptor. A detailed theoretical model of G_α structure (Berlot & Bourne 1992) gives some idea of structure-function relations, but G_α crystallography data is pending. ADP-ribosylation of the $G_{\alpha i}$ family is catalyzed by pertussis toxin at the carboxyl cysteine, disrupting receptor interaction. Some members of both the $G_{\alpha s}$ and $G_{\alpha i}$ families have cholera-toxin mediated ADP ribosylation at a site in the GTPase domain, disrupting hydrolysis of G_α-bound GTP and locking the dissociated subunits in their active states. Some of the G_α subunits are posttranslationally modified. ($G_{\alpha i}$ is myristoylated at the amino terminus.) These fatty acyl modifications help attach the G_α subunit to the membrane but may be even more important in defining protein-protein interactions (Spiegel et al 1991).

The $G_{\beta\gamma}$ subunit is a dimer under native conditions (Neer 1992). The four known mammalian G_β subunits share over 80% amino acid identity; each is 340 amino acids in length. The G_β subunit is composed of a putative α-helical amino terminal region that may participate in coil-coil interactions (Lupas et al 1992), followed by seven repeating units of approximately 40 amino acids, each rich in tryptophan and aspartate [WD40 (Simon et al 1991)]. The G_γ subunit appears to interact with the amino-terminal portion of G_β. At present, seven different G_γ subunits have been identified on the basis of cloning or peptide sequencing of purified proteins (Cali et al 1992). The G_γ subunits are ~75 amino acids long. They are posttranslationally modified at the carboxyl terminus by cleavage of the three carboxyl terminal amino acids and subsequent methylation and prenylation of the carboxy terminal cysteine. The retinal G_γ subunit, $G_{\gamma 1}$, is farnesylated, while $G_{\gamma 2}$ (common in brain) and others are probably geranyl geranylated (Spiegel et al 1991).

G PROTEIN FUNCTION

Although the number of identified G protein G_α and $G_{\beta\gamma}$ subunits has increased rapidly, largely because of advances in molecular cloning techniques, functional identification has been slow. In early studies of G protein function, G_α subunits were named by function, hence $G_{\alpha s}$ for stimulation of adenylyl cyclase, and $G_{\alpha i}$ for inhibition of adenylyl cyclase. Little is known about the detailed molecular interactions in effector stimulation in purified reconstituted systems, and even less about how G protein subunits activate effectors in the cell membrane. Rather than reiterate the evidence for G_α activation of cGMP phosphodiesterase, adenylyl cyclase, and phospholipase Cβ (the three systems in which purified components have been reconstituted by numerous laboratories), a more detailed review of the new evidence for $G_{\beta\gamma}$ activation of effectors follows.

Since the first description of the ability of $G_{\beta\gamma}$ to activate an effector was described in 1987 (Logothetis et al 1987), the list of targets for this protein has grown steadily. Now investigators know of as many or more effectors that are activated by $G_{\beta\gamma}$ as those activated by G_α. That $G_{\beta\gamma}$ can activate some effectors does not change the key role that G_α has in determining the rate of GTP hydrolysis. The formation of a heterotrimer of $G_{\beta\gamma}$ and GDP-liganded G_α subunits inactivates both subunits. Thus, activation by $G_{\beta\gamma}$ can be blocked by many manipulations affecting G_α, for example by pertussis toxin treatment and resulting ADP-ribosylation of the G_α subunit.

The first effector proposed to be activated by $G_{\beta\gamma}$ was the cardiac muscarinic-gated K^+ channel as discussed below. However, soon thereafter Jelsema & Axelrod (1987) suggested that $G_{\beta\gamma}$ subunits activated phospholipase A_2 (PLA$_2$), based on the observed reconstitution of purified retinal $G_{\beta\gamma}$ with G protein–depleted retinal membranes. Subsequently, strong genetic evidence for an active role for $G_{\beta\gamma}$ in cellular signaling came from the yeast pheromone system. Null mutations in the β and γ yeast homologues (STE4 and STE18) disrupted the mating and pheromone response while disruption of the yeast G_α homologue, SCG1 or GPA1, led to a constitutive mating response signal. These data were interpreted to mean that the yeast G_α homologue is a negative regulator of $G_{\beta\gamma}$ and that $G_{\beta\gamma}$ constitutes the direct signaling element (Whiteway et al 1989), to a downstream protein kinase (Leberer et al 1992).

Adenylyl cyclase has at least six cloned isoforms. Purification of the adenylyl cyclase recombinant subtypes allowed Tang & Gilman (1991) to test the effect of individual subunits on defined enzyme preparations. Type I adenylyl cyclase, the calmodulin-sensitive enzyme from the brain, is activated by GTPγS-liganded $G_{\alpha s}$ and inhibited by $G_{\beta\gamma}$. However, $G_{\alpha s}$ and $G_{\beta\gamma}$ synergistically activated type II adenylyl cyclase (Tang & Gilman 1991). Type IV adenylyl cyclase, like type II, is activated by $G_{\beta\gamma}$, but $G_{\beta\gamma}$ neither activates nor inhibits type III (Tang & Gilman 1992). Different $G_{\beta\gamma}$ isoforms have different degrees of effectiveness in activating the type II adenylyl cyclase.

Phospholipase C (PLC), which splits phosphatidylinositol 4,5-bisphosphate (PIP$_2$) into inositol trisphosphate (InsP$_3$) and diacylglycerol, has several isoforms. The $G_{\alpha q}$ class ($G_{\alpha q}$, $G_{\alpha 11}$, $G_{\alpha 16}$) subunits activate the PLCβ1 isoform

(Lee et al 1992, Smrcka et al 1991, Taylor et 1991). Giershik and colleagues (Camps et al 1992) showed that retinal transducin $G_{\beta 1 \gamma 1}$ could activate soluble PLCβ2. Blank et al (1992) demonstrated that a cytosolic PLC enzyme from liver could be activated 50-fold by $G_{\beta \gamma}$, and Camps et al (1992) showed that both PLCβ1 and PLCβ2 could be activated by exogenous transducin $G_{\beta \gamma}$ subunits, although PLCβ2 was much more sensitive to stimulation by $G_{\beta 1 \gamma 1}$ than was PLCβ1. Current evidence suggests that $G_{\beta \gamma}$ activates PLCβ3>PLCβ2>PLCβ1 while $G_{\alpha q}$ activates PLCβ1> PLCβ2>PLCβ3. Only $G_{\alpha q}$ family G_{α} activates PLCβ4 (reviewed in Clapham & Neer 1993).

An enzyme that phosphorylates the muscarinic receptor is activated by brain $G_{\beta \gamma}$ (Haga & Haga 1992). Similarly, the agonist-dependent phosphorylation of purified and reconstituted β_2 adrenergic receptor by β adrenergic receptor kinase (βARK) is enhanced ~10-fold by $G_{\beta \gamma}$ (Pitcher et al 1992). $G_{\beta \gamma}$ associates with the carboxyl terminus of βARK and participates in the translocation of the kinase to the membrane where it can phosphorylate the receptor (Inglese et al 1992).

In conclusion, (a) $G_{\beta \gamma}$ can regulate receptor function directly and indirectly, and (b) both $G_{\beta \gamma}$ and G_{α} independently activate downstream effectors. Although not discussed here, effectors may also enhance G_{α} GTPase activity. A recent review of $G_{\beta \gamma}$ (Clapham & Neer 1993) discusses new models for G protein action.

DEFINITION OF DIRECT ION CHANNEL GATING

Two criteria evince G protein gating of ion channels. First, data should show that G protein–linked receptors regulate the effector and that this regulation can also be evoked by agents that lock the G protein in the active state, even without receptor binding (GTPγS, AlF_4^-). Additional evidence for this point may be provided by pertussis or cholera toxin modification of $G_{\alpha o}$, $G_{\alpha i}$, $G_{\alpha t}$, or $G_{\alpha s}$ subtypes, again bypassing the receptor. However, since all ion channels are also modulated by indirect phosphorylation pathways (Figure 2) and G proteins are often involved in these pathways, direct gating of ion channels by G proteins requires further proof. The second criterion is that the purified channel should be gated by a purified G protein subunit in a simple system. Direct physical association between the G protein and the channel must be demonstrated. This second criterion has not been reproducibly met for any ion channel. Thus proponents of direct G protein gating were forced to use the term membrane-delimited G protein gating, which was coined to summarize the limits of patch clamp measurements of ion channels, a point that requires some explanation.

With patch clamp measurements (Hammill et al 1981), one electrically

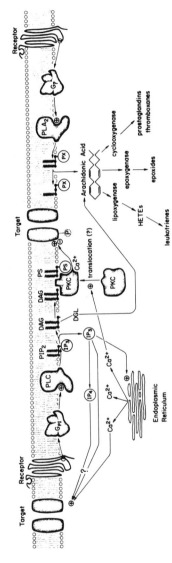

Figure 2 G protein–dependent pathways that may be involved in membrane-delimited channel modulation and thus be difficult to separate from direct G protein activation. PI, phosphatidylinositol; PLA$_2$, phospholipase A$_2$; PS, phosphatidylserine; IP$_3$, inositol (1,4,5)tris phosphate; IP$_4$, inositol (1,3,4,5)tetrakisphosphate; DAG, diacylglycerol; DGL, diacylglycerol lipase; PKC, protein kinase C; PIP$_2$, phosphatidylinositol 4,5-bisphosphate; PLC, phospholipase C, also known as phosphoinositidase C (PIC); HETE, hydroxyeicosatetraenoic acid. From Clapham 1990.

isolates a small patch of membrane that is either part of the intact cell (cell-attached patch) or a piece of membrane excised from the cell (inside-out, outside-out patch). Whole-cell variants of the patch clamp technique allow one to measure from the entire cell surface. This is done while perfusing the cell with the entire contents of the pipette (traditional whole-cell) or with only monovalent ions from the pipette (permeabilized patch). In any of these configurations, the purified G protein is introduced into a complex system of associated membrane structures, native G proteins, and ion channels. Following observations of nonhydrolysable GTP analogue stimulation of the cardiac muscarinic-gated inwardly rectifying K^+ channel ($I_{K.ACh}$) by Breitwieser & Szabo (1985) and Pfaffinger et al (1985), Logothetis et al (1987) tested the action of purified G proteins on the channel. Application of purified G protein subunits to the cytoplasmic surface of an inside-out patch of cardiac atrial membrane dramatically increased ion channel opening ($>$ 500-fold). This behavior is one to two orders of magnitude more robust than the enhancement of channel opening achieved by application of purified G proteins to other ion channels. However, the inside-out patch is not a simple lipid bilayer–ion channel system, although it does rapidly dilute freely diffusable ions such as ATP and GTP from the inside surface of the patch. The patch may vary in size from ~ 1 to 10 μm^2 and frequently contains intracellular surface cytoskeletal elements (Sokabe & Sachs 1990). The ion channel protein itself is only ~ 10 nm in diameter, making it at best 0.1% of the total patch area, which is analogous to a doughnut placed in a football field. Many ion channels are expressed at low densities (e.g. roughly 1 μm^{-2}, or ~ 2000 channels per cell for many 10- to 20-μm-diameter cells). Receptor densities vary widely, but 200,000 receptors per cell (100 μm^{-2}) is not unusual. G protein densities may be even higher, constituting up to 1% (growth cones) of the total membrane protein, with densities greater than 1000 μm^{-2}. Numerous other proteins reside in this same area. Thus, far from being a single ion channel in a field of lipid, the patch is more accurately depicted as a mini-cell with a leakage pathway allowing rapid washout of hydrophilic elements. The ion channel represents a distinct minority in a membrane crowded with signaling components.

Many studies show that ion channels are phosphorylated through protein kinase A and protein kinase C pathways, much like many other proteins containing serine or threonine consensus phosphorylation sites. A smaller, but well-established, body of evidence indicates that arachidonic acid and its metabolites also modify channel gating. Protein kinases A and C, and phospholipase A_2, are activated by receptor- and G protein–stimulated pathways (Figure 2). These levels of channel modification are usually modest, increasing mean channel activity two- to fivefold. Indirect stimulation of ion channels via these phosphorylation and lipophilic mediator routes should not

Table 2 Putative G protein regulated ion channels

Gating action	Ion channel current	Tissue
Probable significant direct G protein gating	$I_{K.ACh}$ (I_{GIRK1})	Cardiac, brain, ATT20, GH3
	I_{CaN}	Neuronal (DRG, sympathetic, NG108)
Predominantly gated by other mechanisms (but may be modulated by direct G protein route)	I_{CaL}	Cardiac, neuronal
	$I_{K.ATP}$	Cardiac
	I_{Na}	Cardiac
Postulated direct G protein gating	Amiloride-sensitive I_{NA}	Kidney epithelia
	I_{Cl}	Skeletal
	I_{Cl} (CFTR)	Airway epithelia
	Cation channel	Kidney collecting duct

be confused with direct G protein stimulation. Although this is a rather straightforward conceptual point, it is not easily established experimentally. With rare exceptions, discussed below, all experiments testing the direct G protein–ion channel hypothesis have done so using complex systems of membranes and proteins. Table 2 lists potential direct G protein–gated ion channels. By the criteria given above, the data are inadequate to conclude that any channels are directly G protein gated, and at least for one, the evidence to the contrary is substantial.

CHANNEL REGULATION

K^+ Channels

$I_{K.ACh}$ Acetylcholine is released by the vagus nerve onto pacemaker cells of the sino- and atrioventricular nodes and the atria. Acetylcholine binds muscarinic type 2 (m2) receptors on these cells to activate a pertussis toxin–sensitive G protein, which in turn gates an inwardly rectifying K^+ channel ($I_{K.ACh}$). Recently $I_{K.ACh}$ was cloned (Kubo et al 1993) and labelled I_{GIRK1}. This and other mechanisms cause heart rate to slow. In cell-attached patches, Noma & Soejima (1984) discovered that $I_{K.ACh}$ was activated when ACh was in the pipette but not when ACh was applied to the bath and isolated from the channel by the pipette glass. He suggested that the second-messenger pathway was limited to the membrane or to a region close to the channel. Pfaffinger et al (1985) and Breitweiser & Szabo (1985) established in whole-cell recordings that $I_{K.ACh}$ was activated on intracellular perfusion with nonhydrolyzable GTP analogues and that this pathway was sensitive to pretreatment by pertussis toxin. This K^+ channel is easily identified in cardiac atrial cells, either in cell-attached or excised inside-out patches of membrane (Sakmann et al 1983). Similarly, many neurons have a pertussis toxin–sensitive, G protein–gated inwardly rectifying K^+ channel. These channels are activated by m2 muscarinic, D2 dopamine, serotonin, histamine, opioid, somatostatin, and α_2 adrenergic receptors to produce inhibitory postsynaptic potentials (North 1989).

Logothetis et al (1987) first proposed that $G_{\beta\gamma}$ subunits had a direct role in effector function when they reported that $G_{\beta\gamma}$ purified from bovine brain activated the cardiac muscarinic-gated K^+ channel ($I_{K.ACh}$). Subsequently, Codina et al (1987) showed that the G_α subunit (G_α–GTP$_\gamma$S) activated the same channel (Mattera et al 1989; see also Yatani et al 1987a). At the time many thought these two findings were contradictory. In particular, Birnbaumer & Brown (1988) argued that $G_{\beta\gamma}$ activation was an artifact caused by detergent, despite the numerous controls on this point previously reported by Logothetis et al (1987). This attempt to simplify G protein activation into an either/or situation was unnecessary. Now several effector systems are known to be activated by both G_α and $G_{\beta\gamma}$, as detailed above. Logothetis et al (1988) demonstrated apparent activation by both subunits (see also Ito et al 1991). However, the in vivo mechanism cannot be determined from any of these studies alone, and artifacts, such as GTP$_\gamma$S from α to activate native G proteins must be eliminated. Also, without reconstitution of purified channel and purified G protein, there is no method to determine if a direct interaction between proteins occurs. Reconstitution of the purified channel can now be attempted since it has been cloned (Kubo et al 1993).

In excised patches, the cytoplasmic face may be perfused with GTP analogues or purified $G_{\beta\gamma}$ or G_α subunits, resulting in large (500-fold) increases in channel activity (Logothetis et al 1987). Activation by G_α and $G_{\beta\gamma}$ is not additive. The $G_{\beta\gamma}$ and G_α subunits activate the channel at different concentrations (Codina et al 1987, Ito et al 1992, Logothetis et al 1987, 1988): G_α activates at lower concentrations but less consistently than $G_{\beta\gamma}$. $G_{\beta\gamma}$ activates 95% of patches while G_α only activates ~30% of patches (Figure 3) (Ito et al 1992, Logothetis et al 1988, Nanavati et al 1990b). Channel activation was observed with $G_{\beta\gamma}$ purified from bovine brain, rat brain, and placenta, and with recombinant $G_{\beta\gamma}$. Concentrations of transducin $G_{\beta\gamma}$ 10- to 100-fold higher were required for activation (Logothetis et al 1988). G_α subunit activated the channel at lower concentrations than $G_{\beta\gamma}$ ($K_d \approx 10–100$ pM). However, the concentration of $G_{\beta\gamma}$ that activates the K^+ channel (10 nM) is similar to the concentration subsequently shown to regulate other effectors. Also, G_α-GTP$_\gamma$S, not the physiologically relevant and hydrolyzable G_α-GTP concentrations, are compared to $G_{\beta\gamma}$ concentrations in published work. Concerns about contamination of $G_{\beta\gamma}$ by G_α subunits, or artifactual

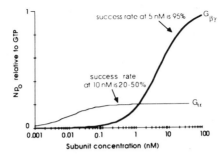

Figure 3   Dose response curves comparing activation by G_α vs $G_{\beta\gamma}$. Not only does G_α not activate $I_{K.ACh}$ as completely (on average) as $G_{\beta\gamma}$, it also only activates ~1/3 of patches in which it is tried. $G_{\beta\gamma}$ activates virtually all patches. Reprinted from Nanavati et al (1990).

activation by detergent (Brown & Birnbaumer 1988), have been addressed several times (Ito et al 1992, Logothetis et al 1988, Nanavati et al 1990b) and found to be groundless (Figure 4). Elegant recent experiments by Kurachi and colleagues (Ito et al 1992) showed that $G_{\beta\gamma}$ preferentially activated $I_{K.ACh}$ over G_α in reconstitutions in intact patches, while G_α preferentially activated another type of K^+ channel, $I_{K.ATP}$, in the same patch. This is an important

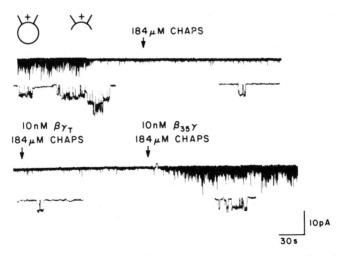

Figure 4   The $I_{K.ACh}$ channel in a neonatal rat atrial patch (cell-attached configuration) was activated by 10 μM ACh in the pipette (+). Channel activity ceased after the patch was excised into an inside-out configuration and cellular GTP diffused away. Addition of detergent (CHAPS), used to suspend $G_{\beta\gamma}$, did not activate the channel. Transducin $G_{\beta\gamma}$ (10 nM βγT) in the presence of CHAPS was also ineffective. Placental βγ(35kD) (10 nM) then activated IK.ACh (as does bovine brain $G_{\beta\gamma}$). Reprinted from Logothetis et al (1988).

result because it showed that both $G_{\beta\gamma}$ and G_α subunits were active while controlling for nonspecific effects.

If an antibody could be made to block the effector site on a G_α subunit without affecting G_α-$G_{\beta\gamma}$ dissociation or GTPase activity, one could test whether muscarinic receptor stimulation depended on G_α activity. In such an experiment, Yatani et al (1988a) found that Mb4A, a monoclonal antibody directed against transducin G_α (Rarick et al 1992), when perfused onto inside-out patches blocked carbachol-stimulated $I_{K.ACh}$ activity. However, Mb4A antibody inhibits the interaction of both G_α and $G_{\beta\gamma}$ with the receptor by simultaneous steric blockade (Navon & Fung 1988) and would thus interfere with normal G protein cycling when the effect depends on carbachol stimulation. This is not a test for G_α-$I_{K.ACh}$ interaction.

$I_{K.ACh}$ is truly a unique channel; it is readily activated by at least eight receptors [muscarinic type 2 (m2), somatostatin, adenosine, α_1, platelet activating factor, neuropeptide Y, cGRP, and endothelin receptors], apparently through a pertussis toxin–sensitive G protein in cardiac atria alone. As shown above, although $G_{\beta\gamma}$ activates the channel more reliably than G_α, both proteins seem to activate it. Furthermore, arachidonic acid, 12-HPETE, 8 HETE, and the leukotrienes, LTD4 and LTC4, also activate the channel (Kim et al 1989, Kurachi et al 1989). Finally, the small G protein *ras*, complexed with GTPase-activating protein (*ras* GAP) inhibited GTP-dependent channel opening. GAP deletion mutant experiments were interpreted to mean that *ras* p21 induced a conformational change in GAP that allowed its SH2-SH3 (src homology domains) to bind to a membrane component and uncouple receptor and G protein. This finding suggested that GAP was downstream of *ras* (Martin et al 1992, Yatani et al 1990b). One can imagine several other more complex models for $I_{K.ACh}$ activation by multiple receptors and G protein subunits given the putative interaction with *ras* GAP. These interesting results still await confirmation in other laboratories.

Kim et al (1989) suggested that $G_{\beta\gamma}$ activated $I_{K.ACh}$ via phospholipase A_2 because an antibody shown to block PLA_2 activity in other systems blocked $G_{\beta\gamma}$ activation of $I_{K.ACh}$. In support of this hypothesis, arachidonic acid and metabolites activated the channel (Kim et al 1989, Kurachi et al 1989). However, arachidonic acid never activated $I_{K.ACh}$ as rapidly or as well as $G_{\beta\gamma}$ (Kim et al 1989). Kurachi et al (1989) found the channel was activated by arachidonic acid only in cell-attached patches and proposed that PLA_2 products merely enhanced heterotrimer dissociation. Scherer & Breitwieser (1990) have shown the LTD4, a product of arachidonic acid, activates the channel and suggest a direct effect on G protein turnover. Still under debate is whether $G_{\beta\gamma}$ activates $I_{K.ACh}$ directly as originally proposed (Logothetis et al 1987), indirectly via PLA_2 pathways (Kim et al 1989), or both. Our current information leads us to hypothesize direct $G_{\beta\gamma}$ activation of $I_{K.ACh}$ and indirect activation of the channel through PLA_2 pathways.

$I_{K.ATP}$ Kirsch et al (1990) studied the regulation of the ATP-sensitive K^+ channel of rat ventricular myocytes by excising patches and applying purified G_α subunits to the intracellular surface. They found that 200 pM human erythrocyte $G_{\alpha i1,2,3}$ mimicked GTPγS while bovine brain $G_{\alpha o}$ and $G_{\alpha s}$ did not. Ito et al (1992) found that 100 pM bovine brain $G_{\alpha i1}$ and $G_{\alpha i2}$ increased channel activity 12- to 15-fold over the background, while $G_{\alpha o}$ increased channel activity in only two of five patches, and $G_{\beta\gamma}$ did not activate the channel. Regulation of $I_{K.ATP}$ is otherwise complex because it is phosphorylated, inhibited by direct ATP binding, and modulated by nucleotide diphosphates (Ashcroft & Ashcroft 1990).

I_{KCa} AND OTHERS   Vaca et al (1992) postulated that bradykinin may activate the calcium-activated channel of vascular endothelium via a G protein–mediated change in the sensitivity of the channel for calcium. No purified G protein subunits were used but GTP sensitivity in excised patches suggested a membrane-delimited pathway.

Four partially characterized neuronal K channels were proposed to be directly activated by purified $G_{\alpha o}$ subunits (VanDongen et al 1988).

Ca^{2+} Channels

A growing number of voltage-dependent calcium channels have been recognized and are defined functionally as L, N, T, and P types. These correspond to the dihyrodropyridine-sensitive, high threshold (more depolarized), slowly inactivating calcium current (L); the high-threshold, ω-conotoxin–sensitive N type channel common in neurons; the low-threshold (more hyperpolarized), fast inactivating or transient (T) type current; and the more recently characterized high threshold ω-Aga-IVA-sensitive calcium current (Mintz et al 1992, Regan et al 1991) common in cerebellar Purkinje fibers (Tsien et al 1991). Several calcium channels have now been cloned, but the expressed clones have not yet been sufficiently characterized to clearly correlate all functional and cloned channels. Various voltage-dependent Ca^{2+} channels appear to use different (but homologous) α_1 subunits as the Ca^{2+}-selective pore. These subunits are variably associated with α_2, β, γ, and δ subunits whose functions are incompletely understood. For the purposes of this review, the evidence for G protein control of Ca^{2+} channels is separated into neuronal high-threshold (both L and N) and cardiac dihydropyridine-sensitive (L) type calcium channels. The field suffers from incomplete identification of the subtype of calcium current in much of the work in which G protein activation of channels has been proposed. Growing evidence indicates that neurons have fast and slow modes of neurotransmitter-mediated high threshold (both N and L type) calcium channel inhibition (Hille 1992). More controversial evidence suggests a fast mode of β-receptor mediated enhancement of the dihydropyridine-sensitive L current in heart cells.

NEURONAL HIGH THRESHOLD CALCIUM CHANNEL INHIBITION   Holz et al (1986) proposed that norepinephrine and GABA receptors activated G proteins to inhibit voltage-dependent calcium channels in embryonic chick dorsal root ganglion (DRG). Phorbol esters mimicked this response, and protein kinase C and a direct G protein pathway were suggested as possible mediators. Bean (1989) found that this inhibition did not result from a decrease in the number of functional channels or voltage dependence of inactivation, but rather from a change in the voltage-dependence of opening. A model in which separate G_α subunits bind to each of the four domains of the α_1 channel pore has been proposed (Boland & Bean 1993). Single channel recordings of frog sympathetic neuron N type channels revealed stabilization of particular patterns of channel opening (called modes) after norepinephrine activation (Delcour & Tsien 1993), which is consistent with alteration of gating behavior by a membrane-delimited pathway. Kasai & Aosaki (1989) suggested that adenosine inhibition of I_{CaN} in chick sensory neurons might involve direct action of G proteins on Ca channels, because phorbol ester, arachidonic acid, and nitroprusside had no effect. Lipscombe et al (1989) proposed that α-adrenergic inhibition of sympathetic neurotransmitter release was mediated by modulation of N-type channels via a membrane-delimited pathway. In NG108-15 cells, a mutant pertussis toxin–resistant $G_{\alpha o}$ subunit was stably expressed and used to rescue Leu-enk and NE receptor–dependent inhibition of the N type Ca^{2+} current. Somatostatin was not recoupled in this approach, and because it was previously established that bradykinin inhibited I_{CaN} through a non-PTX sensitive G protein, at least three different G protein subunit subtypes couple to I_{CaN} (Taussig et al 1992). In the superior cervical ganglion, Hille and coworkers (Beech et al 1992, Bernheim et al 1992, Mathie et al 1992) separated neurotransmitter-activated channel modulation pathways into pertussis toxin–sensitive and –insensitive fast and slow pathways, some of which involve unidentified diffusable second messenger pathways. Although fast mechanisms imply a membrane-delimited route because of diffusion constraints, one should remember that fast kinase routes may be activated near the membrane, as in the $G_{\beta\gamma}$-dependent activation of βARK (Pitcher et al 1992). Also, phosphorylation of a chromaffin cell L type channel apparently can occur in less than 50 ms (Artalejo et al 1990).

    Several groups have attempted to test the possibility of direct G protein interaction. In one of the first experiments reconstituting purified G protein subunits in an electrophysiological assay, Hescheler et al (1987) perfused purified, but inactive, G_α subunits into a pertussis toxin–treated neuroblastoma-glioma cell line and activated the G protein via the native enkephalin receptors. Concentrations effective at reconstituting the native inhibitory response were roughly 0.4 nM for $G_{\alpha o}$ and 4 nM for $G_{\alpha i}$. Unfortunately, the calcium current subtype was not characterized. $G_{\alpha o}$ (10 nM) and GTP perfused into rat dorsal root ganglion cells reconstituted the neuropeptide Y-mediated inhibitory effect on I_{CaL} current, even after PKC was downregulated by pretreatment with phorbol ester (Ewald et al 1988). Other G protein subunits were not tested. Dolphin and coworkers (Dolphin 1991, Dolphin et al 1988, Scott & Dolphin 1987, 1988) found a pertussis toxin–sensitive, GTPγS-induced increase in the rate of activation of I_{CaL}. Furthermore, internal GTPγS enhanced the stimulatory response to calcium channel antagonists. Some have interpreted these results as indicating a direct interaction of the G protein with the calcium channel, perhaps close to or affecting the dihydropyridine binding site, but clearly the exact mechanism is unknown. (Figure 5)

Regardless of the details of the effector pathway, receptor-mediated increases in calcium current show remarkable G protein subtype specificity. In a surprising series of experiments, nuclear microinjection of antisense oligonucleotides specific for G protein subunits was used to eliminate coupling between the muscarinic and somatostatin receptors and inhibition of calcium current in GH_3 cells. The specific subtype of calcium current being studied was not demonstrated. The evidence supports the specific participation of both $G_{\beta\gamma}$ and G_α in this pathway. First, Kleuss et al (1991) showed that $G_{\alpha o1}$ mediated the muscarinic-dependent, and $G_{\alpha o2}$ the somatostatin-dependent,

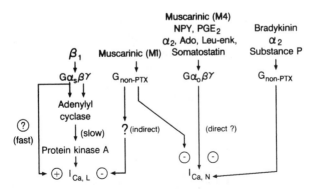

Figure 5 Summary of receptor–G protein–calcium channel pathways. NPY, neuropeptide Y; PGE2, prostaglandin E2; Ado, adenosine; Leu-enk, Leu-enkephalin. Modified after Hille (1992).

inhibition of the calcium current. Kleuss et al (1991) then demonstrated that antisense oligonucleotides for $G_{\beta 1}$ blocked Ca^{2+} channel inhibition by somatostatin receptors, while antisense oligonucleotides for $G_{\beta 3}$ selectively blocked inhibition of Ca^{2+} current by muscarinic receptors (Kleuss et al 1992). These investigators interpreted these results in light of the standard G protein dogma; knocking out G_α blocked the effector pathway while eliminating $G_{\beta\gamma}$ simply disrupted the G protein cycle. However, elimination of either G_α or $G_{\beta\gamma}$ completely blocks the Ca^{2+} channel inhibition, and we cannot conclude which parts of the G protein cycle and effector interactions are disrupted. Different $G_{\beta\gamma}$ subunits may also specify which receptor activates which $G_{\alpha o}$ subtype, effectively presorting subunits to specific receptors. Results of the experiments in intact cells do not allow resolution of these questions.

NEURONAL AND CARDIAC L TYPE CHANNEL ENHANCEMENT   Fenwick et al (1982a,b) found that depolarization beyond 0 mV facilitated subsequent calcium currents in adrenal chromaffin cells. Artalejo et al (1992) showed that a normally silent L type calcium current is activated by strong pre-depolarization steps and that this rapid voltage-dependent facilitation depends on phosphorylation by a protein kinase. Dopamine (D1) receptor activation mimicked the same potentiation. Support for a shift of L type calcium channel gating comes from the data of Pietrobon & Hess (1990), who found that strong depolarization shifts single L type channel behavior into a long opening mode. Yue et al (1990) also found that adrenergic receptor stimulation led to enhancement of the long opening mode.

The initial evidence for direct G protein activation of cardiac I_{CaL} relied largely on patch reconstitution. First, Yatani et al (1987b) showed that $G_{\alpha s}$ could restore activity to rundown patches containing I_{CaL}. Second, a small, fast component of calcium channel activation in response to β adrenergic receptor stimulation in whole-cell recordings was used to justify the direct route on a kinetic basis (Brown 1990). Strictly speaking, these kinds of experiments do not address the issue of whether the G protein interaction is direct or merely membrane delimited. However, two other experiments did attempt to answer this question. First, Yatani and coworkers (Imoto et al 1988) incorporated partially purified cardiac sarcolemmal vesicles in bilayers (channel opening potentiated by BayK8644). $G_{\alpha s}$ potentiated the probability of channel opening four- to sixfold. In similar bilayer reconstitutions with skeletal muscle channels, G_α increased channel activity 13- to 25-fold (Yatani et al 1988b). If pure channel reconstitutions had been possible, this would have proved direct activation in vitro. Finally, partially purified I_{CaL} from skeletal muscle bound purified $G_{\alpha s}$ (Hamilton et al 1991). Unfortunately only one G protein was bound for every 300 dihydropyridine receptors, suggesting a low affinity of binding under these conditions and invalidating firm conclusions. The assertion that G proteins directly activate the dihydropyridine-sensitive (L type) calcium channel (I_{CaL}) has been contested by Hartzell & Fischmeister (1992). The discussion here is not which subunit activates the channel, which has not been well addressed, but rather if phosphorylation accounts for all receptor-mediated stimulation of calcium current. In cardiac cells, from which most of the evidence has been collected, the point is more conceptual than practically significant because investigators generally agree that the majority of calcium-channel modulation (>80%) in heart is via activation of adenylyl cyclase and subsequent phosphorylation of the channel. In tests of the degree to which phosphorylation could account for the receptor-mediated increase in I_{CaL}, Hartzell et al (1991) found that in rat, guinea pig, and frog, isoprenaline-stimulated, cAMP-dependent activation accounted for the entire increase in I_{CaL} within experimental error. Other experimentalists (Shuba et al 1990) found that phosphorylation inhibitors could not fully block receptor-mediated stimulation. An even more complicated alternative, that G protein binding to the channel might prime the channel for phosphorylation, has been suggested by Cavalie et al (1991).

Na^+ Channels

Schubert et al (1989) proposed that G_α mediated a β-adrenergic inhibition of the cardiac fast sodium current, I_{Na}. In whole-cell recordings and single channel patch reconstitutions from neonatal rat ventricle, isoprenaline and $G_{\alpha s}$ subunits decreased sodium current. The reduction (~40%) in whole-cell recordings was most prominent when the cells were held at partially depolarized levels where the sodium current was half-inactivated. These authors demonstrated an ~ −20 mV shift in the inactivation curve [$h_\infty(V)$], suggesting that isoproterenol preferentially induced block of inactivated channels. Preactivated erythrocyte $G_{\alpha s}$ completely eliminated single channel opening at 100–200 pM concentrations in excised patches. The authors proposed both direct and indirect modulation of the sodium current. These rather striking observations of G protein subunit reconstitution have not been repeated in the literature to date, but it would be interesting to know if the single channel inhibition by G protein subunits correlates to the whole-cell current partial inhibition. Reminiscent of the I_{CaL} controversy discussed above, Ono et al (1989) found that the inhibition of the guinea pig ventricular sodium current could be completely accounted for by cAMP-dependent phosphorylation pathways because forskolin and dibutyryl cAMP effects were reversible. They also found that isoproterenol and dibutyryl cAMP induced a negative shift in the inactivation curve of I_{Na}.

The amiloride-sensitive sodium channel of a renal epithelial cell line (A_6) appears to be regulated by a pertussis toxin–sensitive G protein. $G_{\alpha i3}$, but not $G_{\alpha i2}$, revived channel activity in patch reconstitutions (Ausiello et al 1992,

Cantiello et al 1989). Further experimentation suggested that the reconstituted channel activation was via phospholipase A_2 and generation of second messengers (Cantiello et al 1990). Most recently, the partially purified channel reconstituted in liposomes and identified with Na^+ flux measurements was found to copurify with $G_{\alpha i3}$ and other unidentified proteins. Thus the evidence to date suggests that the A_6 amiloride-sensitive sodium channel is regulated via an indirect mechanism involving $G_{\alpha i3}$. Finally, this same group (Light et al 1989) again proposed a direct regulation of a renal inner medullary collecting duct cation channel ($P_{Na} = P_K$) by $G_{\alpha i3}$ (and cGMP), but other G protein subunits were not tried.

Cl^- Channels

Several studies report G protein–dependent activation of various chloride channels. Classification of chloride channel subtypes is rapidly changing and was recently reviewed (Ackerman & Clapham 1993). Schwiebert et al (1990) reported that $G_{\alpha i3}$ activated a large (305-pS) rabbit renal epithelial channel previously shown to be activated by protein kinase C (Schwiebert et al 1990). The subunit induced an approximately fivefold change in activity at picomolar G protein concentrations in inside-out patches. Apparently no other G protein subunits were tried in patch reconstitutions. In cystic fibrosis airway epithelial cells, the same group (Schwiebert et al 1992) proposed that G proteins inhibit cAMP from activating chloride (presumably CFTR) conductance and that pertussis toxin block of the G protein pathway allows cAMP to activate the channel—a rather provocative result that hopefully will be repeated by other groups. Fahlke et al (1992) proposed that a skeletal muscle multistate conductance chloride channel was directly regulated by G proteins. However, this conclusion was based on the fact that the GTPγS-induced current was not blocked by pharmacological agents known to block protein kinases (H8, PKI), phospholipase A_2 (quinacrine), or buffering of internal calcium. No purified G protein reconstitutions were attempted. A stretch-sensitive, GTPγS-induced chloride current in chromaffin cells was mediated by phospholipase A_2 (Doroshenko 1991, Doroshenko & Neher 1992). In membrane vescicles isolated from rat small intestine or a human colon carcinoma cell line, Tilly et al (1991) described a 20-pS chloride conductance that was GTPγS-sensitive and cytosolic ATP-, cAMP-, and Ca^{2+}-insensitive. Finally, Gadsby and colleagues (Hwang et al 1992) found no evidence that the β-adrenergic receptor-activated chloride conductance of guinea pig ventricle (now identified as the time and voltage independent CFTR-cystic fibrosis transmembrane regulator) was directly gated by $G_{\alpha s}$ because all regulation could be inhibited by blocking the phosphorylation pathways. In summary, several chloride channels may be regulated by G proteins, but these channels have not yet been completely characterized.

CONCLUSIONS

Neurons and muscle cells naturally require rapid activation of ion channels. Nicotinic, $GABA_A$, glycine, and glutamate receptors provide fast (millisecond) pathways for depolarization and hyperpolarization because transmitters need only bind these receptors to open them as channels. Cells are replete with slower phosphorylation pathways requiring many seconds for enhancement of channel activity. For second messenger pathways, intermediate time frames are logical and allow the large family of G protein–linked receptors to be adapted to faster modes of signaling. Despite the lack of concrete evidence, direct G protein gating of ion channels seems likely.

Until pure channel and pure G protein subunits are reconstituted in a pure lipid bilayer, the term membrane delimited is more accurately applied to those channels that exhibit evidence of fast G protein–mediated pathways independent of water-soluble diffusable second messengers. Speed of activation alone cannot be used as a criterion for direct interaction since phosphorylation rates vary widely and some may be closely allied to the phosphorylated protein (an extreme example is autophosphorylation of tyrosine kinase receptors). Experiments using agents such as pertussis toxin, cholera toxin, GTPγS, GTPβS, or AlF_4^- merely demonstrate that a G protein may act at some point in the activation pathway. Because phosphorylation pathways are common and often GTP-dependent, these tests cannot be used as arguments for direct activation. G protein reconstitution in inside-out patches eliminates readily diffusable ATP that would support phosphorylation but does not address the role of lipid-soluble second messengers such as arachidonic acid. The use of pharmacologic agents to block pathways only blocks known pathways and cannot prove direct activation by a process of elimination. Several other problematic experimental trends have arisen in the rush to verify direct G protein gating: testing of only one G protein subunit (particularly human erythrocyte $G_{\alpha i3}$) without controls or comparisons to other subunits; the assumption that all pharmacological blockade of known pathways is sufficient to establish a direct G protein interaction; large differences in subunit concentrations (often orders of magnitude) used to activate channels in perfused whole-cell recordings compared to inside-out patch recordings; and assumptions about the simplicity of excised patches. Hopefully, with the availability of cloned channels and G proteins, more direct tests will be applied.

Despite these caveats, the use of patch clamp methods to separate components of signal transduction in the cell and examine the gating of single channel proteins has provided much more information on transmitter-dependent activation of many channels. Single channel recording is a unique method, providing high time resolution recordings of the workings of a single protein in its native environment. Thus while research on channels lags behind work on purified proteins such as adenylyl cyclase and phospholipase C in the assessment of direct binding and mechanism of G protein activation, it may ultimately provide the most detailed information on protein-protein interactions. Given the large number of receptors and G proteins, and the need for fast gating of ion channels in neuronal membranes, it is likely that direct G protein gating of ion channels has evolved as a common mechanism in signal transduction. Direct G protein gating of channels may eventually be proven.

ACKNOWLEDGMENTS

Thanks to Drs. Bruce Bean, Criss Hartzell, and Bertil Hille for discussions on regulation of calcium channels and Yoshi Kurachi for discussions on K^+ channel regulation. The author was supported by an Established Investigator award of the American Heart Association and grants from the NIH.

Literature Cited

Ackerman MJ, Clapham DE. 1993. Cardiac chloride channels. *Trends Cardiovas. Med.* In press

Armstrong DL, White RE. 1992. An enzymatic mechanism for potassium channel stimulation through pertussin-toxin-sensitive G proteins. *Trends Neurosci.* 15:403–8

Artalejo CR, Ariano MA, Perlman RL, Fox AP. 1990. Activation of facilitation calcium channels in chromaffin cells by D_1 dopamine receptors through a cAMP/protein kinase A–dependent mechanism. *Nature* 348:239–42

Artalejo CR, Rossie S, Perlman RL, Fox AP. 1992. Voltage-dependent phosphorylation may recruit Ca^{2+} current facilitation in chromaffin cells. *Nature* 358:63–66

Ashcroft SJH, Ashcroft FM. 1990. Properties and functions of ATP-sensitive K-channels. *Cell. Signal.* 2:197–214

Ausiello DA, Stow JL, Cantiello HF, de Almeida JB, Benos DJ. 1992. Purified epithelial Na^+ channel complex contains the pertussis toxin-sensitive $G\alpha_{i-3}$ protein. *J. Biol. Chem.* 267:4759–65

Bean BP. 1989. Neurotransmitter inhibition of neuronal calcium currents by changes in channel voltage dependence. *Nature* 340:153–56

Beech DJ, Bernheim L, Hille B. 1992. Pertussis toxin and voltage dependence distinguish multiple pathways modulating calcium channels of rat sympathetic neurons. *Neuron* 8:97–106

Berlot CH, Bourne HR. 1992. Identification of effector-activating residues of $G_{s\alpha}$. *Cell* 68:911–22

Bernheim L, Mathie A, Hille B. 1992. Characterization of muscarinic receptor subtypes inhibiting Ca^{2+} current and M current in rat sympathetic neurons. *Proc. Natl. Acad. Sci. USA* 89:9544–48

Blank JL, Brattain KA, Exton JH. 1992. Activation of cytosolic phosphoinositide phospholipase C by G-protein $\beta\gamma$ subunits. *J. Biol. Chem.* 267:23069–75

Boland B, Bean B. 1993. Modulation of N-type calcium channels in bullfrog sypathetic ganglion by luteinizing hormone releasing hormone: kinetics and voltage dependence *J. Neurosci.* 13:516–33

Bourne HR, Sanders DA, McCormick F. 1991. The GTPase superfamily: conserved structure and molecular mechanism. *Nature* 349:117–27

Breitwieser GE. 1991. G protein–mediated ion channel activation. *Hypertension* 17:684–92

Breitwieser GE, Szabo G. 1985. Uncoupling of cardiac muscarinic and β-adrenergic receptors from ion channels by a guanine nucleotide analogue. *Nature* 317:538–40

Brown AM. 1990. Regulation of heartbeat by G protein–coupled ion channels. *Am. J. Physiol.* 259:H1621–28

Brown AM. 1991. A cellular logic for G protein–coupled ion channel pathways. *FASEB J.* 5:2175–79

Brown AM, Birnbaumer L. 1988. Direct G protein gating of ion channels. *Am. J. Physiol.* 254:H401–10

Cali JJ, Balcueva EA, Rybalkin I, Robishaw JD. 1992. Selective tissue distribution of G protein γ subunits, including a new form of the γ subunits identified by cDNA cloning. *J. Biol. Chem.* 267:24203–27

Camps M, Hou C, Sidiropoulos D, Stock JB, Jakobs KH, Giershik P. 1992. Stimulation of phospholipase C by guanine-nucleotide-binding protein βγ subunits. *Eur. J. Biochem.* 206:821–31

Cantiello HF, Pantenaude CR, Ausiello DA. 1989. G protein subunit, α_f-3, activates a pertussis toxin-sensitive Na^+ channel from the epithelial cell line, A6. *J. Biol. Chem.* 264:20867–70

Cantiello HF, Pantenaude CR, Codina J, Birnbaumer L, Ausiello DA. 1990. $G_{\alpha i\text{-}3}$ regulates epithelial Na^+ channels by activation of phospholipase A_2 and lipoxygenase pathways. *J. Biol. Chem.* 265:21624–28

Cavalie A, Allen TJA, Trautwein W. 1991. Role of the GTP-binding protein G_s in the adrenergic modulation of cardiac Ca channels. *Pflugers Arch.* 419:433–43

Clapham DE. 1990. Intracellular regulation of ion channels. In *Cardiac Electrophysiology*, ed. Zipes, Jalife, pp. 85–94

Clapham DE, Neer EJ. 1993. Bifurcating pathways for transmembrane signalling; new roles for G protein βγ subunits. *Nature* 365:403–6

Codina J, Yatani A, Grenet D, Brown AM, Birnbaumer L. 1987. The α subunit of the GTP binding protein G_k opens atrial potassium channels. *Science* 236:442–45

Delcour AH, Tsien R. 1993. Altered prevalence of gating modes in neurotransmitter inhibition of N-type calcium channels. *Science* 259:980–84

Dolphin AC. 1991. Regulation of calcium channel activity by GTP binding proteins and second messengers. *Biochem. Biophys. Acta* 1091:68–80

Dolphin AC, Wootton JF, Scott RH, Trentham DR. 1988. Photoactivation of intracellular guanosine triphosphate analogues reduces the amplitude and slows the kinetics of voltage-activated calcium channel currents in sensory neurones. *Pflugers Arch.* 411: 628–36

Doroshenko P. 1991. Second messengers mediating activation of chloride current by intracellular GTPγS in bovine chromaffin cells. *J. Physiol.* 436:725–38

Doroshenko P, Neher E. 1992. Volume-sensitive chloride conductance in bovine chromaffin cell membrane. *J. Physiol.* 449: 197–218

Dunlap K, Holz GG, Rane SG. 1987. G proteins as regulators of ion channel function. *Trends Neurosci.* 10:241–44

Ewald DA, Sternweis PC, Miller RJ. 1988. Guanine nucleotide-binding protein G_o-induced coupling of neuropeptide Y receptors to Ca^{2+} channels in sensory neurons. *Proc. Natl. Acad. Sci. USA* 85:3633–37

Fahlke C, Zachar E, Haussler U, Rudel R. 1992. Chloride channels in cultured human skeletal muscle are regulated by G proteins. *Pflugers Arch.* 421:566–71

Fenwick EM, Marty A, Neher E. 1982a. A patch-clamp study of bovine chromaffin cells and of their sensitivity to acetylcholine. *J. Physiol.* 331:577–97

Fenwick EM, Marty A, Neher E. 1982b. Sodium and calcium channels in bovine chromaffin cells. *J. Physiol.* 331:599–635

Haga K, Haga T. 1992. Activation by G protein βγ subunits of agonist- or light-dependent phosphorylation of muscarinic acetylcholine receptors and rhodopsin. *J. Biol. Chem.* 267:2222–27

Hamilton SL, Codina J, Hawkes MJ, Yatani A, Sawada T, et al. 1991. Evidence for direct interaction for $G_{s\alpha}$ with Ca^{2+} channel of skeletal muscle. *J. Biol. Chem.* 266: 19528–35

Hammill OP, Marty A, Neher E, Sakmann B, Sigworth FJ. 1981. Improved patch-clamp techniques for high-resolution current recording from cells and cell-free membrane patches. *Pflugers Arch.* 391:85–100

Hartzell HC, Fischmeister R. 1992. Direct regulation of cardiac Ca^{2+} channels by G proteins: neither proven nor necessary? *Trends Pharmacol. Sci.* 13:380–85

Hartzell HC, Mery P-F, Fischmeister R, Szabo G. 1991. Sympathetic regulation of cardiac calcium current is due exclusively to cAMP-dependent phosphorylation. *Nature* 351: 573–76

Hepler JR, Gilman AG. 1992. G proteins. *Trends Biol. Sci.* 17:383–87

Hescheler J, Rosenthal W, Trautwein W, Schultz G. 1987. The GTP-binding protein, G_0, regulates neuronal calcium channels. *Nature* 325:445–47

Hille B. 1992. G protein-coupled mechanisms and nervous signaling. *Neuron* 9:187–95

Holz GGI, Rane SG, Dunlap K. 1986. GTP-binding proteins mediate transmitter inhibition of voltage-dependent calcium channels. *Nature* 319:670–72

Houslay MD. 1987. Ion channels controlled by guanine nucleotide regulatory proteins. *Trends Biol. Sci.* 12:167–68

Hwang T-C, Horie M, Nairn AC, Gadsby DC. 1992. Role of GTP-binding proteins in the regulation of mammalian cardiac chloride conductance. *J. Gen. Physiol.* 99:465–89

Imoto Y, Yatani A, Reeves JP, Codina J, Birnbaumer L, et al. 1988. α-Subunit of G_s directly activates cardiac calcium channels in lipid bilayers. *Am. J. Physiol.* 255:H722–28

Inglese J, Koch WJ, Caron MG, Lefkowitz RJ.

1992. Isoprenylation in regulation of signal transduction by G-protein-coupled receptor kinases. *Nature* 359:147–50

Ito H, Sugimoto T, Kobayashi I, Takahashi K, Katada T, et al. 1991. On the mechanism of basal and agonist-induced activation of the G protein-gated muscarinic K$^+$ channel in atrial myocytes of guinea pig heart. *J. Gen. Physiol.* 98:517–33

Ito H, Tung RT, Sugimoto T, Kobayashi I, Takahashi K, et al. 1992. On the mechanism of G protein βγ subunit activation of the muscarinic K$^+$ channel in guinea pig atrial cell membrane. *J. Gen. Physiol.* 99:961–83

Jelsema CL, Axelrod J. 1987. Stimulation of phospholipase A$_2$ activity in bovine rod outer segments by the βγ subunits of transducin and its inhibition by the α subunit. *Proc. Natl. Acad. Sci. USA* 84:3623–27

Kasai H, Aosaki T. 1989. Modulation of Ca-channel current by an adenosine analog mediated by a GTP-binding protein in chick sensory neurons. *Pflugers Arch.* 414:145–49

Kaziro Y, Itoh H, Kozasa T, Nakafuku M, Satoh T. 1991. Structure and function of signal-transducing GTP-binding proteins. *Annu. Rev. Biochem.* 60:349–400

Kim D, Lewis DL, Graziadei L, Neer EJ, Bar-Sagi D, et al. 1989. G-protein βγ-subunits activate the cardiac muscarinic K$^+$-channel via phospholipase A$_2$. *Nature* 337:557–60

Kirsch GE, Codina J, Birnbaumer L, Brown AM. 1990. Coupling of ATP-sensitive K$^+$ channels to A$_1$ receptors by G proteins in rat ventricular myocytes. *Am. J. Physiol.* 259: H820–26

Kleuss C, Hescheler J, Ewel C, Rosenthal W, Schultz G, et al. 1991. Assignment of G-protein subtypes to specific receptors inducing inhibition of calcium currents. *Nature* 353:43–48

Kleuss C, Scherubl H, Hescheler J, Schultz G, Wittig B. 1992. Different β-subunits determine G-protein interaction with transmembrane receptors. *Nature* 358:424–26

Kubo Y, Reuveny E, Slesinger P, Jan YN, Yan LY. 1993. Primary structure and functional expression of a rat G protein-coupled muscarinic potassium channel. *Nature* 364: 802–06

Kurachi Y, Ito H, Sugimoto T, Shimizu T, Miki I, et al. 1989. Arachidonic acid metabolites as intracellular modulators of the G protein–gated cardiac K$^+$ channel. *Nature* 337:555–57

Kurachi Y, Tung RT, Ito H, Nakajima T. 1992. G protein activation of cardiac muscarinic K$^+$ channels. *Prog. Neurobiol.* 39:229–46

Leberer E, Dignard D, Harcus D, Thomas DY, Whiteway M. 1992. The protein kinase homologue Ste20p is required to link the yeast pheromone response G-protein βγ sub-

units to downstream signalling components. *EMBO J.* 11:4815–24

Lee CH, Park D, Wu D, Rhee SG, Simon MI. 1992. Members of the G$_q$ α subunit gene family activate phospholipase C β isozymes. *J. Biol. Chem.* 267:160444–47

Light DB, Ausiello DA, Stanton BA. 1989. Guanine nucleotide-binding protein, α$_{i-3}$, directly activates a cation channel in rat renal inner medullary collecting duct cells. *J. Clin. Invest.* 84:352–56

Lipscombe D, Kongsamut S, Tsien RW. 1989. α-Adrenergic inhibition of sympathetic neurotransmitter release mediated by modulation of N-type calcium-channel gating. *Nature* 340:639–42

Logothetis DE, Kim D, Northup JK, Neer EJ, Clapham DE. 1988. Specificity of action of guanine nucleotide-binding regulatory protein subunits on the cardiac muscarinic K$^+$ channel. *Proc. Natl. Acad. Sci. USA* 85: 5814–18

Logothetis DE, Kurachi Y, Galper J, Neer EJ, Clapham DE. 1987. The βγ subunits of GTP-binding proteins activate the muscarinic K$^+$ channel in heart. *Nature* 325:321–26

Lupas AN, Lupas JM, Stock JB. 1992. Do G protein subunits associate via a three-stranded coiled coil? *FEBS Lett.* 314:105–8

Martin GA, Yatani A, Clark R, Conroy L, Polakis P, et al. 1992. GAP domains responsible for ras p21-dependent inhibition of muscarinic atrial K$^+$ channel currents. *Science* 255:192–94

Mathie A, Bernheim L, Hille B. 1992. Inhibition of N- and L-type calcium channels by muscarinic receptor activation in rat sympathetic neurons. *Neuron* 8:907

Mattera R, Yatini A, Kirsch GE, Graf R, Okabe K, et al. 1989. Recombinant α$_{r}$3 subunit of G protein activates G$_k$-gated K$^+$ channels. *J. Biol. Chem.* 264:465–71

Mintz IM, Venema VJ, Swiderek KM, Lee TD, Bean BP, et al. 1992. P-type calcium channels blocked the spider toxin ω-Aga-IVA. *Nature* 355:827–29

Nanavati C, Clapham DE, Ito H, Kurachi Y. 1990. A comparison of the roles of purified G protein subunits in the activation of the cardiac muscarinic K$^+$ channel. In *G Proteins and Signal Transduction*, pp. 29–42. New York: Rockefeller Univ. Press

Navon SE, Fung BK-K. 1988. Characterization of transducin from bovine retinal rod outer segments. *J. Biol. Chem.* 263:489–96

Neer EJ. 1992. Subunit interactions of heterotrimeric G proteins. *Handb. Exp. Pharmacol.* In press

Neer EJ, Clapham DE. 1988. Roles of G protein subunits in transmembrane signalling. *Nature* 333:129–34

North RA. 1989. Drug receptors and the inhibition of nerve cells. *Br. J. Pharmacol.* 98:13–28

Ono K, Kiyosue T, Arita M. 1989. Isoproterenol, DBcAMP, and forskolin inhibit cardiac sodium current. *Am. J. Physiol.* 256:C1131–37

Pfaffinger PJ, Martin JM, Hunter DD, Nathanson NM, Hille B. 1985. GTP-binding proteins couple cardiac muscarinic receptors to a K channel. *Nature* 317:536–38

Pietrobon D, Hess P. 1990. Novel mechanism of voltage-dependent gating in L-type calcium channels. *Nature* 346:651–55

Pitcher JA, Inglese J, Higgins JB, Arriza JL, Casey PJ, et al. 1992. Role of βγ subunits of G proteins in targeting the β-adrenergic receptor kinase to membrane-bound receptors. *Science* 257:1264–67

Rarick HM, Artemyev NO, Hamm HE. 1992. A site on rod G protein α subunit that mediates effector activation. *Science* 256: 1031–33

Regan LJ, Sah DWY, Bean BP. 1991. Ca^{2+} channels in rat central and peripheral neurons: high-threshold current resistant to dihydropyridine blockers and ω-conotoxin. *Neuron* 6:269–80

Rosenthal W, Hescheler J, Trautwein W, Schultz G. 1988. Control of voltage-dependent Ca^{2+} channels by G protein-coupled receptors. *FASEB J.* 2:2784–90

Sakmann B, Noma A, Trautwein W. 1983. Acetylcholine activation of single muscarinic K^+ channels in isolated pacemaker cells of the mammalian heart. *Nature* 303:250–53

Scherer RW, Breitwieser GE. 1990. Arachidonic acid metabolites alter G protein-mediated signal transduction in heart. *J. Gen. Physiol.* 96:735–55

Schubert B, VanDongen AMJ, Kirsch GE, Brown AM. 1989. β-Adrenergic inhibition of cardiac sodium channels by dual G-protein pathways. *Science* 245:516–19

Schwiebert EM, Kizer N, Gruenert DC, Stanton BA. 1992. GTP-binding proteins inhibit cAMP activation of chloride channels in cystic fibrosis airway epithelial cells. *Proc. Natl. Acad. Sci. USA* 89:10623–27

Schwiebert EM, Light DB, Fejes-Toth G, Naray-Fejes-Toth A, Stanton BA. 1990. A GTP-binding protein activates chloride channels in a renal epithelium. *J. Biol. Chem.* 265:7725–28

Scott RH, Dolphin AC. 1987. Activation of a G protein promotes agonist responses to calcium channel ligands. *Nature* 330:760–62

Scott RH, Dolphin AC. 1988. The agonist effect of Bay K 8644 on neuronal calcium channel current is promoted by G-protein activation. *Neurosci. Lett.* 89:170–75

Shuba YM, Hesslinger B, Trautwein W, McDonald TF, Pelzer D. 1990. Whole-cell calcium current in guinea-pig ventricular myocytes dialysed with guanine nucleotides. *J. Physiol.* 424:205–28

Simon MI, Strathmann MP, Gautam N. 1991. Diversity of G proteins in signal transduction. *Science* 252:802–8

Smrcka AV, Hepler JR, Brown KO, Sternweis PC. 1991. Regulation of polyphosphoinositide-specific phospholipase C activity by purified G_q. *Science* 251:804–7

Soejima M, Noma A. 1084. Mode of regulation of the ACh-sensitive K-channel by the muscarinic receptor in rabbit atrial cells. *Pflugers Arch.* 400:424–31

Sokabe M, Sachs F. 1990. The structure and dynamics of patch-clamped membranes: a study using differential interference contrast light microscopy. *J. Cell Biol.* 111:599–606

Spiegel AM, Backlund PSJ, Butrynski JE, Jones TLZ, Simonds WF. 1991. The G protein connection: molecular basis of membrane association. *Trends Biol. Sci.* 16:338–41

Spiegel AM, Shenker A, Weinstein LS. 1992. Receptor-effector coupling by G proteins: implications for normal and abnormal signal transduction. *Endocr. Rev.* 13:536–65

Sternweis PC, Pang I.-H. 1990. The G protein-channel connection. *Trends Neurosci.* 13: 122–26

Tang W-J, Gilman AG. 1991. Type-specific regulation of adenylyl cyclase by G protein βγ subunits. *Science* 254:1500–3

Tang W-J, Gilman AG. 1992. Adenylyl cyclases. *Cell* 70:869–72

Taussig R, Sanchez S, Rifo M, Gilman AG, Belardetti F. 1992. Inhibition of the ω-conotoxin-sensitive calcium current by distinct G proteins. *Neuron* 8:799–809

Taylor SJ, Chae HZ, Rhee SG, Exton JH. 1991. Activation of the β1 isozyme of phospholipase C by α subunits of the G_q class of G proteins. *Nature* 350:516–18

Tilly BC, Kansen M, vanGageldonk PGM, van den Berghe N, Galjaard H, et al. 1991. G-proteins mediate intestinal chloride channel activation. *J. Biol. Chem.* 266: 2036–40

Tsien RW, Ellinor PT, Horne WA. 1991. Molecular diversity of voltage-dependent Ca^{2+} channels. *Trends Pharmacol. Sci.* 12: 349–54

Vaca L, Schilling WP, Kunze DL. 1992. G-protein-mediated regulation of a Ca^{2+}-dependent K^+ channel in cultured vascular endothelial cells. *Pflugers Arch.* 422:66–74

VanDongen AMJ, Codina J, Olate J, Mattera R, Joho R, et al. 1988. Newly identified brain potassium channels gated by the guanine nucleotide binding protein G_o. *Science* 242:1433–37

Whiteway M, Hougan L, Dignard D, Thomas DY, Bell L, et al. 1989. The STE4 and STE18 genes of yeast encode potential β and g subunits of the mating factor receptor-coupled G protein. *Cell* 56:467–77

Yatani A, Codina J, Brown AM, Birnbaumer L. 1987a. Direct activation of mammalian atrial muscarinic potassium channels by GTP regulatory protein G_k. *Science* 235: 207–11

Yatani A, Codina J, Imoto Y, Reeves JP, Birnbaumer L, et al. 1987b. A G protein directly regulates mammalian cardiac calcium channels. *Science* 238:1288–92

Yatani A, Hamm H, Codina J, Mazzoni MR, Birnbaumer L, et al. 1988a. A monoclonal antibody to the α subunit of G_k blocks muscarinic activation of atrial K^+ channels. *Science* 241:828–31

Yatani A, Imoto Y, Codina J, Hamilton SL,

Brown AM, et al. 1988b. The stimulatory G protein of adenylyl cyclase, G_s, also stimulates dihydropyridine-sensitive Ca^{2+} channels. *J. Biol. Chem.* 263:9887–95

Yatani A, Okabe K, Codina J, Birnbaumer L, Brown AM. 1990a. Heart rate regulation by G proteins acting on the cardiac pacemaker channel. *Science* 249:1163–66

Yatani A, Okabe K, Polakis P, Halenbeck R, McCormick F, et al. 1990b. ras p21 and GAP inhibit coupling of muscarinic receptors to atrial K^+ channels. *Cell* 61:769–76

Yue DT, Herzig S, Marban E, et al. 1990. β-Adrenergic stimulation of calcium channels occurs by potentiation of high-activity gating modes. *Proc. Natl. Acad. Sci. USA* 87:753–57

Annu. Rev. Neurosci. 1994. 17:465–88

THE ANATOMY AND PHYSIOLOGY OF PRIMATE NEURONS THAT CONTROL RAPID EYE MOVEMENTS

A. K. Moschovakis

Department of Basic Sciences, Division of Medicine, School of Health Sciences, University of Crete, Post Office Box 1393, Iraklion, Crete, Greece

S. M. Highstein

Department of Otolaryngology, Washington University School of Medicine, St. Louis, Missouri 63110

KEYWORDS: saccades, oculomotor system, superior colliculus, vector subtraction hypothesis

INTRODUCTION

Primates are frontal eyed animals possessing a sophisticated retina endowed with a highly differentiated foveal region with which the visual scene can be examined in detail. Vision is very important to these animals and is probably the predominant sensory modality with which they orient themselves and interact with the world. The relationship between the visual system and the motor system that moves the eyes (oculomotor) is intimate. Motion of the visual scene across the retina, particularly the fovea, can degrade visual acuity. Hence there is a necessity for accurate oculomotor control. Not surprisingly, a considerable portion of the brain, including the superior colliculus (SC), the frontal eye field (FEF), the thalamus, and the cerebellum as well as several brainstem nuclei, is entrusted with the control of eye movements (Hepp et al 1989).

The most ubiquitous movements of the eyes are the rapid ones (also called saccadic) that reorient the eyes quickly toward interesting features of the surrounding world. During a saccade, two or more of six extraocular muscles

465

(per eye) contract or relax in proportion to the signals they receive from extraocular motoneurons (MNs). At the MN level, movement direction is specified by the relative activity of MN pools innervating muscles with different pulling directions, while movement amplitude is specified by the total activity of engaged MNs. The discharge of extraocular MNs is in turn due to the direct or indirect influence exerted on them by a large variety of saccade related cells scattered throughout the brain. Some of these neurons exhibit a brief burst of activity before saccades of particular directions. Depending upon the latency of the burst in relation to the onset of the movement, they are called medium lead burst neurons or long lead burst neurons (Luschei & Fuchs 1972, Raybourn & Keller 1977, Hepp & Henn 1983). Other neurons modulate their discharge in proportion to eye position (tonic; Keller 1974), while others pause during saccades irrespective of direction (omnipause; Keller 1974). Generally the discharge characteristics of saccade related neurons specify the metrical and often the dynamical properties of impending saccades (Hepp et al 1988).

To provide a description of the causal chain of neural events that leads to saccade generation, it is necessary to know how saccade related cells are connected to one another. Some of the necessary information has been obtained with the recent use of the intraaxonal recording and injection technique in alert behaving animals; accordingly, the targets of single, identified medium lead burst (Yoshida et al 1982, Strassman et al 1986a,b, Moschovakis et al 1990, Moschovakis et al 1991a,b), omnipause (Strassman et al 1987) and long lead burst (Scudder et al 1989, Moschovakis et al 1988b) neurons have been delineated. The axonal projections of some of these cells are appropriate for distributing the output of the burst generator to extraocular MNs. Other projections can ensure the coordination between distinct burst generators. Finally, some cell types are more appropriate for closing the feedback paths of the control loops involved. Here, the activity and connections of these neurons is reviewed, and their implications for current models of oculomotor control are discussed.

HOW IS THE OUTPUT OF THE BURST GENERATORS DISTRIBUTED TO OCULOMOTONEURONS?

During saccades in the pulling direction of the muscle they innervate, extraocular MNs exhibit a pulse of activity (Robinson 1970). This is due to the input they receive from medium lead burst neurons, several groups of which have been described to date (Cohen & Henn 1972, Luschei & Fuchs 1972, Keller 1974, Henn & Cohen 1976, Buttner et al 1977, King & Fuchs 1979, van Gisbergen et al 1981). As long as the animal is alert, these cells discharge exclusively before (with a latency shorter than 12 ms) and during rapid eye movements (Figure 1*a*). At the level of the medium lead burst neurons, the amplitude and direction of saccades are specified by the amplitude of their horizontal and vertical components. These are computed relatively independently, by distinct groups of medium lead burst neurons.

Taking into consideration their preferred directions (Figure 1*b*), it is possible to delineate three groups of medium lead burst neurons (and therefore three burst generators), within each half of the brain (right side in the case of Figure 1*b*): (i) horizontal (ipsilateral), (ii) upward, and (iii) downward. Rightward (leftward) excitatory burst neurons are located in the right (left) paramedian pontine reticular formation (PPRF), just rostral to the abducens nucleus, and they deal with the horizontal component of saccades (Henn & Cohen 1976, Strassman et al 1986a). Upward and downward medium lead burst neurons are located in the right (left) rostral interstitial nucleus of the medial longitudinal fasciculus (riMLF), and these deal with the vertical component of saccades (Buttner et al 1977, King & Fuchs 1979, Moschovakis et al 1990, 1991a,b).

There are several reasons to think that the horizontal and vertical medium lead burst neurons are causally relevant for the execution of saccades in their on-direction. Firstly, their discharge is both a necessary and a sufficient condition for the occurrence of saccades in their on-direction. For example, records similar to the one illustrated in Figure 1*a* indicate that no downward saccades occur during periods of quiescence of excitatory downward medium

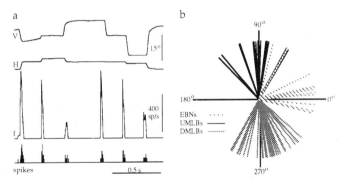

Figure 1 (*a*) Discharge pattern of a downward medium lead burst neuron. Traces illustrate from top to bottom: instantaneous vertical eye position (*V*), instantaneous horizontal eye position (*H*), instantaneous firing rate of the neuron (*f*) and spike train. Upward deflection of the eye position traces indicates right and up movement of the eyes in the horizontal and vertical directions, respectively. Calibration bar of 15 degrees applies to both eye position traces. (*b*) Distribution of the on-directions of 14 upward (UMLBs; adapted from Moschovakis et al 1991a), 29 downward (DMLBs; adapted from Moschovakis et al 1991b), and 11 rightward (EBNs; adapted from Strassman et al 1986a) medium lead burst neurons. To compensate for the fact that vertical neurons were recorded from the left side of the brain, their on-directions are presented as mirror symmetrical (with respect to the vertical plane) to those included in previous reports.

lead burst neurons and that all relatively intense bursts of all downward medium lead burst neurons are accompanied by downward saccades. Figure 1a also illustrates that some off-direction (upward) saccades are also accompanied by downward medium lead burst neuron bursts. In this case, the bigger the off-direction saccade the smaller the burst. This is suggestive of a push-pull coupling between on- and off-direction burst generators (Strassman et al 1986b, Scudder 1988, Moschovakis et al 1991b), an issue discussed in the next section.

Secondly, destruction of medium lead burst neurons leads to a loss of saccades in their on-direction. Thus, PPRF lesions lead to a loss of horizontal saccades while sparing vertical saccades (Henn et al 1984). Conversely, lesions of the riMLF cause paresis of vertical saccades while leaving horizontal saccades intact (Henn et al 1983). The fact that pontine and mesodiencephalic lesions can lead to similar syndromes in human subjects has been known since the previous century (reviewed in Hepp et al 1989).

Finally, axons of medium lead burst neurons convey the output of the burst generators directly to extraocular MNs. Figure 2 summarizes the projections of horizontal excitatory burst neurons that were intracellularly injected with horseradish peroxidase. One of their most prominent projections is to the ipsilateral abducens nucleus (Figure 2a), thus directly providing lateral rectus MNs with the excitatory drive they need during abducting saccades (Strassman et al 1986a). Vertical medium lead burst neurons ramify extensively within the oculomotor complex (i.e. the nuclei from which the third and fourth cranial nerves originate). The precise location within the oculomotor complex where each vertical medium lead burst neuron distributes its terminals varies in relation to the cell's preferred direction (Figure 3). Terminal fields of excitatory upward medium lead burst neurons overlie the territories of motoneurons supplying the superior rectus (SR) and the inferior oblique (IO) muscles (Moschovakis et al 1991a). The oculomotor terminations of downward medium lead burst neurons overlie the territories of motoneurons supplying the inferior rectus (IR) and the superior oblique (SO) muscles (Moschovakis et al 1991b). There is neurophysiological evidence indicating that such connections between excitatory medium lead burst neurons and extraocular MNs with similar preferred directions probably exist in other species as well (e.g. Sasaki & Shimazu 1981, Nakao & Shiraishi 1983, 1985).

Due to their widespread terminations, both upward and downward medium lead burst neurons can establish connections with motoneuron pools innervating yoked muscles of the two eyes (Moschovakis et al 1990). One of the purposes served by such a distribution of the output of the burst generators could be the establishment of eye conjugacy (i.e. so that both eyes move simultaneously in the same direction and by roughly the same amount). According to Hering's (1868) principle of equal innervation, for the eyes to move in this fashion, synergistic muscles of the two eyes must be activated

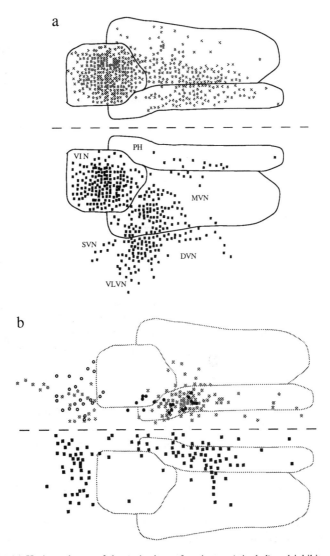

Figure 2 (*a*) Horizontal map of the projections of excitatory (stippled) and inhibitory (solid) horizontal medium lead burst neurons to the abducens, vestibular and prepositus nuclei (adapted from Strassman et al 1986b). The dashed line indicates the midline. (*b*) Horizontal map of the projections of excitatory (stippled) and inhibitory (solid) horizontal medium lead burst neurons to the reticular formation (adapted from Strassman et al 1986b). Open and closed circles represent the location of somata of excitatory and inhibitory neurons, respectively. The borders of the midline and the dorsally situated abducens, vestibular, and prepositus nuclei are indicated by dashed and stippled lines, respectively. Both maps are oriented so that the right-rostral part of the brainstem is to the top-left part of the figure. (Abbreviations: VI N, abducens nucleus; DVN, descending vestibular nucleus; MVN, medial vestibular nucleus; PH, nucleus prepositus hypoglossi; SVN, superior vestibular nucleus; VLVN, ventrolateral vestibular nucleus.)

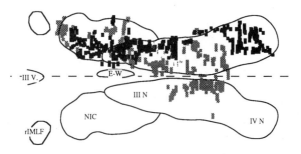

Figure 3   Horizontal map of the projections of one upward (stippled) and one downward (solid) medium lead burst neuron of the riMLF (adapted from Moschovakis et al 1990). The dashed line indicates the midline. Orientation as in Figure 2. To compensate for the fact that the somata of illustrated neurons were recovered in the left side of the brain, their terminal fields are presented as mirror symmetrical (with respect to the sagittal plane) to those included in previous reports. (Abbreviations: III N, oculomotor nucleus; III V, third ventricle; IV N, trochlear nucleus; E-W, nucleus Edinger-Westfal; NIC, interstitial nucleus of Cajal; riMLF, rostral interstitial nucleus of the medial longitudinal fasciculus.)

simultaneously and by roughly the same amount. In the vertical oculomotor system, there are four pairs of yoked muscles: (i) left SR-right IO, (ii) left SO-right IR, (iii) left IO-right SR, (iv) left IR-right SO (pairs 1 & 2 are innervated by MNs in the right oculomotor complex while pairs 3 and 4 are innervated by MNs in the left oculomotor complex). For each one of these four pairs, the establishment of contacts between axonal branches of single excitatory vertical medium lead burst neurons and MNs innervating both of the muscles in a pair could allow a simple neural implementation of Hering's principle. Such an arrangement would also imply that there are two groups of anatomically defined vertical medium lead burst neurons in each side of the brain. Upward medium lead burst neurons located in the left (right) mesencephalon would contact IO and SR MNs in the left (right) oculomotor complex, while downward medium lead burst neurons located in the left (right) mesencephalon would contact IR and SO MNs in the left (right) oculomotor complex. Such a distribution of the output of the vertical burst generators is consistent with the results of selective, reversible lesions of the riMLF, as well as observations concerning the relationship between the activity of vertical medium lead burst neurons and torsional components of vertical eye movements (Hepp et al 1988). The pattern of termination of single downward medium lead burst neurons is also consistent with this idea (Moschovakis et al 1991b). Downward medium lead burst neurons of the left riMLF contact MNs that supply the IR muscle of the left eye and the SO muscle of the right eye (Moschovakis et al 1991b), while downward medium lead burst neurons of the right riMLF contact MNs that supply the IR muscle of the right eye and the SO muscle of the left eye. Since the iso-frequency curves of all MNs that innervate upward pulling muscles are indistinguishable (Hepp & Henn 1985), all upward medium lead burst neurons could contact all upward MNs. Indeed, terminal fields of upward medium lead burst neurons have been shown to overlie the MN pools that supply both the SR and the IO muscles of both eyes (Moschovakis et al 1991a).

Differences in the immediacy with which vertical and horizontal burst generator output is distributed to extraocular MNs could be responsible for the fact that vertical saccades are more strictly conjugate than horizontal ones (Collewijn et al 1988b). In the horizontal system, eye conjugacy is accomplished through the connections that excitatory medium lead burst neurons establish with a second class of neurons, which are also located within the ipsilateral abducens nucleus and project to the contralateral medial rectus (MR) subdivision of the oculomotor complex (Highstein & Baker 1978). These internuclear cells could provide the contralateral MR MNs with a copy of the inputs which they receive and presumably share with abducens MNs. As suggested by their dissimilar appearance (Highstein et al 1982, McCrea et al 1986), the discharge patterns of abducens motor and internuclear neurons are not identical. Internuclear neurons start discharging earlier and have a velocity sensitivity higher than that of abducens MNs (Delgado-Garcia et al 1986a,b, Fuchs et al 1988). These differences are suggestive of a neuronal mechanism (Fuchs et al 1985), introduced to compensate for the additional filtering step introduced in the horizontal system by one additional neuron (the internuclear) and one more synapse (between the internuclear neuron and the MR MN). Its inability to act in a fully compensatory role is indicated by the relative sluggishness of the adducting eye during horizontal saccades (Collewijn et al 1988a).

In addition to the simultaneous activation of MNs innervating synergist muscles of both eyes, extraocular MNs innervating antagonist muscles are inactivated during saccades (Robinson 1970). In the horizontal system, activation of ipsilateral abducens MNs is accompanied by inhibition of contralateral abducens MNs (these innervate the LR muscle of the opposite eye) as well as disfacilitation of ipsilateral MR MNs (these innervate the MR muscle of the ipsilateral eye). Both are largely mediated by inhibitory burst neurons. In 1977, Maciewicz and coworkers demonstrated that neurons located in the dorsomedial gigantocellular tegmental field of the cat project to the contralateral abducens nucleus. Simultaneously, Hikosaka & Kawakami (1977) showed that these neurons monosynaptically inhibited abducens motoneurons and carried a burst signal that encoded the amplitude and direction of saccadic eye movements. In the monkey, the discharge pattern of inhibitory burst neurons is similar to that of excitatory burst neurons (Strassman et al 1986b). This could be due to the fact that the latter drive the former or to inputs shared by both neuronal cell types. Both mechanisms are supported by currently available evidence. Terminal fields of single identified primate tectal long lead burst neurons are distributed within the region of the PPRF that contains horizontal excitatory burst neurons (Scudder et al 1989) while inhibitory burst neurons of the cat respond within monosynaptic latencies to SC stimulation (Hikosaka & Kawakami 1977). On the other hand, terminal fields of identified excitatory burst neurons of the monkey are distributed within the region of the PPRF that contains inhibitory burst neurons (Figure 2b). Also, a monosynaptic connection between excitatory and inhibitory burst neurons has been demonstrated in the cat (Sasaki & Shimazu 1981).

The projections of single identified primate horizontal inhibitory burst neurons have been delineated (Strassman et al 1986b), and as shown in Figure 2, they include the contralateral abducens nucleus. This could account for the pause of MNs located therein during ipsiversive saccades. In addition, their inhibitory termination onto contralateral abducens internuclear neurons, which has been demonstrated in the cat (Hikosaka et al 1978b, 1980), could account for the disfacilitation of ipsilateral MR MNs during abduction of the eyes. In the vertical system, an IPSP also partly underlies the pause of MNs (e.g. downward) during off-direction (e.g. upward) saccades (Baker & Berthoz 1974). This is presumably due to the influence exerted by inhibitory vertical medium lead burst neurons. The existence of both upward and downward inhibitory medium lead burst neurons is supported by the disclosure of monosynaptic IPSPs in both the SO (Nakao & Shiraishi 1985) and the SR (Nakao & Shiraishi 1983) MNs in response to electrical stimulation of the mesodiencephalic junction. The oculomotor terminations of inhibitory vertical medium lead burst neurons are presumably similar to those of excitatory burst neurons with opposite on-directions. Indeed, complete reconstruction of the axonal system of an identified inhibitory upward medium lead burst neuron of the monkey has demonstrated that its oculomotor terminations are entirely contained within the IR-SO subdivisions of the ipsilateral oculomotor complex (Moschovakis et al 1991a).

For saccadic eye movements to be executed, engaged extraocular MNs must be supplied with a static component, proportional to eye position, in addition to the phasic component, related to eye displacement. Only the latter can be provided directly to oculomotoneurons by the excitatory medium lead burst neurons. The static component must be extracted from the medium lead burst neuron activity, through a process akin to mathematical integration (Robinson 1975). In both the cat and the monkey, evidence from lesion studies (Cheron & Godaux 1987, Cannon & Robinson 1987), and from the analysis of discharge of single cells (Lopez-Barneo et al 1982), points to the medial vestibular nucleus and the nucleus prepositus hypoglossi as potential sites where integration concerning the horizontal plane could be neurally implemented. The projections of horizontal excitatory medium lead burst neurons to these structures (Figure 2a) support this notion (Strassman et al 1986a). Also in both species, the analysis of the consequences of lesions (Anderson 1981, Anderson et al 1979, Crawford et al 1991) and of the discharge of single neurons (King et al 1981, Fukushima 1991) has implicated the interstitial nucleus of Cajal (NIC) in the neural implementation of integration in the vertical plane. The fact that all upward (Moschovakis et al 1991a) and downward (Moschovakis et al 1991b) medium lead burst neurons recovered to date project to the NIC (Figure 3) again supports this notion. Inhibitory medium lead burst neurons also project to the neural integrators, thus ensuring their bidirectional modulation. Terminal fields of vertical inhibitory medium lead burst neurons include the NIC (Moschovakis et al 1991a), while those of horizontal inhibitory medium lead burst neurons include the vestibular nuclei and the nucleus prepositus hypoglossi (Figure 2a).

Strassman and coworkers (1986a,b) have suggested that the projections of horizontal excitatory and inhibitory medium lead burst neurons to the vestibular nuclei and the nucleus prepositus hypoglossi could also play a role in the generation of both the bursts and the pauses that neurons of these structures display during saccades or quick phases of nystagmus (Baker et al 1975, Hikosaka et al 1978a, McCrea et al 1980, Tomlinson & Robinson 1984). As documented in these studies, many of the neurons in the vestibular three neuron arc (e.g. tonic-vestibular-pause) pause for saccades in some directions. The same cells may also emit a weak burst of activity for saccades in their on-direction. Since their on-directions are correlated with their patterns of axonal projection within the vestibular nuclei, individual horizontal medium lead burst neurons could convey necessary spatial information to this structure (Strassman et al 1986b). Burst neurons with oblique on-directions project to regions of the vestibular nuclei known to contain vertical vestibular relay neurons. Conversely, burst neurons with more nearly horizontal on-directions project to areas of the vestibular nuclei known to contain interneurons that participate in the horizontal vestibulo-ocular reflex pathway.

HOW IS THE ACTIVITY OF DISTINCT BURST GENERATORS COORDINATED?

The activity of burst generators with different on-directions may also have to be coordinated, to ensure the simultaneous activation of extraocular MNs innervating several synergist muscles and the inactivation of MNs innervating antagonist muscles. For example, horizontal excitatory burst neurons are presumably kept from firing an inappropriate burst during contraversive saccades, due to the input they receive from horizontal inhibitory burst neurons of the opposite side (Strassman et al 1986b). The intense terminations that the inhibitory burst neurons of one side deploy within the area containing the excitatory burst neurons of the opposite side (Figure 2b) offer an anatomical substrate for this phenomenon. Vertical burst generators with opposite on directions would also be coupled in a push-pull fashion, if the recurrent collaterals of inhibitory upward (downward) medium lead burst neurons were directed toward downward (upward) medium lead burst neurons. Encouragingly, one morphological characteristic that distinguishes putative inhibitory upward medium lead burst neurons from putative excitatory ones is the deployment of recurrent collaterals (Moschovakis et al 1991a). A push-pull coupling between burst generators with opposite on-directions also occurs at even higher levels of saccade programming. For example, cells of the feline SC that burst before saccades in one direction are inhibited during saccades in the opposite direction (Infante & Leiva 1986, Peck 1990). The axon collaterals of the tectal long lead burst neurons (Moschovakis et al 1988b), which participate in the presumably inhibitory commissural pathway of the SC (Moschovakis & Karabelas 1982), could provide the anatomical substrate for this phenomenon.

To account for the straight trajectories of oblique saccades, the activity of orthogonal burst generators (vertical and horizontal) must also be coordinated. The vertical and horizontal components of oblique saccades have roughly similar duration despite their dissimilar size (King et al 1986). This could be due to the fact that the discharges of the vertical and horizontal burst generators have roughly similar duration (King & Fuchs 1979). Coupling of orthogonal burst generators through the omnipause neurons can account for the equalization of orthogonal burst generator discharge durations and the straight trajectories of eye movements during oblique saccades (Scudder 1988). Of the many oculomotor related neurons located in the PPRF, it is the omnipause neurons that have been shown to project to vertical medium lead burst neurons, in both the monkey and the cat (Buttner-Ennever & Buttner 1988, Nakao et al 1988). Omnipause neurons also contact horizontal medium lead burst cells in the same species (Nakao et al 1980, Furuya & Markham 1982, Curthoys et al 1984, Strassman et al 1987). Actually, the omnipause projection to the vertical and horizontal medium lead burst neurons could originate from single bifurcating axons (King et al 1980, Strassman et al 1987).

In turn, the discharge of omnipause neurons is influenced by the output of both orthogonal burst generators. Presently available evidence is consistent with the notion that excitatory medium lead burst neurons inhibit omnipause neurons through sign inverting local circuit cells (Strassman et al 1986a,b, Hepp et al 1989). The latter could be embodied by the still elusive "latch" neurons, which are thought to silence the omnipause neurons and thereby enable medium lead burst neurons (van Gisbergen et al 1981). The existence of excitatory burst neuron terminations within the midline region that contains omnipause neurons is consistent with this notion (Strassman et al 1986a). One could expect projections of excitatory vertical medium lead burst neurons also to include the omnipause region. In the cat, perifascicular neurons (including cells within the H field of Forel which is homologous to the primate riMLF) have been backlabeled from the omnipause region (Langer & Kaneko 1984). The alternative notion, that inhibitory medium lead burst neurons contact omnipause neurons directly, is not supported by presently available physiological (Nakao et al 1980, Hikosaka et al 1980), and anatomical (Yoshida et al 1982, Strassman et al 1987), evidence.

A second source of input, shared by both vertical and horizontal burst generators, probably arises from the long lead burst neurons of the SC. The discharge pattern of medium lead burst neurons differs in several respects from that of tectal long lead burst neurons. The activity of both classes of cells has been the object of extensive quantitative study (for recent reviews see Fuchs et al 1985, Sparks & Mays 1990). As a result, it is now known that several of the parameters that characterize the burst of medium lead neurons are strongly correlated with parameters that characterize the saccade (Fuchs et al 1985). The two strongest ones concern the relation between: (a) burst duration and saccade duration, (b) the number of spikes in the burst and component size. Neither relation holds for individual long lead burst neurons of the primate SC (Sparks & Mays 1990). Instead, these neurons discharge before saccades of particular amplitude and direction (collectively defined as the movement field of the cell). The location of tectal long lead burst neurons is such that cells with small movement fields are located in the rostral SC while cells with large movement fields are located more caudally. Also, cells that discharge before upward saccades are located medially in the SC whereas cells that discharge before downward saccades are located laterally (Sparks & Mays 1980, Moschovakis et al 1988b). It would seem, therefore, that commands issuing from the SC are organized in a place code and that they represent the whole range of saccade vectors spanning the entire contralateral hemifield.

These vectorial command signals must then be reorganized into commands along the pulling directions of the extraocular muscles. This process of vector decomposition is partly accomplished already at the level of medium lead burst neurons. The mean on-direction for the population of upward, rightward, and downward medium lead burst neurons does not differ in a statistically significant manner from 90°, 0° and 270°, respectively (Strassman et al 1986a, Moschovakis et al 1991a,b). However, the plot of their on-directions indicates that vector decomposition is not yet complete at the level of individual medium lead burst neurons (Figure 1b). For example, there are horizontal medium lead burst neurons with a considerable vertical component. This suggests that their input comes from neurons that supply upward and downward medium lead burst neurons, as well. Similar conclusions can be reached through an inspection of the on-directions of individual vertical medium lead burst neurons.

The neural events interposed between the tectal long lead and the medium lead burst neurons are still a matter of conjecture. Each tectal presaccadic cell is thought to establish appropriately weighted connections with long lead burst neurons of the midbrain and pons (Scudder 1988); these then synapse with the medium lead burst neurons. Clearly, there is considerable anatomical and physiological evidence to support this scheme (reviewed in Sparks & Mays 1990). On the other hand, the relationship between the discharge of long lead burst neurons and the metrics of saccades is so weak that their involvement in oculomotor control has been questioned (van Gisbergen et al 1981). Furthermore, few of the intraaxonally HRP-injected long lead burst neurons, recovered to date, demonstrate the requisite pattern of axonal projections (Scudder et al 1989).

Alternatively, each tectal presaccadic cell could directly establish appropriately weighted connections with medium lead burst neurons. This scheme is also supported by presently available evidence. In the cat, disynaptic EPSPs have often been recorded from both vertical (Grantyn et al 1982) and horizontal (Grantyn & Grantyn 1976) MNs in response to stimulation of the SC. The proximal leg of the relevant three neuron arc could be embodied by axons of tectal long lead burst neurons. In primates, each one of these axons is known to bifurcate into two major branches (Moschovakis et al 1988b). One of them is uncrossed and ascending (participating in the ventral ascending tectofugal fiber bundle; Grantyn & Grantyn 1982, Moschovakis & Karabelas 1985, Moschovakis et al 1988a). The other is crossed and descending (participating in the predorsal bundle; Grantyn & Grantyn 1982, Moschovakis & Karabelas 1985, Moschovakis et al 1988a). Targets of the former include the riMLF (Harting et al 1980), while those of the latter include a region in the PPRF which contains horizontal medium lead burst neurons (Scudder et al 1989). Although their presence is well supported, connections between tectal long lead and medium lead burst neurons are not necessarily powerful. This is indicated by the fact that single shock electrical stimulation of the SC cannot elicit responses from medium lead burst neurons (Raybourn & Keller 1977).

HOW DOES THE BURST GENERATOR COMPUTE SACCADE AMPLITUDE AND DIRECTION?

Each group of medium lead burst neurons controls component size (left or right, up or down) by adjusting the total number of spikes it emits for a saccade. How is the number of spikes emitted by medium lead burst neurons computed so that saccades land on target? Figure 4*a* illustrates an early answer to this question, in its original form as proposed by Robinson in 1975. Several models with similar premises have appeared since (e.g. Guitton & Volle 1987). In Robinson's model, a copy of eye position (E) is fed back to a comparator (MLB) where it is subtracted from the desired eye position signal (E'). Therefore, the eye is driven until E matches E', when the motor error (Me = E – E') equals zero, the burst stops, and the eye lands on the desired target. In this model, neither the amplitude (number of spikes), nor the duration of the burst are preprogrammed. Instead, they are automatically adjusted for the desired saccade size through the operation of the feedback loop (Robinson 1981). In addition to the direct path connecting the medium lead burst neurons with the MNs, Figure 4*a* illustrates the indirect path through the neural integrator (NI).

This model has shown a remarkable consistency with a large amount of psychophysical data. For example, it is consistent with the fact that some saccades can be modified in midflight, e.g. the slow saccades produced by patients suffering from spinocerebellar atrophy (Zee et al 1976). Also, it incorporates many of the known oculomotor related cells (burst-tonic, tonic, medium lead burst). A potential shortcoming of this model is that it is driven not by retinal error, which is the most accurate estimate of the distance between the eye and the target, but by a neural reconstruction of target position in space (E'). Models that remedy this situation are characterized by their reliance on eye displacement (ΔE) signals for their feedback. Figure 4*b* illustrates a recent version of the eye displacement model (Scudder 1988). In this model, a copy of the output of the burst generator is fed back to the comparator, which this time is embodied by the long lead burst neurons (LLB). Burst generator output is fed back to long lead burst neurons through the inhibitory feedback neurons (IFNs), which are identical to medium lead burst cells in terms of their discharge but different in terms of their connections. Long lead burst neurons integrate their two inputs (symbol of integration) and stop firing when the number of spikes due to their excitatory input equals the number of spikes due to the inhibitory feedback.

The eye displacement model is supported by much of the evidence that supports the eye position model. For example, both models can account for the relationship that exists between the motor error and the instantaneous discharge of medium lead burst neurons. Nevertheless, the two models do not always behave identically. For example, the eye position model replicates the saturation of the discharge of medium lead burst neurons more realistically than the eye displacement model (Scudder 1988). Also, destruction of omnipause neurons should result in saccadic oscillations according to the eye position model and lower saccadic peak velocities according to the eye displacement model. In recent experiments, Kaneko (1990) has lesioned the

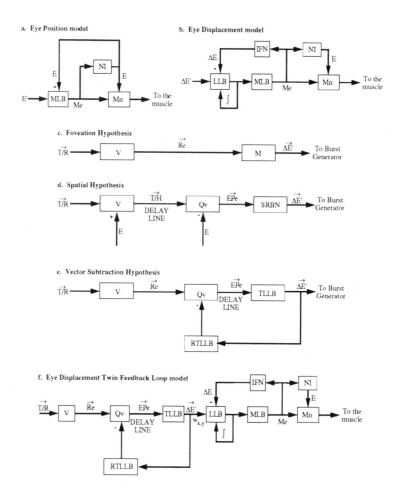

Figure 4 Models of the primate saccadic system. (*a*) Eye position local feedback loop model (adapted from Robinson 1975). (*b*) Eye displacement local feedback loop model (adapted from Scudder 1988). (*c*) Foveation hypothesis (adapted from Schiller & Koerner 1971). (*d*) Spatial hypothesis (adapted from Sparks & Mays 1982). (*e*) Vector subtraction hypothesis (adapted from Moschovakis 1987). (*f*) Eye displacement twin feedback loop model. (Abbreviations: ΔE, eye displacement; ΔE′, desired eye displacement; E, instantaneous eye position; E′, desired eye position; EPe, eye position error; IFN, inhibitory feedback neuron; LLB, long lead burst neuron; M, movement related neuron; MLB, medium-lead burst neuron; Me, motor error; Mn, motoneuron; NI, neural integrator; Qv, quasivisual neuron; Re, retinal error; RTLLB, reticulotectal long lead burst neuron; SRBN, saccade related burst neuron; TLLB, tectal long lead burst neuron; T/H, target location with respect to the head, T/R, target location with respect to the retina, V, visual cell, $w_{x,y}$, synaptic weights adjusted according to neuron's location (in x, y coordinates) in the SC. Vector notation indicates place coding of relevant information. (-) indicates inhibitory connection. All other connections are excitatory.)

omnipause neurons and has found that peak saccadic velocity decreased while saccadic duration was proportionately lengthened. Further, following histo-logically verified omnipause cell lesions, no saccadic oscillations were observed. These findings support the eye displacement model rather than the eye position model.

Experiments that characterize the physiology and anatomy of primate saccade related burst neurons should also shed light on the sort of burst generator implemented in the brain. If the burst generator resembles that embodied by the eye position model, then the existence of neurons firing in relation to eye position and projecting to medium lead burst neurons should be demonstrated. No such neurons have been found to date (Fuchs et al 1985). By contrast, if the eye-displacement model is the one implemented in the brain, it should be possible to document the existence of neurons that: (i) discharge in relation to eye displacement, (ii) are located in a nucleus that receives projections from the medium lead burst neurons, and (iii) project back to nuclei that contain medium lead burst neurons. The existence of such cells has recently been demonstrated (Moschovakis et al 1991b). Thus: (i) The firing pattern of one such neuron (Figure 5c) resembles that of downward medium lead burst neurons, in that it consists of high frequency bursts of spikes emitted shortly before downward saccades. (ii) Its soma was recovered within the interstitial nucleus of Cajal (Figure 5b, arrow), which receives a heavy projection from downward medium lead burst neurons (Moschovakis et al 1991b). (iii) Its axonal terminations targeted the riMLF (Figure 5a), which contains premotoneuronal downward medium lead burst neurons (Moschovakis et al 1991b). Therefore, the presently available morphophysi-ological evidence supports a version of the eye displacement model, at least as concerns the burst generator for downward saccades. Similar evidence obtained from neurons of the upward burst generator also supports a version, albeit a different one, of the eye displacement model (Moschovakis 1993).

The eye position and displacement models also differ in their proposed inputs. Namely, the position model relies upon information about target location in head-centered coordinates (E') while the displacement model needs a command to move the eyes in a certain direction and by a certain amount ($\Delta E'$). No neurons that compute target location in head-centered coordinates have yet been discovered. There are posterior parietal (Zipser & Andersen 1988) and intralaminar thalamic (Schlag et al 1980) neurons whose respon-siveness to visual stimuli is dependent on eye position. However, these cells do not encode the spatial location of targets, and further, their axonal projections are unknown. On the other hand, both the FEF (Segraves & Goldberg 1987) and the SC (Keller 1979, Moschovakis et al 1988b) issue commands to move the eyes in a certain direction and by a certain amount. Furthermore, both areas project to medium lead burst neurons, albeit

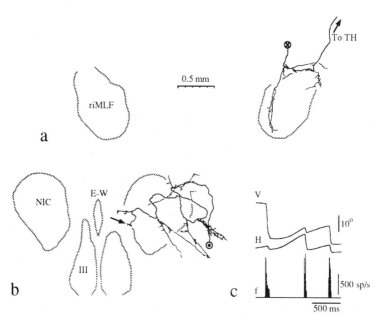

Figure 5 (*a, b*) Camera lucida reconstruction of the axonal system of a putative downward inhibitory feedback neuron intracellularly injected with horseradish peroxidase (adapted from Moschovakis et al 1991b). The rostral (*a*) and caudal (*b*) levels through the brainstem are about 1.5 mm apart. Arrow indicates the continuation of the fiber toward thalamic nuclei (To Th). Encircled dot and x symbols indicate that the axon can be followed in the more rostral or caudal level, respectively. Abbreviations as in Figure 3. (*c*) Discharge pattern of the same neuron (adapted from Moschovakis et al 1991b). Abbreviations as in Figure 1*a*.

indirectly, again providing experimental evidence in support of the displacement model.

HOW IS THE EYE DISPLACEMENT COMMAND THAT LEAVES THE SUPERIOR COLLICULUS COMPUTED?

Figure 4*c* illustrates a classical early answer to this question, known as the "foveation hypothesis" (Schiller & Koerner 1971). According to this hypothesis, the input to the SC is a retinal error signal (T/R) and is represented by the location of visually driven neurons in the superficial tectal layers. The activity of these superficial neurons is in turn thought to activate presaccadic neurons (M) that are located in the subjacent regions of the SC. Because the motor map of the deeper tectal layers, described above, is in register with the visual map of the superficial layers, the saccade that is due to the activation of the presaccadic neurons ultimately results in the displacement of the eye in such a way that the target ends up being projected onto the fovea.

Refutation of the foveation hypothesis was in part based on the absence of a crucial anatomical piece of evidence, namely connections between the superficial and the deeper tectal layers (Sparks 1986). Relevant evidence was not forthcoming despite considerable effort (reviewed in Edwards 1980). It is now known, that the superficial layers of the primate SC contain a particular class of neurons (L neurons) whose axons ramify profusely within the deeper tectal layers (Moschovakis et al 1988a). However, by the time the existence of these cells was established, an impressive array of psychophysical and neurophysiological objections to the foveation hypothesis had been amassed (Sparks 1986). Probably the most dramatic of these was the demonstration that subjects are able to saccade to a target briefly flashed on the fovea during the execution of a previous saccade (Hallett & Lightstone 1976). Under the circumstances of this paradigm, computation of the metrics of a saccade back to the location of the target cannot rely on retinal error alone, because the latter is equal to zero. Also, the foveation hypothesis could not account for the ability of monkeys to compensate for the perturbation of orbital position that results from the electrical stimulation of the SC, by executing saccades to the approximate location of a visual target extinguished before the electrical stimulation (Sparks & Mays 1983). Finally, the foveation hypothesis could not account for the discharge pattern of a particular class of tectal cells, called quasivisual (Qv), which carry a signal proportional to eye position error (i.e. the difference between desired and actual eye position), irrespective of whether this is due to the presence of a target in the cell's receptive field or because of a saccade that moved the eyes in such a way that the now extinct target lies within the cell's field (Sparks & Porter 1983, Mays & Sparks 1980).

To circumvent these objections, a model that relies on nonretinotopic encoding of target location (Figure 4d) was proposed (Sparks & Mays 1982). According to it, the target location in retinotopic coordinates is first combined with eye position information (+E) in order to obtain target location in spatial (i.e. head centered) coordinates (T/H). After a certain delay (i.e. just before saccade launching), the eye position information is subtracted (−E) from this signal, to obtain eye position error information (EPe). The latter is carried by quasivisual neurons in such a way that different amounts of EPe are coded by different sets of quasivisual cells, arranged over the mediolateral and anteroposterior extent of the SC in an organized fashion. EPe is held in memory (a topographic register) and is then used to generate the desired eye displacement signal that leaves the SC (Sparks & Mays 1982).

Despite arguments to the contrary (e.g. Hallett & Lightstone 1976, Scudder 1988, Sparks 1986, Sparks & Mays 1982), the evidential basis of the spatial hypothesis is still compatible with the usage of a purely retinotopic frame of reference (i.e. centered on the eye and moving with it) in the SC. Instead of relying on eye position information to remap target coordinates in a spatial and then back in a motor frame of reference, the SC could use a feedback signal proportional to eye displacement to extract the EPe signal from the Re signal. Conceptually, this process would resemble a vectorial subtraction, in that the vector of eye displacement would be subtracted from the vector of retinal error (Moschovakis 1987).

The flow of information through the SC according to this "vector subtraction" hypothesis is illustrated in Figure 4e. Here, and as in the foveation hypothesis, the input to the SC is Re. Also, in common to the spatial hypothesis, the EPe vector is first computed by the quasivisual cells and then used by tectal long lead burst neurons to compute the vector of desired eye displacement ($\Delta E'$). However, in contrast to both the foveation and the spatial hypotheses, it is a neural replica of $\Delta E'$ that is fed back to the SC, in the vector subtraction hypothesis. Here, to update the EPe vector, the $\Delta E'$ vector is subtracted from the Re vector (Moschovakis et al 1988b). More recently, a similar scheme concerning the FEF, was proposed (Goldberg & Bruce 1990).

Clearly, the vector subtraction and the spatial hypotheses differ considerably in terms of the signals they manipulate. For example, the spatial hypothesis relies on a signal of target location in head centered coordinates for its input. On the other hand, the vector subtraction hypothesis relies on a feedback signal proportional to saccadic eye displacement. Which of the two hypotheses is more consistent with presently available evidence concerning the response properties and connections of nerve cells? As already discussed, there is no evidence that neurons encode target location in head centered coordinates, or that such a signal reaches the SC, as required by the spatial hypothesis. In contrast, if a vector subtraction scheme is implemented in the brain, then it should be possible to demonstrate the existence of neurons that carry a signal proportional to eye displacement, are located in a region of the brain where axons of tectal long lead burst neurons terminate, and deploy terminal fields within the SC. The existence of such neurons has indeed been demonstrated; because of their activity and appearance, they have been termed reticulotectal long lead burst neurons (Moschovakis et al 1988b). Figure 6 provides an illustrative summary of pertinent data. The somata of reticulotectal long lead burst neurons (Figure 6, arrow) are located in a portion of the mesencephalic reticular formation that receives a heavy projection from tectal long lead burst neurons (Moschovakis et al 1988b), and from whence saccades are generated in response to electrical stimulation (Cohen et al 1986). Also, the axonal terminations of reticulotectal long lead burst neurons are confined to the SC (Figure 6). Finally, the activity of reticulotectal long lead burst neurons is related to saccadic occurrence; their rate of discharge starts increasing long before the start of the eye movement and reaches a peak that roughly coincides

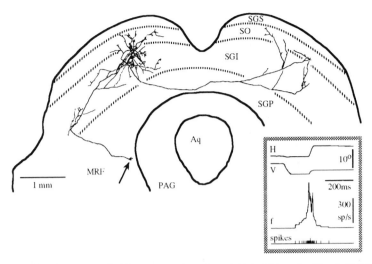

Figure 6   Complete camera lucida reconstruction of a reticulotectal long lead burst neuron, in the frontal plane, and an example of its saccade related activity (inset). (Abbreviations: Aq, aqueduct; MRF, mesencephalic reticular formation; PAG, periaqueductal grey; SGI, stratum griseum intermedium; SGP, stratum griseum profundum; SGS, stratum griseum superficiale; SO, stratum opticum. Other abbreviations as in Figure 1*a*.)

with it (Figure 6, inset). Because the correlation coefficient between the number of spikes in the burst and the amplitude of saccades in their on-direction can be as high as 0.85, the activity of reticulotectal long lead burst neurons provides a good estimate of the metrics of impending saccades. Although their on-directions are often horizontal, reticulotectal long lead burst neurons with vertical and oblique on-directions have also been disclosed (Moschovakis et al 1988b).

As with the spatial hypothesis, the advantages of the vector subtraction over the foveation hypothesis are evident only if the eyes move in the interval that elapses between the presentation of a target and its acquisition (cf objections to the foveation hypothesis, above). Why would a neural system ever have developed if its only function were to generate saccades in such artificial circumstances? Actually, it may have more to do with the need for a system that programs complicated sequences of saccades. Thus, the line of sight can serially visit important features of a complicated image without the need to reprocess visual information after each saccade. An additional advantage of the vector subtraction hypothesis is the ease with which it accounts for the frequently observed "corrective" saccades. This is the term used for small saccades, which follow incorrect saccades at a fraction of their latency (Becker 1976). To see how the vector subtraction hypothesis accounts for corrective saccades, let us assume that an inappropriate eye displacement command exits the SC. This results in the execution of an inappropriate saccade. Due to the existence of the feedback path embodied by reticulotectal long lead burst neurons, a copy of the inappropriate eye displacement command can be compared with the originally computed eye position error. After this comparison, a second, error correcting command can be issued without having to resample retinal error after the end of the first saccade. Consistent with this idea is the relatively short latency of corrective saccades and the fact that their execution does not depend on retinal feedback (Becker & Fuchs 1969).

CONCLUSION

All in all, presently available evidence suggests that the primate nervous system can accurately compute the amplitude and direction of saccadic eye movements thanks to the presence of two feedback loops connected in series (Figure 4f). The first one implements a vector subtraction scheme in the SC to create a command of desired eye displacement; this then controls a second feedback loop, namely an eye displacement type burst generator. Each burst generator contains neurons that can distribute its output to extraocular MNs in such a way as to account for eye conjugacy, and to the neural integrators in such a way as to account for their bidirectional modulation. It also contains neurons that can ensure the push-pull coupling between burst generators with opposite on directions and the two-dimensional coordination between orthogonal burst generators.

This eye-displacement twin feedback loop model is relatively complete (in the sense that it receives retinal information as input and produces motoneuron discharge as output). According to it, target acquisition proceeds without invoking eye position feedback signals. This in no way implies that all eye position information is irrelevant for brain function. Actually, the idea that a feedback signal indicating the position of the eyes in the orbit is crucial for the perception of visual direction has a long tradition (Helmholtz 1910, Holst 1954, Shebilske 1976), and some experimental support (Gauthier et al 1990). However, the fact that one part of the CNS (e.g. the visual system) makes use of certain information does not guarantee that a different part of it (e.g. the oculomotor system) will do the same, despite the apparent parsimony of this arrangement. Such issues can be decided only on the basis of evidence about the signals carried by, and the connections between, real neurons.

ACKNOWLEDGMENTS

Support provided by CODEST grant SCI-CT91-0643 (to AKM) and by National Institutes of Health grant EY-01670 (to SMH) is gratefully acknowledged.

Literature Cited

Anderson JH. 1981. Behavior of the vertical canal VOR in normal and INC-lesioned cats. In *Progress in Oculomotor Research*, ed. A Fuchs, W Becker, pp. 395–401. Amsterdam: Elsevier

Anderson JH, Precht W, Pappas C. 1979. Changes in the vertical vestibulo-ocular reflex due to kainic acid lesions of the interstitial nucleus of Cajal. *Neurosci. Lett.* 14: 259–64

Baker R, Berthoz A. 1974. Organization of vestibular nystagmus in oblique oculomotor system. *J. Neurophysiol.* 37:195–217

Baker R, Gresty M, Berthoz A. 1975. Neuronal activity in the prepositus hypoglossi nucleus correlated with vertical and horizontal eye movement in the cat. *Brain Res.* 101:366–71

Becker W. 1976. Do correction saccades depend exclusively on retinal feedback? A note on the possible role of non-retinal feedback. *Vision Res.* 16:425–7

Becker W, Fuchs AF. 1969. Further properties of the human saccadic system: Eye movements and correction saccades with and without visual fixation points. *Vision Res.* 9:1247–58

Büttner U, Büttner-Ennever JA, Henn V. 1977. Vertical eye movement related activity in the rostral mesencephalic reticular formation of the alert monkey. *Brain Res.* 130:239–52

Büttner-Ennever JA, Büttner U. 1988. The reticular formation. In *Neuroanatomy of the Oculomotor System*, ed. JA Büttner-Ennever, pp. 119–76. Amsterdam: Elsevier

Cannon SC, Robinson DA. 1987. Loss of the neural integrator of the oculomotor system from brain stem lesions in monkey. *J. Neurophysiol.* 57:1383–409

Cheron G, Godaux E. 1987. Disabling of the oculomotor neural integrator by kainic acid injections in the prepositus-vestibular complex of the cat. *J. Physiol.* 394:267–90

Cohen B, Henn V. 1972. Unit activity in the pontine reticular formation associated with eye movements. *Brain Res.* 46:403–10

Cohen B, Waitzman DM, Büttner-Ennever JA, Matsuo, V. 1986. Horizontal saccades and the central mesencephalic reticular formation. In *Progress in Brain Research*, ed. H-J Freund, U Büttner, B Cohen, J Noth, 64: 243–56. Amsterdam: Elsevier

Collewijn H, Erkelens CJ, Steinman RM. 1988a. Binocular co-ordination of human horizontal saccadic eye movements. *J. Physiol.* 404:157–82

Collewijn H, Erkelens CJ, Steinman RM. 1988b. Binocular co-ordination of human vertical saccadic eye movements. *J. Physiol.* 404:183–97

Crawford JD, Cadera W, Vilis T. 1991. Generation of torsional and vertical eye position signals by the interstitial nucleus of Cajal. *Science* 252:1551–53

Curthoys IS, Markham C, Furuya N. 1984. Direct projection of pause neurons to nystagmus-related excitatory burst neurons in the cat pontine reticular formation. *Exp. Neurol.* 83:414–22

Delgado-Garcia JM, del Pozo F, Baker R. 1986a. Behavior of neurons in the abducens nucleus of the alert cat. I. Motoneurons. *Neuroscience* 17:929–52

Delgado-Garcia JM, del Pozo F, Baker R. 1986b. Behavior of neurons in the abducens nucleus of the alert cat. II. Internuclear neurons. *Neuroscience* 17:953–73

Edwards SB. 1980. The deep cell layers of the superior colliculus: Their reticular characteristics and structural organization. In *The Reticular Formation Revisited*, ed. A Hobson, M Brazier, pp. 193–209. New York: Raven

Fuchs AF, Kaneko CRS, Scudder CA. 1985. Brainstem control of saccadic eye movements. *Annu. Rev. Neurosci.* 8:307–37

Fuchs AF, Scudder CA, Kaneko CRS. 1988. Discharge patterns and recruitment order of identified motoneurons and internuclear neurons in the monkey abducens nucleus. *J. Neurophysiol.* 60:1874–95

Fukushima K. 1991. The interstitial nucleus of Cajal in the midbrain reticular formation and vertical eye movement. *Neurosci. Res.* 10: 159–87

Furuya N, Markham CH. 1982. Direct inhibitory synaptic linkage of pause neurons with burst inhibitory neurons. *Brain Res.* 245: 139–43

Gauthier GM, Nommay D, Vercher J-L. 1990. The role of ocular muscle proprioception in visual localization of targets. *Science* 249: 58–61

Goldberg ME, Bruce CJ. 1990. Primate frontal eye fields. III. Maintenance of a spatially accurate saccade signal. *J. Neurophysiol.* 64:489–508

Grantyn A, Grantyn R. 1982. Axonal patterns and sites of termination of cat superior colliculus neurons projecting to the tectobulbo-spinal tract. *Exp. Brain Res.* 46:243–56

Grantyn A, Grantyn R, Berthoz A, Ribas J. 1982. Tectal control of vertical eye movements: A search for underlying neuronal circuits in the mesencephalon. In *Physiological and Pathological Aspects of Eye Movements*, ed. A Roucoux, M Crommelinck, pp. 337–44. The Hague: Junk

Grantyn A, Grantyn R. 1976. Synaptic actions of tectofugal pathways on abducens motoneurons in the cat. *Brain Res.* 105:269–85

Guitton D, Volle M. 1987. Gaze control in humans: eye-head coordination during orienting movements to targets within and beyond the oculomotor range. *J. Neurophysiol.* 58:427–59

Hallett PE, Lightstone AD. 1976. Saccadic eye movements to flashed targets. *Vision Res.* 16:107–14

Harting JK, Huerta MF, Frankfurter AJ, et al. 1980. Ascending pathways from the monkey superior colliculus: an autoradiographic analysis. *J. Comp. Neurol.* 192:852–82

Helmholtz H. 1910. *Treatise on Physiological Optics.* (Transl. JPC Southall, 1924) New York: Dover (From German)

Henn V, Cohen B. 1976. Coding of information about rapid eye movements in the pontine reticular formation of alert monkeys. *Brain Res.* 108:307–25

Henn V, Lang W, Hepp K, Reisine H. 1984. Experimental gaze palsies in monkeys and their relation to human pathology. *Brain* 107:619–36

Henn V, Schnyder H, Hepp K, Reisine H. 1983. Loss of vertical rapid eye movements after kainic acid lesions in the rostral mesencephalon in the rhesus monkey. *Proc. Soc. Neurosci.* 9:749 (Abstr.)

Hepp K, Henn V. 1983. Spatio-temporal recoding of rapid eye movement signals in the monkey paramedian pontine reticular formation (PPRF). *Exp. Brain Res.* 52:105–20

Hepp K, Henn V. 1985. Iso-frequency curves of oculomotor neurons in the rhesus monkey. *Vision Res.* 25:493–9

Hepp K, Henn V, Vilis T, Cohen B. 1989. Brainstem regions related to saccade generation. In *The Neurobiology of Saccadic Eye Movements,* ed. RE Wurtz, ME Goldberg, pp. 105–212. Amsterdam: Elsevier

Hepp K, Vilis T, Henn V. 1988. On the generation of rapid eye movements in three dimensions. In *Representation of Three-dimensional Space in the Vestibular, Oculomotor, and Visual Systems,* ed. B Cohen, V Henn, pp. 140–53. New York: NY Acad. Sci.

Hering E. 1868. *The Theory of Binocular Vision.* (Transl. B. Bridgeman, 1977) New York: Plenum (From German)

Highstein SM, Baker R. 1978. Excitatory synaptic terminations of abducens internuclear neurons on medial rectus motoneurons: relationship to syndrome of internuclear ophthalmoplegia. *J. Neurophysiol.* 41:1647–61

Highstein SM, Karabelas A, Baker R, McCrea RA. 1982. Comparison of the morphology of physiologically identified abducens motor and internuclear neurons in the cat: A light

microscopic study employing the intracellular injection of horseradish peroxidase. *J. Comp. Neurol.* 208:369–81

Hikosaka O, Igusa Y, Imai H. 1978a. Firing pattern of prepositus hypoglossi and adjacent reticular neurons related to vestibular nystagmus in the cat. *Brain Res.* 144:395–403

Hikosaka O, Igusa Y, Imai H. 1980. Inhibitory connections of nystagmus-related reticular burst neurons with neurons in the abducens, prepositus hypoglossi and vestibular nuclei in the cat. *Exp. Brain Res.* 39:301–11

Hikosaka O, Igusa Y, Nakao S, Shimazu H. 1978b. Direct inhibitory synaptic linkage of pontomedullary reticular burst neurons with abducens motoneurons in the cat. *Exp. Brain Res.* 33:337–52

Hikosaka O, Kawakami T. 1977. Inhibitory reticular neurons related to the quick phase of vestibular nystagmus—their location and projection. *Exp. Brain Res.* 27:377–96

Holst E. 1954. Relations between the central nervous system and the peripheral organs. *Br. J. Anim. Behav.* 2:89–94

Infante C, Leiva J. 1986. Simultaneous unitary neuronal activity in both superior colliculi and its relation to eye movements in the cat. *Brain Res.* 381:390–2

Kaneko CRS. 1990. *Tests of two models of the neural saccade generator: saccadic eye movement deficits following ibotenic acid lesions of the nuclei raphe interpositus and prepositus hypoglossi in monkey.* Presented at Satellite Symp. Barany Soc. Meet., 16th, Tokyo

Keller EL. 1974. Participation of medial pontine reticular formation in eye movement generation in monkey. *J. Neurophysiol.* 37:316–32

Keller EL. 1979. Coliculoreticular organization of the oculomotor system. In *Progress in Brain Research,* ed. R Granit, O Pompeiano, 50:725–34. Amsterdam: Elsevier

King WM, Fuchs AF. 1979. Reticular control of vertical saccadic eye movements by mesencephalic burst neurons. *J. Neurophysiol.* 42:861–76

King WM, Fuchs AF, Magnin M. 1981. Vertical eye movement-related responses of neurons in midbrain near interstitial nucleus of Cajal. *J. Neurophysiol.* 46:549–62

King WM, Lisberger SG, Fuchs AF. 1986. Oblique saccadic eye movements of primates. *J. Neurophysiol.* 56:769–84

King WM, Precht W, Dieringer N. 1980. Afferent and efferent connections of cat omnipause neurons. *Exp. Brain Res.* 38:395–403

Langer TP, Kaneko CRS. 1984. Brainstem afferents to the omnipause region in the cat: A horseradish peroxidase study. *J. Comp. Neurol.* 230:444–58

Lopez-Barneo J, Darlot C, Berthoz A, Baker R. 1982. Neuronal activity in prepositus nucleus correlated with eye movement in the alert cat. *J. Neurophysiol.* 47:329–52

Luschei ES, Fuchs AF. 1972. Activity of brain stem neurons during eye movements of alert monkeys. *J. Neurophysiol.* 35:445–61

Maciewicz RJ, Eagen K, Kaneko CRS, Highstein SM. 1977. Vestibular and medullary brain stem afferents to the abducens nucleus in the cat. *Brain Res.* 123:229–40

Mays LE, Sparks DL. 1980. Dissociation of visual and saccade-related responses in superior colliculus neurons. *J. Neurophysiol.* 43:207–32

McCrea RA, Strassman A, Highstein SM. 1986. Morphology and physiology of abducens motoneurons and internuclear neurons intracellularly injected with horseradish peroxidase in alert squirrel monkeys. *J. Comp. Neurol.* 243:291–308

McCrea RA, Yoshida K, Berthoz A, Baker R. 1980. Eye movement related activity and morphology of second order vestibular neurons terminating in the cat abducens nucleus. *Exp. Brain Res.* 40:468–73

Moschovakis AK. 1987. *Observations on the appearance and function of neurons in the primate superior colliculus. An intracellular HRP study.* PhD thesis. Washington Univ., St. Louis. 204 pp

Moschovakis AK. 1993. Neural network simulations of the primate oculomotor system. I. The vertical saccadic burst generator. *Biol. Cybern.* In press

Moschovakis AK, Karabelas AB. 1982. Tectotectal interactions in the cat. *Proc. Soc. Neurosci.* 8:293 (Abstr.)

Moschovakis AK, Karabelas AB. 1985. Observations on the somatodendritic morphology and axonal trajectory of intracellularly HRP-labeled efferent neurons located in the deeper layers of the superior colliculus of the cat. *J. Comp. Neurol.* 239:276–308

Moschovakis AK, Karabelas AB, Highstein SM. 1988a. Structure-function relationships in the primate superior colliculus. I. Morphological classification of efferent neurons. *J. Neurophysiol.* 60:232–62

Moschovakis AK, Karabelas AB, Highstein SM. 1988b. Structure-function relationships in the primate superior colliculus. II. Morphological identity of presaccadic neurons. *J. Neurophysiol.* 60:263–302

Moschovakis AK, Scudder CA, Highstein SM. 1990. A morphological basis for Hering's law: Projections to extraocular motoneurons. *Science* 248:1118–9

Moschovakis AK, Scudder CA, Highstein SM. 1991a. Structure of the primate burst generator. I. Medium-lead burst neurons with upward on-directions. *J. Neurophysiol.* 65:203–17

Moschovakis AK, Scudder CA, Highstein SM, Warren JD. 1991b. Structure of the primate burst generator. II. Medium-lead burst neurons with downward on-directions. *J. Neurophysiol.* 65:218–29

Nakao S, Curthoys I, Markham C. 1980. Direct inhibitory projection of pause neurons to nystagmus-related pontomedullary reticular burst neurons in the cat. *Exp. Brain Res.* 40:283–93

Nakao S, Shiraishi Y. 1983. Excitatory and inhibitory synaptic inputs from the medial mesodiencephalic junction to vertical eye movement-related motoneurons in the cat oculomotor nucleus. *Neurosci. Lett.* 42:125–30

Nakao S, Shiraishi Y. 1985. Direct excitatory and inhibitory synaptic inputs from the medial mesodiencephalic junction to motoneurons innervating extraocular oblique muscles in the cat. *Exp. Brain Res.* 61:62–72

Nakao S, Shiraishi Y, Oda H, Inagaki, M. 1988. Direct inhibitory projection of pontine omnipause neurons to burst neurons in the Forel's field H controlling vertical eye movement-related motoneurons in the cat. *Exp. Brain Res.* 70:632–6

Peck CK. 1990. Neuronal activity related to head and eye movements in cat superior colliculus. *J. Physiol.* 421:79–104

Raybourn MS, Keller EL. 1977. Colliculoreticular organization in primate oculomotor system. *J. Neurophysiol.* 40:861–78

Robinson DA. 1970. Oculomotor unit behavior in the monkey. *J. Neurophysiol.* 33:393–404

Robinson DA. 1975. Oculomotor control signals. In *Basic Mechanisms of Ocular Motility and their Clinical Implications*, ed. P Bach-y-Rita, G Lennerstrand, pp. 337–74. Oxford: Pergamon

Robinson DA. 1981. The use of control systems analysis in the neurophysiology of eye movements. *Annu. Rev. Neurosci.* 4:463–503

Sasaki S, Shimazu H. 1981. Reticulovestibular organization participating in generation of horizontal fast eye movement. *Ann. NY Acad. Sci.* 374:130–43

Schiller PH, Koerner F. 1971. Discharge characteristics of single units in the superior colliculus of the alert rhesus monkey. *J. Neurophysiol.* 34:920–36

Schlag J, Schlag-Rey M, Peck CK, Joseph J-P. 1980. Visual responses of thalamic neurons depending on the direction of gaze and the position of targets in space. *Exp. Brain Res.* 40:170–84

Scudder CA. 1988. A new local feedback model of the saccadic burst generator. *J. Neurophysiol.* 59:1455–75

Scudder CA, Highstein SM, Karabelas TA, Moschovakis AK. 1989. Anatomy and physiology of long-lead burst neurons (LLBNs). *Proc. Soc. Neurosci.* 15:238 (Abstr.)

Segraves MA, Goldberg ME. 1987. Functional properties of corticotectal neurons in the monkey's frontal eye field. *J. Neurophysiol.* 58:1387–419

Shebilske WL. 1976. Extraretinal information in corrective saccades and inflow vs. outflow theories of visual direction constancy. *Vision Res.* 16:621–8

Sparks DL. 1986. Translation of sensory signals into commands for control of saccadic eye movements: role of primate superior colliculus. *Physiol. Rev.* 66:118–71

Sparks DL, Mays LE. 1980. Movement fields of saccade-related burst neurons in the monkey superior colliculus. *Brain Res.* 190:39–50

Sparks DL, Mays LE. 1982. Role of the monkey superior colliculus in the spatial localization of saccade targets. In *Spatially Oriented Behavior,* ed. A Hein, M Jeannerod, pp. 63–85. New York: Springer

Sparks DL, Mays LE. 1983. Spatial localization of saccade targets. I. Compensation for stimulation-induced perturbations in eye position. *J. Neurophysiol.* 49:45–63

Sparks DL, Mays LE. 1990. Signal transformations required for the generation of saccadic eye movements. *Annu. Rev. Neurosci.* 13:309–36

Sparks DL, Porter JD. 1983. Spatial localization of saccade targets. II. Activity of superior colliculus neurons preceding compensatory saccades. *J. Neurophysiol.* 49:64–74

Strassman A, Evinger C, McCrea RA, et al. 1987. Anatomy and physiology of intracellularly labelled omnipause neurons in the cat and squirrel monkey. *Exp. Brain Res.* 67:436–40

Strassman A, Highstein SM, McCrea RA. 1986a. Anatomy and physiology of saccadic burst neurons in the alert squirrel monkey. I. Excitatory burst neurons. *J. Comp. Neurol.* 249:337–57

Strassman A, Highstein SM, McCrea RA. 1986b. Anatomy and physiology of saccadic burst neurons in the alert squirrel monkey. II. Inhibitory burst neurons. *J. Comp. Neurol.* 249:358–80

Tomlinson RD, Robinson DA. 1984. Signals in vestibular nucleus mediating vertical eye movements in the monkey. *J. Neurophysiol.* 51:1121–36

van Gisbergen JAM, Robinson DA, Gielen S. 1981. A quantitative analysis of generation of sacccadic eye movements by burst neurons. *J. Neurophysiol.* 45:417–42

Yoshida K, McCrea R, Berthoz A, Vidal P. 1982. Morphological and physiological characteristics of inhibitory burst neurons controlling horizontal rapid eye movements in the alert cat. *J. Neurophysiol.* 48:761–84

Zee DS, Optican LM, Cook JD, et al. 1976. Slow saccades in spinocerebellar degeneration. *Arch. Neurol.* 33:243–51

Zipser D, Andersen RA. 1988. A back-propagation programmed network that simulates response properties of a subset of posterior parietal neurons. *Nature* 331:679–84

Annu. Rev. Neurosci. 1994. 17:489–517

NORMAL AND ABNORMAL BIOLOGY OF THE β-AMYLOID PRECURSOR PROTEIN

Dennis J. Selkoe

Department of Neurology and Program in Neuroscience, Harvard Medical School and Center for Neurologic Diseases, Brigham and Women's Hospital, Boston, Massachusetts 02115

KEY WORDS: Alzheimer's disease, neurodegeneration, amyloid β-protein, transmembrane proteins, amyloidosis

INTRODUCTION

Few macromolecules have become the object of more intense chemical and biological scrutiny in recent years than has the β-amyloid precursor protein. This highly conserved and widely expressed integral membrane protein is the focus of a rapidly growing number of studies aimed at its characterization by protein chemical, molecular biological, cell biological, and neuroanatomical methods. The principal reason for this attention is clear. The early promise that a 39–43 residue fragment of βAPP, the amyloid β-protein (Aβ), might play a central role in the pathogenesis of Alzheimer's disease has increasingly been borne out. Although there is still substantial debate about exactly when Aβ becomes involved in Alzheimer's disease and how attractive a therapeutic target it represents, accumulating evidence now points to an early and sometimes causative role in the pathogenetic cascade. βAPP has become a compelling example of the phenomenon of research that begins with a strictly disease-oriented focus giving rise to novel insights about the normal biology of a whole family of macromolecules, in this case about the processing of certain cell surface, single membrane-spanning polypeptides.

Here, I review recent information about the normal structure and function of βAPP and how its Aβ fragment contributes to Alzheimer's disease. Such

489

a summary is, of necessity, selective rather than exhaustive, and its perspective may stimulate others to put forward different views of this complex area of applied biology.

A VARIETY Of Aβ-CONTAINING LESIONS OCCUR IN ALZHEIMER'S DISEASE

Because the starting point for βAPP research was the neuropathology of Alzheimer's disease, an updated view of the nature of the structural alterations is in order. Alzheimer himself described the characteristic light microscopic lesions—neurofibrillary tangles and amyloid-bearing senile plaques—in the limbic and cerebral cortices by using the Bielschowsky method of silver impregnation. The neurofibrillary tangles were eventually shown by electron microscopy to be composed of nonmembrane-bound bundles of paired, helically-wound ~10-nm filaments (PHF) in the perinuclear cytoplasm of selected neurons (Kidd 1963, Terry 1963). The senile, or neuritic, plaque is a complex, multicellular lesion, the temporal genesis of which is only imperfectly understood. So-called "classical" neuritic plaques consist of a compacted spherical deposit of extracellular, ~8-nm amyloid filaments surrounded by variable numbers of dilated (dystrophic) neurites, both axonal terminals and dendrites. Many such classical plaques contain apparently activated microglial cells intimately surrounding the amyloid core (Wisniewski et al 1989) as well as reactive fibrous astrocytes around the periphery of the plaque.

Although many plaques with these features can be found in the hippocampus, amygdala, entorhinal cortex, and cerebral association cortex in Alzheimer's disease, the advent of highly sensitive antibodies to purified or synthetic Aβ has shown that they are a minority of all cerebral Aβ deposits. Far more abundant in most Alzheimer brains are amorphous, roughly spherical, and less dense deposits of Aβ referred to as "diffuse" or "pre-amyloid" plaques (see for example, Tagliavini et al 1988, Yamaguchi et al 1988). When observed with electron microscopy, they display very few, if any, structurally altered neurites, astrocytes, or microglial cells, and their Aβ immunoreactivity is not explained by amyloid filaments, which are very sparse or absent (Verga et al 1989; Yamaguchi et al 1989, 1990). Precisely what structural and biochemical form the Aβ of diffuse plaques takes remains unsettled. In Down's syndrome (trisomy 21), small or moderate numbers of diffuse plaques can be found in the cerebral cortex when subjects are in their teens or twenties, and these plaques are unassociated with surrounding neuritic dystrophy, gliosis, or neurofibrillary tangles (Giaccone et al 1989, Mann et al 1990). Indeed, occasional diffuse plaques have been described in the brains of Down's patients less than three years old (Ter-Minassian et al 1992).

of Down's patients less than three years old (Ter-Minassian et al 1992). Because virtually all Down's subjects show classical neuritic plaques and neurofibrillary tangles if they survive beyond age 50, it has been postulated that Aβ deposition can long precede the other structural changes of Alzheimer's disease. In support of this hypothesis are two other findings: (*a*) highly variable numbers of diffuse plaques alone are observed in the brains of some neurologically normal individuals dying after age 60 or so; and (*b*) in Alzheimer's disease, abundant diffuse plaques occur in areas such as striatum and cerebellar molecular cortex in the virtual absence of local neuronal or glial alteration and in the absence of any neurological deficits associated with these brain regions (Joachim et al 1989b, Gearing et al 1993). These findings suggest but do not prove that at least some diffuse plaques may evolve gradually into denser, mostly fibrillar amyloid plaques, perhaps by the accrual of high local concentrations of Aβ. If so, most Aβ deposits have still not undergone this evolution by the end of the disease.

Besides diffuse and compacted plaques, extracellular Aβ deposits often occur in other morphological forms. Long, ribbon-like amorphous-looking deposits are frequently observed in the superficial subpial cortex (neocortical layer I). Some or many meningeal and intracerebral microvessels, particularly arterioles and capillaries, bear fibrillar Aβ deposits in their walls (Mandybur 1975, Vinters et al 1988). Ultrastructurally, this vascular β-amyloid has a clear predilection for the abluminal portion of the vessel basement membrane (Yamaguchi et al 1992), not unlike certain systemic vascular amyloid deposits known to be of circulating origin (e.g. transthyretin and AA amyloid). This parallel with peripheral vascular amyloidosis led to the detection of small amounts of Aβ-immunoreactive material in the walls of occasional microvessels of nonneural tissues such as skin, intestine, and spleen in Alzheimer's disease, Down's syndrome, and normal aging (Joachim et al 1989a, Lemere et al 1991, Ikeda et al 1993). Attempts to extract and biochemically characterize the sparse Aβ-like deposits from samples of Down's syndrome or AD skin have been unsuccessful, whereas electron microscopy has demonstrated amorphous, electron-dense, nonfibrillar material at the site of the Aβ immunoreactivity (H Yamaguchi & D Selkoe, unpublished data). In a different disorder, typical fibrillar β-amyloid, which reacts with the classical amyloid dye, Congo red, has been detected within muscle fibers of patients with inherited or sporadic forms of the progressive muscle disease inclusion body myositis (Askanas et al 1992). This observation provides the first clear example of Aβ deposition intracellularly, and it supports the notion that Aβ can occasionally be deposited outside of the nervous system.

In addition to the altered neurites that intimately surround many fibrillar amyloid plaques, a widespread structural abnormality of cortical neurites,

Kosik 1987), or simply dystrophic neurites, is often present in Alzheimer cerebral cortex. Abnormal neurites that are either associated or unassociated with plaques may contain PHF and antigenically related straight 10- to 15-nm filaments. On the other hand, it appears that not all neurites or perikarya that degenerate in Alzheimer cortex necessarily contain PHF.

STRUCTURE AND EXPRESSION OF THE B-AMYLOID PRECURSOR PROTEIN

In 1984, Glenner & Wong first isolated and purified the subunit protein of the meningovascular amyloid filaments in Alzheimer's disease and determined its amino-terminal sequence (Glenner & Wong 1984). They dubbed this novel $\sim$ 4-kDa protein the amyloid β-peptide (Aβ). Shortly thereafter, compacted amyloid plaque cores were partially purified from AD cerebral cortex by Masters and colleagues (1985) and independently by other laboratories (Gorevic et al 1986, Roher et al 1986, Selkoe et al 1986). These studies demonstrated that the subunit of the amyloid in plaque cores was apparently the same 4-kDa peptide, with an amino acid composition indistinguishable from that of the meningovascular Aβ. However, there was some disagreement among these early studies as to whether the core-derived Aβ could be quantitatively sequenced or rather whether at least some of it had a blocked amino-terminus. Recent studies suggest that both free and blocked N-termini can indeed be found in isolated plaque cores (Mori et al 1992, Miller et al 1993).

The provision of the partial amino acid sequence of Aβ enabled four laboratories independently to clone cDNAs encoding part or all of the precursor of the peptide (Goldgaber et al 1987, Kang et al 1987, Robakis et al 1987, Tanzi et al 1987). The deduced amino acid sequence obtained from a full-length cDNA predicted a large precursor polypeptide of 695 amino acids containing a single hydrophobic stretch near its carboxyl terminus and having the properties of a membrane-spanning domain (Kang et al 1987) (Figure 1). The precursor, now generally designated β-amyloid precursor protein (βAPP) or sometimes simply APP, contained a 17-residue signal peptide for transport into the endoplasmic reticulum and two consensus sequences for N-linked glycosylation (Kang et al 1987). Of particular interest was the observation that the 39- to 43-residue Aβ sequence began 28 residues amino-terminal to the single transmembrane domain and extended 11–15 residues into that domain. Thus, the predicted primary structure of βAPP immediately raised the question of how a hydrophobic, putatively membrane-anchoring portion of the precursor could be cleaved and released from cells, where it accumulated as β-amyloid deposits in the extracellular spaces of the brain and its microvessels.

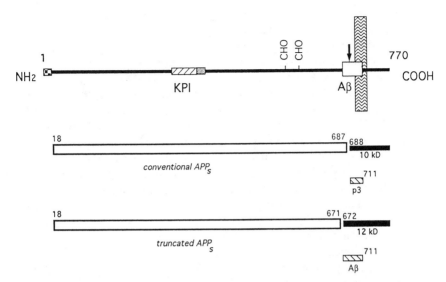

Figure 1   Schematic diagrams of the β-amyloid precursor protein and its principle metabolic derivatives. The upper diagram depicts the largest of the known βAPP alternate transcripts, comprising 770 amino acids. Regions of interest are indicated at their correct relative postions. A 17 residue signal peptide occurs at the N-terminus. Two alternatively spliced exons of 56 and 19 amino acids are inserted at residue 289; the first contains a serine protease inhibitor domain of the Kunitz type (KPI). Two sites of N-glycosylation (CHO) are found at residues 542 and 571. A single membrane-spanning domain at amino acids 700–723 is indicated by the vertical hatched bar. The amyloid β-protein (Aβ) fragment (white box) includes 28 residues just outside the membrane plus the first 11–15 residues of the transmembrane domain. The arrow indicates the site (after residue 687) of a constitutive proteolytic cleavage that enables secretion of the large, soluble ectodomain of βAPP (APPs) into the medium and retention of the 83-residue C-terminal fragment (~10 kD) in the membrane (middle diagram). The lower diagram depicts the alternative proteolytic cleavage after residue 671 that results in the formation of an ~12 kD C-terminal fragment. The latter may serve as an intermediate in the generation of Aβ, whereas the 10 kD fragment is believed to give rise to the p3 peptide.

lated as β-amyloid deposits in the extracellular spaces of the brain and its microvessels.

   Northern analyses and in situ hybridization demonstrated that βAPP was widely expressed in virtually all neural and nonneural mammalian tissues, although the highest levels of expression appeared to be in brain and kidney. Within brain, neurons showed particularly high levels of expression of βAPP$_{695}$. The βAPP gene was detected in a wide range of mammals, and homologous proteins have been identified in *Drosophila* (Rosen et al 1989) and *C. elegans* (Rosoff & Li 1992). In humans, the βAPP gene was found to be localized to the mid-portion of the long arm of chromosome 21. This observation strongly suggested that the virtually invariant development of

The βAPP gene undergoes alternative exon splicing to yield several distinct isoforms. The most widely and abundantly expressed of these is a 751-residue protein that contains a 56–amino acid insert in the middle of the extramembranous region with ~50% homology to the Kunitz family of serine protease inhibitors (Kitaguchi et al 1988, Ponte et al 1988, Tanzi et al 1988). A 770-residue isoform contains this exon plus an immediately adjacent exon encoding a 19–amino acid sequence of unknown functional significance. There is also a soluble, secreted isoform that contains the KPI exon but lacks the Aβ, transmembrane, and cytoplasmic regions (de Sauvage & Octave 1989). The role, if any, of the KPI motif in the proteolytic processing of βAPP itself has not been clearly established. An additional alternatively spliced form has been described in macrophage/microglial-type cells that lack the exon immediately preceding the two exons comprising the Aβ region (Koenig et al 1992). Whether alternative splicing of βAPP or the expression of a particular transcript has any direct bearing on the processing of the precursor to release Aβ remains unknown.

A variety of posttranslational modifications of βAPP have been characterized in cultured cells. Pulse-chase experiments have shown that βAPP undergoes first N-linked and then O-linked glycosylation in the Golgi, following which a variable portion of the molecules becomes inserted at the cell surface. The protein has also been shown to undergo phosphorylation (Oltersdorf et al 1990) and sulfation (Weidemann et al 1989, Oltersdorf et al 1990). Following its N- and O-linked glycosylation, mature βAPP can undergo proteolytic cleavage after residue 612 of βAPP_{695} (residue 16 of Aβ) to release the large, soluble, extramembranous portion (designated APP_s) into the medium (Schubert et al 1989b, Weidemann et al 1989, Esch et al 1990, Sisodia et al 1990). This scission leaves a ~10-kDa carboxyl-terminal fragment containing the transmembrane and cytoplasmic domains within the cell (Selkoe et al 1988). Because such constitutive secretion precludes formation of intact Aβ, it was assumed until recently that release of the Aβ region must involve abnormal proteolytic processing (see below). The percentage of the precursor that undergoes this secretory cleavage appears to vary widely among cell types but seems not to exceed about a third of newly synthesized molecules (Weidemann et al 1989). Much of the remaining polypeptide apparently remains inserted into internal membranes in cells such as primary astrocytes, microglia, and neurons (Haass et al 1991). The various alternatively spliced and posttranslationally modified forms of full-length βAPP just reviewed appear as a complex group of polypeptides migrating at ~100–140 kDa in electrophoretic gels (Selkoe et al 1988). An additional posttranslational modification of βAPP that markedly increases its molecular weight is the apparent addition of glycosaminoglycan side chains to yield a

posttranslational modification of βAPP that markedly increases its molecular weight is the apparent addition of glycosaminoglycan side chains to yield a chondroitin sulfate proteoglycan (Shioi et al 1992). This specialized form has been described in C6 glioma cells and rat brain.

At least two distinct mammalian gene products with a high degree of homology to βAPP have recently been identified. One of these, called amyloid precursor-like protein (APLP) 1, is encoded by a gene on human chromosome 19 and shows a high degree of sequence conservation in the extramembranous and cytoplasmic regions but divergence in the Aβ and transmembrane domains (Wasco et al 1992). The other, APLP 2, has similar features to APLP 1 (Wasco et al 1993). The effects that the expression of these molecules may have on the expression and processing of βAPP and the generation of Aβ are now under study. Thus, βAPP is one member of a highly conserved gene family. The existence of these structurally similar molecules in many cells and tissues complicates the correct biochemical identificiation of βAPP in some kinds of experiments.

CELLULAR PROCESSING OF βAPP AND THE PHYSIOLOGICAL PRODUCTION OF Aβ

Because the first proteolytic processing event defined for βAPP involved a secretory cleavage within the Aβ region (Esch et al 1990), it was concluded that alternative and presumably abnormal processing steps must sometimes occur to liberate the amyloid peptide. Chen et al (1990) called attention to the fact that βAPP contains a conserved asn-pro-xxx-tyr consensus motif in its cytoplasmic tail that might mediate its reinternalization from the cell surface via clathrin-coated vesicles and subsequent targeting to the endosomal/lysosomal system, as previously observed in the low-density lipoprotein (LDL) receptor, the insulin receptor, the EGF receptor, and several other single membrane spanning surface proteins. The presence of this signal, as well as the apparent localization of some βAPP molecules to lysosomes (Benowitz et al 1989, Cole et al 1989), led Younkin and colleagues to develop evidence of lysosomal generation of βAPP fragments. They observed that βAPP-transfected cells treated with certain inhibitors of lysosomal proteolysis, such as leupeptin or ammonium chloride, accumulated several low molecular weight carboxyl-terminal fragments of the precursor, some of which showed a size and Aβ immunoreactivity suggesting that they contained the intact Aβ region (Golde et al 1992). Fragments with these properties were also detected in human postmortem brain by these (Estus et al 1992) and other investigators (Nordstedt et al 1991, Tamaoka et al 1992). This work provided indirect evidence for the involvement of lysosomal processing in the generation of

the reinternalization of βAPP and its targeting to late endosomes/lysosomes; purification of these vesicles confirmed that they contained not only mature βAPP but also an array of low molecular weight C-terminal fragments containing Aβ (Haass et al 1992a).

These experiments thus defined a second normal processing pathway for βAPP that does not involve cleavage of the precursor within the Aβ region. Moreover, at least two secreted APP$_s$ derivatives that do not end at residue 16 of Aβ have been identified. One of these is a longer soluble form that appears to extend at least to residue Aβ$_{28}$ and probably beyond (Anderson et al 1992); whether this derivative contains the entire Aβ region is not yet certain. The other species is a shorter form of APP$_s$ that appears to end at met$_{596}$ of βAPP$_{695}$ (Seubert et al 1993). The existence of this intriguing soluble derivative indicates that normal secretion of βAPP can involve a proteolytic cleavage that creates the amino terminus of the Aβ peptide. This observation shows that secretory cleavage is not necessarily incompatible with the generation of Aβ and that the amyloid fragment may arise in part from secretory processing at or near the cell surface.

These findings suggest that both the exocytic and endocytic pathways for βAPP may be capable of generating Aβ-bearing fragments. In these and all previous studies of βAPP processing, the actual production of Aβ was not specifically searched for. It had been widely assumed that the membrane-anchored Aβ sequence would not be found as a free peptide in the absence of membrane injury that could somehow allow a protease to cleave βAPP at the C-terminus of Aβ. Further, it was assumed that Aβ was only produced in the brain tissue of certain aged mammals, e.g. primates (including humans) and dogs, and in patients with Aβ-type amyloidosis. Moreover, the hydrophobic, self-aggregating nature of Aβ suggested that it would not occur as a soluble, circulating peptide in normal biological fluids. It therefore came as a surprise when three laboratories reported the secretion of soluble Aβ into the conditioned media of a variety of primary or transfected βAPP-expressing cells under normal metabolic conditions (Haass et al 1992b, Seubert et al 1992, Shoji et al 1992). Subconfluent cultures of apparently healthy, intact cells routinely release the ~ 4-kDa Aβ peptide into medium in high picomolar to low nanomolar concentrations. The peptide can readily be immunoprecipitated from the conditioned media of radiolabeled cells and detected autoradiographically (Haass et al 1992b, Shoji et al 1992), and an Aβ-specific sandwich ELISA has been developed to quantitate its levels in culture media and biological fluids (Seubert et al 1992). Aβ was also detected in normal cerebrospinal fluid (CSF) of humans and lower mammals, including rats and guinea pigs (Seubert et al 1992, Shoji et al 1992). Moreover, Aβ-immunoreactivity has been detected by the ELISA in normal human plasma; the

guinea pigs (Seubert et al 1992, Shoji et al 1992). Moreover, Aβ-immunoreactivity has been detected by the ELISA in normal human plasma; the chemical nature of this species has not yet been established (Seubert et al 1992). Initial analysis of CSF by Western blotting showed no major change in the levels of Aβ in a few patients with Alzheimer's disease (Shoji et al 1992). High-speed centrifugation of tissue culture medium has shown that the Aβ is virtually entirely soluble, in contrast to its insoluble, aggregated state in Aβ brain tissue (Haass et al 1992b). Purification of Aβ from both culture medium and human CSF has confirmed the expected Aβ N-terminal sequence, and mass spectrometry has revealed that at least a portion of the secreted peptide is 40 residues long (Seubert et al 1992), just as has been detected in Alzheimer brain parenchyma (Mori et al 1992) and microvessels (Joachim et al 1988, Miller et al 1993, Roher et al).

The discovery that Aβ is continuously produced and secreted by cells under normal metabolic conditions and is present in biological fluids has several implications. First, this finding enables the dynamic study of Aβ production and regulation in vitro and in vivo and the assessment of the effects of a wide range of physiological and pharmacological modulators on this process. Heretofore, Aβ had only been examined biochemically in small amounts after laborious isolation of insoluble, aggregated amyloid deposits from postmortem human brain. Second, the detection of soluble Aβ in cerebrospinal fluid and perhaps in plasma raises the possibility of finding differences in Aβ levels between at least some Aβ patients and controls that might be useful in monitoring the disease process. However, details of the in vivo synthesis, protein binding, and degradation of soluble Aβ; its stability in various fluids; and the effects of a range of physiological variables on its levels (e.g. fasting, age, activity, diurnal variation, drug usage, etc) must be determined before a meaningful analysis of Aβ levels in various Aβ and control patient groups can be carried out.

The third and most important implication of the new findings is the ability to use cellular Aβ production as an in vitro screen to identify compounds that lower Aβ levels. In the past, it was not possible to assay for quantitative changes in Aβ production before and after treatment with a physiological or pharmacological agent. Now, primary or transfected βAPP-expressing cells (including human fetal neurons) can readily be used to determine if particular compounds raise or lower Aβ production, and this can be done both with rationally selected classes of compounds and by random screening. The detailed biochemical mechanism of Aβ generation and the specific proteolytic enzymes involved need not necessarily be understood in order to carry out such drug screening. Following the identification of cell-penetrating compounds that apparently inhibit Aβ production or release without producing

cerebral levels of Aβ are decreased. In short, the new knowledge leads at once to in vitro and in vivo screening systems for identifying and then modifying Aβ inhibitors.

It is unclear by what cellular processing pathway Aβ is generated—secretory, endosomal/lysosomal, a combination of these, or a route not yet defined. However, early studies of the mechanism of Aβ production in vitro have established some of the characteristics of the processing pathway. Treatment of βAPP-expressing cells with agents that alter intracellular pH, such as chloroquine (Shoji et al 1992) or ammonium chloride (Haass et al 1993), leads to a substantial decrease in Aβ production, suggesting the importance of an acidic intracellular compartment for its generation. Treatment with leupeptin, which can inhibit lysosomal thiol proteases, has no significant effect on Aβ production (Haass et al 1993). The use of the monovalent ionophore monensin, which alters ionic gradients across vesicular compartments, results in a failure of full maturation of βAPP and the appearance of an intermediate, apparently partially glycosylated form, as well as the abolition of Aβ production. Exposure to brefeldin A, a compound that collapses the Golgi into the endoplasmic reticulum and prevents the normal maturation of proteins, similarly leads to an immature βAPP species and no Aβ generation. Treating cells with agents that depolymerize microtubules (e.g. colchicine or nocodazole) and should therefore alter trafficking of endosomes to lysosomes has no effect on Aβ production. Purification of late endosomes/lysosomes from βAPP-transfected cells that release substantial amounts of the peptide reveals no Aβ within these structures. Indeed, total cell lysates apparently contain no detectable Aβ. These various findings (Haass et al 1993) suggest that Aβ production requires, among other things, maturation of βAPP through the Golgi and processing in an acidic compartment other than lysosomes. Precisely which vesicular compartments are the sites of the Aβ N- and C-terminal cleavages remains to be elucidated.

Pulse chase experiments show that Aβ and a slightly smaller, ~3-kDa fragment (p3) are secreted in parallel with APP$_s$ following the N- and O-glycosylation of the precursor (Haass et al 1993). Since radiosequencing reveals that the p3 peptide begins principally at lys 17 of Aβ, p3 presumably derives as an amino terminal fragment of the ~10-kDa carboxyl terminal βAPP derivative generated by conventional secretory cleavage. Enhanced production of Aβ in vitro is reportedly accompanied by an increase in an ~12-kDa carboxyl terminal βAPP fragment that begins at or near asp$_1$ of Aβ (Cai et al 1993). This result suggests that the Aβ N-terminal cleavage occurs first, apparently followed by the C-terminal cleavage and subsequent rapid release of Aβ from cells. There is currently no evidence that preexisting membrane injury is required for this release. It is noteworthy that the metabolic

release of Aβ from cells. There is currently no evidence that preexisting membrane injury is required for this release. It is noteworthy that the metabolic fate of membrane-retained cytoplasmic stubs of integral membrane proteins that undergo secretory cleavage of the ectodomain is generally unknown.

The measurement of Aβ in vitro has made it possible to determine the effects of specific transcriptional and posttranslational modifications of βAPP on the generation of the amyloid subunit. Aβ seems to be produced in approximately equal quantities from 293 cells transfected with either $βAPP_{695}$ or $βAPP_{751}$ cDNAs (Haass et al 1992b). This result suggests that the KPI domain does not significantly influence the generation of the peptide, although some experiments have shown a modestly higher production of Aβ—and also of other proteolytic fragments, such as the 10-kDa C-terminal stub (Haass et al 1991)—from $βAPP_{695}$ transfectants. As mentioned above, Aβ production in cultured cells appears to follow the full maturation of βAPP, including its N- and O-glycosylation. However, the actual glycosylation of the precursor may not be a critical prerequisite, since preliminary data indicate that Aβ is still generated from βAPP constructs lacking the middle third of the molecule, including the N-linked glycosylation sites (WQ Qiu and D Selkoe, unpublished data).

The role of phosphorylation in βAPP processing has been the subject of a limited number of studies to date. Addition of phorbol esters to several cell types has been shown to stimulate release of APP_s and to increase the amounts of membrane-retained carboxyl-terminal fragments (Buxbaum et al 1990, Caporaso et al 1992, Gillespie et al 1992). This result suggests that stimulating protein phosphorylation by activating protein kinase C (PKC) leads to increased secretory cleavage of βAPP. Because PKC can phosphorylate a synthetic peptide corresponding to amino acid 645–661 of βAPP (Gandy et al 1988), specifically by adding phosphate to serine 655, and because exogenous PKC appears to phosphorylate this same serine in permeabilized PC12 cells in vitro (Suzuki et al 1992), it has been proposed that direct phosphorylation of the precursor upregulates secretion in vivo. This mode of regulation could be relevant not only to cells undergoing direct activation of PKC by phorbol esters but also to the regulation of APP_s release from neurons in response to a variety of extracellular signals, such as certain first messengers or depolarization. Indeed, human 293 cells and rat PC12 cells stably transfected with muscarinic acetylcholine receptors subtypes m1 or m3 show increased secretion of APP_s upon addition of the cholinergic agonist carbachol, a process that is blocked by addition of the protein kinase inhibitor staurosporine (Buxbaum et al 1992, Nitsch et al 1992). In addition, bradykinin (Nitsch et al 1992) and interleukin-1 (Buxbaum et al 1992) apparently have similar effects on βAPP processing. Electrical stimulation of hippocampal slices in

Although these results provide indirect evidence about the role of protein phosphorylation in βAPP metabolism, direct examination of the phosphorylation state of βAPP itself has only recently been undertaken (Hung et al 1993). In vivo phosphate labeling of intact βAPP-transfected 293 cells has shown βAPP to be phosphorylated solely on serine residues. Surprisingly, the phosphoserines were located exclusively in the ectodomain, resulting in the secretion of phosphorylated APP$_s$. Nevertheless, the stimulation of secretion by phorbol esters occurred independently of direct phosphorylation of βAPP, since such treatment did not increase phosphorylation of the holoprotein. Further support for this conclusion is the finding that PKC-mediated activation of APP$_s$ release still occurs following mutation of one or both cytoplasmic serine residues (positions 655 and 675 of βAPP$_{695}$) and even after deletion of the entire cytoplasmic domain (Hung et al 1993). PKC activation has also been shown to substantially decrease Aβ production, again independent of intracellular βAPP phosphorylation (Hung et al 1993). Taken together, the results suggest that protein phosphorylation plays a critical role in the regulation of APP$_s$ and Aβ secretion and that PKC-mediated pathways may help control amyloid formation. However, in contrast to previous postulates, PKC appears to mediate βAPP secretory processing not directly but via additional intracellular protein messengers. One reasonable candidate for the phosphoacceptor protein is the βAPP secretase itself. The identity of the phosphoprotein(s) regulating βAPP metabolism is now important to establish.

NORMAL FUNCTIONS OF βAPP

The high degree of evolutionary conservation of the cytoplasmic and extramembranous domains of βAPP, its widespread tissue expression, and its abundance in cells such as neurons and glia all support the likelihood of several important functions for the molecule. The first function to be described involved the identification of the KPI-like motif as an inhibitor of certain serine proteases, such as trypsin and chymotrypsin (Kitaguchi et al 1988, Oltersdorf et al 1989, Van Nostrand et al 1989, Sinha et al 1990). An example of how this inhibitory function might occur in vivo is provided by the discovery that a previously described inhibitor of factor XIa of the clotting cascade is in fact protease nexin II, the KPI-containing form of APP$_s$ (Smith et al 1990). Following the model of protease nexin I (an entirely different gene product), it has been speculated that secreted protease nexin II might act by complexing with and inactivating extracellular serine proteases; the cell would then take up the complex via a conformationally altered motif in the bound protease nexin II, which could putatively interact with a serpine-enzyme complex

nexin II, which could putatively interact with a serpine-enzyme complex (SEC) receptor (Van Nostrand & Cunningham 1987, Joslin et al 1992). Such a mode of action has not yet been established.

Another function of βAPP postulated on the basis of cell culture studies is as a growth promoting or autocrine molecule (Saitoh et al 1989). Fibroblasts treated with antisense βAPP oligonucleotides showed a marked decrease in βAPP expression and a concomitant decrease in proliferation. Cell growth could be restored by adding exogenous, purified APP_s. These results obtained whether $βAPP_{695}$ or $βAPP_{751}$ was studied, suggesting the presence of a sequence in the extracellular portion of βAPP other than the KPI motif that mediated these effects. Further studies by Saitoh and colleagues have localized the apparent determinant of this function to residues 328–332 [arg-glu-arg-met-ser (RERMS)] of $βAPP_{695}$, just beyond the point of insertion of the KPI domain (Ninomiya et al 1993). Synthetic peptides bearing this sequence are capable of promoting the growth of fibroblasts, PC12 cells, and human neuroblastoma cells in vitro and may have neurite-promoting properties. A synthetic peptide with the partially overlapping sequence arg-met-ser-gln-/RMSQ ($βAPP$ $APP_{330-333}$) is apparently capable of antagonizing this function (Ninomiya et al 1993). In another study, APP_s has been found to act as a neuroprotective molecule that can downregulate intracellular calcium levels in cultured rat and human neurons (Mattson et al 1993). The putative cell-surface receptor mediating the effects of APP_s has not been identified.

Another possible function for the secreted ectodomain is as an extracellular substrate promoting cell-cell or cell-matrix interactions (Schubert et al 1989a). APP_s has been shown to bind to certain extracellular matrices and to promote attachment of neuronal-like cells (Klier et al 1990). Several studies have provided direct or indirect evidence that APP_s may mediate cell-cell or cell-substrate adhesion in cultures of primary neurons or neuronal-like cell lines (Schubert et al 1989a, Breen et al 1991, Chen & Yankner 1991, Milward et al 1992). There has been much less work on the function of full-length membrane-inserted βAPP. Recently, evidence of neuronal adhesion and neurite-promoting activities of the holoprotein when it is inserted at the cell surface has been presented (Qiu et al 1993). In this work, primary rat hippocampal neurons grown on a monolayer of βAPP-transfected CHO cells showed enhanced neurite outgrowth compared with neurons grown on nontransfected CHO cells. Cell-surface expressed $βAPP_{695}$, $βAPP_{751}$, and $βAPP_{770}$ all could promote neurite outgrowth, although the KPI-containing forms did so more robustly. Control experiments in which only the conditioned media of the CHO transfectants was added to the neuronal cultures showed that secreted APP_s did not account for the neurite-promoting effect. Moreover, βAPP synthetic peptides containing the RERMS growth-promoting or RMSQ growth-inhibiting sequences of APP_s had no effect on the neurons. These data

can interact with an as-yet-unknown molecule on the neuronal surface to promote neuronal adhesion and neurite extension in vitro. Whether βAPP expressed on astrocytes, microglia, or neurons could provide such a function in the intact nervous system remains to be seen.

The original elucidation of the βAPP primary structure by cDNA sequencing (Kang et al 1987) led to the proposal that the molecule itself might act as a cell-surface receptor similar to other single membrane spanning polypeptides, such as the transferrin receptor and the IGF II receptor. Subsequent studies have indicated that the ectodomain of the precursor has specialized regions that are capable of binding heparin (Schubert et al 1989b) and certain metallic cations, including zinc (Busch et al 1993), in vitro. However, a specific soluble ligand that binds to surface-inserted βAPP with receptor-like kinetics has not been identified. Potential support for a receptor-like function of βAPP is the observation that the precursor can associate in vitro with the heterotrimeric GTP-binding protein G_o via a motif in the βAPP cytoplasmic domain (Nishimoto et al 1993). If this observation can be confirmed in intact cells, the effects of the presence or absence of G_o binding on βAPP processing and Aβ production should be investigated.

The possibility that the Aβ fragment may itself have a physiological function throughout life has arisen from the realization that Aβ released from many cell types. Prior to this discovery, work on the biological effects of Aβ revolved largely around its toxic activities, based on the assumption that it was a pathological metabolic product. Now that we know that fetal human neurons and other cells in culture continuously release significant quantities of Aβ, the search for a function is under way. It must be borne in mind, however, that the particular hydrophobic sequence of the fragment could simply make it a semistable proteolytic intermediate during βAPP catabolism, without it necessarily having a specific function.

Evidence that Aβ may have neurotrophic and neurite-promoting activity has come from studies involving the addition of solubilized Aβ synthetic peptides in relatively high doses to aqueous culture media (Whitson et al 1989, 1990; Yankner et al 1990). An alternative way to model the effects of Aβ in the nervous system involves the addition of Aβ peptide to extracellular matrix molecules to determine whether it can serve as a trophic substrate. In such experiments, Aβ immobilized onto nitrocellulose filters could not serve as a substrate for neurite outgrowth from rat dorsal root ganglia (DRG) explants (Koo et al 1993). Furthermore, addition of Aβ to optimal growth-promoting doses of laminin had neither a positive nor a negative effect on neurite outgrowth. However, when low, suboptimal doses of laminin were used as a substrate, DRG neurites grew very poorly or not at all, and growth could be restored to near-normal levels by addition of synthetic $A\beta_{1-40}$ (Koo et al 1993). Similar results were obtained when fibronectin served as the matrix molecule

Similar results were obtained when fibronectin served as the matrix molecule to which Aβ was bound. These findings point to a possible interaction of secreted Aβ with extracellular substrate molecules in the nervous system to promote neurite growth. Such an activity could also provide an inappropriate trophic stimulus when aggregated; particulate Aβ begins to accumulate with heparan sulfate proteolycans and perhaps other matrix molecules during the genesis of diffuse plaques.

Deletion of the βAPP gene product and regions thereof via homologous recombination will be of special interest as an indication of the importance of the molecule and its putative functional domains to the physiology of both neural and nonneural tissues.

GENETIC EVIDENCE IMPLICATES βAPP IN THE PATHOGENESIS OF ALZHEIMER'S DISEASE

Over the past decade, accumulating data has indirectly implicated βAPP in the basic mechanism of Alzheimer's disease. Two salient findings in this regard were: (a) the discovery that the βAPP gene is localized to chromosome 21 and is overexpressed in Down's syndrome, and (b) the realization that amorphous deposits of Aβ appear to precede any other Alzheimer-type brain lesions in both trisomy 21 and normal brain aging. Evidence directly implicating βAPP in the pathogenesis came from the discovery of missense mutations in the βAPP gene that were strongly linked genetically to familial Alzheimer's disease or to the related disorder hereditary, cerebral hemorrhage with amyloidosis of the Dutch type (HCHWA-Dutch) (Hardy 1992). The first such mutation was actually found in HCHWA-Dutch, a substitution of glutamine for glutamate at position 22 of Aβ (residue 693 of βAPP_{770}) that was shown to cause this rare form of cerebral β-amyloidosis (Levy et al 1990, van Broeckhoven et al 1990). This disease is marked by the premature and severe deposition of Aβ peptide in innumerable microvessels of the brain and meninges, resulting in multiple hemorrhages that are eventually fatal. Importantly, these patients also develop large numbers of parenchymal Aβ deposits in cerebral cortex that are indistinguishable from the diffuse plaques of Alzheimer's disease and in Down's syndrome. Precisely why there is such a strong vascular predilection for Aβ peptide bearing this mutation and why the cortical deposits appear not to progress to full-blown neuritic plaques in these nondemented patients are unanswered questions.

Subsequently, at least six distinct missense mutations have been detected within or immediately flanking the Aβ region of βAPP in families with autosomal dominant, early onset Aβ (Goate et al 1991, Hardy 1992) (Figure 2). At least one additional mutation has been found in a living subject with schizophrenia not displaying the Alheimer's disease phenotype. Although

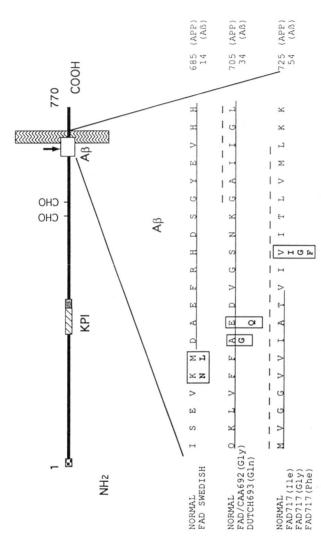

Figure 2 βAPP misssense mutations associated with familial AD or closely related β-amyloidotic disorders. The sequence of βAPP containing the Aβ and transmembrane region is expanded and shown by the single-letter amino acid code. The underlined residues represent the Aβ₁₋₄₃ peptide. The broken line indicates the location of the transmembrane domain. The boxed residues depict the currently known missense mutations identified in patients with familial Alzheimer's disease (FAD) or hereditary cerebral hemorrhage with amyloidosis of the Dutch type (Dutch). CAA indicates that the family with the 692 ala → gly mutation contains individuals with the congophilic amyloid phenotype and/or the FAD phenotype. Three digit numbers refer to the residue number according to the βAPP₇₇₀ isoform. Two digit numbers refer to the residue number according to the Aβ sequence (asp₆₇₂ = 1).

most of the known missense mutations in exons 16 and 17 of βAPP are strongly genetically linked to the occurrence of Alzheimer's disease, the mechanisms by which they produce the phenotype remain largely unresolved. One FAD genotype, a double mutation of 670 lys → asn/671 met → leu (βAPP$_{770}$ numbering) just before the Aβ start site that occurs in a large Swedish family, has been analyzed as regards Aβ production. Transfection of several cell types with constructs bearing one or both of these mutations has demonstrated that the double mutation results in a marked (approximately five- to eightfold) increase in Aβ production in vitro and that the second substitution (met → leu) confers most of this effect (Citron et al 1992, Cai et al 1993). This dramatic increase in Aβ release has not yet been associated with a notable change in APP$_s$ secretion, but the possibility that release of the shorter APP$_s$ form ending at met$_{671}$ (Seubert et al 1993) is increased has not been ruled out. One may conclude that the Swedish met → leu mutation provides a highly favorable substrate for the unidentified enzyme that normally cleaves at the met-asp bond to create the N-terminus of Aβ. Indeed, cells expressing the Swedish mutation can apparently show elevation of the ~12-kDa carboxyl-terminal βAPP fragment that presumably begins at asp-1 of Aβ, suggesting that this N-terminal cleavage event occurs prior to the Aβ C-terminal scission (Cai et al 1993). The mechanisms by which the other known missense mutations in and around Aβ cause markedly enhanced β-amyloid deposition remain to be elucidated but are likely to differ, given the variety of amino acids that are substituted and their loci in the molecule. Thus, the allelic heterogeneity of βAPP-linked familial AD will probably be reflected in somewhat different pathogenetic mechanisms.

Non-allelic heterogeneity in familial AD has also been firmly established (Figure 3). It had been known for some time that a number of large pedigrees with early-onset AD did not link to the βAPP gene or any other locus on chromosome 21. Most of these families have now been linked to an as-yet-unknown gene on the long arm of chromosome 14 (Mullan et al 1992, Schellenberg et al 1992, St. George-Hyslop et al 1992, van Broeckhoven et al 1992). The onset of clinical dementia in the families linked to chromosome

Chromosome	Gene	Onset of AD
21	βAPP	early (~50s)
	Trisomy 21 => very early Aβ deposition	
14	unknown	early (~40s)
19	ApoE	late (ε4 earlier than ε3)

Figure 3 Genetic alterations leading to Alzheimer's disease. Other chromosomal loci are likely but not yet identified.

in the βAPP gene (early 50s) (M Mullan and J Hardy, personal communication). Currently available evidence suggests that the clinical and neuropathological phenotypes are otherwise highly similar, if not indistinguishable, in these two forms of familial AD. The fact that the chromosome 14-linked subjects undergo a very early and severe Aβ deposition suggests that the defective gene product may have some direct or indirect role in the expression, regulation, or processing of βAPP or perhaps in the metabolism of the Aβ fragment itself. Few candidate genes in the mid-region of chromosome 14q have been identified. One, the cFOS gene, has already been ruled out (St. George-Hyslop et al 1992), and apparently a novel member of the heat shock protein gene family is also not implicated. It is already clear that one or more additional genetic loci for other forms of early-onset familial AD must exist. For example, a large kindred, originally of German origin, that emigrated from the Volga River valley shows no linkage to either chromosomes 14 or 21.

The families that have been linked to these two chromosomes represent only a tiny fraction of all cases of AD. The fact that the vast majority of AD subjects experience the onset of clinical disease well after age 65 has greatly hampered efforts to apply conventional linkage analysis to the search for genetic factors in such cases. However, the use of affected pedigree member analysis in a group of late-onset subjects with family histories of AD enabled Pericak-Vance and colleagues (1991) to obtain evidence of an apparent locus on the long arm of chromosome 19. These investigators have subsequently shown that a specific allele (E4) of the gene on chromosome 19 that encodes apolipoprotein E occurs at a substantially higher frequency in late-onset AD subjects (~50%) than in the general population (~50%) (Strittmatter et al 1993). Immunocytochemical data indicates that apolipoprotein E can bind to Aβ, the prion protein, and certain systemic amyloid proteins. It has been suggested that the E4 allele of apolipoprotein E might have a capacity to bind and transport Aβ that is different from that of other alleles. Regardless of the biochemical mechanism, this work indicates that ApoE4 represents an important risk factor for late-onset AD.

Progress in elucidating the genetic heterogeneity that underlies familial forms of AD has again highlighted the longstanding question of what proportion of all AD cases are genetically determined. Although it is commonly stated that some 5–15% of AD subjects have a clearly positive family history consistent with an autosomal dominant mendelian trait, numerous epidemiological reports indicate that a much higher percentage of patients have one or more first degree relatives with a history of a clinical dementia highly similar to that of the patient. Longitudinal studies of a large number of AD subjects have suggested that the likelihood of detecting the AD phenotype in first degree relatives rises steadily with age, achieving a

AD phenotype in first degree relatives rises steadily with age, achieving a figure of approximately 50% if case ascertainment extends to age 90 (Mohs et al 1987). One may speculate that an array of distinct genetic defects on various chromosomes will be found to underlie many if not most cases of AD.

The identification of environmental factors that could precipitate the onset of AD or even represent causative agents has been elusive. The disorder appears to occur in virtually all races and ethnic groups that have been investigated, and no population isolates with either a very high or very low incidence of definite Alzheimer's disease have been clearly identified. Earlier reports suggesting that the concordance rate for Alzheimer's disease in monozygotic twins was relatively low (0.4–0.5) and that therefore environmental factors might be important are not supported by more recent studies, which suggest concordance rates following vigorous ascertainment that approach 0.8 or above.

AMYLOID β PROTEIN AND THE ORIGIN OF ALZHEIMER'S DISEASE

Perhaps the most frequently posed question in current research on AD is whether Aβ deposition is a cause or an effect of the disease. It appears increasingly likely that the answer is both. In the only patients in which a specific molecular cause of AD has been identified to date, that cause is a missense mutation in the amyloid precursor protein. Genetic linkage analyses, as well as in vitro expression of at least one of the mutant βAPP genes, have provided very strong evidence that these mutations can be truly causative. βAPP also appears to be responsible for the AD phenotype in trisomy 21; increased gene dosage and increased tissue expression of βAPP are very likely to underlie the early and progressive formation of diffuse Aβ deposits, years before any other detectable cytological signature of AD appears.

On the other hand, we now know that in many autosomal dominant cases of AD, Aβ deposition cannot be said to be causative in a true molecular sense but rather follows upon primary defects in still unknown genes on chromosome 14 or other autosomes. Even in these cases, the morphology, distribution, immunochemical nature, and fine structure of the β-amyloid lesions in brain are, so far, indistinguishable from those in βAPP-linked cases. The virtually identical neuropathological (and usually clinical) phenotypes observed in the chromosome 14- and βAPP-linked cases make it unlikely that the pathogenetic mechanisms of the premature β-amyloid deposition seen in both these forms are vastly different. Rather, it is reasonable to hypothesize that the chromosome 14 defect involves an altered gene product that, through one or several biochemical steps, leads to accelerated Aβ deposition. Given that Aβ plaques

higher numbers in all AD cases than in age-matched normal subjects, Aβ deposition likely plays an early pathogenetic role in virtually all cases of Alzheimer's disease, as we currently define the disease. The notion that some cases of AD (e.g. the βAPP-linked cases) can be caused by an accelerated β-amyloidosis, whereas in other subjects a phenotypically indistinguishable β-amyloid process occurs as a late secondary event that has little or no important pathogenetic role in the dementia, seems improbable.

Although this debate cannot be settled unequivocally at present, one may begin with the knowledge that at least a few AD cases are caused by βAPP mutations and proceed to examine current information about how premature, accelerated Aβ deposition might gradually lead to neuronal/neuritic degeneration and hence dementia. From the facts that diffuse plaques show little or no ultrastructural evidence of neuritic alteration, that these plaques occur very early in Down's syndrome, and that they also occur in considerable numbers in many clinically normal aged humans, one may conclude that initial Aβ deposition is not sufficient by itself to produce the AD clinical syndrome. Instead, it appears that the amount of Aβ protein per deposit and the number of deposits that accumulate in the limbic and association cortices over time may be determinants of the eventual development of dementia. Perhaps when Aβ reaches a critical local concentration, particularly as regards the longer peptides of 42–43 (or more) residues, a gradual aggregation of monomers followed by oligomers may occur, leading to fibril formation. Dystrophic neurites, reactive astrocytes, and activated microglia tend to be much more associated with fibrillar than with nonfibrillar Aβ plaques. This model does not necessarily imply that the polymeric, fibrillar Aβ is the sole cytotoxic moiety. Rather, gradual local accumulation of Aβ could simultaneously result in increased toxic effects of monomeric or oligomeric Aβ, assembly of some of the oligomeric Aβ into polymeric fibrils, and association of biologically active proteins and other molecules with the maturing fibrillar deposit.

Intensive recent studies on the mechanism of Aβ cytotoxicity (reviewed in Kosik & Coleman 1992) have led to two broad but not mutually exclusive hypotheses: that Aβ in some form may directly injure cells, perhaps by surface membrane effects; and that Aβ perturbs cells more indirectly by enhancing their vulnerability to a range of potentially toxic factors that occur naturally in cerebral tissue. As regards the first hypothesis, evidence has been presented that Aβ can produce apparent direct neurotoxicity that does not require the involvement of factors such as excitatory amino acids or free radicals (Yankner et al 1990, Kowall et al 1991). According to these data, at least some of the effects of Aβ on neurons cannot be reversed by conventional inhibitors of these neurotoxic molecules. In other studies, Aβ has shown little direct toxicity by itself but has increased the susceptibility of cultured neurons to injury by excitatory amino acids (Koh et al 1990, Mattson et al 1992),

injury by excitatory amino acids (Koh et al 1990, Mattson et al 1992), oxidative stress (Saunders et al 1991), or hypoglycemia. At least some of these indirect effects appear to be mediated by an increase in intracellular free calcium that accompanies Aβ treatment (Mattson et al 1992). A mechanism by which this might occur has been proposed by Arispe et al (1993), who reported that synthetic $Aβ_{1-40}$, when packaged in phospholipid vesicles and then applied to phospholipid bilayers in vitro, could induce the formation of channels that conducted calcium and other cations. Whether such an effect occurs in intact cells requires further study.

Irrespective of whether the neurotoxic effects of Aβ are direct, indirect, or both, emerging evidence suggests that Aβ that has been allowed to aggregate into polymers over time ("in vitro peptide aging") may be more toxic to cultured neurons than is monomeric, soluble Aβ. This notion fits well with the observations that Aβ is continuously produced by healthy cells as a soluble, monomeric substance and that local neuronal injury in AD brain appears to be associated more closely with aggregated, compacted deposits of Aβ.

An additional hypothetical mechanism for the gradual cytotoxic effects that apparently follow Aβ deposition involves the so-called β-amyloid associated proteins. In this scenario, Aβ, particularly in its more aggregated forms, can serve as a matrix upon which a variety of soluble molecules that are normally made in small amounts by local cells can accumulate. Some of the proteins that appear to be rather tightly associated with β-amyloid have already been identified: $α_1$-antichymotrypsin (Abraham et al 1988), several components of the classical complement cascade (Eikelenboom & Stam 1982, Rozemuller et al 1990, Rogers et al 1992), heparan sulfate proteoglycans (Snow et al 1988), and the serum amyloid P component (Kalaria et al 1991). Local astrocytes surrounding neuritic plaques may produce not only one or more of these polypeptides but also cytokines and other acute phase proteins that could mediate the local inflammatory response found in and around mature plaques (Griffin et al 1989). The reactive astrocytic and microglial changes that occur within and around the senile plaque are likely to contribute in complex ways to local neuritic alterations. It will be difficult to sort out the temporal sequence in which these molecules become associated with the developing plaque and their importance in local cytotoxicity without the availability of an animal model that develops amyloid plaque formation very similar to that of AD.

A critical result of the complex processes in and immediately surrounding the plaque is the dystrophy and apparent degeneration of neurites of varying neurotransmitter specificities. The existence of widespread neuritic dystrophy in limbic and association cortices in AD that is not intimately associated with the amyloid deposits suggests that the consequences of β-amyloid deposition may gradually lead to more distant effects on both local-circuit cortical neurons

effects on axonal transport and possible transynaptic degeneration. One result of the insult to axons and dendrites that appears to be associated with amyloid plaque formation may be the selective development in both cell bodies and neurites of altered cytoskeletal elements, including PHF. These cytoskeletal changes, which involve in large part a hyperphosphorylated, aggregated form of the microtubule-associated protein tau (for review, see Lee & Trojanowski 1992), are likely to be accompanied by adverse effects on the stability and function of the cytoskeleton. It remains unclear whether PHF formation is a secondary effect of this disorganization of the neuronal cytoskeleton or whether the PHF are themselves directly responsible for neuronal/neuritic dysfunction and subsequent cell death.

Evidently, numerous neuronal proteins besides altered cytoskeletal elements such as PHF accumulate in peri-plaque dystrophic neurites. Plaque neurites can be immunolabeled with antibodies to enzymes that synthesize or degrade the particular neurotransmitters characteristic of the neurites, as well as by antibodies to other intrinsic proteins, such as certain kinases, proteases, and esterases (see for example Walker et al 1988, Cataldo & Nixon 1990, Munoz 1991, Wright et al 1993). Among the many neuronal proteins that are immunodetectable in altered plaque neurites is βAPP itself. Non-Aβ epitopes of βAPP, both from the amino and carboxyl portions of the precursor, can be detected in plaque neurites, and these sometimes colocalize with cytoskeletal markers (Cras et al 1991, Joachim et al 1991). The occurrence of such neurites does not provide evidence that the extracellular Aβ in the plaques arises from βAPP in surrounding neurites, since the large majority of Aβ plaques in AD brain (i.e. diffuse plaques) show no such βAPP-reactive dystrophic neurites in their periphery. Indeed, dystrophic neurites with βAPP immunoreactivity have been described in the areas immediately surrounding cerebral tissue damaged by ischemia in rodents; these sites have no Aβ deposits (Stephenson et al 1992). Moreover, βAPP-reactive neurites have been observed in damaged cortical areas in other human neurodegenerative disorders, e.g. subacute sclerosing panencephalitis, again in the absence of any Aβ deposition (Cochran et al 1991). Thus, βAPP, as one of many neuronal proteins that are transported down axons, can accumulate nonspecifically at sites of neuritic injury without producing extracellular Aβ deposits.

Among the most sensitive protein markers of dystrophic neurites found within and outside of amyloid plaques is ubiquitin. Neurites showing ubiquitin immunoreactivity have been observed to a limited degree in some apparently diffuse plaques of the cerebellum, in the absence of any other immunocytochemically detectable neuritic marker. Again, this ubiquitin reaction is not specific to Aβ plaques and can be seen in other types of neuronal degeneration. The role that the ubiquitination of still unidentified neuritic proteins may play in the neurodegenerative process remains to be clarified.

proteins may play in the neurodegenerative process remains to be clarified. Recently, the protein within the neurofibrillary tangle that undergoes ubiquitination was shown to be a carboxyl-terminal fragment of the PHF subunit protein, tau (Morishima-Kawashima et al 1993).

The multiplicity of macromolecular and cellular alterations occurring in neurons and their processes, astrocytes, and microglia following the deposition of Aβ suggests that the amyloid peptide or molecules that become associated with it may trigger a variety of complex molecular alterations, presumably followed by secondary feedback reactions that further aggravate the initial insult. Which of the many biochemical and cytological changes occurring in and around plaques are particularly important pathogenetically and might thus be targets for therapeutic inhibition in Alzheimer's disease will be difficult to sort out without a dynamic animal model of the process.

An equally difficult issue is the longstanding and unresolved question of the cellular origin of Aβ deposits. Their remarkable predilection for cerebral gray matter and the abundant neuronal expression of βAPP (particularly the 695 residue form) have led to the common assumption that neurons are the most likely source of Aβ in plaques. However, the expression of multiple βAPP isoforms in astrocytes (Siman et al 1989, Haass et al 1991) and the observation of apparent Aβ fibrils inside microglial cells (Wegial & Wisniewski 1990) have also led to hypotheses that either or both of these cells could contribute to the Aβ deposits. The abundance of β-amyloid fibrils localized preferentially to the abluminal basement membranes of meningeal arterioles lacking neuronal or astrocytic elements and of cerebral capillaries has raised the possibility that either vessel wall cells (endothelial and/or smooth muscle) or the plasma itself could serve as a source of Aβ. The recent discovery of soluble Aβ in extracellular fluids, including CSF and apparently plasma (Seubert et al 1992), fulfills one necessary requirement for the hypothesis that the Aβ in cerebral blood vessels and perhaps even in brain tissue could arise in part from an extracellular fluid source (Selkoe 1989). Further circumstantial evidence consistent with this hypothesis is the light and electron microscopic findings that most Aβ plaques show a close spatial association with capillaries (Miyakawa et al 1982, Kawai et al 1992). However, none of these findings provides direct evidence that Aβ could arise from a circulating (CSF or plasma) source. Given that all βAPP-expressing cells seem to be capable of secreting Aβ into the extracellular space, it may be that multiple cell types can contribute to the various morphological types of amyloid deposits. The consistently spherical shape of most Aβ plaques suggests that Aβ diffuses from a point source at the center of the plaques, whether a neural cell or a capillary. The ribbon-like subpial deposits of Aβ found in many AD brains might well arise by diffusion from the CSF into the superficial cortex. Morphologically very similar subpial deposits com-

cases of familial amyloidotic neuropathy caused by point mutations in the transthyretin gene (Ushiyama et al 1991). In the final analysis, the precise cellular origin(s) of the pleiomorphic Aβ deposits found in AD brain may be impossible to ascertain without an animal model that closely recreates the amyloid pathology.

CONCLUSION

The elucidation of the molecular and cellular pathogenesis of Alzheimer's disease is at an intermediate stage, and the significance of many observations cannot yet be determined. The temporal sequence of alterations and their relative importance in producing progressive neuronal/neuritic dysfunction and hence dementia are incompletely understood and thus are the appropriate subjects of lively debate. Nevertheless, the concept that deposition of the Aβ protein plays an early and critical role in some if not most cases of Alzheimer's disease has received increasing acceptance. This amyloid cascade hypothesis of AD has at present the greatest amount of supporting experimental data among the various hypotheses that have been proposed to explain the etiology and mechanism of this disorder. The next few years should bring a steady accrual of information about both the normal function and processing of βAPP and the reasons for the excessive accumulation of the Aβ fragment in areas of the brain important for memory and cognition.

The model of Alzheimer's disease I have reviewed here points to new therapeutic strategies aimed at one or more crucial steps in its molecular progression. First, one could partially inhibit the proteases that liberate Aβ from its precursor. Second, one could decrease the production and release of Aβ by a variety of means other than protease inhibition, for example by pharmacologically diverting some βAPP molecules from an amyloidogenic to a nonamyloidogenic processing pathway. Third, one could retard the apparent maturation of extracellular Aβ deposits into neuritic plaques, perhaps by interfering with the fibrillogenesis that seems to accompany this change. Fourth, one could interfere with the activities of microglia, astrocytes, and other cells that contribute to the chronic inflammatory process around the neuritic plaques. And fifth, one could attempt to block the molecules on the surface of neurons or their intracellular effectors that mediate the apparent neurotoxic effects of Aβ and the proteins intimately associated with it. None of these pharmacological objectives will be easy to reach. But the strength of accumulating data emerging from laboratories worldwide suggests that these are rational therapeutic targets which offer the best hope of slowing or arresting the progression of the fundamental disease process early in its tragic course.

Literature Cited

Abraham CR, Selkoe DJ, Potter H. 1988. Immunochemical identification of the serine protease inhibitor, α₁-antichymotrypsin in the brain amyloid deposits of Alzheimer's disease. *Cell* 52:487–501

Anderson JP, Chen YU, Kim KS, Robakis NK. 1992. An alternative secretase cleavage produces soluble Alzheimer amyloid precursor protein containing a potentially amyloidogenic sequence. *J. Neurochem.* 59:2328–31

Arispe N, Rojas E, Pollard H. 1993. Alzheimer disease amyloid βA protein forms calcium channels in bilayer membranes: blockade by tromethamine and aluminum. *Proc. Natl. Acad. Sci. USA* 90:567–71

Askanas V, Engel WK, Alvarez RB. 1992. Light and electron microscopic localization of β-amyloid protein in muscle biopsies of patients with inclusion-body myositis. *Am. J. Pathol.* 141:31–36

Benowitz LI, Rodriguez W, Paskevich P, et al. 1989. The amyloid precursor protein is concentrated in neuronal lysosomes in normal and Alzheimer disease subjects. *Exp. Neurol.* 106:237–50

Braak H, Braak E, Grundke-Iqbal I, Iqbal K. 1986. Occurrence of neuropil threads in the senile human brains and in Alzheimer's disease: a third localization of paired helical filaments outside of neurofibrillary tangles and neuritic plaques. *Neurosci. Lett.* 65: 351–55

Breen KC, Bruce M, Anderton BH. 1991. Beta amyloid precursor protein mediates neuronal cell-cell and cell-surface adhesion. *J. Neurosci. Res.* 28:90–100

Bush A, Multhaup G, Moir R, et al. 1993. A novel Zinc(II) binding site modulates the function of the βA4 amyloid protein precursor of Alzheimer's disease. *J. Biol. Chem.* In press

Buxbaum JD, Gandy SE, Cicchetti P, et al. 1990. Processing of Alzheimer β/A4 amyloid precursor protein: Modulation by agents that regulate protein phosphorylation. *Proc. Natl. Acad. Sci. USA* 87:6003–6

Buxbaum JD, Oishi M, Chen HI, et al. 1992. Cholinergic agonists and interleukin 1 regulate processing and secretion of the Alzheimer β/A4 amyloid precursor protein. *Proc. Natl. Acad. Sci. USA* 89:3055–59

Cai X-D, Golde TE, Younkin GS. 1993. Release of excess amyloid β protein from a mutant amyloid A protein precursor. *Science* 259:514–16 0

Caporaso GL, Gandy SE, Buxbaum JD, et al. 1992. Protein phosphorylation regulates secretion of Alzheimer β/A4 amyloid precursor protein. *Proc. Natl. Acad. Sci. USA* 89:3055–59

Cataldo AM, Nixon RA. 1990. Enzymatically active lysosomal proteases are associated with amyloid deposits in Alzheimer brain. *Proc. Natl. Acad. Sci. USA* 87:3861–65

Chen M, Yankner BA. 1991. An antibody to β-amyloid and the amyloid precursor protein inhibits cell-substratum adhesion in many mammalian cell types. *Neurosci. Lett.* 125:223–26

Chen W-J, Goldstein JL, Brown MS. 1990. NPXY, a sequence often found in cytoplasmic tails, is required for coated-pit mediated internalization of the low density lipoprotein receptor. *J. Biol. Chem.* 265:3116–23

Citron M, Oltersdorf T, Haass C, et al. 1992. Mutation of the β-amyloid precursor protein in familial Alzheimer's disease increases β-protein production. *Nature* 360:672–74

Cochran E, Bacci B, Chen Y, et al. 1991. Amyloid precursor protein and ubiquitin immunoreactivity in dystrophic axons not unique to Alzheimer's disease. *Am. J. Pathol.* 139:485–89

Cole GM, Huynh TV, Saitoh T. 1989. Evidence for lysosomal processing of beta-amyloid precursor in cultured cells. *Neurochem. Res.* 10:933–39

Cras P, Kawai M, Lowery D, et al. 1991. Senile plaque neurites in Alzheimer disease accumulate amyloid precursor protein. *Proc. Natl. Acad. Sci. USA* 88:7552–56

de Sauvage F, Octave JN. 1989. A novel mRNA of the A4 amyloid precursor gene coding for a possibly secreted protein. *Science* 245:651–53

Eikelenboom P, Stam FC. 1982. Immunoglobulins and complement factors in senile plaques: an immunoperoxidase study. *Acta Neuropathol.* 57:239–42

Esch FS, Keim PS, Beattie EC, et al. 1990. Cleavage of amyloid β-peptide during constitutive processing of its precursor. *Science* 248:1122–24

Estus S, Golde TE, Kunishita T, et al. 1992. Potentially amyloidogenic carboxyl-terminal derivatives of the amyloid protein precursor. *Science* 255:726–28

Farber SA, Nitsch RM, Wurtman RJ. 1992. Alzheimer amyloid precursor protein can be released from hippocampal slices in vitro. *Soc. Neurosci. Abstr.* 18:765

Gandy S, Czernik AJ, Greengard P. 1988. Phosphorylation of Alzheimer disease amyloid precursor peptide by protein kinase C and Ca²⁺/calmodulin-dependent protein kinase II. *Proc. Natl. Acad. Sci. USA* 85: 6218–21

Gearing M, Wilson RW, Unger ER, et al. 1993. Amyloid precursor protein (APP) in the striatum in Alzheimer's disease: An immunohistochemical study. *J. Neuropathol. Exp. Neurol.* 52:22–30

Giaccone G, Tagliavini F, Linoli G, et al. 1989. Down patients: extracellular preamyloid deposits precede neuritic degeneration and senile plaques. *Neurosci. Lett.* 97: 232–39

Gillespie SL, Golde TE, Younkin SG. 1992. Secretory processing of the Alzheimer amyloid β/A4 protein precursor is increased by protein phosphorylation. *Biochem. Biophys. Res. Commun.* 187:1285–90

Glenner GG, Wong CW. 1984. Alzheimer's disease: Initial report of the purification and characterization of a novel cerebrovascular amyloid protein. *Biochem. Biophys. Res. Commun.* 120:885–90

Goate A, Chartier-Harlin M-C, Mullan M, et al. 1991. Segregation of a missense mutation in the amyloid precursor protein gene with familial Alzheimer's disease. *Nature* 349: 704–706

Golde TE, Estus S, Younkin LH, et al. 1992. Processing of the amyloid protein precursor to potentially amyloidogenic carboxyl-terminal derivatives. *Science* 255:728–30

Goldgaber D, Lerman MI, McBridge OW, Saffiot U, Gajdusek DC. 1987. Characterization and chromosomal localization of a cDNA encoding brain amyloid of Alzheimer's disease. *Science* 235:877–80

Gorevic P, Goni F, Pons-Estel B, et al. 1986. Isolation and partial characterization of neurofibrillary tangles and amyloid plaque cores in Alzheimer's disease: Immunohistological studies. *J. Neuropathol. Exp. Neurol.* 45: 647–64

Griffin WST, Stanley LC, Ling C, et al. 1989. Brain interleukin I and S-100 immunoreactivity are elevated in Down syndrome and Alzheimer disease. *Proc. Natl. Acad. Sci. USA* 86:7611–15

Haass C, Hung AY, Schlossmacher MG, et al. 1993. β-amyloid peptide and a 3-kDa fragment are derived by distinct cellular mechanisms. *J. Biol. Chem.* 268:3021–24

Haass C, Hung AY, Selkoe DJ. 1991. Processing of β-amyloid precursor protein in microglia and astrocytes favors a localization in internal vesicles over constitutive secretion. *J. Neurosci.* 11:3783–93

Haass C, Koo EH, Mellon A, et al. 1992a. Targeting of cell-surface βA-amyloid precursor protein to lysosomes: alternative processing into amyloid-bearing fragments. *Nature* 357:500–3

Haass C, Schlossmacher MG, Hung AY, et al. 1992b. Amyloid β-peptide is produced by cultured cells during normal metabolism. *Nature* 359:322–25

Hardy J. 1992. Framing β-amyloid. *Nature Genet.* 1:233–34

Hung A, Munsat T, Selkoe D. 1993. Regulation of β amyloid precursor protein processing into non-amyloidogenic and amyloidogenic derivations by protein kinase C. *Soc. Neurosci. Abstr.* In Press

Ikeda M, Shoji M, Yamaguchi H, et al. 1993. Diagnostic significance of skin immunolabelling with antibody against native cerebral amyloid in Alzheimer's disease. *Neurosci. Lett.* 150:159–61

Joachim CL, Duffy LK, Morris JH, Selkoe DJ. 1988. Protein chemical and immunocytochemical studies of meningovascular β−amyloid protein in Alzheimer's disease and normal aging. *Brain Res.* 474:100–11

Joachim CL, Games D, Morris J, et al. 1991. Antibodies to non-β regions of the β-amyloid precursor protein detect a subset of senile plaques. *Am. J. Pathol.* 138:373–84

Joachim CL, Mori H, Selkoe DJ. 1989a. Amyloid β-protein deposition in tissues other than brain in Alzheimer's disease. *Nature* 341:226–30

Joachim CL, Morris JH, Selkoe DJ. 1989b. Diffuse senile plaques occur commonly in the cerebellum in Alzheimer's disease. *Am. J. Pathol.* 135:309–19

Joslin G, Griffin GL, August AM, et al. 1992. The serpin-enzyme complex (SEC) receptor mediates the neutrophil chemotactic effect of alpha-1 antitrypsin-elastase complexes and amyloid-beta peptide. *J. Clin. Invest.* 90: 1150–54

Kalaria RN, Galloway PG, Perry G. 1991. Widespread amyloid P component immunoreactivity in cortical amyloid deposits of Alzheimer's disease and other degenerative disorders. *Neuropathol. Appl. Neurobiol.* 17:189–201

Kang J, Lemaire H, Unterbeck A, et al. 1987. The precursor of Alzheimer's disease amyloid A4 protein resembles a cell-surface receptor. *Nature* 325:733–36

Kawai M, Cras P, Perry G. 1992. Serial reconstruction of beta-protein amyloid plaques: relationship to microvessels and size distribution. *Brain Res.* 592:278–82

Kidd M. 1963. Paired helical filaments in electron microscopy of Alzheimer's disease. *Nature* 197:192–93

Kitaguchi N, Takahashi Y, Tokushima Y, et al. 1988. Novel precursor of Alzheimer's disease amyloid protein shows protease inhibitory activity. *Nature* 331:530–32

Klier FG, Cole G, Stallcup W, Schubert D. 1990. Amyloid β-protein precursor is associated with extracellular matrix. *Brain Res.* 515:336–42

Koenig G, Monning U, Czech C, et al. 1992. Identification and differential expression of a novel alternative splice isoform of the βA4 amyloid precursor protein (APP) mRNA in leukocytes and brain microglial cells. *J. Biol. Chem.* 267:10804–9

Koh J-Y, Yang L-L, Cotman CW. 1990. β-Amyloid protein increases the vulnerability of cultured cortical neurons to excitotoxic damage. *Brain Res.* 533:315–20

Koo EH, Park L, Selkoe DJ. 1993. Amyloid β-protein as a substrate interacts with extracellular matrix to promote neurite outgrowth. *Proc. Natl. Acad. Sci. USA* In press

Kosik KS, Coleman P. 1992. Is β-amyloid neurotoxic. *Neurobiol. Aging* 13:535–630

Kowall NW, Beal MF, Busciglio J, et al. 1991. An in vivo model for the neurodegenerative effects of β-amyloid and protection by substance P. *Proc. Natl. Acad. Sci. USA* 88:7247–51

Kowall NW, Kosik KS. 1987. The cytoskeletal pathology of Alzheimer's disease is characterized by aberrant tau distribution. *Ann. Neurol.* 22:639–43

Lee VM, Trojanowski JQ. 1992. The disordered neuronal cytoskeleton in Alzheimer's disease. *Curr. Opin. Neurobiol.* 2:653–56

Lemere C, Yamaguchi H, Joahcim C, Selkoe D. 1991. Amyloid β-protein immunoreactivity in spleen, lymph node and other non-neural tissues using a new AβP antiserum. *Soc. Neurosci. Abstr.* 17:1443

Levy E, Carman MD, Fernandez-Madrid IJ, et al. 1990. Mutation of the Alzheimer's disease amyloid gene in hereditary cerebral hemorrhage, Dutch-type. *Science* 248:1124–26

Mandybur TI. 1975. The incidence of cerebral amyloid angiopathy in Alzheimer's disease. *Neurology* 25:120–26

Mann DMA, Jones D, Prinja D, Purkiss MS. 1990. The prevalence of amyloid (A4) protein deposits within the cerebral and cerebellar cortex in Down's syndrome and Alzheimer's disease. *Acta Neuropathol.* 80:318–27

Masters CL, Simms G, Weinman NA, et al. 1985. Amyloid plaque core protein in Alzheimer disease and Down syndrome. *Proc. Natl. Acad. Sci USA* 82:4245–49

Mattson MP, Cheng B, Davis D, et al. 1992. Amyloid peptides destabilize calcium homeostasis and render human cortical neurons vulnerable to excitotoxicity. *J. Neurosci.* 12:379–89

Mattson M, Culwell AR, Esch FS, et al. 1993. Evidence for neuroprotective and intraneuronal calcium-regulating roles for secreted forms of the β-amyloid precursor protein. *Neuron* 10:243–54

Miller DL, Papayannopoulos IA, Styles J, Bobin SA. 1993. Peptide compositions of the cerebrovascular and senile plaque core amyloid deposits of Alzheimer's disease. *Arch. Biochem. Biophys.* 301:41–52

Milward EA, Papadopoulos R, Fuller SJ, et al. 1992. The amyloid protein precursor of Alzheimer's disease is a mediator of the effects of nerve growth factor on neurite outgrowth. *Neuron* 9:129–37

Miyakawa T, Shimoji A, Kuramoto R, Higuchi Y. 1982. The relationship between senile plaques and cerebral blood vessels in Alzheimer's disease and senile dementia: morphological mechanism of senile plaque production. *Virchows Arch. B* 40:121–29

Molls RC, Breitner JC, Silverman JM, Davis KL. 1987. Alzheimer's disease. Morbid risk among first-degree relatives approximates 50% by 90 years of age. *Arch. Gen. Psychiatry* 44:405–8

Mori H, Takio K, Ogawara M, Selkoe DJ. 1992. Mass spectrometry of purified amyloid β protein in Alzheimer's disease. *J. Biol. Chem.* 267:17082–86

Morishima-Kawashima MHM, Takio K, Suzuki M, et al. 1993. Ubiquitin is conjugated with amino-terminally processsed tau in parted helical filaments. *Neuron.* 44:1151–60

Mullan M, Houlden H, Windelspecht M, et al. 1992. A locus for familial early onset Alzheimer's disease on the long arm of chromosome 14, proximal to α₁-antichymotrypsin. *Nature Genet.* 1:340–42

Munoz DG. 1991. Chromogranin A-like immunoreactive neurites are major constituents of senile plaques. *Lab. Invest.* 64:826–32

Ninomiya H, Roch J, Sundsmo MP, et al. 1993. Amino acid sequence RERMS represents the active domain of amyloid β/A4 protein precursor that promotes fibroblast growth. *J. Cell Biol.* 121:879–86

Nishimoto I, Okamoto T, Matsuura Y, et al. 1993. Alzheimer amyloid protein precursor complexes with brain GTP-binding protein G_o. *Nature* 362:75–79

Nitsch RM, Slack BE, Wurtman RJ, Growdon JH. 1992. Release of Alzheimer amyloid precursor derivatives stimulated by activation of muscarinic acetylcholine receptors. *Science* 258:304–7

Nordstedt C, Gandy SE, Alafuzoff I, et al. 1991. Alzheimer β/A4-amyloid precursor protein in human brain: aging-associated increases in holoprotein and in a proteolytic fragment. *Proc. Natl. Acad. Sci. USA* 88:8910–14

Oltersdorf T, Fritz LC, Schenk DB, et al. 1989. The secreted form of the Alzheimer's amyloid precursor protein with the Kunitz domain is protease Nexin-II. *Nature* 341:144–47

Oltersdorf T, Ward PJ, Henriksson T, et al. 1990. The Alzheimer amyloid precursor protein. Identification of a stable intermediate in the biosynthetic/degradative pathway. *J. Biol. Chem.* 265:4492–97

Pericak-Vance MA, Bebout J, Gaskell P. 1991. Linkage studies in familial Alzheimer disease: Evidence for chromosome 19 linkage. *Am. J. Hum. Genet.* 48:1034–50

Ponte P, Gonzalez-DeWhitt P, Schilling J, et al. 1988. A new A4 amyloid mRNA contains a domain homologous to serine proteinase inhibitors. *Nature* 331:525–27

Qiu W, Ferreira A, Miller C, et al. 1993. Cell-surface β-amyloid precursor protein stimulated hippocampal neuronal adhesion and neurite outgrowth in an isoform-dependent manner. *Soc. Neurosci. Abstr.* In Press

Robakis NK, Ramakrishna N, Wolfe G, Wisniewski HM. 1987. Molecular cloning and characterization of a cDNA encoding the cerebrovascular and the neuritic plaque amyloid peptides. *Proc. Natl. Acad. Sci. USA* 84:4190–94

Rogers J, Cooper NR, Websger S, et al. 1992. Complement activation by β-amyloid in Alzheimer disease. *Proc. Natl. Acad. Sci. USA* 89:10016–20

Roher AE, Lowenson JD, Clarke S, et al. 1993. Structural alterations in the peptide backbone of β-amyloid core protein may account for its deposition and stability in Alzheimer's disease. *J. Biol. Chem.* 268:3072–83

Roher A, Wolfe D, Palutke M, KuKuruga D. 1986. Purification, ultrastructure, and chemical analyses of Alzheimer's disease amyloid plaque core protein. *Proc. Natl. Acad. Sci. USA* 83:2662–66

Rosen DR, Martin-Morris L, Luo L, White K. 1989. A *Drosophila* gene encoding a protein resembling the human beta-amyloid precursor protein. *Proc. Natl. Acad. Sci. USA* 86:2478–82

Rosoff M, Li C. 1992. Isolation and characterization of FMRFamide-like peptides from *C. elegans* and mapping of promoter elements for the corresponding gene. *Soc. Neurosci. Abstr.* 18:1090

Rozemuller JM, Stam FC, Eikelenboom P. 990. Acute phase proteins are present in amorphous plaques in the cerebral but not cerebellar cortex of patients with Alzheimer's disease. *Neurosci. Lett.* 119:75–78

Saitoh T, Sunsdmo M, Roch J-M, et al. 1989. Secreted form of amyloid β protein precursor is involved in the growth regulation of fibroblasts. *Cell* 58:615–22

Saunders RD, Luttman CA, Keith PT, Little SP. 1991. Beta-amyloid protein potentiates H_2O_2-induced neuron degeneration in vitro. *Soc. Neursci. Abstr.* 17:1447

Schellenberg GD, Bird TD, Wijsman EM, et al. 992. Genetic linkage evidence for a familial Alzheimer's disease locus on chromosome 14. *Science* 258:668–71

Schubert D, Jin L-W, Saitoh T, Cole G. 1989a. The regulation of amyloid β protein precursor secretion and its modulatory role in cell adhesion. *Neuron* 3:689–94

Schubert D, LaCorbiere M, Saitoh T, Cole G. 1989b. Characterization of an amyloid beta precursor protein that binds heparin and contains tyrosine sulfate. *Proc. Natl. Acad. Sci. USA* 86:2066–69

Selkoe DJ. 1989. Molecular pathology of amyloidogenic proteins and the role of vascular amyloidosis in Alzheimer's disease. *Neurobiol. Aging* 10:387–95

Selkoe DJ, Abraham CR, Podlisny MB, Duffy LK. 1986. Isolation of low-molecular-weight proteins from amyloid plaque fibers in Alzheimer's disease. *J. Neurochem.* 146:1820–34

Selkoe DJ, Podlisny MB, Joachim CL, et al. 1988. β-amyloid precursor protein of Alzheimer disease occurs as 110–135 kilodalton membrane-associated proteins in neural and nonneural tissues. *Proc. Natl. Acad. Sci. USA* 85:7341–45

Seubert P, Oltersdorf T, Lee MG, et al. 1993. Secretion of β-amyloid precursor protein cleaved at the amino-terminus of the β-amyloid peptide. *Nature* 361:260–63

Seubert P, Vigo-Pelfrey C, Esch F, et al. 1992. Isolation and quantitation of soluble Alzheimer's β-peptide from biological fluids. *Nature* 359:325–27

Shioi J, Anderson JP, Ripellino JA, Robakis NK. 1992. Chondroitin sulfate proteoglycan form of the Alzheimer's β-amyloid precursor. *J. Biol. Chem.* 267:13819–22

Shoji M, Golde TE, Ghiso J, et al. 1992. Production of the Alzheimer amyloid β protein by normal proteolytic processing. *Science* 258:126–29

Siman R, Card JP, Nelson RB, Davis LG. 1989. Expression of β-amyloid precursor protein in reactive astrocytes following neuronal damage. *Neuron* 3:275–85

Sinha S, Dovey HF, Seubert P, et al. 1990. The protease inhibitory properties of the Alzheimer's β-amyloid precursor protein. *J. Biol. Chem.* 265:8983–85

Sisodia SS, Koo EH, Bayreuther K, et al. 1990. Evidence that β-amyloid protein in Alzheimer's disease is not derived by normal processing. *Science* 248:492–95

Smith RP, Higuchi DA, Broze GJ Jr. 1990. Platelet coagulation factor XIa-inhibitor, a form of Alzheimer amyloid precursor protein. *Science* 248:1126–28

Snow AD, Mar H, Hochlin D, et al. 1988. The presence of heparan sulfate proteoglycans in the neuritic plaques and congophilic angiopathy in Alzheimer's disease. *Am. J. Pathol.* 133:456–63

St. George-Hyslop P, Haines J, Rogaev E, et al. 1992. Genetic evidence for a novel familial Alzheimer's disease locus on chromosome 14. *Nature Genet.* 2:330–34

Stephenson DT, Rash K, Clemens JA. 1992. Amyloid precursor protein accumulates in regions of neurodegeneration following focal cerebral ischemia in the rat. *Brain Res.* 593:128–35

Strittmatter WJ, Saunders AM, Schmechel D, et al. 1993. Apolipoprotein E: high-avidity binding to β-amyloid and increased frequency of type 4 allele in late-onset familial

Alzheimer disease. *Proc. Natl. Acad. Sci. USA* 90:1977–81

Suzuki T, Nairn AC, Gandy SE, Greengard P. 1992. Phosphorylation of Alzheimer amyloid precursor protein by protein kinase C. *Neuroscience* 48:755–61

Tagliavini F, Giaccone G, Frangione B, Bugiani O. 1988. Preamyloid deposits in the cerebral cortex of patients with Alzheimer's disease and nondemented individuals. *Neurosci. Lett.* 93:191–96

Tamaoka A, Kalaria RN, Lieberburg I, Selkoe DJ. 1992. Identification of a stable fragment of the Alzheimer amyloid precursor containing the β-protein in brain microvessels. *Proc. Natl. Acad. Sci. USA* 89:1345–49

Tanzi RE, Gusella JF, Watkins PC, et al. 1987. Amyloid β-protein gene: cDNA, mRNA distribution, and genetic linkage near the Alzheimer locus. *Science* 235:880–84

Tanzi RE, McClatchey AI, Lamperti ED, et al. 1988. Protease inhibitor domain encoded by an amyloid protein precursor mRNA associated with Alzheimer's disease. *Nature* 331:528–32

Ter-Minassian M, Kowall NW, McKee AC. 1992. Beta amyloid protein immunoreactive senile plaques in infantile Down's syndrome. *Soc. Neurosci. Abstr.* 18:734

Terry RD. 1963. The fine structure of neurofibrillary tangles in Alzheimer's disease. *J. Neuropathol. Exp. Neurol.* 22:629–42

Ushiyama M, Ikeda S, Yanagisawa N. 1991. Transthyretin-type cerebral amyloid angiopathy in type I familial amyloid polyneuropathy. *Acta. Neuropathol.* 81:524–28

van Broeckhoven C, Backhovens H, Cruts M, et al. 1992. Mapping of a gene predisposing to early-onset Alzheimer's disease to chromosome 14q24.3. *Nature Genet.* 2:335–39

van Broeckhoven C, Haan J, Bakker E, et al. 1990. Amyloid β-protein precursor gene and hereditary cerebral hemorrhage with amyloidosis. *Science* 248:1120–22. In Dutch

Van Nostrand WE, Cunningham DD. 1987. Purification of protease nexin-II from human fibroblasts. *J. Biol. Chem.* 262:8508–14

Van Nostrand WE, Wagner SL, Suzuki M, et al. 1989. Protease nexin-II, a potent antichymotrypsin, shows identity to amyloid β-protein precursor. *Nature* 341:546–49

Verga L, Frangione B, Tagliavini F, et al. 1989. Alzheimer patients and Down patients: cerebral preamyloid deposits differ ultrastructurally and histochemically from the amyloid of senile plaques. *Neurosci. Lett.* 105:294–99

Vinters HV, Pardridge WM, Yang J. 1988. Immunohistochemical study of cerebral amyloid angiopathy: Use of an antiserum to a synthetic 28-amino-acid peptide fragment of the Alzheimer's disease amyloid precursor. *Human Pathol.* 19:214–22

Walker LC, Kitt CA, Cork LC, et al. 1988. Multiple transmitter systems contribute neurites to individual senile plaques. *J. Neuropathol. Exp. Neurol.* 47:138–44

Wasco W, Bupp K, Magendantz M, et al. 1992. Identification of a mouse brain cDNA that encodes a protein related to the Alzheimer disease-associated amyloid β-protein precursor. *Proc. Natl. Acad. Sci. USA* 89:10758–62

Wasco W, Gurubhagavatula S, Paradis MD, et al. 1993. Isolation and characterization of the human APLP2 gene encoding a homologue of the Alzheimer's associated amyloid β protein precursor. *Nature Genet.* In Press

Wegial J, Wisniewski HM. 1990. The complex of microglial cells and amyloid star in three-dimensional reconstruction. *Acta Neuropathol.* 81:116–24

Weidemann A, Konig G, Bunke D, et al. 1989. Identification, biogenesis and localization of precursors of Alzheimer's disease A4 amyloid protein. *Cell* 57:115–26

Whitson JS, Glabe CG, Shintani E, et al. 1990. β-amyloid protein promotes neuritic branching in hippocampal cultures. *Neurosci. Lett.* 110:319–24

Whitson JS, Selkoe DJ, Cotman CW. 1989. Amyloid β protein enhances the survival of hippocampal neurons in vitro. *Science* 243:1488–90

Wisniewski HM, Wegiel J, Wang KC, et al. 1989. Ultrastructural studies of the cells forming amyloid fibers in classical plaques. *Can. J. Neurol. Sci.* 16:535–42

Wright CI, Guela C, Mesulam MM. 1993. Protease inhibitors and indoleamines selectively inhibit cholinesterases in the histopathologic structures of Alzheimer disease. *Proc. Natl. Acad. Sci. USA* 90:683–86

Yamaguchi H, Hirai S, Morimatsu M, et al. 1988. Diffuse type of senile plaques in the brains of Alzheimer-type dementia. *Acta Neuropathol.* 77:113–19

Yamaguchi H, Nakazato Y, Hirai S, Shoji M. 1990. Immunoelectron microscopic localization of amyloid β protein in the diffuse plaques of Alzheimer-type dementia. *Brain Res.* 508:320–24

Yamaguchi H, Nakazato Y, Hirai S, et al. 1989. Electron micrograph of diffuse plaques: initial stage of senile plaque formation in the Alzheimer brain. *Am. J. Pathol.* 135:593–97

Yamaguchi H, Yamazaki T, Lemere CA, et al. 1992. Beta amyloid is focally deposited within the outer basement membrane in the amyloid angiopathy of Alzheimer's disease. *Am. J. Pathol.* 141:249–59

Yankner BA, Duffy LK, Kirschner DA. 1990. Neurotrophic and neurotoxic effects of amyloid β protein: reversal by tachykinin neuropeptides. *Science* 250:279–82

Annu. Rev. Neurosci. 1994. 17:519–549

ORGANIZATION OF MEMORY TRACES IN THE MAMMALIAN BRAIN

Richard F. Thompson and David J. Krupa

Program for Neural, Informational, and Behavioral Science, University of Southern California, Los Angeles, California 90089-2520

KEY WORDS: cerebellum, learning and memory, eyeblink conditioning, memory localization, classical conditioning

INTRODUCTION

Perhaps the most fundamental issue in the broad field of neuronal substrates of learning and memory concerns the physical/biological mechanisms underlying long-term memory formation, storage, and retrieval in the mammalian brain. Considerable progress has been made in elucidating memory storage mechanisms in simpler invertebrate systems (Abrams et al 1991, Bergold et al 1990, Clark & Schuman 1991, Crow & Forrester 1990, Nelson & Alkon 1990), but similar advances have yet to be made in understanding the mammalian brain. As Lashley (1929) stressed so many years ago, the overriding problem for understanding memory mechanisms in the mammalian brain is localization of memory storage. Mechanisms of memory storage cannot be analyzed until the memory storage sites, whether localized or distributed, have been identified. Given the nature of the vertebrate brain, memories should, to some degree, be distributed over ensembles of neurons, but whether this is within a localized region, over several regions in a given structure, or over several structures, is not yet known for most forms of memory.

Once the locus, or loci, of a given form of long-term memory has been determined, we should be able to analyze the mechanisms of memory storage. These will likely involve chemical/structural changes at synapses and possibly long-lasting changes in gene expression in the relevant neurons. Determining these mechanisms, however, will not inform us of what the memory is; the content of a neuronal memory store can only be determined by a detailed characterization of the neural networks that subserve the memory.

We now know that different forms or aspects of memory critically involve different neural systems in the brain. The hippocampus plays a key role in relational, contextual, spatial, and olfactory learning in lower mammals (Becker et al 1980, Berger & Orr 1983, Eichenbaum et al 1986, Lynch 1986, Moyer et al 1990, O'Keefe & Nadel 1978, Ross et al 1984, Solomon et al 1986b, Squire 1992). In monkeys, the hippocampus and adjacent cortex, particularly the perirhinal, parahippocampal, and entorhinal areas, are critical for delayed non-matching to sample and for establishing (and/or retrieving) long-term visual memories, which are often termed declarative or working, but do not appear to be the site of long-term storage (Meunier et al 1993, Squire 1992, Zola-Morgan & Squire 1990). The cerebral cortex is the proposed repository of long-term declarative memories, but conclusive evidence supporting this belief is lacking. The amygdala plays a key role in initial learning of conditioned heart rate and blood pressure, conditioned potentiation of startle, conditioned freezing to tone, and instrumental avoidance (Hitchcock & Davis 1986, Iwata et al 1986, Fanselow et al 1991, Liang et al 1982); however, it may not be the site of long-term storage (see Lavond et al 1993, McGaugh 1989). [Interestingly, the hippocampus is critical for initial learned freezing to context (Kim & Fanselow 1992).] Recent evidence suggests that the basal ganglia may play an important role in some aspects of instrumental learning (Packard et al 1989). A region of the cerebellum is necessary for basic delay classical conditioning of discrete motor responses (see below). On the other hand, it is possible that all these memory systems become engaged, to varying degrees, in multiple aspects of learning, with each system playing its separate role (see Lavond et al 1993, Wagner & Brandon 1989).

With the possible exception of declarative or working memory, all these aspects of learning exhibit similar basic parametric features (e.g. massed vs spaced practice, temporal relations of stimuli, stimulus salience, etc), which suggests that at some level common mechanisms are involved. As Rescorla stressed (1988a,b), basic associative learning, which results from exposure to relations among events in the world, is the way organisms, including humans, learn about causal relationships in the world. For both modern Pavlovian and cognitive views of learning and memory, the individual learns a representation of the causal structure of the world and, as a result of experience, then adjusts this representation to bring it in tune with the real causal structure of the world, thus striving to reduce any discrepancies or errors between its internal representation and external reality (see also Dudai 1989, Squire 1987).

THE ISSUE OF LOCALIZATION

Permanent lesions that abolish a learned behavior (i.e. the behavioral expression of the memory) but do not interfere with the ability of the organism to otherwise perform the behavior can serve to identify the brain structures necessary for a given form of memory but cannot illuminate precisely what roles these structures play in memory storage. Recording of neuronal activity that changes in tight correlation with learning can identify structures involved in or influenced by formation of the memory. Such evidence, per se, cannot localize the site(s) of memory formation. But at the very least the locus of memory storage must in some way receive information about the conditioned or signal stimulus and the unconditioned or reinforcing stimulus and/or response and must exhibit learning-related changes in activity. Hence, electrophysiological evidence can serve to both identify putative sites of storage and rule out possible storage sites. Electrical microstimulation, substituting for peripheral stimulation, can help to identify the neuronal circuitry sufficient for a given form of learning. Anatomical and biochemical changes, [in number or microstructure of synapses, in receptor number or affinity, in gene expression, in activation (e.g. cytochrome oxidase), etc] can indicate persisting and localized changes. However, unlike electrophysiological activity, these changes cannot be directly related temporally to learning-induced changes in behavior. Further, unless the locus of the synaptic plasticity coding a memory trace has been determined, as in certain invertebrate preparations, it is difficult to distinguish between changes in neuron activity, per se, and memory storage. Collectively, these methods can serve to identify an essential (necessary and sufficient) memory circuit, but they cannot provide definitive evidence to identify the locus of memory storage. Note that identification of the neural circuitry essential for a given form of learning and memory is also necessary to characterize the content of the memory store.

The methods of reversible inactivation offer promise of localizing sites of memory storage. The logic is straightforward, given that the essential circuit for a given form of memory has been identified. If naive animals are trained during reversible inactivation of a structure or region that is part of the essential circuit, the animals will, by definition, show no expression of learning during inactivation training. If they show no evidence of having learned in post-inactivation training, and learn with no savings as though naive, then the inactivated region is either the site of memory storage or a mandatory afferent to the site of memory storage. If, on the other hand, the animals show asymptotic learned performance from the beginning of postinactivation training, the inactivated region cannot be the site of the memory storage and is efferent from the site. The same logic would seem to apply if the memory trace is distributed over several regions. Degree of savings in post-inactivation training as a function of regions inactivated, compared to non-inactivated controls, can assess the relative contribution of each component of a distributed memory representation.

CLASSICAL CONDITIONING OF DISCRETE RESPONSES

The specific focus of this review is on identification of the essential memory trace circuit and localization of the memory trace within that circuit for a basic form of long-term associative memory, basic delay classical conditioning of discrete behavioral responses. Eyeblink conditioning is the most widely used procedure, but classical conditioning of other discrete responses, particularly limb flexion, provides additional evidence. Perhaps the simplest descriptor of these types of basic associative learning is sensory-motor memory. We will argue that the evidence now strongly favors the hypothesis that the essential memory trace circuit includes the cerebellum and its associated brain stem circuitry and that the memory traces are formed and stored in the cerebellum. For reasons that are not entirely clear, this has been an extremely contentious field. Several researchers appear to have an unshakable a priori belief that associative memory traces cannot possibly be formed and stored in the cerebellum (Bloedel 1992, Welsh & Harvey 1989).[1]

Dating from the classic papers of Brindley (1964), Marr (1969), and Albus (1971), the cerebellum has long been a favored structure for modeling a neuronal learning system. Recent empirical studies reviewed below have been guided by these models and the related views of Eccles et al (1967), Eccles (1977), and Ito (1972, 1984), and results to date constitute a remarkable verification of the spirit of these earlier theories. The highly simplified schematic block diagram of Figure 1 summarizes results to date and may be considered a qualitative model of the role of the cerebellum in classical conditioning of discrete behavioral responses (see also reviews by Lavond et al 1993, Steinmetz et al 1992, Thach et al 1992, Thompson 1986, 1990). Earlier Russian research indicated that removal of the cerebellum impaired or abolished classical conditioning of the leg-flexion response (Karamian et al 1969, see also discussion in Clark et al 1984 and Thompson et al 1983); however, these observations were not pursued.

Based on recent anatomical findings, patterns of neural discharge related to specific behaviors, and the effects of focal lesions, Thach et al (1992) have recently proposed a model of cerebellar function that suggests that the job of

---

[1]For an interesting example of how extreme this bias can be, see the recent chapter by Welsh & Harvey (1992) in which most of the directly relevant evidence is not considered or even cited.

the cerebellum is, among other things, to coordinate elements of movement in its downstream targets and to adjust old movement synergies while learning new ones. Although an in-depth discussion of this literature is beyond the scope of this review, a brief mention of this work is warranted, especially as it pertains to motor learning. The model was inspired, in part, by recent findings that there is a somatotopic mapping of body parts and modes of motor control within the cerebellar deep nuclei (Asanuma et al 1983, Kane et al 1988, Thach et al 1982). This fact, coupled with recent assessments of a much longer length for the cerebellar parallel fibers than previously thought existed (Brand et al 1976, Mugnaini 1983), suggested that beams of Purkinje cells connected by the longer parallel fibers could link actions of different body parts represented within each nucleus and exert control across nuclei into coordinated multijointed movements. This hypothesis is supported, in part, by the finding that reversible or permanent lesions of deep cerebellar nuclei selectively and severely disrupt tasks requiring coordinated activation of multi-jointed movements while having little or no effect upon previously trained single-joint movements (Kane et al 1988, 1989). Although these lesion studies suggest only a minor role for the cerebellum in single-jointed movements, recordings from the deep nuclei during these tasks find marked changes in neural discharge correlated with the parameters of movement (Scheiber & Thach 1985, Thach 1968, 1970). It appears, however, that while performing the single-joint movements, the monkeys also moved many more muscles than those minimally required for the task, and the recorded-unit activity may have been better correlated with the covert multijointed movements than with the single-jointed task (Thach et al 1992, p. 420).

The model Thach et al proposed suggests the cerebellum is involved not only in coordinating multijointed tasks, but that it also learns new tasks through activity-dependent modification of parallel-fiber Purkinje cell synapses (Thach et al 1992, Gilbert & Thach 1977). Evidence in support of this hypothesis includes Purkinje cell recordings, which reveal patterns of activity consistent with a causal involvement in the modification of the coordination between eye position and hand/arm movement (Keating & Thach 1990) and studies in which focal lesions by microinjection of muscimol severely impair adaptation of hand/eye coordination without affecting the performance of the task (Keating & Thach 1991). The implications of this model for eyeblink conditioning are intriguing since, as noted below, the conditioned response is a highly coordinated activation of several muscle groups.

The cerebellum is believed to be critically involved in a number of other learned behaviors. Supple & Leaton (1990a,b) have recently shown that the cerebellar vermis is necessary for classical conditioning of heart rate in both restrained and freely moving rats. A growing body of literature, which is beyond the scope of this review, implicates the cerebellum as critical for a range of instrumental tasks in animals: avoidance learning of the eyeblink in rabbits (Polenchar et al 1985); avoidance but not appetitive learning of bar press in rats (Steinmetz et al 1993); and Morris water maze (invisible platform, i.e. place learning) in mutant pcd (Purkinje cell degeneration) mice (Goodlett et al 1992) and in mutant staggerer mice (Purkinje, granule and inferior olive neuron degeneration) (Lalonde 1987) (see Lalonde & Botez 1990 for review). There is also an increasing amount of research on the human cerebellum that indicates critical cerebellar involvement in complex learning/cognitive tasks (Akshoomoff & Courchesne 1992, Canavan & Homberg 1991, Daum et al 1991, Fiez et al 1992, Grafman et al 1992, Ingvar 1990, Petersen et al 1989; see Leiner et al 1989, Schmahmann 1991, for reviews).

The Nature of the Eyeblink Conditioned Response

Gormezano et al (1962) showed some years ago in separate studies of the rabbit that eyeball retraction, nictitating membrane (NM) extension, and external eyelid closure all had essentially identical acquisition functions (Deaux & Gormezano 1963, Schneiderman et al 1962). Simultaneous recording of NM extension and external eyelid closure (EMG recordings from *orbicularis oculi*) during acquisition and extinction showed that they were, in essence, perfectly correlated, both within trials and over training (McCormick et al 1982c, Lavond et al 1990). Furthermore, some degree of conditioned contraction of facial and neck musculature also developed and was also strongly correlated with NM extension. These observations led to characterization of the conditioned response (CR) as a "synchronous facial 'flinch' centered about closure of the eyelids and extension of the NM" (McCormick et al 1982c, p. 773). Substantial learning-induced increases in neuronal unit activity that correlate very closely with the conditioned NM extension response have been reported in several motor nuclei: oculomotor, trochlear, motor trigeminal, abducens, accessory abducens, and facial (Berthier & Moore 1983, Cegavske et al 1979, Disterhoft et al 1985, McCormick et al 1983). These are all components of the same global conditioned response involving, to the extent studied, essentially perfectly coordinated activity in a number of muscles and associated motor nuclei. The NM extension response is but one component of the conditioned response. The suggestion that different motor nuclei might somehow exhibit different conditioned responses in the eyeblink conditioning paradigm (Delgado-Garcia et al 1990) is not supported by evidence.

The CR and the unconditioned response (UR) are similar in eyeblink conditioning in the sense that to a large extent the same muscles and motor nuclei are engaged. However, the CR and the UR differ fundamentally in a number of respects. The minimum onset latency of the CR to a tone conditioning stimulus (CS) in well trained rabbits, measured as NM extension, is about 90–100 ms; the minimum onset latency of the NM extension UR to a 3 psi corneal airpuff unconditioned stimulus (US) in the rabbit is about 25–40 ms. Perhaps most important, the variables that determine the topographies of the UR and CR are quite different. The topography of the UR is under the control of the properties of the US: for example, stimulus intensity, rise-time, and duration. In marked contrast, the topography of the CR is substantially independent of the properties of the US and is determined primarily by the interstimulus interval (the CS-US onset interval)—the CR peaking at about the onset of the US over a wide range of effective CS-US onset intervals (Coleman & Gormezano 1971, Millenson et al 1977, Schneiderman 1966, Steinmetz 1990a). This key property of the CR cannot be derived from the properties of the US or the UR. The CR and the UR also differ in that certain components of the UR can be elicited separately by appropriate peripheral stimuli but the CR always occurs as a global coordinated response (McCormick et al 1982c). Another important difference is that the CR exhibits much greater plasticity in recovery from lesions of the motor nuclei that impair performance of the UR than does the UR itself (Disterhoft et al 1985, Steinmetz et al 1992).

In sum, the conditioned eyeblink response involves highly coordinated activity in a number of motor nuclei and muscles; it is one global defensive response that is conditioned to a neutral stimulus as a result of associative training. The very small lesion of the interpositus nucleus, which is effective in completely and permanently abolishing the conditioned NM extension response (see below), also completely and permanently abolishes all other components of the conditioned response that have been studied—eyeball retraction, external eyelid closure, orbicularis oculi EMG—without producing any impairment in performance of the reflex response (Steinmetz et al 1992).

Interpositus Nucleus and the CR Pathway

Recordings of neuronal unit activity from the interpositus nucleus during eyeblink conditioning revealed populations of cells that discharged when the conditioning stimuli (CS and US) were presented, and, more importantly, populations of cells that, as a result of training, discharged just prior to execution of the classically conditioned response (Berthier & Moore 1990, Foy et al 1984, McCormick & Thompson 1984a,b; McCormick et al 1981, 1982a). These cells fired in a pattern that was similar to the learned behavioral response, i.e. they formed an amplitude-time course "model" of the learned response that preceded and predicted the occurrence of the behavioral conditioned response within trials and over the trials of training. Recently, Yang & Weisz (1992) recorded activity of single neurons in the dentate and interpositus nuclei, which had been identified as excitatory by antidromic activation from stimulation of the red nucleus, in response to the CS and US in the beginning of training. The great majority of cells showed increased discharge frequency to the US (substantial) and the CS (small but significant). Paired CS-US presentations resulted in marked enhancement of the response to the US in cells in the interpositus nucleus and depression of response to the US in cells in the dentate nucleus. In other animals, small lesions of the region of recording in the anterior interpositus nucleus completely prevented acquisition of the eyeblink CR but lesions of the dentate recording region had no effect at all on learning (see below).

McCormick et al (1981), using a tone CS and corneal airpuff US in the basic delay paradigm (CS 350 ms; US 100 ms; coterminating), initially reported that lesions of the cerebellum ipsilateral to the trained eye (large aspirations and electrolytic lesions of the interpositus nucleus) abolished the eyeblink conditioned response (CR) completely and selectively, i.e. the CR was completely abolished, but the lesion had no effect on the unconditioned response (see also McCormick et al 1982a). The lesions did not prevent learning in the contralateral eye. If the lesion was made before training, animals were completely unable to learn any CRs at all with the eye ipsilateral to the lesion (Lincoln et al 1982). In other studies the same results were obtained with lesions of the superior cerebellar peduncle, the efferent pathway from interpositus to red nucleus (Lavond et al 1981, McCormick et al 1982b, Rosenfield et al 1985).

Electrolytic lesions of the interpositus nucleus ipsilateral to the trained eye again demonstrated that if the lesions completely destroyed the critical region of the interpositus nucleus the CR was abolished, with no effect on the UR (Clark et al 1984). Importantly, if the lesions were incomplete, there was a marked decrease in the amplitude and frequency of occurrence of the CR and a marked increase in CR onset latency that did not recover with post-operative training (see also Welsh & Harvey 1989). Since electrolytic lesions of the interpositus cause retrograde degeneration in the inferior olive, kainic acid lesions of the critical region of the interpositus were made, with identical results, i.e. complete and selective abolition of the CR (Lavond et al 1985). Yeo and associates (1985a) replicated the interpositus lesion result, using light and white noise conditioned stimuli (CSs) and a periorbital shock US, thus extending the generality of the findings. A number of subsequent studies showed identical effects of interpositus lesions, i.e. complete and selective abolition of the ipsilateral eyeblink CR with no effect on the UR (Lavond et al 1984a,b; Lavond et al 1987, McCormick & Thompson 1984a, Polenchar et al 1985, Sears & Steinmetz 1990, Steinmetz et al 1991, Weisz & LoTurco 1988, Woodruff-Pak et al 1985). This effect was extremely localized. Electrolytic interpositus lesioned animals were periodically trained for periods up to 8 months; no CRs ever developed on the side of the lesion (Lavond et al 1984b). Reversible inactivation by microinfusion of nanomolar amounts of neurotransmitter antagonists in the critical region of the interpositus completely and reversibly abolished the CR, in a dose-dependent fashion, with no effect at all on the UR (Mamounas et al 1987).

Recently, the interpositus lesion abolition of the CR was studied in great detail (Steinmetz et al 1992). Appropriate interpositus lesions completely and permanently prevent acquisition and completely and permanently abolish retention of the eyeblink CR, over all conditions of training and measurements that have been used, and have no persisting effects on performance of the UR, regardless of the amount of pre- or postoperative training (see also Steinmetz & Steinmetz 1991). Finally, there are now several human studies showing that appropriate cerebellar damage completely prevents learning of the eyeblink CR (Daum et al 1993, Lye et al 1988, Solomon et al 1989).

Electrical microstimulation of the critical region of the anterior interpositus nucleus evokes an eyeblink response in naive animals, and lesion of the superior cerebellar peduncle abolishes this response; the eyeblink circuit is hard-wired from interpositus nucleus to behavior (McCormick & Thompson 1984a). If the stimulus intensity is set to elicit an eyeblink comparable in amplitude to that elicited by the standard 3 psi corneal airpuff and this interpositus stimulus is used as a US, neither learning of the CR to a tone CS nor maintenance of a CR previously learned to tone with a corneal airpuff US occurs. However, animals given tone-interpositus stimulation training show marked transfer in subsequent tone-airpuff training (Chapman et al 1988).

The region of the contralateral magnocellular red nucleus that receives projection from the region of the anterior interpositus critical for eyeblink conditioning also exhibits a learning-induced pattern of increased unit activity in eyeblink conditioning very similar to that shown by interpositus neurons (Chapman et al 1990). Microstimulation of this region of the red nucleus in naive animals also elicits eyeblink responses. When this is used as a US, neither learning of the CR to a tone CS nor maintenance of a CR previously learned with tone-airpuff training occurs. Further, there is no transfer from tone-red nucleus training to subsequent tone-airpuff training (Chapman et al 1988). If the red nucleus is reversibly inactivated in trained animals, the eyeblink CR is reversibly abolished but the learning-induced neuronal model of the CR in the interpositus nucleus is unaffected. In contrast, when the anterior interpositus nucleus is reversibly inactivated, both the behavioral CR and the learning-induced neuronal model of the CR in the magnocellular red nucleus are completely abolished (Chapman et al 1990, Clark et al 1992, Clark & Lavond 1993).

Small lesions of the appropriate region of the magnocellular red nucleus contralateral to the trained eye cause complete abolition of the CR with no effect on the UR (Chapman et al 1988, Haley et al 1983, Rosenfield et al

1985, Rosenfield & Moore 1983). This same lesion also abolishes the eyeblink response elicited in untrained animals by stimulating the interpositus nucleus. Microinfusions of nanomolar amounts of neurotransmitter antagonists in a very localized region of the magnocellular red nucleus reversibly abolished the conditioned eyeblink response with no effect on the UR (Haley et al 1988). Identical results were obtained for the conditioned limb flexion response; appropriate interpositus lesions abolished the conditioned hindlimb flexion response with no effect on reflex flexion in the rabbit (Donegan et al 1983). Lesions of the red nucleus or descending rubral pathway abolished the conditioned limb flexion response in the cat with no effect on the reflex limb flexion response and no effect on normal behavioral movement control of the limb (Smith 1970, Tsukahara et al 1981, Voneida 1990).

The Issue of Performance

One of the advantages of classical conditioning for neurobiological analysis is that performance of the behavioral response, per se (the UR), can be measured independently of the performance of the learned response (the CR). A consistent finding of the many studies showing interpositus lesion abolition of the CR is that the lesion has no effect on the UR (see above). Welsh & Harvey (1989) claimed that interpositus lesions did affect the UR. However, they did not in fact measure the effects of interpositus lesions on reflex responses in the same animals; instead they compared only some of their postlesion animals separated post hoc into different groups. Importantly, they did not find any effect of interpositus lesions on amplitude of the UR at any US intensity. They reported only very small lesion effects on UR topography at low US intensities. Many factors can influence the UR and there are extreme individual animal differences in properties of the UR. In order to demonstrate effects of lesions on the UR it is essential to compare URs in the same animals before and after lesion. Given the fact that the evidence necessary to demonstrate interpositus lesion effects on performance of the UR does not exist, it is somewhat surprising that the performance argument could have been taken seriously. In any event, possible effects of interpositus lesions (that abolished the CR) on the UR to US alone stimuli were examined in detail over a wide range of US intensities, comparing pre- vs postlesion URs in the same animals. No persisting effects of the lesions were found on any property of the UR (Steinmetz et al 1992).

Welsh & Harvey (1989) asserted that, "when one attempts to equate the CS and the UCS [unconditioned stimulus] as response-eliciting stimuli, the deficits in the CR and the UCR [unconditioned response] become more alike" (p. 309). The only way to evaluate this statement is to equate the CS and the US in terms of response elicitation prior to lesion and then determine the effect of the lesion on the two responses that were "psychophysically equivalent" prior to lesion. Welsh & Harvey did not make this comparison and did not provide any information on the properties of the URs prior to lesion in their animals. When this is done, e.g. when the intensity of the US is reduced so that the UR (US alone trials) is matched in amplitude and percent response to the CR before lesion, the interpositus lesion abolishes the CR and has no effect at all on the prelesion equivalent UR (Steinmetz et al 1992). All these studies used a standard US intensity for training (e.g. ≥3 psi), which typically yields a UR (US alone trials) larger in amplitude than the CR. Recently, animals were trained with a low intensity US just suprathreshold to establish learning (Ivkovich et al 1993). Under these conditions the CR (CS alone trials) and the UR (US alone trials) are equivalent in amplitude (the CR is numerically but not statistically larger than the UR). Interpositus lesions completely abolished the CR and had no effect at all on the UR.

There is in fact a double dissociation in terms of various brain lesion effects on the CR and the UR in eyeblink conditioning. Appropriate partial lesions of the motor nuclei involved in generating the CR and the UR cause immediate abolition of both the CR and the UR; however, with post-operative training the CR recovers almost to the pre-operative level but the UR shows little recovery (Disterhoft et al 1985, Steinmetz et al 1992). Large lesions of appropriate regions of cerebellar cortex that markedly impair or abolish performance of the CR (see below) result in an increase in the amplitude of the UR (Logan 1991, Yeo 1991).

Collectively, this evidence decisively rules out the "performance" argument that interpositus lesion abolition of the CR is somehow due to lesion effects on the UR.

The CS Pathway

The pontine nuclei receive projections from auditory, visual, and somatosensory systems (Brodal 1981, Mower et al 1979). Several regions of the pontine nuclei exhibit short latency evoked unit responses to auditory stimuli (Aitkin & Boyd 1978, Steinmetz et al 1987). Appropriate lesions of the pontine nuclei can abolish the CR established to a tone CS but not a light CS, i.e. can be selective for CS modality (interpositus lesions abolish the CR to all modalities of CS) (Steinmetz et al 1987). Lesions of the regions of the pons receiving projections from the auditory cortex abolish the CR established with electrical stimulation of auditory cortex as a CS (Knowlton & Thompson 1992). Infusion of lidocaine into the pontine nuclei in trained animals reversibly attenuates the CR (tone CS) (Knowlton & Thompson 1988). Extensive lesions of the middle cerebellar peduncle (mcp), which conveys mossy fibers from the pontine nuclei and other sources to the cerebellum, abolish the CR to all modalities of CS (Lewis et al 1987).

Electrical stimulation of the pontine nuclei serves as a "supernormal" CS, yielding more rapid learning than does a tone or light CS (Steinmetz et al 1986). With a pontine stimulation CS, lesion of the middle cerebellar peduncle abolishes the CR, thus ruling out the possibility that the pontine CS is activating noncerebellar pathways, e.g. by stimulation of fibers of passage or antidromic activation of sensory afferents (Solomon et al 1986a). Stimulation of the middle cerebellar peduncle itself is an effective CS and lesion of the interpositus nucleus abolishes the CR established with a pontine or middle peduncle stimulation CS (Steinmetz et al 1986). When animals are trained using electrical stimulation of the pontine nuclei as a CS (corneal airpuff us), some animals show immediate and complete transfer of the behavioral eyeblink CR and the learning-induced neuronal model of the behavioral CR in the interpositus nucleus to a tone CS. These results suggest that the pontine stimulus and tone must activate a large number of memory circuit elements (neurons) in common (Steinmetz 1990b).

Electrical stimulation of the pontine nuclei can elicit a variety of movements, depending on electrode location, but at rather high thresholds. When such stimulation well below movement threshold is used as a CS (corneal airpuff us), the eyeblink CR develops to the CS and the threshold CS to elicit the eyeblink CR becomes very low, orders of magnitude lower than the threshold for the movement (e.g. head turn) elicited before training (Tracy et al 1992). This is true regardless of the nature of the high threshold pontine-elicited movement before training. Extinction training results in a return to the high threshold movement. The eyeblink threshold to stimulation of the critical region of the interpositus nucleus does not change at all over the course of such training and extinction. Very small lesions via the interpositus stimulating electrode abolish the CR to the pontine stimulation CS. Thus, learning results in a marked increase in synaptic efficacy in the cerebellar circuit between the pontine and interpositus electrodes.

These results have interesting implications for the readout from higher brain structures to specific learned behaviors. Stimulation of any region of the pontine nuclei can serve as an effective CS for eyeblink conditioning. The pontine nuclei receive input from virtually all areas of the cerebral cortex (Brodal 1981) and from the hippocampus via the retrosplenial cortex (Berger & Bassett 1992). Thus, stimulation of the auditory cortex as a CS in eyeblink conditioning relays through the pontine nuclei to the cerebellum (Knowlton & Thompson 1992; 1993). The organization of the pontine-mossy fiber projections to the cerebellum is such that activation of virtually any subset of pontine neurons (so long as a minimum number are activated) can become associated with any specific response to be learned. The cerebellum appears able to translate distributed input via the pontine-mossy fibers into a behavioral response that is completely specific to the local sign of the US-UR. The latter is determined, we and others have suggested (see below), by the US-activated climbing fiber projections to the cerebellum (see Marr 1969). This view implies that the convergence of mossy-parallel fibers and climbing fibers is such that virtually any subset of mossy fibers can converge to virtually any subset of climbing fibers, i.e. those activated by a particular aversive US. This convergence may be due in part to the complex, patchy fractured sensory representation in cerebellar cortex (Welker 1987).

Collectively, these data strongly support the view that the essential CS pathway includes the pontine nuclei and mossy fiber projections to the cerebellum.

The US Pathway

Neuronal unit activity recorded in the critical region of the dorsal accessory olive (DAO, see below) exhibits no responses at all to the tone CS and a clear evoked increase in unit activity to the onset of the corneal airpuff US prior to training, which supports the argument that the memory trace is not in the DAO. Interestingly, this US-evoked neuronal activity decreases as animals learn and perform the CR but is still fully present on US alone trials (Sears & Steinmetz 1991), as is true for complex spikes evoked in Purkinje neurons by US onset (see below).

Electrolytic or chemical lesions of the critical region of the inferior olive, which is the face representation in the dorsal accessory olive (DAO), completely prevented learning of the conditioned eyeblink response if made before training and yielded extinction with continued training and hence abolition of the CR if made after training (McCormick et al 1985, Mintz et al 1988; see also Yeo et al 1986). The lesion had no effect at all on the UR. Voneida et al (1990) made chemical lesions of the DAO in cats trained in limb flexion conditioning and found extinction of the CR with continued training. All studies agree that the DAO is necessary for both learning and retention of the CR.

Electrical microstimulation of this region of the DAO elicits eyeblink responses before training; indeed virtually any phasic behavioral response can be so elicited, depending on the locus of the stimulating electrode. When DAO stimulation is used as a US, the exact response elicited by DAO stimulation is learned as a CR to a tone CS (Mauk et al 1986, Steinmetz et al 1989). Control procedures showed that the effective stimulus is to the climbing fibers (Thompson 1989). First, climbing fiber field potentials were recorded in cerebellar cortex during DAO stimulation when implanting the DAO electrodes. Second, interpositus lesions abolished both the CR- and DAO-elicited UR, thus ruling out the possibility of antidromic or current spread activation of reflex afferents (interpositus lesions do not affect reflex URs). Third, in some animals the 2-deoxy-glucose (2DG) technique was used to map the regions of cerebellar cortex activated by the DAO-US— they corresponded to climbing fiber projections from the DAO. Fourth, electrodes just dorsal to the DAO in the reticular formation also elicited movements, but these could not be conditioned to a CS. Fifth, DAO stimulus intensities below movement elicitation threshold can serve as an effective US. Sixth, extensive lesions of the reticular formation just dorsal to the DAO (where low threshold movements can be elicited by stimulation, but such stimulation is not an effective US) have no effect on the CR (Swain 1992). Finally, movements (URs) elicited by electrical stimulation of cerebellar cortex and white matter (US) can also be conditioned to a tone CS, and interpositus lesions abolish both the CR and the cerebellar elicited UR (Swain et al 1992, Swain 1992). So far as can be determined, the DAO-climbing fiber-cerebellar circuit is the only system other than reflex afferents in the brain where discrete specific movements elicited by stimulation can be conditioned to neutral stimuli. These data suggest that this is a system specialized for the learning of discrete movements.

It should be noted that when peripheral stimuli are presented, e.g. tone CS and corneal airpuff US, the CS may evoke some climbing fiber activity and the US will certainly evoke converging climbing fiber and mossy fiber activity to localized regions of the cerebellum. This converging activity may play an important role in learning with peripheral USs but does not appear essential, i.e. with direct stimulation of climbing fibers as a US. Indeed, normal behavioral learning occurs when electrical stimulation of pontine nuclei, which are mossy fibers, is used as a CS and electrical stimulation of the DAO–climbing fibers is used as a US; whatever behavioral response is evoked by the DAO stimulation US, this response is learned to the pontine nuclei CS, which itself does not elicit movements before training (Steinmetz et al 1989).

Collectively, this evidence strongly supports the hypothesis that the DAO and its climbing fiber projections to the cerebellum forms the essential US, or reinforcing, pathway for the learning of discrete responses. This evidence also rules out the possibility that the memory trace is localized to the DAO.

Cerebellar Cortex

Using extracellular single unit recording, it has been found that many Purkinje neurons, particularly in lobule HVI, are responsive to the tone CS and the corneal airpuff US in naive animals. Before training, the majority of Purkinje neurons that are responsive to the tone show variable increases in simple spike discharge frequency in the CS period (Foy & Thompson 1986). After training, the majority show learning-induced decreases in simple spike frequency in the CS period that correlate closely in onset latency with the behavioral CR; however, a significant number show the opposite effect (Berthier & Moore 1986, Donegan et al 1985, Foy et al 1992). These results are consistent with the possibility that a process of long-term depression (LTD) may be a mechanism of synaptic plasticity in cerebellar cortex in eyeblink conditioning (Chen & Thompson 1992, Ito 1989, Schreurs & Alkon 1992) (see below).

Before training, Purkinje neurons that are influenced by the corneal airpuff consistently show an evoked complex spike to US onset. In trained animals, this US evoked complex spike is virtually absent on paired CS-US trials when the animal gives a CR but present and normal on US alone test trials (Foy & Thompson 1986, Krupa et al 1991). This learning-induced reduction in US-evoked complex spikes is consistent with, and accounted for, by the DAO unit recording study noted above (Sears & Steinmetz 1991). These results led to the hypothesis that the DAO-climbing fiber system, a necessary part of the US pathway, could function as the error-correcting algorithm in classical conditioning (Rescorla & Wagner 1972) by way of the direct GABAergic descending pathway from the interpositus to the inferior olive (Nelson & Mugnaini 1989) (see Figure 1) (Donegan et al 1989, Gluck et al 1990, Thompson 1990). Current evidence supports this view: when animals are trained in the "blocking" paradigm, infusion of the GABA antagonist picrotoxin into the DAO during the compound stimulus training phase completely prevents blocking, as revealed in subsequent training to the novel stimulus of the compound (Kim et al 1992).

In initial studies, there appeared to be some disagreement about effects of cerebellar cortical lesions on eyeblink conditioning (McCormick & Thompson 1984a, Yeo et al 1985b). However, there now appears to be a growing consensus. In trained animals, lesions limited largely to HVI cause variable degrees of impairment in the conditioned response but with substantial or complete recovery (Lavond et al 1987, Yeo & Hardiman 1992). Larger lesions can cause greater impairments, and very large lesions (HVI, Crus I, Crus II, paramedian, and anterior lobes) can cause substantial and, in some animals but not others, persisting impairments (Lavond et al 1987, Perrett et al 1993, Yeo 1990, Yeo & Hardiman 1992). If lesions were made before training, lesions limited to HVI slowed acquisition somewhat; larger lesions (HVI, Crus I, Crus II, paramedian lobule and in one study the flocculus and paraflocculus as well) markedly impaired acquisition (Lavond & Steinmetz 1989, Logan 1991); and very large lesions, including some anterior lobe, markedly impaired and in a few cases, but not others, prevented acquisition (Logan 1991). Animals with these very large lesions that do learn and/or retain CRs exhibit disruption of adaptive timing of the CR (Logan 1991, Perrett et al 1993). Such lesions also prevent acquisition of conditioned inhibition (Logan 1991). Experimentally, it is extremely difficult to remove a very large extent of cerebellar cortex without damaging the cerebellar nuclei. It seems clear that cerebellar cortex plays a critically important role in normal learning of the eyeblink CR; whether or not it is essential remains unresolved.

Latency measures are consistent with the cerebellar hypothesis. Under standard training conditions (85 dB tone CS, 3 psi corneal airpuff, 250 ms CS-US onset interval) the mean minimum onset latency of the NM extension CR is about 90–100 ms. The onset of learning-induced increase in unit activity in the interpositus nucleus varies somewhat from animal to animal; it can precede onset of the learned NM response by as much as 60–70 ms and a typical value is about 50 ms. Learning-induced decreases in Purkinje neuron simple spikes can precede the onset of the learned NM response by as much as 60–80 ms (Foy & Thompson 1986). The latency of activation of the cerebellum by peripheral somatosensory stimuli is about 20 ms (Ekerot et al 1987). Finally, onset of the NM extension response to electrical stimulation of the interpositus nucleus is about 50 ms (McCormick & Thompson 1984a). These time delays account for the otherwise puzzling fact that the minimum onset latency of the conditioned NM response is about 90–100 ms, substantially longer than the minimum onset latency of the reflex NM response to corneal airpuff (25 ms) and the even more puzzling fact that no learning occurs if the CS-US onset interval (the interstimulus interval used in training) is shorter than about 80 ms (Gormezano et al 1983, Steinmetz 1990a).

Decerebration

If normal rabbits are trained in eyeblink conditioning and then acutely decerebrated, they exhibit retention of the CR, compared with control animals, if, and only if, the transection is sufficiently rostral not to damage the red nucleus (Mauk & Thompson 1987). Bloedel and associates (Kelly et al 1990) claimed to have established conditioned eyeblink responses in the acute decerebrate, decerebellate rabbit. However, their training procedure was extreme massed practice (9 s intertrial interval); they used an idiosyncratic definition of the CR; they did not measure or control the excitability of their preparations; they did not run any control groups for alpha conditioning, sensitization, or pseudoconditioning; and they reported results from only a very few of the animals they ran. Replication of their procedures in intact rabbits showed that no learning at all occurred (Nordholm et al 1991). Yeo (1991) and Yeo & Hardiman (1992) reported that they could establish eyeblink CRs in the acute decerebrate rabbit, using more standard training procedures. Subsequent removal of the cerebellum in their preparations completely abolished the eyeblink CR. It seems very likely that the eyeblink responses reported by Bloedel and associates (Kelly et al 1990) in the acute decerebrate, decerebellate rabbit were due to nonassociative processes (see discussion in Nordholm et al 1991).

For the sake of argument, let us assume that there is an associative component to the eyeblink responses of the decerebrate-decerebellate preparation, i.e. that with appropriate controls and training conditions and sufficient training, an associative conditioned eyeblink response could be established.

As with any new preparation for the study of learning, a number of control procedures must be done in order to define and characterize the phenomenon (Cohen 1984, Gormezano et al 1962, Rescorla 1988a,b; Schneiderman et al 1962, Carew et al 1983). A case in point is work on classical conditioning of the hindlimb flexion reflex in the acute spinal mammal (Beggs et al 1983, Durkovic 1975, Fitzgerald & Thompson 1967, Patterson et al 1973). Necessary control procedures demonstrate that there is indeed an associative component but that the conditioned increase in flexion reflex differs in certain of its properties from hindlimb flexion conditioning in the intact animal. Indeed, it is clear that the spinal cord alone does not contain the circuits that are necessary and sufficient for limb flexion conditioning in the intact animal. Thus, in the otherwise intact animal, lesions in the cerebellar nuclei, red nucleus, or rubrospinal tract produce complete and specific abolition of the conditioned limb flexion response with no effect on the reflex limb flexion response (Donegan et al 1983, Smith 1970, Voneida 1990). Further, normal animals that undergo leg flexion training prior to spinal transection show no retention or savings of conditioned responses following transection (J Steinmetz, personal communication). The isolated spinal cord is thus capable of mediating a kind of associative neuronal plasticity but does not subserve classical conditioning of the limb flexion response in the intact animal, whereas the cerebellar circuitry does. Spinal conditioning is a useful model to study basic associative plasticity in a simplified neuronal network, but it does not tell us where or how such memories are formed in the intact animal.

The same may well be true for the decerebrate-decerebellate preparation. Once the necessary control procedures have been run, it may prove to be a simplified brainstem model of associative plasticity but with different properties than associative learning in the intact, behaving animal. Indeed, there is some evidence suggesting that it may be possible to establish eyeblink CRs or at least eyeblink responses to tone in the brainstem preparation. Yeo & Hardiman (1992) reported that with extended training following decerebellation in their decerebrate preparations, some small eyeblink responses developed to the CS. However, as they note, they did not run controls for nonassociative processes. An earlier study by Norman et al (1977) with long-term chronic decerebrate cats reported that eyeblink responses to tone could be developed within sessions in some animals, although performance was quite variable. In some of their animals, the transections appeared to be below the level of the red nucleus. No separate control groups were run to evaluate nonassociative processes. In any event this says nothing about where and how the memory traces for classical conditioning of the eyeblink response are formed in the intact animal (see discussion in Thompson et al 1983).

The only way to demonstrate that a significant component of the memory trace for eyelid conditioning in the normal mammal is established in the brainstem would be to demonstrate that animals trained in the normal state show significant retention following decerebration and decerebellation, compared to appropriate control groups for nonassociative processes. Indeed, as noted above, in our study of retention of the conditioned eyeblink response following decerebration, there was retention of the CR if the decerebration was sufficiently high that it did not damage the red nucleus, but there was no retention if the red nucleus (a part of the essential CR pathway from the cerebellum) was damaged (Mauk & Thompson 1987).

Reversible inactivation

The diagram of Figure 1 shows in highly simplified schematic form the essential memory trace circuit for classical conditioning of discrete responses, based on the lesion, recording, and stimulation evidence described above. Interneuron circuits are not shown, only net excitatory or inhibitory actions of projection pathways. Other pathways, known and unknown, may also of course be involved. Many uncertainties still exist, e.g. concerning details of sensory-specific patterns of projection to pontine nuclei and cerebellum (CS pathways), details of red nucleus projections to premotor and motor nuclei (CR pathway), the relative roles of the cerebellar cortex and interpositus nucleus, and the possible roles of recurrent circuits.

Several parts of the circuit have been reversibly inactivated for the duration of training (eyeblink conditioning) in naive animals, indicated by shadings

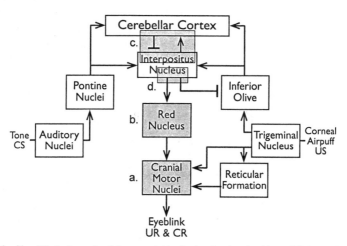

Figure 1   Simplified schematic of the essential brain circuitry involved in eyeblink conditioning. Shadowed boxes represent areas that have been reversibly inactivated during training. (*a*) Inactivation of motor nuclei including facial (seventh) and accessory (sixth). (*b*) Inactivation of magnocellular red nucleus. (*c*) Inactivation of dorsal aspects of the interpositus nucleus and overlying cerebellar cortex. (*d*) Inactivation of white matter ventral to the interpositus.

labeled a, b, c, d in Figure 1. The motor nuclei essential for generating the UR & CR (primarily seventh and accessory sixth) were inactivated by infusion of muscimol (six days) or cooling (five days) during standard tone-airpuff training (*a* in Figure 1) (Thompson et al 1993, Zhang & Lavond 1991). The animals showed no CRs and no URs during this inactivation training; indeed they showed no behavior at all; performance was completely abolished. However, the animals exhibited asymptotic CR performance and normal UR performance from the very beginning of postinactivation training. Thus, performance of the CR and UR are completely unnecessary for normal learning and the motor nuclei make no contribution to formation of the memory trace; they are efferent from the trace.

Inactivation of the magnocellular red nucleus is indicated by *b* in Figure 1. Inactivation by low doses of muscimol for six days of training had no effect on the UR but completely prevented expression of the CR (Krupa et al 1993). Yet animals showed asymptotic learned performance of the CR from the beginning of postinactivation training. Training during cooling of the magnocellular red nucleus gave identical results—animals learned during cooling, as evidenced in postinactivation training, but did not express CRs at all during inactivation training (Clark & Lavond 1993). However, cooling did impair performance of the UR (but the animals learned normally), yet another line of evidence against the performance argument. Consequently, the red nucleus must be efferent from the memory trace.

Inactivation of the dorsal anterior interpositus and overlying cortex (*c* in Figure 1) by low doses of muscimol (six days), by lidocaine (three and six days), and by cooling (five days) resulted in no expression of CRs during inactivation training and no evidence of any learning at all having occurred during inactivation training (Clark et al 1992, Krupa et al 1993, Nordholm et al 1993). In subsequent postinactivation training, animals learned normally as though completely naive; they showed no savings at all relative to noninactivated control animals. None of the methods of inactivation had any effect at all on performance of the UR in US alone trials. In one study (Nordholm et al 1993), cerebellar lidocaine infusions effective in abolishing the CR and preventing learning were subsequently tested on US alone trials over a wide range of US intensities and had no effect on performance of the UR; indeed URs were numerically larger with lidocaine inactivation of CRs than with saline control infusions that had no effect on CRs. The distribution of [^{3}H]-muscimol completely effective in preventing learning included the anterior dorsal interpositus and overlying cortex of lobule HVI, a volume approximately 2% of the total volume of the cerebellum (Krupa et al 1993). The region of the cerebellum essential for learning this task is extremely localized.

Welsh & Harvey (1991) gave rabbits extensive training to a light CS; then gave them transfer training to a tone CS with interspersed light CSs (one session) with lidocaine infusion in the anterior interpositus; and then trained them to the tone without infusion. Control animals were treated identically except given saline infusions during the transfer training session. Welsh & Harvey reported that CRs to the light CS were prevented during lidocaine infusion but that the animals exhibited virtually asymptotic performance to the tone CS in post-infusion training. As noted above, if naive animals are given tone CS training during lidocaine infusion (three or six days) they do not learn at all during infusion and subsequently learn with no savings in postinfusion training. The simplest explanation of the Welsh & Harvey (1991) results is that substantial transfer of training occurred; indeed, their control animals showed very substantial transfer compared to naive animals (see also Schreurs & Kehoe 1987). Cannula location may also be a factor. When lidocaine was infused in the white matter ventral to the interpositus to inactivate the efferent projections from the interpositus (see below and d in Figure 1), normal learning occurred although no CRs were expressed during infusion training (Nordholm et al 1993), a result analogous to the result reported by Welsh & Harvey (1991). In any event, all studies in which naive animals were trained during inactivation of the anterior interpositus and overlying cortex (cooling, muscimol, lidocaine) agree in showing that no learning at all occurs during inactivation training.

The white matter ventral to the interpositus nucleus includes the efferent projections conveying information from this portion of the cerebellum to other brain structures. Animals were trained (three days) during lidocaine inactivation of this fiber region (d in Figure 1) (Nordholm et al 1993). They showed no evidence of CRs during inactivation training but exhibited asymptotic CR performance from the beginning of postinactivation training. These ventral infusions had no effect on performance of the UR. This result argues that interpositus projections to other brain regions play no essential role in this learning and memory. The only alternative is that efferent fibers from the interpositus project to structure X, which is critical for formation of the memory trace, and X in turn projects back to the cerebellum. The only way this could occur is if the lidocaine infusion ventral to the interpositus does not inactivate these efferent fibers, since the animals learn. However, this hypothetical pathway can play no role in expression of the CR since the ventral infusion, although not preventing learning, completely prevents expression of the CR. Although this seems most unlikely, it is possible. This is not to say, however, that such recurrent feedback circuits play no role in learning (see p. 533). A recent connectionist-level computational model of this circuit (Gluck et al 1993) suggests that feedback to the cerebellum from the output of the interpositus may also play a key role in adaptive timing of the CR.

Over a wide CS-US onset interval range, the peak of the CR occurs at about the time of onset of the US.

Collectively, these data strongly support the hypothesis that the memory trace is formed and stored in a localized region of the cerebellum (anterior interpositus and overlying cortex). Inactivation of this region (Figure 1c) during training completely prevents learning but inactivation of the output pathway from the region (Figure 1d) and its necessary (for the CR) efferent target, the red nucleus (Figure 1b) do not prevent learning at all. In no case do the drug inactivations have any effect at all on performance of the reflex response on US alone trials. If even a part of the essential memory trace were formed prior to the cerebellum in the essential circuit, then following cerebellar inactivation training, the animals would have to show savings, and they show none at all. Similarly, if a part of the essential memory trace were formed in the red nucleus or other efferent targets of the interpositus, then following red nucleus (Figure 1b) or interpositus efferent (Figure 1d) inactivation training, animals could not show asymptotic CR performance but they do.

These results would seem conclusively to rule out the possibility, as hypothesized by Bloedel (1992) and Welsh & Harvey (1989, 1991), that the essential memory trace is formed and stored in the brainstem reflex pathways. The brainstem hypothesis has not been elaborated much beyond the assertion that the memory trace occurs in the brainstem and there is at present no evidence to support it. There are perhaps two alternative possibilities: (a) the trace is established in the brainstem largely independent of the cerebellum, but excitatory drive from the cerebellum is necessary for its behavioral expression, or (b) the trace is established in the brainstem at least in part as a result of the actions of the cerebellum on the brainstem circuitry. Insofar as expression of the CR is concerned, lesions of the interpositus nucleus, its efferent projection to the red nucleus via the superior cerebellar peduncle, the target region of the magnocellular red nucleus, or the descending rubral pathway projecting to the brainstem and motor nuclei completely prevent and abolish expression of the CR with no effect on the UR (see above). Consequently, the cerebellum must exert its actions on the brainstem and motor nuclei, regarding the eyeblink CR, via its efferent projection to the red nucleus and rubral projections to the brainstem. So if the cerebellum–red nucleus circuit simply facilitates expression of the brainstem CR, then inactivation of the critical region of the cerebellum during training could not possibly prevent learning, yet it does. On the other hand, if the cerebellum–red nucleus circuit actually establishes the brainstem CR, reversible inactivation of either structure, each of which completely prevents expression of the CR, must have the same effect on learning of the CR in naive animals. However, as noted, inactivation of the critical cerebellar region completely prevents learning, but inactivation of the red nucleus region does not prevent learning at all, even though in both cases the necessary cerebellar actions on the brainstem are completely prevented.

PUTATIVE MECHANISMS

The evidence for localization of memory traces for classical conditioning of discrete responses to the cerebellum is now sufficiently compelling that a focus on cerebellar cellular mechanisms of plasticity is warranted. Indeed, since much of the essential circuitry has been identified, this system may provide the first instance in the mammalian brain where the entire circuitry from memory trace to learned behavior, the read out of memory, and the neuronal content of the memory store, can be analyzed.

Although the question of whether or not the cerebellar cortex is essential has not yet been resolved, it is clear that it is critically important for normal learning (see above). The fact that the majority of Purkinje neurons exhibiting learning-related changes show decreases in simple spike responses in the CS period, which would of course result in disinhibition of interpositus neurons, is consistent with a mechanism of long-term depression (LTD). A possible counter-example was reported by Schreurs et al (1991). They took slices of cerebellar cortex from trained and untrained rabbits and reported that the majority of Purkinje neurons from trained animals showed decreases in afterhyperpolarizations to intracellular depolarization relative to controls, implying an increase in excitability. (This of course does not indicate possible changes in synaptic efficacy.) However, there is a sampling problem; many Purkinje neurons in HVI, which is the tissue they used, do not show any learning-related changes and a significant number show increases in simple spike discharges in the CS period (see above). It is possible that the two types of learning-induced changes are from different microzones in cerebellar cortex (see Ito 1984). Indeed, microstimulation through the recording microelectrode evokes small eyelid closure movements when the recording is from Purkinje neurons that show decreases during the CS period, but no such responses are evoked by stimulation when recording is from Purkinje neurons that do not exhibit this pattern of simple spike response (DJ Krupa & RF Thompson, unpublished observations).

Mechanisms of LTD have been reviewed elsewhere in this volume and earlier (Ito 1989). In brief, conjoint activation of mossy-parallel fiber and climbing fiber synapses on Purkinje neurons produces a prolonged depression of the parallel fiber-Purkinje dendrite synapses (AMPA receptors). Current evidence suggests that glutamate activation of AMPA and metabotropic receptors together with increased intracellular calcium, which is produced by climbing fiber activation, yields the persisting decrease in AMPA receptor function (see also Ito & Karachot 1990, Linden & Connor 1991). LTD has been proposed as the mechanism underlying synaptic plasticity in the flocculus that subserves adaptation of the vestibulo-ocular reflex (Ito 1989).

In behavioral studies, classical conditioning of discrete responses, for example, eyeblink and head turn, evoked by stimulation of the DAO-climbing fibers occurs when paired with stimulation of the mossy fibers as a CS (Steinmetz et al 1989), the exact procedure used in the initial studies of LTD (Ito et al 1982). However, the temporal properties of LTD and classical conditioning appear to differ: most studies of LTD have used near-simultaneous activation of parallel and climbing fibers, but in classical conditioning, mossy-parallel–fiber stimulation must precede climbing fiber stimulation by about 100 ms and best learning occurs with an interval of about 250 ms; simultaneous activation results in extinction of the CR so induced (Steinmetz et al 1989). Virtually all current studies of LTD have used the cerebellar slice or tissue culture and have used GABA blockers. In recent work, parallel fiber-Purkinje cell field potentials in cerebellar slice were used to assess the temporal properties of LTD (Chen & Thompson 1992, Chen 1993). Simultaneous stimulation of parallel fibers and climbing fibers (100 pairings) yielded LTD in the presence of bicuculline, as in other studies, but did not yield LTD in the absence of bicuculline; however if parallel fiber stimulation preceded climbing fiber stimulation by 250 ms, robust LTD developed in the absence of bicuculline, suggesting that GABA, i.e. inhibitory interneurons, may play a key role in determining the temporal properties of the synaptic plasticity underlying LTD. This possibility is not inconsistent with current views of the mechanisms of LTD, e.g. G protein activation of intracellular cascades (Ito 1994, Ito & Karachot 1990, Linden & Connor 1991). It is perhaps relevant that very large cerebellar cortical lesions severely disrupt adaptive timing of the conditioned eyeblink response (see above; Logan 1991, Perret et al 1993). Nitric oxide (NO) has also been implicated in the establishment of LTD in cerebellar cortex (Ito & Karachot 1990) but there are complications; for example, there is apparently no NO synthase in Purkinje neurons (see Ito 1994). Interestingly, Chapman et al (1992) reported that systemic injection of an NO synthase inhibitor impaired acquisition of the conditioned eyeblink response.

The possibility of learning-induced synaptic plasticity in the interpositus nucleus has not been much explored. One study (in vivo anesthetized rat) reported that tetanus of white matter yielded long-term potentiation of the stimulus-evoked field potential in the interpositus (Racine et al 1986). Given the critical importance of the anterior interpositus nucleus for classical conditioning of discrete responses, more work must be done on possible processes of synaptic plasticity in this structure.

Insofar as classical conditioning is concerned, and granting that the memory trace is formed and stored in the cerebellum, evidence at present precludes exclusion of either cerebellar cortex or interpositus as the site of memory storage. The most conservative hypothesis is that there are multiple sites of storage in cortex and interpositus (Thompson 1986). [Interestingly, recent modeling of the VOR circuitry (Lisberger & Sejnowski 1992a,b) suggests that adaptation of the VOR involves synaptic plasticity both in cerebellar cortex and vestibular nuclei, the latter analogous to the interpositus nucleus.] Because the neurons in the interpositus nucleus send substantial direct projections to cerebellar cortex (Batini et al 1989, Buisseret-Delmas & Angaut 1989, Chan-Palay 1977), inactivation limited to the critical region of the interpositus during training that completely prevents learning would still not rule out involvement of the cerebellar cortex since such inactivation would also of course inactivate these nucleo-cortical projections. The *experimentum crucium* would seem to be complete inactivation of the ipsilateral cerebellar cortex but not the interpositus during training; this experiment has not yet been achieved.

ACKNOWLEDGMENT

This paper was supported by the National Institute on Aging grant AF05142, the Office of Naval Research contract N00014-91-J-0112, the National Science Foundation (IBN 9215069), and Sankyo.

Literature Cited

Abrams TW, Kari KA, Kandel ER. 1991. Biochemical studies of stimulus convergence during classical conditioning in Aplysia: dual regulation of adenylate cyclase by Ca^{2+} calmodulin and transmitter. *J. Neurosci.* 11:2655–65

Aitkin LM, Boyd J. 1978. Acoustic input to the lateral pontine nuclei. *Hearing Res.* 1: 67–77

Akshoomoff NA, Courchesne E. 1992. A new role for the cerebellum in cognitive operations. *Behav. Neurosci.* 106:731–38

Albus JS. 1971. A theory of cerebellar function. *Math Biosci.* 10:25–61

Asanuma C, Thach WT, Jones EG. 1983. Anatomical evidence for segregated focal groupings of efferent cells and their terminal ramifications in the cerebellothalamic pathway of the monkey. *Brain Res. Rev.* 5:267–99

Batini C, Buisseret-Delmas C, Compoint C, Daniel H. 1989. The GABAergic neurones of the cerebellar nuclei in the rat: projections to the cerebellar cortex. *Neurosci. Lett.* 99: 251–56

Becker JT, Walker JA, Olton DS. 1980. Neuroanatomical bases of spatial memory. *Brain Res.* 200:307–20

Beggs AL, Steinmetz JE, Romano AG, Patterson MM. 1983. Extinction and retention of a classically conditioned flexor nerve response in acute spinal cat. *Behav. Neurosci.* 97:530–40

Berger TW, Bassett JL. 1992. System properties of the hippocampus. In *Learning and Memory: the Biological Substrates*, ed. I Gormezano, EA Wasserman. pp. 275–320. Hillsdale, NJ: Lawrence Erlbaum

Berger TW, Orr WB. 1983. Hippocampectomy selectively disrupts discrimination reversal conditioning of the rabbit nictitating membrane response. *Behav. Brain Res.* 210:411–17

Bergold PJ, Sweatt JD, Winicov I, Weiss KR, Kandel ER, Schwartz JH. 1990. Protein synthesis during acquisition of long-term facilitation is needed for the persistent loss of regulatory subunits of the Aplysia cAMP-dependent protein kinase. *Proc. Natl. Acad. Sci. USA* 87:3788–91

Berthier NE, Moore JW. 1983. The nictitating membrane response: an electrophysiological study of the abducens nerve and nucleus and the accessory abducens nucleus in rabbit. *Brain Res.* 258:201–10

Berthier NE, Moore JW. 1986. Cerebellar Purkinje cell activity related to the classically conditioned nictitating membrane response. *Exp. Brain Res.* 63:341–50

Berthier NW, Moore JW. 1990. Activity of deep cerebellar nuclear cells during classical conditioning of nictitating membrane extension in rabbits. *Exp. Brain Res.* 83: 44–54

Bloedel JR. 1992. Functional heterogeneity with structural homogeneity: how does the cerebellum operate? *Behav. Brain Sci.* 15: 666–78

Brand S, Dahl A-L, Mugnaini E. 1976. The length of parallel fibers in the cat cerebellar cortex. An experimental light and electron microscopic study. *Exp. Brain Res.* 26:39–58

Brindley CS. 1964. The use made by the cerebellum of the information that it receives from sense organs. *IBRO Bull.* 3:80

Brodal A. 1981. *Neurological Anatomy*, New York: Oxford Univ. Press

Buisseret-Delmas C, Angaut P. 1989. Anatomical mapping of the cerebellar nucleocortical projections in the rat: a retrograde labeling study. *J. Comp. Neurol.* 288:297–310

Canavan AGM, Homberg V. 1991. Visuomotor associative learning in cerebellar disease. *IBRO World Congress Neurosci., 3rd, Abstr.*

Carew TJ, Hawkins RD, Kandel ER. 1983. Differential classical conditioning of a defensive withdrawal reflex in *Aplysia californica. Science* 219:397–400

Cegavske CF, Patterson MM, Thompson RF. 1979. Neuronal unit activity in the abducens nucleus during classical conditioning of the nictitating membrane response in the rabbit (*Oryctolagus cuniculus*). *J. Comp. Physiol. Psychol.* 93:595–609

Chan-Palay V. 1977. *Cerebellar Dentate Nucleus* Berlin: Springer-Verlag

Chapman PF, Atkins CM, Allen MT, Haley JE, Steinmetz JE. 1992. Inhibition of nitric oxide synthesis impairs two different forms of learning. *Neuroreport* 3:567–70

Chapman PF, Steinmetz JE, Thompson RF. 1988. Classical conditioning does not occur when direct stimulation of the red nucleus or cerebellar nuclei is the unconditioned stimulus. *Brain Res.* 442:97–104

Chapman PF, Steinmetz JE, Sears LL, Thompson RF. 1990. Effects of lidocaine injection in the interpositus nucleus and red nucleus on conditioned behavioral and neuronal responses. *Brain Res.* 537:149–56

Chen C. 1993. *Temporal specificity of cerebellar long-term depression.* PhD dissertation. Univ. Southern Calif., Los Angeles, CA

Chen C, Thompson RF. 1992. Associative long-term depression revealed by field potential recording in rat cerebellar slice. *Soc. Neurosci. Abstr.* 18:1215

Clark GA, McCormick DA, Lavond DG, Thompson RF. 1984. Effects of lesions of cerebellar nuclei on conditioned behavioral and hippocampal neuronal responses. *Brain Res.* 291:125–36

Clark GA, Schuman E.M. 1991. Snails' tales: initial comparisons of synaptic plasticity underlying learning in *Hermissenda and Aplysia.* In *Neuropsychology of Memory,* ed. LR Squire, N Butters. New York: Guildford

Clark RE, Lavond DG. 1993. Reversible lesions of the red nucleus during acquisition and retention of a classically conditioned behavior in rabbit. *Behav. Neurosci.* 107: 264–70

Clark RE, Zhang AA, Lavond DG. 1992. Reversible lesions of the cerebellar interpositus nucleus during acquisition and retention of a classically conditioned behavior. *Behav. Neurosci.* 106:879–88

Cohen DH. 1984. Identification of vertebrate neurons modified during learning analysis of sensory pathways. In *Primary Neural Substrates of Learning and Behavioral Change,* ed. DL Alkon, J Farley. London: Cambridge Univ. Press

Coleman SR, Gormezano I. 1971. Classical conditioning of the rabbit's (*Oryctolagus cuniculus*) nictitating membrane response under symmetrical CS-US interval shifts. *J. Comp. Physiol. Psychol.* 77:447–55

Crow T, Forrester J. 1990. Inhibition of protein synthesis blocks long-term enhancement of generator potentials produced by one-trial in vivo conditioning in *Hermissenda. Proc. Natl. Acad. Sci. USA* 87:4490–94

Daum I, Channon S, Polkey CE, Gray JA. 1991. Classical conditioning after temporal lobe lesions in man: impairment in conditional discrimination. *Behav. Neurosci.* 105: 396–408

Daum I, Schugens MM, Ackermann H, Lutzenberger W, Dichgans J, Birbaumer N. 1993. Classical conditioning after cerebellar lesions in human. *Behav. Neurosci.* In press

Deaux EG, Gormezano I. 1963. Eyeball retraction: classical conditioning and extinction in the albino rabbit. *Science* 141:630–31

Delgado-Garcia JM, Evinger C, Escudero M, Baker R. 1990. Behavior of accessory abducens and abducens motoneurons during eye retraction and rotation in the alert cat. *J. Neurophysiol.* 64:413–22

Disterhoft JF, Quinn KJ, Weiss C, Shipley MT. 1985. Accessory abducens nucleus and conditioned eye retraction/nictitating membrane extensions in rabbit. *J. Neurosci.* 5: 941–50

Donegan NH, Foy MR, Thompson RF. 1985. Neuronal responses of the rabbit cerebellar cortex during performance of the classically conditioned eyelid response. *Soc. Neurosci. Abstr.* 11:835

Donegan NH, Gluck MA, Thompson RF. 1989. Integrating behavioral and biological models of classical conditioning. In *Psychology of Learning and Motivation,* ed. RD Hawkins, GH Bower, pp.109–56. New York: Academic

Donegan NH, Lowry RW, Thompson RF. 1983. Effects of lesioning cerebellar nuclei on conditioned leg-flexion responses. *Soc. Neurosci. Abstr.* 9:331

Dudai Y. 1989. *The Neurobiology of Memory: Concepts, Findings, Trends.* New York: Oxford Univ. Press

Durkovic R. 1975. Classical conditioning, sensitization and habituation in the spinal cat. *Physiol. Behav.* 14:297–304

Eccles JC. 1977. An instruction-selection theory of learning in the cerebellar cortex. *Brain Res.* 127:327–52

Eccles JC, Ito M, Szentagothai J. 1967. *The Cerebellum as a Neuronal Machine.* Berlin: Springer

Eichenbaum H, Fagan A, Cohen NJ. 1986. Normal olfactory discrimination learning set and facilitation of reversal learning after medial-temporal damage in rats: implications for an account of preserved learning abilities in amnesia. *J. Neurosci.* 6:1876–84

Ekerot CF, Gustavsson P, Oscarsson O, Schouenborg J. 1987. Climbing fibres projecting to cat cerebellar anterior lobe activated by cutaneous A and C fibres. *J. Physiol.* 386:529–38

Fanselow MS, Kim JJ, Landeira-Fernandez J. 1991. Anatomically selective blockade of Pavlovian fear conditioning by application of an NMDA antagonist to the amygdala and periaqueductal gray. *Soc. Neurosci. Abstr.* 17:659

Fiez JA, Petersen SE, Cheney MK, Raichle ME. 1992. Impaired non-motor learning and error detection associated with cerebellar damage. *Brain* 115:155–78

Fitzgerald LA, Thompson RF. 1967. Classical conditioning of the hindlimb flexion reflex in the acute spinal cat. *Psychon. Sci.* 9:511–12

Foy MR, Krupa DJ, Tracy J, Thompson RF. 1992. Analysis of single unit recordings from cerebellar cortex of classically conditioned rabbits. *Soc. Neurosci. Abstr.* 18:1215

Foy MR, Steinmetz JE, Thompson RF. 1984. Single unit analysis of the cerebellum during classically conditioned eyelid responses. *Soc. Neurosci. Abstr.* 10:122

Foy MR, Thompson RF. 1986. Single unit analysis of Purkinje cell discharge in classically conditioned and untrained rabbits. *Soc. Neurosci. Abstr.* 12:518

Gilbert PFC, Thach WT. 1977. Purkinje cell activity during motor learning. *Brain Res.* 128:309–28

Gluck MA, Reifsnider, Thompson RF. 1990. Adaptive signal processing and the cerebellum: models of classical conditioning and VOR adaptation. In *Neuroscience and Connectionist Models,* ed. MA Gluck, DE Rumelhart, pp. 131–85. New Jersey: Lawrence Erlbaum

Gluck MA, Goren O, Myers C, Thompson RF. 1993. A higher order recurrent network model of the cerebellar substrates of response timing in motor-reflex conditioning. *J. Cog. Neurosci.* In press

Goodlett CR, Hamre KM, West JR. 1992. Dissociation of spatial navigation and visual guidance performance in Purkinje cell degeneration (pcd) mutant mice. *Behav. Brain Res.* 47:129–41

Gormezano I, Schneiderman N, Deaux EG, Fuentes J. 1962. Nictitating membrane classical conditioning and extinction in the albino rabbit. *Science* 138:33–34

Gormezano I, Kehoe EJ, Marshall-Goodell BS. 1983. Twenty years of classical conditioning research with the rabbit. In *Progress in Physiological Psychology,* ed. JM Sprague, AN Epstein, pp. 197–275. New York: Academic

Grafman J, Litvan I, Massauoioi S, Steward M, Sirigu A, Hallet M. 1992. Cognitive planning deficit in patients with cerebellar atrophy. *Neurology* 42:1493–96

Haley DA, Lavond DG, Thompson RF. 1983. Effects of contralateral red nuclear lesions on retention of the classically conditioned nictitating membrane/eyelid response. *Soc. Neurosci. Abstr.* 9:643

Haley DA, Thompson RF, Madden J, IV. 1988. Pharmacological analysis of the magnocellular red nucleus during classical conditioning of the rabbit nictitating membrane response. *Brain Res.* 454:131–39

Hitchcock JM, Davis M. 1986. Lesions of the amygdala, but not of the cerebellum or red nucleus, block conditioned fear as measured with the potentiated startle paradigm. *Behav. Neurosci.* 100:11–22

Ito M. 1972. Neural design of the cerebellar motor control system. *Brain Res.* 40:81–84

Ito M. 1984. *The Cerebellum and Neural Control.* New York: Raven Press

Ito M. 1989. Long-term depression. *Annu. Rev. Neurosci.* 12:85–102

Ito M. 1994. Cerebellar mechanisms of long-term depression. In *Synaptic Plasticity: Molecular and Functional Aspects,* ed. M Baudry, JL Davis, RF Thompson, Cambridge, MA: MIT Press. In press

Ito M, Karachot L. 1990. Messenger mediating long-term desensitization in cerebellar Purkinje cells. *Neuroreport* 1:129–32

Ito M, Sakurai M, Tongroach P. 1982. Climbing fibre induced depression of both mossy fibre responsiveness and glutamate sensitivity of cerebellar Purkinje cells. *J. Physiol. London,* 324:113–34

Ivkovich D, Lockard JM, Thompson RF. 1993. Interpositus lesion abolition of the eyeblink CR is not due to effects on performance. *Behav. Neurosci.* 107:530–32

Iwata J, LeDoux JE, Meeley MP, Arneric S, Reis DJ. 1986. Intrinsic neurons in the amygdaloid field projected to by the medial geniculate body mediate emotional responses conditioned to acoustic stimuli. *Brain Res.* 383:195–214

Kane SA, Mink JW, Thach WT. 1988. Fastigial, interposed, and dentate cerebellar nuclei: somatotopic organization and the movements differentially controlled by each. *Soc. Neurosci. Abstr.* 14:954

Kane SA, Goodkin HP, Keating JG, Thach WT. 1989. Incoordination in attempted reaching and pinching after inactivation of cerebellar dentate nucleus. *Soc. Neurosci. Abstr.* 15:52

Karamian AI, Fanardijian VV, Kosareva AA. 1969. The functional and morphological evolution of the cerebellum and its role in behavior. In *Neurobiology of Cerebellar Evolution and Development, First International Symposium,* ed. R Llinas, pp. 639–673.. Chicago: Am. Med. Assoc.

Keating JG, Thach WT. 1990. Cerebellar motor learning: quantitation of movement adaptation and performance in rhesus monkeys and humans implicates cortex as the site of adaptation. *Soc. Neurosci. Abstr.* 16:762

Keating JG, Thach WT. 1991. The cerebellar cortical area required for adaptation of monkey's "jump" task is lateral, localized, and small. *Soc. Neurosci. Abstr.* 17:1381

Kelly TM, Zuo C, Bloedel JR. 1990. Classical conditioning of the eyeblink reflex in the decerebrate-decerebellate rabbit. *Behav. Brain Res.* 38:7–18

Kim JJ, Fanselow MS. 1992. Modality-specific retrograde amnesia of fear. *Science* 256:675–77

Kim JJ, Krupa DJ, Thompson RF 1992. Intraolivary infusions of picrotoxin prevent "blocking" of rabbit conditioned eyeblink response. *Soc. Neurosci. Abstr.* 18:1562

Knowlton BJ, Thompson RF. 1988. Microinjections of local anesthetic into the pontine nuclei reduce the amplitude of the classically conditioned eyeblink response. *Physiol. Behav.* 43:855–57

Knowlton BJ, Thompson RF. 1992. Conditioning using a cerebral cortical CS is dependent on the cerebellum and brainstem circuitry. *Behav. Neurosci.* 106:509–17

Knowlton BJ, Thompson JK, Thompson RF. 1993. Projections from the auditory cortex to the pontine nuclei in the rabbit. *Behav. Brain Res.* 56:23–30

Krupa DJ, Thompson JK, Thompson RF. 1993. Localization of a memory trace in the mammalian brain. *Science* 260:989–91

Krupa DJ, Weiss C, Thompson RF. 1991. Air puff evoked Purkinje cell complex spike activity is diminished during conditioned responses in eyeblink conditioned rabbits. *Soc. Neurosci. Abstr.* 17:322

Lalonde R. 1987. Exploration and spatial learning in staggerer mutant mice. *J. Neurogenet.* 4:285–92

Lalonde R, Botez MI. 1990. The cerebellum and learning processes in animals. *Brain Res. Rev.* 15:325–32

Lashley KS. 1929. *Brain Mechanisms and Intelligence.* Chicago: Univ. Chicago Press

Lavond DG, Steinmetz JE. 1989. Acquisition of classical conditioning without cerebellar cortex. *Behav. Brain Res.* 33:113–64

Lavond DG, Hembree TL, Thompson RF. 1985. Effects of kainic acid lesions of the cerebellar interpositus nucleus on eyelid conditioning in the rabbit. *Brain Res.* 326:179–82

Lavond DG, Kim JJ, Thompson RF. 1993. Mammalian brain substrates of aversive classical conditioning. *Annu. Rev. Psychol.* 44:317–42

Lavond DG, Lincoln JS, McCormick DA, Thompson RF. 1984a. Effect of bilateral lesions of the dentate and interpositus cerebellar nuclei on conditioning of heart-rate and nictitating membrane/eyelid responses in the rabbit. *Brain Res.* 305:323–30

Lavond DG, Logan CG, Sohn JH, Garner WDA, Kanzawa SA. 1990. Lesions of the cerebellar interpositus nucleus abolish both nictitating membrane and eyelid EMG conditioned response. *Brain Res.* 514:238–48

Lavond DG, McCormick DA, Clark GA, Holmes DT, Thompson RF. 1981. Effects of ipsilateral rostral pontine reticular lesions on retention of classically conditioned nictitating membrane and eyelid response. *Physiol. Psychol.* 9:335–39

Lavond DG, McCormick DA, Thompson RF. 1984b. A nonrecoverable learning deficit. *Physiol. Psychol.* 12:103–10

Lavond DG, Steinmetz JE, Yokaitis MH, Thompson RF. 1987. Reacquisition of classical conditioning after removal of cerebellar cortex. *Exp. Brain Res.* 67:569–93

Leiner HC, Leiner AL, Dow RS. 1989. Reappraising the cerebellum: what does the hindbrain contribute to the forebrain? *Behav. Neurosci.* 103:998–1008

Lewis JL, LoTurco JJ, Solomon PR. 1987.

Lesions of the middle cerebellar peduncle disrupt acquisition and retention of the rabbit's classically conditioned nictitating membrane response. *Behav. Neurosci.* 101: 151–57

Liang KC, McGaugh JL, Martinez JL, Jensen RA Jr, Vasquez BJ, Messing RB. 1982. Post-training amygdaloid lesions impair retention of an inhibitory avoidance response. *Behav. Brain Res.* 4:237–49

Lincoln JS, McCormick DA, Thompson RF. 1982. Ipsilateral cerebellar lesions prevent learning of the classically conditioned nictitating membrane/eyelid response of the rabbit. *Brain Res.* 242:190–93

Linden DJ, Connor JA. 1991. Participation of postsynaptic PKC in cerebellar long-term depression in culture. *Science* 254:1656–59

Lisberger SG, Sejnowski TJ. 1992a. Motor learning in a recurrent network model based on the vestibulo-ocular reflex. *Nature* 360: 159–61

Lisberger SG, Sejnowksi TJ. 1992b. Computational analysis suggests a new hypothesis for motor learning in the vestibulo-ocular reflex. *Tech. Rep. INC-9201*

Logan CG. 1991. *Cerebellar cortical involvement in excitatory and inhibitory classical conditioning.* PhD dissertation, Stanford Univ., Stanford, CA

Lye RH, O'Boyle DJ, Ramsden RT, Schady W. 1988. Effects of a unilateral cerebellar lesion on the acquisition of eye-blink conditioning in man. *J. Physiol.* 403:58

Lynch G. 1986. *Synapses, Circuits, and the Beginnings of Memory.* Cambridge, MA: MIT Press

Mamounas LA, Thompson RF, Madden, J IV. 1987. Cerebellar GABAergic processes: evidence for critical involvement in a form of simple associative learning in the rabbit. *Proc. Natl. Acad. Sci. USA* 84:2101–5

Marr D. 1969. A theory of cerebellar cortex. *J. Physiol.* 202:437–70

Mauk MD, Thompson RF. 1987. Retention of classically conditioned eyelid responses following acute decerebration. *Brain Res.* 403: 89–95

Mauk MD, Steinmetz JE, Thompson RF. 1986. Classical conditioning using stimulation of the inferior olive as the unconditioned stimulus. *Proc. Natl. Acad. Sci. USA* 83:5349–53

McCormick DA, Thompson RF. 1984a. Cerebellum: essential involvement in the classically conditioned eyelid response. *Science* 223:296–99

McCormick DA, Thompson RF. 1984b. Neuronal responses of the rabbit cerebellum during acquisition and performance of a classically conditioned nictitating membrane-eyelid response. *J. Neurosci.* 4:2811–22

McCormick DA, Clark GA, Lavond DG, Thompson RF. 1982a. Initial localization of the memory trace for a basic form of learning. *Proc. Natl. Acad. Sci. USA* 79:2731–42

McCormick DA, Guyer PE, Thompson RF. 1982b. Superior cerebellar peduncle lesions selectively abolish the ipsilateral classically conditioned nictitating membrane/eyelid response of the rabbit. *Brain Res.* 244:347–50

McCormick DA, Lavond DG, Clark GA, Kettner RE, Rising CE, Thompson RF. 1981. The engram found? Role of the cerebellum in classical conditioning of nictitating membrane and eyelid responses. *Bull. Psychon. Soc.* 18:103–5

McCormick DA, Lavond DG, Thompson RF. 1982c. Concomitant classical conditioning of the rabbit nictitating membrane and eyelid responses: correlations and implications. *Physiol. Behav.* 28:769–75

McCormick DA, Lavond DG, Thompson RF. 1983. Neuronal responses of the rabbit brainstem during performance of the classically conditioned nictitating membrane (NM)/eyelid response. *Brain Res.* 271:73–88

McCormick DA, Steinmetz JE, Thompson RF. 1985. Lesions of the inferior olivary complex cause extinction of the classically conditioned eyeblink response. *Brain Res.* 359: 120–30

McGaugh J. 1989. Involvement of hormone and neuromodulatory systems in the regulation of memory storage. *Annu. Rev. Neurosci.* 12:255–87

Meunier M, Bachevalier J, Mishkin M, Murray EA. 1993. Effects on visual recognition of combined and separate lesions of the entorhinal perirhinal cortex in rhesus monkeys. *J. Neurosci.* In press

Millenson JR, Kehoe EJ, Gormezano I. 1977. Classical conditioning of the rabbit's nictitating membrane response under fixed and mixed CS-US intervals. *Learn. Motiv.* 8: 351–66

Mintz M, Yun Y, Lavond DG, Thompson RF. 1988. Unilateral inferior olive NMDA lesion leads to unilateral deficit in acquisition of NMR classical conditioning. *Soc. Neurosci. Abstr.* 14:783

Mower G, Gibson A, Glickstein M. 1979. Tectopontine pathway in the cat: laminar distribution of cells of origin and visual properties of target cells in dorsolateral pontine nucleus. *J. Neurophysiol.* 42:1–15

Moyer JR Jr, Deyo RA, Disterhoft JF. 1990. Hippocampectomy disrupts trace eyeblink conditioning in rabbits. *Behav. Neurosci.* 104:243–52

Mugnaini E. 1983. The length of cerebellar parallel fibers in chicken and Rhesus monkey. *J. Comp. Neurol.* 220:7–15

Nelson B, Alkon DL. 1990. Specific high molecular weight mRNAs induced by associative learning in *Hermissenda*. *Proc. Natl. Acad. Sci. USA* 87:269–73

Nelson B, Mugnaini E. 1989. GABAergic innervation of the inferior olivary complex and experimental evidence for its origin. In *The Olivocerebellar System in Motor Control*, ed. P Strata, pp. 86–107. New York: Springer-Verlag

Nordholm AF, Lavond DG, Thompson RF. 1991. Are eyeblink responses to tone in the decerebrate, decerebellate rabbit conditioned responses? *Behav. Brain Res.* 44:27–34

Nordholm AF, Thompson JK, Dersarkissian C, Thompson RF. 1993. Lidocaine infusion in a critical region of cerebellum completely prevents learning of the conditioned eyeblink response. *Behav. Neurosci.* In press

Norman RJ, Buchwald JS, Villablanca JR. 1977. Classical conditioning with auditory discrimination of the eyeblink in decerebrate cats. *Science* 196:551–53

O'Keefe J, Nadel L. 1978. *The Hippocampus as a Cognitive Map*. London: Oxford Univ. Press

Packard MG, Hirsh R, White NM. 1989. Differential effects of fornix and caudate nucleus lesions on two radial maze tasks: evidence for multiple memory systems. *J. Neurosci.* 9:1465–72

Patterson MM, Cegavske CE, Thompson RF. 1973. Effects of a classical conditioning paradigm on hindlimb flexor nerve response in immobilized spinal cat. *J. Comp. Physiol. Psychol.* 84:88–97

Perrett SP, Ruiz BP, Mauk MD. 1993. Cerebellar cortex lesions disrupt the timing of conditioned eyelid responses. *J. Neurosci.* 13:1708–18

Petersen SE, Fox PT, Posner MI, Minto M, Raichle ME. 1989. Positive emission tomographic studies of the processing of single words. *J. Cog. Neurosci.* 1:153–70

Polenchar BE, Patterson MM, Lavond DG, Thompson RF. 1985. Cerebellar lesions abolish an avoidance response in rabbit. *Behav. Neural Biol.* 44:221–27

Racine RJ, Wilson DA, Gingell R, Sutherland D. 1986. Long-term potentiation in the interpositus and vestibular nuclei in the rat. *Exp. Brain Res.* 63:158–62

Rescorla RA. 1988a. Behavioral studies of Pavlovian conditioning. *Annu. Rev. Neurosci.* 11:329–52

Rescorla RA. 1988b. Pavlovian conditioning: it's not what you think it is. *Am. Psychol.* 43:151–60

Rescorla R, Wagner A. 1972. A theory of Pavlovian conditioning: variations in the effectiveness of reinforcement. In *Classical Conditioning II: Current Research Theory*, ed. AH Black, WF Prokasy, pp. 64–99. New York: Appleton-Century-Crofts

Rosenfield ME, Moore JW. 1983. Red nucleus lesions disrupt the classically conditioned nictitating membrane response in rabbits. *Behav. Brain Res.* 10:393–98

Rosenfield ME, Dovydaitis A, Moore JW. 1985. Brachium conjunctivum and rubrobulbar tract: brainstem projections of red nucleus essential for the conditioned nictitating membrane response. *Physiol. Behav.* 34:751–59

Ross RT, Orr WB, Holland PC, Berger TW. 1984. Hippocampectomy disrupts acquisition and retention of learned conditional responding. *Behav. Neurosci.* 98:211–25

Scheiber MH, Thach WT. 1985. Trained slow tracking. II. Bidirectional discharge patterns of cerebellar nuclear, motor cortex, and spindle afferent neurons. *J. Neurophysiol.* 55:1228–70

Schmahmann JD. 1991. An emerging concept. The cerebellar contribution to higher function. *Arch. Neurol.* 48:1178–87

Schneiderman N. 1966. Interstimulus interval function of the nictitating membrane response of the rabbit under delay versus trace conditioning. *J. Comp. Physiol. Psychol.* 62:397–402

Schneiderman N, Fuentes I, Gormezano I. 1962. Acquisition and extinction of the classically conditioned eyelid response in the albino rabbit. *Science* 136:650–52

Schreurs BG, Alkon DL. 1992. Long-term depression and classical conditioning of the rabbit nictitating membrane response: an assessment using the rabbit cerebellar slice. *Soc. Neurosci. Abstr.* 18:337

Schreurs BG, Kehoe EJ. 1987. Cross-model transfer as a function of initial training level in classical conditioning with the rabbit. *Animal Learn. Behav.* 15:47–54

Schreurs BG, Sanchez-Andres JV, Alkon DL. 1991. Learning-specific differences in Purkinje cell dendrites of lobule HVI (lobulus simplex): intracellular recording in a rabbit cerebellar slice. *Brain Res.* 548:18–22

Sears LL, Steinmetz JE. 1990. Acquisition of classical conditioned-related activity in the hippocampus is affected by lesions of the cerebellar interpositus nucleus. *Behav. Neurosci.* 104:681–92

Sears LL, Steinmetz JE. 1991. Dorsal accessory inferior olive activity diminishes during acquisition of the rabbit classically conditioned eyelid response. *Brain Res.* 545:114–22

Smith AM. 1970. The effects of rubral lesions and stimulations on conditioned forelimb flexion responses in the cat. *Physiol. Behav.* 5:1121–26

Solomon PR, Lewis JL, LoTurco JJ, Steinmetz JE, Thompson RF. 1986a. The role of the middle cerebellar peduncle in acquisition and retention of the rabbit's classically conditioned nictitating membrane response. *Bull. Psychon. Soc.* 24:75–78

Solomon PR, Stowe GT, Pendlebury WW. 1989. Disrupted eyelid conditioning in a patient with damage to cerebellar afferents. *Behav. Neurosci.* 103:898–902

Solomon PR, Vander Schaaf ER, Thompson RF, Weisz DJ. 1986b. Hippocampus and trace conditioning of the rabbit's classically conditioned nictitating membrane response. *Behav. Neurosci.* 100:729–44

Squire LR. 1987. *Memory and Brain.* New York: Oxford Univ. Press

Squire LR. 1992. Memory and the hippocampus: a synthesis from findings with rats, monkeys, and humans. *Psychol. Rev.* 99: 195–231

Steinmetz JE. 1990a. Classical nictitating membrane conditioning in rabbits with varying interstimulus intervals and direct activation of cerebellar mossy fibers as the CS. *Behav. Brain Res.* 38:91–108

Steinmetz JE. 1990b. Neuronal activity in the rabbit interpositus nucleus during classical NM-conditioning with a pontine-nucleus-stimulation CS. *Psychol. Sci.* 1:378–82

Steinmetz JE, Steinmetz SS. 1991. Rabbit classically conditioned eyelid responses fail to reappear after interpositus lesions and extended post lesion training. *Soc. Neurosci. Abstr.* 17:323

Steinmetz JE, Lavond DG, Thompson RF. 1989. Classical conditioning in rabbits using pontine nucleus stimulation as a conditioned stimulus and inferior olive stimulation as an unconditioned stimulus. *Synapse* 3:225–32

Steinmetz JE, Lavond DG, Ivkovich D, Logan CG, Thompson RF. 1992. Disruption of classical eyelid conditioning after cerebellar lesions: damage to a memory trace system or a simple performance deficit? *J. Neurosci.* 12:4403–26

Steinmetz JE, Logan CG, Rosen DJ, Thompson JK, Lavond DG, Thompson RF. 1987. Initial localization of the acoustic conditioned stimulus projection system to the cerebellum essential for classical eyelid conditioning. *Proc. Natl. Acad. Sci. USA* 84: 3531–35

Steinmetz JE, Logue SF, Miller DP. 1993. Using signaled bar-pressing tasks to study the neural substrates of appetitive and aversive learning in rats: behavioral manipulations and cerebellar lesions. *Behav. Neurosci.* In press

Steinmetz JE, Rosen DJ, Chapman PF, Lavond DG, Thompson RF. 1986. Classical conditioning of the rabbit eyelid response with a mossy fiber stimulation CS. I. Pontine nuclei and middle cerebellar peduncle stimulation. *Behav. Neurosci.* 100:871–80

Steinmetz JE, Sears LL, Gabriel M, Kubota Y, Poremba A, Kang E. 1991. Cerebellar inter-

positus nucleus lesions disrupt classical nictitating membrane conditioning but not discriminative avoidance learning in rabbits. *Behav. Brain Res.* 45:71–80

Supple WF Jr, Leaton RN. 1990a. Lesions of the cerebellar vermis and cerebellar hemispheres: effects on heart rate conditioning in rats. *Behav. Neurosci.* 104:934–47

Supple WF Jr, Leaton RN. 1990b. Cerebellar vermis: essential for classical conditioned bradycardia in rats. *Brain Res.* 509:17–23

Swain RA. 1992. *Cerebellar cortical stimulation and classical conditioning of discrete motor movements.* PhD dissertation, Univ. Southern Calif., Los Angeles, CA

Swain RA, Shinkman PG, Nordholm AF, Thompson RF. 1992. Cerebellar stimulation as an unconditioned stimulus in classical conditioning. *Behav. Neurosci.* 106:739–50

Thach WT. 1968. Discharge of Purkinje and cerebellar nuclear neurons during rapidly alternating arm movements in the monkey. *J. Neurophysiol.* 33:527–36

Thach WT. 1970. Discharge of cerebellar neurons related to two maintained postures and two prompt movements. I. Nuclear cell output. *J. Neurophysiol.* 33:527–36

Thach WT, Perry JG, Scheiber MH. 1982. Cerebellar output: body maps and muscle spindles. In *The Cerebellum—New Vistas,* ed. SL Palay, V Chan-Palay, pp. 440–54. New York: Springer-Verlag

Thach WT, Goodkin HG, Keating JG. 1992. The cerebellum and the adaptive coordination of movement. *Annu. Rev. Neurosci.* 15:403–42

Thompson JK, Krupa DJ, Weng J, Thompson RF. 1993. Inactivation of motor nuclei blocks expression but not acquisition of rabbit's classically conditioned eyeblink response. *Soc. Neurosci. Abstr.* 19:999

Thompson RF. 1986. The neurobiology of learning and memory. *Science* 233:941–47

Thompson RF. 1989. Role of inferior olive in classical conditioning. In *The Olivocerebellar System in Motor Control,* ed. P Strata, pp. 347–62. New York: Springer-Verlag

Thompson RF. 1990. Neural mechanisms of classical conditioning in mammals. *Philos. Trans. R. Soc. London Ser. B* 329:161–70

Thompson RF, Berger TW, Madden J IV. 1983. Cellular processes of learning and memory in the mammalian CNS. *Annu. Rev. Neurosci.* 6:447–91

Tracy J, Krupa DJ, Bernasconi NL, Grethe JS, Thompson RF. 1992. Changes in stimulation thresholds for DLPN-elicited eyeblinks reflects CR-behavior in the rabbit. *Soc. Neurosci. Abstr.* 18:1561

Tsukahara N, Oda T, Notsu T. 1981. Classical conditioning mediated by the red nucleus in the cat. *J. Neurosci.* 1:72–79

Voneida TJ. 1990. The effect of rubrospinal

tractotomy on a conditioned limb response in the cat. *Soc. Neurosci. Abstr.* 16:270

Voneida T, Christie D, Boganski R, Chopko B. 1990. Changes in instrumentally and classically conditioned limb-flexion responses following inferior olivary lesions and olivocerebellar tractotomy in the cat. *J. Neurosci.* 10:3583–93

Wagner AR, Brandon SE. 1989. Evolution of structured connectionist model of Pavlovian conditioning (AESOP). In *Contemporary Learning Theories: Pavlovian Conditioning and the Status of Traditional Learning Theory*, ed. SB Klein, RR Mowrer, pp. 149–90. Hillsdale NJ: Lawrence Erlbaum.

Weisz DJ, LoTurco JJ. 1988. Reflex facilitation of the nictitating membrane response remains after cerebellar lesions. *Behav. Neurosci.* 102:203–9

Welker WI. 1987. Spatial organization of somatosensory projection to granule cell cerebellar cortex: functional and connectional implications of fractured somatotopy (summary of Wisconsin studies). In *New Concepts in Cerebellar Neurobiology*, ed. JS King, pp. 239–80. New York: Liss

Welsh JP, Harvey JA. 1989. Cerebellar lesions and the nictitating membrane reflex: performance deficits of the conditioned and unconditioned response. *J. Neurosci.* 9:299–311

Welsh JP, Harvey JA. 1991. Pavlovian conditioning in the rabbit during inactivation of the interpositus nucleus. *J. Physiol.* 444: 459–80

Welsh JP, Harvey JA. 1992. The role of the cerebellum in voluntary and reflexive movements: history and current status. In *The Cerebellum Revisited*, ed. R Llinas, C Sotelo, pp. 301–34. New York: Springer-Verlag

Woodruff-Pak DS, Lavond DG, Thompson RF. 1985. Trace conditioning: abolished by cerebellar nuclear lesions but not lateral cerebellar cortex aspirations. *Brain Res.* 348: 249–60

Yang B-Y, Weisz DJ. 1992. An auditory conditioned stimulus modulates unconditioned stimulus-elicited neuronal activity in the cerebellar anterior interpositus and dentate nuclei during nictitating membrane response conditioning in rabbits. *Behav. Neurosci.* 106:889–99

Yeo CH. 1991. Cerebellum and classical conditioning of motor response. *Ann. NY Acad. Sci.* 627:292–304

Yeo CH, Hardiman MJ. 1992. Cerebellar cortex and eyeblink conditioning: a reexamination. *Exp. Brain Res.* 88:623–38

Yeo CH, Hardiman MJ, Glickstein M. 1985a. Classical conditioning of the nictitating membrane response of the rabbit I. Lesions of the cerebellar nuclei. *Exp. Brain Res.* 60:87–98

Yeo CH, Hardiman MJ, Glickstein M. 1985b. Classical conditioning of the nictitating membrane response of the rabbit II. Lesions of the cerebellar cortex. *Exp. Brain Res.* 60:99–113

Yeo CH, Hardiman MJ, Glickstein M. 1986. Classical conditioning of the nictitating membrane response of the rabbit IV. Lesions of the inferior olive. *Exp. Brain Res.* 63:81–92

Zhang AA, Lavond DG. 1991. Effects of reversible lesions of reticular or facial neurons during eyeblink conditioning *Soc. Neurosci. Abstr.* 17:869

Zola-Morgan SM, Squire LR. 1990. The primate hippocampal formation: evidence for a time-limited role in memory storage. *Science* 250:288–90

Annu. Rev. Neurosci. 1994. 17:551–67

THE PROTEIN KINASE C FAMILY FOR NEURONAL SIGNALING

Chikako Tanaka and Yasutomi Nishizuka*

Departments of Pharmacology and Biochemistry*, Kobe University School of Medicine, and Biosignal Research Center*, Kobe University, Kobe 650 Japan

KEYWORDS: heterogeneity, postsynaptic role, presynaptic role, neuronal plasticity, neuronal disorders

INTRODUCTION

Protein kinase C (PKC) is present in high concentrations in neuronal tissues and has been implicated in a broad spectrum of neuronal functions. Nerve cells can transmit signals over long distances by means of electrical impulses. Opening of voltage-gated Ca^{2+} channels following depolarization of the presynaptic membrane by an action potential, normally translates the electrical signal into several chemical messages. The influx of Ca^{2+} triggers an exocytotic release of a variety of neurotransmitters from synaptic vesicles. The chemical messages are then reverted back to electrical form through channel-linked receptors such as nicotinic, glutamate, and $GABA_A$ receptors located on postsynaptic membranes. Many proteins related to these processes of synaptic transmission may be the prime targets of PKC action. Activation of this enzyme in nerve cells is frequently associated with the modulation of ion channels (Shearman et al 1989), desensitization of receptors (Huganir & Greengard 1990), and enhancement of neurotransmitter release (Robinson 1992). The PKC pathway may modulate the efficacy of synaptic transmission, thus providing a basis for some forms of memory.

On the other hand, non-channel-linked receptors respond to agonists, thereby initiating a cascade of enzymatic reactions. The first step in this cascade is activation of G protein, which may either interact directly with ion channels or control the production of intracellular second messengers. When phospholipases are activated via G protein-linked receptors, PKC is activated by increased amounts of diacylglycerol (DAG) in membranes, as a result of

551

agonist-induced hydrolysis of inositol phospholipids (PI) by phospholipase C (PLC) (Nishizuka 1984). Upon cell stimulation, DAG is detected in various intracellular compartments at different times during the cellular responses. The early peak of DAG is transient and reverts back to basal line within seconds, at most several minutes, temporally corresponding to the formation of inositol 1,4,5-trisphosphate and to the rise in intracellular Ca^{2+} concentration. At a relatively later phase of cellular responses, the formation of DAG has a slow onset but is more sustained. It is most likely derived from the hydrolysis of major constituents of the phospholipid bilayer such as phosphatidylcholine (PC) by phospholipase D (PLD), in a signal-dependent manner, yielding phosphatidic acid (PA), which is then dephosphorylated to DAG (Exton 1990). The activation of PLC and PLD is often accompanied by a signal-dependent release of arachidonic acid (AA) through phospholipase A2(PLA2)-catalysed hydrolysis. The reaction products of PC hydrolysis by PLA2, cis-unsaturated fatty acid (FFA) and lysophosphatidylcholine (lysoPC), are both enhancer molecules of PKC activation (Asaoka et al 1992, Nishizuka 1992). It is plausible that the agonist-induced cascade of degradation of various membrane phospholipids is necessary for transducing full information from extracellular signals across the membrane.

In this article, evidence is summarized for the molecular and biochemical heterogeneity and differential spatial expression of the multiple members of the PKC family within the neuron and its potential roles in neuronal signaling.

HETEROGENEITY OF MOLECULAR STRUCTURES AND MODE OF ACTIVATION

Ten isoforms (α, βI, βII, γ, δ, ϵ, η, θ, ζ, and λ) have so far been identified in mammalian tissues (Table 1). These isoforms show subtly different enzymological properties, differential tissue expression, and specific intracellular localization. Some of these isoforms in their mode of activation do not show typical characteristics of classical PKC enzymes. Four PKC isoforms, α, βI, βII, and γ (cPKC), originally emerged from the initial screening (Nishizuka 1988, Bell & Bums 1991). Four additional PKC isoforms, δ, ϵ, η(L), and θ (nPKC) were subsequently identified (Ono et al 1988, Osada et al 1990, 1992, Liyanage et al 1992). Two atypical PKC isoforms, ζ and λ (aPKC), were more recently characterized (Ono et al 1989, Nakanishi & Exton 1992).

The cPKC, the α, βI, βII, and γ isoforms, have four regions of conserved sequence (C1-C4) with five regions of variable sequence (V1-V5). These isoforms are activated by DAG and Ca^{2+} and are further enhanced by FFA and lysoPC. Tumor-promoting phorbol esters act as an analog of DAG, a physiological activator of PKC, and they activate PKC in vitro. The repeat of the cysteine-rich sequence present in the C1 region is essential for phorbol ester binding. The C2 region appears to be related to Ca^{2+} sensitivity of the

Table 1 PKC isoforms in mammalian tissues[a]

Isoforms	Activators	Tissue expression	Intracellular localization (in neurons)
cPKC			
α	Ca^{2+}, DAG, PS, FFA, LysoPC	Universal	Golgi complex, dendrites, periphery of perikarya
βI	Ca^{2+}, DAG, PS, FFA, LysoPC	Some tissues	Periphery of perikarya
βII	Ca^{2+}, DAG, PS, FFA, LysoPC	Many tissues	Golgi complex, dendrites
γ	Ca^{2+}, DAG, PS, FFA, LysoPC	Brain, spinal cord	Throughout cytoplasm; nucleus, plasma membrane
nPKC			
δ	DAG, PS	Universal	
ϵ	DAG, PS, FFA	Brain and others	Golgi complex, axons, nerve terminals
η	?	Lung, skin, heart	
θ	?	Skeletal muscle	
aPKC			
ζ	PS, FFA	Universal	
λ	?	Ovary, testis, and others	

[a]Abbreviations: DAG, diacylglycerol; PS, phosphatidykserine; FFA, cis-unsaturated fatty acid; Lyso PC, lysophosphatidylcholine

enzyme. The C3 region includes the catalytic site, and the C4 region seems to be necessary for recognition of the substrate to be phosphorylated. The α, βI, βII, and γ isoforms were identified in isolated fractions of type III, II, and I, from the soluble fraction of rat brain on a hydroxyapatite column (Huang et al 1986, Kikkawa et al 1987). It is worth noting that the cPKC isoforms exhibit nearly full enzymatic activity at the basal level of Ca^{2+} when both DAG and FFA are available (Nishizuka 1992).

The nPKC, δ, ϵ, η, and θ isoforms, which lack the C2 region, do not require Ca^{2+} and exhibit their enzymatic activities in the presence of phosphatidylserine (PS) and DAG or phorbol esters (Ono et al 1988, 1989, Konno et al 1989, Bacher et al 1991). The δ and ϵ isoforms were purified from the soluble fraction of rat brain by successive chromatographic fractionation (Koide et al 1992, Ogita et al 1992). The δ and ϵ isoforms exist in phosphorylated forms in native tissues and reveal doublet bands upon electrophoresis. Presumably this group of enzymes may be activated by DAG even after the CA^{2+} concentration returns to basal levels. The ϵ but not δ isoform is activated by FFA. It is possible that nPKC enzymes are integrated directly or indirectly in a protein kinase cascade that is initiated by activation of growth factor receptors, eventually leading to the regulation of nuclear events such as cell cycle control (Ahn et al 1990).

The aPKC, ζ, and λ isoforms, which lack the C2 region and have only one cysteine-rich zinc finger-like motif, are dependent on PS but are not affected by DAG, phorbol esters, nor Ca^{2+} for their activation (Ono et al 1989, Nakanishi & Exton 1992). The ζ isoform is activated significantly by FFA in the presence of PS. At present, the signal activating the aPKC enzymes remains unknown. The members of the PKC family so far examined respond perhaps differently to various combinations of Ca^{2+}, DAG, and other phospholipid degradation products, and thereby they produce distinct activation patterns with respect to the extent, duration, and intracellular localization.

DIFFERENTIAL EXPRESSION OF PKC ISOFORMS

The α, βI, βII, γ, ϵ, δ, and ζ isoforms and their mRNAs have been identified in the brain and spinal cord using Western and Northern blot analysis and in situ hybridization. Immunohistochemical analysis using isoform-specific antibodies revealed the differential distribution of the PKC isoforms in the mammalian central nervous system (Tanaka & Saito 1992). PKC isoform containing neurons were mapped in the rat brain, and typical examples are summarized in Table 2. The γ isoform is apparently expressed solely in the brain and spinal cord (Saito et al 1988) and has not been found in any other tissues. In the rat brain, the γ isoform is very low at birth and progressively increases up to 2–3 weeks of age (Hashimoto et al 1988, Hirata et al 1991); it is localized mainly in excitatory and inhibitory amino acid neurons such as cortical pyramidal cells, hippocampal pyramidal and granule cells, cerebellar Purkinje cells, and thalamic neurons. The βI and βII isoforms generated by alternative splicing of a common primary transcript are expressed in different ratios and in different neurons at a distinct stage of development (Hosoda et al 1989, Saito et al 1989, Hirata et al 1991). The βI isoform is already expressed at birth, mainly in the brain stem, and it gradually increases with increasing age. The βII isoform is poorly expressed in the forebrain at birth but rapidly increases up to 2–3 weeks of age. The βII isoform is localized mainly in amino acid neurons including cortical and hippocampal pyramidal cells and striatal neurons. The α and δ isoforms are universally distributed in all tissues and cell types so far examined; however, the α isoform is distributed unevenly in the brain (Ito et al 1990). The ϵ isoform is expressed mainly in the forebrain, spinal cord, and primary sensory neuron (Saito et al 1993).

The differential expression of the cPKC in the cerebellum has been most extensively analyzed. In the cerebellar cortex (Table 2), the γ isoform is present in the Purkinje and Golgi cells with GABA (Hashimoto et al 1988, Hidaka et al 1988, Huang et al 1988, Saito et al 1988), the α isoform is localized in the climbing fiber (Ito et al 1990), and the βI isoform is contained

Table 2 The α, βI, βII, γ, and ε PKC isoform-containing neurons in the neocortex, hippocampus, striatum, substantia nigra, and cerebellum of rat

	α	βI	βII	γ	ε
Neocortex	Round cell (I) Pyramidal cell (III) Nonpyramidal (IV–VI)	Horizontal cell (I) Nonpyramidal (II–VI, GABA)	Pyramidal cell (II, III, V)	Pyramidal cell (II, V, VI)	Nerve terminal
Hippocampal formation	Pyramidal cell (CA1–CA3) Interneuron	—	Pyramidal cell (CA1)	Pyramidal cell (CA1–CA3) Granule cell (DG)	Pyramidal cell (CA3) Mossy fiber Schaffer colateral Perforant pathway
Straitum	ACh neuron	GABA neuron (intrinsic)	GABA neuron (projecting) GABA terminal	Medium-sized cell (projecting) Nerve terminal	Nerve terminal
Substantia nigra	Dopamine neuron (pars compacta)	Nondopamine (pars reticulata)		Nerve terminal	Nerve terminal
Cerebellar cortex	Climbing fiber Basket cell Stellate cell	Mossy fiber Basket cell Stellate cell	Parallel fiber	Purkinje cell Basket cell Stellate cell Golgi cell	Parallel fiber
Deep cerebellar nuclei	—	—	Purkinje cell terminal	—	—

in the mossy fibers (Hosoda et al 1989), respectively. However, the βII and ε isoforms are co-localized in parallel fibers and granule cells (Saito et al 1989, Saito et al, 1993) and the α, βI, and γ isoforms are co-localized in basket and stellate cells. Another striking example of the differential expression of these isoforms is seen in the striatal and nigral neurons. Four isoforms of the cPKC are expressed solely within different neurons with a specific neurotransmitter, in the striatum and substantia nigra (Yoshihara et al 1991) (Table 2). The glutamate neurons in the hippocampal formation display multiple PKC isoforms (Table 2). The α, βII, and γ isoforms are contained in pyramidal cells, and the α and γ isoform in granule cells (Ito et al 1990, Kose et al 1990). The ε isoform is present predominantly in nerve terminals rather than the perikarya (Saito et al 1993).

Electron microscopic analysis revealed that each isoform is localized in distinct intracellular compartments in various brain regions, including cerebellum, hippocampal formation, neocortex, and spinal cord (Table 1) (Kose et al 1988, Kose et al 1990, Mori et al 1990, Stichel et al 1990, Tsujino et al 1990, Saito et al 1993). The γ isoform is localized predominantly in the cytoplasm associated with most ribosomes and outer membranes of cell organelles and with a weaker density in the nucleoplasm. This isoform is localized postsynaptically in both dendritic shaft synapses and spine synapses but poorly in presynaptic terminals, with the exception of the terminals of Purkinje cells. In most brain regions, the α isoform is present in the periphery of the perikarya and is also scattered sparsely in the perinuclear area. The βI isoform is also present clustered in the cytoplasm just adjacent to the plasma membrane, whereas the βII isoform is concentrated around the trans-face of the Golgi complex and the proximal dendrite, with the exception of dendritic spines. The cellular and intracellular distribution patterns of individual PKC isoforms yields further insight into the role of this enzyme in nerve function, depending on where each isoform is located with what kind of substrate.

POTENTIAL ROLES OF PKC IN POSTSYNAPTIC PROCESSES

Modulation of the sensitivity of channels and receptors is a major role of the PKC family on postsynaptic membrane (Shearman et al 1989, Huganir & Greengard 1990). Storage of information in the brain appears to involve persistent, use-dependent alteration in the efficacy of synaptic transmission. One such model is the changes in the glutamatergic synapses such as long-term potentiation (LTP) in the visual cortex and hippocampal formation and long-term depression (LTD) in the cerebellum. In the cerebellum, co-activation of climbing fiber and parallel fiber input to the Purkinje cell leads to a LTD of the parallel fiber input. This phenomenon is mimicked by phorbol ester and is prevented by PKC inhibitors (Ito 1989, Linden & Connor 1991).

Cerebellar LTD is caused postsynaptically by desensitization of the α-amino-3-hydroxy-5-methyl-4-isoxazolepropinate (AMPA) type of glutamate receptors (Ito 1989, Linden & Connor 1991). Desensitization of AMPA receptors is induced by activation of metabotropic receptors, recently identified as linked to the phosphoinositide messenger pathway (Schoepp et al 1990). Both LTD and desensitization of AMPA receptors in cerebellar Purkinje cells are induced by activation of PKC. Most results appear to indicate that the γ isoform is involved in the LTD of glutamatergic transmission in the dendritic spine of Purkinje cells. This hypothesis, however, is supported presently only by anatomical evidence that the γ isoform is present in the dendritic spines that form synaptic contacts with parallel fibers (Kose et al 1988) and that metabotropic and AMPA types of glutamate receptors are expressed in Purkinje cells (Masu et al 1991).

The mechanisms underlying LTP have been most extensively studied in the dentate gyrus and in the CA1 region of the hippocampal formation (Bliss & Collingridge 1993). During the induction phase of about 30 seconds, a cascade of events, starting with activation of N-methyl-D-asparate (NMDA) receptor channels and resulting in Ca^{2+} influx into the postsynaptic cell, triggers enhancement of the synaptic response. Thereafter, the response declines and reaches a stable plateau, and during the maintenance phase another cascade of events results in enhancement of the synaptic response, which is maintained for hours and even days. The phosphorylation cascade induced by Ca^{2+} sensitive enzymes was thought to be possibly involved in converting the initial signal, the rise in $[Ca^{2+}]$, to persistent enhancement of synaptic efficacy. Evidence is accumulating that PKC activation is required to convert short-term potentiation (STP), which is induced by weak stimuli to LTP. Several chemicals related to PKC activation such as phorbol esters and FFA potentiate a synaptic transmission that resembles LTP in the hippocampal slices (Linden et al 1986, Malenka et al 1986, Williams & Bliss 1989). Injection of PKC into the postsynaptic cell in CA1 also mimics LTP (Hu et al 1987). Extracellular application of PKC inhibitors, which bind to either the regulatory or the catalytic domain of PKC, prevents induction of LTP but not fully established LTP. An elegant study has demonstrated that an intracellular injection of the selective 19–31 PKC inhibitory peptide prevents the induction of LTP, and that after its establishment, LTP appears unresponsive to postsynaptic H7, a nonselective PKC inhibitor, although it remains sensitive to externally applied H7 (Malinow et al 1989). Thus, the activation of postsynaptic PKC may possibly be required for induction of LTP, and that of presynaptic PKC appears to be necessary for the maintenance of LTP.

Postsynaptic PKC is activated for less than a few minutes following tetanus, whereas the presynaptic PKC is activated for long period (Huang et al 1992). It is plausible that glutamate receptors are a possible target of the postsynaptic PKC action during LTP. Phorbol ester positively modulates the activity of the

NMDA receptor $\zeta 1$ subunits expressed in *Xenopus* oocytes (Yamazaki et al 1992). The reduction of the Mg^+ block of the channels by PKC appears to be responsible for sustained potentiation of NMDA responses (Chen & Huang 1992). The metabotropic glutamate receptor could be coupled through G protein to PLC, PLA2 or adenylate cyclase, resulting in generation of inositol trisphosphate (IP_3), DAG, AA, lysophospholipid, and cAMP (Aramori & Nakanishi 1992). In the CA1 hippocampal neuron, the activation of metabotropic glutamate receptors enhances the amplitude of NMDA currents, and facilitates induction of LTP (Aniksztejn et al 1992). This up-regulation of NMDA receptor channels and facilitation of LTP are reproduced by the intracellular injection of PKC activators and are prevented by PKC inhibitors, thereby suggesting that the induction of LTP can be modulated by a positive feedback action exerted by PKC on NMDA receptor channels. During induction of LTP, AMPA receptor response is gradually increased, and this up-regulation can be prevented by the PKC inhibitor, K252b (Reymann et al 1990). AMPA receptors are also up-regulated by cAMP-dependent protein kinase (Greengard et al 1991, Wang et al 1991, Yamazaki et al 1992). Direct phosphorylation of glutamate channel receptors appears to be the most likely event linked to the postsynaptic component of LTP. Because the γ isoform seems to be localized in the postsynaptic side of the glutamatergic synapse related to LTP or LTD, including hippocampal pyramidal cells, granule cells in the dentate gyrus, pyramidal cells in the visual cortex, cerebellar Purkinje cells, and neurons in the cochlear nucleus and superior colliculus, this PKC isoform is a likely candidate for involvement in postsynaptic regulation of glutamate receptors.

SUSTAINED PKC ACTIVATION AND RETROGRADE MESSENGER

The sustained activation of PKC is needed for neuronal plasticity and nerve growth. The sustained elevation of DAG, which may come from PC hydrolysis, is often seen in response to various long-lasting signals that can activate PKC. Several mechanisms have been postulated for signal-induced formation of DAG from PC. PC is hydrolyzed by PLD in a signal-dependent fashion to produce phosphatidic acid, which is then dephosphorylated to DAG. PLD was detected in the rat brain (Saito & Kafner 1975), and PLD from rat synaptosomes is activated in the presence of low concentration of Ca^{2+} (Chalifa et al 1990). Because PLD is also activated by phorbol ester or membrane-permeant DAG, sometimes synergistically with the Ca^{2+} ionophore, PKC, once activated initially by PI hydrolysis, may play a role in producing DAG from PC for prolonging PKC activation. It is possible that upon stimulation of the cell, PC is first converted to PI by exchange of the choline moiety with free inositol, and the resulting PI is subsequently hydrolyzed by PLC to produce DAG and inositol phosphate (Nishizuka 1992). This apparent transphosphatydylation reaction is activated by phorbol esters. The occurrence of PC-reactive PLC has been suggested, but the evidence available is insufficient.

On the other hand, PLA2, which hydrolyzes phospholipids to liberate FFA and lysophospholipid, is present ubiquitously in mammalian tissues, including brain; receptor-mediated activation of PLA2 has been proposed (Axelrod et al 1988). Several FFA, including oleic, linoleic, linolenic, arachidonic, and docosahexaenic acids also enhance DAG-dependent activation of PKC, particularly at lower Ca^{2+} concentration (Shinomura et al 1991). The potentiation of cellular responses by FFA in a neuronal system has been noted for DAG-induced reduction of K^+ channel conductance in Hermissenda photoreceptor cells (Lester et al 1991) and for the glutamate exocytosis induced by metabotropic presynaptic receptor stimulation (Herrero et al 1992). The activation of PLC and PLD is often accompanied by signal dependent release of AA through PLA2-catalysed hydrolysis. Two forms of PLA2 are found in the cytosolic fraction of rat brain (Yoshihara & Watanabe 1990). One is a Ca^{2+}-independent high molecular weight enzyme. The other is a low molecular weight enzyme requiring Ca^{2+} which appears to translocate to membranes in response to agonists that provoke PI hydrolysis and mobilize Ca^{2+}. The endogenous PLA2 activity of brain synaptic vesicles is CA^{2+}-dependent, and stimulation of the enzyme is correlated with the induction of vesicle-vesicle aggregation (Moskowitz et al 1982). The secretory (group II) PLA2 is found in cultured astrocytes and in synaptosomes from rat brain and is active in the presence of Ca^{2+} (Oka & Arita 1991). If group II PLA2 itself would be released into the synaptic cleft, FAA might be generated in the extracellular space and diffuse back to presynaptic terminals. PKC may play a role in the signal-induced activation of PLA2, because PKC activators induce a release of AA, sometimes in synergy with physiological agonists (Rehfeld et al 1991).

The maintenance of LTP has been attributed in part to presynaptic mechanisms such as an increase in presynaptic transmitter release. The sustained increase in glutamate release was noted in LTP produced by stimulation of the perforant path-granule cell synapse in the dentate gyrus (PP-DG) in vivo (Dolphin et al 1982) and in slices (Lynch & Bliss 1986). The increased glutamate release in LTP is also enhanced by phorbol esters (Malenka et al 1987). However, a plausible mechanism for inducing a sustained increase in glutamate release after tetanic stimulus has not been identified. AA was proposed to be a candidate for the retrograde messenger of LTP (Williams & Bliss 1989). AA would be produced on the postsynaptic side, then diffuse back across to the presynaptic terminal to activate PKC and induce a sustained increase in transmitter release. AA is released from cultured neurons by activation of NMDA receptors (Dumuis et al 1988), and its efflux is increased in LTP. Application of mepacrine, a PLA2 inhibitor, immediately before LTP induction, produces the blockade of maintenance of LTP, which is prevented by prior application of FFA (Williams & Bliss 1989). AA satisfies many but not all of the criteria for a retrograde messenger at the perforant pathway to granule cell synapses in the dentate gyrus (PP-DG), but not at the Schaffer collateral synapse onto CA1 pyramidal cell (SS-CA1).

The presynaptic glutamate receptor has been identified as being responsible for activation of PKC linking to K^+ channel inhibition, but only in the presence of AA (Herrero et al 1992). Activation of the metabotropic receptor by agonist produces an enhancement of glutamate release from cerebrocortical nerve terminals. AA can be produced in neurons by PLA2 activated by Ca^{2+} entering through NMDA receptor channels (Dumuis et al 1988). Although it remains to be determined if the activation of metabotropic receptors mediates glutamate release from the hippocampus during LTP, the presynaptic glutamate receptor may have a physiological role in the maintenance of LTP.

LTP maintenance is blocked by inhibition of PLA2 and this blockade can be reversed by a prior application of FFA (Linden et al 1987). Potentiation of the frequency of miniature synaptic currents is associated with sustained enhancement of evoked transmission and is induced by mechanisms similar to those underlying LTP. The sustained potentiation of the frequency of the miniature synaptic currents is expressed even in the absence of Ca^{2+} entry into presynaptic terminals (Malgaroli & Tsien 1992). This Ca^{2+}-independent neurosecretory modification may be regulated by CA^{2+}-independent PKC isoforms. In the absence of Ca^{2+} and PS, FFA activates purified PKC (Murakami & Routtenberg 1985). The ε-PKC can be activated by FFA in the absence of Ca^{2+}, and it exists in phosphorylated forms in nature (Koide et al 1992). The ε-PKC is mainly present in presynaptic terminals which make contact with dendrites of the pyramidal cells in CA1, in a giant terminal of the mossy fiber in CA3 and in axon terminals in the dentate gyrus (Saito et al 1993). AA released from postsynaptic cells may diffuse to the presynaptic membrane to sensitize PKC and couple with the metabotropic glutamate receptor to PKC through PLC and DAG. In this putative model of retrograde messenger, AA seems to sensitize the ε-PKC and to result in an enhanced release of glutamate at the maintenance stage of LTP.

Nitric acid (NO) may be another candidate for a retrograde messenger in the maintenance of LTP at SS-CA1 synapses. Injection of an inhibitor of NO synthase into the postsynaptic cell as well as the extracellular application of hemoglobin, a chelator of extracellular NO, prevents LTP at the SS-CA1 synapse (O'Dell et al 1991). Although NO synthase is phosphorylated by PKC and by other kinases, this enzyme has not been shown to be co-localized with PKC in CA1 pyramidal cells (Brecht & Snyder 1992). The effects of

NO synthase inhibitors on LTP are presently controversial (Bliss & Collingridge 1993).

ADDITIONAL POSSIBLE ROLES OF PKC IN PRESYNAPTIC PROCESSES

Several lines of evidence indicate that neurotransmitter release is linked to PKC activation. Ca^{2+}–dependent release of neurotransmitter evoked by chemical or electrical depolarization is normally potentiated by PKC activators such as phorbol ester, membrane permeant DAG and FFA including AA, and is antagonized by PKC inhibitors (Tanaka et al 1984, 1986; Zurgil & Zisapel 1985, Nichols et al 1987, Versteeg & Ulenkate 1987, Allgaier et al 1988, Shuntoh et al 1988). For example, the presynaptic role of the γ isoform has been predicted from the pure anatomical observations that this isoform is present throughout the cytoplasm of nerve terminals of Purkinje cells in deep cerebellar nuclei. The high K^+ evoked release of GABA from deep cerebellar nuclei is potentiated by activation of PKC by AA as well as by phorbol ester, and the enhancement of the evoked release is antagonized by PKC inhibitors (Shuntoh et al 1989, Taniyama et al 1990). Positive feedback control of GABA release in the terminals of Purkinje cell is most likely due to activation of the γ isoform. Similarly, activation of PKC potentiated the GABA release from substantia nigra where βII-PKC is located in the nerve terminals of striato-nigral GABAergic neuron. In the GABAergic terminals of striato-nigral pathway, a GABA transporter (GAT1) is abundantly expressed, and the activity of the presynaptic GABA transporter is suppressed by Ca^{2+} dependent activation of PKC. The increase in GABA release from the substantia nigra may be due to the inhibitory modulation of GABA transporter as well as the facilitation of GABA exocytosis from the terminals by βII-PKC. Whereas acetylcholine (ACh) release from the striatal cholinergic neuron may be modulated by the α isoform which is co-localized with choline acetyltransferase in the striatal neuron (Tanaka et al 1986, Yoshihara et al 1991). It is likely that the uptake and release of neurotransmitter are modulated by each PKC isoform present in these nerve terminals.

The growth associated protein GAP-43 (F1, B50) (Coggins & Zweirs 1991) and dephosphin (Robinson 1992) are presynaptic substrates specific to PKC. It has been postulated that GAP-43 is involved in diverse neuronal functions including neurotransmitter release, neuronal development and regeneration, synaptogenesis, synaptic plasticity and memory formation, and other cognitive behaviors. GAP-43 is specifically phosphorylated by PKC during exocytosis, depolarization, axogenesis, and LTP (Dekker et al 1989, 1990, Nelson et al 1989, Meiri et al 1991). This neuron-specific PKC substrate apparently exists as an exclusively presynaptic component (Gispen et al 1985). An immunohistochemical study using antibodies that recognize selectively the phosphorylated and nonphosphorylated form of GAP-43, has shown that the phosphorylated GAP-43 appears only when axogenesis has begun and is spatially restricted to the distal axon and growth cone (Meiri et al 1991), thereby suggesting that the phosphorylation of GAP-43 by PKC is not required for axon outgrowth or growth cone function per se. Antibodies against this protein can block glutamate release (Dekker et al 1989). An increase in the phosphorylation of GAP-43 may be one molecular mechanism underlying LTP. The presynaptic localization of the α, βII, but not γ isoform has been noted in synaptosomes prepared from the rat hippocampus (Shearman et al 1991). Since the βII isoform phosphorylates GAP-43 to a greater extent than the γ isoform, it has been suggested that the βII isoform is a candidate for the enzyme responsible for phosphorylating GAP-43, which may facilitate release of neurotransmitter (Sheu et al 1990).

Dephosphin and its phosphorylation by PKC may also be associated with evoked release (Robinson 1992). This protein is a synaptic phosphoprotein in intact nerve terminals that is rapidly dephosphorylated upon depolarization and Ca^{2+} influx (Robinson 1992). Upon repolarization and decline in the rise of intracellular Ca^{2+}, dephosphin is phosphorylated again by PKC. PKC mediated phosphorylation of this protein is proposed to be an obligatory step that initiates the release of neurotransmitter.

PKC AND NEURONAL DISORDERS

PKC seems to be involved in the survival of neurons, as many neuronal trophic factors function through PKC activation (Montz et al 1985, Hama et al 1986). A substantial body of evidence has implicated excitotoxicity as a mechanism of both acute and chronic neurologic diseases. During hypoxia and ischemia, the glutamate concentration at synapses is increased for sustained periods. Persistent activation of the glutamate receptor increases neuronal Ca^{2+} influx. The cascade of events including the rise in intracellular Ca^{2+} and irreversible destabilization of Ca^{2+} homeostasis, triggers the formation of AA and of NO, the activation and translocation of PKC and the activation of Ca^{2+}-dependent enzymes, the result being neuronal cell death (Choi & Rothman 1990, Manev et al 1990). The sustained PKC activation may be due to the generation of DAG from PC and to the enhancement by FFA, lysoPC, and possibly other metabolites which are produced by a cascade of hydrolysis of various membrane phospholipids (Nishizuka 1992). A brief anoxic episode is found to induce in CA1 a persistent potentiation of the synaptic NMDA component (Ben-Ari et al 1992). PKC activation is known to enhance NMDA currents by reducing the Mg^+ block of the channels (Chen & Huang 1992). Enhancement of the NMDA receptor activity induced by PKC activation probably underlies the persistent destabilization of $[CA^{2+}]i$ and the subsequent neurotoxicity. The excitotoxicity may also contribute to some of the many neurological diseases associated with early neurological degeneration, dementia, seizures, and motor disturbances, although it is presently difficult to perceive how a similar mechanism could occur in chronic disease.

Reduced PKC levels determined by phorbol ester binding in the particulate fraction and a trend toward elevated PKC levels has been found in autopsied brains from patients with Alzheimer's disease (Cole et al 1988). An immunocytochemical study revealed that the βII isoform is reduced in cortical neurons, and that the βI and βII isoforms accumulate in neurite plaques (Masliah et al 1990). Such PKC-related alteration correlates with neuritic plaque formation and not with neurofibrillary tangle formation. There is a hypothesis that abnormal protein phosphorylation may play a role in development of the cerebral amyloidosis that accompanies Alzheimer's disease, and PKC has been postulated to take part in this process. PKC phosphorylates the P/4 amyloid precursor protein (APP) in vitro (Gandy et al 1988). Activation of PKC and/or protein phosphatase 1 and 2A results in an increase in proteolytic processing (Buxbaum et al 1990) and secretion (Caporaso et al 1992) of APP, in intact cells. PKC also regulates APP mRNA expression (Goldgaber et al 1989). Intact Golgi function is necessary for APP maturation and processing, and the proteolysis and secretory cleavage of APP might occur in the endoplasmic reticulum or proximal Golgi (Caporaso et al 1992). These biochemical findings suggest a hypothetical connection between processing and secretion of APP and the βII PKC, which is located around the trans face of the Golgi complex.

A reduction in phorbol ester binding in the striatum and substantia nigra has been found in patients with Parkinson's disease (Nishino et al 1989) and Huntington's disease (Hashimoto et al 1992). The decreased phorbol ester binding in Parkinson's disease may be attributed to both the decreased α isoform in dopaminergic neurons and the decreased βII isoform in the GABAergic nerve terminals due to transsynaptic degeneration of striatal neurons. Huntington's disease is characterized by degeneration of striato-nigral GABAergic neurons. Western blotting using PKC isoform–specific antibodies revealed that the βII but not γ isoform was decreased in the striatum from patients with Huntington's disease (Hashimoto et al 1992). This is consistent with anatomical evidence that the βII isoform is selectively localized in the rat striato-nigral pathway as described above. These results suggest a differential expression of PKC isoform in human brain as well as rat brain, and involvement of PKC isoforms in neurodegenerative diseases, especially the βII isoform which is present in cerebral pyramidal cells, hippocampal pyramidal cells, and striato-nigral neurons. While it remains to be determined whether these alterations in PKC simply reflect the loss of neurons or actually induce neuronal death, a direct relation to neuronal dysfunctions seems likely.

ACKNOWLEDGMENTS

This work was supported in part by research grants from the Special Research Fund of Ministry of Education, Science, and Culture, Japan. We thank Mariko Ohara for assistance with the manuscript.

Literature Cited

Ahn NG, Weiel JE, Chan CP, Krebs EG. 1990. Identification of multiple epidermal growth factor-stimulated protein serine/threonine kinases from Swiss 3T3 cells. *J. Biol. Chem.* 265:11487–94

Allgaier C, Daschmann B, Huang H-Y, Hertting G. 1988. Protein kinase C and presynaptic modulation of acetylcholine release in rabbit hippocampus. *Br. J. Pharmacol.* 93: 525–34

Aniksztejn L, Otani S, Ben-Ari Y. 1992. Quisqualate metabotropic receptors modulate NMDA currents and facilitate induction of long-term potentiation through protein kinase C. *Eur. J. Neurosci.* 4:500–5

Aramori I, Nakanishi S. 1992. Signal transduction and pharmacological characteristics of a metabotropic glutamate receptor, mGluR1, in transfected CHO cells. *Neuron* 8:757–65

Asaoka Y, Nakamura S, Yoshida K, Nishizuka Y. 1992. Protein kinase C, calcium and phospholipid degradation. *Trends Biochem. Sci.* 17:414–17

Axelrod J, Burch RM, Jelsema CL. 1988. Receptor-mediated activation of phospholipase A2 via GTP-binding proteins: arachidonic acid and its metabolites as second messengers. *Trends Neurosci.* 1988: 117–23

Bacher N, Zisman Y, Berent E, Livneh E. 1991. Isolation and characterization of PKC-L, a new member of the protein kinase C-related gene family specifically expressed in lung, skin, and heart. *Mol. Cell. Biol.* 11:126–33

Bell RM, Bums DJ. 1991. Lipid activation of protein kinase C. *J. Biol. Chem.* 266:4661–64

Ben-Ari Y, Aniksztejn L, Bregestovski P. 1992. Protein kinase C modulation of NMDA currents: an important link for LTP induction. *Trend. Neurosci.* 15:333–39

Bliss TVP, Collingridge GL. 1993. A synaptic model of memory:long-term potentiation in the hippocampus. *Nature* 361:31–39

Brecht DS, Snyder SH. 1992. Nitric oxide, a novel neuronal messenger. *Neuron* 8:3–11

Buxbaum JD, Gandy SE, Cicchetti P, et al. 1990. Processing of Alzheimer β/A4 amyloid precursor protein: Modulation by agents that regulate protein phosphorylation. *Proc. Natl. Acad. Sci. USA* 87:6003–6

Caporaso GL, Gandy SE, Buxbaum JD, Greengard P. 1992. Chloroquine inhibits intracellular degradation but not secretion of Alzheimer β/A4 amyloid precursor protein. *Proc. Natl. Acad. Sci. USA* 89:2252–56

Chalifa V, Möhn H, Liscovitch M. 1990. A neutral phospholipase D activity from rat brain synaptic plasma membranes. Identification and partial characterization. *J. Biol. Chem.* 265:17512–19

Chen L, Huang L-YM. 1992. Protein kinase C reduces Mg^{2+} block of NMDA-receptor channels as a mechanism of modulation. *Nature* 356:521–23

Choi DW, Rothman SM. 1990. The role of glutamate neurotoxicity in hypoxic-ischemic neuronal death. *Annu. Rev. Neurosci.* 13: 171–182

Coggins PJ, Zwiers H. 1991. B-50(GAP-43): Biochemistry and functional neurochemistry of a neuron-specific phosphorylation. *J. Neurochem.* 56:1095–1106

Cole G, Dobkins KR, Hansen LA, et al. 1988. Decreased levels of protein kinase C in Alzheimer brain. *Brain Res.* 452:165–70

Dekker LV, DeGraan PN, Oestreicher AB. 1989. Inhibition of noradreneline release by antibodies to B-50. *Nature* 342:74–76

Dekker LV, DeGraan PN, Oestreicher AB. 1990. Depolarization-induced phosphorylation of the protein kinase C substrate B-50 (GAP-43) in rat cortical synaptosomes. *J. Neurochem.* 54:1645–52

Dolphin AC, Errington ML, Bliss TVP. 1982. Long-term potentiation of the perforant path in vivo is associated with increased glutamate release. *Nature* 297:496–98

Dumuis A, Sebben M, Haynes L, et al. 1988. NMDA receptors activate the arachidonic acid cascade system in striatal neurons. *Nature* 336:68–70

Exton JH. 1990. Signaling through phosphatidylcholine breakdown. *J. Biol. Chem.* 265: 1–4

Gandy S, Czernik AN, Greengard P. 1988. Phosphorylation of Alzheimer disease amyloid precursor peptide by protein kinase C and CA^{2+}/calmodulin-dependent protein kinase II. *Proc. Natl. Acad. Sci. USA* 83: 6218–621

Gispen WH, Leunissen JLM, Oestreiter AB, et al. 1985. Presynaptic localization of B-50 phosphoprotein:the ACTH-sensitive protein kinase C substrate involved in rat brain phosphoinositide metabolism. *Brain Res.* 28: 381–85

Goldgaber D, Garris HW, Hla T, et al. 1989. Interleukin I regulates synthesis of amyloid β-protein precursor mRNA in human endothelial cells. *Proc. Natl. Acad. Sci. USA* 86:7606–10

Greengard P, Jen J, Nairn AC, Stevens CF. 1991. Enhancement of the glutamate response by cAMP-dependent protein kinase in hippocampal neurons. *Science* 253:1135–38

Hama T, Huang K-P, Guroff G. 1986. Protein kinase C as a component of a nerve growth factor-sensitive phosphorylation system in PC12 cells. *Proc. Natl. Acad. Sci. USA* 83:2353–57

Hashimoto T, Ase K, Sawamura S, et al. 1988. Postnatal development of a brain specific subspecies of protein kinase C in rat. *J. Neurosci.* 8:1678–83

Hashimoto T, Kitamura N, Saito N, et al. 1992. The loss of βII-protein kinase C in the striatum from patients with Hungtington's disease. *Brain Res.* 585:303–306

Herrero I, Miras-Portugal MT, Sanchez-Pietro J. 1992. Positive feedback of glutamate exocytosis by metabotropic presynaptic receptor stimulation. *Nature* 360:163–66

Hidaka H, Tanaka T, Onoda K, et al. 1988. Cell type-specific expression of protein kinase C isozymes in the rabbit cerebellum. *J. Biol. Chem.* 263:4523–26

Hirata M, Saito N, Kono M, Tanaka C. 1991. Differential expression of the βI- and βII-PKC subspecies in the postnatal developing rat brain; an immunocytochemical study. *Dev. Brain. Res.* 62:229–38

Hosoda K, Saito N, Kose A, et al. 1989. Immunocytochemical localization of βI-subspecies of protein kinase C in rat brain. *Proc. Natl. Acad. Sci. USA* 86:1393–97

Hu G-Y, Hvalby O, Walaas SI, et al. 1987. Protein kinase C injection into hippocampal pyramidal cells elicits features of long term potentiation. *Nature* 328:426–29

Huang FL, Yoshida Y, Nakabayashi H, et al. 1988. Immunocytochemical localization of protein kinase C isozymes in rat brain. *J. Neurosci.* 8:4734–44

Huang KP, Nakabayashi H, Huang FL. 1986. Isozymic forms of rat brain Ca^{2+}-activated and phospholipid dependent protein kinase. *Proc. Natl. Acad. Sci. USA* 84:8535–39

Huang YY, Colley PA, Routtenberg A. 1992. Postsynaptic then presynaptic protein kinase C activity may be necessary for long ten-n potentiation. *Neuroscience* 49:819–27

Huganir RL, Greengard P. 1990. Regulation of neurotransmitter receptor densensitization by protein phosphorylation. *Neuron* 5:555–67

Ito A, Saito N, Hirata M, et al. 1990. Immunocytochemical localization of the α subspecies of protein kinase C in rat brain. *Proc. Natl. Acad. Sci. USA* 87:3195–99

Ito M. 1989. Long term depression. *Annu. Rev. Neurosci.* 12:85–102

Kikkawa U, Ono Y, Ogita K, et al. 1987. Identification of the structures of multiple subspecies of protein kinase C expressed in rat brain. *FEBS Lett.* 217:227–31

Koide H, Ogita K, Kikkawa U, Nishizuka Y. 1992. Isolation and characterization of the ε-subspecies of protein kinase C from rat brain. *Proc. Natl. Acad. Sci. USA* 89:1149–53

Konno Y, Ohno S, Akita Y, et al. 1989. Enzymatic properties of a novel phorbol ester/protein kinase, nPKC. *J. Biochem.* 106:673–78

Kose A, Ito A, Saito N, Tanaka C. 1990. Electron microscopic localization of γ- and βII-subspecies of protein kinase C in rat hippocampus. *Brain Res.* 518:209–17

Kose A, Saito N, Ito H, et al. 1988. Electron microscopic localization of type I protein kinase C in rat Purkinje cells. *J. Neurosci.* 8:4262–68

Lester DS, Collin C, Etcheberrigaray R, Alkon DL. 1991. Arachidonic acid and diacylglycerol act synergistically to activate protein kinase C in vitro and in vivo. *Biochem. Biophys. Chem. Commun.* 179: 1522–28

Linden DJ, Connor JA. 1991. Participation of postsynaptic PKC in cerebellar long-term depression in culture. *Science* 254:1656–59

Linden DJ, Murakami K, Routtenberg A. 1986. A newly discovered protein kinase C activator (oleic acid) enhances long term potentiation in the intact hippocampus. *Brain Res.* 379:358–63

Linden DJ, Sheu F-S, Murakami K, Routtenberg A. 1987. Enhancement of long-term potentiation by cis-unsaturated fatty acid: relation to protein kinase C and phospholipase A2. *J. Neurosci.* 7:3783–92

Liyanage M, Frith D, Livneh E, Stabel S. 1992. Protein kinase C group B members PKC-δ, -ε-, -ζ and PKC-L (η). *Biochem. J.* 283:781–787

Lynch MA, Bliss TVP. 1986. On the mechanism of enhanced release of [14C] glutamate in hippocampal long-term potentiation. *Brain Res.* 369:405–8

Malenka RC, Ayoub GS, Nicoll RA. 1987. Phorbol esters enhance transmitter release in rat hippocampal slices. *Brain Res.* 403:198–203

Malenka RC, Madison DV, Nicoll RA. 1986. Potentiation of synaptic transmission in the hippocampus by phorbol esters. *Nature* 321: 175–77

Malgaroli A, Tsien RW. 1992. Glutamate-induced long-term potentiation of the frequency of miniature synaptic currents in cultured hippocampal neurons. *Nature* 357: 134–39

Malinow R, Schulman H, Tsien RW. 1989. Inhibition of postsynaptic PKC or CaMKII blocks induction but not expression of LTP. *Science* 245:862–66

Manev H, Costa E, Wroblewoki JT, Guidotti A. 1990. Abusive stimulation of excitatory aminoacid receptors: a strategy to limit neurotoxicity. *FASEB J.* 4:2788–97

Masliah E, Cole G, Shimohama S, et al. 1990. Differential involvement of protein kinase C isozymes in Alzheimer's disease. *J. Neurosci.* 10:2113–24

Masu M, Tanabe Y, Tsuchida K, et al. 1991. Sequence and expression of a metabotropic glutamate receptor. *Nature* 349:760–65

Meiri KF, Bickerstaff LE, Schwob JE. 1991. Monoclonal antibodies show that kinase C phosphorylation of GAP-43 during axogenesis is both spatially and temporally restricted in vivo. *J. Cell Biol.* 112:991–1005

Montz HPM, Davis GE, Skaper SD, et al. 1985. Tumor promoting phorbol diester mimics two distinct neuronotrophic factors. *Dev. Brain Res.* 23:150–54

Mori M, Kose A, Tsujino T, Tanaka C. 1990. Immunocytochemical localization of protein kinase C subspecies in the rat spinal cord: light and electron microscopic study. *J. Comp. Neurol.* 299:167–177

Moskowitz N, Schook W, Puszkin S. 1982. Interaction of brain synaptic vesicles induced by endogenous Ca^{2+}-dependent phospholipase A_2. *Science* 216:305–7

Murakami K, Routtenberg A. 1985. Direct stimulation of purified protein kinase C by unsaturated fatty acids (oleate, arachidonate) in the absence of phospholipids and calcium. *FEBS Lett.* 192:189–93

Nakanishi H, Exton JH. 1992. Purification and characterization of the ζ-isoform of protein kinase C from bovine kidney. *J. Biol. Chem.* 267:16347–54

Nelson RB, Linden DJ, Hyman C, et al. 1989. The two major phosphoproteins in growth cones are probably identical; to two protein kinase C substrates correlated with the persistence of long term potentiation. *J. Neurosci.* 9:381–89

Nichols RA, Haycock JW, Wang JKT, Greengard P. 1987. Phorbol ester enhancement of neurotransmitter release from rat brain synaptosomes. *J. Neurochem.* 48:615–21

Nishino N, Kitamura N, Nakai T, et al. 1989. Phorbol ester binding sites in human brain: characterization, regional distribution, age-correlation, and alterations in Parkinson's disease. *J. Mol. Neurosci.* 1:19–26

Nishizuka Y. 1984. The role of protein kinase C in cell surface signal transduction and tumour promotion. *Nature* 308:693–98

Nishizuka Y. 1988. The molecular heterogeneity of protein kinase C and implications for cellular regulation. *Nature* 334:661–65

Nishizuka Y. 1992. Intracellular signalling by hydrolysis of phospholipids and activation of protein kinase C. *Science* 258:607–14

O'Dell TJ, Hawkins RD, Kandel ER, Arancio O. 1991. Tests of the roles of the two diffusible substances in long-term potentiation: Evidence for nitric oxide as a possible early retrograde messenger. *Proc. Natl. Acad. Sci. USA* 88:11285–89

Ogita K, Miyamoto S, Yamaguchi K, et al. 1992. Isolation and characterization of δ-subspecies of protein kinase C from rat brain. *Proc. Natl. Acad. Sci. USA* 89:1592–96

Oka S, Arita H. 1991. Inflammatory factors stimulate expression of group II phospholipase A_2 in rat cultured astrocytes. *J. Biol. Chem.* 266:9956–60

Ono Y, Fujii T, Igarashi K, et al. 1989. Phorbol ester binding to protein kinase C requires a cystein-rich zinc-finger like sequence. *Proc. Natl. Acad. Sci. USA* 86:4868–71

Ono Y, Fujii T, Ogita K, et al. 1988. The structure, expression, and properties of additional members of the protein kinase C family. *J. Biol. Chem.* 263:6927–32

Ono Y, Fujii T, Ogita K, et al. 1989. Protein kinase C ζ subspecies from rat brain: its structure, expression and properties. *Proc. Natl. Acad. Sci. USA* 86:3099–3103

Osada S, Mizuno K, Saido TC, et al. 1990. A phorbol ester receptor/protein kinase, nPKC η, a new member of the protein kinase C family predominantly expressed in lung and skin. *J. Biol. Chem.* 265:22434–40

Osada S-I, Mizuno K, Saido TC, et al. 1992. A new member of the protein kinase C family, nPKCθ, predominantly expressed in skeletal muscle. *Mol. Cell. Biol.* 12:3930–38

Rehfeld W, Hass R, Goppelt-Struebe M. 1991. Characterization of phospholipase A_2 in monocytic cell lines. *Biochem. J.* 276:631–36

Reymann KG, Davies SN, Matthies H, et al. 1990. Activation of a K-252b-sensitive protein-kinase is necessary for a postsynaptic phase of long-term potentiation in area of CA1 of rat hippocampus. *Eur. J. Neurosci.* 2:481–86

Robinson PJ. 1992. The role of protein kinase C and its neuronal substrates dephosphin, B-50, and MARCKS in neurotransmitter release. *Mol. Neurobiol.* 5:87–130

Saito M, Kafner J. 1975. Phosphatidohydrolase

activity in a solubilized preparation from rat brain particulate fraction. *Arch. Biochem. Biophys.* 169:318–23

Saito N, Itouji A, Totani Y, et al. 1993. Cellular and intracellular localization of ε-subspecies of protein kinase C in the rat brain; presynaptic localization of ε-subspecies. *Brain Res.* 609:241–48

Saito N, Kikkawa U, Nishizuka Y, Tanaka C. 1988. Distribution of protein kinase C-like immunoreactive neurons in rat brain. *J. Neurosci.* 8:369–82

Saito N, Kose A, Ito A, et al. 1989. Immunocytochemical localization of βII subspecies of protein kinase C in rat brain. *Proc. Natl. Acad. Sci. USA* 86:3409–13

Schoepp D, Bockaert J, Sladeczek F. 1990. Pharmacological and functional characteristics of metabotropic excitatory amino acid receptors. *Trends Pharmacol. Sci.* 11:508–15

Shearman MS, Sekiguchi K, Nishizuka Y. 1989. Modulation of ion channel activity: a key function of the protein kinase C enzyme family. *Pharmacol. Rev.* 41:211–37

Shearman MS, Shinomura T, Oda T, Nishizuka Y. 1991. Synaptosomal protein kinase C subspecies: A dynamic change in the hippocampus and cerebellar cortex concomitant with synaptogenesis. *J. Neurochem.* 56:1565–72

Sheu F-S, Marias RM, Parher PJ, et al. 1990. Neuron-specific protein F1/GAP-43 shows substrate specificity for the beta subtype of protein kinase C. *Biochem. Biophys. Res. Commun.* 171:1236–43

Shinomura T, Asaoka Y, Oka M, et al. 1991. Synergistic action of diacylglycerol and unsaturated fatty acid for protein kinase C activation:its possible implication. *Proc. Natl. Acad. Sci. USA* 88:5149–53

Shuntoh H, Taniyama K, Fukuzaki H, Tanaka C. 1988. Inhibition by cyclic AMP of phorbol ester-potentiated norepinephrine release from guinea pig brain cortical synaptosomes. *J. Neurochem.* 51:1565–72

Shuntoh H, Taniyama K, Tanaka C. 1989. Involvement of protein kinase C in the Ca^{2+}-dependent vesicular release of GABA from central and enteric neurons of the guinea pig. *Brain Res.* 483:384–88

Stichel CC, Singer W, Zilles K. 1990. Ultrastructure of PkC(II/III)-immunopositive structures in rat promary visual cortex. *Exp. Brain Res.* 82:575–84

Tanaka C, Fujiwara H, Fujii Y. 1986. Acetylcholine release from guinea pig caudate slices evoked by phorbol ester and calcium. *FEBS Lett.* 195:129–34

Tanaka C, Saito N. 1992. Localization of subspecies of protein kinase C in the mammalian central nervous system. *Neurochem. Int.* 21:499–512

Tanaka C, Taniyama K, Kusunoki M. 1984. A phorbol ester and A23187 act synergistically to release acetylcholine from the guinea pig ileum. *FEBS Lett.* 175:165–69

Taniyama K, Saito N, Kose A, et al. 1990. Involvement of γ-subtype of protein kinase C in GABA release from the cerebellum. In *Advances in Second Messenger and Phosphoprotein Research*, ed. Y Nishizuka, M Endo, C Tanaka, pp. 399–404. 24. New York:Raven

Tsujino T, Kose A, Saito N, Tanaka C. 1990. Light and electron microscopic localization of βI-, βII-, and γ-subspecies of protein kinase C in rat cerebral neocortex. *J. Neurosci.* 10:870–84

Versteeg DHG, Ulenkate HJLM. 1987. Basal electrically stimulated release of [³H]noradrenaline and [³H]dopamine from rat amygdala slices in vitro:effects c,f 4β-phorbol 12,13-dibutyrate, 4α-phorbol 12,13-didecanoate and polymyxin B. *Brain Res.* 416:343–48

Wang L-Y, Salter MW, MacDonald JF. 1991. Regulation of kinate receptors by cAMP-dependent protein kinase and phosphatases. *Science* 253:1132–35

Williams JH, Bliss TVP. 1989. An in vitro study of the effect of lipoxygenase and cyclo-oxygenase inhibitors of arachidonic acid on the induction and maintenance of long-term potentiation in the hippocampus. *Neurosci. Lett.* 107:301–6

Yamazaki M, Mori H, Araki K, et al. 1992. Cloning, expression and modulation of a mouse NMDA receptor subunit. *FEBS Lett.* 300:39–45

Yoshihara C, Saito N, Taniyama K, Tanaka C. 1991. Differential localization of four subspecies of protein kinase C in the rat striatum and substantia nigra. *J. Neurosci.* 11:690–700

Yoshihara Y, Watanabe Y. 1990. Translocation of phospholipase A_2 from cytosol to membranes in rat brain induced by calcium ions. *Biochem. Biophys. Res. Commun.* 170:484–90

Zurgil N, Zisapel N. 1985. Phorbol ester and calcium act synergistically to enhance neurotransmitter release by brain neurons in culture. *FEBS Lett.* 185:257–61

Annu. Rev. Neurosci. 1994. 17:569–602

GABA$_A$ RECEPTOR CHANNELS

Robert L. Macdonald

Departments of Neurology and Physiology, University of Michigan Medical Center, Ann Arbor, Michigan 48104

Richard W. Olsen

Department of Pharmacology, UCLA School of Medicine, Los Angeles, California 90024

KEY WORDS: benzodiazepine receptors, neurosteroids, barbituates, phosphorylation

INTRODUCTION

The neutral amino acid γ-aminobutyric acid (GABA) is the major inhibitory neurotransmitter in the central nervous system. GABA is released from GABAergic neurons and binds to both GABA$_A$ receptors (GABARs) and GABA$_B$ receptors. GABA$_B$ receptors are coupled to calcium or potassium ion channels via GTP binding proteins. Only GABARs will be discussed in this review. The GABAR is a macromolecular protein that contains specific binding sites at least for GABA, picrotoxin, barbiturates, benzodiazepines, and the anesthetic steroids, and forms a chloride ion-selective channel (Twyman & Macdonald 1991, DeLorey & Olsen 1992, Macdonald & Twyman 1992) (Table 1).

GABAR AGONIST BINDING SITES

GABARs are activated by GABA and structural analogues of GABA, such as muscimol, a natural product from the hallucinogenic mushroom *Amanita muscaria,* and synthetic analogues including 4,5,6,7-tetrahydroisoxazolo-pyridin-3-ol (THIP), 3-aminopropanesulfonate, piperidine-4-sulfonate, and isoguvacine (Johnston 1986, Ticku 1991). GABA binds to GABARs to regulate gating (opening and closing) of the chloride ion channel. GABA

569

Table 1 GABA$_A$ Receptors

Selective agonists	Muscimol, GABA
Competitive antagonist	Bicuculline
Non-competitive antagonists	Picrotoxin, PTZ, TBPS
Channel blocker	Penicillin
Benzodiazepine receptor agonists	Diazepam, clonazepam
Benzodiazepine receptor inverse agonists	DMCM, βCCM
Benzodiazepine receptor antagonist	Flumazenil
Barbitrate receptor agonists	Pentobarbital, phenobarbital
"Anesthetic steroid receptor" agonist	Alphaxalone
Channel ion selectivity	Chloride ions

concentration response curves are sigmoidal and have Hill numbers of about two, suggesting that at least two molecules of GABA must bind to the GABAR for full activation of the native receptor channel (Sakmann et al 1983).

The GABA binding sites on the GABAR complex in mammalian brain have been characterized in cell-free homogenates by using specific radioligands (Zukin et al 1974). Vigorous homogenization, multiple wash steps, the use of sodium-free buffer, and cooling to 0°C serves to remove endogenous GABA (and other potential endogenous ligands) and to disable the GABA transport system. GABAR-specific binding is defined as sodium-independent binding of GABA that is inhibited by the antagonist bicuculline and the agonist muscimol and that shows multiple high affinity sites (Kd < 1 μM); however, the affinities of these sites are one to three orders of magnitude lower than the concentrations needed (> 1 μM) to activate Cl$^-$ channels (Barker & Owen 1986, Yang & Olsen 1987). The use of more physiological conditions for the binding assays has only partially alleviated the problem. For example, binding can be studied with synaptoneurosomes, which are homogenized membrane particles containing membrane-enclosed pre- and postsynaptic elements. Synaptoneurosomes have the ability to maintain a membrane potential and to respond functionally to GABA agonists measured with radiotracer ion ^{36}Cl$^-$ influx or efflux (DeLorey & Brown 1992, Edgar & Schwartz 1992). Rapid kinetic measurements of ligand binding and ion flux on native membrane (Kardos & Cash 1990) or reconstituted GABARs (Dunn et al 1989) should be useful for quantitative comparisons of biochemical and physiological data. The growing diversity and complexity of our knowledge of this receptor system underscores the necessity to resolve contradictions between pharmacologically and physiologically relevant events.

GABAR BENZODIAZEPINE, BARBITURATE, PICROTOXIN AND STEROID BINDING SITES

GABAR currents are universally reduced by the plant convulsant picrotoxin and generally enhanced by three classes of CNS depressant drugs: the benzodiazepines, the barbiturates, and the anesthetic steroids (Olsen 1981, 1987; Olsen et al 1986). The mechanisms for enhancement or reduction of GABAR currents by these drugs will be discussed in later sections. The properties of the GABAR drug binding sites, including the picrotoxin/convulsant sites (Ticku et al 1978, Squires et al 1983) and the benzodiazepine sites (Squires & Braestrup 1977, Möhler & Okada 1977), have been characterized by using radioligands. However, no radioligands for the steroid or barbiturate sites have been described.

High and low affinity GABA binding sites copurify with each other and with benzodiazepine binding activity. Cloning and expression of benzodiazepine-modulated GABA-regulated chloride channels have proved that the benzodiazepine receptor is one and the same with the GABAR complex and that this protein demonstrates both high affinity (nM) binding of GABA and low affinity (μM) activation by GABA of chloride channel function (Sigel et al 1983, Schofield et al 1987, Stauber et al 1987). Whether the high and low affinity GABA binding sites are distinct or interconvertible has not been established. Barbiturates and related CNS depressants, including anesthetic steroids, allosterically modulate the binding of radiolabeled GABA,

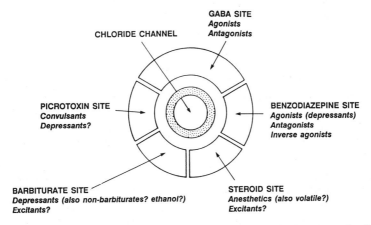

Figure 1   Schematic "donut" model of the GABA$_A$ receptor–chloride ion channel complex. Each of the sections represents a distinct functional binding domain for the drug class indicated, plus the chloride channel itself in the middle. No implications about the protein subunit structure are made in this model. From Olsen et al (1991b), with permission.

benzodiazepines, and picrotoxin (Olsen 1981, 1987; Ticku 1991). TBPS/ picrotoxin binding is allosterically modulated by GABA, barbiturates, benzodiazepines, and steroids, and copurifies with the $GABA_A$-benzodiazepine receptor complex (Sigel & Barnard 1984, King et al 1987). This supports the concept of the supramolecular GABAR complex schematized in Figure 1 (Olsen 1981).

GABAR SUBUNIT MOLECULAR STRUCTURE

A major advance in determining the molecular structure of the GABAR was the discovery that the receptor protein could be photoaffinity labeled by the benzodiazepine [^{3}H]flunitrazepam (Möhler et al 1980). A single polypeptide was specifically labeled in crude brain homogenates, with a molecular weight of 51 kDa on polyacrylamide gel electrophoresis in sodium dodecyl sulfate (SDS-PAGE). Subsequently, microheterogeneity was shown for the benzodiazepine photolabeling, with minor bands at 53, 55, and 59 kDa. These additional benzodiazepine binding polypeptides were shown to vary in brain regional abundance, ontogenetic development, one-dimensional peptide mapping of proteolytic fragments, and binding specificity (Sieghart & Drexler 1983), consistent with the presence of a family of gene products. These findings have been subsequently confirmed by molecular cloning.

A variety of mild detergents have been used to solubilize GABAR binding activity from brain membranes. High solubilization yields are obtained with deoxycholate. Although GABA and benzodiazepine binding are reasonably stable in Triton X-100 solution, barbiturate modulation, bicuculline, and TBPS binding are well preserved with the zwitterionic detergent 3[(3-cholamido-propyl) dimethyl ammonio]-1-propane-sulfonate (CHAPS) plus protease inhibitors (Sigel et al 1983, DeLorey & Olsen 1992). The solubilized GABA, benzodiazepine, and TBPS binding activities copurified on benzodiazepine affinity columns, leading to a homogeneous preparation purified over 1000-fold (Sigel et al 1983, Stauber et al 1987). The purified receptor contained (apparently) two major polypeptides on SDS-PAGE (α at 51 kDa and β at 56 kDa); a hetero-oligomeric complex of about 220–300 kDa was proposed (Sigel et al 1983). The purified receptor was photoaffinity labeled on the 51-kDa α band with [^{3}H]flunitrazepam, whereas [^{3}H]muscimol photoaffinity labeled the 56-kDa β band (Deng et al 1986, Stauber et al 1987). Subsequently, microheterogeneity of both bands was observed by protein staining, photoaffinity labeling, and immunoblotting (Bureau & Olsen 1990, 1993; Fuchs et al 1990; Park & deBlas 1991). What originally appeared to be two polypeptide bands actually contained about a dozen polypeptides of similar size (Table 2).

Based on sequence similarity, five different GABAR subunit families have

Table 2[a]   GABAR subunit biochemistry

Subunits	α[b]		β[c]		γ[d]		δ[e]		ρ[f]	
Number of subtypes	6		4		3		1		2	
Number of splice variants	0		1		1		0		0	
Size range (kDa)										
AA Sequence	48–64		51		48		48		52	
SDS-PAGE[g]										
α1	51		β1	?	γ1	?	δ	54	ρ1	?
α2	53		β2	56	γ2	45			ρ2	?
α3	58		β3	58	γ3	?				
α4	?		β4	?						
α5	55									
α6	57									
% AA homology (intrafamily)	70–80		70–80		70–80		NA		70–80	
% AA homology (interfamily)	30–40		30–40		30–40		30–40		30–40	
Consensus sequence sites for phosphorylation	α4, α6: PKA PKC		β1–β4: PKA PKC		γ1, γ3: PTK γ2S/L: PTK PKC		?		ρ1, ρ2: PKC	

[a] The table was modified from one developed by Dr. Timothy P. Angelotti (personal communication).

[b] Data for the α subunits were obtained from Schofield et al (1987), Levitan et al (1988), Wisden et al (1991), Kato (1990), Luddens et al (1990), and Pritchett & Seeburg (1990).

[c] Data for the β subunits were obtained from Schofield et al (1987) and Ymer et al (1989).

[d] Data for the γ subunits were obtained from Pritchett et al (1989b), Ymer et al (1990), and Knoflach et al (1991).

[e] Data for the δ subunit was obtained from Shivers et al (1989).

[f] Data for the ρ subunits were obtained from Cutting et al (1991, 1992).

[g] SDS PAGE data for rat subunits were obtained from Stephenson et al (1989, 1990), Fuchs et al (1990), Benke et al (1991a,b), Buchstaller et al (1991), McKernan et al (1991), Olsen et al (1991a), Pollard et al (1991), DeLorey et al (1992), Wieland et al (1992), Endo & Olsen (1993), and Machu et al (1993).

been identified and have been named α, β, γ, δ, and ρ subunits. There is 30–40% sequence identity among the subunit families. About 20–30% sequence homology exists among all GABAR subunit candidates and other gene products of the superfamily (Schofield et al 1987, Olsen & Tobin 1990, DeLorey & Olsen 1992). Most of the subunit families have multiple subtypes[1] (α1–6, β1–4, γ1–3, δ, and ρ1–2), and all of the sequences within each subunit family are homologous, with about 70–80% amino acid sequence identity (Table 2). Additional diversity arises from RNA splice variants, described so far for γ2 (Whiting et al 1990, Kofuji et al 1991) and β4 (Bateson et al 1991, Lasham et al 1991) subtypes. Each GABAR subunit cDNA encodes for a

[1] We have used the term subunit to refer to families of GABAR proteins such as an α subunit, the term subtype to refer to members of a subunit "family" such as an α1 subtype, and the term isoform to refer to GABARs composed of different combinations of subunit subtypes such as an α1β1γ2 GABAR isoform.

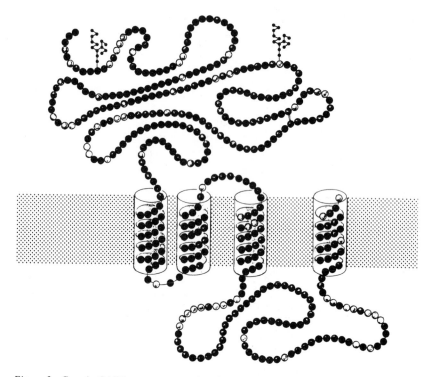

Figure 2 Generic GABA_A receptor protein subunit sequence and topological structure. The N-terminal half of the polypeptide is suggested to be extracellular, with probable sites for asparagine-glycosylation shown by polymers attached at positions 10 and 110 (*black circles*) and the conserved cystine-bridge indicated at positions 138–152. Four putative membrane-spanning domains are shown as α-helical cylinders within the cell membrane (*stippled*), with the C-terminal at the extracellular end of the fourth membrane-spanning region. Between the third and fourth membrane-spanning regions is a large putative intracellular loop. Modified from Olsen & Tobin (1990), adapted from Schofield et al (1987).

polypeptide of about 50 kDa, with putative N-glycosylation sites, and four α-helical hydrophobic membrane-spanning regions (Figure 2; Schofield et al 1987, Olsen & Tobin 1990). Between the third and fourth membrane-spanning regions is a hydrophilic putative cytoplasmic region of highly variable sequence involved in intracellular regulatory mechanisms such as phosphorylation.

DIVERSITY OF THE GABAR-ION CHANNEL COMPLEX

The current understanding of the molecular structure of the GABAR-ion channel complex is that it is a heteropentameric glycoprotein of about 275 kDa,

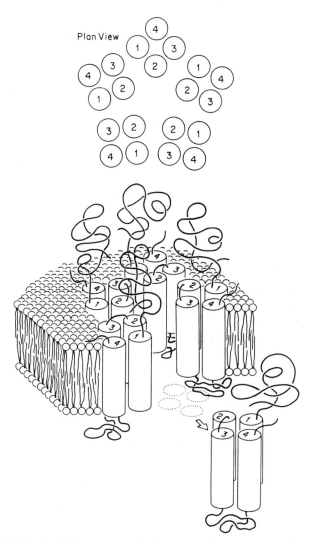

Figure 3   Model of the GABA_A receptor-chloride ion channel protein complex. The ligand-gated ion channel is proposed to be a heterooligomer composed of five subunits of the type shown in Figure 2, including some of the α, β, γ, δ, and ρ polypeptides. The exact subunit composition/ stoichiometry is not known. Each subunit has four putative membrane-spanning domains numbered 1–4 (*cylinders*), one or more of which contribute to the wall of the ion channel. The structure is patterned closely after the well-characterized nACHR, another member of the same gene superfamily (Schofield et al 1987, DeLorey & Olsen 1992); modified from Olsen & Tobin (1990), with permission.

composed of combinations of multiple polypeptide subunits. The subunits form a quasisymmetric structure around the ion channel, each subunit contributing to the wall of the channel (Figure 3; Olsen & Tobin 1990). The model is based heavily on analogy with the nicotinic acetylcholine receptor (nACHR), another member of the ligand-gated ion channel gene superfamily. The existence of 15 or so subunit subtype candidates leads to several questions about the molecular structure of GABAR isoforms: (*a*) How many oligomeric GABAR isoforms exist in nature? (*b*) What is the subunit subtype composition of each GABAR isoform? (*c*) For what purpose do they exist? (*d*) What aspects of pharmacological and physiological heterogeneity can be related to these structural GABAR isoforms? It is likely that different combinations of subunit subtypes are present in different neuronal populations. Localization of the GABAR gene products, including mRNA by in situ hybridization and polypeptides by immunocytochemistry, provides information about where each is found and the possible overlap or coexistence of more than one product in the same cell. The distribution of mRNAs in the central nervous system determined by in situ hybridization is very different for each subunit subtype. For example, the ρ subunit is expressed only in retina (Cutting et al 1991), whereas various α, β, and γ subtypes and the one δ subunit show very different regional as well as developmental distributions (e.g. Shivers et al 1989, Zhang et al 1991, MacLennan et al 1991, Richards et al 1991, Poulter et al 1992, Wisden et al 1992). Partial colocalization suggests some tentative major oligomeric assemblies; for instance, Wisden et al (1992) propose the likely existence of at least five combinations: $\alpha 1 \beta 2 \gamma 2$, $\alpha 2 \beta 3 \gamma x$, $\alpha 5 \beta 1 \gamma x$, $\alpha 1 \alpha 4 \beta 2 \delta$, and $\alpha 1 \alpha 6 \beta 2 \delta$. They also note that the $\gamma 1$ subtype is limited to the limbic regions of amygdala, hypothalamus, and septum, and the $\alpha 2$ and $\alpha 3$ subtypes also have a limited distribution. A given α, β, or γ subtype may well combine with multiple different subunit subtypes, thus creating numerous GABAR isoforms.

An alternative approach to determining the subunit subtype composition of GABARs has been to use GABAR subunit subtype-specific antibodies. Antibodies to synthetic peptides based on subunit subtype-specific clone sequences have been used to identify the subunit on Western blots, to localize in brain sections the presence of subunit subtypes, and to analyze subunit composition by immunoprecipitation and immunoaffinity chromatography. Antibodies directed against sequences of the original α ($\alpha 1$) and β ($\beta 1$) subunits were found to stain several polypeptides by cross-reaction in purified receptor preparations, now believed to represent other subtypes of these subunits (Kirkness & Turner 1988, Fuchs et al 1990). Gene products were identified by Western blotting as $\alpha 1$ (51 kDa: Duggan & Stephenson 1990, Stephenson et al 1990, Olsen et al 1991a), $\alpha 2$ (53 kDa: Fuchs et al 1990, Stephenson et al 1990, Olsen et al 1991a), $\alpha 3$ (58 kDa: Stephenson et al 1989, Fuchs et al 1990, Olsen et al 1991a, Endo & Olsen 1993), $\alpha 5$ [55 kDa in rat: McKernan et al 1991; or 57 kDa in cow (termed $\alpha4$): Endo & Olsen 1993]; $\beta2$ (57 kDa in cow, 56 kDa in rat: DeLorey et al 1992, Machu et al 1993); $\beta3$ (54–56 kDa in cow: DeLorey et al 1992; 57–59 kDa in rat: Pollard et al 1991, Buchstaller et al 1991, DeLorey et al 1992); $\gamma2$ (43–47 kDa: Stephenson et al 1990, Benke et al 1991a, Khan et al 1993); δ (54 kDa: Benke et al 1991b). The $\alpha6$ subtype is believed to represent the 57-kDa polypeptide band photolabeled in cerebellum by [^{3}H]Ro 15-4513 (Wieland et al 1992). Microsequencing (e.g. Olsen et al 1991a) has confirmed the identification of some stained bands in the purified receptor preparations. Table 2 summarizes the subunit subtypes and the relative positions on SDS-PAGE of those subunit subtypes that have been identified.

Antibodies specific for α subtypes were shown to immunoprecipitate variable amounts of receptor, depending on the subunit subtype and brain region. Oligomers pelleted by subunit subtype-specific antibodies contained primarily a single α subunit (Duggan & Stephenson 1990), although some receptors apparently contained multiple subtypes (Duggan et al 1991, McKernan et al 1991, Endo & Olsen 1993, Pollard et al 1993). Antibodies recognizing all β subtypes can immunoprecipitate virtually all benzodiazepine affinity-purified receptors from bovine cortex (Endo & Olsen 1992), whereas antibodies to the $\beta3$ subtype immunoprecipitate about 30% (Pollard et al 1991) to 50% (Huh et al 1992). Antibodies to the $\gamma2$ subtype immunoprecipitate 40% (Benke et al 1991a) to 75% (Stephenson et al 1990) of purified receptors, and antibodies to the δ subunit precipitate about 20–30% (Benke et al 1991b, Mertens et al 1993). Early studies on immunohistochemical localization of subunit subtypes confirmed the complex distribution pattern in the CNS. For example, Fritschy et al (1992), using four subunit subtype-selective antisera, concluded that there exist at least five oligomeric receptor isoforms to explain the differential localization of $\alpha1$, $\alpha3$, $\beta2/3$, and $\gamma2$ immunoreactivities, not to mention the other nine less abundant clones. Thus one or two dozen oligomeric GABAR isoforms tentatively have been identified in the CNS. Some GABAR isoforms may coexist in a single neuronal cell type, making even more difficult the determination of a given receptor isoform. Further, GABARs appear to exist in glial cells, and heterogeneity of receptor isoforms occurs for Bergmann glia, oligodendrocytes, and astrocytes (e.g. MacVicar et al 1989). The exact GABAR subunit subtype composition and stoichiometries need to be determined, as well as the localization, physiological functions, and pharmacological profiles of these receptor isoforms.

Identification of multiple GABAR subunit subtype mRNAs in specific cell types or of receptors immunoprecipitated from localized brain regions is necessary but not sufficient for identification of functionally expressed GABARs. Determination of the subunit subtype composition of specific GABARs in cells containing multiple GABAR subunit subtype mRNAs requires answering several questions: (*a*) In what cell type is the mRNA expressed? (*b*) Do all combinations of subunit subtypes translated from multiple GABAR mRNAs assemble into GABARs? (*c*) Do all GABARs assembled become inserted into the cell membrane? (*d*) Are all inserted GABARs functional, i.e. do they pass chloride current following application of GABA? Assuming that the GABAR subunits assemble randomly in a pentamer, Burt & Kamatchi (1991) determined that one, two, or three GABAR subunits expressed in a cell could assemble into one, eight, or fifty-one unique configurations, respectively. It is unclear how many of these different pentamers actually assemble to produce functional receptors and if they have similar or different pharmacological or biophysical properties. Alternatively, a single subunit subtype configuration may exist as an exclusive or preferred form. The structure of the muscle nACHR has been extensively studied and may be used to predict the likely GABAR structure. The nACHR consists of a pentamer composed of α, β, γ or ϵ, and δ subunits, which could assemble randomly into 208 different configurations. However, it forms two almost exclusive pentameric structures, $\alpha2\beta\gamma\delta$ or $\alpha2\beta\epsilon\delta$ with a fixed arrangement within the pentamer (Toyoshima & Unwin 1990). Furthermore, there is ordered assembly of the nACHR, which proceeds from the formation of $\alpha\gamma$ and $\alpha\delta$ subunits to the assembly of the mature receptor (Blount et al 1990, Paulson et al 1991, Saedi et al 1991). It is uncertain if assembly of the GABAR is random, as suggested by some expression studies, or is ordered and a unique form(s) of the receptor is expressed.

Although most of the above questions remain unanswered, heterologous expression of GABAR subunits in *Xenopus* oocytes and mammalian cells has been utilized to determine the specific properties of GABARs assembled from different subunit subtypes (Levitan et al 1988, Sigel et al 1990, Verdoorn et al 1990, Smart et al 1991, Knoflach et al 1991, Angelotti et al 1993a). Different functional GABAR isoforms can be produced from single, double, or triple GABAR subunit combinations, depending upon the cell type used for expression (Schofield et al 1987, Blair et al 1988, Shivers et al 1989, Sigel et al 1990, Verdoorn et al 1990, Angelotti et al 1993a). Expression of α, β, and γ GABAR subunits in *Xenopus* oocytes has suggested that receptors form from monomeric, dimeric, and trimeric combinations, although with different levels of expression (Schofield et al 1987, Blair et al 1988, Shivers et al 1989, Sigel et al 1990). In contrast, expression of $\alpha1$, $\beta2$, and $\gamma2S$ subtypes in HEK 293 fibroblasts has demonstrated that functional GABARs were produced only with $\alpha1\beta2$, $\alpha1\gamma2$, and $\alpha1\beta2\gamma2S$ subtype combinations (Verdoorn et al 1990). The $\alpha1\beta2$ GABARs were potently blocked by Zn^{2+}, but $\alpha1\beta2\gamma2S$ GABARs were insensitive to Zn^{2+}. However, Zn^{2+} sensitive receptors were rarely formed when $\alpha1$, $\beta2$, and $\gamma2S$ subtypes were coexpressed, suggesting that only Zn^{2+} insensitive $\alpha1\beta2\gamma2S$ GABARs were expressed (Draguhn et al 1990). The possibility that multiple distinct Zn^{2+} insensitive GABARs were formed, however, could not be excluded. Recently, the issue of functional GABAR subunit assembly has been studied in L929 fibroblasts (Angelotti et al 1993a, Angelotti & Macdonald 1993). In these cells, only $\alpha1\beta1$ and $\alpha1\beta1\gamma2S$ GABARs were formed when different combinations of GABAR subunit subtype cDNAs were transfected into the L929 cells (Angelotti et al 1993a). Furthermore, single-channel recordings obtained from cells transfected with $\alpha1$, $\beta1$, and $\gamma2S$ cDNAs revealed that $\alpha1\beta1\gamma2S$ GABAR channels were expressed, $\alpha1\beta1$ GABARs were not expressed, and the $\alpha1\beta1\gamma2S$ GABAR channels expressed had uniform biophysical properties (Angelotti & Macdonald 1993). These heterologous cell expression studies need to be evaluated carefully, however, since the subunit combinations tested may never occur in nature. Furthermore, the injected/transfected cells may be capable of synthesizing endogenous GABAR subunits, e.g. $\beta3$ mRNA is detectable in the commonly used HEK293 cells (Kirkness & Fraser 1993). Apparent expression of homooligomeric channels from other subunits in these cells, therefore, actually might reflect more complex structures. Taken together, these data suggest that there is ordered assembly of GABAR subunits into a preferred form of functional GABAR channel, that not all subunits can combine randomly to produce functional GABARs in the same cell, and that not all cells assemble the same subunit subtype combinations into functional receptors. Hadingham et al (1992) have recently described the stable expression of $\alpha1\beta1\gamma2L$ triple subtype combinations in mouse L cells by using a steroid-inducible promoter, which should lead to further information on complex subunit subtype combinations.

By comparing the single-channel properties of GABARs produced with coexpression of two or more GABAR subunits in a cell, it is possible, therefore, to determine how many variant configurations of the receptor could be assembled and if there is a preferred final form of the GABAR. Thus biochemical, immunochemical, binding, and functional analysis of naturally occurring receptors should be combined with heterologous cell expression to determine which GABAR subunit subtype combinations occur in nature, whether or not they are functional, and what their pharmacological and biophysical properties are.

PHARMACOLOGICAL PROPERTIES OF GABAR SUBUNIT SUBTYPES

Role of GABAR α Subunit Subtypes

Expression of various combinations of recombinant subunit subtypes in oocytes or cultured mammalian cell lines can suggest the GABAR subunit subtype combinations likely responsible for a given pharmacological property. For example, benzodiazepine sensitivity is conferred by the presence of the γ2 subtype (Pritchett et al 1989b). Varying the α subtype in combination with constant β and γ subtypes suggested differences in benzodiazepine pharmacology, GABA-benzodiazepine interaction, and steroid modulation of GABA responses, but not barbiturate, picrotoxin, or bicuculline sensitivity. Previous photoaffinity and autoradiography binding studies had defined apparent benzodiazepine receptor subtypes differing in affinity for certain ligands. For example, binding sites with high or low affinity for certain ligands such as CL 218,872 were designated the BZ1 and BZ2 types, respectively (Sieghart & Drexler 1983). The α1 subtype expressed with β and γ subunits gives GABARs with a benzodiazepine binding site that has BZ1 receptor specificity, namely, high affinity for CL 218,872 (Pritchett et al 1989a). Immunoprecipitation of receptor binding activity with antibodies specific for α1 subtype sequences showed that the α1-containing oligomers in the pellet were enriched in BZ1 type binding whereas the supernatant was enriched in BZ2 type binding (McKernan et al 1991). Recombinant α2, α3, and α5 (previously also called α4; Khrestchatisky et al 1989) subtypes all showed lower affinity for BZ1-selective ligands and thus may produce subvariants of BZ2 receptors, obviously a very heterogeneous group. For example, the BZ1-selective imidazopyridine hypnotic drug, zolpidem, has a relatively higher affinity for recombinant GABARs containing α1 subtypes but varies in affinity for GABARs containing other α subtypes. Affinity for zolpidem binding ranges from slightly lower to GABARs containing the α2 subtype to no binding to GABARs containing the α5 subtype (Pritchett et al 1989a, Puia et al 1991). In the brain, however, receptor oligomers containing the α5 subtype immunoprecipitated with α subtype-selective antisera bound zolpidem with an affinity that varied with brain region (Mertens et al 1993). This suggests that subunit subtypes other than the α5 subtype, possibly including other α subtypes, contribute to the benzodiazepine (zolpidem) binding specificity, and that several receptor isoforms occur. The nature of the α subunit appears to affect the efficacy as well as the affinity of benzodiazepine ligands (Sigel et al 1990, Verdoorn et al 1990, von Blankenfeld 1990, Puia et al 1991). Likewise, the α6 subtype, which is only expressed in cerebellar granular cells, appears to be responsible for a novel GABAR isoform (Wieland et al 1992) that binds the partial inverse agonist ligand [3H]Ro 15-4513 but not the classical benzodiazepines such as diazepam. The α4 subtype may also be a constituent of GABARs that lacks benzodiazepine binding activity (Wieland et al 1992).

Site-directed mutagenesis has confirmed the α sequence specificity for benzodiazepine ligand binding. Comparisons of sequences for α1, α2, and a3 subtypes revealed differences in the N-terminal putative extracellular domain. The synthesis of chimeric cDNAs with mixed domains of α1 and α3 subtypes led to the identification of a single residue (α1 subtype amino acid 201) that determined the affinity of GABARs for BZ1 and BZ2 agonists. In the α1 subtype that confers BZ1-binding, amino acid 201 was Gly, whereas in the α2, α3, or α5 subtypes that confer BZ2-binding, the amino acid in homologous positions was Glu (Pritchett & Seeburg 1991). Similarly, comparisons of α6 and α1 subtype sequences identified a single residue that confers the ability to bind [^{3}H]flunitrazepam: whereas the α6 subtype expressed with β and γ subunits produces receptors that bind Ro 15-4513 in a diazepam insensitive manner and do not bind flunitrazepam, mutation of the α6 subtype residue Arg-100 to His-100 as found in α1, α2, α3, and α5 subtypes (His-102 in α1) "restored" the flunitrazepam binding (Wieland et al 1992). The same residue was found to vary in allelic fashion in two strains of rats differing in sensitivity to the pharmacological actions of alcohol: Arg 100 gives the benzodazepine insensitive binding (normal α6 subtype), whereas Gln in that position produces diazepam sensitive [^{3}H]Ro 15-4513 binding and diazepam enhancement of GABA function. Nevertheless, neither of the two α6 subtype alleles, when expressed with any β and the γ2L subtypes, produce GABARs whose function is enhanced by ethanol (Korpi et al 1993).

Microsequencing of photoaffinity labeled receptor has localized residues involved in ligand binding, primarily the benzodiazepine site. Stephenson & Duggan (1989) showed that [^{3}H]flunitrazepam-photolabeled receptor produced CNBr fragments that could be identified on SDS-PAGE by radiolabel and immunoblotting with antibodies to the N- and C-terminus, thus tentatively narrowing down the possible site of attachment of the photolabel to residues 59–148 of the α1 subtype. However, Smith & Olsen (1992) found that proteolytic fragments (*S. aureus* V8 digest) of photolabeled preparations could be purified to yield partial amino acid sequences starting with residue 8 (31-kDa fragment), and a 17-kDa fragment starting at 298 which was not labeled, limiting the label to residues 8–297 (Olsen et al 1991a). Trypsin digestion of photolabeled gel-purified subunit subtypes and purification of labeled fragments produced microsequence starting with residue 223 of the bovine α1 subtype, with label present at aromatic residues Phe-226 and Tyr-231, at the beginning of the first membrane-spanning region (Smith & Olsen 1992).

The nature of the α subunit(s) can also affect the GABA site properties. The affinity for GABA activation of chloride channels in recombinant α1β2γ2S GABARs was shown to be altered by mutation of Phe-64 of the α1 (or α5) but not β2 or γ2S subtypes, indicating a role for this α subunit residue in GABA binding (Sigel et al 1992). This is consistent with the observation that Phe-64, present in all subunits (except ρ; Cutting et al 1991), is covalently modified in several α subtypes by affinity labeling with [³H]muscimol and microsequencing of chymotryptic fragments (G. B. Smith & R. W. Olsen, manuscript in preparation).

Role of GABAR β and γ Subunit Subtypes

The complexity of GABAR pharmacology is increased further by the heterogeneity of the β and γ subunits, the nature of which can affect pharmacological specificity. The nature of the β subunit affects channel properties (Verdoorn et al 1990) and benzodiazepine efficacy (Sigel et al 1990, von Blankenfeld et al 1990). Affinity-purified receptors from rat brain show variable affinities of different β polypeptides photoaffinity labeled with [³H]muscimol for GABA analogues and variable efficacies for allosteric modulation by barbiturates and steroids (Bureau & Olsen 1990, 1993). The β2 subtype showed a higher affinity for the GABA analogue taurine, lower affinity for THIP, and more enhancement by barbiturates and steroids than the β3 subtype (Huh et al 1992, Bureau & Olsen 1993). This is consistent with heterogeneity for these same ligands across brain regions as demonstrated by radioligand binding autoradiography (Bureau & Olsen 1993), presumably a reflection of regional heterogeneity in the functional properties of GABARs caused by receptor isoforms with differing pharmacological properties. The nature of the γ subunit affects the benzodiazepine sites, that is, the γ3 subtype replaces the γ2 subtype in producing benzodiazepine sensitivity of recombinant receptors, whereas the γ1 subtype leads to lower affinities and/or efficacies (von Blankenfeld et al 1990, Knoflach et al 1991, Puia et al 1991).

BIOPHYSICAL PROPERTIES OF GABAR CHANNELS

GABAR Conductance Levels

Development of the single-channel recording technique permitted direct study of GABAR channels on neurons (Hamill et al 1983, Bormann et al 1987, Macdonald et al 1989a). When GABA is applied to outside-out patches obtained from mouse spinal cord neurons in cell culture, the GABAR channel opens and closes rapidly so that relatively square current pulses are recorded (Figure 4). The GABAR channel opens to multiple conductance levels. A 27–30 pS conductance level is the predominant or main-conductance level (Figure 4C, *double asterisks*) and conductance levels of 17–19 pS (Figure 4C, *single asterisk*) and 11–12 pS occur less frequently. Although the GABAR channel opens to multiple conductance levels, the main-conductance level is responsible for more than 95% of the current through the channel. Based on

GABA 2μM

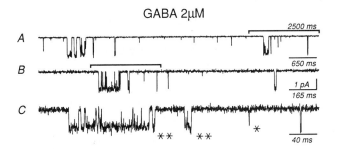

Figure 4   Single-channel currents are shown at increasing time resolution (*A–C*) for GABA (2 μM). GABA evoked bursting single-channel inward (down-going) currents with at least two current amplitudes when outside-out patches were voltage-clamped at −75 mV. The larger 2.04-pA (27-pS) channel (*double asterisk*) occurred more frequently compared with a smaller 1.48-pA (20-pS) channel (*single asterisk*). The portion outlined for each tracing is presented expanded in time in the tracing beneath it. Attenuated brief channel openings seen at lower temporal resolution may be seen at true amplitudes at higher temporal resolution. The time calibration for each trace is shown on the right, below the trace. Current calibration applies throughout. Modified from Twyman et al (1989b).

the permeability sequence for large polyatomic anions, the main-conductance level of the GABAR has an effective pore diameter of 5.6 nm (Bormann et al 1987). The effective pore diameters of the other conductance levels have not been determined.

The basis for the multiple conductance levels remains unknown, but the multiple levels may reflect the configuration or combination of different receptor subunits or the distribution of charges within the ion channel. For example, the main-conductance level of the GABAR has been shown to vary with subunit subtype composition. In a Chinese hamster ovary (CHO) cell line stably transfected and in mammalian fibroblast L929 cell acutely transfected with cDNAs for α1 and β1 subtypes of the receptor, single-channel recordings demonstrated that the main-conductance level of α1β1 GABAR channels was small (15–18 pS) (Figure 5, *top*) (Moss et al 1990, Angelotti & Macdonald 1993, Porter et al 1993). In contrast, acute transfection of L929 cells with α1β1γ2S cDNAs produced GABAR channels with a main-conductance level of 29 pS (Figure 5, *middle*), similar to the main-conductance level found in neuronal GABAR channels (Angelotti & Macdonald 1993), and transfection of L929 cells with α6β1γ2 cDNAs produced GABAR channels with a larger conductance level of 33 pS (Angelotti et al 1992). Similarly, in human embryonic kidney (HEK) 293 cells acutely transfected with different GABAR subunit subtype cDNAs, α1β2 receptors have a main-conductance

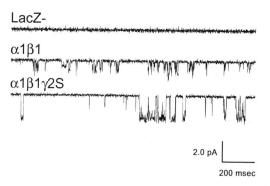

LacZ-

α1β1

α1β1γ2S

2.0 pA

200 msec

Figure 5 GABA (3 μm) was applied to excised outside-out patches held at a holding potential of −75 mV from (*top*) nontransfected LacZ⁻, (*middle*) α1β1 transfected LacZ⁺, and (*bottom*) α1β1γ2S transfected LacZ⁺ L929 fibroblast cells. The amplitude and time scale applies throughout. From Angelotti & Macdonald (1993).

level of 11 pS, whereas α1β2γ2 and α1γ2 receptors have a main-conductance level of 30 pS (Verdoorn et al 1990). Which combination of subunit subtypes is expressed in vivo remains uncertain.

GABAR Equilibrium Single-Channel Gating Properties

The equilibrium single-channel gating properties of the main-conductance level of the GABAR in murine spinal cord neurons in culture have been characterized (Sakmann et al 1983, Macdonald et al 1989a, Weiss & Magleby 1989, Twyman et al 1990, Twyman & Macdonald 1992). The GABAR opens into three different open states with mean durations of 0.5, 2.6, and 7.6 ms. The average open duration increases with increasing GABA concentration owing to a shift in the type of opening. At low GABA concentrations, primarily brief openings are recorded. At higher GABA concentrations, openings of the two longer open states are most frequent. The GABAR enters multiple closed states, including two brief closed states with mean durations of 0.2 and 1.4 ms and at least three closed states with longer mean durations. Bursts of openings are interrupted by brief closures (Figure 4). Bursts may be defined as openings or groups of openings separated by relatively long closed periods (Colquhoun & Sigworth 1983). Bursts are composed of repeated openings into the same open states, and the longer the mean duration of the open state, the more mean openings per burst occur. Thus, bursts of brief openings have only a few openings per burst whereas bursts of long openings have many openings per burst (Macdonald et al 1989a, Twyman et al 1990).

To explain the complex gating behavior described above, the single-channel

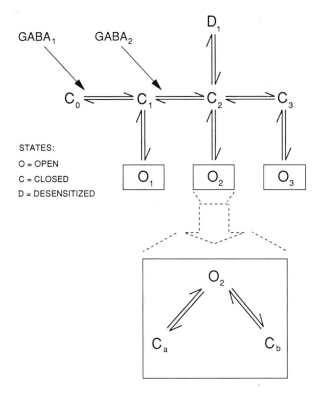

Figure 6 Microscopic reaction scheme for the GABAR main-conductance level shows binding sites for GABA. Two independent but kinetically similar closed states (C_a and C_b, enclosed in the box) are connected to each open state. See text for a discussion of the reaction scheme. From Macdonald & Twyman (1992).

activity of the main-conductance level at equilibrium has been modelled with a reaction scheme that incorporates two sequential GABA binding sites, three open states, ten closed states, and one desensitized state (Figure 6) (Macdonald et al 1989a, Twyman et al 1990, Twyman & Macdonald 1992).

The gating properties of recombinant GABAR single channels vary with subunit subtype combinations. When expressed in mammalian cells, paired combinations of GABAR subunit subtypes assemble with different efficiencies. In L929 fibroblasts, α1β1, but not α1γ2 or β1γ2, subtypes assemble to produce benzodiazepine insensitive GABAR channels (Angelotti et al 1993a). The α1β1 GABAR channels have only two open states (Angelotti & Macdonald 1993, Porter et al 1993). In contrast, α1β1γ2 GABAR channels have gating properties that are similar to those of neuronal GABAR channels

(Angelotti & Macdonald 1993), whereas $\alpha6\beta1\gamma2$ GABARs have different gating properties (Angelotti et al 1992).

Modification of Neuronal GABA$_A$ Receptor Channels by Drugs

A number of drugs have been shown to modify GABAR current (Olsen 1987, Macdonald & Twyman 1992). Barbiturates and benzodiazepines enhance GABAR current (Choi et al 1977; Macdonald & Barker 1978b, 1979), but through different allosteric regulatory sites on the GABAR (Study & Barker 1981; Twyman et al 1989a; Olsen 1981, 1987). The steroids, including anesthetic steroids and progesterone metabolites, also enhance GABAR current (Majewska et al 1986; Callachan et al 1987a,b). The drugs bicuculline, picrotoxin, methyl 6,7-dimethoxy-4-ethyl-β-carboline-3-carbox-ylate (DMCM), and penicillin all reduce GABA-mediated inhibition, but each drug interacts with a different site on the GABAR. To enhance GABAR current, a drug may increase channel conductance, increase channel open and burst frequencies, and/or increase channel open and burst durations. Conversely, to reduce GABAR current, a drug may decrease channel conductance, decrease channel open and burst frequencies, and/or decrease channel open and burst durations. By determining the drug-induced alter-ations produced in the open, closed, and burst properties of GABAR single-channel currents, the mechanism(s) of action of the drugs on GABAR channels can be determined.

Bicuculline

GABARs are selectively competitively antagonized by the plant convulsant bicuculline, and also in vitro by the quaternary nitrogen analogues of bicuculline. Other antagonists include pitrazepin, the amidine steroid RU5135, and a series of pyridazinyl-GABA derivatives typified by SR-95531 (Johnston 1991). Bicuculline reduces GABAR current by decreasing open frequency and mean duration (Macdonald et al 1989a). Although detailed kinetic studies have not been published, bicuculline probably produces a competitive antagonism of GABAR currents by competing with GABA for binding to the receptor (Figure 7). Whether bicuculline binds to one or both of the GABA binding sites remains uncertain.

Barbiturates

Sedative-hypnotic, anesthetic, and anticonvulsant barbiturates, as well as related nonbarbiturates such as etazolate, etomidate, and LY81067, enhance GABAR current measured by electrophysiological methods or by radiotracer ion flux. Barbiturates both enhance GABA responses and mimic GABA by opening the GABAR channel in the absence of GABA. The active barbiturates allosterically inhibit the in vitro binding of picrotoxin (Ticku et al 1978).

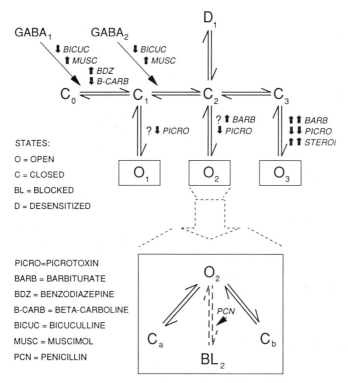

Figure 7 Microscopic reaction scheme for the GABAR main-conductance level shows binding sites for GABA and proposed sites of action of regulatory drugs. See text for a discussion of the reaction scheme. To enhance GABAR current, barbiturates (BARB) appear primarily to increase the opening transition rate into a long duration open state (O_3) relative to opening transition rates to short duration open states (O_1 and O_2), thereby prolonging the time spent in open states. The convulsant picrotoxin (PICRO) acts in a reciprocal fashion to the barbiturates. Benzodiazepines (BDZ) modify transition rates or the affinity of the first GABA binding site ($GABA_1$) to increase GABAR channel opening frequency and thus do not alter average open and burst durations. Convulsant β-carbolines (BCARB) reduce GABAR currents by a mechanism reciprocally related to anticonvulsant benzodiazepines. The convulsant penicillin (PCN) blocks GABA-evoked openings and introduces a new blocked state distal to each of the three open states (shown as BL_2 for O_2). From Macdonald & Twyman (1992).

Barbiturates likewise allosterically enhance the binding of GABA and benzodiazepine agonists, while inhibiting the binding of GABA antagonists and benzodiazepine inverse agonists (Olsen et al 1986). Barbiturates enhance the affinity of GABA binding to low and intermediate affinity sites (Yang & Olsen 1987). The relationship of this binding enhancement to channel activation is unclear. These interactions are dependent upon the presence of physiological concentrations of chloride or other anions that can permeate the

GABAR channel. A long list of compounds including stereoisomers shows excellent correlation between the structure-activity relationships for enhancement of GABAR function and the allosteric binding interactions (Olsen 1981, 1987; Ticku 1991). Good correlation with the in vivo pharmacological effects of barbiturates is also found, suggesting that the enhancement of GABAR-mediated inhibition is an excellent candidate mechanism for these drug effects.

Results from fluctuation analysis suggest that phenobarbital and pentobarbital increase the average open duration of GABAR single-channel currents without altering channel conductance (Barker & McBurney 1979, Study & Barker 1981). Single-channel recordings of barbiturate-enhanced single GABAR currents directly demonstrated that barbiturates increase average channel open duration but do not alter receptor conductance or opening frequency (Mathers & Barker 1981, Jackson et al 1982, Macdonald et al 1989b, Twyman et al 1989a). On the other hand, analysis of open duration frequency histograms in the presence of clinically relevant free-serum therapeutic concentrations of phenobarbital and pentobarbital revealed that the barbiturates do not alter the open duration time constants (Macdonald et al 1989b). Rather, they reduce the relative proportion of openings with short durations (states 1 and 2) and increase the proportion with the longer durations (state 3). Thus, the mean durations of each of the GABAR open states are unchanged in the presence of the barbiturates, but the average open duration of all openings of the channel is increased. These findings suggest that barbiturates alter the intrinsic gating of the channel once GABA is bound, so that the rate of opening of the receptor to the longest duration state 3 is increased relative to the rates of opening into states 1 and 2 (Figure 7). Whether the opening rates to open states 1 and 2 are also increased to a lesser extent, decreased, or remain unchanged is uncertain.

Picrotoxin

GABAR currents are inhibited by the plant convulsant picrotoxin (the active ingredient is picrotoxinin). This nonnitrogenous compound is not a structural analogue of GABA and exerts a noncompetitive inhibition via a site distinct from that for GABA itself (Olsen 1981). Drugs active on the picrotoxin site include the natural product tutin and the experimental synthetic polycyclic convulsants, such as pentylenetetrazol and the bicyclophosphate "cage" compounds such as TBPS. Certain benzodiazepines (e.g. Ro5-3663, Ro5-4864) are convulsant by virtue of their activity at this site rather than at the benzodiazepine site. All these compounds competitively inhibit radiolabeled picrotoxin or TBPS binding and block GABAR channel activation by GABA in the same manner as picrotoxin (Olsen 1981, 1987; Squires et al 1983; Ticku 1991; King et al 1987).

Picrotoxin binding either allosterically prevents the conformational change of the GABAR protein that opens the chloride channel upon GABA binding, and/or physically blocks the channel. The nature of the picrotoxin binding site has not been identified biochemically. GABAR channels produced from any combination of recombinant subunit subtypes are blocked by picrotoxin. Picrotoxin closes GABAR channels that are spontaneously open, as well as those opened by barbiturates, steroids, or avermectin (Olsen & Tobin 1990, Ticku 1991, DeLorey & Olsen 1992). Consistent with the hypothesis that the picrotoxin site is the channel itself are two observations: First, binding of the picrotoxin site ligand [^{35}S]TBPS cannot be detected in Triton X-100-solubilized receptor preparations that can bind radiolabeled GABA and benzodiazepine ligand, but picrotoxin binding is present in CHAPS-solubilized receptors. This detergent preserves allosteric interactions between the various sites on the GABAR complex, presumably requiring subunit-subunit interactions of the heterooligomeric protein. Second, the target size of the [^{35}S]TBPS binding activity determined by high energy irradiation inactivation is significantly larger than that of GABA and benzodiazepine binding: the latter can be interpreted as requiring a minimal-size protein molecule equal to a single subunit of about 50 kDa, thus the TBPS binding activity requires a multi-subunit structure (King et al 1987).

Single-channel recordings revealed that picrotoxin reduced GABA-evoked average open duration and burst duration (Twyman et al 1989b). These findings are not consistent with a fast open channel block mechanism and suggest that picrotoxin either (*a*) produces a slow open channel block (high affinity binding to the channel with slow off rate) or (*b*) reduces opening transition rates of the bound receptor (entry into the longer open states 2 and 3 appears to be reduced more than entry into the briefest open state 1). The kinetic analysis is more consistent with the latter alternative. These findings suggest that picrotoxin may alter the intrinsic gating of the channel once GABA is bound, so that the rate of opening of the receptor to the longest duration open state 3 is decreased relative to the rates of opening into open states 1 and 2 (Figure 7). Whether the opening rates to open states 1 and 2 are also decreased to a lesser extent or remain unchanged is uncertain.

Thus the barbiturates and picrotoxin both seem to act on the same process, the gating open of the GABAR channel, but their effects on opening rate constants appear to be opposite—barbiturates favor opening of long lasting open states whereas picrotoxin favors opening of brief duration open states.

Benzodiazepines and β-carbolines

Benzodiazepines, used clinically for antianxiety, antiepileptic, muscle relaxant, and hypnotic activity, enhance GABAR currents. Some analogues of benzodiazepines and related substances have antagonist (inhibiting benzo-diazepine agonists but not GABA) or inverse agonist (inhibiting GABAR function) efficacy. The binding of benzodiazepine agonists in vitro is allosterically enhanced by GABA agonists, whereas inverse agonists are inhibited and antagonists are unaffected (Richards et al 1991), consistent with coupling of the GABA and benzodiazepine receptor sites in the GABAR complex. Benzodiazepines do not enhance the high and intermediate affinity binding of GABA agonists (Olsen 1981, 1987; Olsen et al 1986); they may enhance the binding of the low affinity GABA sites thought to be more closely associated with channel activation (Skerritt et al 1982, Concas et al 1985), but this is difficult to detect.

Benzodiazepines increase GABAR current (Choi et al 1977, Macdonald & Barker 1978b). Results from fluctuation analysis suggest that the benzodiazep-ine diazepam increases GABAR current by increasing opening frequency without altering channel conductance or open duration (Study & Barker 1981). Single-channel recordings have confirmed that benzodiazepines increase receptor opening frequency without altering mean open time or conductance (Figure 8) (Vicini et al 1987; Rogers et al 1988, 1989, In press; Twyman et al 1989a). If benzodiazepine enhancement of the GABAR current is due purely to increased affinity of the receptor for GABA, the single-channel kinetic properties should change with increasing concentrations of benzodiazepine in a manner similar to that obtained with increased concentrations of GABA: channel open and burst frequencies and average channel open and burst durations would be expected to increase in the presence of a benzodiazepine. Analysis of single-channel kinetic properties did not support this expectation (Rogers et al 1988, In press). At clinically relevant concentrations of diazepam (< 100 nM), channel open and burst frequencies increase, but average open and burst durations are not altered. These results contrast with the increase in burst duration with little effect on burst frequency seen in the presence of phenobarbital (Twyman et al 1989a). For diazepam, these results could be explained by an increased affinity of the GABAR at one, but not both, of the GABA binding sites (Figure 7). More specifically, the increased open and burst frequencies with no change in open and burst durations could be explained by an increased association rate or a decreased dissociation rate only at the first binding site (Figure 7). Alteration of these rates for the second binding site would significantly alter the open and burst durations. Another explanation is that benzodiazepines could reduce the rate of entry into a desensitized state without altering the gating of the bound GABAR channel.

Reduction of GABAR currents by an inverse agonist for the benzodiazepine receptor is produced by a mechanism opposite to the action of benzodiazepine receptor agonists. The inverse agonist β-carboline, DMCM, does not alter GABAR conductance or average open and burst durations (Rogers et al 1989,

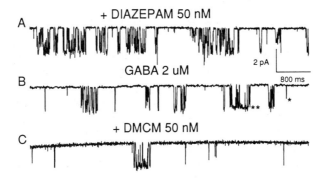

Figure 8   (*A*) GABA (2 μm) with DZ (50 nm) resulted in increased opening frequency. Time and current calibration bars are applicable to all traces shown. (*B*) GABA opened chloride channels, resulting in single and bursting inward currents. (*C*) GABA with DMCM (50 nm) resulted in decreased opening frequency. Data were obtained from different excised outside-out patches. From Rogers et al (In press).

In press) but does reduce open and burst frequencies (Figure 8). These results suggest that modulation of GABAR single-channel kinetics by DMCM could be explained by a reduction of the affinity of GABA binding at the first, but not second, GABA binding site (Figure 7). Again, an alternative interpretation is that β-carbolines increase the rate of entry into a desensitized state without altering the gating of the bound GABAR.

Steroids

The steroid female sex hormone progesterone has sedative activity, and synthetic analogues sharing this action led to the development of a steroid intravenous general anesthetic, alphaxalone. The active naturally-occurring steroids are primarily metabolites of hormones that lack activity on the hormone receptors. These compounds are found to enhance GABAR function and binding in a manner that resembles the barbiturates (Harrison & Simmonds 1984; Majewska et al 1986; Callachan et al 1987a,b; Barker et al 1987; Cottrell et al 1987). Synergistic or additive effects of steroids and barbiturates indicate that they have distinct sites of action on the GABAR complex (Figure 1) (Callachan et al 1987b, Gee et al 1988, Turner et al 1989, Morrow et al 1990). Direct GABAR activation by high concentrations of steroids was further modulated by low concentrations of barbiturate (Callachan et al 1987a). These steroids exert a direct nongenomic effect on the neuronal cell membrane. Metabolites of progesterone, corticosterone, and testosterone are candidates for physiological modulators of the GABAR system (Majewska et al 1986, Ticku 1991).

Structurally different steroids also potentiate GABAR responses (Mienville & Vicini 1989). Initial reports of modulation of single GABAR channel currents by steroids demonstrated that the conductance of the receptor is unaltered (Figure 9) (Callachan et al 1987a). Prolongation of average channel open time was inferred by fluctuation analysis, and marked prolongation of single-channel burst duration was reported (Barker et al 1987, Callachan et al 1987a).

Modification of the single-channel kinetic properties of GABARs by the steroids androsterone (5α-androstan-3α-ol-17-one) and pregnanolone (5β-pregnan-3α-ol-20-one) has been reported (Twyman & Macdonald 1992). Both steroids increase the average GABAR channel open and burst durations. The basis for the prolonged average GABAR channel open durations is an increase in opening frequency and an increase in the frequency of occurrence of the two longer open states 2 and 3 relative to the briefest open state 1. The increased average burst duration is steroid concentration-dependent and results from a shift in the proportion of bursts with shorter durations versus bursts with longer durations. The steroids do not alter intrinsic burst properties of the GABAR channel, but rather increase the likelihood of longer bursts, which are comprised primarily of longer duration openings. The steroid enhancement of GABAR current results from an increase in channel opening frequency and an increase in the probability of opening of longer openings, with no change in the durations of the open states of the GABAR channel. The mechanism for prolongation of average open and burst durations is similar to that of barbiturates, but the increase in channel opening frequency produced by the steroids has not been reported for barbiturates (Macdonald et al 1989b). Although it has been suggested that steroids and barbiturates bind to different sites (and their differential effect on opening frequency seems to corroborate

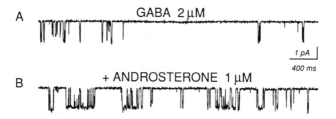

Figure 9   The neurosteriod anderosterone (100 μM) increases GABAR current by increasing frequency and mean open time of the GABA-evoked openings. From Macdonald & Twyman (1992).

this), the mechanism for the observed prolongation of the GABAR channel is similar to that described for barbiturates (Figure 7), suggesting that these steroids and barbiturates may regulate the GABAR channel through at least one common effector mechanism.

Penicillin

The convulsant penicillin G inhibits GABAR current (Raichle et al 1971); it inhibits picrotoxin binding, but only at concentrations higher than those reducing GABAR current. Its site and mechanism of action are biochemically unknown. GABA-evoked responses in the presence of penicillin are reduced in amplitude and prolonged in duration (Macdonald & Barker 1978a), and there is evidence for open-channel blockade of the GABAR channel by penicillin (Chow & Mathers 1986, Twyman et al 1991). Open-channel blockers enter open ion channels and physically block current flow, usually completely occluding it when the channel is blocked; when the channel is unblocked, the current flow is unaltered. Penicillin reduces average channel open duration and increases average burst duration without altering single-channel conductance (Figure 10). Single-channel kinetic analysis revealed that the reduction of open state duration and prolongation of burst duration are consistent with open channel block of the GABAR (Twyman et al 1991). In the GABAR kinetic scheme, penicillin introduces a distal blocked closed state for each of the three open states (Figure 7). Penicillin is a negatively charged molecule at physiological pH, and therefore it must interact intermittently with positively charged amino acids within the channel to occlude the flow of chloride ions through the channel.

Summary of Drug Modification of GABAR Channels

Virtually all steps in the binding and gating processes of the GABAR channel have been shown to be subject to modification by drugs (Figure 7). The

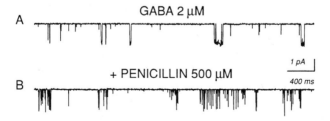

Figure 10   The convulsant drug penicillin (500 μM) decreases GABAR current by decreasing mean channel open time, increasing burst duration, and increasing the number of openings per burst. These alterations in single-channel currents suggest that penicillin produces a simple open channel block of the GABAR channel. From Macdonald & Twyman (1992).

binding steps appear to be regulated by benzodiazepines, β-carbolines, bicuculline, and possibly steroids. The gating process is apparently regulated by steroids, barbiturates, and picrotoxin. The open states of the channel can be occluded by penicillin. Whether or not the final state(s), the desensitization state(s), is regulated remains to be determined but seems likely. This form of regulation of a neurotransmitter receptor channel is likely to be similar to the regulation of another neurotransmitter receptor channel, the NMDA receptor channel.

Future work will concentrate on the physical bases for this modification by drugs. Although the specific subunits and subunit subtypes composing the GABAR are known, it is not clear what receptor isoform(s) is expressed in mouse spinal cord neurons. Very likely, however, these receptors are composed of α, β, and γ receptor subunits, since they have the full GABAR pharmacology. Study of receptors expressed in oocytes and eucaryotic cells and the use of site-directed mutagenesis will allow determination of the portions of the GABAR that are subject to binding and modification. The combination of single-channel kinetic techniques and molecular biological techniques should allow a detailed elucidation of the allosteric mechanisms for regulation of the GABAR channel.

MODIFICATION OF NEURONAL GABA$_A$ RECEPTOR CHANNELS BY PHOSPHORYLATION

GABAR channel function may be modified by treatment with agents that increase protein phosphorylation. Although the GABAR has not been demonstrated to be phosphorylated in vivo, several GABAR subunit subtypes contain consensus kinase substrate sequences in the cytoplasmic loop between the third and fourth transmembrane region for the cyclic AMP-dependent protein kinase (PKA), Ca^{2+}-phospholipid-dependent protein kinase C (PKC), and tyrosine protein kinase (Table 2). The γ2 subtype has an mRNA splice variant, γ2L, which contains an 8–amino-acid insert with a consensus substrate sequence for phosphorylation by PKC (Whiting et al 1990). This insert, and presumably receptor phosphorylation as well, may be necessary for modulation of GABAR function by ethanol (Wafford et al 1991). The GABAR protein can be phosphorylated in vitro by purified PKC and PKA, but not by calmodulin-dependent protein kinase (Browning et al 1990). Interestingly, the best substrates for both PKA and PKC are β subunits. The in vitro phosphorylation by both kinases could be completely prevented by preincubation with an antibody prepared against a synthetic peptide corresponding to the consensus substrate for PKA in the β subunits. Further, the major phosphorylation substrate for PKA in the receptor is the β3 subtype. Tryptic digestion and microsequencing revealed that the phosphorylated residue is in a fragment containing the cytoplasmic loop (Browning et al 1993).

Alteration of GABARs by PKA

Cyclic AMP-elevation decreases GABAR function in rat cortical synaptoneurosomes (Heuschneider & Schwartz 1989) and in cultured chick cortical neurons (Tehrani et al 1989). An effect of PKA-mediated phosphorylation was demonstrated by injection of the catalytic subunit of PKA (cPKA) into mouse spinal cord neurons (Porter et al 1990), which showed that GABAR currents are reduced by cPKA. The reduction in current is due to a decrease in opening frequency. cPKA also reduces GABAR Cl$^-$ flux into lysed and resealed rat synaptoneurosomes and increases phosphorylation of a polypeptide that could be immunoprecipitated with antibodies specific for GABARs (Leidenheimer et al 1991). Recombinant receptor channels expressed in HEK 293 cells with $\alpha1\beta1$ and $\alpha1\beta1\gamma2$ subtype combinations are also reduced by cAMP, and this functional modulation was prevented by site-directed mutagenesis of residue Ser-409 on the $\beta1$ subtype, indicating that acute phosphorylation of this residue was responsible for the reduction of GABAR current (Moss et al 1992b).

In contrast to the above studies, in cerebellar Purkinje cells application of 8-bromo cAMP increased GABAR current, mimicking the physiological effect of norepinephrine innervation, which is known to be mediated by β-adrenergic receptors coupled to PKA (Cheun & Yeh 1993). Furthermore, expression of $\alpha1\beta1\gamma2$ subtype combinations in mammalian fibroblast L929 cells stably transfected with cPKA cDNA resulted in enhanced GABAR currents when compared with expression of $\alpha1\beta1\gamma2$ subtype combinations in the parent L929 cell line (Angelotti et al, submitted). The enhancement in GABAR current was abolished by mutation of the PKA phosphorylation site on the β subunit. These experiments suggest that chronic PKA phosphorylation enhances assembly or decreases degradation of GABARs.

Alteration of GABARs by PKC

PKC phosphorylates recombinant $\beta1$ and $\gamma2S$ as well as $\gamma2L$ subtypes on intracellular serine residues, as shown by site-directed mutagenesis; no functional consequences were determined in this study (Moss et al 1992a). In oocytes injected with brain or GABAR subunit mRNA, phorbol esters reduced GABAR function via activation of PKC; site-directed mutagenesis of the Ser-410 residue in the $\beta2$ subtype and the Ser-327 in the $\gamma2S$ subtype demonstrated that phosphorylation of these residues in the GABAR by phorbol-stimulated kinase was responsible for the inhibition of function (Kellenberger et al 1992). The phorbol ester inhibition of GABAR function expressed in oocytes also could be shown in brain microsacs, where PKC activation appeared to inhibit selectively the fraction of GABAR flux that was not rapidly desensitized by prolonged (several-second) exposure to agonist (Leidenheimer et al 1992).

Alteration of GABARs by Unknown Kinases

GABAR current also appears to be "maintained" by phosphorylation (involving an unknown kinase and unknown substrate): the activity "runs down" in some cells when cytoplasmic contents are dialyzed during whole cell recording and can be maintained by addition of ATP-Mg^{2+} (Stelzer et al 1988). Evidence suggests that intracellular messengers, including Ca^{2+}, can modulate GABAR function, possibly via phosphorylation events.

Summary of Phosphorylation of GABAR Channels

The physiological significance and specific consequences of phosphorylation of the GABAR continue to be uncertain, but the initial studies suggest that phosphorylation of the GABAR may be of importance for both short-term and long-term regulation of GABAR function and expression.

ACKNOWLEDGMENTS

We wish to thank Drs. M. Bureau, T. DeLorey, S. Endo, K. Huh, G. Smith, R. Tyndale, R. Twyman, and T. Angelotti for their contributions to studies published by the authors; the National Institutes of Health for funding our research (NS28772 to RWO and NS19613 to RLM); and the Markey Foundation for funding (RLM).

Literature Cited

Angelotti TP, Macdonald RL. 1993. Assembly of GABA$_A$ receptor subunits: $\alpha1\beta2$ and $\alpha1\beta1\gamma2S$ subunits produce unique ion channels with dissimilar single-channel properties. J. Neurosci. 13:1429–40

Angelotti TP, Tan F, Macdonald RL. 1992. Molecular and electrophysiological characterization of an allelic variant of the rat $\alpha6$ GABA$_A$ receptor subunit. Mol. Brain Res. 16:173–78

Angelotti TP, Uhler MD, Macdonald RL. 1993a. Assembly of GABA$_A$ receptor subunits: Analysis of transient single cell expression utilizing a fluorescent substrate/marker gene combination. J. Neurosci. 13: 1418–28

Angelotti TP, Uhler MD, Macdonald RL. 1993b. Enhancement of recombinant $\alpha1\beta1\gamma2S$ GABA$_A$ receptor currents by chronic elevation of cAMP-dependent protein kinase is mediated by phosphorylation of the $\beta1$ subunit. Mol. Pharmacol. In press

Barker JL, Harrison NL, Lange GD, Owen DG. 1987. Potentiation of γ-aminobutyric-acid-activated chloride conductance by a steroid anesthetic in cultured rat spinal neurones. J. Physiol. (London) 386:485–501

Barker JL, McBurney RM. 1979. Phenobarbitone modulation of postsynaptic GABA receptor function on cultured neurons. Proc. R. Soc. London Ser. B 206:319–27

Barker JL, Owen DG. 1986. Electrophysiological pharmacology of GABA and diazepam in cultured CNS neurons. In Benzodiazepine/GABA Receptors and Chloride Channels: Structural and Functional Properties, ed. RW Olsen, JC Venter, pp. 135–65. New York: Alan R. Liss

Bateson AL, Lasham A, Darlison, MG. 1991. γ-Aminobutyric acid$_A$ receptor heterogeneity is increased by alternative splicing of a novel β-subunit gene transcript. *J. Neurochem.* 56:1437–42

Benke D, Mertens S, Trzeciak A, et al. 1991a. GABA$_A$ receptors display association of γ2-subunit with α1- and β2/β3-subunits. *J. Biol. Chem.* 266:4478–83

Benke D, Mertens S, Trzeciak A, et al. 1991b. Identification and immunohistochemical mapping of GABA$_A$ receptor subtypes containing the δ-subunit in rat brain. *FEBS Lett.* 283:145–49

Blair LA, Levitan ES, Marshall J, et al. 1988. Single subunits of GABA$_A$ receptor form ion channels with properties of the native receptor. *Science* 242:577–79

Blount P, Smith MM, Merlie JP. 1990. Assembly intermediates of the mouse muscle nicotinic acetylcholine receptor in stably transfected fibroblasts. *J. Cell Biol.* 111:2601–11

Bormann J, Hamill OP, Sakmann, B. 1987. Mechanism of anion permeation through channels gated by glycine and γ-aminobutyric acid in mouse cultured spinal neurones. *J. Physiol. (London)* 385:243–86

Browning MD, Bureau M, Dudek EM, Olsen RW. 1990. Protein kinase C and cAMP-dependent protein kinase phosphorylate the β-subunit of the purified GABA$_A$ receptor. *Proc. Natl. Acad. Sci. USA* 87:1315–18

Browning MD, Endo S, Smith G, et al. 1993. Phosphorylation of the GABA$_A$ receptor by cAMP-dependent protein kinase and protein kinase C: analysis of the substrate domain. *Neurochem. Res.* 18:95–100

Buchstaller A, Adamiker D, Fuchs K, Sieghart W. 1991. N-Deglycosylation and immunological identification indicates the existence of β-subunit isoforms of the rat GABA$_A$ receptor. *FEBS Lett.* 287:27–30

Bureau M, Olsen RW. 1990. Multiple distinct subunits of the γ-aminobutyric acid-A receptor protein show different ligand-binding affinities. *Mol. Pharmacol.* 37:497–502

Bureau MH, Olsen RW. 1993. GABA$_A$ receptor subtypes: ligand binding heterogeneity demonstrated by photoaffinity labeling and autoradiography. *J. Neurochem.* In press

Burt DR, Kamatchi GL. 1991. GABA$_A$ receptor subtypes—from pharmacology to molecular biology. *FASEB J.* 5:2916–23

Callachan H, Cottrell GA, Hather NY, et al. 1987a. Modulation of the GABAa receptor by progesterone metabolites. *Proc. R. Soc. London* B231:359–89

Callachan H, Lambert JJ, Peters JA. 1987b. Modulation of the GABA$_A$ receptor by barbiturates and steroids. *Neurosci. Lett. Suppl.* 29:S21

Cheun JE, Yeh HH. 1993. Modulation of GABA$_A$ receptor-activated current by nor-epinephrine in cerebellar Purkinje cells. *Neuroscience* 51:951–60

Choi DW, Farb DH, Fischbach GD. 1977. Chlordiazepoxide selectively augments GABA action in spinal cord cell cultures. *Nature* 269:342–44

Chow P, Mathers D. 1986. Convulsant doses of penicillin shorten the lifetime of GABA-induced channels in cultured central neurones. *Br. J. Pharmacol.* 88:541–47

Colquhoun D, Sigworth FJ. 1983. Fitting and statistical analysis of single-channel records. In *Single-Channel Recordings*, ed. B Sakmann, E Neher, pp. 191–263. New York: Plenum Press

Concas A, Seera M, Crisponi G, et al. 1985. Changes in the characteristics of low affinity GABA binding sites elicited by Ro 15-1788. *Life Sci.* 36:329–37

Cottrell GA, Lambert JJ, Perters JA. 1987. Modulation of GABA$_A$ receptor activity by alphaxalone. *Br. J. Pharmacol.* 98:491–500

Cutting GR, Curristin RS, Zoghbi H, et al. 1992. Identification of a putative gamma-aminobutyric acid (GABA) receptor subunit ρ2 cDNA and colcalization of the genes encoding ρ2 (GABRR2) and ρ1 (GABRRI) to human chromosome 6q14–q21 and mouse chromosome-4. *Genomics* 12:801–6

Cutting GR, Lu L, O'Hara BF, et al. 1991. Cloning of the gamma-aminobutyric acid (GABA) rho1 cDNA: a GABA receptor subunit highly expressed in the retina. *Proc. Natl. Acad. Sci. USA* 88:2673–77

DeLorey TM, Brown GB. 1992. γ-Aminobutyric acid$_A$ receptor pharmacology in rat cerebral cortical synaptoneurosomes. *J. Neurochem.* 58:2162–69

DeLorey TM, Endo S, Machu TK, et al. 1992. Identification of β2 and β3 subunits of the GABA$_A$ receptor from rat brain with subunit subtype-specific antibodies. *Soc. Neurosci. Abstr.* 18:263

DeLorey TM, Olsen RW. 1992. γ-Aminobutyric acid$_A$ receptor structure and function. *J. Biol. Chem.* 267:16747–50

Deng L, Ransom RW, Olsen RW. 1986. [^{3}H]Muscimol photolabels the γ-aminobutyric acid receptor binding site on a peptide subunit distinct from that labeled with benzodiazepines. *Biochem. Biophys. Res. Commun.* 138:1308–14

Draguhn A, Verdorn TA, Ewert M, Seeburg PH, Sakmann B. 1990. Function and molecular distinction between recombinant rat GABA$_A$ receptor subtypes by ZN^{2+}. *Neuron* 5:781–88

Duggan MJ, Pollard S, Stephenson FA. 1991. Immunoaffinity purification of GABA$_A$ receptor α-subunit iso-oligomers. Demonstration of receptor populations containing α1α2, α1α3 and α2α3 subunit pairs. *J. Biol. Chem.* 266:24778–84

598   MACDONALD & OLSEN

Duggan MJ, Stephenson FA. 1990. Biochemical evidence for the existence of γ-aminobutyrate$_A$ receptor iso-oligomers. *J. Biol. Chem.* 265:3831–35

Dunn SMJ, Martin CR, Agey MW, Miyazaki R. 1989. Functional reconstitution of the bovine brain GABA$_A$ receptor from solubilized components. *Biochemistry* 28:2545–51

Edgar PP, Schwartz RD. 1992. Functionally relevant γ-aminobutyric acid$_A$ receptors: equivalence between receptor affinity (Kd) and potency (EC50)? *Mol. Pharmacol.* 41:1124–29

Endo S, Olsen RW. 1992. Preparation of antibodies to beta subunits of GABA$_A$ receptors. *J. Neurochem.* 59:1444–51

Endo S, Olsen RW. 1993. Antibodies specific for α subunit subtypes of GABA$_A$ receptors reveal brain regional heterogeneity. *J. Neurochem.* 60:1388–98

Fritschy JM, Benke D, Mertens S, et al. 1992. Five subtypes of type A γ-aminobutyric acid receptors identifed in neurons by double and triple immunofluorescence staining with subunit-specific antibodies. *Proc. Natl. Acad. Sci. USA* 89:6726–30

Fuchs K, Adamiker D, Sieghart W. 1990. Identification of α2- and α3-subunits of the GABA$_A$-benzodiazepine receptor complex purified from the brains of young rats. *FEBS Lett.* 261:52–54

Gee KW, Bolger MB, Brinton RE, et al. 1988. Steroid modulation of the chloride iontophore in rat brain: structure-activity requirements, regional dependence and mechanism of action. *J. Pharmacol. Exp. Ther.* 246:803–12

Hadingham KL, Harkness PC, McKernan RM, et al. 1992. Stable expression of mammalian type A γ-aminobutyric acid receptors in mouse cells: Demonstration of functional assembly of benzodiazepine-responsive sites. *Proc. Natl. Acad. Sci. USA* 89:6378–82

Hamill OP, Bormann J, Sakmann B. 1983. Activation of multiple-conductance state chloride channels in spinal neurones by glycine and GABA. *Nature* 305:805–8

Harrison NL, Simmonds MA. 1984. Modulation of the GABA receptor complex by a steroid anaesthetic. *Brain Res.* 323:287–92

Heuschneider G, Schwartz RD. 1989. cAMP and forskolin decrease gamma-aminobutyric acid-gated chloride flux in rat brain synaptoneurosomes. *Proc. Natl. Acad. Sci. USA* 86:2938–42

Huh KH, DeLorey TM, Endo S, Olsen RW. 1992. Characterization of structural and functional subtypes of GABA$_A$ receptors with antibodies to the β3 subunit. *Soc. Neurosci. Abstr.* 18:263

Jackson MB, Lecar H, Mathers DA, Barker JL. 1982. Single channel currents activated by γ-aminobutyric acid, muscimol, and (−)pentobarbital in cultured mouse spinal neurons. *J. Neurosci.* 2:889–94

Johnston GAR. 1986. Multiplicity of GABA receptors. In *Benzodiazepines/GABA Receptors and Chloride Channels: Structural and Functional Properties Isoguvacine*, ed. RW Olsen, JC Venter, pp. 57–71. New York: Alan R. Liss

Johnston GAR. 1991. GABA$_A$ antagonists. *Semin. Neurosci.* 3:205–10

Kardos J, Cash DJ. 1990. Transmembrane ^{36}Cl$^-$ flux measurements and desensitization of the γ-aminobutyric acid$_A$ receptor. *J. Neurochem.* 55:1095–99

Kato K. 1990. Novel GABA$_A$ receptor a subunit is expressed only in cerebellar granule cells. *J. Mol. Biol.* 214:619–24

Kellenberger S, Malherbe P, Sigel E. 1992. Function of the α1β2γ2S γ-aminobutyric acid type A receptor is modulated by protein kinase C via multiple phosphorylation sites. *J. Biol. Chem.* 267:25660–63

Khan AU, Fernando LP, Escriba P, et al. 1993. Antibodies to the human γ2 subunit of the γ-aminobutyric acid$_A$/benzodiazepine receptor. *J. Neurochem.* 60:961–71

Khrestchatisky M, MacLennan AJ, Chiang MY, et al. 1989. A novel alpha-subunit in rat brain GABA$_A$ receptors. *Neuron* 3:745–53

King RG, Nielsen M, Stauber GB, Olsen RW. 1987. Convulsant/barbiturate activities on the soluble GABA/benzodiazepine receptor complex. *Eur. J. Biochem.* 169:555–62

Kirkness EF, Fraser CM. 1993. A strong promoter element is located between alternative exons of a gene encoding the human GABAR β3 subunit (GABRB3). *J. Biol. Chem.* 268:4420–28

Kirkness EF, Turner AJ. 1988. Antibodies directed against a nonapeptide sequence of the γ-aminobutyrate (GABA)/benzodiazepine receptor α-subunit. *Biochem. J.* 256:291–94

Knoflach F, Rhyner T, Villa M, et al. 1991. The γ3-subunit of the GABA$_A$-receptor confers sensitivity to benzodiazepine receptor ligands. *FEBS Lett.* 293:191–94

Kofuji P, Wang JB, Moss SJ, et al. 1991. Generation of two forms of the γ-aminobutyric acid$_A$ receptor γ2-subunit in mice by alternative splicing. *J. Neurochem.* 56:713–15

Korpi ER, Kleingoor C, Kettenmann H, Seeburg PH. 1993. Benzodiazepine-induced motor impairment linked to point mutation in cerebellar GABA$_A$ receptor. *Nature* 361:356–59

Lasham A, Vreugdenhil E, Bateson AN, et al. 1991. Conserved organization of γ-aminobutyric acid$_A$ receptor genes: cloning and analysis of the chicken β4-subunit gene. *J. Neurochem.* 57:352–55

Leidenheimer NJ, Machu TK, Endo S, et al. 1991. Cyclic AMP-dependent protein kinase decreases γ-aminobutyric acid$_A$ receptor-mediated ^{36}Cl$^-$ uptake by brain microsacs. *J. Neurochem.* 57:722–25

Leidenheimer NJ, McQuilken SJ, Hahner LD, et al. 1992. Activation of protein kinase C selectively inhibits the γ-aminobutyric acid$_A$ receptor: role of desensitization. *Mol. Pharmacol.* 41:1116–23

Levitan ES, Blair LAC, Dionne VE, Barnard EA. 1988. Biophysical and pharmacological properties of cloned GABA$_A$ receptor subunits expressed in *Xenopus* oocytes. *Neuron* 1:773–81

Luddens H, Pritchett DB, Kohler M, et al. 1990. Cerebellar GABA$_A$ receptor selective for a behavioral alcohol antagonist. *Nature* 346:648–51

Macdonald RL, Barker JL. 1978a. Specific antagonism of GABA-mediated postsynaptic inhibition in cultured mammalian neurons: a common mode of anticonvulsant action. *Neurology* 28:325–30

Macdonald RL, Barker JL. 1978b. Benzodiazepines specifically modulate GABA-mediated postsynaptic inhibition in cultured mammalian neurones. *Nature* 271: 563–64

Macdonald RL, Barker JL. 1979. Anticonvulsant and anesthetic barbiturates: different post-synaptic actions in cultured mammalian neurons. *Neurology* 29:432–47

Macdonald RL, Rogers CJ, Twyman RE. 1989a. Kinetic properties of the GABA$_A$ receptor main-conductance state of mouse spinal cord neurons in culture. *J. Physiol. (London)* 410:479–99

Macdonald RL, Rogers CJ, Twyman RE. 1989b. Barbiturate modulation of kinetic properties of GABA$_A$ receptor channels in mouse spinal neurons in culture. *J. Physiol. (London)* 417:483–500

Macdonald RL, Twyman RE. 1992. Kinetic properties and regulation of GABA$_A$ receptor channels. In *Ion Channels,* ed. T Narahashi, 4:315–43. New York: Plenum

Machu TK, Olsen RW, Browning MB. 1993. Immunochemical characterization of the β2 subunit of the GABA$_A$-receptor. *J. Neurochem.* In press

MacLennan AJ, Brecha N, Khrestchatisky M, et al. 1991. Independent cellular and ontogenetic expression of mRNAs encoding three α polypeptides of the rat GABA$_A$ receptor. *Neuroscience* 43:369–80

MacVicar BA, Tse FWY, Crackton SE, Kettenmann H. 1989. GABA activated Cl-channels in astrocytes of hippocampal slices. *J. Neurosci.* 9:3577–83

Majewska MD, Harrison NL, Schwartz RD, et al. 1986. Steroid hormone metabolites are barbiturate-like modulators of the GABA receptor. *Science* 232:1004–7

Mathers DA, Barker JL. 1981. GABA and muscimol open ion channels of different lifetimes on cultured mouse spinal cord cells. *Brain Res.* 204:242–47

McKernan RM, Quirk K, Prince R, et al. 1991. GABA$_A$ receptor subtypes immunopurified from rat brain with a subunit-specific antibodies have unique pharmacological properties. *Neuron* 7:667–76

Mertens S, Benke D, Mohler H. 1993. GABA$_A$ receptor populations with novel subunit combinations and drug binding profiles identified in brain by α5- and δ-subunit-specific immunopurification. *J. Biol. Chem.* 268:5965–73

Mienville JM, Vicini S. 1989. Pregnenolone sulfate antagonizes GABA$_A$ receptor-mediated currents via a reduction of channel opening frequency. *Brain Res.* 489:190–94

Möhler H, Battersby MK, Richards JG. 1980. Benzodiazepine receptor protein identified and visualized in brain tissue by a photoaffinity label. *Proc. Natl. Acad. Sci. USA* 77: 1666–70

Möhler H, Okada T. 1977. Benzodiazepine receptors: demonstration in the central nervous system. *Science* 198:849–51

Morrow AL, Pace JR, Purdy RH, Paul SM. 1990. Characterizations of steroid interactions with the GABA receptor-gated ion channel: evidence for multiple steroid recognition sites. *Mol. Pharmacol.* 37:263–70

Moss SJ, Doherty CA, Huganir RL. 1992a. Identification of the cAMP-dependent protein kinase and protein kinase C phosphorylation sites within the major intracellular domains of the β1, γ2S, and γ2L subunits of the γ-aminobutyric acid type A receptor. *J. Biol. Chem.* 267:14470–76

Moss SJ, Smart TA, Porter NM, et al. 1990. Cloned GABA receptors are maintained in a stable cell line: allosteric and channel properties. *Eur. J. Pharmacol.* 189:77–88

Moss SJ, Smart TG, Blackstone CD, Huganir RL. 1992b. Functional modulation of GABA$_A$ receptors by cAMP-dependent protein phosphorylation. *Science* 257:661–65

Olsen RW. 1981. GABA-benzodiazepine-barbiturate receptor interactions. *J. Neurochem.* 37:1–13

Olsen RW. 1987. The γ-aminobutyric acid/benzodiazepine/barbiturate receptor chloride ion channel complex of mammalian brain. In *Synaptic Function,* ed. GM Edelman, WE Gall, WM Cowan, pp. 257–71. Neuroscience Research Foundation/Wiley

Olsen RW, Bureau MH, Endo S, Smith G. 1991a. The GABA$_A$ receptor family in the mammalian brain. *Neurochem. Res.* 16:317–25

Olsen RW, Sapp DM, Bureau MH, et al. 1991b. Allosteric actions of CNS depressants including anesthetics on subtypes of the inhibitory GABA$_A$ receptor-chloride channel complex. In *Molecular and Cellular Mechanisms of Alcohol and Anesthetics*, ed. E Rubin, KW Miller, SH Roth, 625:145–54. New York: Ann. NY Acad. Sci.

Olsen RW, Tobin AJ. 1990. Molecular biology of GABA$_A$ receptors. *FASEB J.* 4:1469–80

Olsen RW, Yang J, King RG, et al. 1986. Barbiturate and benzodiazepine modulation of GABA receptor binding and function. *Life Sci.* 39:1969–76

Park D, deBlas AL. 1991. Peptide subunits of γ-aminobutyric acid$_A$/benzodiazepine receptors from bovine cerebral cortex. *J. Neurochem.* 56:1972–79

Paulson HL, Ross AF, Green WN, Claudio T. 1991. Analysis of early events in acetylcholine receptor assembly. *J. Cell Biol.* 113: 1371–84

Pollard S, Duggan MJ, Stephenson FA. 1991. Promiscuity of GABA$_A$ receptor β3 subunits as demonstrated by their presence in α1, α2 and α3 subunit-containing receptor subpopulations. *FEBS Lett.* 295:81–83

Pollard S, Duggan MJ, Stephenson FA. 1993. Further evidence for the existence of a subunit heterogeneity within discrete γ-aminobutyric acid$_A$ receptor subpopulations. *J. Biol. Chem.* 268:3753–57

Porter NM, Angelotti TP, Twyman RE, Macdonald RL. 1993. Kinetic properties of α1β1 γ-aminobutyric acid$_A$ receptor channels expressed in Chinese hamster ovary cells: regulation by pentobarbital and picrotoxin. *Mol. Pharmacol.* 42:872–81

Porter NM, Twyman RE, Uhler MD, Macdonald RL. 1990. Cyclic AMP-dependent protein kinase decreases GABA$_A$ receptor current in mouse spinal neurons. *Neuron* 5:789–96

Poulter MO, Barker JL, O'Carroll AM, et al. 1992. Differential and transient expression of GABA$_A$ receptor α-subunit mRNAs in the developing rat CNS. *J. Neurosci.* 12:2888–2900

Pritchett DB, Lüddens H, Seeburg P. 1989a. Type I and type II GABA$_A$-benzodiazepine receptor produced in transfected cells. *Science* 245:1389–92

Pritchett DB, Seeburg PH. 1990. Gamma-aminobutyric acid$_A$ receptor α5-subunit creates novel type II benxodiazepine receptor pharmacology. *J. Neurochem.* 54:1802–4

Pritchett DB, Seeburg PH. 1991. γ-Aminobutyric acid type A receptor point mutation increases the affinity of compounds for the benzodiazepine site. *Proc. Natl. Acad. Sci. USA* 88:1421–25

Pritchett DB, Sontheimer H, Shivers BD, et al. 1989b. Importance of a novel GABA$_A$ receptor subunit for benzodiazepine pharmacology. *Nature* 338:582–85

Puia G, Vicini S, Seeburg PH, Costa E. 1991. Influence of recombinant γ-aminobutyric acid$_A$ receptor subunit compositions on the action of allosteric modulators of γ-aminobutyric acid-gated Cl$^-$ currents. *Mol. Pharmacol.* 39:691–96

Raichle ME, Kult H, Louis S, McDowell F. 1971. Neurotoxicity of intravenously administered penicillin G. *Arch. Neurol. (Chicago)* 25:232–39

Richards JG, Schoch P, Haefely W. 1991. Benzodiazepine receptors: new vistas. *Semin. Neurosci.* 3:191–203

Rogers CJ, Twyman RE, Macdonald RL. 1988. Diazepam does not alter the gating kinetics of GABA receptor channels. *Soc. Neurosci. Abstr.* 14:642

Rogers CJ, Twyman RE, Macdonald RL. 1989. The benzodiazepine diazepam and the beta-carboline DM.CM modulate GABA$_A$ receptor currents by opposite mechanisms. *Soc. Neurosci. Abstr.* 15:1150

Rogers CJ, Twyman RE, Macdonald RL. 1993. Benzodiazepine and beta-carboline regulation of single GABA$_A$-receptor channels of mouse spinal neurons in culture. *J. Physiol. (London)* In press

Saedi MS, Conroy WG, Lindstrom J. 1991. Assembly of *Torpedo* acetylcholine receptors in *Xenopus* oocytes. *J. Cell Biol.* 112: 1007–15

Sakmann B, Hamill OP, Bormann J. 1983. Patch-clamp measurements of elementary chloride currents activated by the putative inhibitory transmitters GABA and glycine in mammalian spinal neurons. *J. Neural Transmission Suppl.* 18:83–95

Schofield PR, Darlison MG, Fujita N, et al. 1987. Sequence and functional expression of the GABA-A receptor shows a ligand-gated receptor superfamily. *Nature* 328:221–27

Shivers BD, Killisch I, Sprengel SR. 1989. Two novel GABA$_A$ receptor subunits exist in distinct neuronal subpopulations. *Neuron* 3:327–37

Sieghart W, Drexler G. 1983. Irreversible binding of [^{3}H]flunitrazepam to different proteins in various brain regions. *J. Neurochem.* 41:47–55

Sigel E, Barnard EA. 1984. A γ-aminobutyric acid/benzodiazepine receptor complex from bovine cerebral cortex. Improved purification with preservation of regulatory sites and their interactions. *J. Biol. Chem.* 259:7219–

Sigel E, Baur R, Kellenberger S, Malherbe P. 1992. Point mutations affecting antagonist and agonist dependent gating of GABA$_A$ receptor channels. *EMBO J.* 11:2017–23

Sigel E, Baur R, Trube G, et al. 1990. The effect of subunit composition of rat brain GABA$_A$ receptors on channel function. *Neuron* 5:703–11

Sigel E, Stephenson FA, Mamalaki C, Barnard EA. 1983. A γ-aminobutyric acid/benzodiazepine receptor complex of bovine cerebral cortex: Purification and partial characterization. *J. Biol. Chem.* 258:6965–71

Smart TG, Moss SJ, Xie X, Huganir RL. 1991. GABA$_A$ receptors are differentially sensitive to zinc-dependence on subunit comparison. *Brit. J. Pharmacol.* 103:1837–39

Skerritt JH, Willow M, Johnston GAR. 1982. Diazepam enhancement of low affinity GABA binding to rat brain membranes. *Neurosci. Lett.* 29:63–66

Smith GB, Olsen RW. 1992. Identification of a [³H]flunitrazepam photoaffinity substrate in bovine GABA-A receptor. *Soc. Neurosci. Abstr.* 18:264

Squires RF, Braestrup C. 1977. Benzodiazepine receptors in rat brain. *Nature* 266:732–34

Squires RF, Casida JE, Richardson M, Saederup E. 1983. [35S]t-Butyl bicyclophosphorothionate binds with high affinity to brain-specific sites coupled to γ-aminobutyric acid-A and ion recognition sites. *Mol. Pharmacol.* 23:326–36

Stauber GB, Ransom RW, Dilber AI, Olsen RW. 1987. The γ-aminobutyric acid-benzodiazepine receptor protein from rat brain: large-scale purification and preparation of antibodies. *Eur. J. Biochem.* 167: 125–33

Stelzer A, Kay AR, Wong RKS. 1988. GABA$_A$-Receptor function in hippocampal cells is maintained by phosphorylation factors. *Science* 241:339–41

Stephenson FA, Duggan MJ. 1989. Mapping the benzodiazepine photoaffinity-labelling site with sequence-specific γ-aminobutyric acid$_A$-receptor antibodies. *Biochem. J.* 264: 199–206

Stephenson FA, Duggan MJ, Casalotti SO. 1989. Identification of the α3-subunit in the GABA$_A$ receptor purified from bovine brain. *FEBS Lett.* 243:358–62

Stephenson FA, Duggan MJ, Pollard S. 1990. The γ2 subunit is an integral component of the γ-aminobutyric acid$_A$ receptor but the α1 polypeptide is the principal site of the agonist benzodiazepine photoaffinity labeling reaction. *J. Biol. Chem.* 265:21160–65

Study RE, Barker JL. 1981. Diazepam and (+/-) pentobarbital: Fluctuation analysis reveals different mechanisms for potentiation of γ-aminobutyric acid responses in cultured central neurons. *Proc. Natl. Acad. Sci. USA* 78:7180–84

Tehrani MHJ, Hablitz JJ, Barnes EM Jr. 1989. cAMP increases the rate of GABA-A receptor desensitization in chick cortical neurons. *Synapse* 4:126–31

Ticku MK. 1991. Drug modulation of GABA$_A$-mediated transmission. *Semin. Neurosci.* 3:211–18

Ticku MK, Ban M, Olsen RW. 1978. Binding of [³H]α-dihydropicrotoxinin, a γ-aminobutyric acid synaptic antagonist, to rat brain membranes. *Mol. Pharmacol.* 14:391–402

Toyoshima C, Unwin N. 1990. Three-dimensional structure of the acetylcholine receptor by cryoelectron microscopy and helical image reconstruction. *J. Cell Biol.* 111: 2623–35

Turner DM, Ransom RW, Yang JS, Olsen RW. 1989. Steroid anesthetics and naturally occurring analogs modulate the γ-aminobutyric acid receptor complex at a site distinct from barbiturates. *J. Pharmacol. Exp. Ther.* 248:960–66

Twyman RE, Green RM, Macdonald RL. 1991. Kinetics of open channel block of single GABA$_A$ receptor channels by penicillin. *Biophys. J.* 59:256a

Twyman RE, Macdonald RL. 1991. Antiepileptic drug regulation of GABA$_A$ receptor channels. In *GABA Mechanisms in Epilepsy*, ed. G Tunnicliff, BU Raess, Chapter 4, pp. 89–104. New York: Wiley Liss

Twyman RE, Macdonald RL. 1992. Neurosteroid regulation of GABA$_A$ receptor single channel kinetic properties. *J. Physiol. (London)* 456:215–45

Twyman RE, Rogers CJ, Macdonald RL. 1989a. Differential mechanisms for enhancement of GABA by diazepam and phenobarbital: a single channel study. *Ann. Neurol.* 25:213–20

Twyman RE, Rogers CJ, Macdonald RL. 1989b. Pentobarbital and picrotoxin have reciprocal actions on single GABA-Cl⁻ channels. *Neurosci. Lett.* 96:89–95

Twyman RE, Rogers CJ, Macdonald RL. 1990. Intraburst kinetic properties of the GABA$_A$ receptor main conductance level of mouse spinal cord neurons in culture. *J. Physiol. (London)* 423:193–220

Verdoorn TA, Draguhn A, Ymer S, et al. 1990. Functional properties of recombinant rat GABA$_A$ receptors depend upon subunit composition. *Neuron* 4:919–28

Vicini S, Mienville JM, Costa E. 1987. Actions of benzodiazepine and beta-carboline derivatives on GABA-activated Cl⁻ channels recorded from membrane patches of neonatal rat cortical neurons in culture. *J. Pharmacol. Exp. Ther.* 243:1195–1201

von Blankenfeld G, Ymer S, Pritchett DB, et al. 1990. Differential benzodiazepine pharmacology of mammalian recombinant GABA$_A$ receptors. *Neurosci. Lett.* 115:269–73

Wafford KA, Burnett DM, Leidenheimer NJ, et al. 1991. Ethanol sensitivity of the GABA$_A$ receptor expressed in *Xenopus* oocytes requires 8 amino acids contained in the γ2L subunit. *Neuron* 7:27–33

Weiss DS, Magleby K. 1989. Gating scheme for single GABA-activated Cl^- channels determined from stability plots, dwell-time distributions, and adjacent-interval durations. *J. Neurosci.* 9:1314–24

Whiting P, McKernan RM, Iversen LL. 1990. Another mechanism for creating diversity in γ-aminobutyrate type A receptors: RNA splicing directs expression of two forms of γ2 subunit, one of which contains a protein kinase C phosphorylation site. *Proc. Natl. Acad. Sci. USA* 87:9966–70

Wieland HA, Lüddens H, Seeburg PH. 1992. A single histidine in GABA$_A$ receptors is essential for benzodiazepine agonist binding. *J. Biol. Chem.* 267:1426–29

Wisden W, Herb A, Wieland H, et al. 1991. Cloning, pharmacological characteristics and expression pattern of the rat GABA$_A$ receptor α4 subunit. *FEBS Lett.* 289:227–30

Wisden W, Laurie DJ, Monyer H, Seeburg PH. 1992. The distribution of 13 GABA$_A$ receptor subunit mRNAs in the rat brain I. Telencephalon, diencephalon, mesencephalon. *J. Neurosci.* 12:1040–62

Yang JS, Olsen RW. 1987. γ-Aminobutyric acid receptor binding in fresh mouse brain membranes at 22°C: ligand-induced changes in affinity. *Mol. Pharmacol.* 32:266–77

Ymer SA, Draguhn W, Wisden P, et al. 1990. Structural and functional characterization of the γ1 subunit of GABA$_A$ benzodiazepine receptors. *EMBO J.* 9:3261–67

Ymer S, Schofield PR, Draghun A, Werner P, Kohler M, Seeburg PH. 1989. GABA$_A$ receptor beta subunit heterogeneity: functional expression of cloned cDNAs. *EMBO J.* 8:1665–70

Zhang JH, Sato M, Tohyama M. 1991. The region-specific expression of the mRNAs encoding β subunits (β1, β2 and β3) of GABA$_A$ receptor in the rat brain. *J. Comp. Neurol.* 303:637–57

Zukin SR, Young AB, Snyder SH. 1974. Gamma-aminobutyric acid binding to receptor sites in the rat central nervous system. *Proc. Natl. Acad. Sci. USA* 71:4802–7

SUBJECT INDEX

A

N-Acetylapartylglutamate
(NAAG), 80
Acetylcholine, 449–53
Achaete-scute complex (ASC),
18
ADP-ribosylation
nitric oxide–stimulated, 160–
61, 169
ADP-ribosyltransferase
(ADPRT), 169–70
ω-Aga-IVA, 453
Alphaxalone, 570
Alz-50, 208
Alzheimer's disease
amyloid cascade hypothesis
of, 507–12
amyloid β-protein and, 489–
512
glutamate receptor system
and, 32, 98–99
neuronal polarity and
alterations in, 300–1
protein kinase C and, 563
α-Amidating enzyme, 226
γ-Aminobutyric acid (See
GABA)
α-Amino-3-hydroxy-5-methyl-4-
isoxazole propionate receptor (See AMPA receptor)
AMPA receptor, 31, 41–62, 97–
100
cloning of, 41–42
divalent cation permeability
of, 54–56
mutagenesis experiments
on, 54–55
expression of receptor RNA,
56
flip and flop splice variants
of, 47, 53, 98
functional properties of, 48–
49, 53–54
gene structure and chromosomal localization, 60–62
long-term potentiation and,
47, 60
mechanics of creating receptor diversity, 47–48
recombinant receptors
protein-biochemical characterization of, 56–60
sequence homologies with
other genes, 46–47
structural features of, 42–46
N-glycosylation and, 42
signal peptides and, 42, 44
transmembrane domains
and, 42, 44–46
Amphiphysin, 224, 230
Amygdala, 520
Amyloid plaques
formation of, 332–33

prion diseases and, 325, 327,
332–33
Amyloid β-protein
Alzheimer's disease and
pathogenesis of, 489–
512
Down's syndrome and, 490–
91
Amyloidosis, 489–512
Amytrophic lateral sclerosis
(ALS)
GluR5 and, 67
glutamate receptor system
and, 32, 67
Anesthetic steroids, 569, 571–
72, 586, 591–93
GABA receptor channels by
modification of, 586, 591–
93
Arachidonic acid
long-term potentiation and,
166–67
Argiotoxin, 76
Astrocytes, 133–49
assemblies of, 134, 138
brain homeostatic functions
of, 134–37
brain injury and, 133–42
carbon dioxide metabolism
and, 136
functions of, 134–37
gap junction syncytium of,
134, 136, 139
glial fibrillary acidic protein
(GFAP) and, 139–41
hypertrophy of, 140
interaction with endothelial
cells in brain, 145
major histocompatibility
genes and, 140
migration of, 141
neurotransmitter metabolism
and, 136–37
GABA, 136–37
glutamate, 136–38
NMDA, 137
platelet-derived growth factor
and, 145
proliferation of, 140–41
responses to injury
early, 134, 139–40, 148
intitial (immediate), 134,
137–39, 148
late, 134, 140–41, 148
swelling of, 135–38, 147
Ataxia telangiectasia, 61
Autoimmunity
multiple sclerosis, 247, 256–
62
Axons, 267–301
differences from dendrites,
268–301
dimensions of, 272–73
morphology of

cell culture studies and,
268–71
targeting, 419–35

B

Barbituates
GABA$_A$ receptor channels
and, 569, 571–71, 586–
88, 594
Barrel differentiation, 421–24
rodent primary somatosensory
cortex and, 421–22
ventrobasal thalamic afferents
and, 421–23
Benzodiazepines, 569, 571–72,
586
modification of neuronal
GABA$_A$ channels by,
586, 589–91, 594
Bicuculline, 586, 594
Blood-brain barrier, 134, 143–45
endothelial injury and, 144–45
vascular endothelial cells and,
143–44
Bovine spongiform encephalopathy, 312
Brain homeostasis, 134–49
astrocytes and, 133–42, 145–
49
carbon dioxide metabolism
and, 136
neurotransmitter metabolism
and, 136–37
potassium homeostasis and,
135–36, 148–49
Brain injury, 32, 98, 100, 133–
49
astrocytic function and, 133–
42, 145–49
NGF and, 142
potassium homeostasis and,
135–36, 146–49
potassium siphoning and,
135, 138
astrocytic hypertrophy and,
140
astrocytic proliferation and,
140–41
astrocytic responses to
early, 134, 139–40, 148
initial, 134, 137–39, 148
late, 134, 140–41, 148
axon growth and
failure of, 148–49
brain macrophages and, 142–
43
signaling by, 143
endothelial cells and, 143–46
ion homeostasis and permeability changes and,
144–45
signaling by, 145–46
extracellular matrix and, 146

603

CUMULATIVE INDEXES

CONTRIBUTING AUTHORS, VOLUMES 11–17

CHAPTER TITLES, VOLUMES 11–17

617

❑ *Annual Review of* **CELL BIOLOGY**
Vols. 1-7   (1985-91)...................$41 / $46
Vols. 8-9   (1992-93)...................$46 / $51
Vol.   10   (avail. Nov. 1994).... ....$49 / $54 Vol(s). _____ Vol. _____

❑ *Annual Review of* **COMPUTER SCIENCE** (Series suspended)
Vols. 1-2   (1986-87)...................$41 / $46
Vols. 3-4   (1988-89/90)...........................$47 / $52 Vol(s). _____
Special package price for
Vols. 1-4   (if ordered together)....... ....$100 / $115 ❑ Send all four volumes.

❑ *Annual Review of* **EARTH AND PLANETARY SCIENCES**
Vols. 1-6, 8-19 (1973-78, 80-91)................$55 / $60
Vols. 20-21   (1992-93)...................$59 / $64
Vol.   22   (avail. May 1994).... ....$62 / $67 Vol(s). _____ Vol. _____

❑ *Annual Review of* **ECOLOGY AND SYSTEMATICS**
Vols. 2-12, 14-17, 19-22..(1971-81, 83-86, 88-91)...$40 / $45
Vols. 23-24   (1992-93).....................$44 / $49
Vol.   25   (avail. Nov. 1994).... ....$47 / $52 Vol(s). _____ Vol. _____

❑ *Annual Review of* **ENERGY AND THE ENVIRONMENT**
Vols. 1-16   (1976-91).....................$64 / $69
Vols. 17-18   (1992-93).....................$68 / $73
Vol.   19   (avail. Oct. 1994).... ....$71 / $76 Vol(s). _____ Vol. _____

❑ *Annual Review of* **ENTOMOLOGY**
Vols. 10-16, 18, 20-36 (1965-71, 73, 75-91)...........$40 / $45
Vols. 37-38   (1992-93) ....$44 / $49
Vol.   39   (avail. January 1994).... ....$47 / $52 Vol(s). _____ Vol. _____

❑ *Annual Review of* **FLUID MECHANICS**
Vols. 2-4, 7   (1970-72, 75)
      9-11, 16-23 (1977-79, 84-91).......................$40 / $45
Vols. 24-25   (1992-93) ....$44 / $49
Vol.   26   (avail. January 1994).... ....$47 / $52 Vol(s). _____ Vol. _____

❑ *Annual Review of* **GENETICS**
Vols. 1-12, 14-25 (1967-78, 80-91)...................$40 / $45
Vols. 26-27   (1992-93).....................$44 / $49
Vol.   28   (avail. Dec. 1994).... ....$47 / $52 Vol(s). _____ Vol. _____

❑ *Annual Review of* **IMMUNOLOGY**
Vols. 1-9   (1983-91) ....$41 / $46
Vols. 10-11   (1992-93).....................$45 / $50
Vol.   12   (avail. April 1994).... ....$48 / $53 Vol(s). _____ Vol. _____

❑ *Annual Review of* **MATERIALS SCIENCE**
Vols. 1, 3-19 (1971, 73-89) ....$68 / $73
Vols. 20-23   (1990-93).....................$72 / $77
Vol.   24   (avail. August 1994).... ....$75 / $80 Vol(s). _____ Vol. _____

❑ *Annual Review of* **MEDICINE: Selected Topics in the Clinical Sciences**
Vols. 9, 11-15, 17-42 (1958, 60-64, 66-42) ....$40 / $45
Vols. 43-44   (1992-93).....................$44 / $49
Vol.   45   (avail. April 1994).... ....$47 / $52 Vol(s). _____ Vol. _____